ENVIRONMENTAL ENGINEERING – 2

WASTEWATER ENGINEERING

(INCLUDING AIR POLLUTION)

[CONTAINING 23 CHAPTERS]

ENVIRONMENTAL ENGINEERING–II

WASTE WATER ENGINEERING

[INCLUDING AIR POLLUTION]

By

Dr. B.C. PUNMIA

Formerly,
Professor and Head, Deptt. of Civil Engineering, &
Dean, Faculty of Engineering
M.B.M. Engineering College,
Jodhpur, Rajasthan

Er. ASHOK KUMAR JAIN

Director,
Arihant Consultants,
Jodhpur Rajasthan

Dr. ARUN KUMAR JAIN

Assistant Professor
M.B.M. Engineering College,
Jodhpur, Rajasthan

IN SI UNITS

LAXMI PUBLICATIONS (P) LTD

(An ISO 9001:2015 Company)

BENGALURU • CHENNAI • GUWAHATI • HYDERABAD • JALANDHAR
KOCHI • KOLKATA • LUCKNOW • MUMBAI • RANCHI
NEW DELHI

ENVIRONMENTAL ENGINEERING–II
WASTE WATER ENGINEERING (INCLUDING AIR POLLUTION)

Printed and bound in India
First Edition : July 1996, *Second Edition* : July 1998, *Reprint* : Sept. 1999, Sept. 2000 Aug. 2001, Jan. 2003, Aug. 2003, Feb. 2005, Nov. 2005, June 2006, Jan. 2007, Aug. 2007 Feb. 2008, Aug. 2008, Jan. 2009, July 2010, Jan. 2011, May 2011, Nov. 2011, Aug. 2012 April 2013, Dec. 2013, April 2014, Dec. 2015, April 2017, 2018, 2020, *Edition* : 2022
ISBN : 978-81-318-0596-1

Branches

Bengaluru	080-26 75 69 30
Chennai	044-24 34 47 26
Guwahati	0361-254 36 69
Hyderabad	040-27 55 53 83
Jalandhar	0181-222 12 72
Kochi	0484-405 13 03
Kolkata	033-40 04 77 79
Lucknow	0522-430 36 13
Ranchi	0651-224 24 64

PUBLISHED IN INDIA BY

Laxmi Publications (P) Ltd.
(An ISO 9001:2015 Company)
113, GOLDEN HOUSE, GURUDWARA ROAD,
DARYAGANJ, NEW DELHI - 110002, INDIA
Telephone : 91-11-4353 2500, 4353 2501
www.laxmipublications.com
info@laxmipublications.com

C—R/022/011
Printed at: Ajit Printing Press, Delhi.

DEDICATED

TO

LATE SHRI NIHAL CHAND JAIN

WHO TAUGHT US

THE VALUES OF LIFE

Preface

Every Community produces both liquid as well as solid wastes. If the waste matter, created and given out by the human and animal life, and also by industries etc., is allowed to accumulate, it will get decomposed and will contaminate or pollute air, water and food. Hence it is essential to remove the contaminants from the wastewater through appropriate treatment methods. The ultimate disposal of treated wastewater, solid and semisolid residuals and concentrated contaminants removed by the treatment continues to be one of the most difficult and expensive problems in the field of wastewater engineering.

The need for direct and deliberate reuse of reclaimed sewage effluents is increasing in many areas of the World. The location of several cities on a single river/stream leads to unavoidable water reuse when each city uses the water body for water supply and receiving body for wastewater. Wastewater collection, treatment and disposal is now confronted with myriads of problems due to additional pollutional loads of industrial wastes. This book is an attempt to present essential principles and modern practices and techniques in the field of wastewater engineering through twenty three chapters. While chapter 1 is introductory, chapter 2 deals with collection and conveyance of wastewater. Chapter 3 discusses wastewater flow rates while chapter 4 deals with hydraulics and design of sewers. Chapter 5 deals with contruction of sewers and chapter 6 describes sewer appurtenances and structures. Wastewater pumping has been given in chapter 7. Chapter 8 deals with various characteristics of wastewater. Chapter 9 deals with natural disposal methods while chapter 10 introduces various unit operations and treatment methods used for wastewater. The primary treatment of wastewater has been given in chapter 11, while sedimentation and chemical clarification has been given in chapter 12. Chapter 13 is devoted to sewage filtration while activated sludge process has been elaborately discussed in chapter 14. Chapter 15 gives miscellaneus methods of biological treatment. Chapter 16 is devoted to the treatment and disposal of sludge. Chapter 17 deals with the design and working of septic and Imhoff tanks. Chapter 18 is devoted to advanced wastewater treatment while chapter 19 is on industrial wastewater treatment. House drainage and rural sanitation have been discussed in chapters 20 and 21 respectively. Chapter 22 is devoted to solid waste disposal. Lastly air pollution, one of the important aspects of environmental pollution, has been introduced in chapter 23.

The book is not a treatise or a design manual. Instead, it is a text book written for engineering students at undergraduate and postgraduate levels. A large portion of the material presented in this book has been derived from the work of others. Their contribution is greatly acknowledged. The recommendations of various Indian Standards on the subject, along with those of Manual on Sewerage and Sewage Treatment, prepared by the Central Public Health and Environmantal Engineering Organisation under the Ministry of Urban Development, have been closely followed. Each chapter begins with clear statement of pertinent definitions, design principles, and theories and the design procedures. The basic principles are supplemented with numerous design examples and illustrations. Further suggestions on the improvement of the book are welcome. The authors are thankful to Shri Mool Singh Gahlot and Shri Prem Singh Sankhla for the fine laser typesetting done by them.

JODHPUR **B.C. PUNMIA**

Ist July, 1996 **ASHOK K. JAIN**

Preface to the Second Edition

In the Second edition, the subject matter has been revised and updated. The following new Tables have been added in the appendix : (*i*) Properties of water (In FPS units), (*ii*) Conversion factors from customory units to SI metric, (*iii*) Saturation values of dissolved oxygen in water exposed to water-saturated air containing 20.90% oxygen under a pressure of 760 mm of mercury and (*iv*) Table of relative atomic weights. Account has been taken throughout of suggestions offered by the many users of the book and grateful acknowledgement is made to them. Further suggestions will be greatly appreciated.

JODHPUR **B.C. PUNMIA**

Ist July, 1998 **ASHOK K. JAIN**

CONTENTS

CHAPTER 13. BIOLOGICAL TREATMENT I : SEWAGE FILTRATION

CHAPTER 14. BIOLOGICAL TREATMENT II : ACTIVATED SLUDGE PROCESS

CHAPTER 15. BIOLOGICAL TREATMENT III : MISCELLANEOUS METHODS

1

Introduction

1.1. SANITARY ENGINEERING

Sanitary Engineering is the branch of Public Health Engineering which deals with the preservation and maintenance of health of the individual and the community, by preventing communicable deseases. It consists of scientific and methodical collection, conveyance, treatment and disposal of the waste matter, so that public health can be protected from the offensive and injurious substances. *Sanitation* is the prevention of the sporadic outbreak of diseases, and can be achieved by either controlling or eliminating such environmental factors that contribute in some form or the other to the transmission of diseases.

If the waste matter, created and given out by the human and animal life, and also by industries etc., is allowed to accumulate, it will get decomposed and will contaminate or pollute air, water and food. Hence sanitary disposal of the waste, either in the solid form or in the liquid form, is most essential. The *sanitary sewage* includes excreta (i.e. waste matter eliminated from the body), *domestic sewage* (i.e. used water from home or community which includes toilet, bath, laundry, lavatory and kitchen-sink wastes) and *industrial wastes*. The improper disposal of human excreta and sewage is the major factor threatening the health and comfort of individuals where satisfactory sewerage systems are not available.

1.2. HISTORICAL DEVELOPMENTS

The need for proper drainage of wastewater seems to have been felt even during the ancient times. Among the first historical traces of sanitary engineering construction, dating back to about 3750

B.C., is the sewer arch at Nippur in India. A sewer running under an important street in Tell Asmar, near Bagdad, was probably constructed during the twenty sixth century B.C. Excavations in the environs of Bagdad reveal sewers constructed of bricks, with laterals connected to water flushed latrines, date from 2500 B.C. Curt Merckel, an antiquarian of Engineering, was able to find an old Babylonian seal cylinder. Layard's explorations revealed arched severs in Nineveh and Babylon dating from seventh century B.C. Schick and Warren have unearthed considerable informations about the sewers of Jerusalem. The works of this class in Grecian cities are fairly well known, and the great underground drains of Rome have repeatedly been described. The Cloaca Maxima, an ancient arched channel which drained the Roman Forum is still is use in Rome. It is believed that these early channels were installed as ground and surface drains and not for the water carriage of human faecal matter. Following Roman practice, early sewers elsewhere were constructed originally for the removal of storm water. All human excreta were excluded from the sewers of London until 1815, from those of Boston until 1833 and from those of Paris until 1880.

There was no marked progress of sewerage from first century A.D. to the beginning of nineteenth century. However, wastewater farming was practiced in Germany in 1550 and in England in 1700. In England, chemical precipitation of wastewater was tried in 1762. The subject of wastewater treatment and disposal received only occasional local attention in England until the construction of sewerage systems after the cholera epidemics of the mid–1800 s. The early sanitary conditions in London, Paris and United States were the worst possible. John Phillips was the first engineer to make a comprehensive study of metropolitan sewerage needs in an official capacity in 1847. In the years 1848, 1849, 1852, 1853 and 1854, there was an out break in cholera epidemic in England, resulting in a total mortality of 25000 lives. The developments in the sewerage works was the result of the awakening of the people by a succession of cholera epidemics. The actual works started in 1859 under the supervision of J.W. Bazalgette. The sewerage system of Paris, like that of London, was inaugurated as a result of cholera epidemic. In United States, J. W. Adams first applied the engineering knowledge to design the sewer system for Brooklyn. E.S. Chesbrough designed the sewer system for Chicago in 1858. In 1880, Waring system of separate sewers was installed in Memphis. Early Studies in sewage treatment were made in the United States through the establishment of the Lawrence Experimental Station in 1887 by the Massachusetts State Board of Health. The influence of the research done there has been profound and far reaching. The intensive methods of treatment developed at Lawrence are still in use today. In the early stages, both England and United States suffered from the improper design of separate systems of sewerage in which the house sewerage and rain water were led separately. It is not definitely known as to who designed the first system of sewers for removing house sewage separately, but the principle was strongly advocated by Chadwick, the father of sanitation in England. Excreta was legally admitted to sewers of London, Boston and Paris in 1815, 1833 and 1880 respectively. It was in 1847 that connections of houses and cesspools to the sewers were made compulsory by the law in London.

The discharge of wastewater to environments (i.e. land and streams) caused adverse condition, and this led to the development of intensive methods of sewage treatment. Sedimentation and chemical precipitation (1762) were one of the first processes used for wastewater treatment. In 1865, early experiments on microbiology of sludge digestion were conducted in England. In 1868, early experiments on intermittent filtration of wastewater were conducted, while in 1870 early experinents on intermittent sand filtration were made in England. In 1876, first septic tank was developed in the United States. In 1882, first experiments on aeration took place in England. United States was the first to use bar racks in 1884. In United States first chemical precipitation treatment plant was installed in 1887. In 1889, filtration in contact beds was tried at the Lawrence Experiment station, Massachusetts. In 1891, the method of sludge digestion in lagoons was developed in Germany. In 1895, methane gas was collected from the septic tanks and used for plant lighting in England. The first rotary sprinklers for rotary filters were developed in 1898. The first grit chambers were developed in the United States in 1904.

The offensive character of the sludge produced by sedimentation led to the use of septic tanks in which the solids were rendered more or less inoffensive, but difficulties of various kinds led to the general adaptation of Travis two-story septic (hydrolytic) tank in England in 1904, and the Imhoff tank which was patented in Germany in 1904. The chlorination of wasterwater for disinfection was demonstrated by Phelps in the United States, in 1906. The first municipal installation of a trickling filler was made in United States 1908, and at the same time laws of disinfection were formulated by Chick in United States. In United States, the first Imhoff tanks were constructed in 1911. Simultaneously, in 1911, separate sludge digestion was adopted in United states, to distinguish it from the two-storey tank process. In 1912–13, aeration of wasterwater in tanks containing slate was carried out at Lawrence Experiment station. In 1914, experiments were conducted by Ardern and Lockett that led to the development of the activated sludge process, wherein a high degree of purification is achieved. The process was first applied in a municipal plant for treating sewage at San Marces, Tex., in 1916. In 1925, contact aerators were developed by Buswell in United States.

The changing characteristics of wasterwater, due to discharge of many contaminants, are responsible for the many changes that are taking place today in the wastewater treatment. More than half a million types of organic compounds have been synthesized since the turn of the century and many of these compounds find their way in the wastewater from many cities. Most of these can be treated effectively, but the number of those compounds which are not amenable to treatment is also increasing. The number of industries that discharge wastes to domestic sewers has increased significantly in the past 3 to 4 decades. In the future, it is anticipated that rather than combining industrial and domestic waste, separate treatment facilities will be provided for the industrial wastes before they can be discharged to the domestic sewers.

1.3. IMPORTANT TERMS AND DEFINITIONS

1. Refuse : Refuse is a general term used to indicate what is rejected or left out as worthless. It may be in liquid, semi-solid or solid form, and many be divided into six categories : (*i*) garbage (*ii*) rubbish (*iii*) sullage (*iv*) sewage (*v*) subsoil water and (*vi*) storm water.

2. Garbage : Garbage indicates *dry refuse.* It includes waste paper, decayed fruits and vegetables, grass and leaves, and sweepings from streets, markets and other public places. Thus, garbage contains large amounts of organic and putritying matter.

3. Rubbish : Rubbish indicates sundry solid wastes from offiees, residences and other buildings. It also includes waste building materials, brocken furniture, paper, rags etc. Generally, rubbish is dry and is of combustible nature.

4. Sullage : Sullage is a term used to indicate the wastewater from bath rooms, kitchens, washing places and wash basins etc. It does not create bad smell since organic matter in it is either absent or is of negligible amount.

5. Sewage : Sewage indicates the liquid waste from the community. It includes sullage, discharge from laterines, urinals, stables industrial waste and also the ground surface and storm water that may be admitted into the sewer. It is extremely putrescible ; its decomposition produces large quantities of malodorous gases, and it may contain numerous pathogenic or disease producing bacteria.

6. Sub-Soil water : It is the ground water that finds it entry into sewers through leaks.

7. Storm water : It indicates the rain water of the locality.

8. Sanitary sewage : Sanitary sewage or domestic sewage indicates sewage mainly derived from the residential building and industrial establishments. It is extremely foul in nature. Sanitary sewage may be classified as (*i*) domestic sewage and (*ii*) industrial sewage.

9. Domestic sewage : It is the sewage obtained from the lavatory basins, urinals and water closets of residential buildings office buildings, theatres and other institutions. Since it contains human excreta and urine, it is extremely foul in nature.

10. Industrial sewage : It is wastewater obtained from the industrial and commercial establishments. It may contain objectionable organic compounds that may not be amenable to conventional treatment processes.

11. Night soil : It is a term used to indicate the human and animal excreta.

12. Sewer : It is an under-ground conduit or drain through which sewage is carried to a point of discharge or disposal. *Separate sewers* are those which carry the house hold and industrial wastes only. *Storm water drains* are those which carry rain water from the roofs and street surfaces. *Combined sewers* are those which carry both sewage and storm water. *House sewer (or drain)* is a pipe carrying away the sewage from a building to a street sewer. *Main sewer* or *trunk sewer* is a sewer that receives sewage from many tributary branches and sewers, serving as an outlet for a large territory. *Branch sewer or submain sewer* is a sewer which receives sewage from a relatively small area, usually a few laterals, and discharge into a main sewer. *Lateral sewer* is a sewer which collects sewage directly from the houses. It indicates the first stage of sewage collection. *Depressed sewer* is a section of sewer constructed lower than adjacent sections to pass beneath an obstacle or obstruction. It runs full under the force of gravity and at greater than atmospheric pressure, the sewage entering and leaving at atmospheric pressure. *Intercepting sewer* is a sewer laid transversely to general sewer system to intercept the dry-weather flow of sewage and such additional surface and storm water an may be desirable. An intercepting sewer is usually a large sewer, flowing parallel to a natural drainage channel, into which a number of main or out fall sewers discharge. *Out fall sewer* is a sewer that receives the sewage from the collecting system and conducts it to a point of final discharge or to a disposal plant.

Relief sewer or overflow sewer is a sewer built to carry the flow in excess of the capacity of an existing sewer.

13. Sewerage : The term *sewerage*, as a noun, means the structures, device, equipment and appurtenances intended for the collection, transportation and pumping of sewage and liquid wastes, but excluding works for the treatment of sewage. The term sewerage also includes the entire science of collecting and carrying sewage by water carriage system through sewers.

14. Wastewater : The term *wastewater* is now increasingly used is place of *sewage*. Wastewater includes both organic and mineral matter carried through liquid media. While the organic portion of the wastewater is putrescible, undergoing biological decomposition, the mineral matter may combine with water to form dissolved solids, may form unsightly sludge deposits and may contribute to the hardness of the water in the effluent.

1.4. WASTEWATER TREATMENT SYSTEMS

The wastewater treatment or sewage treatment is a broad term that applies to any process, operation or combination of processes and operations, that can reduce the objectionable properties of water-carried waste and render it less dangerous and repulsive to man. Thus, the waterwater should be treated before its ultimate disposal in order to :

(*a*) reduce the spread of communicable diseases caused by the pathogenic organisms in the sewage and

(*b*) prevent the pollution of surface and ground water.

Wastewater treatment is a combination of physical, chemical and biological processes.

Methods of treatment in which the application of physical forces predominate are known as *unit operations*, while the methods of treatment in which chemical or biological activity are involved are known as *unit processes*. There are three types of unit operations and processes.

1. Physical unit operations
2. Chemical unit processes

and 3. Biological unit processes.

Physical unit operations are those in which application of physical forces predominate. They consist of screening, mixing, flocculation, sedimentation, flotation etc. *Chemical unit processes* are those in which removal of contaminants are brought about by chemical activity. Common examples of chemical unit processes are : chemical precipitation, gas transfer, adsorption, ion exchange, electrodialysis etc. *Biological unit processes* are these in which the removal of contaminants are brought about by biological activity. Common examples of such processes are : activated sludge process, tricking filtration, sludge digestion etc.

Conventional treatment is the term used to describe the standard method of sewage treatment in temperate climates. It comprises of the following stages of treatment :

1. Preliminary processes/treatment
2. Primary treatment

and 3. Secondary (or biological) treatment.

The *preliminary processes* consist of removing floating materials as well as heavy settleable inorganic solids; these also include pumping and flow measurement. The *primary treatment* is aimed at the *sedimentation* of organic suspended solids. In *secondary treatment*, biological processes are used to remove most of the organic matter.

Sometimes, the wastewater is also subjected to *tertiary treatment* (also known as *advanced treatment*) with the aim of removing the *pollutants* not removed in primary and secondary treatment. These pollutants may include soluble inorganic compounds (such as phosphorus and nitrogen) remaining organic material, pathogens, colloidal solids, soluble minerals and other toxic substances.

In addition to the various stages of treatment of wastewater mentioned above, the processing and disposal of *sludge* obtained from the above treatment methods is of equal importance. Disposal of the accumulated waste sludge is a major economic factor since the cost of its processing is about one third of that involved in the treatment plants. The basic features and characteristics of these unit operation/processes and treatment methods have been described in Chapter 10.

1.5. WASTEWATER MANAGEMENT

There are three constituents and interrelated aspects of wastewater management :

1. Collection
2. Treatment
3. Disposal/Re-use.

Collection of domestic and industrial wastewater is best achieved by a fully developed sewerage or water carriage system. The planning and design of wastewater collection facilities involves the determination of wastewater flow rates; the hydraulic design of sewers, large conduits, and junction and diversion structures ; the selection of sewer appurtenances and pumping stations. Most sewers are laid under roads to avoid interference with private property when connections and repairs have to be made. Sewers are commonly laid in straight lines, man holes being provided at all changes of direction, gradient and diameter. With the modern developments in other fields, many recent changes have been made in wastewater collection – notable amongst them being the photogrammetric and computer techniques to the design of sewers, the improvement of construction materials, and the application of computers in the control of storm sewers. The topics on the wasterwater collection have been discussed in chapters 2 through 7.

Treatment of wastewater is essential in order to reduce the spread of communicable diseases caused by the pathogenic organisms in the sewage and to prevent the pollution of surface and ground water. The treatment of wastewater is carried out by a combination of physical unit operations and chemical and biological unit processes, before the end products can be safely disposed off. The form of treatment normally adopted consists of providing an environment in which natural processes of decay can be intensified and controlled so as to take place in the least objectionable manner. The various methods of wastewater treatment have been discussed in chapter 9 through 20.

After the treatment, the problems of *disposal* arises. The most important recent trend in the field of disposal is the establishment of increasingly stringent requirements to protect the environment. Currently used effluent disposal methods are (*i*) dilution in streams and rivers (*ii*) land application and (*iii*) re-use in aquaculture and crop irrigation. The most striking example of agricultural re-use of domestic wastes is in China in which over 90% of national night soil production is, after treatment, applied to the land and it represents about one third of the nutrients actually used by the crops. The sludge and concentrated pollutants are disposed of in lagoons, sludge drying beds or sanitary land fills, for small treatment plants, and by vacuum filtration followed by heat drying and incineration in large treatment plants.

Performance criteria for wastewater management system

An ideal wastewater management system, consisting of collection, treatment and disposal/reuse, should satisfy the following criteria.

1. ***Health criteria*** The system should be so designed that pathogenic organisms do not spread either by direct contact with the night soil or sewage or indirectly via. soil, water or food. The treatment method chosen should be such that high degree of pathogen destruction is achieved and other harmful chemical pollutants are removed or neutralized to the largest extent possible.

2. ***Reuse criteria.*** The treatment processes used should be such that the treatment effluent is safe to be reused for aquaculture, agriculture, industrial cooling and ground water recharge.

3. Ecological criteria. If the treated effluent is not re-used, but is disposed of in the receiving waster, its characteristics should be such that the aquatic environment is protected and the self purification capacity of the recipient water is not exceeded. For this, a high degree of removal of organic matter (BOD) is required in order to prevent pollution of the receiving watercourse by oxygen depletion.

4. Nuisance criteria. The treatment system should be so designed that either it does not release objectional odour or the degree of odour released must be below the nuisance thresh hold. No part of the system should be aesthetically offensive.

5. Cultural criteria. The methods chosen for collection, treatment, disposal and reuse should be such that they match with local habits, social customs and religious practices of the people of the area.

6. Operational criteria. The system should be such that it may operate with minimum difficulties, and greater degree of spill is not required. It should be such that operation is possible with locally available staff, with minimum possible training.

7. Cost criteria. The system should have minimum possible capital and running costs, which can be easily paid by the community. The return of the reuse of the end products is an important factor. If the end products are not reused, their disposal should be achieved with minimum possible cost.

2

Collection and Conveyance of Sewage

2.1. SYSTEMS OF WASTE AND SEWAGE DISPOSAL

Wastes or *refuse* are of two types : solid waste and liquid waste. *Night soil* consists of human or animal excreta and ureas, originating at privies, water-closets, urinals and stables. Garbage indicates dry refuse from a town, and includes sweepings from houses, streets, markets and such other public places, waste paper, leaves, grass, parings from vegetables, decaying fruit etc. constitute garbage. *Sullage* indicates waste waster from bath rooms, kitchens, washing places and wash basins etc. *Sewage* indicates the liquid waste from the community. It includes sullage, discharge from latrines, urinals, stables, and the industrial waste. It is extremely putrescible; its decomposition produces large quantities of malodorous gases, and it may contain numerous pathogenic or disease producing bacteria. The refuse, consisting of all above item, formed in any sanitary system should be rapidly and safely carried to its disposal site so as to maintain a clean environment. Excreta and sewage is satisfactorily disposed of if the following conditions are satisfied.

1. The *waste* does not pollute the ground surface, nor is it exposed to atmosphere when in adequately treated.
2. It should also not be accessible to children or household pets.
3. It does not pollute or contaminate drinking water supply.
4. It does not give rise to odour nuisance.

5. It does not give unsightly appearance.

6. It does not give rise to mosquito nuisance. It should also not be accessible to insects and rodents.

7. It does not pollute or contaminate the waters of bathing beach or streams used for domestic water supply.

Depending upon the type of *waste*, two systems may be employed for its collection, conveyance and disposal :

(*a*) Conservancy system

(*b*) Water carriage system.

2.2. CONSERVANCY SYSTEM

This is an old system in which various types of wastes, such as night soil, garbage etc. are collected separatety in vessels or deposited in pools or pits and then removed periodically atleast once in 24 hours. On account of this method of handling independently the different types of refuse from a place, the term conservancy system has been derived. The system is also known as the *dry system*. The following are the methods of collection of various types of wastes in the system.

*1. **Night soil**.* Night soils or human excreta in latrines, privies or cesspools etc. is collected separately in pans or pails and carried on heads of sweepers to a central place from where it is transported in bullock carts or motor vans to a place away from the town for its final disposal. Normally, it is buried into ground, in trenches, to give excellent manure in one or two years.

2. ***Garbage.*** Garbage is collected separately, in dust bins and conveyed on hand carts or motor van once or twice is a day. It may consist of waste matter of both non-combustible as well as combustible type. The two are therefore sorted out. Garbage disposal methods include the open dump, hog feeding, incineration, dumping into sanitary fill, fermentation or biological digestion. Incineration, if properly controlled, is satisfactory for burning combustible refuse.

*3. **Sullage and storm water.*** Sullage and storm water are collected and conveyed separately in closed or open gutters. The liquid and semi-liquid mass of filth which frequently overflow the receptacles in privies is swept away by the sweepers to drain from the privies, which carry it to drains carrying sullage and storm water, along the public lanes or streets.

There was a time when the conservancy system was favoured and the water carriage system was opposed by European chemists, physicians and agriculturists because of a fear of contamination of the soil by leakage from the sewers and the possible pollution of bodies of water receiving the sewage, and possible nuisance where sewage was distributed on land. However, in India, the conservancy system is still on vogue in all the villages and small towns. Only a few cities have the water carriage system.

Disadvantages of conservancy system

The conservancy system has the following disadvantages :

1. Hygiene and sanitary aspect : The consevancy system is highly unhygienic and cause insanitary conditions since the excreta starts decomposing within few hours of its production. Even if it is assumed that cleaning will take place twice in a day, the excreta remaining in the previes will emit bad smell and will give rise to fly nuisance.

2. Transportation aspect : Transportation of night soil takes place in open carts through streets and other crowded localities. This is highly undesirable.

3. Labour aspect : The working of the system depends entirely on the mercy of labour (sweepers). If they go on strike even for one day for any reason whatsoever, the previes can not be used because of foul smell. The whole locality will smell very badly.

4. Building design aspect : The lavatories or previes are to be located outside the house and slightly away from the main building. The compact design is therefore not possible.

5. Conditions of drains : Insanitation may be there due to carriage of sullage through open drains laid in the streets.

6. Human aspect : In the present day world, when man has progressed much, it is highly humiliating to ask human beings to transport night soil in pails on their heads.

7. Risk of epidemic : Due to improper or careless disposal of night soil, there are more chances of outbreak of epidemic.

8. Pollution problems : The liquid wastes from lavatories etc., during their washing, may soak in the ground, thus contaminating the soil. If the ground water is at a shallow depth, it may also be polluted due to percolation of waste water.

9. Cost consideration : Though the system is quite cheap in the beginning, its maintenance and establishment costs (i.e. recurring expenditure) are very high.

10. Disposal land requirement : The system requires considerable land for the disposal of sewage.

2.3. WATER CARRIAGE SYSTEM

In this system, the collection, conveyance and disposal of various type of wastes are carried out with the help of water. Thus, water is used as medium to convey the waste from its point of production to the point of its treatment or final disposal. Sufficient quantity of water is required to be mixed with the wastes so that dilution ratio is so great that the mixture may flow just like water.

In this system, specially designed latrines, called *water closets* (W.C.) are used which are flushed with 5 to 10 litres of water after its use by every person. The human excreta is thus flushed away and led to suitable designed and maintained *sewers*. The wastes from kitchens, baths, wash basins etc. are also led to the sewers. The sewers are the underground closed pipes which are laid on suitable longitudinal gradient so that flow takes under gravity and proper flow velocity is maintained to keep the sewer clean. The sewers lead the *sewarage* so collected, to a suitable site where it is treated suitably and then is disposed off by irrigation or by dilution. It should be noted than the *garbage* is collected separately and conveyed in the same manner as is done in the case of conservancy system. If garbage is permitted in the sewers, they may be clogged.

The system requires large initial cost of installation and it requires large quantity of water also to create efficient flow conditions. If the financial conditions of people are poor, it may be difficult to adopt this system. However, this is the most efficient and hygienic system of sewage disposal, and may be adopted in stages if sufficient funds are not available in the beginning.

Advantages of water carriage system The water carriage system is the most modern system of drainage, and has the following advantages:

1. Hygienes and sanitary aspect : The system is very hygienic since the night soil and other waste water is conveyed through closed conduits which are not directly exposed to the atmosphere. There is no bad smell because of continuous flow.

2. Epidemic aspect : There are no chances of outbreak of edpidemic because flies and other insects do not have direct access to the sewage.

3. Pollution aspect : The liquid wastes etc. are directly conveyed through the sewers, and therefore there are no changes of the waste water being soaked in the ground thus contaminating the soil. The waste water does not percolate down to join the ground water. There are no chances of pollution of water of wells in individual houses if any.

4. Compactness in design : Since the latrines are flushed after every use, excreta does not remain and there are no foul smells. The latrines can therefore be attached to the living and bed rooms. This permits a compact design. The lavatories can be accommodated in any part of the house.

5. Labour aspect: The labour required for the operation and maintenance is extremely small. In fact, the functioning of the system is practically automatic, except for the operation of certain pumps etc. Therefore, there is no labour problem. In the individual houses, the latrines/lavatories can be coveniently cleaned by occupants themselves.

6. ***Treatment aspect :*** The system permits the use of modern methods of treatment of the sewerage collected through the sewers. The treated waste water and sewage can be safely disposed off without any risk.

7. ***Land disposal requirements :*** Because of treatment facilities, the land required for the disposal of the treated wastewater is very much smaller than that required for the conservancy system.

8. ***Cost consideration*** : Though the initial cost of installation of the system are very high, the running costs are very small since manual labour is very much reduced.

2.4. COMPARISON AND CHOICE OF THE TWO SYSTEMS

The water carrigage system is the most modern and hygienic system of sweage disposal, and should be invariabley adopted wherever sufficient funds are available. Table 2.1 gives comparison between the two systems.

2.5. CLASSIFICATION OF WATER CARRIAGE SYSTEM

The water carriage system can be divided into the following types :

(1) Separate system
(2) Combined system
(3) Partially separate system

The above three types are commonly referred to as the three *systems of sewerage.*

1. Separate system

The separate system provides two separate systems of sewers—the one intended for the conveyance of foul sewage only, such as faecal matter, domestic wastewaters, the washings, and drainings of places such as slaughter houses, laundries, stables, and the waste waters derieved from the manufacturing processes ; and the other for the rain water, including the surface washing from certain streets, overflow from public baths and foundations etc. The sewage from the first system of sewers can be led to the treatment works, while the flow from the second system of sewers can be discharged directly to natural streams etc. without any treatment.

Advantages

The separate system has the following advantages :

1. The cost of installation is low. The storm water can be

TABLE 2.1 COMPARISON BETWEEN CONSERVANCY AND WATER CARRIAGE SYSTEM

S.N.	Conservancy System	Water Carriage System
1.	The system is unhygienic since every thing is visible.	The system is hygienic. Sewers are laid below the ground and hence excreta etc. is not visible.
2.	Due to putrefication, there is a lot of foul smell.	No chances of putrefication, and hence no foul smell
3.	Compact house design is not possible	Compact design is possible
4.	Large labour force is required	Labour force is negligibly small
5.	Water consumption is small	Requires high water consumption
6.	Initial cost is small, though the running costs are high	High initial cost. Running costs small
7.	No technical persons required	Technical persons required for operation and maintenance
8.	Acute pollution problems	Pollution problems are rare.
9.	Risk of spread of epidemic	No such risk
10.	Large land required for the disposal of untreated sewage.	Small land required for the disposal of treated sludge
11.	Final disposal into streams etc not free from risks	Final disposal easier because of treatment works.
12.	Good quality manure available from the end products	The sludge has small manure value. However, treated waste water can be used for irrigation etc.
13.	The system is more suitable for rural conditions	The system is better suited for urban conditions.

disposed off through the open channels along the road sides. Old sewers may also be suitably converted to carry rain water. Thus, the actual sewers carrying foul sewage will be of smaller size.

2. The load on the treatment units will be lowered, since only the foul sewage carried by the separate sewers need be treated.

3. If there is necessity of lifting the sewage mechanically, the system will prove to be economical both from the point of view of capital costs, as well as from the point of view of running costs.

4. The sewages in the separate system will be of more unifrom character, and so will lend itself more easily to putrification.

5. There is no necessity of providing automatic flushing tanks, for use in dry weather, because the flow in a sewer of smaller section is much more efficient.

6. Sewers of smaller section can be easily ventilated than those of larger section.

7. The night flow will be comparatively small this may facilirate operations at the outfall works.

8. Rain water can be discharged into streams or rivers without any treatment.

Disadvantages

1. Since the sewers are of small size, it is difficult to clean them.

2. They are likely to get chocked.

3. Two sets of sewers may ultimately prove to be costly.

4. There is a likelyhood of connections being wrongly made through a confusion of the systems.

5. Storm water sewers or drains comes in use only during the rainy season. During other part of the year, these may serve as dumping place for garbage, and may get chocked.

6. Because of lesser air contact in small size sewers, foul smell may be there due to the sewage gas formed.

2. Combined system

The combined system provides only one sewer to carry both the foul sewage as well as the rain water. The sewage and rain water are carried to the sewage treatment plant, before its final disposal. The combined system is advocated on the ground that the street surface washings are as impure as the sewage itself, and should therefore be suitably treated before being allowed to enter the natural stream.

Advantages

1. The system requires only one set of sewers. Hence the maintenance costs are reduced.

2. The sewers are of larger size, and therefore the chances of their choking are rare. Also, it is easy to clean them.

3. The strength of the sewage is reduced by dilution.

4. There is more air in the larger sewers than in smaller ones of the separate system. Hence the sewer gas that may be formed gets diluted. Thus the chances of foul smell are reduced.

Disadvantages

1. The cost of construction are very high because of large dimensions of the sewers to be constructed at sufficient depth to receive the sewage from the basement.

2. Beasue of large size of sewers, their handling and transportation is difficult.

3. Due to the inclusion of the storm water, the load on the treatment plant increases.

4. The system is uneconomical in the circumstances when pumping is required for lifting of sewage.

5. During heavy rains, the sewers may overflow, and may thus create unhygienic conditions and cause pollution problems.

6. Storm water is unnecessarily polluted.

7. The large sewers get easily silted if not properly designed. They may become foul in dry weather, when rain water, is not available.

8. Large sewers are more difficult to be ventilated than the smaller ones.

3. Partially combined system

In this system, only one set of underground sewers is laid. These sewers admit the foul sewage as well as the early washings by rains. As soon as the quantity of storm water exceeds a certain limit, the storm water overflows, and is thus collected and conveyed in open drains to the natural streams. The foul sewage, however, continues to flow in the sewers.

Advantages

1. The sewers are of reasonable size. Their cleaning is therefore not very difficult.
2. It combines the advantage of both the separate as well as the combined systems.
3. The storm water permitted in the sewers eliminates its chances of chocking. The sewers are completely cleaned during rainy season.
4. The problem of disposing off storm water from homes is simplified.

Disadvantages

1. During the dry weather, when there is no rain water, the velocity of flow will be low. Thus self cleansing velocity may not be achieved.
2. The storm water increases the load on treatment units.
3. The storm water also increases the cost of pumping.

Factors governing choice of separate system The separate system is adopted under the following conditions:

1. Financial aspect : If sufficient funds are not available in the beginning, sewers may be constructed to carry only domestic sewage, and the rain water may be conveyed thro' the open drains. These drains can be converted into regular sewers later when sufficient funds are available.

2. Flat topography : If the country is flat, the sewers will have to be laid at a depth in order to achieve reasonable gradient. In such circumstances, separate sewers are more economical. A combined sewer of a bigger size, laid at a great depth will be costlier.

3. Rainfall pattern : If the rainfall is there for a shorter duration, and does not take place throughout the year, it is more economical to adopt separate system.

4. Outlet conditions : If a river or natural stream is available nearby, separate system is best suited. The storm water can be directly conveyed to the stream, while the foul sewage can be led to the treatment plant.

5. Pumping aspects : Separate system is best suited under the conditions when the sewage has to be lifted up by pumping. The separate sewers for the storm water will reduce the load on the pumps.

6. Steep topography : If the ground has steep slopes, it is easier to convert storm water through open drains, to the natural stream.

7. Subsoil conditions : If the subsoil is hard, it is difficult and costlier to lay combined sewer of larger size.

8. Development pattern : If the sewers are laid before the area is developed, it is preferable to adopt separate system.

Factors governing choice of combined system

A combined system is adopted under the following conditions:

1. Space considerations : Combined system is preferred when space available for laying the sewers is restricted.

2. Integrated development : If the sewers are laid along with the overall development of the area, a combined system is preferred.

3. Even rainfall pattern : Combined system is preferred if the rainfall is evenly distributed through the year, and also when the rainfall during the monsoon season is not that intensive.

4. Conversion of existing storm water drains : The combined system is preferred if an existing storm water drain is being converted into a combined sewer. This is possible only if the quantity of sewage is small.

5. Pumping requirements : If the ground slopes are such that it is necessary to lift both the sewage as well as the storm water, it is preferable to use the combined system.

Choice of the system The factors governing the choice of any system are so vast and varied that no generalisation can be done regarding the final choice. The conditions vary from place to place. If the *availability of funds* is the main factor, then one may conclude that separate system may be adopted if sufficient funds are not available in the beginning. In that case, sewers may be designed to carry the foul sewage and rain water may flow through the open drains along the roads and streets. This suggestion may be more appropriate for small cities. For large metropolitan cities, however, a combined system is highly desirable. It is possible to design combined sewers such that reasonable velocities are maintained in them through the year. Alternatively, a partially separate system may be adopted for these big metropolitan cities.

PROBLEMS

1. Describe (*i*) conservancy system, and (*ii*) water carriage system. What are the relative advantages and disadvantages of the two systems ?
2. What are the conditions that are to be satisfied for the satisfactory disposal of sewerage.
3. Describe in brief the methods of collection of various types of wastes. in the conservancy system.
4. Describe in brief various types of water carriage system, stating advantages and disadvantages of each.
5. Discuss the relative merits of the separate and combined system of sewage, and give the conditions favourable for the adoption of each one of them.

3

Wastewater Flow Rates

3.1. GENERAL

In order to determine the section of the sewer, it is essential to know the total quantity of wastewater or sewage that would flow through the sewer. The total wastewater flow can be divided two components :

(*a*) Dry weather flow (D.W.F.)

and (*b*) Storm water flow

The *dry weather flow* is the flow through the sewers that would normally be available during non-rainfall periods. It consists mainly of (*i*) *Domestic sewage and* (*ii*) *Industrial wastewater.*

The *storm water flow* is the additional flow that would occur during the rainy season. It consists of runoff available from roofs, streets, yards, open spaces etc. during rainfall.

3.2. DRY WEATHER FLOW

It is that quantity of wastewater that flows through a sewer in dry weather when no storm water is in the sewer. The dry weather flow is also sometimes called as 'sanitary sewage', and is obtained from the following sources :

(i) Domestic wastewater : A certain quantity of water is being supplied daily by the Water Works Department, for domestic use. This water gets consumed in various ways, and therefore all of it does not reach the sewer. The wastewater reaching the sewers will be that part of water which is used for flushing water closets (W.C.), urinals, washing clothes, bathing, washing utensils etc.

(ii) Industrial wastewater : This is the wastewater generated by the industry after consuming water for its manufacturing processes.

(iii) ***Waste water from public facilities*** : This is the wastewater from schools, cinemas, hotels, railway stations street washings.

(iv) ***Ground water infiltration*** : This is quantity of ground water that infiltrates into the sewers. It depends upon the nature of soil, material of sewers, nature and condition of sewer joints, depth of sewer and position of water table.

3.3. FACTORS AFFECTING DRY WEATHER FLOW

The dry weather flow or the quantity of sanitary sewage depends upon the following factors :

(*a*) Rate of water supply.
(*b*) Population growth.
(*c*) Type of area served
(*d*) Infiltration of ground water

(*a*) **Rate of water supply** The quantity of wastewater produced from a community would naturally depend upon the rate of water supply per capita per day. The quantity of wastewater entering the sewers would be less than the total quantity of water supplied. This is because of the fact that water is lost in domestic consumption (*i.e.* cooking etc.), evaporation, lawn sprinkling, fire fighting, industrial consumption etc. However, private source of water supply (*i.e.* water from domestic wells etc.) and infiltration of sub-soil water in the sewers increase the wastewater flow rate. This extra water that enters the sewers can be assumed to approximately equal to the water lost in consumption etc. On this reasoning, one can assume the wastewater flow rates *equal* to the rates of water supply by the municipal authorities. If however, one is sure that no extra water enters the sewers, the wastewater quantity may be assumed to be 80% of the quantity of water supply. The sewers should be designed for a minimum of 150 litres per capita per day.

(*b*) **Population growth** The sewerage system is designed for the quantity of wastewater not only of the present population but also of the population a few year hence. The population forecast may be done by the following methods :

1. Arithmetical increase method
2. Geometrical increase method
3. Incremental increase method
4. Decreased rate of growth method
5. Graphical extension method.
6. Graphical comparison method.
7. Zoning method or master plan method.
8. Ratio and correlation method.
9. Growth composition analysis method.

These methods have been discussed at length in Vol. I (Water Supply Engineering) of the book. Out of the various methods mentioned above, the most suitable approach is to base the estimation either on *anticipated ultimate density of population* or on *floor space index* (FSI).

In case the desired information on population is not available in the Master Plan of the town, the densities given in Table 3.1 may be adopted as suggested by the Manual on Sewage and Sewage Treatment prepared by the Public Health and Environmental Engineering Organisation.

TABLE 3.1 POPULATION DENSITIES

	Size of Town (Population)	*Density of population per hectare*
1.	Upto 5000	75 – 150
2.	5000 – 20,000	150 – 250
3.	20,000 – 50,000	250 – 300
4.	50,000 to 100,000	300 – 350
5.	Above 100,000	350 – 1000

In cities where *floor space index* (FSI) or *floor area ratio* (FAR) limits are fixed by the municipal authorities, population density may be easily worked out as illustrated below. Let us assume that a particular *development plan rules* provide for the following reservations for different *land uses* :

Roads	:	....	20%
Gardens	:	....	15%
Schools	:	....	6%
Markets	:	....	3%
Hospitals and dispensaries	:	....	2%
			46%

∴ Area available for residential development= 100 − 46 = 54%

Let us consider an area of 1 hectare ($= 10^4 \text{ m}^2$)

∴ Actual total floor area residential development

$$= (0.54 \times 10^4) \times \text{FSI}.$$

Let us assume FSI of 0.5.

∴ Area available for residences $= 0.54 \times 10^4 \times 0.5 = 2700 \text{ m}^2$

Assuming a floor area requirement of 9 m^2 per person,

$$\text{Density of population/hectare } \frac{2700}{9} = 300$$

Design period: The branches and main sewers are designed for the population which may occur at the end of one generation of 30 years. This period of 30 years is called the *design period*. However, the pumping plants etc. are designed for a design period of 5 to 10 years only, since additional pumps can be installed whenever needed. The treatment units are designed for 10 to 30 year period.

The rate of use of water and the consequent rate of sewage production increases with increase in population. This is due to the fact that the increase in the population of the town is mainly due to improved facilities. Data in U.S.A. reveal that the percentage increase in per capita water use is equal to 5% of the percentage increase in population. Though no such figures are available for Indian conditions, Table 3.2 shows how water consumption and sewage production increases with the increase in population.

TABLE 3.2. EFFECT OF GROWTH OF POPULATION

Population	*Rate of water supply lit./capita/day*	*Sewage production lit./capita/day*
Below 2 lacs	130—160	110—120
2 to 5 lacs	160—180	130—140
5 to 10 lacs	180—200	150—170
More than 10 lacs	200—250	180—200

(*c*) **Type of area served.** The quantity of wastewater produced depends upon whether the area to be served is residential, commercial, or industrial. The wastewater from the residential area directly depends upon the rate of water supply. If there is no infiltration of water in the sewers, and if there are no private sources of supply, the wastewater produced from the residential area may be assumed to be equal to 70 to 80% of the water supplied through the public supply system. The amount of waste water produced from the industrial locality depends upon the types of industries and their corresponding industrial processes. The same is the case with the commercial undertakings. The sewage flow rates can be determined after careful study of the various process involved.

(*d*) **Infiltration of sub-soil water.** Ground water or subsoil water may infiltrate into the sewers through the leaky joints. *Exfiltration* is the reverse process which indicats the flow of wastewater from the sewer into the ground. While due to the infiltration the quantity of flow through sewer increases, exfiltration results in decrease in the flow and consequent increase in the pollution of ground water. Both infiltration as well as exfiltration are undesirable and take place due to imperfect joints. However, infiltration is much more important from the point of sewer design. Also, infiltration unnecessarily increases the load on the treatment works.

Infiltration of water into the sewer depends upon the following factors :

(*i*) Depth of sewer below ground water level
(*ii*) Size of sewer
(*iii*) Length of sewer through infiltration prone zone
(*iv*) Nature and type of soil through which sewer is laid
(*v*) Type of joints, workmanship etc.
(*vi*) Sewer material.

Infitration of water into the sewer line may be expressed by the following methods.

(*i*) ***Area basis*** : In this, the rate of infiltration is expressed as litres per hectare of area per day, i.e. 5000 litres per hectare per day. Thus, if the area sered by the sewer line is equal to 10 hectares, the total infiltration volume per day will be 5000 × 10 = 50000 litres. The infiltration may range from 0.2 to 28 m^3/hectare/day. In absence of any other data, the infiltration flow rates shown in Fig. 3.1 may be adopted.

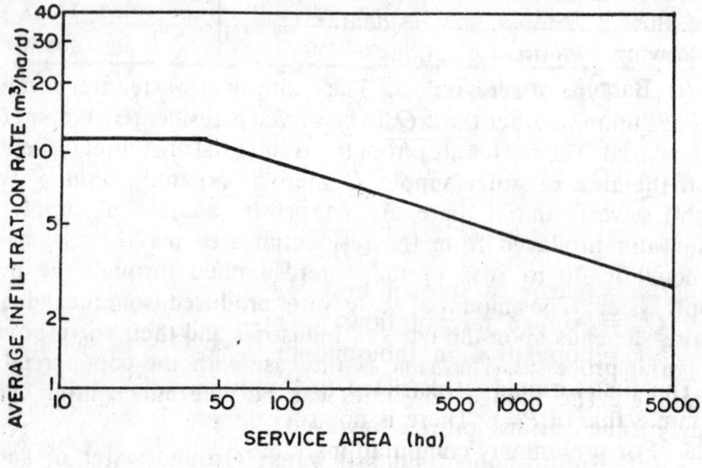

FIG. 3.1. AVERAGE INFILTRATION RATE CURVE

(*ii*) ***Length basis*** : In this, the rate of infiltration is expressed as litres per unit length per day, i.e. 10,000 litres per km length of sewer per day. If the length of sewer through the ground water zone is 5 km, the total infiltration volume will be 10000 × 5 = 50000 litres/day, at the above rate. However the rate of infiltration may vary between wide limits to 10,000 to 50,000 litres per km/day.

(*iii*) ***Diameter-length basis*** : This is the most rational method in which infiltration rate is expressed in litres per cm diameter of sewer per km length per day. The rate of infiltration will increase with the increase in the diameter of the sewer.

3.4. VARIATIONS IN RATE OF SEWAGE

The rate of sewage flow is not constant. It varies with every hour of the day, every day of the season and with every season of the year. The designer is concerned with both maximum and minimum rate of flow. The sewers should be so designed that it can take the maximum load. At the same time, they should be laid on such longitudinal gradient that there are no deposits in the sewers at the minimum flow. Both the maximum and minimum flows are expressed as percentage of the average rate of flow. The variations in the flow may be due to several factors, such as habits of the people, working hours of the office, climatic conditions, timings of water supply, type of industries etc.

In the residential areas, the sewage flow rises by about 6 AM and reaches the maximum between 8 and 10 AM. Then it falls gradually till minimum flow is reached between 1 to 4 P.M. The rate of flow then rises again. The relation between the maximum and the average rate of flow of sewage for residential areas can be expressed by the following empirical relations :

(*i*) Babbit's formula :

$$Q_{max} = \frac{5\,Q_{av}}{P^{0.2}} \qquad \text{...(3.1)}$$

(*ii*) Harmon's formula

$$Q_{max} = \left(1 + \frac{14}{4 + P^{0.5}}\right) Q_{av} \qquad \text{....(3.2)}$$

where $Q_{max.}$ = maximum rate of flow

Q_{av} = average rate of flow

P = population in thousands

The first formula is restricted to maximum value of P=1000 and a min. value of P=1. There is no such limitation in the second formula. For preliminary computations, the ratio of average to minimum flow may be taken to be the same as the ratio of maximum to average rate of flow. For commercial and industrial undertakings the variables are so many that no definite formulations can be made for the relation between maximum average and minimum flows.

The hourly, daily and seasonal variations of the sewage flow (D.W.F.) of a town affects the working of not only the sewers, but also of the pumping station and treatment plants. The hourly rate of wastewater flow is affected by the hourly variations in water consumptions. In general, the wastewater-discharge curve closely parallels the water consumption curve, but with a time lag of few hours.

Fig. 3.2 shows the hourly variation of domestic sewage. Fig. 3.3 shows the relation between hourly variation of water consumption and sewage flow.

Temperature and climatic conditions are the principal factors affecting the seasonal rate of sewage flow. For Indian conditions, average rate of sewage flow may be assumed to occur in Oct./Nov. and March/ April. Minimum rate of flow may be obtained during

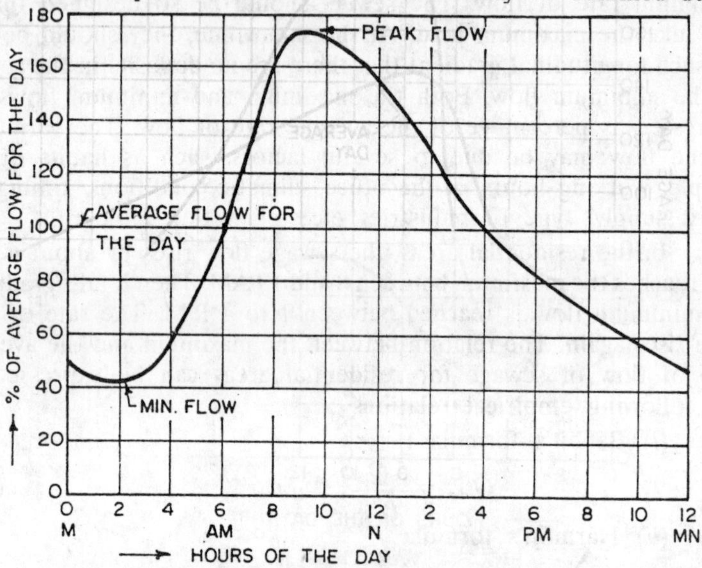

FIG. 3.2. HOURLY VARIATIONS OF SEWAGE FLOW

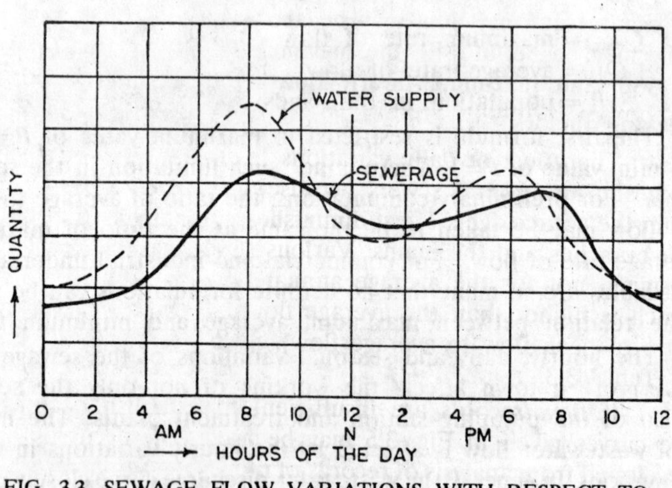

FIG. 3.3. SEWAGE FLOW VARIATIONS WITH RESPECT TO WATER SUPPLY

the Dec./Jan. while max. rate may be obtained during June/July. Fig. 3.4 shows daily variations in sewage flow during minimum–flow–day, average-flow-day and maximum-flow-day.

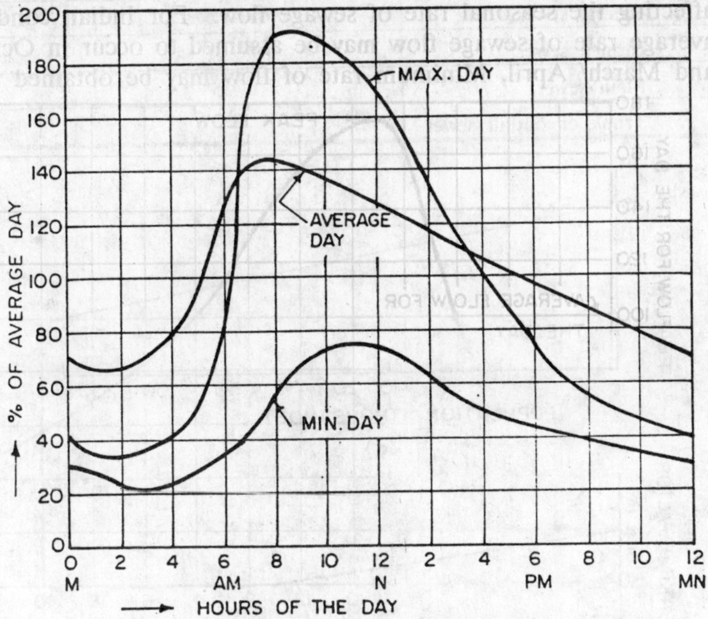

FIG. 3.4. VARIATION IN DAILY FLOW

If the annual average rate of flow is taken as 100, then the maximum seasonal flow may be about 120, the maximum monthly flow may be about 140, the maximum day flow may be about 150 to 180, and maximum hourly flow may be 200 to 300.

Peaking factors

The effect of daily variations in sewage flow is maximum on domestic and lateral sewers because they receive the flow directly from the source. This effect diminishes gradually as the flow reaches the branches and the mains. Various sewers in a sewer-network are designed not for the average annual flow rate, but for a flow rate which is higher than the average flow rate by a peaking factor *i.e.* ratio of peak flow to average flow). Such peaking factors are given in Table 3.3.

If flow records are insufficient to establish peaking factors, the curves given in Fig. 3.5 may be used. These curves have been developed from analysis of records of numerous communities throughout United States.

TABLE 3.3. PEAKING FACTORS OF SEWER DESIGN

	Sewer	*Factor to multiply average rate of annular flow*
1.	Domestic sewer	6
2.	Lateral sewer	4–6
3.	Branch sewer	3
4.	Main sewer	2.5
5.	Trunk or outfall sewer	2

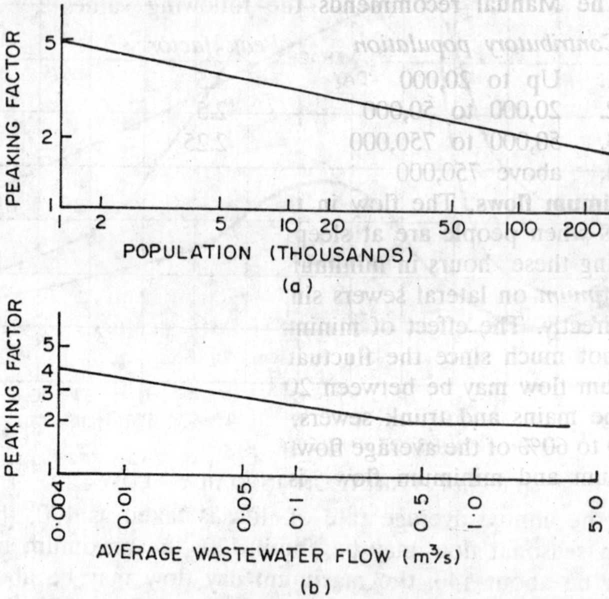

FIG. 3.5. PEAKING FACTORS FOR DOMESTIC SEWAGE.

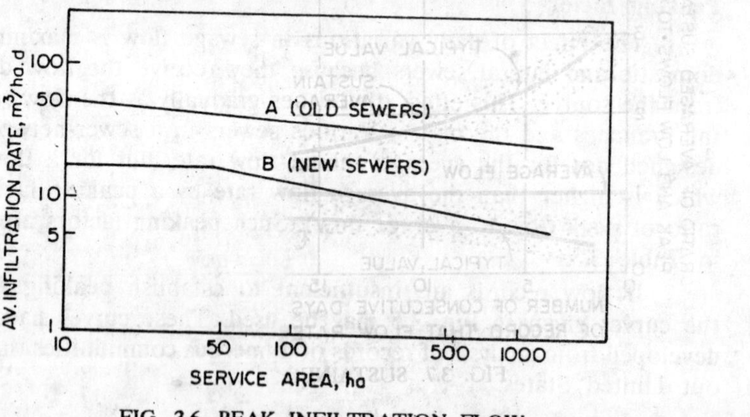

FIG. 3.6. PEAK INFILTRATION FLOW

For determining *peak infiltration flows*, curves shown in Fig. 3.6 may be used. Curve *A* is for old sewers while curve *B* is for new sewers. In addition to newly designed sewers, *new sewers* include vitrified-clay pipe with premolded gasket joints, asbestos cement pipe with coupling and rubber gasket joints, and reinforced concrete pipes with ring or compression rubber gasket joints. Peaking factors of infiltration are derived from flow measurements ; common values range from 1.5 to 2.0.

The peak factor (i.e. the ratio of maximum to average flows) also depends upon the contributing populations, as indicated in Fig. 3.5. (*a*). The Manual recommends the following values :

Contributory population	*Peak factor*
1. Up to 20,000	3.5
2. 20,000 to 50,000	2.5
3. 50,000 to 750,000	2.25
4. above 750,000	2.0

Minimum flows. The flow in the sewers is minimum during night hours when people are at sleep. The use of flushes and bath rooms during these hours in minimum. The effect of this minimum flow is *maximum* on lateral sewers since these sewers get the house drainage directly. The effect of minimum flow on mains and trunk sewers is not much since the fluctuations get damped. In laterals, the minimum flow may be between 20 to 30% of the average flow, while in the mains and trunk sewers, the minimum flow may vary between 40 to 60% of the average flow. The overall variation between the maximum and minimum flow is maximum in lateral sewers,

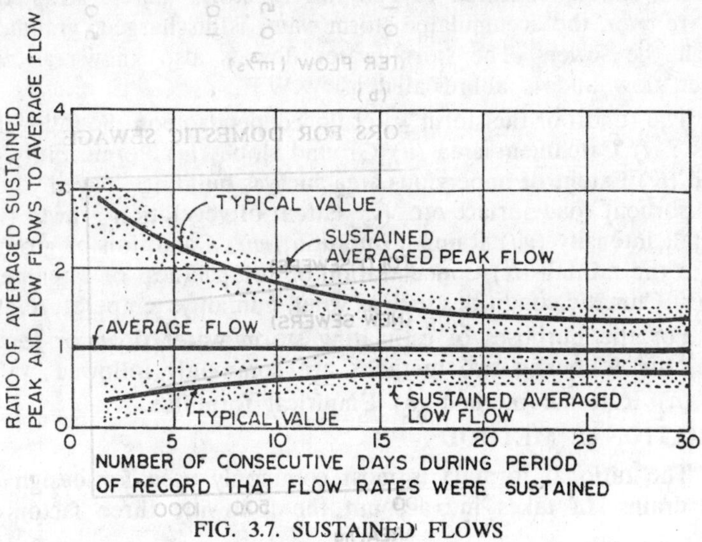

FIG. 3.7. SUSTAINED FLOWS

and minimum in mains and trunk sewers. The ratio of maximum to minimum flow may vary between 6 to 8 for laterals, while it may vary between 2 to 4 for the mains and trunk sewers.

Sustained flows. Apart from information about expected peak flows, information about extended *sustained* flows is equally important. *Sustained flows are flows that persist for various time durations* (say, 2 hours or larger). Sustained flows that are higher than the average flow, or those, lower than the average flow are of importance in sewer design. Fig. 3.7 shows typical *sustained averaged peak flows* and *sustained average low flows* derived from available records for a long durations of 1 to 8 hours :

3.5. STORM WATER FLOW

When rainfall takes place, a part of it infiltrates or percolates into the ground surface while the remaining flows over the land depending upon permeability of the ground, its surface slope and many other factors. The amount of water flowing over the ground surface, pavements, house roofs etc. is commonly known as 'runoff' or the *storm water*. This storm water is ultimately drained through the sewers, otherwise the streets, roads etc. would be flooded. The amount of peak storm water flow may be several times more than D.W.F. In the case of combined system, the sewers are normally not designed for the peak storm water flow, otherwise the size of the sewers would be alarmingly large. During ordinary or normal rain, the sewers run full of storm water. However, during heavy rains, the peak storm water is allowed to accumulate for sometime, while the sewers continue to run full of storm water. After the rains are over, the accumulated storm water is discharged gradually through the sewers. The storm water flow is also known as wet weather flow and is abbreviated as W.W.F.

The runoff or the storm water flow depends upon the following factors : (*i*) Catchment area (*ii*) Ground slope (*iii*) Permeability of ground (*iv*) Extent of impervious area such as buildings, paved yards, non-absorbent road surface etc. (*v*) Extent of vegetation growth (*vi*) Rain fall intensity (*vii*) Rainfall duration (*viii*) Condition of ground prior to the rainfall (*ix*) Concentration or compactness of catchment area. (*x*) Climatic conditions such as wind, humidity, temperature etc.

For the purposes of estimating storm water flow for sewer design, the following two methods are commonly followed

(*a*) Rational method (*b*) Empirical formulae.

3.6. RATIONAL METHOD

The rational formula is most commonly used for design of storm drains. In takes into account the following three factors :

(*i*) Catchment area A (*ii*) Impermeability factor (I) of the catchment area. (*iii*) Intensity of rainfall (R)

The rational formula, can be expressed as follows in its generalised form :

$$Q = K . A . I . R_i \qquad \text{...(3.3)}$$

where Q = run off or storm water flow

K = constant which permits the expression of the factors A, I and R in convenient units.

Let Q = runoff in cubic metres per second (cumec)

A = catchment area in hectares

A_i = impervious area $= A \times I$

R_i = Intensity of rainfall in mm per hour.

In order to find corresponding value of factor K, we have

$$Q = (10^4 A)(I)\left(\frac{R_i}{1000 \times 3600}\right)$$

or

$$Q = \frac{AIR_i}{360} = \frac{A_i R_i}{360} \qquad \text{...(3.4)}$$

In other words, $K = \dfrac{1}{360}$

The rational formula (Eq. 3.4) is also sometimes expressed in the following form :

$$Q = \frac{1}{360}\ C.R_i A. \qquad \text{...(3.4 a)}$$

where C is called the *coefficient of runoff*.

If, however Q is expressed litres per second, A in hectare, and R_i in cm per hour, we have

$$Q = (10^4 A)(I)\left(\frac{R_i}{100 \times 3600}\right) \times 1000$$

or

$$Q = \frac{250}{9} AIR_i \approx 28\, AIR_i \qquad \text{...(3.5)}$$

(1) Catchment area. The three factors operative in Eqs. 3.3, 3.4 and 3.5 are : catchment area, impermeability factor (I) and the intenisity of rainfall (R_i). The catchement area served by a given storm water sewer can be found directly form the map of the town showing the positions of streets, houses, play grounds, sewers etc. However, since the factor I depends upon the type of the surface, the catchment areas of different types of surfaces, as indicated in Table 3.4, should be found separately.

(2) Impermeability factor. The storm water flow depends upon the imperviousness of the surface over which rainfall takes place. If the ground is relatively impervious, more runoff takes place. The percentage of rain water that is available in the form of runoff is known as *impermeability factor* or runoff coefficient. Table 3.4

gives Kuichling's impermeability factors, while Table 3.5 gives Fruhling's impermeability factors for various types of surfaces. For design purposes the factor I must be estimated not as it is at present but what it will be after development at the end of 30 years design period.

TABLE 3.4. KUICHLING'S IMPERMEABILITY FACTORS

	Type of surface	*Factor I*
1.	Water tight roof surface	0.7 to 0.95
2.	Asphaltic pavement in good order	0.85 to 0.90
3.	Stone, brick and wood block pavements with tightly cemented joints	0.75 to 0.85
4.	Same as above, with open or uncemented joints	0.50 to 0.70
5.	Inferior block pavements with open joints	0.40 to 0.50
6.	Macadamized road ways	0.25 to 0.60
7.	Gravel road ways and walks	0.15 to 0.30
8.	Unpaved surfaces, rail road yards, and vacant lots	0.10 to 0.30
9.	Parks, gardens, lawns, and meadows, depending on surface slope and characteristics of subsoil	0.05 to 0.25
10.	Wooded area or forest land, depending upon surface slope and charcteristics of subsoil	0.01 to 0.20
11.	Most densely populated or built-up portion of a city	0.70 to 0.90

TABLE 3.5. FRUHLINGS FACTORS

	Type of Locality	*Factor I*
1.	Business areas	0.85
2.	Areas closely built up	0.75
3.	Areas with 50% attached houses and 50% detached houses	0.65
4.	Suburban area with detached houses	0.45 to 0.55
5.	Extremely suburban areas with 20 to 60% parking and widely detached houses	0.35

Average Impermeability Factor. As stated earlier, a given catchment area may consist of various types of surfaces for which different impermeability factors are applicable. The average impermeability factor for the whole area can be computed as follows :

Let $A_1, A_2 \ldots\ldots A_n$ = areas of the different surface of the catchment area.

$I_1, I_2 \ldots\ldots I_n$ = corresponding impermeability factors for the different surfaces. Then,

$$\text{Impermeable area} = A_1 I_1 + A_2 I_2 + \ldots\ldots A_n I_n = \Sigma A \cdot I$$

Hence $$I_{av} = \frac{A_1 I_1 + A_2 I_2 ..... A_n I_n}{A_1 + A_2 + ... A_n} = \frac{\Sigma A I}{\Sigma A} \quad ...(3.6)$$

(3) Intensity of rainfall. The value of factor R_i *i.e.* intensity of rainfall in mm/hour (or cm/hour) can be worked out from the rainfall records of the area. However R also depends upon *frequency* and the *duration* of the storm.

(i) Storm frequency : The frequency of storm for which the sewers are to be designed depends upon the importance of the drainage area. Commercial and highly priced areas should be subject to less frequent flooding than the residential area while the other unimportant areas can be subject to more frequent flooding. Manual suggests the following values of frequency :

(*a*) *Residential Areas:*

(*i*) Peripheral areas : Twice a year

(*ii*) Central and comparatively area priced high : One a year

(*b*) *Commercial and high priced areas* : One in two years.

(ii) Duration of storm : The *duration of storm* is taken equal to the *time of concentration*. It is the time required for the run-off to be contributed to the point of concentration from the entire area.

Fig. 3.8 shows typical rainfall intensity curves for various frequencies of storm. The value of the rainfall intensity (R_i) can be determined for any given time of concentration (t_c), using these curves. Where rainfall records are not available, the following empirical formula may be used for computing R_i in mm/hour. In all the formulae, R_i is inversely proportional to t, i.e. shorter the duration of rainfall, greater will be the intensity during that period.

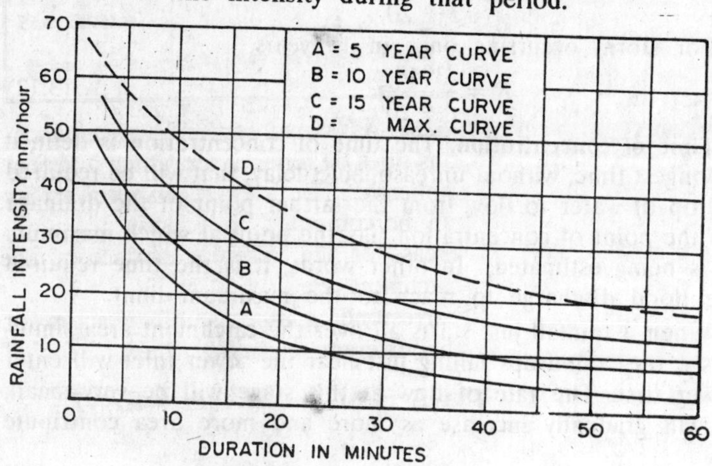

FIG. 3.8. RAINFALL INTENSITY CURVES.

(a) General formula :

$$R_i = \frac{25.4\,a}{t + b} \quad ...(3.7)$$

where R_i = rainfall intensity in mm/hour.

t = duration of storm in minutes≈time of concentration.

a, b = constants.

The United States Ministry of Health recommend the following values of constants a and b

Duration of storm	*constant a*	*constant b*
5 to 20 min.	30	10
20 to 100 min.	40	20

(b) For areas of heavy and frequent rainfall

$$R_i = \frac{3430}{t + 18} \quad ...(3.8)$$

The above formula gives rainfall intensity of frequency once in 5 years.

(c) For storm occurring once in a year

$$R = \frac{150}{t^{0.625}} \quad ...(3.9)$$

(d) For storm occurring once in 10 years.

$$R = \frac{380}{t^{0.5}} \quad ...(3.10)$$

(e) Kuichling's formulae

For storm occurring once in 10 years,

$$R = \frac{2667}{t + 20} \quad ...(3.11)$$

For storm occurring once in 15 years,

$$R = \frac{3048}{t + 20} \quad ...(3.12)$$

Time of concentration. The time of concentration is defined as the longest time, without unreasonable delay, that will be required for a drop of water to flow from the farther point of the drainage area to the point of concentration (i.e. the point at which maximum runoff is being estimated). In other words, it is the time required for the flood discharge to reach to the maximum limit.

When a rainfall just starts all over the catchment area simultaneously, the raindrops falling just near the sewer inlet will enter the sewer first. The rate of flow at this stage will be very small, but it will gradually increase as more and more area contribute to the flow. Finally, when the whole area is contributing, maximum rate of runoff will be obtained, which will be equal to rate of precipitation over the whole of the impevious area. The time required from the beginning of rainfall to the one corresponding to the achievement of maximum rate of runoff is called the *time of concentration*. This maximum rate of runoff will continue till the rainfall stops. After that, the runoff will gradually decrease.

The time of concentration consists of two components : (*a*) time of entry, t_e and (*b*) time of flow, t_f.

The *time of entry* or the *inlet time* is the time taken by the rainfall to run from the most distant point of the watershed to the inlet of the sewer. The areas tributary to most of the sewers are usually small. Hence it is customary to assume suitable value of time of entry (t_e) based on experience under similar conditions. The time of entry decreases with increase in slope and imperviousness of the ground, while it increases with distance and storage conditions of the ground. A normal practice is to assume it between 3 to 20 minutes, the lower value being adopted for the steepest slope while higher value is adopted for grassy plots.

The *time of flow* is the time required for the flow of water in the sewer to the point under consideration. It is computed by assuming the velocity of flow in the drain and measuring the length of drain or sewer from inlet point to the point under consideration, *i.e.* $t_f =$ distance/velocity. The time so computed is usually less than the actual time for the flood crest to reach the point of concentration from the sewer inlet because of the time required to fill the sewer. The time required to fill the sewer is known as the *time of storage*. This time may be appreciable, but is usually neglected.

Thus the total time of concentration ($t_c = t_e + t_f$) can be computed. It is to be noted that larger the catchment, the longer will be the time of concentration. *It is generally considered that the maximum runoff occurs when the duration of storm is equal to the time of concentration.* This is so because if the duration of storm is shorter than the time of concentration, the whole of the catchment will not be contributing water, though the intensity of rainfall will be more than that obtainable for a storm of longer duration. The result will be that lesser runoff will be obtained. If, on the other hand, the duration of storm is longer than the time of concentration, the rainfall intensity will be less, resulting in lesser runoff.

Based on the above discussions, Lloyed Davis arrived at the following conclusions :

(*i*) The total volume of storm water is proportional to the maximum rate of flow. (*ii*) The total discharge of the storm water is proportional to the maximum rainfall during the time of concentration. (*iii*) The storm water discharge of any particular area is proportional to the % of impermeable area expressed as a proportion to the total area. (*iv*) The maximum rate of flow would be obtained when maximum precipitation occurs in the minimum time of concentration, i.e. when the duration of storm is equal to the time of concentration.

These conclusions were utilized by Lloyed Davis in evolving the *rational formula* in the following form :

$$Q = \frac{1}{360}\left(\frac{60\,r}{t_c}\right) A\,I \ \ \text{m}^3/\text{sec.}$$

or

$$Q = \frac{r}{6\,t_c} . AI \qquad ...(3.15)$$

or

$$Q = \frac{r}{6\,t_c} . P \ \text{cumecs} \qquad ...(3.16)$$

where $\frac{60\,r}{t_c} = R_i$ = intensity of rain fall in mm/hour.

t_c = time of concentration in minutes

r = total rainfall in mm, during the time of concentration

A = catchment area in hectares

I = impermeability factor.

P = impermeable area in hectares, out of the total catchment area $= A\,I$

Effect of storm duration on imperviousness. Gregony has stated that the impermeability factor (given in Table 3.4) is a function of duration of storm. If the duration of storm is *less* than time of concentration (t_c), in minutes,

$$I = 0.175\,t_c^{1/3} \qquad ...(3.13)$$

If, however, duration of storm T (in minutes) is greater than the time of concentration,

$$I = \frac{0.175}{t_c}\left(T^{4/3} - t_c^{4/3}\right) \qquad ...(3.14)$$

However, such refinements are normally not used in actual practice.

Time Area Graph : A time area graph is obtained by plotting the time (in minutes) after commencement of storm along the x–axis and the impervious area (in hectares) contributing to the sewer, such as the one shown in Fig. 3.9 (*a*). For the drainage area shown in Fig. 3.9(*a*), let P be the point of measurement. At the commencement of the rainfall (or storm), rain water will start arriving at P only from area A_1, which is in immediate vicinity of point P. As the time increase, water from other areas A_2 , A_3 and A_4 will start arriving.

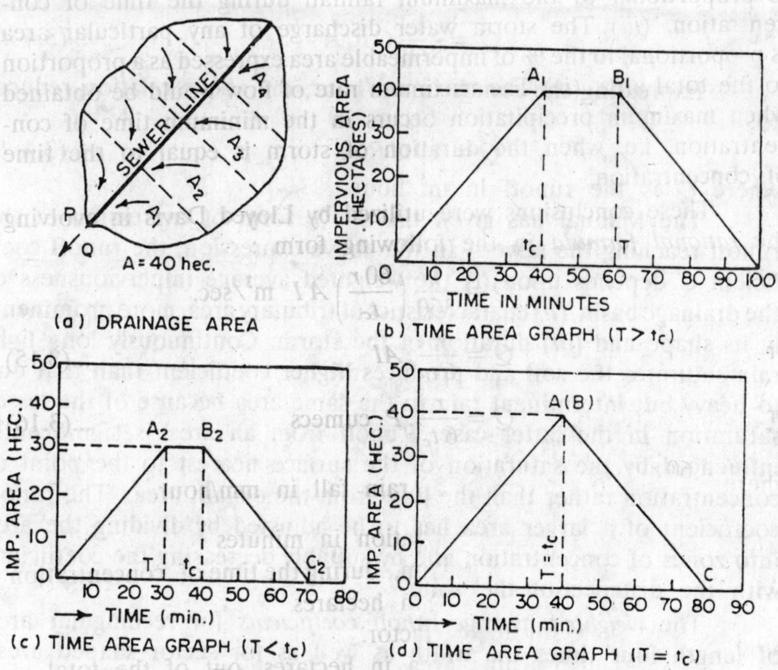

FIG. 3.9. TIME-AREA GRAPHS.

At the time of concentration (t_c) the whole of the area $A (= A_1 + A_2 + A_3 + A_4)$ to the sewer at P, provided the period of rainfall (T) is greater than t_c. Thus, at point A_1 of the graph, the rate of runoff will reach its maximum value and if $T > t_c$, the rate of flow will continue at constant rate upto point B_1 corresponding to the time T of the rainfall. At this point of time, the rainfall stops and hence flow will decrease, as indicated by line, $B_1 C_1$ where point C_1 is such that $OC_1 = T + t_c$.

Fig. 3.9 (c) shows the time-area graph for a situation when the time of rainfall T is less than t_c. For such a case, the storm stops before the whole area starts contributing. Hence rate of runoff (corresponding to point A_2) for such a case will be much less. The flow remains constant from point A_2 (at time T) to point B_2 (at time t_c), and thereafter, it will decrease as indicated by line $B_2 C_2$ when point C_2 is such that $OC_2 = t_c + T$. Fig. 3.9 (d) shows the *ideal case* when the time of rainfall (T) is equal to time of concentration. As indicated earlier, maximum runoff will result for such a case. The point C is such that $OC = t_c + t_c = 2\,t_c$.

Horner's runoff coefficients. As stated earlier, the rational formula can also be expressed in the following form :

$$Q = \frac{1}{360} C R_i A \qquad \text{...(3.4 a)}$$

Expressing the runoff in m³/hour, the above relation reduces to

$$Q = 10\, C R_i A \qquad \text{...(3.4 b)}$$

where Q is the runoff in m³/hour.

The Manual has given the above form of expression of the runoff reaching the sewer. In the above expression, the runoff coefficient C depends upon (*i*) the weighted average imperviousness of the drainage basin, (*ii*) characteristics of tributary area, more prominently its shape and (*iii*) duration of the storm. Continuously long light rain saturates the soil and produces higher coefficient than that due to heavy but interminent rain in the same area because of the lesser saturation in the latter case. Runoff from an area is significantly influenced by the saturation of the surface nearest to the point of concentration rather than the flow from the distant area. The runoff coefficient of a larger area has to be adjusted by dividing the area into zones of concentration and by suitably decreasing the coefficient with the distance of the zone.

The *weighted average runoff coefficients* for rectangular area of length four times the width as well as for sector shaped areas with varying percentages of impervious surface for different times of concentration (t_c) are given in Table 3.6. These values can also be applied to other areas generally encounterd, since the resulting errors will be within the limits of accuracy of the rational method and the assumptions on which it is based.

3.7. EMPIRICAL FORMULAE

Rational formula is used only when the area draining water into the sewer is small, say smaller than 400 hectares. For larger area, empirical formulae are more suitable, since great care and judgment are required for suitably selecting I and R_i in the rational formula. However, the empirical formulae given below are applicable under certain specific conditions. In other words, they are local in nature and they should be used only when certain specific conditions are fulfilled. The various empirical formulae use the following variables:

(*i*) catchment area A, in hectares (*ii*) rate or intensity of rainfall, in mm per hour over entire area (*iii*) relative imperviousness (*iv*) slope of ground, S in metres per 1000 m length

(*i*) **Burkli-Zeiglar formula.** This is the earliest form of empirical formula, originally given by a Swiss engineer for local conditions, but later adopted in U.S.A.

$$Q = 296\, C R_i A \left(\frac{S}{A}\right)^{1/4} \qquad \text{...(3.17)}$$

TABLE 3.6.
WEIGHTED AVERAGE RUNOFF COEFFICIENTS C (AFTER HORNER)

Duration t → *(minutes)*	*10*	*20*	*30*	*45*	*60*	*75*	*90*	*100*	*120*	*135*	*150*	*180*
1. *Sector concentrating in stated time*												
(*i*) Impervious	0.525	0.588	0.642	0.700	0.740	0.771	0.795	0.813	0.828	0.840	0.850	0.865
(*ii*) 60% impervious	0.365	0.427	0.477	0.531	0.569	0.598	0.622	0.641	0.656	0.670	0.682	0.701
(*iii*) 40% impervious	0.285	0.346	0.395	0.446	0.482	0.512	0.535	0.554	0.571	0.585	0.597	0.618
(*iv*) Pervious	0.125	0.185	0.230	0.277	0.312	0.330	0.362	0.382	0.399	0.414	0.429	0.454
2. *Rectangle (Length = 4 × Width) concentrating in stated time*												
(*i*) Impervious	0.550	0.648	0.711	0.768	0.808	0.837	0.856	0.869	0.879	0.887	0.892	0.903
(*ii*) 60% impervious	0.350	0.442	0.499	0.551	0.590	0.618	0.639	0.657	0.671	0.683	0.694	0.713
(*iii*) 40% impervious	0.269	0.360	0.414	0.464	0.502	0.530	0.552	0.572	0.588	0.601	0.604	0.636
(*iv*) Pervious	0.149	0.236	0.287	0.334	0.371	0.398	0.422	0.445	0.463	0.479	0.495	0.522

where Q = runoff in litres per second.
R_i = maximum rate of rainfall in mm/hour over the entire catchment
A = catchment area in hectares
S = slope of ground surface in metres per 1000 m.
C = a constant expressing the nature of the ground surface or relative imperviousness. It may vary between 0.5 to 0.9, though an average value of 0.7 is recommended.

***(ii)* McMath's formula.** This formula was devised for local conditions of St. Louis in U.S.A. using the same notations as above:

$$Q = 292\, C R_i A \left(\frac{S}{A}\right)^{1/5} \quad ...(3.18)$$

The value of C may vary between 0.3 to 0.9, though an average value of 0.75 is recommended.

***(iii)* Metcalf and Eddy's formula** :

$$Q = 28.32\left[\frac{25000}{2.47A + 125} + 15\right] \quad ...(3.19)$$

The above formula contains only the expression for catchment area.

The above three formulae give satisfactory results only when the catchment area is of limiting size of, say, 800 hectares. For larger catchment areas, the following formulae are generally used.

***(iv)* Fanning's Formula**

$$Q = 3125\, A_k^{5/8} \quad ...(3.20)$$

where A_k = area is square km.

***(v)* Talbot's Formula**

$$Q = 87000\, A_k^{1/4} \quad ...(3.21)$$

***(vi)* Dicken's Formula.** This formula is best suited for conditions obtainable in North India.

$$Q = 14\, C A_k^{3/4} \quad ...(3.22)$$

The value of constant C are as follows :

C = 250 for very large areas
= 850 for areas of average size, with annual rainfall varying between 600 to 1200 mm.
= 1600 for small areas.

***(vii)* Ryve's formula.** This is suitable for conditions prevailing in South India.

$$Q = 15\, C A_k^{2/3} \quad ...(3.23)$$

The value of constant C may vary between 450 to 625.

(viii) Inglis's formula. $Q = \frac{123100\, A_k}{\sqrt{A_k + 10.4}}$ ...(3.24)

This formula was adopted for old Bombay Presidency area.

Example 3.1. *A certain district of a city has a projected population of 50,000 residing over an area of 40 hectares. Find the design discharge for the sewer line, for the following data :*

(i) Rate of water supply = 200 litres per capita per day.

(ii) Average impermeability coefficient for the entire area = 0.3

(iii) Time of concentration = 50 minutes

The sewer line is to be designed for a flow equivalent to the wet weather flow (W.W.F.) plus twice the dry weather flow (D.W.F.). Use U.S. ministry of Health formula. Assume that 75% of water supply reaches in sewer as wastewater.

Solution :

The sewage flow is equal to 75% of rate of water supply. Hence sewage flow will be equal to $0.75 \times 200 = 150$ litres/capita/day.

$\therefore$ Sewage flow (D.W.F.) $= \frac{50000 \times 150}{24 \times 60 \times 60} = 86.8$ litres/seconds.

The rainfall intensity is given by

$$R_i = \frac{25.4\, a}{t + b} \text{ mm/hour} \quad ...(3.7)$$

Here $t = 50$ minutes ; $a = 40$; $b = 20$

$$\therefore \quad R_i = \frac{25.4 \times 40}{50 + 20} = 14.5 \text{ mm/hour} = 1.45 \text{ cm/hour}$$

The W.W.F. is given by

$$Q = 28\, A.I.R_i$$

$$= 28 \times 40 \times 0.3\, (1.45) = 487.2 \text{ litres/sec.}$$

Hence the design discharge is given by

$$Q = 2\,(\text{D.W.F.}) + \text{W.W.F.}$$

$$= 2\,(86.8) + 487.2 \approx \mathbf{661 \text{ litres/second}}$$

Comment : Ratio of D.W.F. and W.W.F. $= \frac{86.8}{487.2} = \frac{1}{5.6}$.

Since this ratio is not very large, it is preferable to use a combined sewer system.

Example 3.2. *A sewer has a catchment area of 50 hectares. Estimate the storm water flow corresponding to a rainfall of 40 mm during a time of concentration of 30 minutes. Assume that the impervious area is equal to 55% of the total catchment area. Use Lloyed Davis formula.*

Solution.

$$\text{W.W.F} = \frac{r}{6\,t_c} P \text{ cumecs} \qquad ...(3.16)$$

where r = total rainfall, in mm, during time of concentration = 40 mm

t_c = time of concentration = 30 min.

P = impervious area in hectares = 0.55 × 50 = 27.5 hectares.

$$\therefore \quad \text{W.W.F.} = \frac{40}{6 \times 30} \times 27.5 \text{ m}^3\text{/sec.}$$

$$= 6.1 \text{ cumecs} \approx \mathbf{6100\ litres/sec.}$$

Example 3.3. *The following data is available regarding various types of area and the corresponding impermeability factors of a town:*

	Type	*% Area*	*Impermeability coefficient*
1.	*Roofs*	*15%*	*0.9*
2.	*Pavements*	*20%*	*0.8*
3.	*Lawns, gardens, vegetation*	*40%*	*0.15*
4.	*Unpaved*	*15%*	*0.20*
5.	*Wooded*	*10%*	*0.05*

Determine the average coefficient of runoff.

If the total area of the district is 20 hectares, determine the maximum storm water flow for a rainfall intensity of 50 mm/hour having a frequency of once in five years. Use rational formula.

Solution

The average impermeability factor I_{av} is given by Eq. 3.6

$$I_{av} = \frac{\Sigma A\,.I}{\Sigma A} \qquad ...(3.6)$$

where

$$A_1 I_1 = (0.15\,A)\,0.9 = 0.135\,A$$
$$A_2 I_2 = (0.20\,A)\,0.8 = 0.160\,A$$
$$A_3 I_3 = (0.40\,A)\,0.15 = 0.060\,A$$
$$A_4 I_4 = (0.15\,A)\,0.20 = 0.030\,A$$
$$A_5 I_5 = (0.10\,A) \times 0.05 = 0.005\,A$$

$$\therefore \quad \Sigma AI = (0.135 + 0.160 + 0.060 + 0.030 + 0.005)\,A = 0.390\,A$$

$$\therefore \quad I_{av} = \frac{0.390\,A}{A} = \mathbf{0.39}$$

The storm discharge is given by Eq 3.4

$$Q = \frac{A.I.R_i}{360} \text{ m}^3\text{/sec.}$$

where R_i = intensity of rainfall in mm/hour = 50 mm/hour

I = average impermeability factor = 0.39.

$$A = \text{catchment area in hectares} = 20 \text{ hectares.}$$

$$\therefore \quad Q = \frac{20 \times 0.39 \times 50}{360} = \mathbf{1.08 \ m^3/sec = 1080 \ litres/sec.}$$

Example 3.4. *The drainage area of one sector of a town is 20 hectares. The classification of the surface of this area is as follows.*

Percentage of total surface area	*Type of surface*	*Runoff coefficient*
25%	Hard pavement	0.85
25%	Roof surface	0.80
15%	Unpaved street	0.30
25%	Gardens and lawns	0.15
10%	Wooded area	0.10

If the time of concentration for the area is 30 minutes, find the maximum runoff. Use the following formula for intensity of rainfall:

$$R_i = \frac{900}{t + 60}$$

Solution

$$C_1A_1 = 0.85 \times 0.25\,A = 0.2125\,A$$
$$C_2A_2 = 0.80 \times 0.25\,A = 0.2000\,A$$
$$C_3A_3 = 0.30 \times 0.15\,A = 0.0450\,A$$
$$C_4A_4 = 0.15 \times 0.25\,A = 0.0375\,A$$
$$C_5A_5 = 0.10 \times 0.10\,A = 0.0100\,A$$

$$\text{sum } 0.5050\,A$$

$$\therefore \quad C_{av} = \frac{\Sigma CA}{A} = \frac{0.5050A}{A} = 0.505$$

Now $$R_i = \frac{900}{t+60}, \text{ where } R_i \text{ is in mm/hour}$$

$$= \frac{900}{30 + 60} = 10 \text{ mm/hr.}$$

Hence from Eq. 3.4 (*a*)

$$Q = \frac{1}{360} C R_i A.$$

$$= \frac{1}{360} \times 0.505 \times 10 \times 20 = \mathbf{0.2806} \textit{ cumecs}$$

Example 3.5. *An impervious area of 20 hectares is to be served by a sewer line. This area is subject to three storms P_1, P_2 and P_3 with equal frequency of occurrence. The duration of the three storms is 20 minutes, 30 minutes and 35 minutes respectively. Taking the time* *of concentration equal to 30 minutes, determine the maximum runoff from each storm.*

Solution

(a) Storm P_1 of 20 minutes duration :

The intensity is rainfall is given by Eq. 3.7 :

$$R_i = \frac{25.4\,a}{t+b}$$

For storm of 5 to 20 minutes duration, $a = 30$ and $b = 10$. Also, $t = 20$ minutes for the present storm.

$$\therefore \quad R_i = \frac{25.4 \times 30}{20+10} = 25.4 \text{ mm/hour.}$$

Now, from Eq. 3.4,

$$Q = \frac{AIR_i}{360} = \frac{A_i R_i}{360}$$

where A_i = impervious area $= A \times I$ (hectares).

$$\therefore \quad Q = \frac{20 \times 25.4}{360} = 1.411 \text{ cumecs.}$$

However, the storm lasts only for 20 minutes (*i.e.* $T = 20$ minutes) while the time of concentration (t_c) is 30 minutes. Hence the storm stops much before the whole area starts contributing.

$\therefore$ Max. flow from storm $P_1 = 1.411 \times \frac{20}{30} =$ **0.941 cumecs** ...(*i*)

(b) Storm P_2 of $T_2 = 30$ minutes $(= t_c)$

For storms of 20 to 100 minutes duration, $a = 40$ and $b=20$

$$\therefore \quad R_i = \frac{25.4 \times 40}{30+20} = 20.32 \text{ mm/hour}$$

$$\therefore \quad Q = \frac{A_i R_i}{360} = \frac{20 \times 20.32}{360} = \mathbf{1.129 \text{ cumecs}} \qquad ...(ii)$$

Since $T = t_c$, the whole area starts contributing.

(c) Storm P_3 of $T_3 = 35$ minutes

$$R_i = \frac{25.4 \times 40}{35+20} = 18.473 \text{ mm/hour}$$

$$Q = \frac{A_i R_i}{360} = \frac{20 \times 18.473}{360} = \mathbf{1.026 \text{ cumecs}} \qquad ...(iii)$$

From (*i*), (*ii*) and (*iii*), we find that $Q_{max} = 1.129$ cumecs, and this occurs when the duration of storm is equal to the time of concentration. The results for the three storms are illustrated graphically in Fig. 3.10.

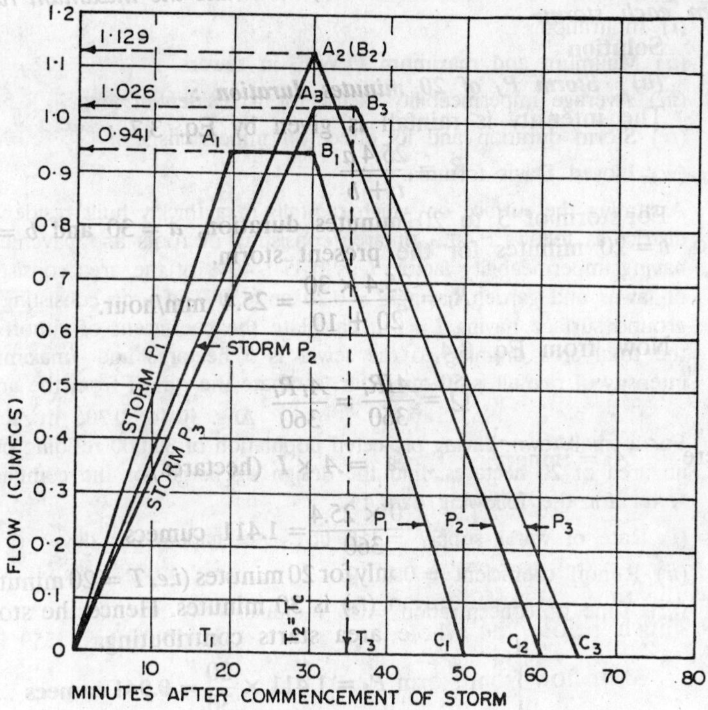

FIG. 3.10.

PROBLEMS

1. What do you understand by dry weather flow. ? Discuss in brief various factors affecting the dry weather flow.
2. (*a*) Discuss the variations of dry weather flow.

   (*b*) Discuss the variation in rate of sewage. What are its effects on the design of a sewer ?
3. State the factors on which the storm water flow on an area depends.
4. Explain the rational method of estimating storm water flow. Discuss the methods of determining various parameters used in the formula.
5. What do you understand by intensity of rainfall ? How do you determine this ?
6. Write a note on 'time of concentration' and its significance in determining the storm water flow.
7. Give various empirical formulae for determining storm water flow in sewers, explaining their applicability or suitability.

8. Write short notes on

(*i*) Infiltration

(*ii*) Minimum and maximum D.W.F. in sewers.

(*iii*) Average impermeability factor for a residential area.

(*iv*) Storm duration and its effect on imperviousness.

(*vi*) Lloyed Davis formula.

9. Assuming the surface on which rainfalls in a thickly built residential district as follows : 40% of area consisting of roofs and pavements having impermeability factor, I as 0.85 ; 40% of the area consisting of lawns and garden having $I = 0.20$ and 20% of area consisting of ground surface having $I = 0.1$, calculate the coefficient of runoff. If the total area draining to the sewer is 5 hectares and maximum intensity of rainfall is 50 mm/hour, estimate the runoff from the area.

Ans. [0.44; 0.306 m^3/sec]

10. For a small town, having projected population of 30,000 residing over an area of 20 hectares, find the design discharge for the combined sewer for the following data :

(*i*) Rate of water supply = 150 litres per capital per day.

(*ii*) Runoff coefficient = 0.4

(*iii*) Time of concentration = 30 minutes.

Ans. [559 lps]

4

Hydraulic Design of Sewers

4.1. INTRODUCTION

The sewage, to be transported through sewers, is mostly liquid (*i.e.* water), containing hardly 0.1 to 0.2 percent of solid matter in the form of organic matter, sediments and minerals. Hence the general approach for the design of sewers is similar to the design of water mains. However, there are two differences in the designs of the sewers and of the water mains :

(*i*) ***Presence of solid matter*** : Water flowing through the water mains is practically free from solid matter, while the sewage flowing thro' sewers contain particles of solid matter (both organic as well as inorganic). These solid particles settle at the bottom and have to be dragged during the sewage transport. In order that the sewers are not clogged, they are to be laid at such a gradient that *self cleansing velocity* is achieved, at all value of discharges. Also the inner surface of the sewers must be resistant to the abrasive action of these solid particles.

(*ii*) ***Pressure*** : Water in the water mains flow under pressure. Hence the water mains can be carried, within certain limits, up and down the hill or gradient. The hydraulic gradient line lies very much above the pipe surface. On the other hand in most cases, sewers may be considered as open channels, wherein the sewage runs under gravity. The sewers seldom run full, and the H.G. line falls within the sewer. Hence the sewers must be laid at continous downward gradient. Sewers run under pressure only when they are designed as force mains and as inverted siphons.

4.2. HYDRAULIC FORMULAE

The design of sewers is done of the basis of the following empirical hydraulic formulae : (1) Chezy's formula (2) Kutter's formula (3) Bazin's formula (4) Manning's formula (5) Crimp and Bruge's formula (6) Hazen and Willian's formula.

Apart from these formulae, there are various charts, diagrams, graphs and tables with the help of which the hydraulic design of sewers can be done. The factors that influence the flow of sewage in the sewers are : (1) Slope of sewer (2) Geometry of sewer (3) Roughness of interiors surface of sewer (4) Bends, transitions, obstructions etc. (5) Flow conditions and (6) Characteristics of sewage.

1. Chezy's formula. Chezy's (1775) gave the following formula:

$$V = C\sqrt{R.S} \qquad ...(4.1)$$

where V = velocity of flow (m/sec.)
S = hydraulic gradient or slope of the sewer
R = hydraulic mean radius (m) $= A/P$
A = area of cross-section (m^2)
P = wetted perimeter.
C = Chezy's constant

The Chezy's constant C is very complex in nature, and it depends upon several factors, such as roughness of inner surface of sewer, hydraulic mean radius, size and shape of sewer, slope etc. Generally, the value of Chezey's constant C is found either by Kutter's formula or by Bazin's formula. Knowing the velocity of flow V from Eq. 4.1, the channel section is designed by the general formula :

$$Q = A \times V$$

where Q = discharge in m^3/sec.

2. Kutter's formula. Kutter and Ganguillet (1869), two Swiss engineers, gave the following expression for Chezy's coefficient C:

$$C = \frac{23 + \dfrac{0.00155}{S} + \dfrac{1}{N}}{1 + \left(23 + \dfrac{0.00155}{S}\right)\dfrac{N}{\sqrt{R}}} \qquad ...(4.2)$$

where R = hydraulic mean radius ; S = slope.
N = rugosity coefficient the value of which depends upon the nature of inside surface of the sewer(Table 4.1)

Usually, the values corresponding to *fair condition* of the interiors surface are taken in the design. A reduction in the value of N has been reported with in the increase in the diameter of the conduit. For cement concrete pipes of dia. 600 mm and above, a value of $N = 0.013$ may be adopted.

TABLE 4.1. VALUES OF KUTTER'S OR MANNING'S COEFFICIENT N

	Conduit Material	*Condition of interior surface*	
		Good	***Fair***
1.	Salt glazed stoneware	0.012	0.014
2.	Cement concrete	0.013	0.015
3.	Cast Iron	0.012	0.013
4.	Brick, unglazed	0.013	0.015
5.	Asbestos cement	0.011	0.012
6.	Plastic smooth	0.011	0.011

3. Bazin's formula. Bazin (1897) gave the following formula for Chezy's constant:

$$C = \frac{157.6}{1.81 + \frac{K}{\sqrt{R}}} \quad ...(4.3)$$

where K = Bazin's constant, the value of which may be taken from Table 4.2.

TABLE 4.2 BAZIN'S CONSTANT K

S.N.	*Type or Nature of inside surface of sewer or drain*	*K*
1.	Very smooth surface	0.109
2.	Smooth brick or concrete surface	0.29
3.	Rough brick or concrete surface	0.833
4.	Smooth rubble masonry surface	1.54
5.	Good earthen channels	0.50
6.	Rough earthen channel	3.17

4. Manning's Formula. Manning (1890) gave the following expression for velocity of flow. The formula is widely used in U.S.A. as well as in India :

$$V = \frac{1}{N} R^{2/3} S^{1/2} \quad ...(4.4)$$

where V, N, R, S have the same meanings, as above. The value of rugosity coefficient N is the same as suggested by Kutter, and may be taken from Table 4.1.

5. Crimp and Bruge's formula. This formula is commonly used in England

$$V = 83.47 R^{2/3} S^{1/2} \quad ...(4.5)$$

where V, R and S have the same meanings, as defined earlier. Comparing this with Manning's formula, we have.

$$V = 83.47 R^{2/3} S^{1/2} = \frac{R^{2/3} S^{1/2}}{N}$$

which gives $N = 1/83.47 = 0.012$. Hence Manning's formula becomes Crimp and Bruge's formula when $N = 0.012$.

For a circular pipe, $R = \frac{A}{P} = \frac{\frac{\pi}{4}D^2}{\pi D} = \frac{D}{4}$.

$$\therefore \quad V = 83.47 \left(\frac{D}{4}\right)^{2/3} S^{1/2}$$

Now, $$Q = A \times V = \frac{\pi}{4} D^2 \times 83.47 \left(\frac{D}{4}\right)^{2/3} S^{1/2}$$

or $$Q = 26.02\, D^{8/3}\, S^{1/2} \qquad ...(4.6)$$

6. Hazen and William's formula. Allen Hazen and G.S. Williams (1902) gave the following formula, which is mostly used for flow under pressure :

$$V = 0.85\, C\, R^{0.63}\, S^{0.54} \qquad ...(4.7)$$

The coefficient C may be taken from Table 4.3.

TABLE 4.3. HAZEN AND WILLIAM'S COEFFICIENT C

S.N.	*Type of material*	*C*
1.	Steel pipe under future conditions	95
2.	Old C.I. pipes; brick sewers in good condition	100
3.	Stoneware pipes in good condition	110
4.	Cement lined pipes	110
5.	New riveted steel pipe	110
6.	Wood stave pipe	120
7.	New C. I. pipes	130
8.	Pipes with very smooth inside surface	140
9.	Asbestos cement pipes	140

4.3. NOMOGRAMS / TABLES FOR HYDRAULIC COMPUTATIONS

The various flow formulae described in the previous article are quite complex and the computations are quite lengthy. The computational work becomes stil' more lengthy while designing a full fledged sewerage scheme. In order to reduce computational work, ready made charts, nomograms and tables are available, based on various flow formulae discussed in the previous article.

1. Nomograms based on Mannings formula. Fig. 4.1, 4.2 and 4.3 shows the nomograms for solution of Mannings formula for circular pipes running full, for $N = 0.013$ and for discharge ranges of 17, 170 and 1700 cumecs respectively. The method of use of the nomograms is explained on Fig. 4.1. For example, let it be required to carry a discharge of 4.0 m^3/min with an average slope of 0.002 (*i.e.* 2 in 1000). A line *AB* is then drawn by placing the straight edge on point A corresponding to $Q = 4\ m^3/min$ on the quantity scale and on point B corresponding to $S = 0.002$ on the slope scale. The straight line AB cuts the diameter scale at 350 mm and the velocity scale at 0.68 m/s. Thus, for the given data of $Q = 4\ m^3/min$ and $s = 0.002$, the diameter of the pipe will be 350 mm and the resulting velocity of flow will be 0.68 m/s when flowing full.

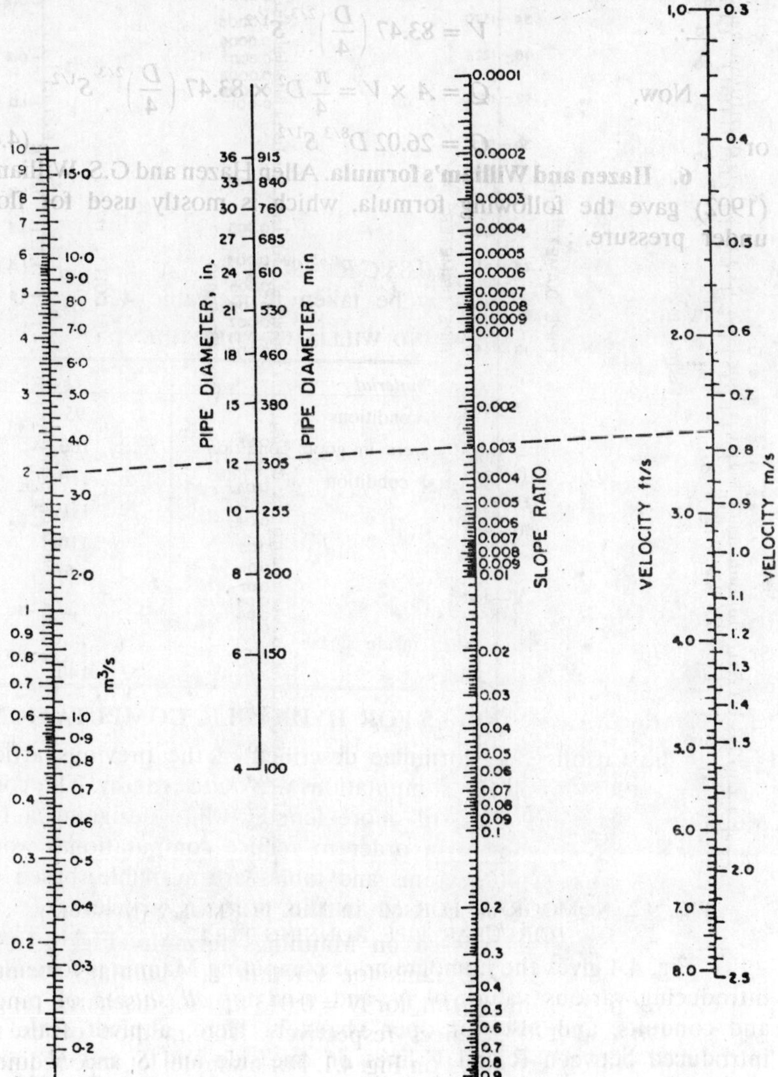

FIG. 4.1. NOMOGRAM FOR MANNINGS FORMULA ($N = 0.013$)
(CIRCULAR PIPE RUNNING FULL)

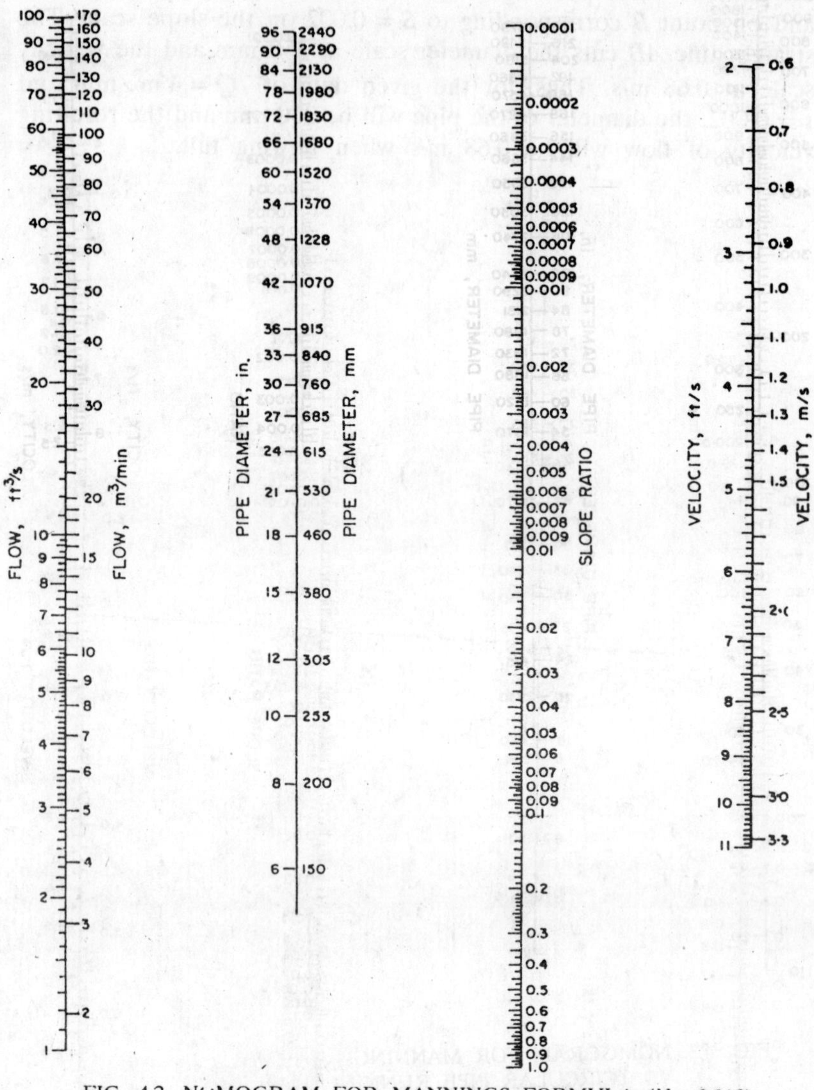

FIG. 4.2. NOMOGRAM FOR MANNINGS FORMULA ($N = 0.013$) (CIRCULAR PIPE RUNNING FULL)

Fig. 4.4 gives the nomogram for computing Mannings formula, introducing various values of N, and used for *all shapes* of pipes and conduits, and also for open channels. Here, a pivot scale is introduced between R and V lines on one side and S and N lines of on the other side. The method of use is illustrated on the diagram itself. Generally, out of the four parameters (*i.e.* R, V, S and N),

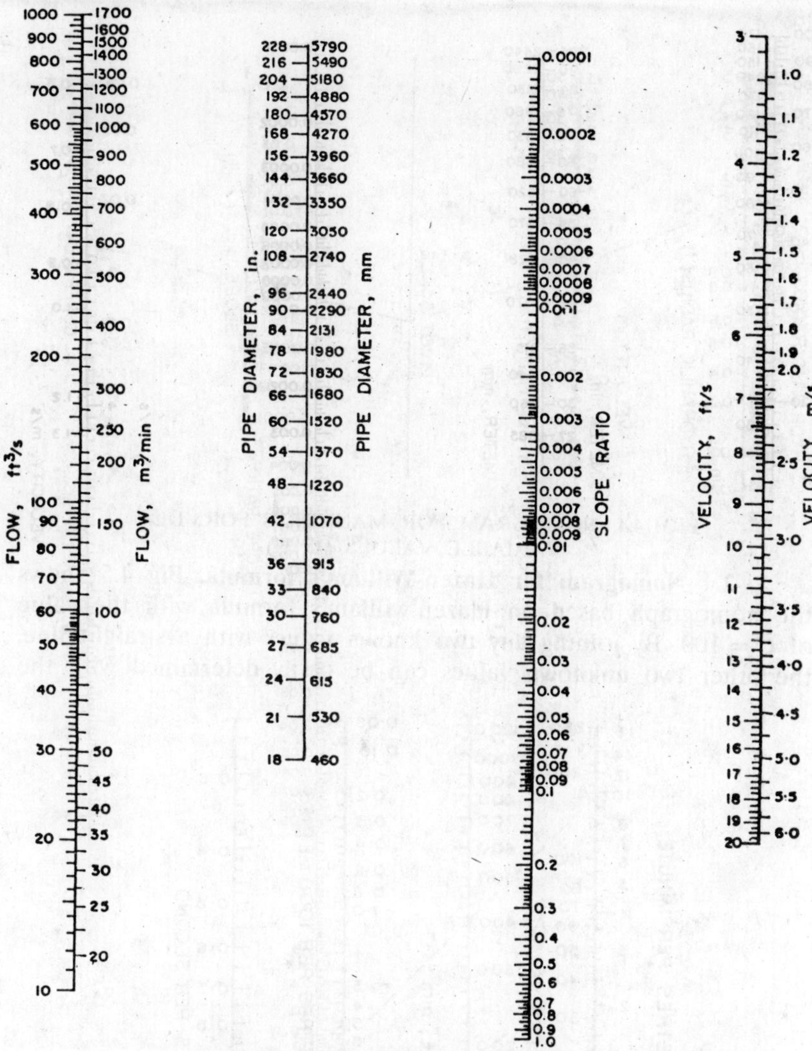

FIG. 4.3. NOMOGRAM FOR MANNING'S FORMULA (N=0.013) (CIRCULAR PIPE RUNNING FULL)

two are known (or assumed) and the other parameters are found from the nomogram. For example, let $N = 0.015$ and $S = 0.002$. Draw a line AB, keeping B corresponding to $S = 0.002$ and A corresponding to $N = 0.015$. This line will cut the pivotal scale at point P. Let the diameter of the conduit be 2 m for which $R = 0.5$. Hence join point P to point C corresponding to $R = 0.5$ m. This will intersect V line at point D which gives $V = 1.75$ m/s. Hence the corresponding discharge will be 5.5 m^3/s when pipe is running full.

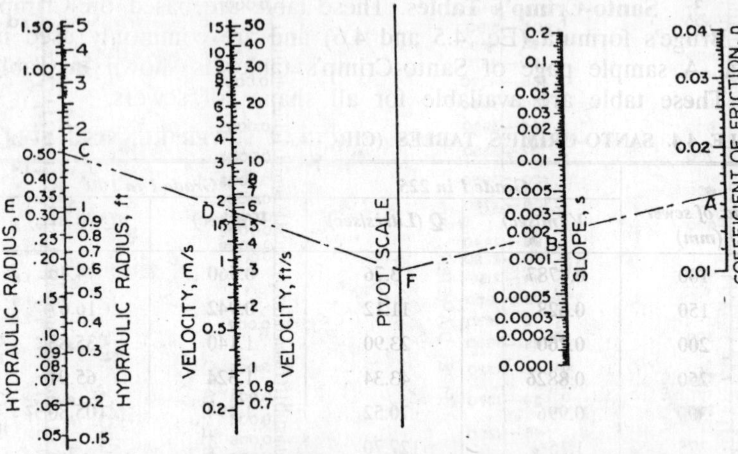

FIG. 4.4. NOMOGRAM FOR MANNING'S FORMULA (VARIABLE VALUES OF N)

2. Nomogram for Hazen-William's formula. Fig 4.5 shows the nomograph based on Hazen-william's formula with the value of $C = 100$. By joining any two known values with a straight line, the other two unknown values can be easily determined with the

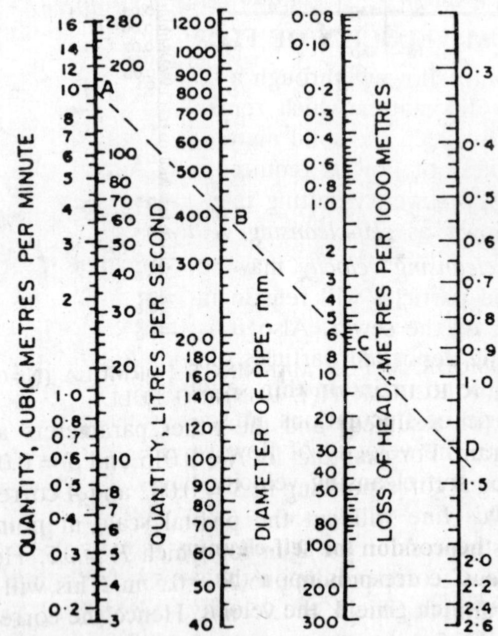

FIG. 4.5. NOMOGRAM FOR HAZEN-WILLIAM'S FORMULA (C=100)

help of the nomograph. For any other value of C, take the value from the chart and multiply it by $C/100$.

3. Santo-Crimp's Tables. These table are based on Crimps and Bruge's formula (Eq. 4.5 and 4.6) and are commonly used in India. A sample page of Santo-Crimp's tables is shown in Table 4.4. These table are available for all shapes of sewers.

TABLE 4.4. SANTO-CRIMP'S TABLES (CIRCULAR SEWER RUNNING FULL)

Dia. of sewer (mm)	*Grade 1 in 225*		*Grade 1 in 100*	
	V (m/sec)	*Q (Litres/sec)*	*V(m/sec)*	*Q(lit./sec.)*
100	0.4787	3.76	0.560	4.40
150	0.628	11.12	0.942	16.67
200	0.760	23.90	1.140	35.83
250	0.8826	43.34	1.324	65.01
300	0.996	70.52	1.494	105.80
375	1.156	127.70	1.735	191.70
450	1.306	208.00	1.960	312.30
525	1.447	313.30	2.170	470.10
600	1.582	448.1	2.373	672.0
675	1.711	611.2	2.566	918.5
750	1.837	812.5	2.754	1219.0
900	2.073	1320.0	3.200	2038.0

4.4. MINIMUM VELOCITY OF FLOW

The sewage flowing through a sewer contains organic as well as inorganic solid matter which remains suspended as the sewage flow. In order to keep the solid matter in suspended form, a certain minimum velocity of flow is required, otherwise the solid particles will settle in the sewer, resulting in its clogging. Such a minimum velocity is known as *self-cleansing velocity.*

A *self cleansing velocity* may be defined as that velocity at which the solid particles will remain in suspension, without settling at the bottom of the sewer. Also it is that velocity at which even the scour of the deposited particles of a given size will taken place. It is not possible to maintain thin self-cleansing velocity throughout the day because of fluctuations in sewage flow. During minimum flow of sewage, the velocity of flow is less than the self cleansing velocity. Hence self cleansing velocity should be maintained atleast once in a day.

Shield's expression for self-cleaning velocity. The self-cleansing property of sewage depends upon the scouring action of the flowing sewage. According to Shield, the velocity required to transport water-

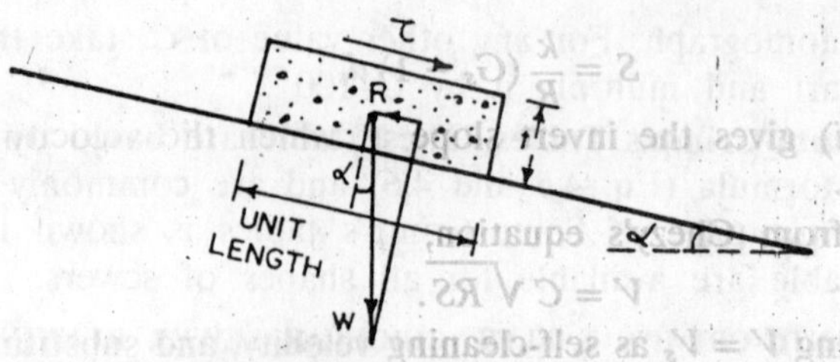

FIG. 4.6. FORCES ACTING ON SEDIMENT.

borne solids is the *self cleansing velocity* which can be determined with reference to Fig. 4.6.

Consider a layer of sediment of unit length, unit width and thickness t, deposited at the invert of the sewer. Let the slope of the sewer be α. The drag force or the intensity of tractive force (τ) exerted by the flowing water on a channel is given by :

$$\tau = \gamma_w . R . S \quad ...(i) \quad ...(4.8\,a)$$

where γ_w = unit weight of water at the prevailing temperature
R = hydraulic radius of filled section.
S = slope of the invert, or head loss per unit length of sewer.

The submerged unit weight of the sediment is given by

$$\gamma_{sub} = \gamma_w \left(\frac{G_s - 1}{1 + e} \right) = \gamma_w (G_s - 1)(1 - n) \quad ...(ii)$$

where G_s = specific gravity of sediment
e = voids ratio of sediment deposit.

$\therefore$ Weight of sediment of unit length and unit width is given by

$$W = \gamma_{sub} \times 1 \times 1 \times t$$

$$\therefore \quad W = \gamma_w (G_s - 1)(1 - n)\, t \quad ...(iii)$$

The frictional resistance R is given by

$$R \approx W \sin \alpha = \gamma_w (G_s - 1)(1 - n)\, t . \sin \alpha \quad ...(iv) \quad ...(4.8\,b)$$

where α is the friction angle.

When the sediment is just on the point of sliding,

$$\tau = R$$

$$\therefore \quad \gamma_w . R . S = \gamma_w (G_s - 1)(1 - n)\, t . \sin \alpha$$

Putting $(1 - n) \sin \alpha = k$, where k is the important characteristic of the sediment, to be determined by experiments, we get

$$S = \frac{k}{R} (G_s - 1)\, t \quad ...(v)$$

For single grains, the volume per unit area t becomes the function of the diameter of the grains d_s as an inverse measure of the surface area of the individual grains exposed to drag or friction.

$$\therefore \qquad S = \frac{k}{R}(G_s - 1)\, d_s \qquad ...(vi)$$

Eq. (*vi*) gives the invert slope at which the velocity will be self-cleansing.

Now, from Chezy's equation,

$$V = C\sqrt{RS}.$$

Denoting $V = V_s$ as self-cleaning velocity, and substituting the value of S from Eq. (*vi*), we get.

$$V_s = C\sqrt{R}\,\sqrt{\frac{k}{R}(G_s - 1)\, d_s}$$

or $$V_s = C\sqrt{k\,(G_s - 1)\, d_s} \qquad ...(vii)$$

In order to get the value of Chezy's constant C, let us use Darcy-Weisbatch, formula for head loss H_L :

$$H_L = \frac{fLV^2}{2gD} \quad \text{(where } D = \text{ pipe dia.)}$$

$$\therefore \qquad S = \frac{H_L}{L} = \frac{fV^2}{2gD}$$

Now $\quad V = C\sqrt{RS}$

$$\therefore \quad C\sqrt{R} = \frac{V}{\sqrt{S}} = \sqrt{\frac{2gD}{f}}$$

For circular pipes running full, $R = D/4$

$$\therefore \qquad C\sqrt{\frac{D}{4}} = \sqrt{\frac{2gD}{f}}$$

or $$C = \sqrt{\frac{8g}{f}} \qquad ...(viii)$$

Substituting this value of C, in Eq. (*vii*),

$$V_s = \sqrt{\frac{8\beta}{f}(G_s - 1)\, gd_s} \qquad ...(4.8)$$

where V_s= self cleansing velocity.

β= Characteristics of solids flowing in the sewage, in suspension. This value of β may be taken as 0.04 for initiating scour of clean grit to 0.8 for full removal of sticky grit. Actual magnitude of k has to be found from experiments.

f= Darcy-Weisbatch friction factor, the common value of which may be taken as 0.03.

G_s=specific gravily of sediments/solids flowing in the sewage. Its value may range from 2.65 for inorganic sediments to 1.2 to for organic sediments.

g= gravitational acceleration constant.

d_s=diameter of solid particles, to be carried by the liquid.

From Eq. 4.8, it is evident that heavier particles and sticky particles require higher velocity for their transport than lighter and cleaner particles. Table 4.5 gives the value of self-cleansing velocities for various types of particles, as recommended by Beardmore. Similarly, Table 4.6 gives the self-classing velocities for sewers of various diameters, as recommended by Badwin Latham.

TABLE 4.5. SELF-CLEANSING VELOCITIES.

S.N.	*Nature of material*	*Self cleansing velocity (cm/sec)*
1.	Angular stones	100
2.	Round pebbles (12 mm to 25 mm dia)	50-60
3.	Fine gravel	30
4.	Coarse sand	20
5.	Fine sand and clay	15
6.	Fine clay and silt	7.5

TABLE 4.6. SELF-CLEANSING VELOCITIES.

S.N.	*Diameter of sewer (cm)*	*Self cleansing velocity (cm/sec)*
1.	15 to 25	100
2.	30 to 60	75
3.	Above 60	60

4.5. MAXIMUM VELOCITY OF FLOW

Though the minimum velocity of flow of sewage should be equal to the self-cleansing velocity so that particles do not settle and stick to the invert, there is also some upper limit of velocity of flow so that the interior surface of the sewer is not damaged due to wear. At higher velocity, the flow becomes turbulent, resulting in continuous abrasion of the interior surface of the sewer, by the suspended particles. Hence maximum velocity of flow is also limited. The maximum velocity at which no such scouring action or abrasion takes place is known as *non-scouring velocity*. Evidently such a velocity depends upon the material used for the construction of sewers. Of the ceramic materials used in sewers, vitrified tiles and glazed bricks are more resistant to wear while building bricks and concrete are less resistant to wear. Also, abrasion is maximum at the bottom of the sewer, because the grit, sand etc. are heavy and travel along the invert. Due to this reason, the bottom of large sewers of brick or concrete are protected by lining them with vitrified tile, glazed bricks or granite blocks. Table 4.7. gives the non-scouring velocity for various materials used for sewer construction.

TABLE 4.7 NON-SCOURING VELOCITIES.

S.N.	Material of sewer	Non scouring velocity (cm/sec)
1.	Earth channels	60 to 120
2.	Ordinary brick-lined sewers	150—250
3.	Cement Concrete sewers	250—300
4.	Stone ware sewers.	300—450
5.	Cast Iron sewer pipes	350—450
6.	Vitrified tile and glazed bricks	450—500

Effect of variations of discharge on velocity in sewers. As stated earlier, the discharge in a sewer does not remain constant at all times. It varies from time to time. Due to the variation in discharge, the hydraulic mean depth (R) also varies. Since the velocity of flow is a function of $R^{2/3}$, the velocity of flow also varies as the sewage discharge varies. This is more prominent in the case of a combined sewer or in a partially combined sewer.

As the flow decreases in the sewer, the velocity of flow also decreases. When the sewer becomes less than half full, (assumed at one third the average flow), it is essential to check that the velocity of flow is atleast equal to 40 cm/sec. At the same time, the designer should ensure that a velocity of about 90 cm/sec is developed outleast at the time of maximum flow. While designing the sewers, the following points should be observed in connection with the self-cleansing velocity and non-scouring velocity.

1. Before the sewer design is done, the discharge is known. Hence the velocity of flow and gradient of the sewers are to be appropriately determined and correlated, to achieve the desired results.

2. For sewers in flat country, the design of sewers should be done in such a way that self-cleansing velocity is obtained at maximum discharge. However, the section of sewer should be such that even at minimum discharge, the velocity is at least equal to 40 cm/sec.

3. For sewers in roughs country, the design of sewers should be done in such a way that self-cleansing velocity is obtained at minimum discharge while non-scouring velocity is obtained at maximum discharge. If due to steep slopes, the velocity is exceeded during maximum discharge, drop man holes should be provided to bring down the velocity within the non-scouring value.

4. In the case of combined sewer, it may be difficult to achieve self-clearing velocity during minimum flow (D.W.F.). In that case, special form of sewers (such as the one shown in Fig. 4.10) should be adopted.

4.6. HYDRAULIC ELEMENTS OF CIRCULAR SEWERS

Sewers of circular cross-section are more commonly used. However 'egg-shaped' sewers are also used and their hydraulic characteristics have been discussed in § 4.7.

Circular sewers offer the following advantages :

1. They are very easily manufactured. 2. A circular section gives the maximum area for a given perimeter, and thus gives the greatest H.M.D. when running full or half full. It is therefore the most efficient section at these flow conditions. 3. It is the most economical section since it utilises minimum quantities of the material. 4. The circular section has uniform curvature all round, and hence it offers less opportunities for deposits.

A circular sewer may run either full or partially full. However the advantage at S.N. 2 above is obtained when the sewer runs *atleast half-full*. This advantage is lost if the depth of flow becomes less than half full, since both the velocity as well as discharge reduce considerably with the reduction in the depth of flow. It will be seen in the later part of this article that minimum grades are enough so long as circular sewers flow more than half full. However, when flow drops to 0.2 depth, grades must be doubled for equal self-cleansing.

(a) Circular section running full. Let D be the internal diameter of circular sewer.

Area of cross-section $A = \frac{\pi}{4} D^2$ ...(4.9 a)

Wetted perimeter $P = \pi D$ ...(4.9 b)

Hydraulic mean depth $R = A/P = D/4$ ...(4.9 c)

(b) Circular sewer running partially full. Fig 4.7 shows a circular sewer running *partially full*. Let d be the depth at partial flow and let α be the central angle subtended as shown in Fig. 4.7.

Let a = area of cross–section.

p = wetted perimeter

r = H.M.D.

and v = velocity of flow

Central angle is given by

$\cos \frac{1}{2}\theta = 1 - 2\,d/D$ ...(4.10)

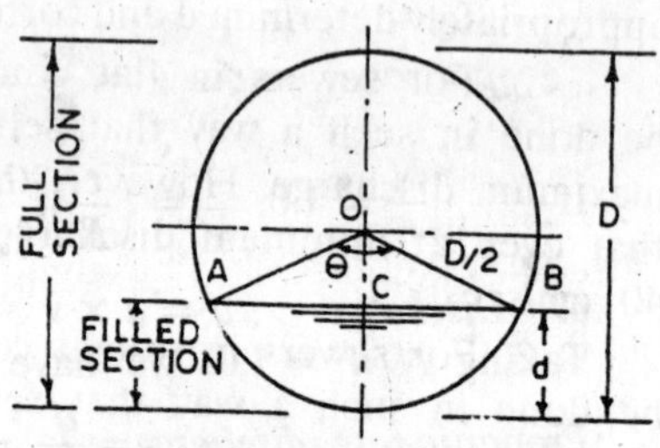

FIG. 4.7

1. Depth $d = \frac{D}{2} - \frac{D}{2}\cos\frac{\theta}{2} = \frac{D}{2}\left(1 - \cos\frac{\theta}{2}\right)$ ...(4.11 a)

$\therefore$ *Proportional depth* $= \frac{d}{D} = \frac{1}{2}\left(1 - \cos\frac{\theta}{2}\right)$ ...(4.11)

2. *Area* $$a = \frac{\pi}{4} D^2 \times \frac{\theta}{360°} - \frac{D}{2}\cos\frac{\theta}{2} \cdot \frac{D}{2}\sin\frac{\theta}{2}$$

or $$a = \frac{\pi D^2}{4}\left[\frac{\theta}{360°} - \frac{\sin\theta}{2\pi}\right] \quad ...(4.12\ a)$$

$\therefore$ *Proportional area* $$= \frac{a}{A}\left[\frac{\theta}{360°} - \frac{\sin\theta}{2\pi}\right] \quad ...(4.12)$$

3. *Wetted perimeter* $$p = \pi D \cdot \frac{\theta}{360} \quad ...(4.13\ a)$$

$\therefore$ *Poportional perimeter* $$= \frac{p}{P} = \frac{\theta}{360°} \quad ...(4.13)$$

4. *H.M.D.* $$r = \frac{a}{p} = \frac{\frac{\pi D^2}{4}\left[\frac{\theta}{360°} - \frac{\sin\theta}{2\pi}\right]}{\pi D \frac{\theta}{360°}}$$

or $$r = \frac{D}{4}\left[1 - \frac{360° \sin\theta}{2\pi\theta}\right] \quad ...(4.14\ a)$$

$\therefore$ *Proportional H.M.D.* $$\frac{r}{R} = \left[1 - \frac{360° \sin\theta}{2\pi\theta}\right] \quad ...(4.14)$$

5. *Velocity of flow*

$$v = \frac{1}{n} r^{2/3} S^{1/2} \quad \text{(Manning's)}$$

where n = Mannings rugosity coefficient applicable for partial flow condition.

$\therefore$ *Proportional velocity* $$= \frac{v}{V} = \frac{N}{n} \cdot \left(\frac{r}{R}\right)^{2/3} \quad ...(4.15\ b)$$

If $\frac{N}{n} = 1.0$, $$\frac{v}{V} = \left(\frac{r}{R}\right)^{2/3} = \left[1 - \frac{360° \sin\theta}{2\pi\theta}\right]^{2/3} \quad ...(4.15)$$

6. *Discharge* $q = a \times v$

Taking $N/n = 1.0$, we have

Proportionate discharge $$= \frac{q}{Q} = \frac{a.v}{AV} = \frac{a}{A} \times \left(\frac{r}{R}\right)^{2/3} \quad ...(4.16)$$

$$\therefore \frac{q}{Q} = \left[\frac{\theta}{360°} - \frac{\sin\theta}{2\pi}\right]\left[1 - \frac{360° \sin\theta}{2\pi\theta}\right]^{2/3} = \frac{\theta}{360}\left[1 - \frac{360° \sin\theta}{2\pi\theta}\right]^{5/3}$$

...(4.16 a)

For variable values of $\frac{N}{n}$, we get

$$\frac{q}{Q} = \frac{N}{n}\left(\frac{a}{A}\right)\left(\frac{r}{R}\right)^{2/3} \quad ...(4.17)$$

It is to be noted that in all the equations derived above [Eqs. 4.12 to 4.16), there is only one variable i.e. angle θ which depends upon d/D ratio as indicated in Eq. 4.10. Hence by taking proportional depth d/D as reference variable, all other quantities can be computed. Table 4.8 gives the computed values of all the hydraulic elements of circular sewers running partially full. The value of N/n given in column (7) of the above taken from 'Water Pollution Control Fed. Manual of Practice, No. 9 (1960).

TABLE 4.8 HYDRAULIC ELEMENTS OF CIRCULAR SEWERS RUNNING PARTIALLY FULL

$\frac{d}{D}$	$\frac{a}{A}$	$\frac{p}{P}$	$\frac{r}{R}$	*For N/n = 1.0*		$\frac{N}{n}$	*For variable N/n*	
				v/V	q/Q		v/V	q/Q
(1)	(2)	(3)	(4)	(5)	(6)	(7)	(8)	(9)
1.00	1.000	1.000	1.000	1.000	1.000	1.00	1.000	1.000
0.90	0.949	0.795	1.192	1.124	1.066	0.94	1.057	1.002
0.80	0.858	0.705	1.217	1.140	0.988	0.88	1.003	0.869
0.70	0.748	0.631	1.185	1.120	0.838	0.85	0.952	0.712
0.60	0.626	0.564	1.110	1.072	0.671	0.83	0.890	0.557
0.50	0.500	0.500	1.000	1.000	0.500	0.81	0.810	0.405
0.40	0.373	0.436	0.857	0.902	0.337	0.79	0.713	0.266
0.30	0.252	0.369	0.684	0.776	0.196	0.78	0.605	0.153
0.20	0.143	0.295	0.482	0.615	0.088	0.79	0.486	0.070
0.10	0.052	0.205	0.254	0.401	0.021	0.82	0.329	0.017
0.00	0.000	—	—	—	0.000	—	—	—

Fig. 4.8 shows the variation of ratios of these hydraulic elements with the depth ratio d/D. The two sets of curves included for v/V and q/Q mark the influence of a variable ratio of N/n. It is interesting to note that if rugosity coefficient (N) is not considered to vary with the depth of flow, the velocities in partialy filled circular section exceed those in full sections whenever sewers flow more than half full, and that maximum velocity is obtained when depth of flow is about 0.81 times the full depth and is about 12.5 % greater than that when running full. However, where changes in the value of N are taken into account, velocities equal to or greater than those in full sections are confined to the upper 20% depth only. Similarly, maximum discharge is obtained when the depth of flow is about 0.95 times the full depth, and this max. discharge is about 7% greater than that when running full.

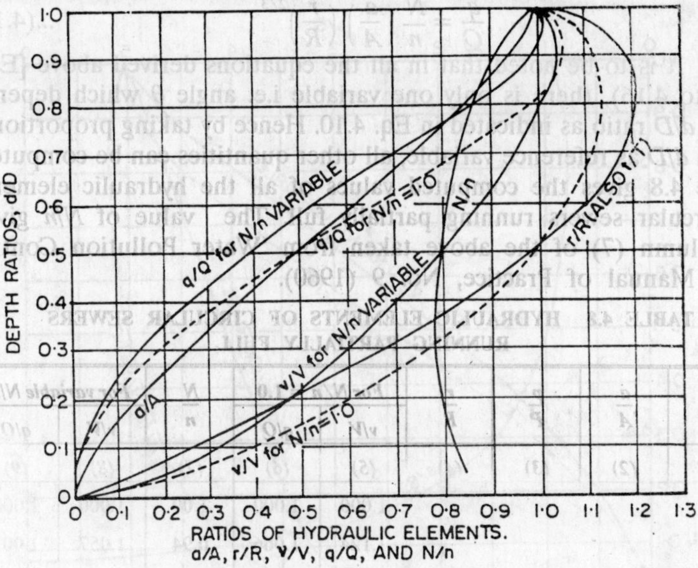

FIG. 4.8. VARIATION OF RATIOS OF HYDRAULIC ELEMENTS

Partial depth self cleansing flow. Sewers flowing between 0.5 and 0.8 full need not be placed on steeper grades to be as self-cleansing as sewers running full. This is so because the velocity (v) and the discharge (q) are functions of the tractive force intensity (τ) which depends upon friction coefficient as well as the velocity. Using subscript s denoting self cleansing equaling that obtained in full section, the needed ratios v_s/v, q_s/Q and s_s/S can be computed with the help of Eq. 4.8 (*a*) on the assumption that equality of tractive force intensity implies equality of cleanings, or

$$\tau = T$$

or $$\gamma_w \,.\, r \,.\, s = \gamma_w RS \quad \text{(where } s = s_s\text{)}$$

Hence $$s_s = \left(\frac{R}{r}\right) S \qquad \text{...(4.18)}$$

or $$\frac{s_s}{S} = \frac{R}{r} \qquad \text{...(4.18 a)}$$

Therefore, $$\frac{v_s}{V} = \frac{N}{n}\left(\frac{r}{R}\right)^{2/3}\left(\frac{s_s}{S}\right)^{1/2} = \frac{N}{n}\left(\frac{r}{R}\right)^{1/6} \qquad \text{...(4.19)}$$

and $$\frac{q_s}{Q} = \left(\frac{N}{n}\right)\left(\frac{a}{A}\right)\left(\frac{r}{R}\right)^{1/6} \qquad \text{...(4.20)}$$

The variation of $\frac{v_s}{V}$ and $\frac{q_s}{Q}$ with d/D ratio are shown in Fig. 4.9, both for $N/n = 1.0$ as well as for N/n variable. The curve for variation of s_s/S with d/D is also shown. *From curve for s_s/S in Fig. 4.9, it is confirmed that minimum grades (corresponding to that*

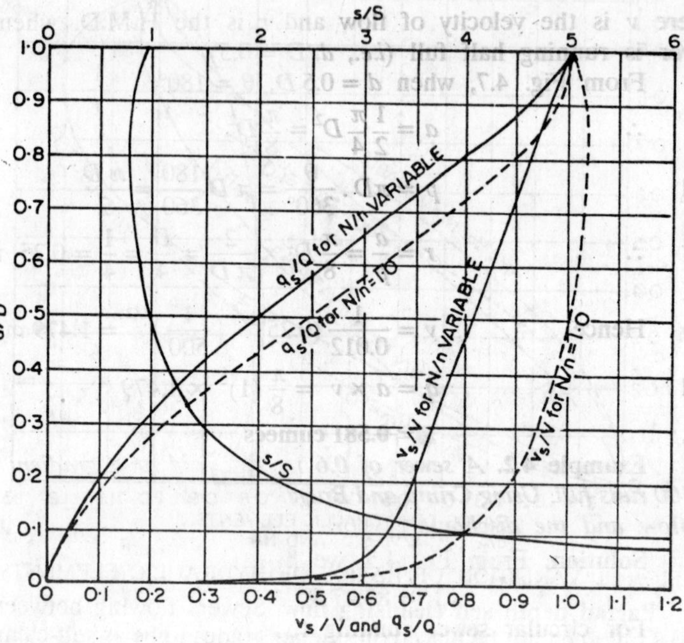

FIG. 4.9. HYDRAULIC ELEMENTS $\frac{v_s}{V}$ and $\frac{q_s}{Q}$ FOR CIRCULAR SEWERS WITH EQUAL SELF-CLEANSING PROPERTIES AT ALL DEPTHS

for running full) are enough so long as circular sewer flow more than half full since they will remain equal self cleaning. However, the curve for s_s/S indicates that when the depth of flow d drops to $0.2\,D$, the grade must be doubled for equal self-cleaning, when for $d = 0.1\,D$, the grade must be four times that required for running full (*i.e.* $s_s = 4S$).

Limitations on depth of flow due to ventilation considerations. Because of ventilation considerations, sewers are not designed to run full even at ultimate peak flows. Small size sewers (upto 0.4 m dia.) are generally desired to flow at 1/2 depth, sewers of size between 0.4 m to 0.9 m are designed to flow at 2/3 depth while larger sewers are generally designed to flow at 3/4 depth, at ultimate peak discharge.

Example 4.1. *Calculate the velocity of flow, and discharge in a sewer of circular section having diameter of 1 m, laid at a gradient of 1 in 500. Use Mannings formula taking N = 0.012. Assume that the sewer is running half full.*

Solution. Using Eq. 4.4 (Mannings formula)

$$v = \frac{1}{N} r^{2/3} s^{1/2}, \text{ (where } N = 0.012\text{).}$$

where v is the velocity of flow and r is the H.M.D. when the sewer is running half full (*i.e.,* $d/D = 0.5$)

From Fig. 4.7, when $d = 0.5\,D$, $\theta = 180°$

$$\therefore \quad a = \frac{1}{2}\frac{\pi}{4}D^2 = \frac{\pi}{8}D^2.$$

and $$p = \pi D \cdot \frac{\theta}{360°} = \pi D\frac{180}{360} = \frac{\pi D}{2}$$

$$\therefore \quad r = \frac{a}{p} = \frac{\pi}{8}D^2 \times \frac{2}{\pi D} = \frac{D}{4} = \frac{1}{4} = 0.25 \text{ m}$$

Hence $$v = \frac{1}{0.012}(0.25)^{2/3}\left(\frac{1}{500}\right)^{1/2} = \mathbf{1.479\ m/s}$$

and $$q = a \times v = \frac{\pi}{8}(1)^2 \times 1.479$$

$$= \mathbf{0.581\ cumecs}$$

Example 4.2. *A sewer of 0.6 m dia., laid at a gradient of 1 in 400 runs full. Using Crimp and Bruge's formula, calculate the velocity of flow and the discharge.*

Solution. From Eq. 4.5, we have

$$V = 83.47\,R^{2/3}\,S^{1/2}$$

For circular sewer running full, $R = D/4 = 0.6/4 = 0.15$ m

$$\therefore \quad V = 83.47\,(0.15)^{2/3}\left(\frac{1}{400}\right)^{1/2} = \mathbf{1.178\ m/sec}$$

and $$Q = \frac{\pi}{4}(0.6)^2 \times 1.178 = \mathbf{0.333\ cumecs}$$

$$= 333 \text{ litres/sec.}$$

Example 4.3. *A stone-ware sewer, 30 cm in diameter is laid at a gradient of 1 in 100. Using N=0.013 in Manning's formula, calculate the velocity, discharge and Chezy's coefficient when the sewer is running full.*

Solution : Manning's formula, given by Eq. 4.5 is

$$V = \frac{1}{N}R^{2/3}\,S^{1/2}, \text{ (where } N = 0.013)$$

where $$R = D/4 = 0.3/4 = 0.075 \text{ m}$$

$$\therefore \quad V = \frac{1}{0.013}(0.075)^{2/3}\left(\frac{1}{100}\right)^{1/2} = \mathbf{1.368\ m/sec}$$

$$Q = \frac{\pi}{4}(0.3)^2 \times 1.368 = \mathbf{0.0967} \text{ cumes}$$

Now Chezy's formula (Eq. 4.1) is

$$V = C\sqrt{RS}$$

$$\therefore \quad 1.368 = C\sqrt{0.075 \times \frac{1}{100}}$$

From which $C \approx \mathbf{50}$

Example 4.4. *Calculate the velocity of flow and corresponding discharge in a sewer of circular section having diameter equal to 1 m, laid at a gradient of 1 in 500. The sewer runs at 0.6 depth. Use Mannings formula taking N = 0.012.*

Solution : Refer Fig. 4.7.

Here $\frac{d}{D} = 0.6$

$\therefore \quad \cos\frac{\theta}{2} = 1 - 2\frac{d}{D} = 1 - 2 \times 0.6 = -0.2$

$\theta = 203.07° \; ; \; \sin\theta = -0.3919$

For sewer running full, $A = \frac{\pi}{4}(1)^2 = 0.7854 \text{ m}^2$

$P = \pi D = \pi \times 1 = 3.1416 \text{ m}$

and $R = D/4 = 1/4 = 0.25$ m.

For partially full running,

$$\frac{a}{A} = \left[\frac{\theta}{360°} - \frac{\sin\theta}{2\pi}\right] = \left[\frac{203.07}{360} + \frac{0.3919}{2\pi}\right]$$

$= 0.6265$

$\therefore \quad a = 0.6265\,A = 0.6265 \times 0.7854 = 0.492 \text{ m}^2$

$$\frac{r}{R} = \left[1 - \frac{360° \sin\theta}{2\pi\theta}\right] = \left[1 + \frac{360 \times 0.3919}{2\pi\,(203.07)}\right]$$

$= 1.1106$

$\therefore \quad r = 1.1106\,R = 1.1106 \times 0.25 = 0.2776$

Now $v = \frac{1}{N} r^{2/3} S^{1/2} = \frac{1}{0.012}(0.2776)^{2/3}\left(\frac{1}{500}\right)^{1/2}$

$= \mathbf{1.586}$ m/sec

Also, $q = a \times v = 0.492 \times 1.586 \approx$ **0.78 cumes**

Note : If the sewer were to run full, the corresponding values would be :

$$V = \frac{1}{0.012}(0.25)^{2/3}\left(\frac{1}{500}\right)^{1/2} = 1.479 \text{ m/sec}$$

and $Q = A \times V = 0.7854 \times 1.479 = 1.162$ cumecs

Hence $\frac{v}{V} = \frac{1.586}{1.479} = 1.072$ and $\frac{q}{Q} = \frac{0.78}{1.162} = 0.671.$

Note that these values are exactly the same as given in Table 4.8.

Example 4.5. *Find the minimum velocity and gradient required to transport coarse sand through a sewer of 60 cm dia. with sand particles of 1 mm dia. and specific gravity 2.66. Assume β = 0.06 and f = 0.02. Assume the sewer to run half full. Take N = 0.012.*

Solution

The minimum velocity (or self cleansing velocity) is given by Eq. 4.8.

$$V_s = \sqrt{\frac{8\beta}{f}(G_s - 1)\,gd_s}$$

Here $\beta = 0.06$; $f = 0.02$; $G_s = 2.66$ and $d_s = 1$ mm $= 0.001$ m

$$\therefore \quad V_s = \sqrt{\frac{8 \times 0.06}{0.02}(2.66 - 1) \times 9.81 \times 0.001}$$

$$= \mathbf{0.625} \text{ m/sec.}$$

Here $D = 60$ cm $= 0.6$ m.

When sewer runs half full, $d/D = 0.5$

Hence $d = 0.5\,D = 0.3$ m and $\theta = 180°$

$$a = \frac{1}{2}\frac{\pi}{4}D^2 = \frac{\pi D^2}{8}$$

$$p = \pi D \frac{\theta}{360°} = \pi D \times \frac{180}{360} = \frac{\pi D}{2}$$

$$\therefore \quad r = \frac{a}{p} = \frac{\pi}{8}D^2 \times \frac{2}{\pi D} = \frac{D}{4} = \frac{0.6}{4} = 0.125 \text{ m}$$

Now $v = \frac{1}{n} r^{2/3} S^{1/2}$

where $v = V_s = 0.625$ m/sec.

Taking $n = N = 0.012$, we have

$$0.625 = \frac{1}{0.012}(0.125)^{2/3} S^{1/2}$$

From which $S = \dfrac{1}{1111}$

Hence the required gradient = 1 in 1111.

Example 4.6. *A main combined sewer is to be designed to serve an area of 12 sq. km with a population density of 250 persons/hectare. The average rate of sewage flow is 250 litres/capita/day. The maximum flow is 100% in excess of average together with the rainfall equivalent of 15 mm in 24 hours, all of which are runoff. Determine the capacity of the sewer. Taking the maximum velocity of flow as 3 m/sec., determine the size of the circular sewer.*

Solution. Total population $= (12 \times 10^2)$ ha $\times$ 250 persons/ha

$= 300000$ persons

Average flow $= 300000 \times 250 = 75 \times 10^6$ litres/day

$$= \frac{75 \times 10^6}{24 \times 60 \times 60} \times \frac{1}{1000} = 0.868 \text{ cumecs.}$$

Max. sewage flow $= 2 \times$ average flow $= 2 \times 0.868$

$= 1.736$ cumecs.

Storm water flow $= 12 \times 10^6 \times \left(\dfrac{15}{1000}\right)\dfrac{1}{24 \times 60 \times 60}$ cumecs

$= 2.083$ cumecs

$\therefore$ Total flow for the combined sewer $= 1.736 + 2.083$
$\approx$ **3.82 cumecs**

This is evidently the capacity of the sewer.

Now $\quad Q = A \times V_{max.} = \frac{\pi}{4} D^2 . V_{max.}$

$$\therefore \quad D = \sqrt{\frac{4Q}{\pi V_{max.}}} = \sqrt{\frac{4 \times 3.82}{\pi \times 3.0}} = 1.273 \text{ m}$$

Hence provide a circular sewer of **130 cm dia.**

Example 4.7. *Calculate the velocity and discharge through a rectangular concrete lined smooth channel 2 m wide and 1 m deep built to a slope of 1 in 200, when running completely full. Use Bazin's coefficient in Chezy's formula as* $C = \dfrac{157.6}{1.81 + \dfrac{K}{\sqrt{R}}}$

where K= 0.3 for smooth concrete lined surface.

Solution

$$A = 2 \times 1 = 2 \text{ m}^2$$

When the rectangular sewer runs full, the roof will also form part of wetted perimeter

$$\therefore \quad P = 2(2 + 1) = 6 \text{ m}$$

$$\therefore \quad R = \frac{A}{P} = \frac{2}{6} = 0.333 \text{ m}$$

$$\text{Hence} \quad C = \frac{157.6}{1.81 + \dfrac{0.3}{\sqrt{0.3333}}} = 67.65$$

From Chezy's formula,

$$V = C\sqrt{RS}$$

$$= 67.65\sqrt{0.3333 \times \frac{1}{200}} = \mathbf{2.762 \text{ m/sec.}}$$

Hence $\quad Q = A \times V = 2 \times 2.762 =$ **5.523 cumecs**

Example 4.8. *For a circular sewer and a rectangular sewer to be hydraulically equivalent, find the relation between the depth of the rectangular sewer and the diameter of the circular sewer. Take the width of rectangular sewer as 1.5 times the depth and assume that only three sides of the rectangular sewer are wetted.*

Solution

(a) ***Rectangular Section***

Let $\quad D =$ Height of rectangular section

$\therefore$ Width $= B = 1.5\,D$

$$A = D \times 1.5\,D = 1.5\,D^2$$

$$P = D + 1.5\,D + D = 3.5\,D$$

$$R = \frac{A}{P} = \frac{1.5\,D^2}{3.5\,D} = 0.4286\,D$$

$$Q = A \times V = A \times \frac{1}{N} R^{2/3} S^{1/2}$$

or $$Q = (1.5\,D^2) \times \frac{1}{N} (0.4286\,D)^{2/3} S^{1/2}$$

or $$Q = 0.8527\,D^{8/3} \frac{S^{1/2}}{N} \qquad ...(i)$$

(b) Circular section

Let d = diameter $\therefore$ $A = \frac{\pi}{4} d^2$

$$P = \pi d$$

$$\therefore \quad R = \frac{A}{P} = \frac{d}{4}$$

Now $$Q = A \times V = A \times \frac{1}{N} R^{2/3} S^{1/2}$$

or $$Q = \frac{\pi}{4} d^2 \times \frac{1}{N} \left(\frac{d}{4}\right)^{2/3} S^{1/2}$$

or $$Q = 0.3117\,d^{8/3} \frac{S^{1/2}}{N} \qquad ...(ii)$$

For the two sewers to be hydraulically equivalent, Q, N and S are the same. Hence from (*i*) and (*ii*), we get

$$0.8527\,D^{8/3} = 0.3117\,d^{8/3}$$

or $$\left(\frac{D}{d}\right)^{8/3} = 0.3655$$

From which $$\frac{D}{d} = 0.6856$$

Hence $$D = \mathbf{0.6856\,d}$$

Example 4.9. *Calculate the diameter and discharge of a circular sewer laid at a slope of 1 in 500 when running half full, and with a velocity of 2 m/sec. Take N = 0.012 in Manning's formula.*

Solution : Refer Fig 4.7. When the sewer runs half full,

$$d = 0.5\,D \text{ and } \theta = 180°$$

$$\therefore \quad a = \tfrac{1}{2} \times \frac{\pi}{4} D^2 = \frac{\pi}{8} D^2$$

$$p = \pi D \frac{\theta}{360°} = \pi D \frac{180}{360} = \frac{\pi D}{2}$$

and $$r = \frac{\pi}{8} D^2 \times \frac{2}{\pi D} = \frac{D}{4}$$

Now $$v = \frac{1}{N} r^{2/3} S^{1/2}$$

$$\therefore \quad 2 = \frac{1}{0.012} \left(\frac{D}{4}\right)^{2/3} \left(\frac{1}{500}\right)^{1/2} = 1.4789\,D^{2/3}$$

From which $D = \mathbf{1.573\ m}$

Hence $Q = a \times v = \frac{\pi}{8}(1.573)^2 \times 2$

$= \mathbf{1.942\ cumecs}$

Example 4.10. *Determine the size of a circular sewer for a discharge of 500 litres per second running half full. Assume S = 0.0001 and N = 0.015.*

Solution : Refer Fig. 4.7. Given $d = 0.5\,D$

$$\therefore \quad \cos\frac{\theta}{2} = 1 - 2\frac{d}{D} = 0$$

From which $\frac{\theta}{2} = 90°$, $\theta = 180°$ and $\sin\theta = 0$

From Eq. 4.12 (*a*)

$$a = \frac{\pi}{4}D^2\left[\frac{\theta}{360} - \frac{\sin\theta}{2\pi}\right] = \frac{\pi}{4}D^2\left[\frac{1}{2} - 0\right] = \frac{\pi}{8}D^2$$

From Eq. 4.13 (*a*)

$$p = \pi D\frac{\theta}{360} = \pi D\frac{180}{360} = \frac{\pi D}{2}$$

$$\therefore \quad r = \frac{\pi}{8}D^2 \times \frac{2}{\pi D} = \frac{D}{4}$$

Now, $q = \frac{1}{N}a \,.\, r^{2/3}\, S^{1/2}$

Here $q = 500$ litres/sec $= 0.5$ cumecs

$$\therefore \quad 0.5 = \frac{1}{0.015} \times \frac{\pi}{8}D^2\left(\frac{D}{4}\right)^{2/3} \times (0.0001)^{1/2}$$

or $D^{8/3} = 4.8128$

From which $D \approx \mathbf{1.80\ m}$

Example 4.11. *A town has a population of 100,000 persons with a per capita water supply of 200 litres/day. Design a sewer running 0.7 times full at maximum discharge. Take a constant value of N = 0.013 at all depths of flow. The sewer is to be laid at a slope of 1 in 500. Take a peaking factor of 3.*

Solution. Water supplied $= 100{,}000 \times 200 = 20 \times 10^6$ litres/day

$$= \frac{20 \times 10^6}{10^3 \times 24 \times 3600} = 0.2315 \text{ cumecs.}$$

Assuming that 80% of the water supplied to the town appears as sewage, we have average discharge in the sewer

$= 0.8 \times 0.2315 = 0.1852$ cumecs.

At a peacking factor of 3,

Max. discharge $= 3 \times 0.1852 = 0.5556$ cumecs

Since the sewer is to be designed as running 0.7 times the full depth, $d/D = 0.7$ and $q = 0.5556$ cumecs.

For a sewer running partially full (Fig. 4.7)

$$\cos\frac{\theta}{2} = 1 - 2\frac{d}{D} = 1 - 2 \times 0.7 = -0.4$$

$$\therefore \frac{\theta}{2} = 113.58° ; \theta = 227.16° ; \sin\theta = -0.7332$$

From Eq. 4.12 (*a*)

$$a = \frac{\pi}{4}D^2\left[\frac{\theta}{360} - \frac{\sin\theta}{2\pi}\right] = \frac{\pi}{4}D^2\left[\frac{227.16}{360} + \frac{0.7332}{2\pi}\right]$$

$$= 0.5872\,D^2 \quad \text{...}(i)$$

$$p = \pi D\frac{\theta}{360} = \pi D\frac{227.16}{360} = 1.9823\,D \quad \text{...}(ii)$$

$$r = \frac{a}{p} = \frac{0.5872\,D^2}{1.9823} = 0.2962\,D \quad \text{...}(iii)$$

Now, $q = \frac{1}{N}a\,r^{2/3}\,S^{1/2}$

$$\therefore\ 0.5556 = \frac{1}{0.013} \times 0.5872\,D^2\,(0.2962\,D)^{2/3}\left(\frac{1}{500}\right)^{1/2}$$

or $D^{8/3} = 0.6190$

From which $D = \mathbf{0.835\ m}$

Check for self cleansing velocity at max. discharge

$$r = 0.2962\,D = 0.2962 \times 0.835 = 0.2474 \text{ m}$$

$$\therefore\ v = \frac{1}{N}r^{2/3}\,S^{1/2} = \frac{1}{0.013}(0.2474)^{2/3}\left(\frac{1}{500}\right)^{1/2}$$

$$= 1.356 \text{ m/sec.}$$

This is much more than the self cleansing velocity of 60 cm/sec (Table 4.6).

Check for self clearing velocity at min. discharge. Let us assume minimum flow equal to $\frac{1}{3}$ times the average flow.

$$\therefore\ q_{min.} = (1/3) \times 0.1852$$

Also, max. flow $Q = 3 \times 0.1852$

$$\therefore\ \frac{q_{min.}}{Q} = \frac{1}{9} \approx 0.11$$

From Fig 4.8, for $\frac{q}{Q} = 0.11$, we get $\frac{d}{D} = 0.22$, assuming $n = N$.

Alternatively, we may get these values from Table 4.8 by interpolation.

$$\therefore\ \frac{d_{min.}}{D} = 0.22$$

Corresponding to this depth ratio, $\frac{v_{min.}}{V} \approx 0.647$

$$\therefore\ v_{min.} = 0.647 \times 1.356 = 0.88 \text{ m/sec.}$$

This is more than the required value of 40 cm/sec.

Note : For analytical solution for this part of the problem, see example 4.12.

Example 4.12. *A circular sewer of 1 m dia. carries a certain discharge while running full. What will be the depth of flow and the corresponding velocity ratio if the discharge in the sewer is reduced to one fourth the value ? Assume that N remains constant at all states of flow.*

Solution. When the sewer runs, full, $R = D/4$

$$A = \frac{\pi}{4}D^2 \;;\quad P = \pi D$$

Now at partial flow, $\frac{q}{Q} = \frac{1}{4}$ (given).

$$\frac{q}{Q} = \left[\frac{\theta}{360} - \frac{\sin\theta}{2\pi}\right]\left[1 - \frac{360° \sin\theta}{2\pi\theta}\right]^{2/3} = 0.25$$

or $\therefore$

$$\frac{\theta}{360°}\left[1 - \frac{360° \sin\theta}{2\pi\theta}\right]\left[1 - \frac{360° \sin\theta}{2\pi\theta}\right]^{2/3} = 0.25$$

or

$$\frac{\theta}{360°}\left[1 - \frac{360° \sin\theta}{2\pi\theta}\right]^{5/3} = 0.25$$

This equation contains only one variable θ. Solving this by trial and error, we get $\theta \approx 143°$.

From Eq. 4.10, $\cos\frac{\theta}{2} = 1 - 2\frac{d}{D}$

or

$$\frac{d}{D} = \frac{1}{2}\left(1 - \cos\frac{\theta}{2}\right) = 0.341$$

$\therefore$ $d = 0.341\,D = \mathbf{0.341\ m}$

From Eq. 4.15, $\frac{v}{V} = \left[1 - \frac{360° \sin\theta}{2\pi\theta}\right]^{2/3}$

$$= \left[1 - \frac{360° \sin 143°}{2\pi \times 143°}\right]^{2/3} = \mathbf{0.832}$$

Example 4.13. *Design an outfall circular sewer of the separate system for a town with a population of 1,50,000 persons with a water supply at 200 litres per head per day. The sewer is to be laid at a slope of 1 in 800 with N = 0.012. A self cleansing velocity of 0.75 m/sec is to be developed. The dry weather flow may be taken as 1/3 of the maximum discharge. Given the following table for partial flow conditions.*

Proportionate depth	*Proportionate velocity*	*Proportionate discharge*
0.37	0.8675	0.2981
0.39	0.8909	0.3217
0.40	0.9022	0.3370
0.42	0.9299	0.3682

Solution

Average rate of water supply $= 150000 \times 200 = 30 \times 10^6$ litres/day

$$= \frac{30 \times 10^6}{10^3 \times 24 \times 60 \times 60} = 0.3472 \text{ cumecs.}$$

$\therefore$ D.W.F. = 0.3472 cumecs

Max. $Q = 3 \times 0.3472 \approx 1.042$ cumecs

Now $Q = \frac{1}{N} A R^{2/3} S^{1/2}$

$$\therefore \quad 1.042 = \frac{1}{0.012}\left(\frac{\pi}{4}D^2\right)\left(\frac{D}{4}\right)^{2/3}\left(\frac{1}{800}\right)^{1/2}$$

or $D^{8/3} = 1.1347$

From which $D = 1.049 \text{ m} \approx 1.05$ m.

$$A = \frac{\pi}{4}(1.05)^2 = 0.8659 \text{ m}^2$$

$$V = \frac{Q}{A} = \frac{1.042}{0.8659} = 1.203 \text{ m/sec.}$$

This is more than the self cleansing velocity.

Now, at D.W.F., $\frac{q}{Q} = \frac{1}{3} = 0.3333$

Hence from the above table, for $\frac{q}{Q} = 0.333$, we get $\frac{d}{D} \approx 0.398$

and hence corresponding $\frac{v}{V} \approx 0.8995$ (by extrapolation)

$$\therefore \quad v = 0.8995 \times 1.203 = 1.082 \text{ m/sec.}$$

This is also more than the self cleansing velocity of 0.75 m/sec.

Hence provide a sewer of 1.05 m dia.

Example 4.14. *A circular sewer is to be designed to carry maximum flow of sewage while flowing 70% full at a velocity of 1 m/sec. If the ratio of maximum/average flow is 3 and average/minimum is also 3, find (i) proportionate depth of flow, and (ii) velocities generated, both at the average flow as well as minimum flow. Neglect the variation of N with depth of flow.*

Solution (Refer Fig. 4.7)

(a) At maximum flow

$$\frac{d_{max.}}{D} = 0.7$$

$$\cos\frac{\theta}{2} = 1 - 2\frac{d}{D} = 1 - 2 \times 0.7 = -0.4$$

$$\therefore \quad \frac{\theta}{2} = 113.58°\,; \theta = 227.16°\,; \sin\theta = -0.7332$$

$$\frac{v}{V} = \left[1 - \frac{360° \sin\theta}{2\pi\theta}\right]^{2/3} = \left[1 + \frac{360 \times 0.7332}{2\pi(227.16)}\right]^{2/3}$$

$$= 1.1198$$

$$\frac{a}{A} = \left[\frac{\theta}{360} - \frac{\sin\theta}{2\pi}\right] = \left[\frac{227.16}{360} + \frac{0.7332}{2\pi}\right] = 0.7478$$

$$\therefore \quad \frac{q_{max.}}{Q} = \frac{v}{V} \times \frac{q}{A} = 1.1198 \times 0.7478 = 0.8373$$

$$\therefore \quad q_{max.} = 0.8373\,Q \qquad ...(i)$$

where Q is the discharge when the sewer is running full.

Also, $\frac{v_{max.}}{V} = 1.1198$ (as obtained above).

$$\therefore \quad V = \frac{v_{max.}}{1.1198} = \frac{1}{1.1198} = 0.893 \text{ m/sec.}$$

(b) At average flow

$$q_{av} = \frac{1}{3} q_{max.} = \frac{1}{3}(0.8373\,Q) = 0.2791\,Q$$

$$\therefore \quad \frac{q_{av}}{Q} = 0.2791 = \frac{\theta}{360°}\left[1 - \frac{360° \sin\theta}{2\pi\theta}\right]^{5/3}$$

Solving this by trial and error, we get $\theta = 147.5°$

$$\therefore \quad \frac{d}{D} = \frac{1}{2}\left(1 - \cos\frac{\theta}{2}\right) = \frac{1}{2}\left(1 - \cos\frac{147.5}{2}\right] = \mathbf{0.36}$$

(Alternatively, we can get the value of d/D either from Fig. 4.8 or from Table 4.8, by extrapolation, for $q/Q = 0.279 \approx 0.28$)

$$\text{Also,} \quad \frac{v_{av.}}{V} = \left[1 - \frac{360° \sin 147.5°}{2\pi\,(147.5)}\right]^{2/3} = 0.8554$$

$$\therefore \quad v_{av.} = 0.8554\,V = 0.8554 \times 0.893 = \mathbf{0.794 \text{ m/sec}}$$

(c) At minimum flow

$$\text{Given :} \quad \frac{q_{av}}{q_{min.}} = 3$$

$$\therefore \quad q_{min.} = \frac{q_{av}}{3} = \frac{0.2791\,Q}{3} \approx 0.093\,Q$$

$$\therefore \quad \frac{q_{min.}}{Q} = 0.093$$

$$\text{Hence} \quad \frac{\theta}{360°}\left[1 - \frac{360° \sin\theta}{2\pi\theta}\right]^{5/3} = 0.093$$

Solving this by trial, we get $\theta = 108°$

$$\therefore \quad \frac{d}{D} = \frac{1}{2}\left(1 - \cos\frac{\theta}{2}\right) = \frac{1}{2}\left(1 - \cos\frac{108°}{2}\right) = \mathbf{0.206}$$

(Alternatively, we can get the value of d/D either from Fig. 4.8 or from Table 4.8, by extrapolation, for $q/Q = 0.093$)

$$\text{Also,} \quad \frac{v_{min.}}{V} = \left[1 - \frac{360 \sin 108°}{2\pi\,(108)}\right]^{2/3} = 0.6261$$

$$\therefore \quad v_{min.} = 0.6261\,V = 0.6261 \times 0.893 \approx \mathbf{0.559 \text{ m/sec}}$$

Example 4.15. *A 60 cm diameter sewer is required to flow at 0.4 depth on a grade ensuring a degree of self cleansing equivalent to that obtained at full depth at a velocity of 85 cm/sec. Find the required grade, associated velocities and rates of discharge at full/depth and 0.4 depth. Take a uniform value of N = 0.015 at all depth of flow.*

Solution

(*a*) ***For the sewer running full.*** $V = \frac{1}{N} R^{2/3} S^{1/2}$

Here, $V = 0.85$ m/sec ; $N = 0.015$

$$R = \frac{D}{4} = \frac{0.6}{4} = 0.15 \text{ m}$$

$$\therefore \quad 0.85 = \frac{1}{0.015} (0.15)^{2/3} S^{1/2}$$

From which $S = 0.00204 = \frac{1}{490.2}$

Also, $Q = \frac{\pi}{4} (0.6)^2 \times 0.85 = 0.2403$ cumecs.

(*b*) ***For partial depth flow***

$$\frac{d}{D} = 0.4 \text{ (given)}$$

$$\cos \frac{\theta}{2} = 1 - 2\frac{d}{D} = 1 - 2 \times 0.4 = 0.2$$

$$\therefore \quad \frac{\theta}{2} = 78.46° ; \theta = 156.93° ; \sin \theta = 0.3919$$

Now $$\frac{a}{A} = \left(\frac{\theta}{360°} - \frac{\sin \theta}{2\pi} \right) \quad ...(4.12\ a)$$

$$= \left(\frac{156.93}{360} - \frac{0.3919}{2\pi} \right) = 0.3735$$

and $$\frac{r}{R} = \left[1 - \frac{360° \sin \theta}{2\pi\theta} \right] \quad ...(4.14)$$

$$= \left[1 - \frac{360 \times 0.3919}{2\pi\,(156.93)} \right] = 0.8568$$

(Alternatively, the value of a/A and r/R can be found from Table 4.8, for $d/D = 0.4$, if Table 4.8 is available)

Hence at 0.4 depth and a grade of 0.00204,

$$\frac{v}{V} = \left(\frac{r}{R} \right)^{2/3} = (0.8568)^{2/3} = 0.9021$$

$$\therefore \quad v = 0.9021 \times 0.85 = 0.767 \text{ m/sec}$$

Also, $$\frac{q}{Q} = \left(\frac{a}{A} \right) \times \left(\frac{v}{V} \right) = 0.3735 \times 0.9021 = 0.3369$$

$$\therefore \quad q = 0.3369 \times 0.2403 = 0.081 \text{ cumecs.}$$

(c) ***For self-cleansing flow*** : For the sewer, flowing at 0.4 depth, to ensure a degree of self cleansing equivalent to that obtained at full depth, it has to have a gradient s_s given by Eq. 4.18.

$$s_s = \left(\frac{R}{r}\right) S$$

$$\therefore \quad s_s = \frac{1}{0.8568} \times 0.00204 = 0.00238$$

$$\text{or} \quad s_s = \frac{\mathbf{1}}{\mathbf{420}}$$

Also, from Eq. 4.19,

$$v_s = \frac{N}{n}\left(\frac{r}{R}\right)^{1/6} V = 1\,(0.8568)^{1/6} \times 0.85$$

$$= \mathbf{0.828\ m/sec}$$

The discharge q_s is given by Eq. 4.20

$$q_s = \left(\frac{N}{n}\right)\left(\frac{a}{A}\right)\left(\frac{r}{R}\right)^{1/6} Q$$

$$= 1 \times 0.3735\,(0.8568)^{1/6} \times 0.2403$$

$$\approx \mathbf{0.0875\ cumecs}$$

Example 4.16. *Solve example 4.15 if $N = 0.015$ at full depth flow and $\frac{N}{n} = 0.79$ for flow at 0.4 depth.*

Solution

(a) ***For full depth flow*** : $V = 0.85$ m/sec.

As determined in example 4.15,

$$S = 0.00204 \; ; \qquad Q = 0.2403 \text{ cumecs}$$

(b) ***For partial depth flow*** : $d/D = 0.4\,;\, N/n = 0.79$

As determined in example 4.15 ,

$$\frac{a}{A} = 0.3735 \text{ and } \frac{r}{R} = 0.8568.$$

Now $$\frac{v}{V} = \frac{N}{n}\left(\frac{r}{R}\right)^{2/3}$$

$$\therefore \quad v = 0.79\,(0.8568)^{2/3} \times 0.85 = \mathbf{0.6058\ m/sec}$$

Also, $$\frac{q}{Q} = \frac{N}{n}\left(\frac{a}{A}\right)\left(\frac{r}{R}\right)^{2/3}$$

$$\therefore \quad q = 0.79\,(0.3735)\,(0.8568)^{2/3} \times 0.2403$$

$$= \mathbf{0.064\ cumecs}$$

(c) ***For self-cleansing flow***

$$s_s = \left(\frac{R}{r}\right) S = \frac{1}{0.8568} \times 0.00204$$

$$= 0.00238 = \frac{\mathbf{1}}{\mathbf{420}}$$

$$\frac{v_s}{V} = \frac{N}{n}\left(\frac{r}{R}\right)^{1/6}$$

$\therefore \quad v_s = 0.79\,(0.8568)^{1/6} \times 0.85 = \mathbf{0.654\ m/sec}$

Also, $\dfrac{q_s}{Q} = \left(\dfrac{N}{n}\right)\left(\dfrac{a}{A}\right)\left(\dfrac{r}{R}\right)^{1/6}$

$\therefore \quad q_s = 0.79\,(0.3735)\,(0.8568)^{1/6} \times 0.2403$
$= \mathbf{0.0691}$ cumecs.

Example 4.17. *A 60 cm diameter sewer is to discharge 0.07 cumecs at a velocity as self-cleansing as a sewer flowing full at 0.85 m/sec. Find the depth and velocity of flow and the required slope. Take uniform value of N = 0.015.*

Solution

(a) For sewer running full

$$V = \frac{1}{N} R^{2/3} S^{1/2}$$

Here $\quad N = 0.015 \,;\, V = 0.85$ cm/sec.

$R = D/4 = 0.6/4 = 0.15$ m

$\therefore \quad 0.85 = \dfrac{1}{0.015}(0.15)^{2/3} S^{1/2}$

From which $\quad S = 0.00204 = \dfrac{1}{490.2}$

And, $\quad Q = \dfrac{\pi}{4}(0.6)^2\, 0.85 = 0.2403$ cumecs.

(b) For partial depth self cleansing flow

$q_s = 0.07$ cumecs (given)

$\therefore \quad \dfrac{q_s}{Q} = \dfrac{0.07}{0.2403} = 0.2913$

But $\quad \dfrac{q_s}{Q} = \dfrac{N}{n}\left(\dfrac{a}{A}\right)\left(\dfrac{r}{R}\right)^{1/6}$ ...(4.20)

Also $\quad \dfrac{a}{A} = \left(\dfrac{\theta}{360} - \dfrac{\sin\theta}{2\pi}\right) = \dfrac{\theta}{360}\left(1 - \dfrac{360° \sin\theta}{2\pi\theta}\right)$ ...(4.12 *a*)

and $\quad \dfrac{r}{R} = \left[1 - \dfrac{360° \sin\theta}{2\pi\theta}\right]$

$$\therefore \quad \frac{q_s}{Q} = \frac{N}{n} \times \frac{\theta}{360}\left(1 - \frac{360° \sin\theta}{2\pi\theta}\right)\left[1 - \frac{360° \sin\theta}{2\pi\theta}\right]^{1/6}$$

or $\quad 0.2913 = \dfrac{1 \times \theta}{360}\left[1 - \dfrac{360° \sin\theta}{2\pi\theta}\right]^{7/6}$

Solving this by trial and error, by assuming several values of θ, we get $\theta \approx 143.5°$

$\therefore \quad \dfrac{d}{D} = \dfrac{1}{2}\left(1 - \cos\dfrac{\theta}{2}\right) = 0.3434$

Also, $\frac{r}{R} = \left[1 - \frac{360 \sin \theta}{2\pi\theta}\right] = \left[1 - \frac{360 \times \sin 143.5}{2\pi\,(143.5)}\right]$

$= 0.7624$

Now $\frac{v_s}{V} = \frac{N}{n}\left(\frac{r}{R}\right)^{1/6}$ ...(4.19)

$\therefore \quad \frac{v_s}{V} = 1\,(0.7624)^{1/6} = 0.9558 \text{ m/sec.}$

$\therefore \quad v_s = 0.9558 \times 0.85 = \mathbf{0.812 \text{ m/sec}}$

and $\frac{s_s}{S} = \left(\frac{R}{r}\right) = \frac{1}{0.7624} = 1.3116$

$\therefore \quad s_s = 1.3116 \times 0.00204 = 0.00268$

$= \mathbf{\frac{1}{373.7}}$

(Alternatively, the values of d/D, v_s/V and s_s/S can be obtained from Fig. 4.9 for $q_s/Q = 0.2913$ if Fig. 4.9 is available)

Example 4.18. *A 40 cm dia. sewer is to flow at 0.3 depth on a grade ensuring a degree of self-cleansing equivalent to that obtained at full depth at a velocity of 80 cm/sec. Find.*

(i) the required grade
(ii) associated velocity
(iii) the rate of discharge at this depth.
Given :
(i) Mannings rugosity coefficient = 0.014
(ii) Proportionate area = 0.252
(iv) Proportionate HMD (r/R) = 0.684

Solution

(a) For sewer running full

$$V = \frac{1}{N} R^{2/3} S^{1/2}$$

Here $V = 0.8 \text{ m/sec}\,;\ R = D/4 = 0.4/4 = 0.1 \text{ m}$

$\therefore \quad 0.8 = \frac{1}{0.014}(0.1)^{2/3} S^{1/2}$

From which $S = 0.0027 \approx \mathbf{\frac{1}{370}}$

Also, $Q = \frac{\pi}{4}(0.4)^2 \times 0.8 = \mathbf{0.1005 \text{ cumecs.}}$

(b) For partial depth self cleansing flow

At 0.4 depth, $\frac{d}{D} = 0.4$

Also, $\frac{a}{A} = 0.252$ and $\frac{r}{R} = 0.684$

From Eq. 4.18,

$$\frac{s_s}{S} = \frac{R}{r} = \frac{1}{0.684} = 1.462$$

$$\therefore \quad s_s = 1.462 \times 0.0027 = 0.00395 = \frac{\mathbf{1}}{\mathbf{253.3}}$$

Also, $$\frac{v_s}{V} = \frac{N}{n}\left(\frac{r}{R}\right)^{1/6} = 1(0.684)^{1/6} = 0.9387$$

$$\therefore \quad v_s = 0.9387 \times 0.8 = \mathbf{0.751\ m/sec}$$

and $$\frac{q_s}{Q} = \left(\frac{N}{n}\right)\left(\frac{a}{A}\right)\left(\frac{r}{R}\right)^{1/6} = 0.252 \times 0.9387$$

$$\therefore \quad q_s = 0.252 \times 0.9387 \times 0.1005 = \mathbf{0.0238\ cumes}$$

Alternatively, $$q_s = a \cdot v_s = \frac{a}{A} \cdot A\, v_s$$

$$= 0.252 \times \frac{\pi}{4}(0.4)^2 \times 0.751 = 0.0238$$

4.7. EGG SHAPED SEWERS

Circular sewers are suitable only where variation in discharge is not large. They are, therefore, suitable with separate *sewerage system.* In thc *combined system*, the discharge is subject to great variation. The sewer is heavily taxed in the rainy season while the D.W.F. during the summer may not be even 5 to 10 percent of the combined discharge. Hence the circular sewer, if provided for a combined system, will run with very low depths. For *combined system*, therefore, egg shaped sewers, (also called ovoid sewers) are more suitable. Their principal advantage is the slightly higher velocity for low flows over the circular sewer of equal capacity. However, egg shaped sewers have the following *disadvantages* :

(1.) An egg-shaped sewer is some what unstable because the small end of the egg is down and it has to support the weight of the upper broader section. (2.) It is more difficult to construct. (3.) It is expensive, as more material is required and as the cost of construction is also high. (4.) In the absence of adequate gradient, it is not self-cleansing.

Egg-shaped sewers were formerly used more than at present. Because of the disadvantages mentioned above, these are becoming obsolete these days.

There are several forms of egg-shaped or ovoid sewers, but the following two forms are very common (Fig. 4.10)

(*i*) Standard or metropolitan section.

(*ii*) New shaped section.

The geometrical characteristics of these two sections are given in Table 4.9, in terms of radius *b* of the broader portion of the

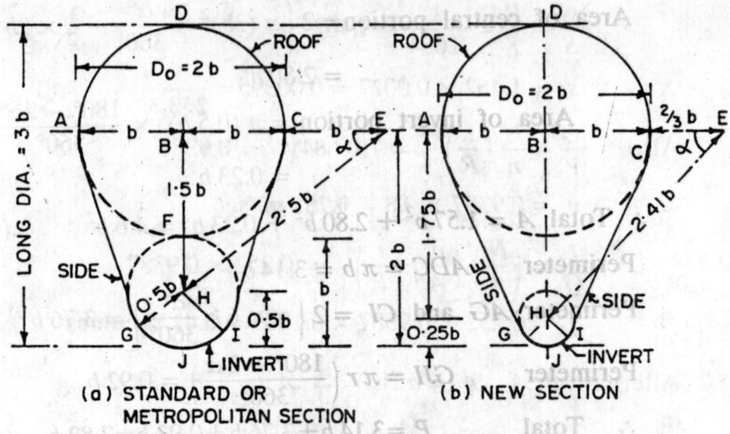

FIG. 4.10 EGG-SHAPED SEWER SECTIONS.

section. The upper portion of the ovoid is known as the *arch* while the lower portion is known as the *invert*.

TABLE 4.9 GEOMETRICAL PARAMETERS OF OVOID SECTION

Parameter	*Standard section*	*New section*
Height	3 *b*	3 *b*
Width	2 *b*	2 b
Radius of arch	*b*	b
Radius of invert	0.5 *b*	0.25 b
Radius of side	3 *b*	$2\frac{2}{3}b$

Hydraulic elements of the standard (or metropolitan) form

The standard (or metropolitan) form is also sometime known as the *old form*. As shown in Fig 4.10 (*a*), the standard form has for its upper portion a semi-circle of a radius (*b*) equal to one-third the depth or the long diameter and an invert curved on a radius of one-sixth of the long diameter, with the sides having radius equal to the long diameter.

(a) Sewer running full. When running full, the total cross sectional area is composed of : (*i*) semi-circular crown (or roof) portion *ADCA* (*ii*) Intermediate (or central) portion *ABCIHGA* and (*iii*) Invert portion *GHIG*

$$\sin\alpha = \frac{1.5\,b}{2.5\,b} = 0.6, \text{ from which } \theta = 36°52'$$

$$\text{Area of crown portion} = \frac{\pi\,b^2}{2} = 1.57\,b^2 \qquad ...(i)$$

$$\text{Area of central portion} = 2\left\{\pi\,(3b)^2 \times \frac{\alpha^\circ}{360^\circ} - \tfrac{1}{2} \times 2b\,(1.5\,b)\right\}$$

$$= 2.80\,b^2 \qquad \text{...(ii)}$$

$$\text{Area of invert portion} = \pi\,(0.5\,b)^2 \times \frac{180^\circ - 2\alpha^\circ}{360^\circ}$$

$$= 0.23\,b^2 \qquad \text{...(iii)}$$

$$\therefore \text{ Total } A = 1.57\,b^2 + 2.80\,b^2 + 0.23\,b^2 = 4.6\,b^2 \qquad \text{...(4.21 a)}$$

Perimeter $ADC = \pi\,b = 3.14\,b$

Perimeter AG and $CI = 2\left\{\pi \times 6\,b\,\frac{\alpha^\circ}{360^\circ}\right\} = 3.76\,b$

Perimeter $GJI = \pi\,r\left(\frac{180^\circ - 2\,\alpha}{360^\circ}\right) = 0.92\,b$

$$\therefore \text{ Total } P = 3.14\,b + 3.76\,b + 0.92\,b = 7.82\,b \qquad \text{...(4.21 b)}$$

$$\therefore \text{ H.M.D. } R = \frac{A}{P} = \frac{4.6\,b^2}{7.82\,b} = 0.58\,b \qquad \text{...(4.21 c)}$$

(b) Sewer running 2/3 full

Sewer will run upto diameter ABC.

Area of central portion $= 2.80\,b^2$

Area of invert portion $= 0.23\,b^2$

$$\therefore \text{ Total } A = 2.80\,b^2 + 0.23\,b^2 = 3.03\,b^2 \qquad \text{...(4.22 a)}$$

Perimeter AG and $CI = 3.76\,b$

Perimeter $GJI = 0.92\,b$

$$\therefore \text{ Total } P = 3.76\,b + 0.92\,b = 4.68\,b \qquad \text{...(4.22 b)}$$

$$\therefore \text{ H. M. D. } R = \frac{3.03\,b^2}{4.68\,b} = 0.64\,b \qquad \text{...(4.22 c)}$$

(c) Sewer running half-full and one third full

On the same lines, it can be shown that

At running half-full, $R = 0.54\,b$ ...(4.23)

At running one third full, $R = 0.41\,b$ ...(4.24)

Hydraulic elements of the new form (Fig. 4.10 *b*)

(a) sewer running full : When running full, the total cross-sections area is composed of : (*i*) semi-circular crown (or roof) portion $ADCA$ (*ii*) Intermediate or central portion $ABCIHGA$ and (*iii*) Invert portion $GHIG$

$$\sin\alpha = \frac{1.75\,b}{2.41\,b} = 0.7261 \;;\; \text{from which } \theta = 46^\circ\,34'$$

$$\text{Area of crown portion} = \frac{\pi b^2}{2} = 1.57\,b^2$$

Area of central portion

$$= 2\left\{\pi\,(2.66\,b)^2 \times \frac{\alpha^\circ}{360^\circ} - \frac{1}{2}\,(1.66\,b)\,(1.75\,b)\right\}$$

$$= 2.86\,b^2$$

Area of invert portion $= \pi\,(0.25\,b)^2 \times \dfrac{180^\circ - 2\alpha^\circ}{360^\circ} = 0.03\,b^2$

$\therefore$ Total $A = 1.57\,b^2 + 2.86\,b^2 + 0.03\,b^2 = 4.46\,b^2$ ...(4.25 *a*)

Perimeter $ADC = \pi\,b = 3.14\,b$

Perimeter AG and $CI = 2\left[\pi \times 5.32\,b \times \dfrac{\alpha^\circ}{360^\circ}\right] = 4.30\,b$

Perimeter $GJI = 0.5\pi\,b\left\{\dfrac{180^\circ - 2\alpha^\circ}{360^\circ}\right\} = 0.38\,b$

$\therefore$ Total $P = 3.14\,b + 4.30\,b + 0.38\,b = 7.82\,b$ ...(4.25 *b*)

$\therefore$ H.M.D. $= R = \dfrac{4.46\,b^2}{7.82\,b} = 0.57\,b$ ...(4.25 *c*)

(*b*) ***sewer running*** $\frac{2}{3}$***rd full*** : The sewer, while running $\frac{2}{3}$rd full, will run upto diameter *ABC*.

Area of central portion $= 2.86\,b^2$

Area of invert portion $= 0.03\,b^2$

$\therefore$ Total $A = 2.86\,b^2 + 0.03\,b^2 = 2.89\,b^2$ ...(4.26 *a*)

Perimeter AG and $CI = 4.30\,b$

Perimeter $GJI = 0.38\,b$

$\therefore$ Total $P = 4.30\,b + 0.38\,b = 4.68\,b$ ...(4.26 *b*)

$\therefore$ H.M.D. $R = \dfrac{2.89\,b^2}{4.68\,b} = 0.62\,b$

(*c*) ***section running one third full*** : On the similar lines, it can be shown that when the section is running one third full,

$R = 0.38\,b$ ...(4.27)

Fig. 4.11 shows the curves of variations of various hydraulic elements such as proportionate dischange (q/Q), proportionate area (a/A), proportionate H.M.D. (r/R) and proportionate velocity (v/V) for various values of depth ratios, for standard ovoid sewers, computed by Crimp and Bruge's formula.

Hydraulically equivalent circular sewers. The computations of various hydraulic elements (such as area, wetted perimeter, hydraulic mean depth etc.) of egg-shaped sewers involve complicated mathematical calculations. Therefore, while designing ovoid sewers, it is the usual practice to calculate the approximate diameter of *hydraulically equivalent* circular sewer first, which would give the same discharge while running full, when laid at the same gradient, and then to

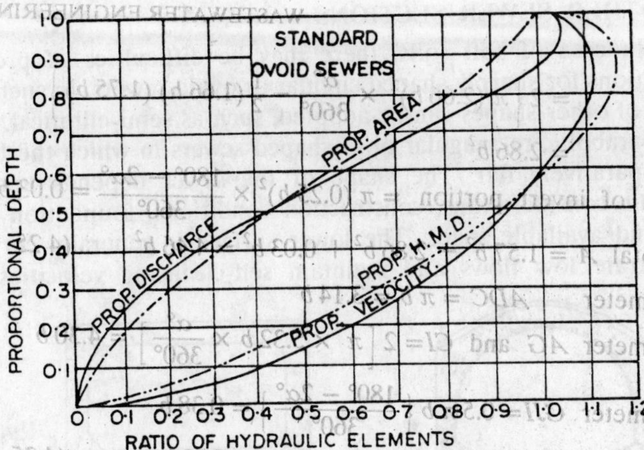

FIG. 4.11 VARIATION OF RATIOS OF HYDRAULIC ELEMENTS OF STANDARD OVOID SEWERS

convert it into dimensions of an egg-shaped section having an equal area.

Let D = diameter of hydraulically equivalent circular section.

D_0 = top horizontal diameter of egg shaped section.

Using suffix 1 for the old (or standard or metropolitan section) and 2 for the new section, we have.

$$D = 1.209\, D_{01} \quad \text{...(4.28 } a)$$

and

$$D = 1.191\, D_{02} \quad \text{...(4.28 } b)$$

For rouch calculations

$$D \approx 1.20\, D_0 \quad \text{...(4.28)}$$

Thus, knowing D, the horizontal diameter D_0 can be very easily worked out. Table 4.10 gives the comparison of proportionate velocities at small depths in circular and standard ovoid sections. From the table, it will be seen that as the depth goes below 0.25 times the full depth, the proportionate velocity in egg shaped sewers become higher than that in hydraulically equivalent circular section.

TABLE 4.10. COMPRESSION OF v/V IN CIRCULAR AND OVOID SEWERS

d/D	v/V	
	Circular sewer	*Standard ovoid sewer*
0.25	0.701	0.698
0.20	0.615	0.627
0.15	0.517	0.544
0.10	0.401	0.440
0.05	0.257	0.295

4.8. OTHER SEWER SECTIONS

In case of soft soils, there may be difficulties of providing foundations for sharply-shaped circular or ovoid sewers. In such cases, sewers of other shapes can be adopted, such as semi-elliptical, horse-shoe, parabolic, rectangular or U-shaped sewers in which the bottom is comparatively flat. The shape of the sewer depends upon the hydraulic considerations, construction conditions, foundation conditions and available space. The lower surface is generally curved to concentrate low flows and maintain self-cleansing velocities.

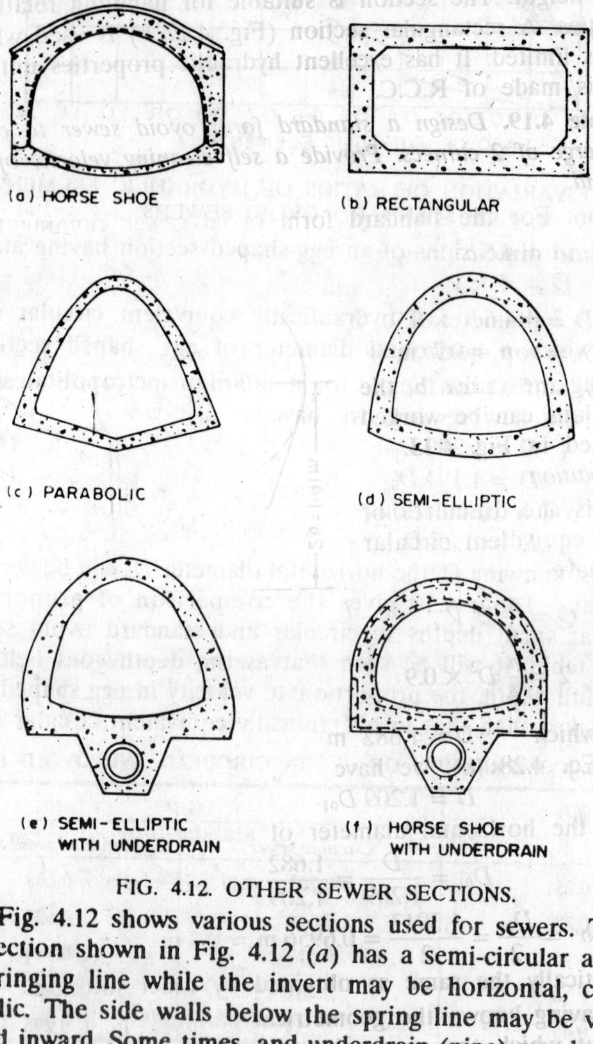

FIG. 4.12. OTHER SEWER SECTIONS.

Fig. 4.12 shows various sections used for sewers. The horse shoe section shown in Fig. 4.12 (*a*) has a semi-circular arch above the springing line while the invert may be horizontal, circular or parabolic. The side walls below the spring line may be vertical or inclined inward. Some times, and underdrain (pipe) may be provided below the horse shoe section, for draining infiltration water, as shown in Fig. 4.12 (*e*). The semi-elliptic section (Fig. 4.12 *d*) is made up of either three circular arcs or of a true semi-ellipse. For sewers having width more than 2 m, this section is more suitable. As shown in Fig. 4.12 (*e*), an underdrain may also be provided for draining infiltration water. The parabolic or delta section, shown in Fig. 4.12 (*c*) has large carrying capacity as compared with a circular section of the same height. The section is suitable for handling relatively small quantities. A rectangular section (Fig. 4.12 *b*) is used where head room is limited. It has excellent hydraulic properties until it is filled. It is made of R.C.C.

Example 4.19. *Design a standard form ovoid sewer to carry a peak discharge of 2 cumecs. Provide a self cleaning velocity of 90 cm per second.*

Solution. For the standard form ovoid sewer running full,

$$A = 4.6\, b^2$$

Now, $Q = A \times V$

or $2 = 4.6\, b^2 \times 0.9$

From which $b \approx 0.7$ m

Knowing the radius b, the other dimensions can be worked out, as marked in Fig. 4.13.

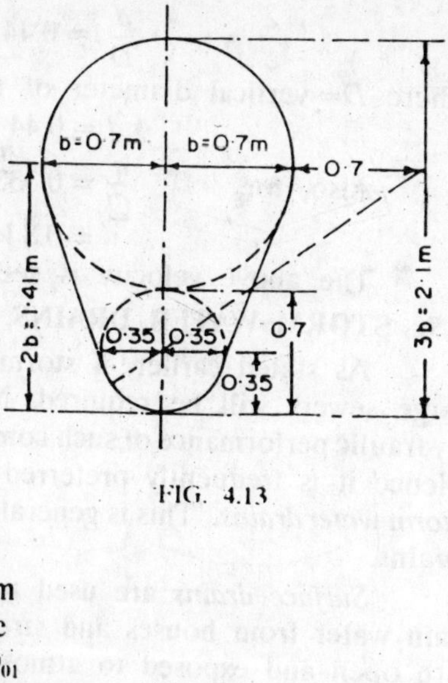

FIG. 4.13

Alternative solution

If D is the diameter of hydraulically equivalent circular section, we have

$$Q = \frac{\pi}{4} D^2 V$$

$$\therefore \quad 2 = \frac{\pi}{4} D^2 \times 0.9$$

From which $D = 1.682$ m

From Eq. 4.28 (*a*), we have

$$D = 1.209\, D_{01}$$

where D_{01} is the horizontal diameter of standard ovoid section.

$$\therefore \quad D_{01} = \frac{D}{1.209} = \frac{1.682}{1.209} = 1.3913 \text{ m}$$

$$\text{Hence } b = \frac{D_{01}}{2} = \frac{1.3913}{2} = 0.6956 \text{ m} \approx 0.7 \text{ m}$$

which is practically the same as obtained by exact solution.

After having known the geometrical parameters, let us find the gradient at which the sewer should be laid.

$$V = \frac{1}{N} R^{2/3} S^{1/2}$$

Here $V = 0.9$ m/sec. ; $R = 0.58\ b$ ...(4.21 c)

$$= 0.58 \times 0.7 = 0.406 \text{ m}$$

$$N = 0.014 \text{ (say)}$$

$$\therefore \quad 0.9 = \frac{1}{0.014}(0.406)^{2/3} S^{1/2}$$

which gives $S = 5.28 \times 10^{-4} = \frac{1}{1894}$

Example 4.20. *If in the sewer designed in Example 4.19, the actual flow in the dry weather (i.e. D.W.F.) is only 1/3 of the peak discharge, what will be the depth of flow and velocity of flow in the standard ovoid sewer ?*

Solution Given : $\frac{q}{Q} = \frac{1}{3}$

From the graphs of Fig. 4.11, for $\frac{q}{Q} = 0.333$, we get

$$\frac{d}{D} \approx 0.44$$

where D= vertical diameter of the sewer $= 3b = 2.1$ m

$$\therefore \quad d = 0.44 \times 2.1 = \mathbf{0.924\ m}$$

Also, for $\frac{q}{Q} = 0.333$, we get $\frac{v}{V} \approx 0.8$

$$\therefore \quad v = 0.8\,V = 0.8 \times 0.9 = 0.72 \text{ m/sec.}$$

The above velocity is self cleansing.

4.9. STORM WATER DRAINS

As stated earlier, if storm water is mixed with sewage flow, large sewers will be required. Not only that, at the D.W.F., the hydraulic performance of such combined sewers will not be satisfactory. Hence it is frequently preferred to carry the storm water through *storm water drains*. This is generally accomplished by providing surface drains.

Surface drains are used to carry both the sullage as well as rain water from houses and streets. They are less hygienic as they are open and exposed to atmosphere. Surface drains are normally laid along either side of the streets facing boundary walls of houses and buildings. They require frequent cleaning as bigger size particles float in due to difficulty in ensuring proper self-cleansing velocity. They carry gravity discharge. For efficient working, the surface drains should fulfill the following requirements : (1.) The inner surface of the drain should be smooth, and this can be achieved by applying plaster to the inner surface. (2.) It should have sufficient carrying capacity, and should have reasonable free board. (3.) It should be laid on a grade so as to achieve self-cleansing velocity even with minimum dry weather flow. (4.) All its joints should be properly and neatly finished. (5.) It should be structurally safe and stable. (6.) It should be constructed with non-corrosive material and should have sufficient resistance to erosion.

Shapes of surface drains. The following shapes are commonly used for surface drains : (*i*) Rectangular section (Fig. 4.14 *a*), (*ii*) Trapezoidal section (Fig. 4.14 *b*), (*iii*) Semi-circular section (Fig. 4.14 *c*), (*iv*) U-section (Fig. 4.14 *d*) and (*v*) V-section (Fig. 4.14 *e*)

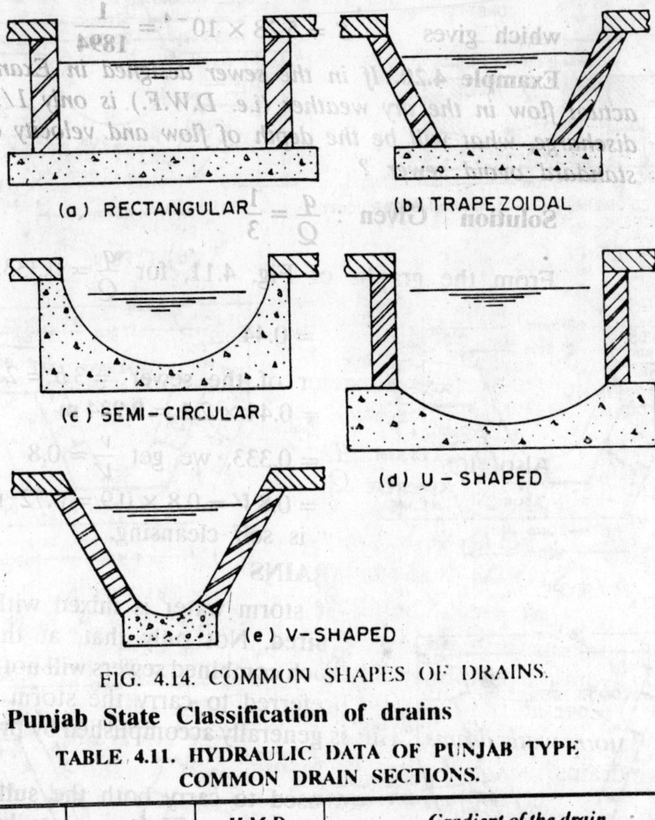

FIG. 4.14. COMMON SHAPES OF DRAINS.

Punjab State Classification of drains

TABLE 4.11. HYDRAULIC DATA OF PUNJAB TYPE COMMON DRAIN SECTIONS.

Type of drain	*A* (m^2)	*H.M.D.* *R* (*m*)	*Gradient of the drain* For V = 0.6 m/sec	For V = 0.75 m/sec	For V = 0.9 m/sec.
I	0.0091	0.00381	1 in 250	1 in 150	1 in 100
II	0.0251	0.06401	1 in 450	1 in 300	1 in 200
III	0.0464	0.08230	1 in 650	1 in 400	1 in 300
IV	0.0706	0.09754	1 in 800	1 in 550	1 in 350
V	0.0947	0.11582	—	1 in 650	1 in 450
VI	0.1375	0.13106	—	1 in 800	1 in 550

Depending upon their cross–section and discharge carrying capacities, Punjab state has classified surface drains into 12 types. Out of these, Type I to type VI are commonly used, and hence these are shown in Fig. 4.15 The hydraulic data of these drain sections, computed by Crimp and Bruges formula, are given in Table 4. 11. The discharges are computed on the basis of self cleansing velocity from 0.6 to 0.9 m/sec.

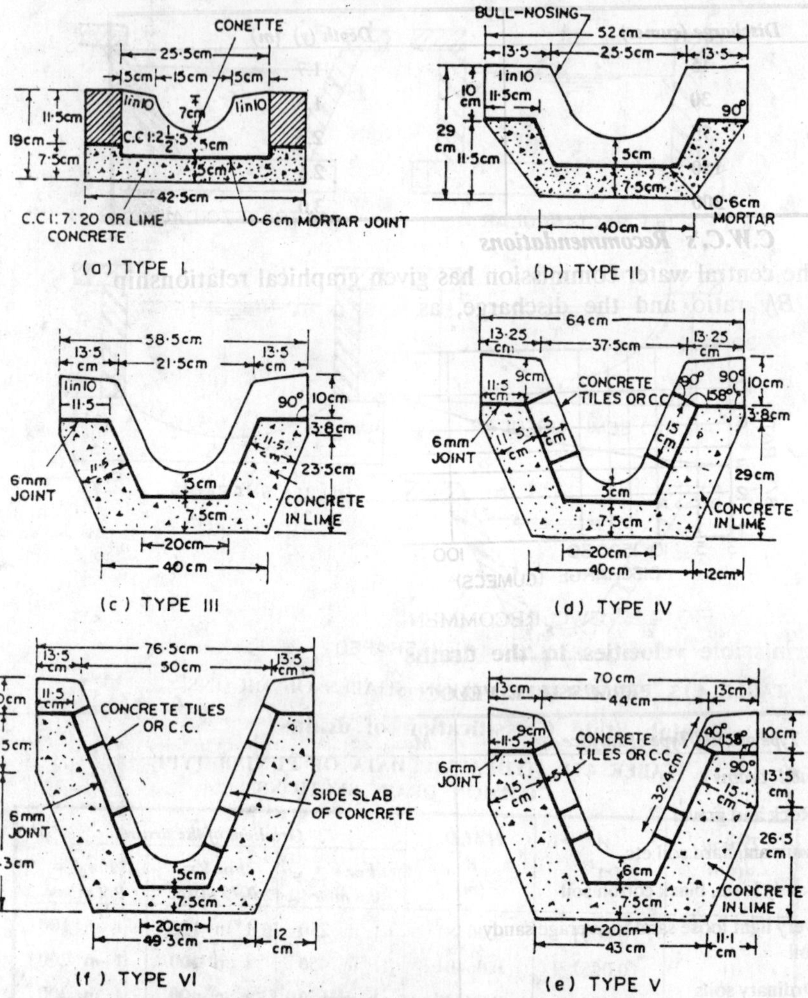

FIG. 4.15. PUNJAB STATE TYPE DRAINS.

Fixation of depth of large open drains. Large open drains are either rectangular or trapezoidal section. The depth of these drains is selected on the basis of certain empirical formulae/curves, based on experimental results.

1. *For drains having discharge upto 15 cumecs*

$$y = 0.5\sqrt{B} \quad ...(4.29)$$

where B = width of the drain

2. *For drains having discharge between 15 to 300 cumecs.*

The suggested values are given in Table 4.12

TABLE 4.12

Discharge (cumes)	*Depth (y) (m)*
15	1.7
30	1.8
75	2.3
150	2.6
300	3.0

3. *C.W.C.'s Recommendations*

The central water commission has given graphical relationship between *B/y* ratio and the discharge, as shown in Fig. 4.16.

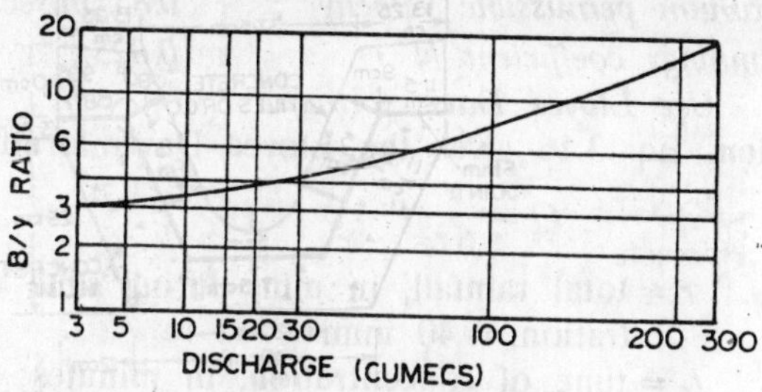

FIG. 4.16. CWC RECOMMENDATIONS.

Permissible velocities in the drains

TABLE 4.13. PERMISSIBLE VELOCITIES IN DRAINS

Type of soil / Type of surface	*Max. Permissible velocity (m/sec)*
(a) Unlined drains	
1. Rock and gravel	1.5
2. Murram, hard soil etc.	1.0 to 1.1
3. Sandy loam, black cotton soil	0.6 to 0.90
4. very light loose sand to average sandy soil	0.3 to 0.6
5. Ordinary soils	0.6 to 0.9
(b) Lined drains	
1. Stone pitched	1.5
2. Burnt clay tile lined	1.8
3. Cement concrete lined	2 to 2.5

The minimum velocity in a drain should be such that the sediments in the flow do not settle. In other words, it should have self-cleansing velocity. At the same time the maximum velocity should be such that it does not erode the bed and sides of the drain. Such a velocity is known as *non-scouring velocity.* Table 4.13 gives values of permissible velocities in lined as well as unlined drains.

Example 4.21. *Design an unlined trapezoidal section for the outfall reach of a storm water drain collecting storm water from a catchment of 50 hectares. The following information is available :*

(i) Inlet time	:	*12 minutes*
(ii) Time of flow in the upper reaches of the drain	:	*18 minutes*
(iii) Rainfall during time of concentration	:	*40 mm*
(iv) Imperviousness factor	:	*0.55*
(v) Design water surface slope	:	*1 in 2000*
(vi) Maximum permissible velocity	:	*0.85 m/sec.*
(vii) Mannings coefficient N	:	*0.025*

Use Lloyed Davis formula.

Solution. Eq. 3.16 gives the Lloyed–Davis formula :

$$Q = \frac{r}{6\,t_c} P$$

where r = total rainfall, in mm, during time of concentration = 40 mm.

t_c = time of concentration, in minutes
$= 12 + 18 = 30$ minutes

P = impervious area in hectares
$= 0.55 \times 50 = 27.5$ hectares

$$\therefore \quad Q = \frac{40}{6 \times 30} \times 27.5 = 6.111 \text{ cumecs.}$$

Let us keep a water depth of 1.2 m. Hence from Eq. 4.29

$$y = 0.5\sqrt{B}$$

or

$$1.2 = 0.5\sqrt{B}$$

From which $B = 5.76$ m. Hence keep $y = 1.2$ m and $B = 5.8$ m.

The drain will be in cutting all along. Provide side slopes of 1:1 and a F.B. of 0.3 m. The section of the drain is shown in Fig. 4.17. Longitudinal slope of the drain is 1 in 2000.

Now $A = (B + y)\,y = (5.8 + 1.2)\,1.2 = 8.4 \text{ m}^2$

$P = B + 2\sqrt{2}\,y = 5.8 + 2\sqrt{2}\,(1.2) = 9.194$ m

$$\therefore \quad R = \frac{A}{P} = \frac{8.4}{9.194} = 0.9136 \text{ m}$$

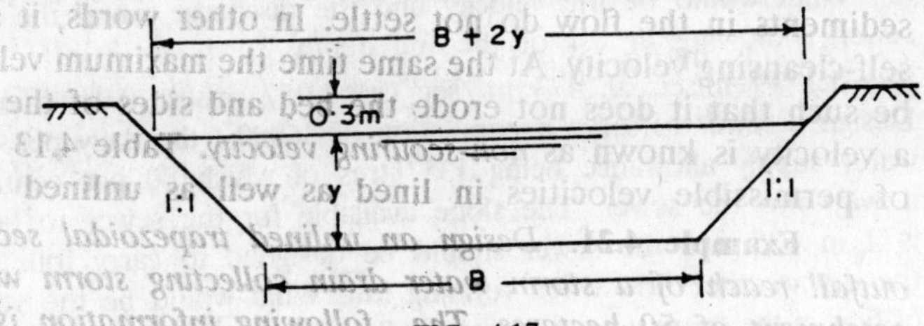

FIG. 4.17.

Now $Q = \frac{1}{N} A R^{2/3} \sqrt{S}$

$= \frac{1}{0.025} \times 8.4\,(0.9136)^{2/3} \sqrt{1/2000} = 7.07$ cumecs

Actual discharge= 6.111 cumecs only.

Hence keep a revised section of $y = 1.2$ m and $B = 5$ m

$$A = (5 + 1.2)\,1.2 = 7.44 \text{ m}^2$$

$$P = 5 + 2\sqrt{2}\,(1.2) = 8.394 \text{ m}$$

$$R = 7.44/8.394 = 0.886 \text{ m}$$

$\therefore$ $Q = \frac{1}{0.025} \times 7.44\,(0.886)^{2/3} \sqrt{1/2000} = 6.139$ cumecs.

Hence OK.

Now $V = \frac{Q}{A} = \frac{6.111}{7.44} = 0.82$ m/sec.

This is less than permissible value of 0.85 m/sec. Hence the designed section is satisfactory.

PROBLEMS

1. (*a*) What do you understand by the terms 'self-cleansing velocity' and 'limiting velocity' in sewers ?

   (*b*) Derive an expression for the self cleansing velocity in the form:

$$V_s = \sqrt{\frac{8\beta}{f}(G_s - 1)\,g\,d_s}$$

2. A circular sewer, 50 cm in diameter is laid at a gradient of 1 in 200. Using $N = 0.0125$ in Manning's formula calculate velocity, disharge and Chezy's coefficent when the sewer is running full.

   Ans. [1.414 m/sec; 0.2777 m^3/sec; 56.56]

3. Circular sewer of 0.5 m diameter is laid at a gradient of 1 in 300. Calculate the velocity of flow and discharge in the sewer when the sewer is running half fall. Use Manning's formula taking N=0.0125.

   Ans. [1.155 m/sec; 0.1134 m^3/sec]

4. (*a*) A 40 cm sewer having invert slope of 1 in 200 was flowing full. What would the velocity of flow and discharge, if $N = 0.013$? Is the velocity self-cleansing ?

(*b*) What would be the velocity and the discharge when the above sewer is flowing at 0.4 of the full depth ?

Ans. [(*a*) 1.1719 m/sec; 0.1473 m^3/sec (*b*) 1.057 m/sec; 0.0497 m^3/sec]

5. Design a sewer to serve a population of 36000 ; the daily per capita water supply allowance being 135 litres of which 80 per cent finds it way into the sewer. The slope available for the sewer to be laid is 1 in 625 and the sewer should be designed to carry four times the dry weather flow when running full. What would be the velocity of flow in the sewer when running full ? Assume $N = 0.012$ in Manning's formula.

(UPSC Engg. Services, 1970)

Ans. [$D = 0.31$ m ; $V = 2.39$ m/sec]

6. Find the minimum velocity and hydraulic gradient required to transport sand through a sewer of 1 m dia. with sand particles of 0.8 mm dia and specific gravity 2.65. Assume value of $\beta = 0.04$ and $f = 0.02$. Assume sewer to run full. What will be the gradient if the sewer runs half full ? Take $N = 0.012$.

Ans. [0.455; 1/5283; 1/5283]

7. Design a outfall circular sewer of the separate system for a town with a population of 1,00,000 persons with a water supply at 180 litres per head per day. The sewer can be laid at a slope of 10 in 10,000 with $n = 0.012$. A self-cleansing velocity of 0.75 m/sec is to be developed. The dry weather flow may be taken as $\frac{1}{3}$ of the maximum discharge. Given the following table :

Proportionate depth	*Proportionate velocity*	*Proportionate discharge*
0.31	0.7901	0.2086
0.35	0.8430	0.2629
0.37	0.8675	0.2981
0.39	0.8909	0.3217
0.40	0.9022	0.3370
0.41	0.9299	0.3682

(U.P.S.C. Engg. Services, 1975) **Ans.** [$D = 0.915$ m]

8. A circular sanitary sewer is designed to carry the maximum flow of sewage while flowing 70% full at a velocity of 0.95 m/sec. The ratio of maximum/average and average/minimum flows are 2.5 and 2 respectively. Find out : (*i*) Proportionate depth of flow, and (*ii*) Velocities of flow generated at the time of (*a*) average and (*b*) minimum flow. Neglect the variation in the values of *N*.

Ans. [(*a*) $d/D = 0.44$; $v = 0.898$ m/sec.
(*b*) $d/D = 0.30$; $v = 0.739$ m/sec.]

9. Write notes on :

(*a*) Self-cleansing velocity (*b*) Non-scouring velocity (*c*) Hydraulically equivalent section (*d*) Circular and egg shaped sewer sections.

10. Find the relation between the side a square section of one sewer and the diameter of a circular section of another sewer when both are hydraulically equivalent.

(AMIE : 1974)

Ans. [$D = 1.094\ B$]

11. Determine the size of a circular sewer for a discharge of 1.2 cumecs running half full. Assume a grade of 1 in 2000 and $N = 0.013$.

Ans. [$D = 1.754$ m]

12. A circular sewer of 1 m dia. carries a certain discharge while running full. What will be the depth of flow and the corresponding velocity ratio if the discharge in the sewer is reduced to $\frac{1}{3}$ rd the value ? Assume that N remains constant with depth.

Ans. [0.3978; 0.8997]

13. A rectangular sewer with width 1.5 times it depth is hydraulically equivalent to a circular one. Find the relation between the width of the rectangular sewer and the diameter of the circular sewer.

(A.M.I.E. 1975)

Ans. [$D = 1.043\ B$]

14. Design an out fall sewer in the separate system for a town with a population of 90,000 persons with water supply of 200 litres per head per day. The sewer is to be of brickwork rendered smooth with cement mortar ($N = 0.012$) . The permissible slope is 1 in 1000. A self-cleansing velocity of 0.75 metres per second is to be developed. The D.W.F. may be taken as $\frac{1}{3}$ rd of the maximum discharge.

What is the equivalent egg-shaped section of this circular section? Calculate the velocity in the new section when flowing full and at D.W.F.

Ans. [$D = 0.825$ m ; $b = 67.5$ m ; 1.1 m/sec.]

15. Develop a relation between the diameter of circular section of a drain and a side of a rectangular section having width as twice its depth. The rectangular drain is open from the top.

Ans. [$H = 0.6\ D$]

16. A 50 cm dia. sewer is required to flow at half–depth on a grade ensuring a degree of self cleansing equivalent to that obtained at full depth of a velocity of 0.9 m/sec. Find the required grade, associated velocities and discharges at full depth and half depth. Take a uniform value of $N = 0.015$.

Ans. [(*a*) 1/343; 0.9 m/sec; 0.1767 cumecs
(*b*) 1/343; 0.9 m/sec; 0.8835 cumecs.]

17. A 50 cm dia. sewer is to discharge 0.1 cumecs at a velocity as self-cleansing as a sewer running full at 0.9 m/sec. Find the depth and velocity of flow and the required slope. Take uniform value of $N = 0.013$.

Ans. [0.274 m; 0.908 m/sec; 1/482.8]

5

Construction of Sewers

5.1. INTRODUCTION : TYPES OF SEWERS

A *sewer* is an under ground conduit or drain through which sewage is conveyed to the point of discharge or disposal.

Main sewer or trunk sewer is a sewer that receives sewage from many tributary branches and sewers, serving as an outlet for a large territory.

Branch sewer or *submain sewer* is a sewer which receives sewage from a relatively small area, usually a few laterals, and dishcarge into a main sewer.

Lateral sewer is a sewer which collects sewage directly from the houses. It indicates the first stage of sewage collection.

Separate sewers are those which carry the house hold and industrial wastes only.

Storm water drains are those which carry rain water from roofs and street surfaces.

Combined sewers are those which carry both sewage and storm water.

House sewer (a drain) is a pipe carrying away the sewage from a building to a street sewer.

Depressed sewer is a section of sewer contructed lower than adjacent sections, to pass beneath an obstacle or obstruction. It runs full under the force of gravity.

Intercepting sewer is a sewer laid transversely to a general sewer system to intercept the dry weather flow (D.W.F.) of sewage and such additional surface and storm water as may be desirable. An intercepting sewer is usually a large sewer, flowing parallel to a natural drainage channel, into which a number of main or outfall sewers discharge.

Outfall sewer is a sewer that receives the sewage from the collecting systems and conducts it to a point of final discharge or to a disposal plant.

Relief sewer or *overflow sewer* is a sewer built to carry the flow in excess of the capacity of existing sewer.

5.2. MATERIALS FOR SEWERS

Sewers may be made of one of the following materials :

(1) Asbestos cement (2) Plain or reinforced cement concrete (3) Vitrified clay or stoneware (4) Brick (5) Cast iron (6) Steel (7) Plastic

While selecting a particular material of a sewer, the following factors must be carefully considered :

1. *Hydraulic efficiency*: The material selected should be such that value of Manning's coefficient N is as low as possible so that a hyraulically efficient surface is available.

2. *Resistance to abrasion*: The flowing sewage may contain a lot of solids such as grit and sand particles. At a high velocity at the sewage invert, erosion of sewer material may take place due to abrasion. For a longer life, the sewer material must possess high degree of resistance to abrasion.

3. *Resistance to corrosion* : The composition of sewage may be such that the sewer material may get corroded. For longer life, the sewer material should possess enough resistence to corrosion.

4. *Strength* : The sewers are generally laid underground and hence they are subjected to a number of external forces, including the weight of overburden and the vehicular/live loads. The material should of strong enough to resist all such external forces.

5. *Durability* : The sewer material should be durable so that expenditure due to its frequent replacement is minimum.

6. *Cost*: The cost of the material should be less so that overall economy is achieved in its construction.

7. *Weight*: The material should be light in weight so that it can be easily handled and transportation costs are also less.

8. *Imperviousness*: The sewer material should be imperivous and should not allow seepage of the sewage from the sewer.

1. Asbestos cement sewers. These sewer pipes are made from a mixture of asbestos fibre, cement and silica. The asbestos fibre serves as a reinforcing material. The mixture of these constituents are converted into a dense homogeneous material under pressure. These pipes are available in various sizes ranging from 75 mm to

500 mm in diameter and 3 to 4 metres in length. These pipes possess the following *advantages*.

(*i*) These pipes possess considerable strength against internal pressure. (*ii*) They are light in weight, due to which these can be easily handled. (*iii*) They can be easily cut and easily joined. (*iv*) These pipes offer good resistance to salts, and other corrosive materials normally present in the sewers. (*v*) The inside surface is relatively smooth. Hence they are hydraulically more efficient.

However these pipes possess the following *disadvantages* : (*i*) They are brittle. They are weak against impact forces. Hence careless handling may result in their breaking. (*ii*) They have less structural strength against heavy external forces. (*iii*) These are susceptible to *sulphide corrosion*. Due to these disadvantages they are used only as verticals.

2. Plain or reinforced cement concrete. Cement concrete pipes can either be plain (for small diameters) or reinforced. Plain cement concrete pipes are used from 80 mm to 450 mm, with their thickness varying from 25 to 35 mm. For bigger diameters, these are refinforced, consisting of longitudinal and transverse reinforcement in the form of rings or elliptical cage, as shown in Fig. 5.1.

The reinforcement pattern shown in Fig.5.1(*a*) is suitable upto diameter of 800 mm, and when pipe is subjected to internal pressure only, while the one shown in Fig. 5.1(*b*) is suitable for greater diameter pipes specially when they are subjected to internal as well as external pressures. The pattern shown in Fig.5.1(*c*) uses elliptical cage (in place of rings) and is suitable for large diameter pipes which are subjected to external pressure only, as is the normal case for sewers.

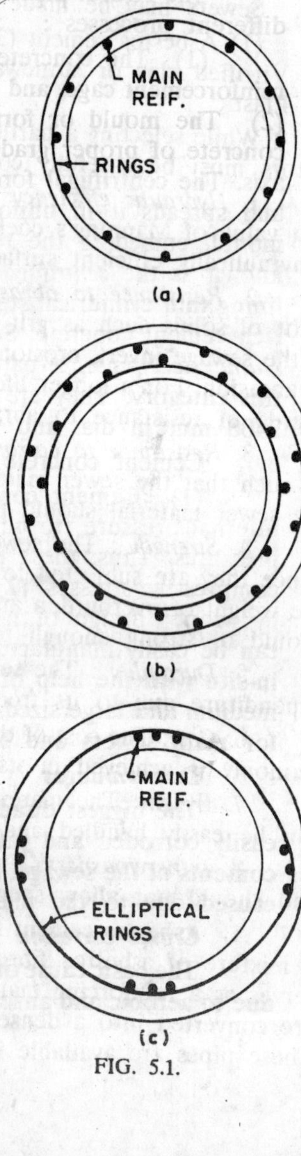

FIG. 5.1.

In order to achieve adequate strength and imperviousness, rich quality cement concrete should be used. The usual mix is $1:1\frac{1}{2}:3$. The grading and mixing of the aggregate should be proper, with the max. size of aggregates limited to 6 mm. The water-cement ratio usually varies between 0.5 to 0.7 depending upon the thickness of the pipe and equipment used for placing and compacting the concrete. The placing of cement concrete should take place immediately after its preparation. The concrete should be placed in such a way that it completely surrounds the circumferential reinforcement.

The R.C.C. pipes can be manufactured by the following three different processes :

(1) The concrete is poured by ordinary method around the reinforcement cage, and tamped. This is known as *stationary method.* (2) The mould or form containing fabricated reinforcement and concrete of proper grade is made to rotate rapidly about the pipe axis. The centrifugal force so created throws the concrete outwards and spreads it in uniform layer over the internal surface of the mould, embeding the reinforcement. Such a pipe is some times known as the *centrifugal type pipe.* (3) The pipes may be made by lining thin cylindrical steel shells, both internally and externally, with rich cement concrete. Such pipes are known as *cylinder types pipes. Hume steel pipes,* a patented form of RCC pipes, also falls under this category. They are available in size varying from 100 mm to 2400 mm in dia. and 900 mm to 2400 mm in length.

Cement concrete pipes possess the following *advantages*

(1) Cement concrete pipes are equally strong, both under internal pressure (to be used as pressure sewer) as well as external pressures. Thus, these pipes can withstand both tensile as well as compressive stresses. (2) These can be made of any desired strength by proper design of mix, thickness and reinforcement. (3) These can be easily manufactured even at the site. They can also be cast in-situ with the help of special forms. (4) They are economical for medium and large sized installations. They are, therfore, widely used for main sewers and branch sewers.

Disadvantages

The biggest disadvantage of concrete sewers are that they get easily corroded and pitted because of the corrosive action of the contents of the sewage. Such a corrosion is known as ***crown corrosion*** caused due to the phenomenon explained below.

Crown corrosion

The basic cause of *crown corrosion* is the formation of H_2S gas due to aerobic and anaerobic conditions prevalent in the lower layers

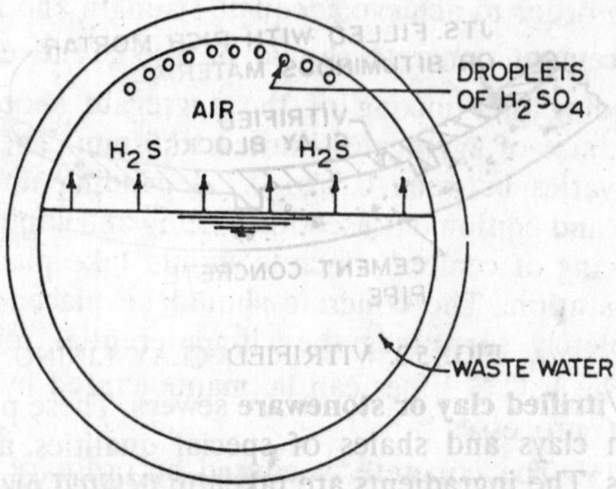

FIG. 5.2. CROWN CORROSION

of wastewater (sewage) which contain nitrogen, calcium, phosphorous, chlorides, sulphates etc. Because of bacterial action, the sulphates are reduced to sulphides and to hydrogen sulphide as under :

$$SO_4^{--} + \text{ organic mater} \xrightarrow{\text{Bacteria}} S^{--} + H_2O + CO_2$$

$$S^{--} + 2H^+ \longrightarrow H_2S$$

The hydrogen sulphide gas so produced is malodourous and inflammable, having odour like rotten eggs, due to which blackening of sewage takes place. Thiobacillus thioxidans bacteria convert the H_2S gas into H_2SO_4 as under :

$$H_2S + 2O_2 \xrightarrow{\text{Bacteria}} H_2SO_4$$

The H_2SO_4 so formed is collected into the droplet form at the crown of the sewer, and acts with concrete to form $CaSO_4$, resulting in *crown corrosion.* This not only reduces the life span of sewers but also reduces their carrying capacity with time. Apart from crown-corrosion, concrete pipes are also susceptical to erosion due to high velocity of sewage containing too much silt and grit.

In order to protect concrete pipes against such erosion and crown corrosion, their interiors are lined with *vitrified clay lining* shown in Fig. 15.3. The blocks of vitrified clay, provided with projections, are fixed in cement concrete. The joints between the adjacent blocks are filled either with rich cement mortar or with bituminous compounds.

Methods of reducing crown corrosion

(1) Lining the concrete pipes by vitrified clay blocks. (2) Ventilating the sewers. (3) Making the sewers run full. (4) Pretreating the sewage so as to reduce sulphate contents. (5) Prohibiting the entry of wastes containing sulphides. (6) Aerating and chlorinating sewage. (7) Neutralising sulphides by addition of chemicals.

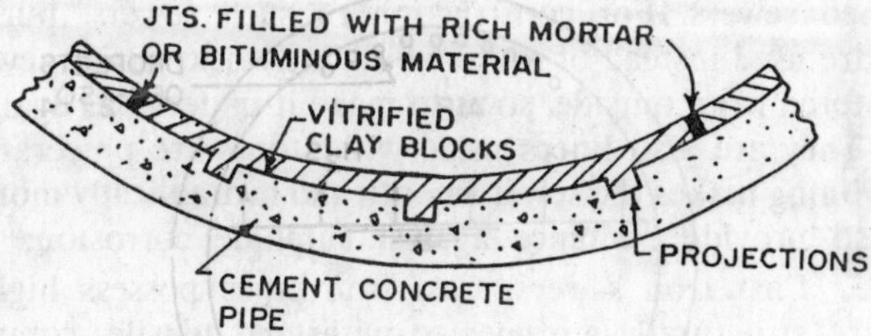

FIG. 5.3. VITRIFIED CLAY LINING.

3. Vitrified clay or stoneware sewers. These pipes are manufactured from clays and shales of special qualities and grades, using hot process. The ingradients are taken in desired proportions, ground to desired fineness and mixed together with water. Pipes are then formed in pipe press, at a specified pressure (about 0.85 N/mm^2) in desired diameters and lengths. The pipes so formed are first dried in warm air and then burnt into a kiln through a well regulated temperature gradually raised from 150°C in the beginning to about 700°C at the end. The vitrification is achieved at a higher temperature of about 1200°C, during which the surface of pipe is glazed by adding small quantity of sodium chloride salt to the kiln. However, in order to make the joints water tight, the exterior surface of spigot and the interior surface of socket are not glazed.

Advantages of vitrified pipes : (1) They are highly resistant to sulphide corrosion. (2) They possess enough resistance to erosion due to grit and silt. (3) Because of smooth interior, they are hydraulically efficient. (4) They are highly imperious. (5) Though weak in tension, they possess high compressive strength. Hence they are quite suitable for withstanding external loads such as back fill pressure, superimposed load etc. They are suitable for use even upto a depth of 4.5 to 5 m. (6) They are cheap and easily available.

Disadvantages : (1) They are weak in tension. Hence these cannot be used for locations where sewage flows under pressure. (2) They are brittle in nature, and are therefore likely to be damaged during transportation. (3) They are quite bulky and heavy. Hence their handling, laying and transportation is difficult.

Because of the above difficulties, they are used only for house drainage and for laterals. They are manufactured only in short lengths ranging from 0.3 m to 0.9 m, and in small diameters, ranging from 0.1 m to 0.3 m.

4. Brick Sewers. Bricks are being used for constructing sewers since olden days. The earliest forms of sewers were made from bricks, though these have now been replaced with the advent of cement concrete. They are still preferred for constructing at site large sized combined sewers. For construction of such sewers, purpose-made bricks are used instead of routine building bricks. Brick sewers should be plastered from outside, so as to make it impervious against ground water. They are also lined inside with stoneware or ceramic blocks. Such a lining makes the sewer smooth and hydraulically more efficient, and also provide resitance against sulphide corrosion.

5. Cast iron sewers. Cast iron pipes possess high strength. They are structurally stronger to withstand tensile, compressive as well as bending stresses. They are available in diameter ranging from 150 mm to 750 mm and lengths upto 3 to 3.5 m. They can be easily acted upon by the acids present in the sewage, unless properly protected by coating it with paint or cement concrete. However, they are relatively costlier, and are therefore adopted only under the following circumstances:

(*i*) *Heavy external loads* : They are suitable at places where heavy external loads on sewers exist, such as sewer lines below railway lines, roads, foundation walls, high overburden of soil etc.

(*ii*) *High internal pressure* : They are preferred when sewage flows under pressure. Such situations arise for conveying sewage through pumping statious and treatment works.

(*iii*) *Crosssing low level areas* : They are used at locations when sewers are to be carried over trestles or pipes, while crossing low lying areas.

(*iv*) *Under expensive road surface* : They are preferred for sewers laid under cement concrete roads where it is not possible to disturb the road surface for the purpose of frequent repairs of the sewers.

(*v*) *Protection against contamination* : They are used where these is danger of contamination of ground water due to leakage of sewage through walls of sewers. Cast iron pipes are 100% leak proof. They are also preferred when sewers are to be laid under or over water supply lines.

(*vi*) *Temperature variations* : Cast iron sewers are highly suitable where there is large variation in temperature during a day or during the year.

(*vii*) *Virbrations* : C.I. sewers are suitable for running through the ground which is susceptible due to heavy movements and vibrations. C.I. sewers are considered to be quite rigid.

(*viii*) *Wet ground conditions* : Because of their imperviousness, they are quite suitable for wet ground conditions, without fear of intiltration.

6. Steel Sewers. Sewers made of steel pipes are used at those locations where high external and or internal pressures are encountered. They are perfectly impervious. They are light in weight. Because of their flexibility, they can absorb vibrations and shock loads. They are used for mains, outfall and trunk sewers having large diameters. They are made corrosion resistant either by heavy galvanisation or by bituminous coatings. Sometimes, special corrosion resistant steel plates may be used for manufacturing steel pipes. The greatest advantage is that these sewers can be easily welded. However, the cost of such sewers is quite high in comparison to C.I. sewers.

7. Plastic sewers. The use of plastics for sewer line is of recent origin, and is still in experimental stage. Use of uPVC pipes for sewer lines started in countries like Netherlands, Scandinavia, France etc, and is not spreading in other countries with the advent of improved technology in the manufacture of such pipes. In India, their use has recently started in the form of domestic sewers. Such pipes have high hydraulic efficiency, thus permitting flatter slopes, because of very low coefficient of friction. They are corrosion resistant. They are available in longer lengths, can be jointed easily and permit cold negotiation of bends.

5.3. JOINTS IN SEWERS

Sewer joints can be of the following types :

(1) Bell and spigot joint. (2) Collar joints. (3) Simplex joints. (4) Flexible or bituminous joints. (5) Mechanical joints. (6) Open joints.

1. Bell and spigot joint. Such joints are quite common in plain or reinforced concrete pipes. Each pipe has a spigot end and a bell or socket end. Cement mortar of proportion 1 : 1 or 1 : 2 is inserted between the space of the bell end and spigot end, as shown in Fig 5.4. Gasket or jute packing may be placed in the inner side, to maintain the alignment of the sewer. The mortar caulked joint is finished at about 45° on the outer face.

2. Collar joints. For such a joint, the plain ends of the consecutive lengths of pipes are kept near each other and a collar of slightly

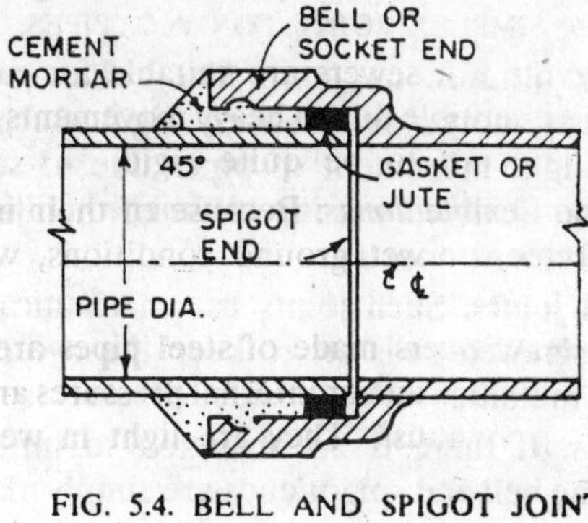

FIG. 5.4. BELL AND SPIGOT JOINT

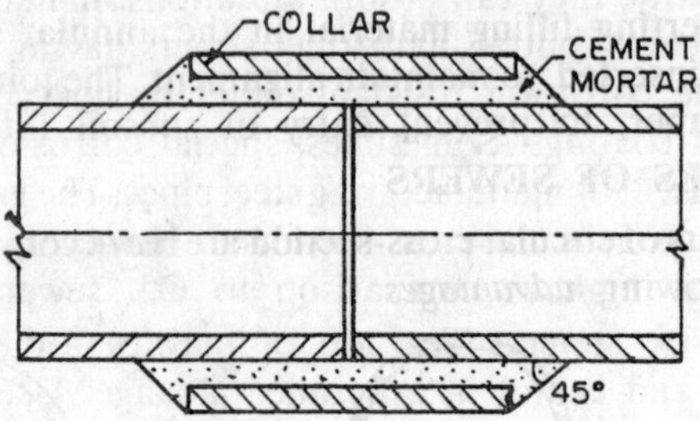

FIG. 5.5. COLLAR JOINT

bigger diameter is placed around. The annular space between the collar and the ends of the pipes is then filled with cement mortar of 1 : 1 proportion. Such joints are used for concrete pipes of larger diameters.

3. Simplex joint. Simplex joint, also known as *ring tie coupling*, is similar to collar joints, and are used for asbestos cement pipes. The joint, shown in Fig. 5.6, consists of a pipe sleeve or coupling of asbestos cement and two rubber rings which are compressed between the exteror of the pipes and the interior of the sleeve. Such a joint is quite flexible.

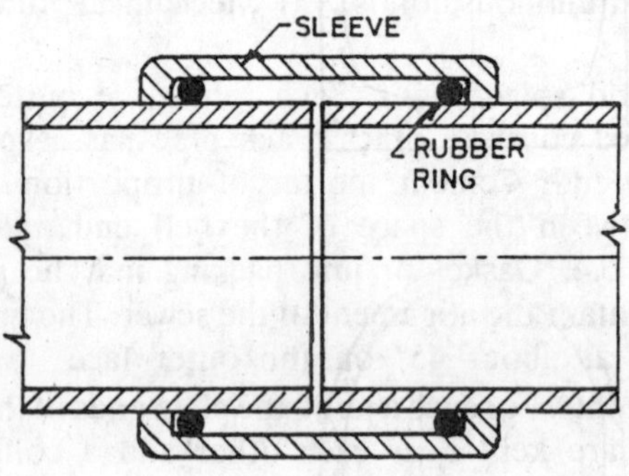

FIG. 5.6. SIMPLEX JOINT FOR A.C. PIPES.

4. Flexible or bituminous joint. The collar joint shown in Fig. 5.5, using cement mortar is relatively rigid. Such joints crack easily due to even slight movement of pipes due to settlement etc. These joints are made flexible by using bitumen or bituminous compounds instead of cement mortar.

5. Mechanical joints. Such joints use mechanical devices like flanged rings, bolts, screwed ends etc to keep the two ends together. They are used for metallic sewers made of cast iron, steel etc.

6. Open joints. If there is no objection to infiltration, open joints are adopted. The bell and spigot ends are simply placed together, without inserting filling material in the annular space. Gasket may however be inserted, to maintain alignment. The joint is merely covered with tar paper, to prevent entry of subsoil into the sewer.

5.4. SHAPES OF SEWERS

Sewers of circular cross-section are more commonly used because of the following *advantages* :

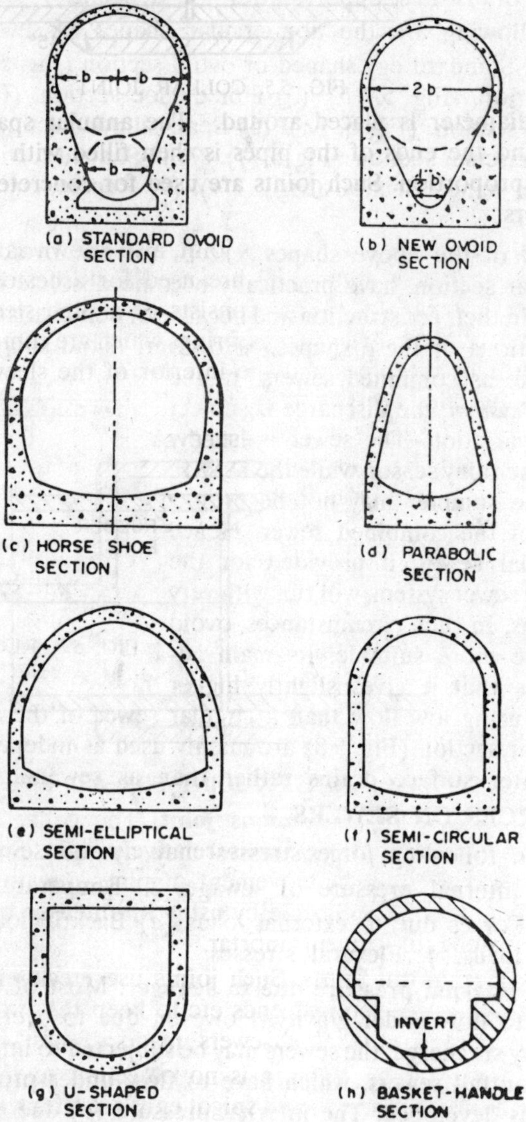

FIG. 5.7. SHAPES OF NON-CIRCULAR SECTIONS.

(1) Circular sewers are easily manufactured. (2) A circular sewer gives the maximum area for a given perimeter and thus gives the greatest H.M.D. when running full or half full. It is therefore the most efficient section at these flow conditions. (3) It is the most economical section since is utilises minimum quantities of the material. (4) Circular section has uniform curvature all round and hence it offers less opportunities for deposits.

Following are the non-circular shapes of sewers (Fig. 5.7)

(1) Standard egg shaped or ovoid section (Fig. 5.7 *a*). (2) New ovoid section. (Fig. 5.7 *b*). (3) Horse shoe section. (Fig. 5.7 *c*). (4) Parabolic section. (Fig. 5..7 *d*). (5) Semi-elliptic section. (Fig. 5.7 *e*). (6) Semi-circular section. (Fig. 5.7 *f*). (7) U-Shaped section. (Fig 5.7 *g*). (8) Basket handle section (Fig 5.7 *h*). (9) Rectangular section. (Fig. 5.7 *i*)

All of the above shapes, except for the ovoid sections and rectangular section, have practically become obsolete because of the difficulty in their construction and because of non-availability of factory made sections of these shapes. However, ovoid shaped sewers are still in use as combined sewers. In the *combined system*, the discharge is subject to great variation. The sewer is heavily taxed in the rainy season while the D.W.F. during the summer may not be even 5 to 10% of the combined sewer. Hence the circular sewer, if provided for the combined sewer system, will run with very low depths. In such circumstances, ovoid sewers are more suitable. Its main advantage is that it gives slightly higher velocity during low flow than a circular sewer of the same capacity. Rectanular section (Fig. 5.8) are mainly used as independant covered storm water surface drains rather than as sewers.

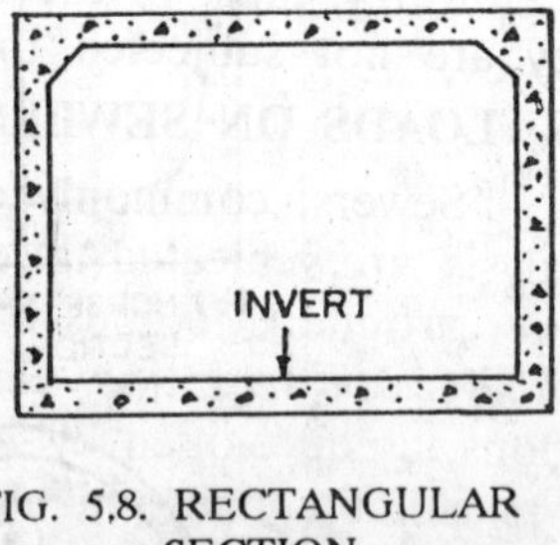

FIG. 5.8. RECTANGULAR SECTION

5.5. FORCES ON SEWERS

The following forces/stresses may act on sewers :

1. Internal pressure of sewage. 2. Temperature stresses

3. Forces due to external loads (*a*) Backfill loads (*b*) Superimposed loads. 4. Flexural stresses.

1. Internal pressure due to sewage : Most of the sewers are designed to flow under gravity. However, due to overflow of sewers and chancy surcharge, the sewers may be subjected to internal pressure. Also, in outfall sewers, which have to flow under pressure, internal pressure is developed. The internal pressure (p) due to sewage flow causes hoop stress (tensile) the magnitude of which is given by

$$p_t = \frac{p\,D}{2\,t} \quad \text{... (5.1)}$$

where p_t = tensile hoop stress ; p = internal pressure
D = diameter of sewer ; t = thickness of shell.

Hence pressure sewers must be designed to withstand the above stress.

2. Temperature stresses. The sewers are subjected to temperature stresses if they are laid above the ground and exposed to atmosphere. The change in the temperature gives rise to temperature stresses, specially when the sewers are not free to expand. The magnitude of tempeature stress is given by :

$$f = E_p \,.\, \alpha\, t \quad \text{..(5.2)}$$

where f = temperature stress induced.
α = coefficient of thermal expansion for the pipe material
t = change in the temperature
E_p = modulus of elasticity of the pipe matterial.

In order to counteract or reduce temperature stresses, special types of expansion joints are provided at an interval of 20 to 30 m. However, most of the sewers are buried underground and hence they are not subjected to temperature stresses.

5.6. LOADS ON SEWERS DUE TO BACK FILL

Sewers, commonly called conduits for the perposes of load analysis, are subjected to loads due to backfill, when they are buried under ground. Methods of determining the vertical load on buried conduits due to gravity earth forces in commonly encountered conditions, as developed by A. Marston, are generally accepted as the most suitable and reliable for computations. Theoretically stated, the load on buried conduit is equal to the weight of prism of soil directly over the conduct, plus or minus the frictional shear forces transferred to the prism by the adjacent prism of soil.

The general form of Marston formula is :

$$W_c = C \,.\, \gamma\, B^2 \quad \text{...(5.3)}$$

where W_c = vertical load per unit length acting on the conduit due to backfill
γ = unit weight of backfill
B = width of the ditch or trench,
C = load coefficient.

Underground conduits (sewers) may fall under the following three categories depending upon the types of installations or construction conditions :

1. Trench ditch condition 2. Embankment condition, and 3. Tunnel condition.

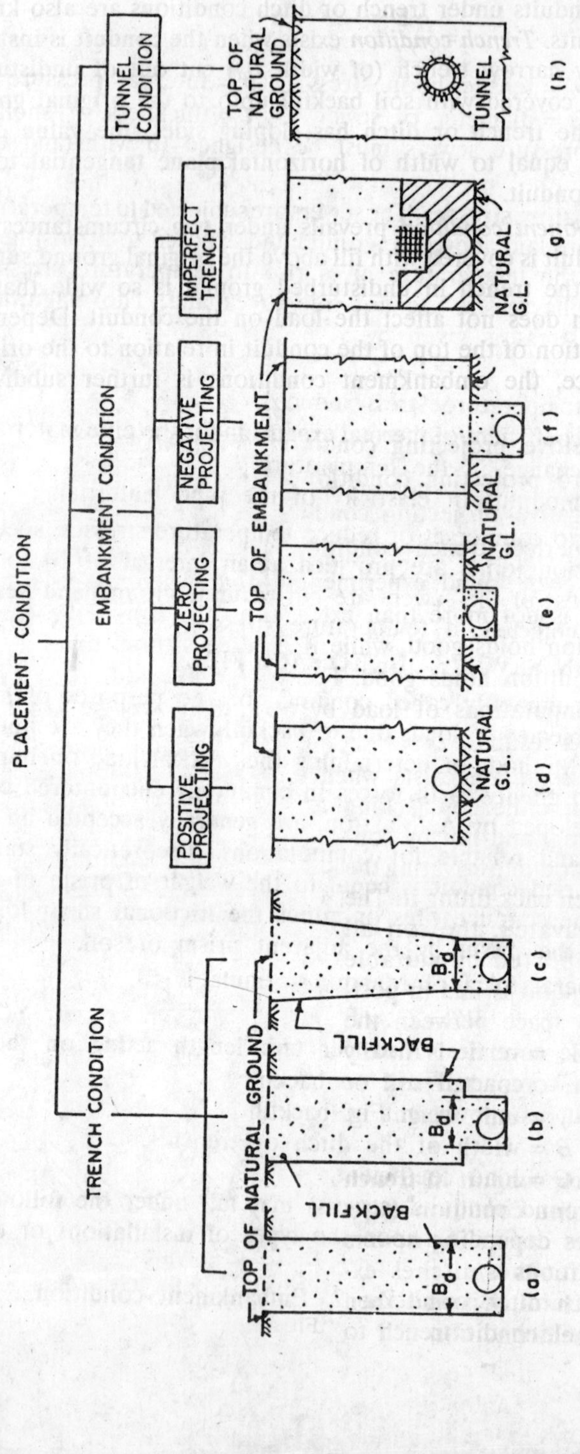

FIG. 5.9. CLASSIFICATION OF CONDUITS ON THE BASIS OF PLACEMENT CONDITIONS

The conduits under trench or ditch conditions are also known as *ditch* conduits. *Trench condition* exists when the conduit is installed in a relatively narrow trench (of width B_d) cut out of undisturbed soil and then covered with soil backfill upto to the original ground surface. If the trench or ditch has sloping sides, the value of B_d will be taken equal to width of horizontal plane tangential to the top of the conduit.

Embankment condition prevails under two circumstances : (*i*) when the conduit is covered with fill above the original ground surface, or (*ii*) when the trench in undisturbed ground is so wide that the trench friction does not affect the load on the conduit. Depending upon the position of the top of the conduit in relation to the original ground surface, the embankment conditions is further subdivided as follows :

(*a*) Positive projecting condition (Fig.5.9*d*)
(*b*) Zero projecting condition (Fig.5.9*e*)
(*c*) Negative projecting condition. (Fig.5.9*f*)
(*d*) Imperfect trench condition. (Fig.5.9*g*)

It has been found experimentally that when the width (B_d) of the trench is not more than external width (B_c) of the conduit, trench condition holds good, while if B_d exceds three times B_c, embankment condition holds good. For transition width $B_d = 2\,B_c$ to $B_d = 3\,B_c$, computations of load by adopting both the conditions give the same results.

Tunnel condition exists when the sewer is laid more than 9 to 12 m deep either by jacking or by tunnelling, and not by conventional method of excavating the trench and then back filling it. The tunnel is excavated, the soil all round is supported by suitable means, the conduit is laid in position, and the space between the tunnel and the conduit is finally filled up with compacted soil or concrete grout, as shown in Fig. 5.10.

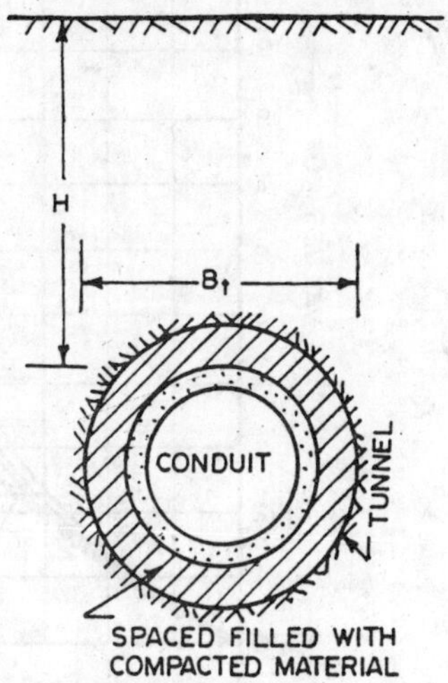

FIG. 5.10. CONDUIT IN TUNNEL.

Loads on conduit in trench condition. Such a condition exists in most of the cases. The sewers are laid in trenches/ditches excavated in natural soil and then covered by refilling the trench to its orginal ground level, as shown in Figs. 5.9 (*a*), (*b*), (*c*).

The load on *rigid conduit* laid in trench condition is given by Marston's formula :

$$W_c = C_d \,.\, \gamma \,.\, B_d^2 \qquad ...(5.4)$$

where B_d = width of the trench at the top of the conduit

C_d = load coefficient which is a function of the ratio of height (H) of the fill to width (B_d) of the trench, and the friction coefficient (μ') between the backfill and the sides ofthe trench, and its value is given by

$$C_d = \left[\frac{1 - e^{-2K\mu' H/B_d}}{2K\mu'}\right] \qquad ...(5.4\,a)$$

where K is the ratio of lateral unit pressure and vertical unit pressure.

The value of the coefficient C_d can be obtained from curves of Fig. 5.11, for various values of H/B_d plotted for several kinds of filling materials having different values of $K\mu'$ (Spanglar, 1963). For curve (*a*), $K\mu'$ = 0.1924 (minimum for granular materials without

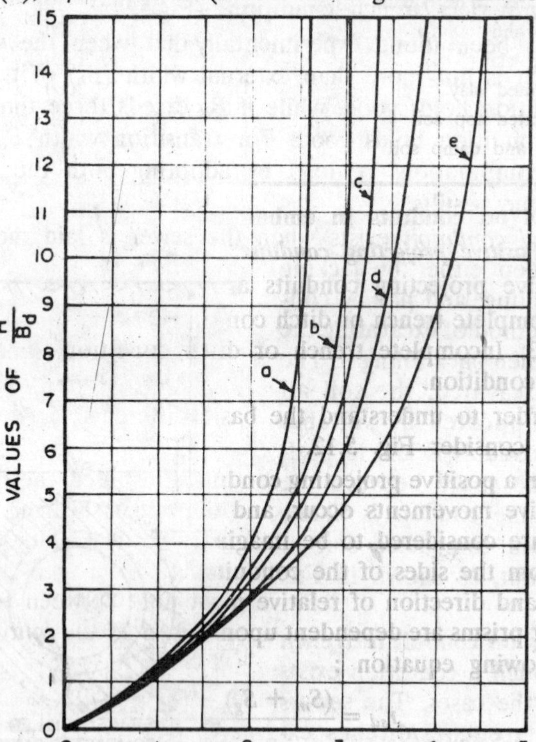

FIG. 5.11. CURVES FOR C_d FOR DITCH CONDUIT

cohesion); for curve (*b*), $K\mu' = 0.165$ (maximum for sand and gravel); for curve (*c*), $K\mu' = 0.150$ *(maximum for saturated top soil), for curve (d)*, $K\mu' = 0.130$ (ordinary maximum for clay), and for curve (*e*), $K\mu' = 0.110$ (max. for saturated clay). In cases of excavations with sloping sides (possible in undeveloped areas), the provision of sub-trench (Fig. 5.9 *b, c*) minimises the load on the pipe by reducing the value of B_d where B_d will be equal to the width of horizontal plane tangential to the top of the conduit.

The load on *flexible conduit* laid in trench condition is given by the following expression

$$W_c = C_d \gamma B_c . B_d \qquad ...(5.5)$$

where B_c = outside width (diameter) of the condult.

The value of unit weight γ for common backfill materials may be taken from Table 5.1.

TABLE 5.1. WEIGHT OF COMMON BACKFILL MATERIALS.

	Type of backfill	*Unit weight γ (kg/m^3)*
1.	Dry sand	1600
2.	Ordinary (damp) sand	1840
3.	Wet sand	1920
4.	Damp clay	1920
5.	Saturated clay	2080
6.	Saturated top soil	1840
7.	Sand and damp soil	1600

Loads on conduits in embankment condition

(a) Positive projecting conduits

Positive projecting conduits are of four types :

1. Complete trench or ditch condition. 2. Complete projection condition. 3. Incomplete trench or ditch condition. 4. Incomplete projection condition.

In order to understand the basic difference between all the four types, consider Fig. 5.12.

When a positive projecting conduit is installed, the plane along which relative movements occur, and on which shearing forces are generated are considered to be imaginary vertical planes extending upwards from the sides of the conduits as shown in Fig. 5.12. The magnitude and direction of relative movement between the interior and exterior prisms are dependent upon the *settlement ratio* r_{sd} defined by the following equation :

$$r_{sd} = \frac{(S_m + S_g) - (S_f + d_c)}{S_m} \qquad ...(5.6)$$

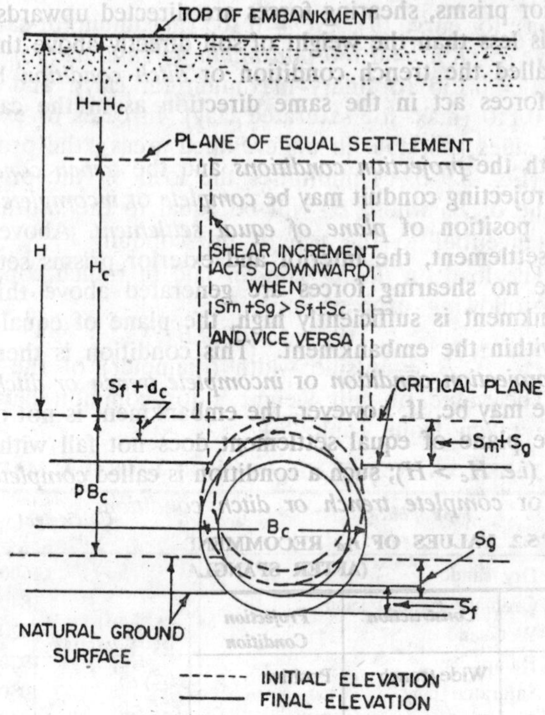

FIG. 5.12. POSITIVE PROJECTING CONDUIT.

where S_m = compression strain of side prisms of soil of height $p \cdot B_c$

p = projection ratio= ratio of the distance from the natural ground surface to the top of the conduit to its width B_c.

S_g = settlment of natural ground surface adjacent to the conduit.

S_f = settlement of the conduit into its foundation.

d_c = shortening of vertical height of the conduit.

The *critical plane* is a horizontal plane, tangential to the top of the conduit. If the critical plane settles more than the top of the conduits, *i.e.* if $(S_m + S_g)$ is more than $(S_f + d_c)$, the settlement ratio is *positive*. Hence the exterior prisms move downward relative to the interior prism, the shearing forces are directed downwards and hence the load on the conduit is *increased*. This is known as *projection condition*. If, however, the critical plane settles less than the top of the conduit, *i.e.* of $(S_m + S_g) < (S_f + d_c)$, the settlement ratio is *negative*. The interior prism moves downwards relative to the exterior prisms, shearing forces are directed upwards and hence the load is less than the weight of soil directly above the structure. This is called the trench condition or *ditch condition* because the shearing forces act in the same direction as in the case of ditch condition.

Both the *projection conditions* and the *trench conditions* of a positive projecting conduit may be *complete* or *incomplete*, depending upon the position of *plane of equal settlement*. Above the plane of equal settlement, the interior and exterior prisms settle equally, and hence no shearing forces are generated above this plane. If the embankment is sufficiently high, the plane of equal settlement may be within the embankment. This condition is then called *incomplete projection condition* or *incomplete trench or ditch condition*, as the case may be. If, however, the embankment is not of sufficient height, the plane of equal settlement does not fall within the embankment (*i.e.* $H_c > H$); such a condition is called *complete projection condition* or *complete trench or ditch condition*.

TABLE 5.2. VALUES OF r_{sd} RECOMMENDED FOR USE IN DESIGN (AFTER SPANGLAR)

Conduit	*Construction*	*Projection Condition*	*Foundation Soil Condition*	*Recommended Value*
Rigid	Wide trench	Positive	Any	+ 1.0
	Embankment	Positive	Rock or hard unyielding soil	+ 1.0
			Ordinary soil	0.5 to 0.8
			Yielding soil	0 to + 5
	Any	Positive	Pipe line supported on piles in soft soil	Uncertain (use complete projection value of C_c)
	Wide trench	Negative	Any	– 0.3
	Embankment	Negative	Any	– 0.3
	Induced trench	Negative	Any	0.0 to – 0.3
Flexible	Wide trench	Positive	Any with poorly compacted side fill	0 to – 0.4
	Embankment	Positive	Any with well compacted side fill	0
	Wide trench	Negative	Any with well compacted side fill	0
	Embankment	Negative	"	0
	Induced trench	"	"	0

The product $r_{sd} \times p$ gives the relative height of plane of equal settlement and hence the magnitude of the shear component of the load. When $r_{sd} \times p = 0$, the plane of equal settlement coincides with the critical plane and there are no shearing forces and the load is equal to the weight of the central prism. It is not practicable to predetermine the value of r_{sd}. However, recommended design values of r_{sd} are given in Table 5.2. Marston gave the following expression for the load in positive projecting conduits (both rigid and flexible):

$$W_c = C_c \cdot \gamma B_c^2 \qquad ...(5.7)$$

where W_c = load on conduit in kg per metre length.

γ = unit weight of soil (kg/m^3)

B_c = outside width of conduit (m)

C_c = load coefficient which is a function of the product of projection ratio and the settlement ratio and the height of fill above the top of the conduit to the out side width of the conduit (H/B_c). It is also influenced by the product of the coefficient of internal friction (μ) of the backfill materal and the Rankine's ratio of lateral pressure to vertical pressure (K). Suggested values of $K\mu$ for positive and negative settlement ratios are 0.19 and 0.13 respectively.

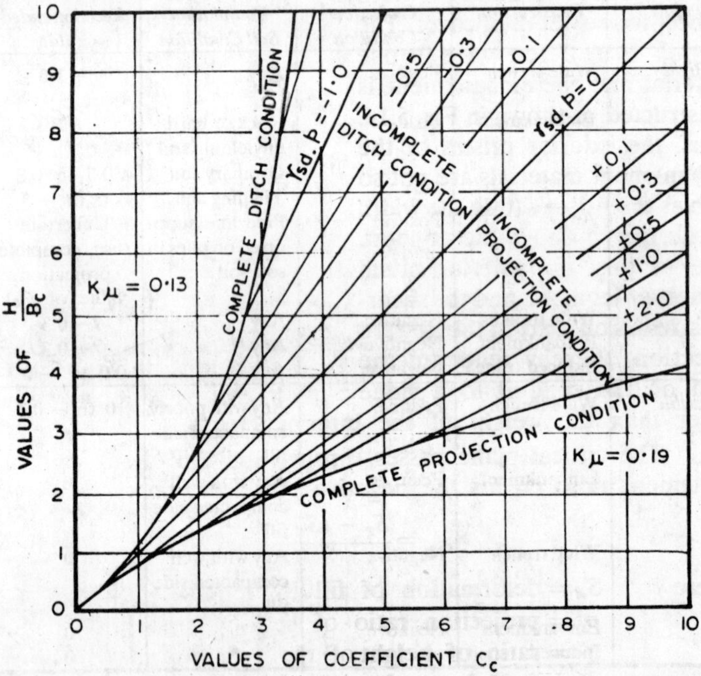

FIG. 5.13. CURVES FOR DETERMINATION OF C_c FOR POSITIVE PROJECTING CONDUITS (SPANGLAR, 1963)

The values of C_c can be obtained from the curves of Fig.5.13. Alternatively, C_c can be computed from the following equation.

$$C_c = \frac{e^{\pm 2K\mu (H/B_c)} - 1}{\pm 2K\mu} \quad ...(5.7\ a)$$

using + sign for complete projection and – sign for complete trench condition.

For incomplete trench condition or in complete projection condition.

$$C_c = \frac{e^{\pm K\mu (H/B_c)} - 1}{\pm 2K\mu} + \left(\frac{H}{B_c} - \frac{H_e}{B_c}\right) e^{\pm 2K\mu (H_e/B_c)} \quad ...(5.7\ a)$$

using + sign for incomplete projection conduit and minus sign incomplete trench conduit.

(*b*) ***Negative projecting conduits.*** As shown in Fig. 5.9 (*f*), a conduit is said to be laid in a negative projecting condition when the trench in which it is laid is narrow with respect to the size of the pipe and shallow with respect to the depth of cover. The trench is refilled with loose material and the embankment is constructed, as shown in Fig. 5.14. Thus, the exterior prisms of the embankment materials are not so high as the prism within the width of the trench above the conduit. Due to this, the interior prism always settles more, and the shearing stresses on it act in the upward direction, thereby reducing the load on the conduit to a value lesser than the weight of the interior prism.

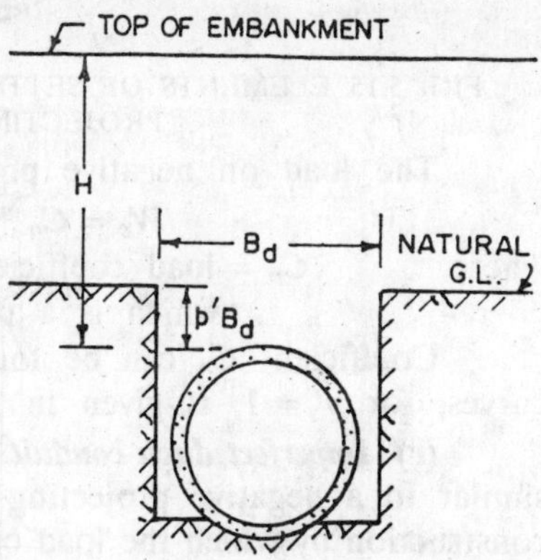

FIG. 5.14. NEGATIVE PROJECTING CONDUIT

The critical plane passes through the top of the trench. The settlement ratio r_{sd} for the negative projecting conduit is given by

$$r_{sd} = \frac{S_g - (S_d + S_f + d_c)}{S_d} \quad ...(5.8)$$

where S_d = deformation of fill in trench within height $p' B_d$

p' = projection ratio of negative projecting conduit.

= ratio of height of the ditch above the top of the conduit to the width of the ditch.

The settlement ratio for a negative projecting conduit is always *negative* since the settlement of the critical plane is more than the settlement of the natural ground. Hence $r_{sd} \cdot p'$ is also negative.

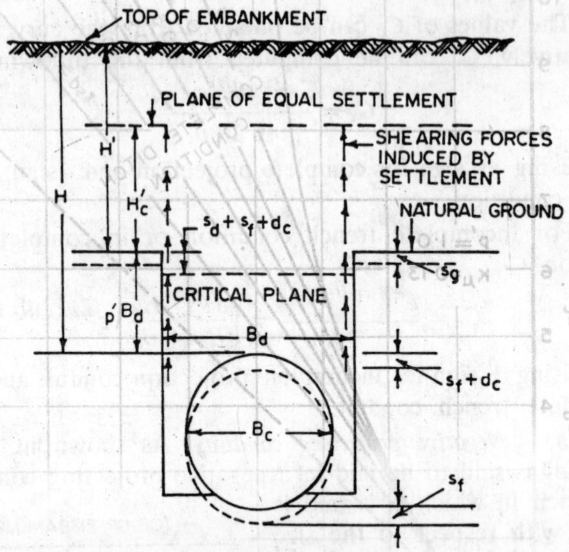

FIG. 5.15 ELEMENTS OF SETTLEMENT RATIO IN A NEGATIVE PROJECTING CONDUIT

The load on negative projecting conduit is given by

$$W_c = C_n \, . \, \gamma \, B_d^2 \qquad ...(5.9)$$

where C_n = load coefficient for negative projecting conduit which is a function of H/B_d ratio, p' and r_{sd}.

Coefficient C_n can be found from curves. One such set of curves, for $p' = 1$, is given in Fig. 5.16.

(c) ***Imperfect ditch conduit.*** Imperfect ditch conduit, somewhat similar to a negative projecting conduit, is a special technique of construction by which the load on the conduit is very much reduced so that the conduit can support a greater height of embankment. Fig. 5.17 shows two stages of construction of such a conduit.

In the first stage of construction, the conduit is installed as a positive projection conduit, and a fill is constructed over it upto a height of $p' \, B_c$ (1 to $1\frac{1}{2} B_c$). The fill is well compacted. In the second stage a trench of width B_c is dug above the conduit, down to the top of the conduit. The trench is refilled with material placed in the loosest possible manner upto the top and the rest of the embankment is constructed as usual. Marston recommended the use of hay, straw, cornstalks etc. to be filled in the trench, in order to increase the compressibility of this prism of soil and thus decrease the load on the conduit. This method of construction was used in Atlanta, Georgia, where the initial embtankment of 35 ft. above a sewer constructed in 1937 was proposed to be raised to about 96 feet in 1955. The

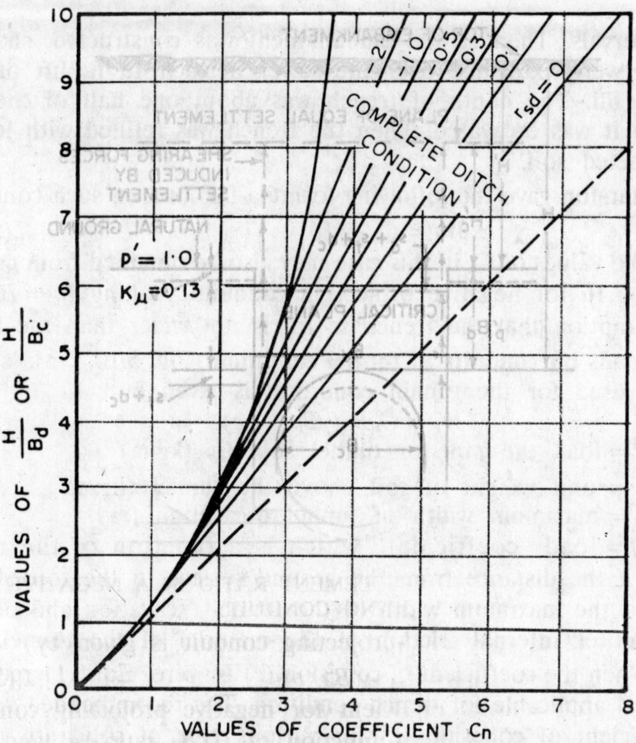

FIG. 5.16. CURVES FOR DETERMINATION OF C_n

sewer was protected from the increased load, by excavating a trench to a depth of 15 ft. below the natural ground, directly over the pipe. The trench was refilled with loose soil in which three 1ft layers of tree leaves and pine straw were incorporated at approximately

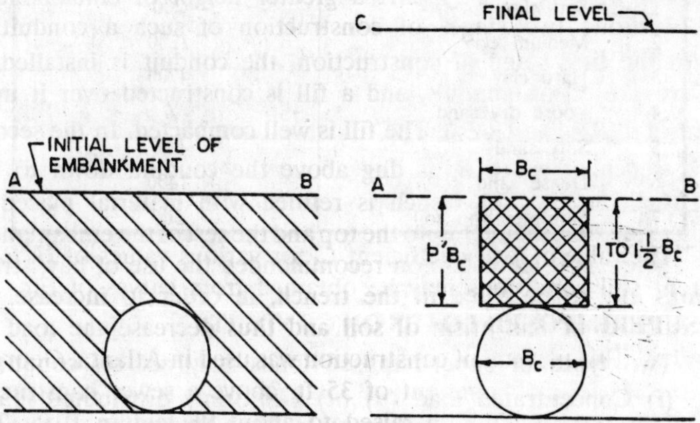

FIG. 5.17. CONSTRUCTION OF IMPERFECT DITCH CONDUIT

5 ft. intervals. Then as the embankment was constructed, shallow trenches were excavated over the sewer in each 15 ft. lift of the compact fill. The depth of trench was about one half of the lift in which it was excavated. Then the trench was refilled with loose, uncompacted soil.

Marston gave the following formula for load on such conduits

$$W_c = C_n \cdot \gamma B_c^2 \qquad ...(5.10)$$

The values of C_n in this case may also be obtained from curves for Fig. 5.16 for negative projecting conduits, taking $B_c = B_d$ on the assumption that the trench in fill is not wider than the pipe.

Loads on conduits in tunnel condition (Fig. 5.10). Marston's formula used for the tunnel condition is given by

$$W_t = C_t B_t (\gamma B_t - 2c) \qquad ..(5.11)$$

where W_t= load the pipe or tunnel support (kg/m)

γ = unit weight of soil above the tunnel (kg/m^3)

B_t = maximum width of tunnel excavation (m)

C_t = load coefficient which is a function of the ratio (H/B_t) of the distance from the ground surface to the top of the tunnel to the maximum width of the tunnel excavation and of the coefficient of internal friction of the material of the tunnel.

When the coefficient of cohesion (c) is zero, Eq. 5.11 reduces to Eq. 5.4 applicable for trench condition. The recommended values of coefficient of cohesion (c) for various type of solid are given in Table 5.3.

TABLE 5.3. RECOMMENDED VALUES OF COEFFICIENT OF COHESION (c)

	Type of soil	*Coefficient of cohesion (c) (kg/m^2)*
1.	Soft clay	200
2.	Medium clay	1200
3.	Hard clay	4700
4.	Loose dry sand	0
5.	Silty sand	500
6.	Dense sand	1400
7.	Saturated top soil	500

The value of load coefficient C_t for various values of H/B_t and different soil conditions may be obtained from curves of Fig. 5.18.

5.7. SUPERIMPOSED LOADS ON CONDUITS

Two types of super-imposed loads may act on conduits :

(*i*) Concentrated load (*p*) (*ii*) Uniformly distributed load of intensity *p* over a rectangular area of width *N* and length *M*.

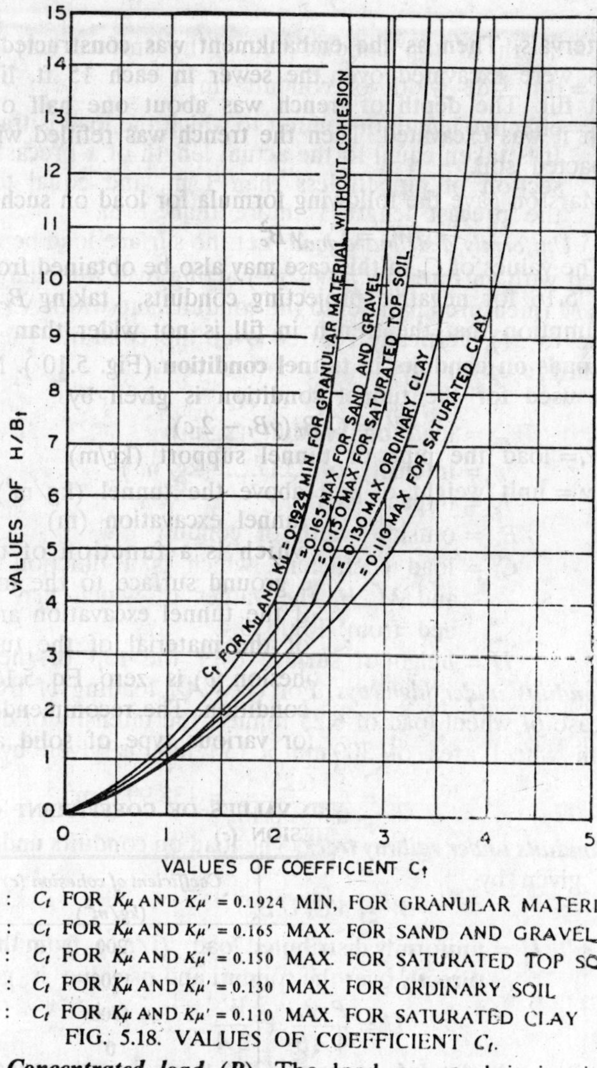

a : C_t FOR $K\mu$ AND $K\mu'$ = 0.1924 MIN, FOR GRANULAR MATERIALS
b : C_t FOR $K\mu$ AND $K\mu'$ = 0.165 MAX. FOR SAND AND GRAVEL
c : C_t FOR $K\mu$ AND $K\mu'$ = 0.150 MAX. FOR SATURATED TOP SOIL
d : C_t FOR $K\mu$ AND $K\mu'$ = 0.130 MAX. FOR ORDINARY SOIL
e : C_t FOR $K\mu$ AND $K\mu'$ = 0.110 MAX. FOR SATURATED CLAY

FIG. 5.18. VALUES OF COEFFICIENT C_t.

(*a*) ***Concentrated load*** (*P*). The load on conduit is given by

$$W_{sc} = C_s \cdot \frac{P I_e}{L_c} \quad ...(5.12)$$

where W_{sc} = load on the conduit (kg/m)

P = concentrated load acting on the surface (kg)

I_e = impact factor, the value of which may be taken as 1.0 for air field runways, 1.5 for highway traffic and air filled taxiways and 1.75 for railway traffic.

C_s = load coefficient, which is a function of $B_c/2H$ and $L_c/2H$. The values of C_s may be taken from Table 5.4.

H = depth of top conduit below the surface (m)

B_c = out side width of conduit (m)

L_c = effective length of conduit to which the load is transmitted. It is taken equal to the actual length of a precast segment section of pipe if less than 1 m, and equal to 1 m is the precast length is more than 1 m.

(*b*) ***Uniformly distributed load***. Let the surface load be uniformly distributed with an intensity p (kg/m^2) acting over a rectangular area of length M (measured parallel to the conduit) and width N (measured transverse to M), vertically centred over the conduit. The load on the conduit is given by

$$W_{sd} = C_s p \cdot B_c \cdot I_e \qquad ...(5.13)$$

where W_{sd} = load on conduit (kg/m)

p = intensity of U.D.L. (kg/m^2)

I_e = impact factor.

B_c = outside width of conduit

C_s = load coefficient, which is a function of $N/2H$ and $M/2H$, the values of which may be obtained from Table 5.4.

H = height of surface over the top of the conduit.

Conduits under highways. For class *AA* loading of IRC, in the critical case of wheel load of 6.25 tonnes, the intensity of distributed load with wheel area of 300 mm × 150 mm is given by

$$p = \frac{6.25}{0.3 \times 0.15} = 138.99 \text{ t/m}^2$$

Conduits under railway track. The load on conduits under railway track is given by

$$W = 4\,C_s\,U\,B_c \qquad ...(5.14)$$

where U = uniformly distributed load, (t/m^2) from the surface directly over the conduit and its value is given by

$$U = \frac{P \cdot I_e}{4AB} + \frac{W_t}{2A} \qquad ...(5.15)$$

where P = axle load in tonnes (22.5 t for broad gauge)

I_e = impact factor (= 1.75 for railroad)

$2A$ = length of the sleeper (= 2.7 m for broad gauge)

$2B$ = distance between the two axles (=1.84 m for broad gauge)

W_t = weight of track structure (= 0.3 t/m for broad gauge)

C_s = Load coefficient which depends on the height of top of sleeper from the top of conduit

B_c = width of conduit (m)

For broad gauge track, Eq. 5.14 reduces to

$$W = 32.14\,C_s \cdot B_c \qquad ...(5.16)$$

TABLE 5.4. VALUES OF LOAD COEFFICIENT C_s

$N/2H$ or $B_c/2H$	$M/2H$ or $L/2H$: 0.1	0.2	0.3	0.4	0.5	0.6	0.7	0.8	0.9	1.0	1.2	1.5	2.0	5.0
0.1	0.019	0.037	0.053	0.067	0.079	0.089	0.097	0.103	0.108	0.112	0.117	0.121	0.124	0.128
0.2	0.037	0.072	0.103	0.131	0.155	0.174	0.189	0.202	0.211	0.219	0.229	0.238	0.244	0.248
0.3	0.053	0.103	0.149	0.190	0.224	0.252	0.274	0.292	0.306	0.318	0.333	0.345	0.355	0.360
0.4	0.067	0.131	0.190	0.241	0.284	0.320	0.349	0.373	0.391	0.405	0.425	0.440	0.454	0.460
0.5	0.079	0.155	0.224	0.284	0.336	0.379	0.414	0.441	0.463	0.481	0.505	0.525	0.540	0.548
0.6	0.089	0.174	0.252	0.320	0.379	0.428	0.467	0.499	0.524	0.544	0.572	0.596	0.613	0.624
0.7	0.097	0.189	0.274	0.349	0.414	0.467	0.511	0.546	0.584	0.597	0.628	0.650	0.674	0.688
0.8	0.103	0.202	0.292	0.373	0.441	0.499	0.546	0.584	0.615	0.639	0.674	0.703	0.725	0.740
0.9	0.108	0.211	0.306	0.391	0.463	0.524	0.574	0.615	0.647	0.673	0.711	0.742	0.766	0.784
1.0	0.112	0.219	0.318	0.405	0.481	0.544	0.597	0.639	0.673	0.701	0.740	0.774	0.800	0.816
1.2	0.117	0.229	0.333	0.425	0.505	0.572	0.628	0.674	0.711	0.740	0.783	0.820	0.849	0.868
1.5	0.121	0.238	0.345	0.440	0.525	0.596	0.650	0.703	0.742	0.774	0.820	0.861	0.894	0.916
2.0	0.124	0.244	0.355	0.454	0.540	0.613	0.674	0.725	0.766	0.800	0.849	0.894	0.930	0.956

5.8. SAFE SUPPORTING STRENGTH AND BEDDING CONDITIONS

The magnitude of the supporting strength of a conduit depends upon the following : (*i*) Distribution of vertical load (*ii*) Bedding reaction (*iii*) Lateral pressure acting on the side of the conduit (*iv*) Inherent strength of conduit.

The inherent strength of a rigid conduit is usually expressed in terms of *three edge bearing test* conducted in the laboratory, using a suitable *load factor. Load factor* is defined as the ratio of the strength of pipe under any stated conditions of loading and bedding, to its strength measured by the three edge bearing test. It should be carefully noted that load factor does not contain a factor of safety.

The *field supporting strength* of a rigid conduit is the maximum load per unit length which the conduit will support while retaining complete serviceability. It is evidently equal to the three-edge bearing strength (measured in the laboratory) multiplied by the load factor.

The *safe supporting strength* (*i.e.* maximum allowable field load) is evidently equal to the field supporting strength divided by a suitable factor of safety :

$$\therefore \text{ Safe supporting strength } = \frac{\text{Field supporting strength}}{\text{Factor of safely}}$$

$$= \frac{\text{Load factor} \times \text{three edge bearing strength}}{\text{Factor of safety}}$$

Hence, Required three-edge bearing strength

$$= \frac{\text{max. allowable field load} \times \text{Factor of safety}}{\text{Load factor}}$$

A factor of safely of atleast 1.5 should be applied. The load factor depends upon the type of bedding specified for various conditions of conduit installations, along with their recommended bedding conditions described below.

(*a*) ***Conduits in trench conditions***. Fig. 5.19 shows *four classes, A, B, C* and *D* of bedding most often used for pipes in trenches. In *class A bedding*, either a concrete cradle is used or a concrete arch is used, both being made of plain cement concrete ; however the concrete cradle, and concrete arch can be reinforced. *Class B bedding* has either a shaped bottom or compacted granular bedding, with a carefully compacted backfill. *Class C* consists of either an ordinary bedding having a shaped bottom (which is not recommended) or compacted granular bedding, but with a lightly compacted backfill. Lastly, *Class D* bedding consists of a flat bottom trench with no care being taken to secure compaction of backfill at the sides and immediately over the pipe and hence is not recommended. Class *B* or *C* bedding with compacted granular bedding is generally recommended.

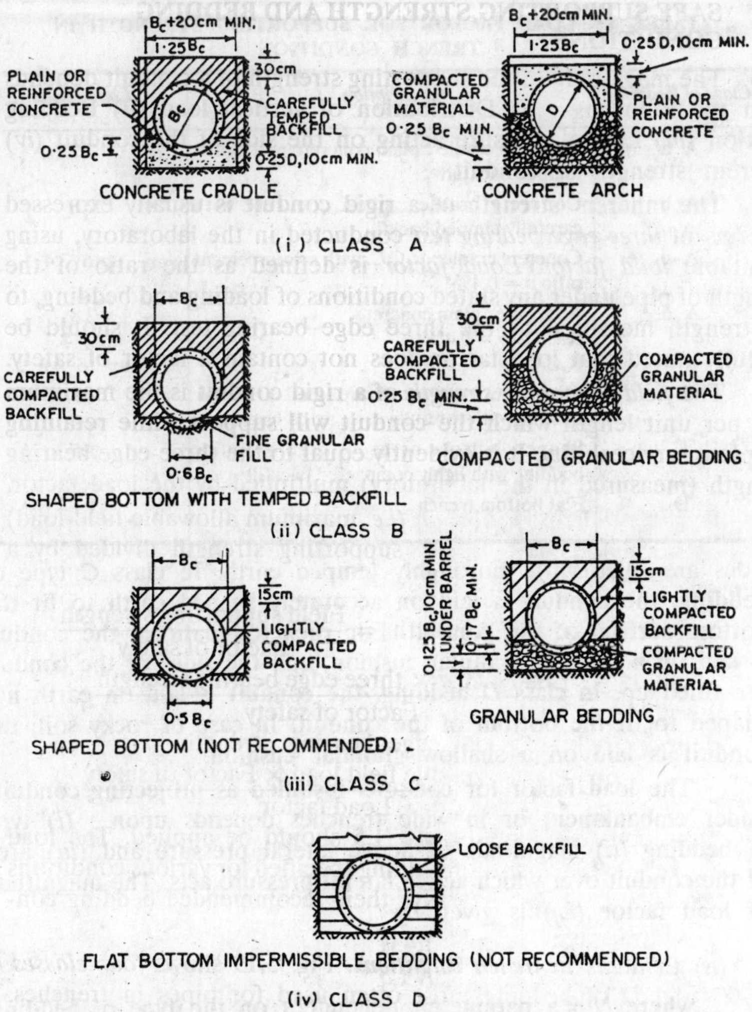

FIG. 5.19. CLASSES OF BEDDING FOR CONDUITS IN TRENCH

Table 5.5 gives the load factors for the above classes of bedding.

(*b*) ***Conduits in embankment conditions***. The soil pressure against the sides of a pipe placed in an embankment may be significant in resisting the vertical load on the structure. Fig. 5.20 shows class *A*, class *B*, class *C* and class *D* type beddings which are generally adopted for projecting conduits laid under the embankment conditions of installation. In class *A* type bedding, the conduit is laid on a mat of concrete. In class *B* type bedding, the conduit is laid on an accurately shaped earth to fit the bottom of the pipe and the

TABLE 5.5. LOAD FACTOR FOR SUPPORTING STRENGTH IN TRENCH CONDITION

Class of Bedding	*Description*	*Load factor*
A a	Concrete cradle : plain concrete and lightly tamped backfull	2.2
A b	Concrete cradle : plain concrete with carefully tamped backfill	2.8
A c	Concrete cradle : R.C.C. with reinforcement ratio $p = 0.4\%$	upto 3.4
A d	Arch type : plain concrete R.C.C. ($p = 0.4\%$) R.C.C. ($p = 1.0\%$)	2.8 upto 3.4 up to 4.8
B	Shaped bottom or compacted granular bedding with carefully compacted backfill	1.9
C	Shaped bottom or co[illegible]cted granuler bedding with lighly compact[illegible] backfill	1.5
D	Flat bottom trench	1.1

sides are filled with thoroughly temped earth. In class *C* type of bedding, the conduit is laid on accurately shaped earth to fit the bottom surface of the conduit. For rock foundations, the conduit is laid on a layer of granular cushion, and the sides of the conduit are filled up. In class *D* bedding, the conduit is laid on earth not shaped to fit the bottom of the conduit. In case of rocky soil, the conduit is laid on a shallow granular cushion.

The load factor for conduits installed as projecting conduits under embankment or in wide trenches depends upon : (*i*) type of bedding (*ii*) magnitude of active lateral pressure and (*iii*) area of the conduit over which active lateral pressure acts. The magnitude of load factor (L_f) is given by

$$L_f = \frac{1.431}{N x q} \quad ...(5.17)$$

where N = a parameter, dependent on the type of bedding, and its value for circular pipe may be taken from Table 5.6.
x = a parameter dependent upon the area over which the lateral pressure acts effectively, and its value may be taken from Table 5.7.
q = ratio of total lateral pressure to total vertical load on the pipe.

For positive projecting conduits, q is given by

$$q = \frac{mk}{C_c}\left[\frac{H}{B_c} + \frac{m}{2}\right] \quad ...(5.18)$$

where k = Rankine's ratio which may be taken as 0.33.
m = fraction of conduit on which lateral pressure acts, its value varying between 0 to 1.

TABLE 5.6. VALUES OF N

	Class of bedding	Value of N
1.	Class A : Reinforced concrete cradle	0.42 to 0.51
2.	Class A : Plain concrete cradle	0.51 to 0.64
3.	Class B	0.71
4.	Class C	0.84
5.	Class D	1.31

TABLE 5.7. VALUES OF PARAMETER x

Fraction (m) of conduit on which lateral pressure act	Value of x for	
	Class A bedding	Other beddings
0	0.150	0
0.3	0.743	0.217
0.5	0.846	0.423
0.7	0.811	0.594
0.9	0.678	0.665
1.0	0.638	0.638

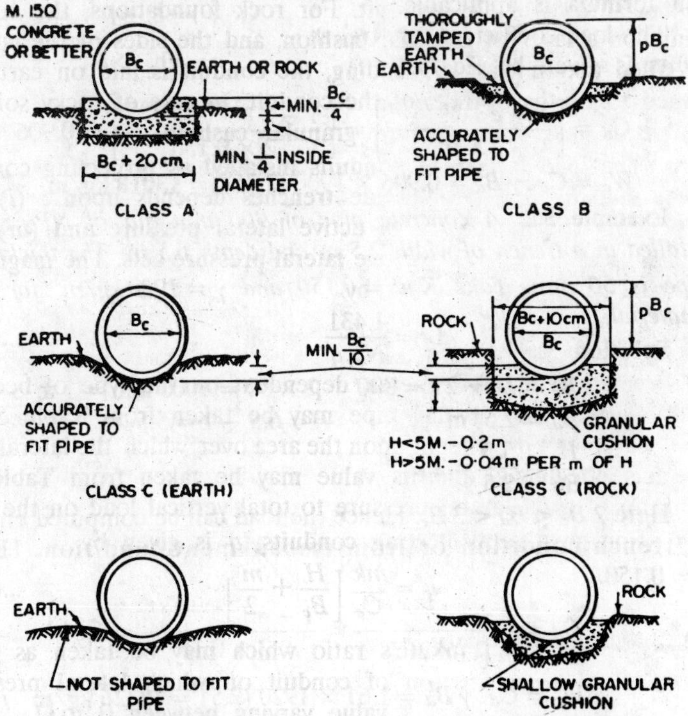

FIG. 5.20. CLASSES OF BEDDING FOR PROJECTING CONDUITS

The load factor for *negative projecting conduits* may also be computed from Eq. 5.17 and 5.18, with a value of Rankine's ratio $k = 0.15$, provided the side fills are well compacted. The equations for positive projecting conditions (*i.e.* Eqs.5.17 and 5.18 with $k=0.33$) will also hold good of conduits in imperfect trench condition.

Example 5.1. *A concrete pipe of 1 5 m dia. is installed in a trench of width 2.75 m and depth 4.5 m. Thickness of pipe is 80 mm. Determine the load on the pipe due to soil fill which consists of clay for which K μ' may be taken as 0.130 and unit weight is 1900 kg/m³.*

Solution : This is the case of a conduit in trench condition.

Here $B_c = D + 2t$

$= 1.5 + 2 \times 0.08 = 1.66$ m

$B_d = 2.75$ m (given)

$H = 4.5 - 1.66 = 2.84$ m

$$\therefore \quad \frac{H}{B_d} = \frac{2.84}{2.75} = 1.033$$

Since $\dfrac{H}{B_d} < 2B_c$,

trench formula is applicable.

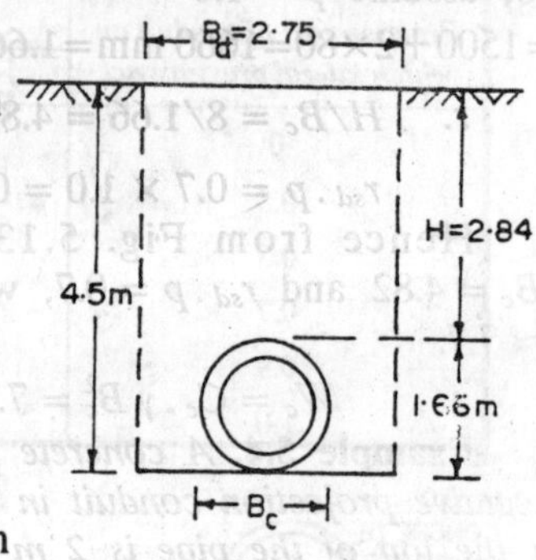

FIG. 5.21.

The load coefficient for trench conduits is given by Eq. 5.4 (*a*).

$$C_d = \left[\frac{1 - e^{-2K\mu' H/B_d}}{2K\mu'}\right] = \left[\frac{1 - e^{-2 \times 0.130 \times 1.033}}{2 \times 0.130}\right] \approx 0.906$$

$$\therefore \quad W_c = C_d \,.\, \gamma B_d^2 = 0.906 \times 1900\,(2.75)^2 = \mathbf{13018\ kg/m}$$

Example 5.2. *A concrete pipe of 900 mm dia of NP₂ class is installed in a trench of width 2.5 m and depth 6.5 m. The thickness of pipe is 50 mm. Take K μ' = 0.150 and γ = 1920 kg/m³ for the fill material.*

Solution

$B_c = D + 2t = 900 + 2 \times 50 = 1000$ mm $= 1$ m

$B_d = 2.5$ m ; $\quad H = 6.5 - 1 = 5.5$ m

$$\frac{H}{B_d} = \frac{5.5}{2.5} = 2.2$$

Here $2B_c < B_d < 3B_c$. Hence the load can be computed either from trench condition or from embankment condition. Here $K\mu' = 0.150$.

$$\therefore \quad C_d = \frac{1 - e^{-2K\mu' H/B_d}}{2K\mu'} = \frac{1 - e^{-2 \times 0.150 \times 2.2}}{2 \times 0.15} = 1.61$$

$$\therefore \quad W_c = C_d \,.\, \gamma B_d^2 = 1.61 \times 1920\,(2.5)^2 = \mathbf{19326\ kg/m}$$

Example 5.3. *A concrete pipe of 1500 mm dia. is installed as a positive projecting conduit under a fill of 8 m height above the top of the pipe. The wall thickness of the pipe is 80 mm. Determine the fill load on the pipe, if the unit weight of the fill material is 1950 kg/m³.*

Solution. From Table 5.2 we find that r_{sd} = ranges from 0.5 to 0.8 for ordinary soil, while its value is 1.0 for unyielding soil. Let us assume $r_{sd} = 0.7$. Also, assume $p = 1.0$

$B_c = 1500 + 2 \times 80 = 1660$ mm $= 1.66$ m

$\therefore \quad H/B_c = 8/1.66 = 4.82$

$r_{sd} \cdot p = 0.7 \times 1.0 = 0.7$

Hence from Fig. 5.13, for $H/B_c = 4.82$ and $r_{sd} \cdot p = 0.7$, we get $C_c \approx 7.7$.

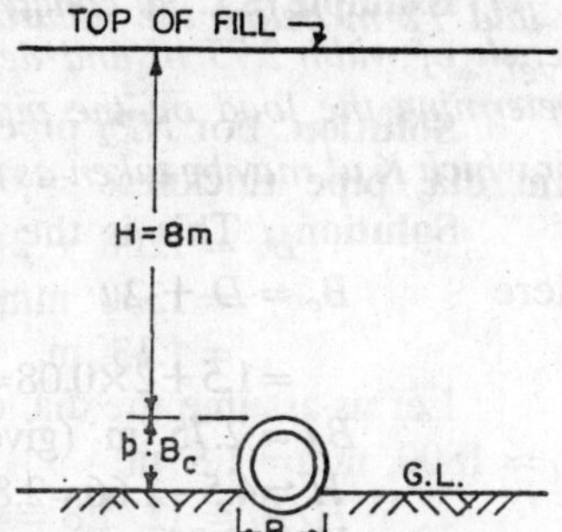

FIG. 5.22

$$\therefore \quad W_c = C_c \cdot \gamma B_c^2 = 7.7 \times 1950\,(1.66)^2 = \mathbf{41375\ kg/m}$$

Example 5.4. *A concrete pipe of 1500 mm dia is installed as a negative projection conduit in a trench, the depth of which is such that the top of the pipe is 2 m below the surface of natural ground in which the trench is dug. The width of trench is 2.0 m. Determine the fill load on the pipe if the height of fill over the top of the pipe is 10 m. The thickness of pipe is 80 mm. Take unit weight of fill equal to 1900 kg/m³.*

Solution

Here $H = 10$ m ; $B_d = 2.0$ m

$p' B_d = 2$ m

$\therefore \quad p' = 2/2 = 1.0$

For negative projection conduits, r_{sd} is always negative.

For the present case, take

$r_{sd} = -0.3$ (Table 5.2).

Also, $H/B_d = 10/2 = 5$

Hence for $p' = 1.0, r_{sd} = -0.3$ and $H/B_d = 5$, we get, from Fig. 5.16, $C_n \approx 3.2$

$$\therefore \quad W_c = C_n \gamma B_d^2$$

$$= 3.2 \times 1900(2)^2$$

$$= \mathbf{24320\ kg/m}$$

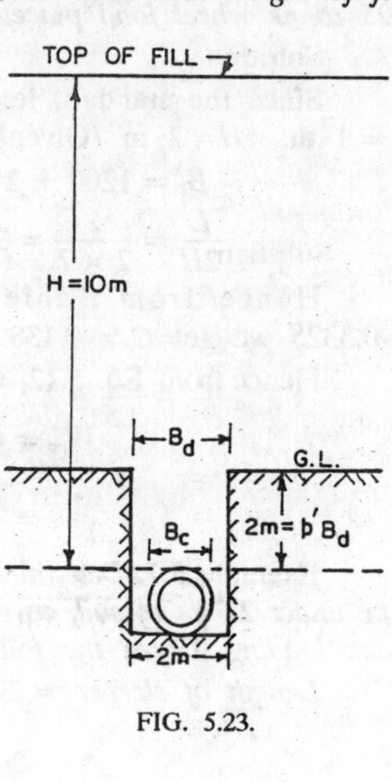

FIG. 5.23.

Example 5.5. *A 1200 mm dia. NP_2 class concrete pipe is installed in tunnel condition, 15 m deep in clay having coefficient of cohesion equal to 900 kg/m², unit weight of 1800 kg/m³ and $K\mu = 0.130$. Determine the load on the pipe if the top of the pipe is laid 12 m below the natural ground level.*

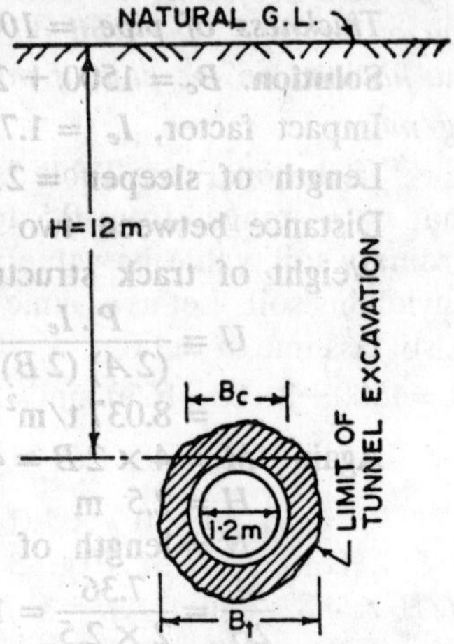

FIG. 5.24.

Solution. For NP_2 pipe of 1200 mm dia, pipe thickness = 65 mm.

$$\therefore \quad B_c = 1200 + 2 \times 65 = 1330 \text{ mm} = 1.33 \text{ m}$$

Let us assume the dia. of tunnel $B_t \approx 1800$ mm = 1.8 m.

$$\therefore \quad H/B_t = 12/1.8 = 6.67$$

From Fig. 5.18, for $H/B_t = 6.67$ and $K\mu = 0.130$, we get $C_t \approx 3.15$.

$$\therefore \quad W_t = C_t B_t (\gamma B_t - 2c) = 3.15 \times 1.8 (1800 \times 1.8 - 2 \times 900) \approx \mathbf{8165 \text{ kg/m}}$$

Example 5.6. *A 1200 mm dia. NP_2 class concrete pipe, having thickness of 65 mm is laid below the ground, with 2 m cover over the top of the pipe. Determine the load on the pipe caused due to 6.25 tonne wheel load placed directly above the centre of the pipe.*

Solution

Since the standard length of pipe is more than 1 m, take $L = 1$ m. $H = 2$ m (Given)

$$B_c = 1200 + 2 \times 65 = 1330 \text{ mm} = 1.33 \text{ m}$$

$$\therefore \quad \frac{L}{2H} = \frac{1}{2 \times 2} = 0.25 \, ; \quad \frac{B_c}{2H} = \frac{1.33}{2 \times 2} = 0.3325$$

Hence from Table 5.4, for $L/2H = 0.25$ and $B_c/2H = 0.3325$ we get $C_s \approx 0.138$.

Hence from Eq. 5.12, taking the impact factor $I_e = 1.5$, we get

$$W_{sc} = C_s \frac{P \cdot I_e}{L_c} = \frac{0.138 \times 6.25 \times 1.5}{1} \approx \mathbf{1.29 \text{ t/m}}$$

Example 5.7. *Determine the load on a 1500 mm dia., concrete pipe under 2.5 m of soil, cover, resulting from a broad gauge railway track loading. Given the following data :*

Length of sleeper = 2.7 m ; Distance between axles = 1.84 m

No. of axles = 4 ; Axle load = 22.5 tonnes
Weight of track structure = 0.3 t/m
Thickness of pipe = 100 mm.

Solution. $B_c = 1500 + 2 \times 100 = 1700$ mm $= 1.7$ m.

Impact factor, $I_e = 1.75$ (for railways)

Length of sleeper $= 2A = 2.7$ m

Distance between two axles $= 2B = 1.84$ m.

Weight of track structure, $W_t = 0.3$ t/m.

$$\therefore \quad U = \frac{P \cdot I_e}{(2A)(2B)} + \frac{W_t}{2A} = \frac{22.5 \times 1.75}{2.7 \times 1.84} + \frac{0.3}{2.7}$$
$$= 8.037 \text{ t/m}^2$$

Again $M = 4 \times 2B = 4 \times 1.84 = 7.36$ m.
$H = 2.5$ m
N = length of sleeper $= 2A = 2.7$ m.

$$\therefore \quad \frac{M}{2H} = \frac{7.36}{2 \times 2.5} = 1.472 \ ; \qquad \frac{N}{2H} = \frac{2.7}{2 \times 2.5} = 0.54$$

Hence from Table 5.4, for $M/2H = 1.472$ and $N/2H = 0.54$, we get load coefficient $C_s = 0.561$.

$$\therefore \quad W = 4\, C_s \,.\, UB_c = 4 \times 0.561 \times 8.037 \times 1.7$$
$$= \mathbf{30.66 \text{ t/m}.}$$

Alternatively, from Eq. 5.16 applicable for broad gauge,

$$W = 32.14\, C_s B_c = 32.14 \times 0.561 \times 1.7 = 30.65 \text{ t/m}.$$

Example 5.8. *Design the structural requirement for a 1200 mm dia. NP₂ class sewer pipe which is to be laid in 6.5 m deep trench of 2.5 m width, assuming that the total vertical load will account for concentrated surcharge of 5 t applied at the centre of the pipe. Assume Ab type of bedding having load factor of 2.8. Use a factor of safety of 1.5 for the saturated top soil, take unit weight as 2000 kg/m² and K μ' = 0.150. Consider water load also, assuming the sewer to run 75% full. The three edge bearing strength for 1200 mm dia. NP₂ pipe is 4320 kg/m.*

Solution.

Thickness of NP_2 class conduit of 1200 mm dia. $= 65$ mm

$$B_c = 1200 + 2 \times 65 = 1330 \text{ mm} = 1.33 \text{ m}$$

$$\therefore \quad H = 6.5 - 1.33 = 5.17 \text{ m}; \qquad B_d = 2.5$$

$$\therefore \quad H/B_d = 5.17/2.5 = 2.068$$

Since $H/B_d < 2\,B_c$, trench formula is applicable.

The load coefficient for trench conduit is given by Eq. 5.4(a)

$$C_d = \left[\frac{1 - e^{-2K\mu'(H/B_d)}}{2K\mu'} \right]$$

$$= \frac{1 - e^{-2 \times 0.150 \times 2.068}}{2 \times 0.150} = 1.54$$

$$\therefore \quad W_c = C_d \cdot \gamma B_d^2 = 1.54 \times 2000\,(2.5)^2 = 19250 \text{ kg/m} \quad ...(i)$$

$$\text{Weight of water, } W_w = \left[\frac{\pi}{4} D^2 \times L\right] w \times 0.75$$

$$= \frac{\pi}{4}(1.2)^2 \times 1 \times 1000 \times 0.75 = 848 \text{ kg/m} \quad ...(ii)$$

Again, due to superimposed load,

$$W_{sc} = C_s \frac{P\,I_e}{L_c}$$

Here $L = 1$ m ; $H = 5.17$ m ; $L/2H = 1/(2 \times 2.17) \approx 0.1$ and $B_c/2H = 1.33/(2 \times 5.17) \approx 0.129$. Hence from Table 5.4, we get $C_s \approx 0.024$.

Taking an impact factor of 1.5 and $L_c = 1$ m

$$W_{sc} = \frac{0.024 \times 5000 \times 1.5}{1} = 180 \text{ kg/m.}$$

Total $W = W_c + W_w + W_{sc} = 19250 + 848 + 180 = 20278$ kg/m

Safe supporting strength of 1200 mm dia. NP_2 pipe

= Three edge bearing strength $\times L_f$

= 4320 × 2.8 = 12096 kg/m.

Since the actual load (= 20278 kg/m) exceeds the safe supporting strength, the pipe is *not safe* unless it is encased in concrete.

5.9. PLANNING OF SEWERAGE SYSTEM

Survey and investigations are prerequisites for the preparation of a detailed sewerage project. Informations on physical, developmental, fiscal and other aspects has to be collected. Out of these, *physical aspects* are very essential for design purposes. The information about the following physical aspects is to be collected.

1. Contour plan of the area to obtain topography or elevation differences needed for the design of sewers and location of out fall and disposal works.

2. Location of streets and adjoining areas likely to be merged or annexed.

3. General arrangement of system-alignment of various laterals, branches and main sewers to be marked on the contour plan.

4. Marking of population or catchment area, to be drained by each sewer.

5. Determination of subsoil conditions, such as types of strata likely to be encountered, depth of ground water table and its fluctuations.

6. Information about underground structures like storm drains and appurtenances, city survey stones and utility services like house

connections for water supply and sewerage, electric and telephone cables and gas lines, if any.

Depending upon whether it is a separate system or a combined system, the sewage discharge likely to flow in each sewer is determined. The required sizes of the sewer sections and their gradients are then computed, so as to enable each sewer to pass the maximum sewage that may enter into it. Separte L–sections are drawn for each sewer line, whether it is a lateral or branch sewer or main sewer, showing invert levels at different R.D'S (reduced distances). Typical cross-sections are also drawn. Manholes are located at all sewer intersections, changes in horizontal directions, major changes in slopes, changes in size of sewer, and at intervals along straight runs. Manhole spacings generally do not exceed 100 m, and should never exceed 150 m. These are numbered and the numbers identify the individual sewer lines as well, since these run from manhole to manhole. The L-section or vertical profile, drawn from survey notes for each sewer line, should also show tentative manhole locations, rock/soil strata indicated by borings, location of bore holes, all under ground structures, cellar elevations, utility services and cross-streets.

5.10. LAYOUT AND CONSTRUCTION OF SEWER LINES

The various steps involved in the layout and construction of sewer lines are :

(1) Setting out (2) Alignment and gradient (3) Excavation of trenches, timbering and dewatering (4) Laying and jointing (5) Testing and (6) Backfilling.

1. Setting out. From the longitudinal section of the sewer line, the positions of manholes are located on the ground. It is general practice to lay sewer line between two manholes at a time. The setting out in the sewerage work is carried out, starting from the tail end or out-fall end, and proceeding upwards. The advantage of starting the layout and construction work from the tail end is the utilization of tail sewers even during the initial period of construction. If on the contrary, this is done from the head-end, the function of the sewerage scheme has to wait till the completion of the entire scheme.

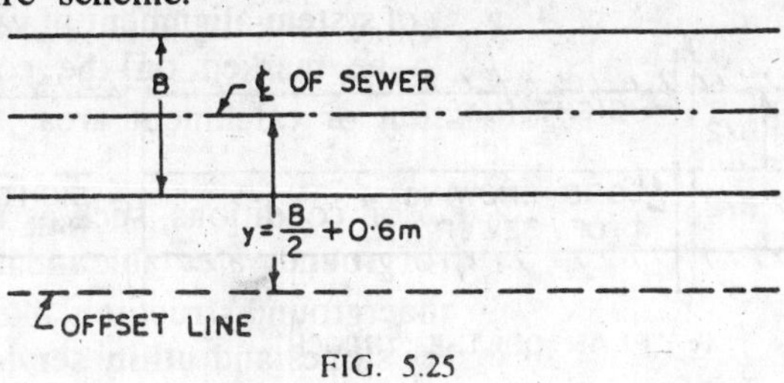

FIG. 5.25

The centre line of the sewer line should properly be maintained during the construction. For this purpose, there are two methods.

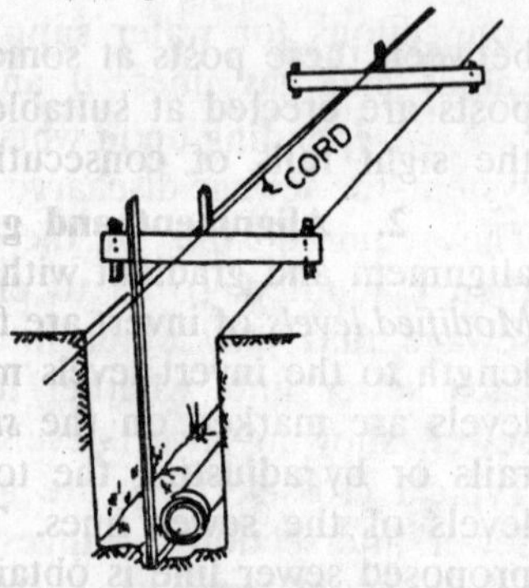

FIG. 5.26

In the *first method*, a line parallel to the centre line of the sewer is drawn and marked ; this line is known as *offset line,* which may be set out at a distance $y = \frac{1}{2}B + 0.6$ m, where B is the width of the trench. Pegs are driven along this line, at an interval of 7.5 to 15 m. Temporary bench marks should be established on this offset line, at suitable interval to carry forward the levels.

In the *second method*, shown in Fig. 5. 26 , two vertical posts are driven into the ground at a known distance (say y) from the centre line peg and one horizontal rail, known as *sight rail* is fixed

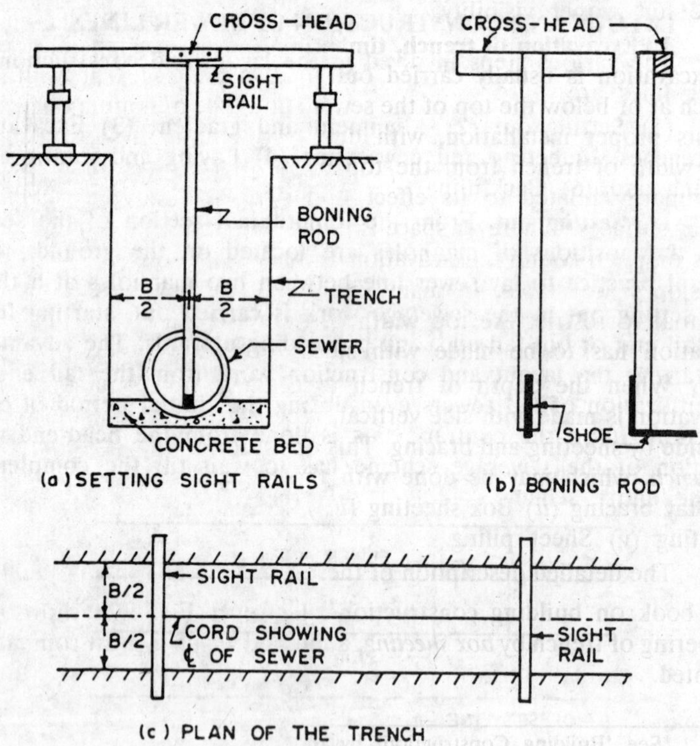

FIG. 5.27.

between these posts at some convenient height above ground. Such posts are erected at suitable interval, and a *cord* is drawn between the sight rails of consecutive posts.

2. Alignment and gradient. The sewers are laid to correct alignment and gradient with the help of boning rods and sight rails. *Modified levels* of invert are first obtained by adding a suitable vertical length to the invert levels marked on the L-section. These modified levels are marked on the *sight rails* either by fixing nails on sight rails or by adjusting the top of sight rails to the modified invert levels of the sewer lines. Thus an imaginary line parallel to the proposed sewer line is obtained on the ground (Fig.5.26 and 5.27*c*).

A *boning rod* (Fig. 5.27 *a*) or a *traveller* is a vertical wooden post suitably shod with shoe iron and fitted with a *cross-head* or *tee*. Its length is made equal to the height of sight rail above the invert line of the sewer. The boning rod can move to and fro in the trench so as to give invert-line on the prepared bed of the sewer. Both the sight rail as well as boning rod have their centre lines accurately marked with thin saw-cut and painted black and white for proper visibility.

3. Excavation of trench, timbering and dewatering. The work of excavation is usually carried out in open cutting. The width of trench at or below the top of the sewer should be minimum necessary for its proper installation, with due consideration to its bedding. The width of trench from the top of the sewer to ground surface is primarily related to its effect upon the adjoining services and nearby surfaces. Wherever space is available, specially in undeveloped areas or open country, exacavation is made with side slopes so that the slopes are stable. In many circumstances, however, it may be essential to restrict the top width of the trench, and hence the ex-acavation has to be made with sides vertical.

When the depth of trench exceeds 1.5 to 2 m, and when excavation is made with side vertical, it becomes necessary to support the side by sheeting and bracing. This operation is known as *timbering of trench*, which may be done with the help of following methods: (*i*) Stay bracing (*ii*) Box sheeting (*iii*) Vertical sheeting (*iv*) Runner sheeting (*v*) Sheet piling.

The detailed description of these may be found in any standard text book on building construction*. However Fig. 5.28 shows the timbering of trench by *box sheeting*, a method which is most commonly adopted.

*See 'Building Construction' by the Authors.

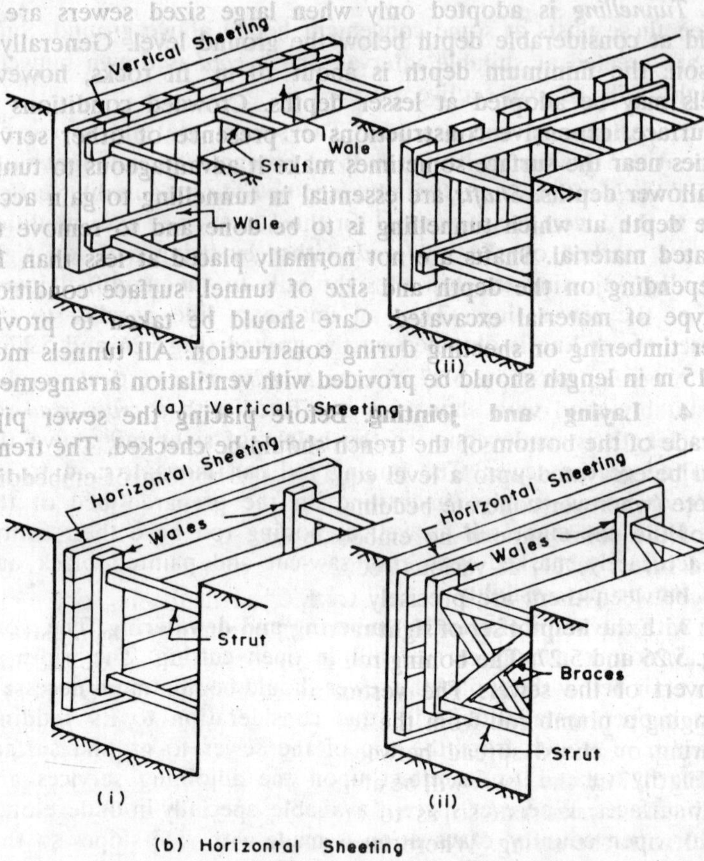

FIG. 5.28. TIMBERING BY BOX SHEETING

Excavation of trenches in ground having high water table, or in water-logged area pose great problems because of water oozing in the trench from sides, bringing with it the soil from the sides. The timbering, if provided, would become loose and collapse. Excavations can be carried out by dewatering the sub-soil water. This can be done by any of the following methods :

(*i*) Ditches and sumps. (*ii*) Well point system (*iii*) Shallow well system (*iv*) Deep well method. (*v*) Vacuum method. or (*vi*) Electro-osmosis method.

A detailed description of these methods can be found in any standard text book on Soil Mechanics*

*See 'Soil Mechanics and Foundations' by the Authors.

Tunnelling is adopted only when large sized sewers are to be laid at considerable depth below the ground level. Generally in soft soil, the minimum depth is about 10 m. In rocks, however, tunnels may be adopted at lesser depths. Crowded conditions on the surface, expensive constructions or presence of other service facilities near the surface sometimes make it advantageous to tunnel at shallower depths. *Shafts* are essential in tunnelling to gain access to the depth at which tunnelling is to be done and to remove the excavated material. Shafts are not normally placed at less than 150 m, depending on the depth and size of tunnel, surface conditions and type of material excavated. Care should be taken to provide proper timbering or sheeting during construction. All tunnels more than 15 m in length should be provided with ventilation arrangement.

4. Laying and jointing. Before placing the sewer pipe, the grade of the bottom of the trench should be checked. The trench should be excavated upto a level equal to the bottom of embedding concrete (wherever concrete bedding is required) or upto the invert level of the sewer pipe if no embedding concrete is provided. The desired invert levels at various reduced distances and the designed slopes between them are precisely transferred to the bottom of the trench with the help of set of sight rails and boning rods, as illustrated in Fig. 5.26 and 5.27. The boning rod is placed with its shoe touching the invert of the sewer. The verticality of boning rod is checked by hanging a plumb rod from the top. If the boning rod is touching the string or chord spread between two consecutive sets of sight rails (Fig 5.27 *c*), the sewer will be at proper gradient. If not, excavation or filling is carried out so as to bring the top of the boning rod in level with the string. Where large sewer lines are laid or where sloped trench walls result in top-of-trench width too great for practical use of sight rails or where soils are unstable, stakes set in the trench bottom itself on the sewer line, as rough grade for the sewer is completed, would server the purpose. A newer technique for maintaining both line and grade employs a *laser beam* generated at a manhole and directed down the pipe as its is placed. Accuracy of line and grade can be held to within 0.01 percent over a range of 300 m by the laser beam technique.

Stoneware pipes are laid with sockets facing up the gradient, on desired bedding. Special bedding, haunching or encasing may be required where conditions so demand (see Example 5.8). The R.C.C. pipes are laid in position over proper bedding, the type of which may be determined in advance, the abutting faces of the pipe being coated by means of a brush with bitumen in liquid condition. The wedge shaped groove in the end of the pipe is filled with sufficient quantity of either special bituminous compound or sufficient quantity of cement mortar in grade M 100 or M-250. The collar is then slipped over the end of the pipe and the next pipe butted well against the plastic ring by appliances so as to compress roughly the plastic ring or cement mortar into the grooves, care being taken to see that eccentricity of the pipe and the levels are not disturbed during the operation.

In the case of *stoneware pipes*, the joints should be caulked with tarred gasket in one length for each joint and sufficiently long to entirely surround the spigot end of the pipe. The gasket should be caulked lightly home but not so as to occupy more than a quarter of the socket depth. The socket is then filled with a mixture of one part of cement and one part of clean fine sand mixed with just sufficient quantily of water to have a consistency of semi-dry condition and a fillet should be formed round the joint with a trovel forming an angle 45° with the barrel of pipe, as specified in IS : 4127. Rubber gaskets may also be used for jointing.

In the case of *stoneware pipes*, collars are placed symmetrically over the end two pipes and the annular space between the inside of the collar and outside of pipe is filled with hemp yarn soaked in tar or cement slurry tamped with just sufficient quantity of water to have a consistency of semi-dry condition, well packed and thoroughly rammed with caulking tools and then filled with cement mortar 1:2. The joint is finished off with a fillet sloping at 45° to the surface for the pipe. The finished joint should be protected and cured for atleast 24 hours. Spigot and socket R.C.C. pipes are laid and jointed in a manner similar to stone ware spigot and socket pipes.

5. Hydraulic testing of pipes. Before backfilling, the pipe are hydraulically tested. The method of testing are explained in the next article.

6. Backfilling the trench. Trenches should be backfilled immediately after the pipe is laid and tested, unless Class *A* bedding is used in which case the backfilling is delayed until the concrete has set up. No water should be permitted to rise in the unbackfilled trenches. The work of backfilling should be carried out with due care, particularly the selection of the soil used for backfilling around the sewer, so as to ensure the future safety of the sewer. The filling in the haunches and upto about 0.6 to 0.9 m above the crown or soffit of the sewer should be made in the finest selected material free of brush, debris, frozen material, large rocks and junk, placed carefully in layers of 15 cm thickness, watered and evenly rammed. After this, the excavated top soil, turf, pavement or road metal are replaced as the top filling material rammed and satisfactorily maintained till the surface has been got reinstated. At each stage the filling should be well rammed, consolidated and completely saturated with water and then only further line should be continued. Before and during the backfilling of a trench, precautions should be taken against the flotation of the pipe line due to large quantities of water into the trench causing an uplift of the empty or the partly filled pipe line. Where trenches are in fields, the backfilling above the 600 mm level is not tamped. All the earth is replaced, and the resulting mound is allowed to settle naturally.

5.11. TESTING OF SEWER LINES

Sewers are generally subjected to the following tests before they are put into service. : (1.) Test for straightness and obstruction. (2.) Water test. (3.) Smoke test (4.) Air test.

1. Test for straightness of alignment and obtruction. As soon as a stretch of sewer is laid, it is tested for straightness and obstruction, by two methods : (a) At the high end of the sewer (or drain), a smooth ball of diameter 13 mm less that the pipe bore is inserted. If there is no obstruction in the form of yarn or mortar projecting through the joint, the ball will roll down the invert of the pipe and emerge at the lower end. Alternatively a double disc or solid or closed cylinder 75 mm less in dimension than the internal dimension of the sewer may be run through the stretch of the sewer to ensure that it is free from any obstruction. (*b*) A mirror is placed at one end of the sewer line and lamp at the other end. If the pipe line is straight, the full circle of light will be observed. If the pipe line is not straight, this would be apparent. The mirror will also indicate any obstruction in the pipe-barrel.

2. Water test. Water test is carried out to find out the water tightness of the joints. This is carried out after giving sufficient time for the joints to set. In case of concrete and stoneware pipes with cement mortar joints, pipes should be tested three days after the cement mortar joints have been made. If is necessary that the pipelines are filled with water for about a week before commencing the application of pressure to allow for the absorption by the pipe wall.

The sewers are tested by plugging the upper end with a provision for an air outlet pipe with stop cock. The water is filled through a funnel connected at the lower end provided with a plug. After the air has been expelled through the air outlet, the stop cock is closed and water level in the funnel is noted after 30 minutes and the quantity of water required to restore the original water level in the tunnel is determined. The pipe line under pressure is then inspected while the funnel is still in position. There shall not be any leaks in the pipe or the joints, though small sweating on the pipe surface is permitted. Leakage in 30 minutes determined by measuring the replenished water in the funnel shall not exceed 15 ml in the smaller diameter and 60 ml in the larger dia. per cm dia. of pipe for 100 mm length. Any sewer or part there of that does not meet the test shall be emptied and repaired or relaid as required and tested again. For concrete, R.C.C. and asbestos cement pipes of more than 600 mm dia, the quantity of water inflow can be increased by 10% for each additional 100 mm of pipe dia. For brick sewers, regardless of their diameter, the permissible leakage of water shall not exceed 10 m^3/24 hours per km length of the sewer.

3. Smoke test. Smoke test is carried out for drainage pipes located in buildings. The smoke is produced by burning oil waste tar paper etc. in the combustion chamber of a smoke machine. The pipes are approved gas-tight by the smoke test conducted under pressure of 25 mm of water, maintained for 15 minutes after all trap seals have been filled with water.

4. Air testing. Air testing becomes necessary, particularly in large dia. pipes when the required quantity of water is not available for testing. It is done by subjecting the stretch of pipe to an air pressure of 100 mm of water by means of a hand pump. It the pressure is maintained at 75 mm, the joints shall be assumed to be water tight. In case the drop is more than 25 mm, the leaking joints shall be traced and suitably treated to ensure water tightness. The exact point of leakage can be detected by applying soap solution to all the joints in the live and looking for air bubbles.

5.12. CLEANING AND MAINTENANCE OF SEWERS

Maintenance of sewers in general relates to the work of keeping any installed sewerage facility in a working condition for the benefit of the people for whom it is intended. Maintenance of sewers conists mainly of the removal or prevention of stoppages, cleaning of sewers and other sewer appurtenances, and repair works. Sewers should be properly cleaned and maintained in good working condition. Sewer maintenance function are two often treated as necessary evil. Maintenance may be *preventive or routine maintenance* which constitutes work executed and precaution taken to prevent any breakdown of sewerage facilities or *corrective maintenance* which constitutes work of repairs after a breakdown has occurred. Preventive maintenance is more economical and provides for reliability in operation of sewerage facilities ; nevertheless, corrective maintenance will also have to be provided for, as the breakdowns are possible inspite of preventive maintenance.

Problems in sewer maintenance. Main problems which are faced in sewer maintenance are : (*a*) Clogging of sewers, and (*b*) Hazards.

(*a*) ***Clogging of sewers*** : sewer may be clogged because of the following factors :

1. Grit and other detritus may get deposited in the sewer. This will create stagnation resulting in the putrefaction of organic matter present in the sewerage, giving rise to odours and poisonous gases. The grease from hot liquid wastes from kitchen, finding entry into sewers, may get deposited on the sides, on cooling, and thus leading to clogging.

2. Roots from plants and nearby trees may penetrate through the joints or cracks in the sewers, resulting in the eventual clogging of the sewers.

3. The growth of fungi results in the formation of a network of tendrils which start floating, and thus offering an obstruction to the flow in side the sewers.

4. Tarry materials may get deposited. This assists in binding the detritus and leading to their growth.

5. Due to improper working of pumping units, sewerage become stagnant in the sewer, resulting in the settlement of grit and other material and dumping of solid wastes in the man holes indiscriminately.

(*b*) ***Hazards*** : Persons engaged in operation and maintenance of sewerage systems, including those at sewage pumping stations, are exposed to different types of occupational hazards like (i) physical injuries (*ii*) injuries caused by chemicals and radio-active wastes (*iii*) infections caused by pathogenic organisms in sewage, and (*iv*) dangers inherent with explosive or noxious vapors and oxygen deficiency. To guard against human error and carelessness, proper *job instructions* and adequate effective supervision by competent personnel are most essential. The most hazardous element during inspection is the presence of gas in sewers. '*Sewer gas*' is a mixture of gases in sewers and manholes containing abnormally high percentage of carbon dioxide, varying amounts of methane, hydrogen, hydrogen sulphide and low percentage of oxygen caused by septic action through the accumulation of organic matter in the sewer. The actual hazard is due to the presence of high level of methane forming and explosive mixture or the oxygen deficiency or hydrogen sulphide in excess of permissible levels. Some trade wastes may also contribute to other gases like chlorine, ammonia, sulphur dioxide etc.

Safety precautions against gas hazards. While entering with man holes, the following precautions should be observed :

1. Traffic warning signs should be erected.

2. No smoking or open flames should be allowed and sparks should be guarded against.

3. Only safety, explosion-proof electric lighting equipment or mirrors for reflection of light should be used.

4. The atmosphere should be tested for noxious gas and oxygen deficiency.

5. If the atmosphere is normal, the worker may enter in the manhole or sewer with safety harness attached with two men available at the top.

6. If any noxious gas or oxygen deficiency is found, forced ventilation should be resorted to using portable blower.

7. Frequent tests for noxious gases and oxygen deficiency should be conducted even if initial tests are satisfatory, as conditions may change during the period workers are inside the manhole or sewers.

8. If forced ventilation is not possible or is not satisfactory and men have to enter urgently, such as in saving a life, a gas mask should be worn and extreme care taken to avoid all sources of ignition if inflammable gas is present. Only permissible safety light (not ordinary flash light), rubbers or non-sparking shoes and non-sparking tools should be used.

9. Only personal experienced in line and familiar with the dangers involved in working in a noxious gas environment and fully equipped with the proper protective safety equipment should be allowed to enter.

Safety equipments. Following are various safety equipments normally required in sewer maintenance work :

(*i*) Gas mask (*ii*) Oxygen breathing apparatus (*iii*) Portable lighting equipment (*iv*) Portable air blowers (*v*) Non-sparking tools (*vi*) Safety belts (*vii*) Inhalators.

(*i*) ***Gas mask*** : Gas masks provide respiratory protection from low and moderately high concentrations of all types of toxic gases and vapors present in the atmosphere in which there is insufficient oxygen to support life. A gas mask consists of face piece, a canister containing purifying chemical, a timer for showing duration of service and a harness for support. Gas masks can not be used in oxygen deficient atmosphere or in unventilated locations or area where large concentration of poisonous gases exist.

(*ii*) ***Oxygen breathing apparatus*** : This apparatus gives respiratory protection from the atmosphere containing toxic gases or that which is deficient in oxygen. This can be either an air hose respirator type or pure oxygen respirator.

(*iii*) ***Portable lighting equipment*** : These consists of permissible type portable electric hand lamps, electric cap lamps and explosion proofs flash lights, for doing work inside the sewer lines.

(iv) ***Portable air blowers*** : These blowers, powered by enclosed explosion prof electric motors provide forced ventilation of manholes. Trailer mounted blowers having a capacity of 210 m^3/min. can ventilate easily many metres of medium sized sewers.

(v) ***Non-sparking tools*** : These are made of an alloy containing atleast 80% of copper, that will not spark, when struck against other objects and metals, and yet have sufficient strength and resistance to wear and tear.

(vi) ***Safety belts*** : This consists of a body belt with a buckle and a shoulder harness. It is used along with a life line and both of these should be tested by lifting the wearer clear of ground before each day's work. A life line consists of a high grade spliced manila rope or a steel cable anchored with rings, on each side of the belt and provided with safety straps for anchoring or securing to a stable support. The life line should be about 15 m in length and the overall assembly should withstand a tensile load of 2000 kg.

(viii) ***Inhalators*** : Inhalators, employing mixture of oxygen and carbon dioxide are used for resuscitating victims of gas collapse, drowning or electric shock. Artificial respiration should be started at once on the patient and an inhalater face piece attached to the victim's mouth as soon as the equipment can be made ready. The carbon dioxide used in small percentages simulates deep breathing so that more oxygen can be inhaled. Pure oxygen should be used only when irritant gases such as hydrogen sulphide or chlorine have caused the victim's collapse.

Sewer clearing equipment and devices. Following equipment is used for sewer cleaning work :

(1) Portable pump set (2) Sectional sewer rods. (3) Flexible sewer rod. (4) Ferret used in conjunction with a fire hose. (5) Sewer cleaning bucket machine. (6) Dredger (7) Rodding machine with flexible sewer rods. (8) Scraper. (9) Hydraulically propelled devices such as *(i)* flush bags *(ii)* sewer balls, *(iii)* sewer scooter.

1. *Portable pump set* : Pumping is resorted to where sewers are blocked completely and sewage has accumulated in manholes. In that case, the collected sewage has to be pumped out to tackle the sewer blockage.

2. *Sectional sewer rods* : Sectional sewer rods are made of bamboo or teakwood or light metal, and are usually of 1 m length having coupling at end. This coupling remains intact in the sewer but can be easily disjointed in the manhole. Sections of the rods are pushed down in the sewer until the obstruction is reached and dislodged. The front or advancing end of the sewer rod is fitted with a cutting edge to cut and dislodge the obstructions. Sewer rods are used to clean small sewers. They may also be used to locate obstruction in big sewers.

3. Flexible sewer rods : These are used for routine cleansing of sewers. The rod is made by sandwiching a manila rope between bamboo strips and typing at short interval.

4. Ferret used in conjunction with a fire hose : In this case, a ferret of sufficient strength is attached to a fire hose of sufficient length to reach at least the next manhole. It is used for breaking and removing sand stoppages. The high velocity jet stream of water is used from the those connected sides of the sewer. The forward stream loosens the accumulated debris ahead of the tool and the rear jets of the ferret admit water to wash the sand back down-stream, where it can be removed from the manhole manually.

5. Sewer cleaning bucket machine : The machine consists of two powered winches with cable in between. The winches are centred over two adjacent manholes of the sewer section to be cleaned. The cable is threaded through the sewer line by means of sewer rods. The cable from the drum of each winch is fastened to the barrel on each end of an expansion sewer bucket fitted with closing device, so that the bucket can be pulled in either direction by machine on the appropriate end.

6. Dredger : A dredger, used to clean larger manholes, consists of a crane and a pulley with the help of which a grab bucket is lowered. The grab bucket scrapes the bottom deposits and brings it to the ground. The disadvantage in this system is that it can not clean the corners of the catchpits of manholes.

7. Rodding machine with flexible sewer rods : Rather than operating the flexible sewer rods manually, these can be operated mechanically. The equipment consists of a machine which rotates a flexible rod to which is attached the cleansing tool such as auger, cork screw or hedgehog and sand cups. The flexible rod consists of a series of steel rods with screw couplings. The rod is pulled in and out in quick succession when the tool is engaging the obstruction, so as to dislodge or loosen it.

8. Scraper : Scraper, used for clearing sewers of dia. larger than 750 mm, consists of an assembly of wooden planks of slightly smaller size than the sewer to be cleaned. The scraper chains, being attached to a control chain in the manhole where it is lowered, is connected to a winch on the next downstream manhole by means of chains. The heading up of the flow behind the scraper will also assist in pushing it in the forward direction. This ensures that the bottom and the side of the sewer are cleaned thoroughly.

*9. **Automatic flushing tanks*** : These tanks employ an outside water source with a controlled flow so that required quantity of water is released at predetermined interval. These are used for routine cleaning of sewers.

*10. **Hydraulically propelled devices*** : Hydraulically propelled devices, such as *flush bags, sewer balls* or *sewer scooter* take advantage of force of empounded water to effectively clean sewers.

(i) Flush bag : A flush bag or sewer flushed is a canvas bag or rubber bag equipped with a fire hose coupler at one end and a reducer at the other end. The flusher is connected to the fire hose and placed in the downstream end from the point where a choke is located. The bag is allowed to fill up until it expands and seals the sewer. The upstream pressure built up due to this damming effect breaks loose the obstructions.

(ii) Sewer balls, and sewer pills : Sewer balls are simple elastic pneumatic type rubber balls, available in different sizes, which can be blown up to varying degree of inflation. The inflated ball of appropriate size is wrapped in a canvass cloth, the edges of which are sewed together, and a trial line, a little longer than the distance between the man holes is attached securely to the covering. Im mediately the ball is thrust into the sewer, sewage commences to back up in the manhole and continues to rise until such time as its pressure is great enough to force sewage under the ball and moving it down stream through the pipe. Acting as a compressible floating plug, it affords enough obstruction, so that a continuous high velocity just spurts under and around to some extent around the ball, thereby sluicing all the movable material ahead of the next manhole. A *sewer pill*, which is a wooden ball, can also be used for this purpose, specially for cleaning large outfall sewers.

(iii) Sewer scooter : This is an improved version of the scraper and consists of two jacks, a controlling rope and the scooter with a tight fitting shield. The scooter attached to the control rope is lowered into the manhole and then into the downstream sever line. The downstream manhole jack is lowered into place from the road and the upper manhole jack set across the top of the manhole. When the scooter is lowered into the sever line, it stops the flow of sewage thus building up a head behind the shield. The resulting pressure causes the scooter to move through the sewer until it accumulates enough debris to stop its movement. The head is then allowed to build up approximately 1 m before the control rope is pulled, causing the shield to fold back, thus allowing the accumulated sewage to gush into the sewer downstream flushing the debris ahead to the next manhole from where it is removed. The control rope

is released, clearing the shield against the sewage and causing the scooter to advance again until the debris stops its movement This process is repeated till the scooter reaches the downstream manhole.

5.13. VENTILATION OF SEWERS

Sewers are properly ventilated to fulfill the following purposes:

(i) Continuous flow : Ventilation relieves air pressure above or below atmospheric caused due to sudden rise or fall of sewage. Due to this, *air locks* will not be formed and continuous flow of sewage inside the sewer will be ensured.

(ii) Disposal of sewer gases : Ventilation prevents accumulation of explosive, corrosive or poisonous gases and vapors. The sewer gases include ammonia, carbon monoxide, carbon dioxide, methane (CH_4), nitrogen, sulphurreted hydrogen (H_2S), petrol vapor etc. Methane gas, being explosive, may even blow off the manhole covers, if it is not removed by proper ventilation. These gases also interfere with the natural flow of sewage.

(iii) Prevention of unpleasant odours : Ventilation prevents concentration if unpleasant odours, causing nuisance.

Methods of ventilation. Ventilation may be done by adopting the following methods.

1. Man holes with gratings : The man hole covers are provided with perforations or gratings through which sewer gases escape. However, this method can not be adopted in residential locality since it will spread foul smell and will cause air pollution. It is adopted only in isolated localities. Another disadvantage of this method is that it permits road dust, large quantities of storm water and other debris.

2. Ventilating columns or shafts : Ventilation columns or shafts are provided at the upper end of every branch sewer and also at every change in size of sewers, apart from their regular placement at the interval of about 200 to 300 m along the sewer lines.

Ventilating column (Fig. 5.29) consists of a vertical shaft made by joining cast iron or steel pipe lengths. A foundation block is provided at the bottom end to

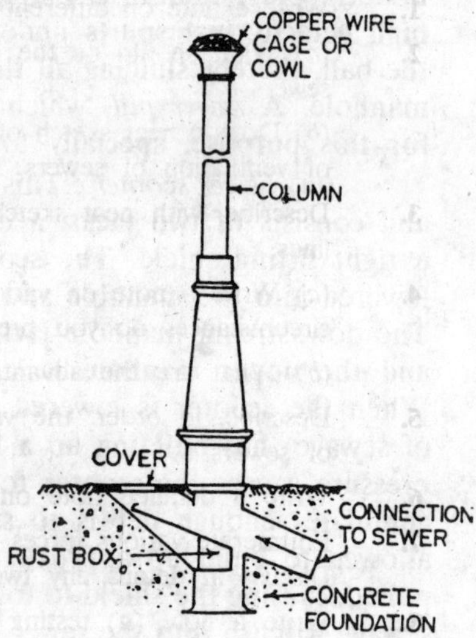

FIG. 5.29.

keep it steady in the vertical position. At the top, copper wire dome or cowl prevent blockage by nesting birds. Foul gases escape from the cowl. The internal diameter of the column is preferably kept one third the diameter of the sewer served by it. The joints of pipes forming the column should be made air tight. The height of the shaft should be sufficient to effectively discharge the foul gases and to be clear of the flat roof of the nearby buildings. The location of the shaft should be such that it obtains sunshine for the major portion of the day.

*3. **Proper design and construction of sewer*** : Sewer should be so designed that they run half or two third full. The space above the flow level will provide ventilation. Also, the flow velocity should be self-cleansing. This can be achieved by laying the sewers at proper gradients.

*4. **Proper house drainage system*** : House vent and soil pipes may be used with advantage to ventilate house drain and the lateral sewers into which they drain, particularly where interceptors are not provided on the sewers connecting houses and buildings.

*5. **Unobstructed outlets*** : Unobstructed outlets provide partial ventilation to storm water drains and sewers.

*6. **Use of mechanical devices*** : Forced draft is provided by exhaust fans to expel out foul gases from sewers.

PROBLEMS

1. Write a note on different types of sewer and the functions of each.
2. (*a*) What should be the characteristics of materials to be used for sewers ?

   (*b*) Draw a neat sketch of a sewage ventilator. What is the necessity of ventilation of sewers ?
3. Describe, with neat sketches, various types of joints used in sewer lines.
4. (*a*) Write a note on various shapes of sewer sections. Under what circumstances do you prefer ovoid sewer ?

   (*b*) What are the advantages circular sewer section ?
5. Describe, in order, the various stages followed in the construction of sewers.
6. Write a detailed note on maintenance of sewers.
7. Enumerate various forces that act on a sewer pipe laid underground. Discuss in details any two of these forces.
8. Write a note (*a*) testing of sewers, (*b*) ventilation of sewers.
9. Describe the procedure for laying and testing of sewers.

10. Explain the phenomenon of crown corrosion with the help of a neat sketch. How can this be avoided.
11. Differentiate clearly between the standard ovoid section and new ovoid section. When are sewers of ovoid shapes adopted ?
12. Discuss, with the help of neat diagrams the classification of sewer conduits based on the placement conditions.
13. How do you determine the loads on conduits due to super-imposed concentrated load and super-imposed distributed load ?
14. Explain how do you determine the load on conduit laid below a railway line.
15. Write a note on self-supporting strength of a conduit.
16. Describe various types of beddings used for sewer lines, both in the trench conditions as well as in embankment condition.
17. What do you understand by the term 'load factor' in connection with the structural design of sewer conduits ? How do you determine it?
18. Determine the fill load a 1200 mm dia. NP_2 class concrete pipe installed in a trench of 2.1 m width and depth of 3.5 m. Take thickness of pipe as 65 mm, unit weight of fill as 1850 kg/m^3 and $K\mu'$ as 0.1924.
Ans [6956 kg/m]
19. Determine the fill load on a 1200 mm dia. NP_2 class concrete pipe installed as a positive projecting conduit under a fill load of 6 m height above top of the pipe. The pipe wall thickness is 65 mm and unit weight of fill is 1850 kg/m^3.
Ans. [22907 kg/m]
20. Determine the fill load on a 1200 mm dia. NP_2 class concrete pipe installed as a negative projection conduit in a trench the depth of which is such that the top of the pipe is 2.5 m below the surface of natural ground is which the trench is dug. The height of fill over the top of the pipe is 12 m. Take width of trench as 2 m and unit weight of soil fill as 1900 kg/m^3.
Ans [28880 kg/m]
21. Determine the load on 1500 mm dia. conduit in tunnel condition 15 m deep in a soil of silty sand for which the coefficient of cohesion is 500 kg/m^2, $K\mu = 0.150$ and unit weight as 1800 kg/m^3. Take max. width of excavation as 1950 mm.
Ans [14680 kg/m]
22. Determine the load on a 600 mm dia. NP_2 class pipe, having thickness = 40 mm, under 2 m cover caused due to 6.25 tonnes wheel load applied directly above the centre of the pipe. Use an impact factor of 1.5.
Ans. [2325 kg/m]

6

Sewer Appurtenances

6.1. INTRODUCTION

Sewer appurtenances are those structures of the sewerage system which are constructed at suitable interval and other locations along a sewer line, to assist in efficient operation and maintenance of the system. Following are important sewer appurtenances.

(1) Inlets (2) Catch basins (3) Clean outs (4) Manholes (5) Drop manholes (6) Lamps holes (7) Flushing tanks (8) Grease and oil traps. (9) Inverted siphons (10) Storm regulators.

6.2. INLETS

An inlet a device meant to admit the storm water or surface wash and convey it into a storm sewer or a combined sewer. Inlets are not necessary in case of separate sewers. Inlets are of three types : (*i*) curb inlet or vertical inlet, and (*ii*) gutter inlet or horizontal inlet and (*iii*) combination inlet, each being either depressed or flush depending upon their elevation with respect to the pavement surface. Maximum spacing of inlets would depend upon various conditions of road surface, size and type of inlet and rainfall. A maximum spacing of 30 m is recommended on straight roads. Inlets are also located at street intersections, as shown in Fig. 6.1, and are so placed that cross walks will not be flooded. Street corners are avoided for this purpose. Inlets are connected to nearby manholes by pipe lines. The arrows in Fig. 6.1 show the direction of surface flow in the streets.

Fig 6.2 shows both the vertical inlets as well as horizontal inlets. Both these types are concrete boxes having gratings in vertical or horizontal direction. A curb inlet does not interface with the

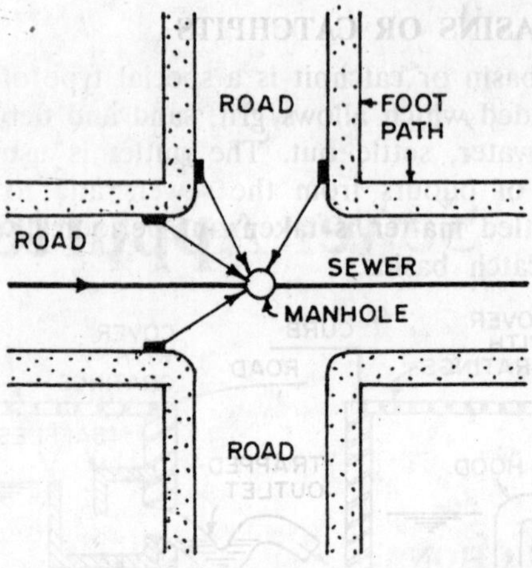

FIG. 6.1 LOCATION OF INLETS AT STREET INTER-SECTION.

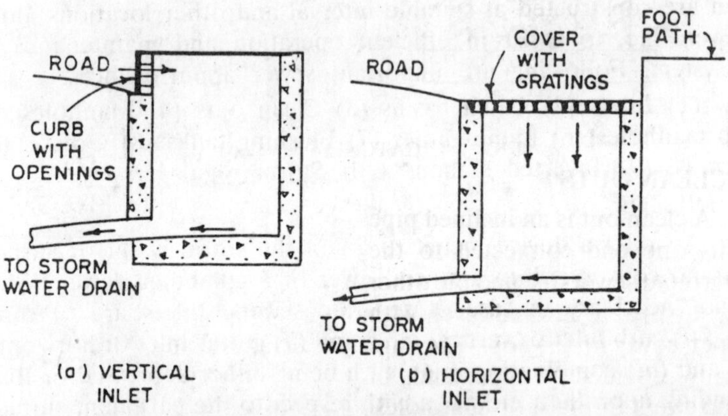

FIG. 6.2 INLETS.

flow of traffic. Combination inlets are composed of a curb and gutter inlet acting as a single unit. Normally, the gutter inlet is placed right in front of the curb inlet but it may be displaced in an overlapping or end-to-end position, as shown in Fig. 6.3.

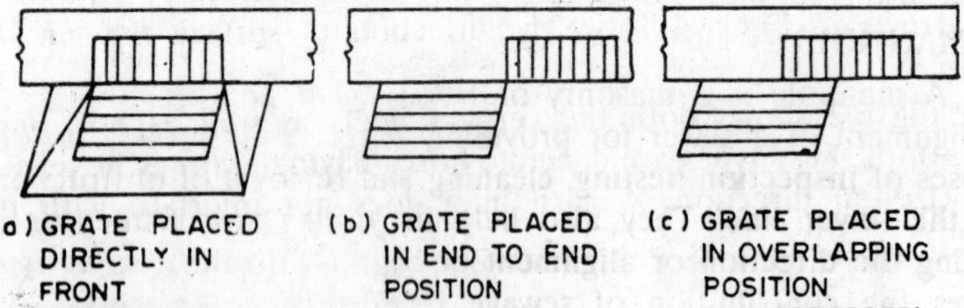

FIG. 6.3 COMBINATION INLETS.

6.3. CATCH BASINS OR CATCHPITS

A catch basin or catchpit is a special type of inlet, in which a basin is provided which allows grit, sand and debris etc., flowing in with storm water, settle out. The outlet is usually trapped to prevent escape of odours from the sewers and to retain floating matter. The settled matter is taken out periodically. Fig 6.4 shows two forms of catch basins.

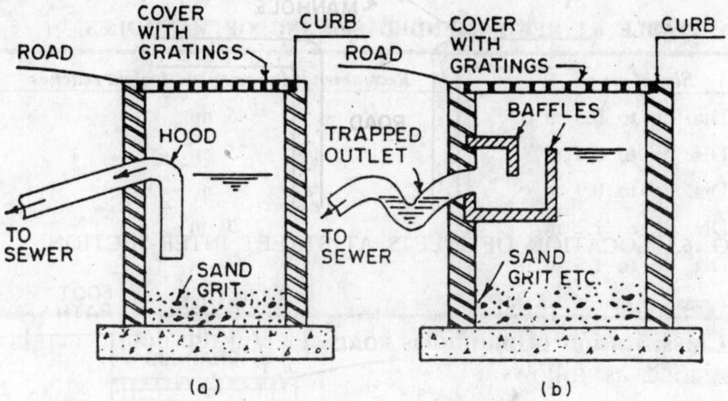

FIG. 6.4. CATCH BASINS.

6.4. CLEAN OUTS

A clean out is an inclined pipe with its one end connected to the underground sewer line and the other end brought upto ground level, with a proper cover at the top. They are generally provided at upper ends of lateral sewers in place of man holes. They are meant for clearing out the lateral sewers. For cleaning purposes, the cover is taken out and water is forced into the clean out pipe. For removing larger obstruction, flexible rod may be inserted or other methods, described in the previous chapter are adopted.

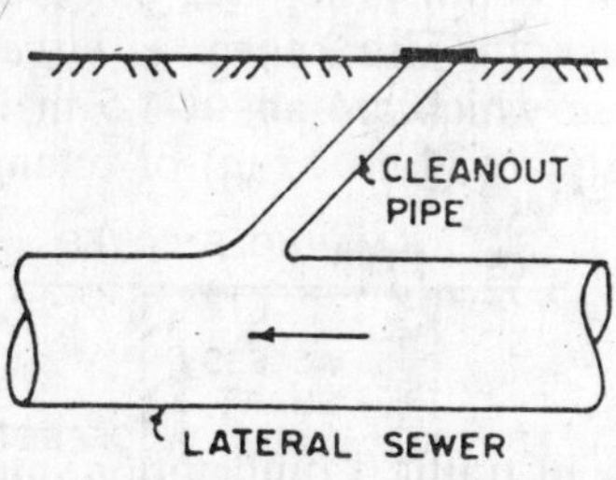

FIG. 6.5. CLEAN OUT

6.5. MANHOLES

A manhole is a masonry or R.C.C. chamber constructed on the alignment of a sewer for providing access to the sewer for the purposes of inspection, testing, cleaning and removal of obstructions from the sewer line. They also help in joining sewer lines or in changing the direction or alignment or both. A manhole sometimes receives the contribution of sewage from sewers of various sizes and coming from various directions.

Location of manholes : Manholes are provided at every bend, junction, change of gradient or change of diameter. On straight reaches, manholes are provided at convenient spacings which depend upon the size of sewers. The larger the diameter of the sewer, the greater may be the spacing between two manholes. The spacing between the manholes will also depend upon the nature of sewer cleaning devices in use. Table 6.1 gives the spacing of manholes as recommended by the Indian Standard IS: 1742–1960.

TABLE 6.1 RECOMMENDED SPACING OF MANHOLES

Size of sewer	*Recommended spacing on straight reaches*
Dia. up to 0.3 m	45 m
Dia. up to 0.6 m	75 m
Dia. up to 0.9 m	90 m
Dia. up to 1.2 m	120 m
Dia. up to 1.5 m	250 m
Dia. greater than 1.5 m	300 m

Classification of manholes : Based upon the depth, manholes are classified as follows :

(*i*) Shallow manholes (*ii*) Normal manholes (*iii*) Deep manholes

Shallow manholes are the one which are about 0.75 to 0.9 m in depth. They are constructed at the start of a branch sewer. These are also known as *inspection chambers*. *Normal manholes* are those which are about 1.5 m in depth. They are constructed either in square (1 m × 1 m) or retangular (0.8 m × 1.2 m) in cross-section.

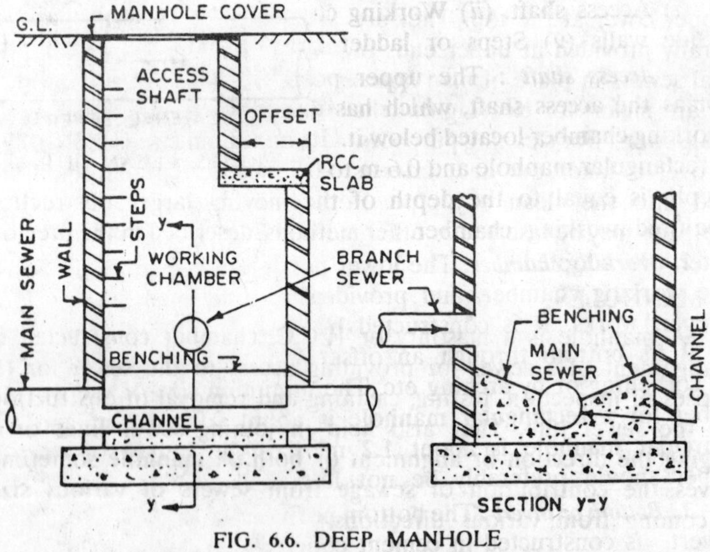

FIG. 6.6. DEEP MANHOLE

The section of such manholes is not changed with depth. It is provided with heavy cover at its top.

Deep manholes are those which are deeper than 1.5 m. The size of such a manhole is larger at the bottom, which is reduced at the top to reduce the size of manhole cover. The reduction in size is achieved by providing an offset of RCC, as shown in Fig. 6.6. Steps are also provided on one vertical wall of the manholes to enable the workers to go upto the bottom. A heavy manholes cover is provided at the top, with suitable C.I. frame.

Dimensions of manholes : Table 6.2 gives minimum internal dimensions for manhole chambers as recommended by Indian Standard IS 1742–1960.

TABLE 6.2

S.N.	*Depth*	*Min. size specified*
1.	0.8 m or less	0.75 m × 0.75 m
2.	0.8 m and 2.1 m	1.2 m × 0.9 m
3.	Greater than 2.1 m	Circular chambers with min. dia. of 1.4 m ; or retanglar chamber with min. dimension of 1.2 × 0.9 m
	Min. wall thickness	
	(*a*) 1.5 m depth	20 cm
	(*b*) >1.5 m depth	30 cm

Component parts of a deep manhole. A manhole typically consists of following component parts.

(*i*) Access shaft. (*ii*) Working chamber. (*iii*) Bottom or invert (*iv*) Side walls (*v*) Steps or ladder (*vi*) Top cover.

1. Access shaft : The upper portion of a deep manhole is known as the access shaft, which has dimension lesser than those of the working chamber located below it. Its minimum size is 0.6×0.75 m for a rectangular manhole and 0.6 m to 0.75 dia. for circular manholes. Its depth is equal to the depth of the manhole minus the height of required working chamber.

2. Working chamber: The lower portion of a manhole is known as the working chamber, and provides working space to carry out the repair works. It is constructed by enlarging the size of access shaft at its bottom through an offset by providing a R.C.C. slab, or by corbelling or by arching etc. The minimum size of the working chamber for a rectangular manhole is about 0.9 × 1.2 m and that for circular manhole is about 1.2 m dia. The height of working chamber should preferably be not less than 1.8 m.

3. Bottom or invert : The bottom portion, also known as benching or invert, is constructed in cement concrete. It consists of a semi-circular or U-shaped main channel and sides are made to slope towards it. The construction of such *benching* facilitates the entry of sewage from branch sewers into main sewer. If the branch sewer and main sewer meet practically at the same level, at the bottom of manhole, channels connected with each other are constructed with easy curves.

*4. **Side walls*** : The side walls of a manhole are constructed of either brick or stone masonry or of R.C.C. They should be structurally strong to withstand earth pressure from the sides, specially in the case of deep manholes which are not circular in shape. Brick masonry walls in cement mortar are quite common since these can be constructed with ease and speed. The minimum specified thickness is 22.5 cm (one brick thick). The appropriate thickness may be determined from the following thumb rule :

$$t = 10 + 4d \qquad \text{...(6.1)}$$

where t = thickness of wall in cm, subject to a min. of 22.5 cm.
d = depth of manhole in metres.

The inside and outside of the brickwork should be plastered 20 mm thick with 1:2 cement plaster, so that a minimum total thickness of 250 mm is obtained.

*5. **Steps or ladder*** : Steps of manholes are made of malleable cast iron. The steps are placed staggered at a horizontal centre to centre distance of 20 cm. If, however, wider steps are provided, such staggering is not needed. The vertical distance (*i.e.* rise) between the steps may be kept upto 30 cm. These steps should be firmly embeded into the walls so that they do not overturn. In very deep manholes, ladders may be provided.

*6. **Top cover and frame*** : Manholes are provided with heavy C.I. covers set in suitable C.I. frame. The depth of the frame is kept 20 to 25 cm, and its width is kept 10 cm. The cover rests in the groove which is kept inside the frame. The frame should rest on concrete and be set in appropriate concrete mix so that the space between the top of the manhole masonry and the bottom flange of the frame should be completely filled and made water tight. A thick ring of mortar extending to the outer edge of the masonry should be placed all around the bottom flange. The frame and top cover should be so set that it should not form a source of inconvenience for road traffic. The cover should be flat and it should be in the plane of the pavement. Covers are generally standardized so that in case of theft or breakage, they can be easily replaced.

6.6. DROP MANHOLE

A drop manhole is a special type of manhole on a sewer line which is constructed to provide a connection between a high level branch sewer to a low level main sewer. When a branch sewer enters a manhole by more than 0.5 to 0.6 m above the main sewer, the sewage is not allowed to fall directly into the manhole, to avoid the possibilities of sewage being thrown on persons entering the working chamber of manhole. Instead, the sewage of the branch sewer is brought into the manhole of the main sewer either by inclined pipe (inclined at 45°) or through a down pipe (vertical pipe). In the former case, it is known a *ramp*, while in the latter case it is known as *a drop manhole*. The construction of drop manhole avoids unnecessary steep gradient of branch sewer thus reducing the amount of earth work.

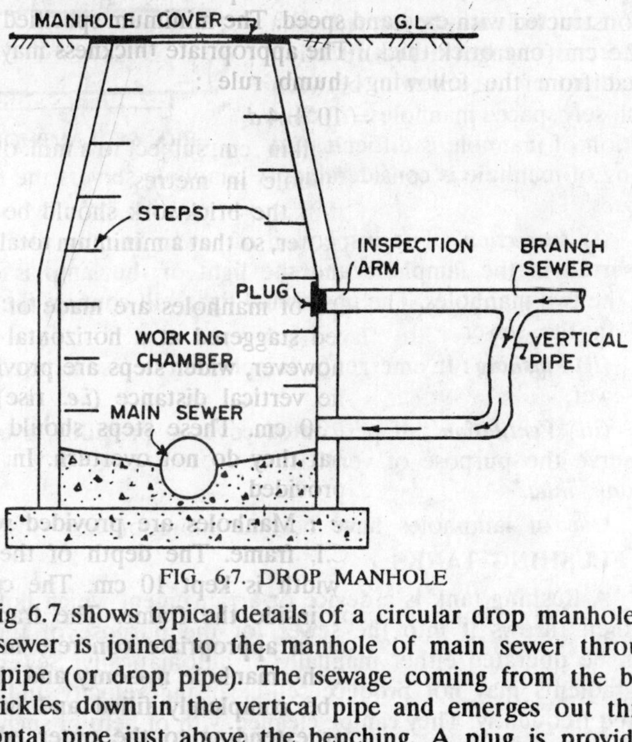

FIG. 6.7. DROP MANHOLE

Fig 6.7 shows typical details of a circular drop manhole. The branch sewer is joined to the manhole of main sewer through a vertical pipe (or drop pipe). The sewage coming from the branch sewer trickles down in the vertical pipe and emerges out through a horizontal pipe just above the benching. A plug is provided at the point where the branch sewer, if prolonged straight, intersects with the vertical wall of the manhole. The prolonged length of branch sewer, beyond the vertical pipe, serves as the inspection arm; after opening the plug, it can be used for inspecting or cleaning the branch sewer.

6.7. LAMPHOLES

Lampholes are special openings, constructed to join a hole in the sewer and the ground, for the purpose of lowering a lamp inside the sewer. A lamphole may consist of either a vertical concrete or stoneware pipe connected to the sewer through a Tee-junction. A manhole cover with suitable supporting frame is provided on its top, to take up the load of the traffic. Fig 6.8 shows typical details of a lamp hole. The vertical pipe is surrounded by concrete to make it stable and give it strength.

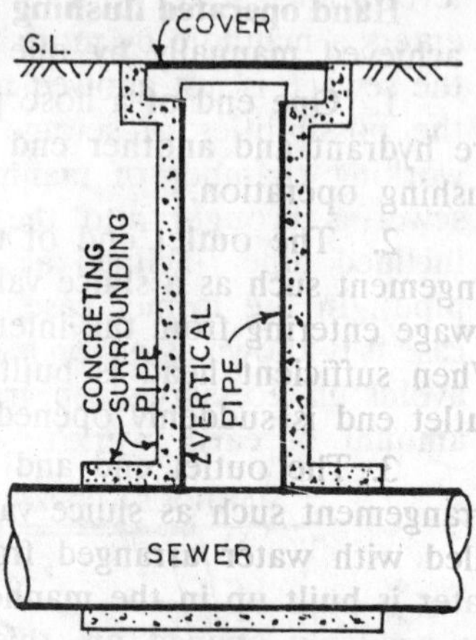

FIG. 6.8. LAMPHOLE

Lamphole may be *located* when (*i*) there is change in the direction or gradient of the sewer in between two closely spaced manholes. (*ii*) construction of manhole is difficult. (*iii*) spacing of manhole is considerable. A lamphole servers the following *purposes*.

(*i*) ***Inspection*** : For inspection of sewer line, an electric lamp is inserted in the lamphole and the light of the lamp is observed from the two manholes. The obstructed light will confirm the obstructions in the sewer.

(*ii*) ***Flushing*** : In emergency, lampholes can also be used flushing the sewer.

(*iii*) ***Ventilation*** : If perforated cover is provided, a lamphole, will serve the purpose of ventilation. In that case, it is known as *fresh air inlet.*

Use of lampholes have become more or less obsolete.

6.8. FLUSHING TANKS

A flushing tank is a device or arrangement which holds water and then throws it into the sewer for the purpose of flushing it. It can be operated either manually or automatically. Sewer laid on flat gradients may not produce self-cleansing velocity and may get blocked frequently. They can be cleaned with of help of such flushing tanks. Apart from this, flushing tanks are also provided near the dead ends of sewers. The quantity of water added in one flush is about 1600 litres. Flushing tanks should have sufficient capacity to hold water temporarily, to serve the required purpose. Generally, the capacity of flushing tank is kept equal to one tenth of the cubical contents of the sewer line served by it.

Flushing may achieved by two methods : 1. Hand operated flushing operation. 2. Use of automatic flushing tanks.

Hand operated flushing operation : In this, the flushing action is achieved manually by one of the following ways :

1. One end of a hose pipe may be connected to the nearest fire hydrant and another end is placed in the manhole to achieve flushing operation.

2. The outlet end of the manhole is closed by suitable arrangement such as a sluice valve or gate. Due to this, the incoming sewage entering from the inlet end starts collecting in the manhole. When sufficient head is built up, the sluice valve or gate of the outlet end is suddenly opened up, thus getting the flushing action.

3. The outlet end and the inlet ends are closed by suitable arrangement such as sluice valve or gate, and the manhole is then filled with water arranged from outside. When sufficient head of water is built up in the manhole, the sluices/gates of both the ends are suddenly opened up, this getting the flushing action.

Automatic flushing tanks : Flushing action may be achieved with flushing tanks, automatically at regular intervals. The entry of water is so regulated that the tank is just filled upto the discharge point, in a period equal to the fluhing interval. An overflow pipe is also provided to drain away excess water in case the tank overflows before the flushing action starts. Fig. 6.9 shows an automatic flushing tank.

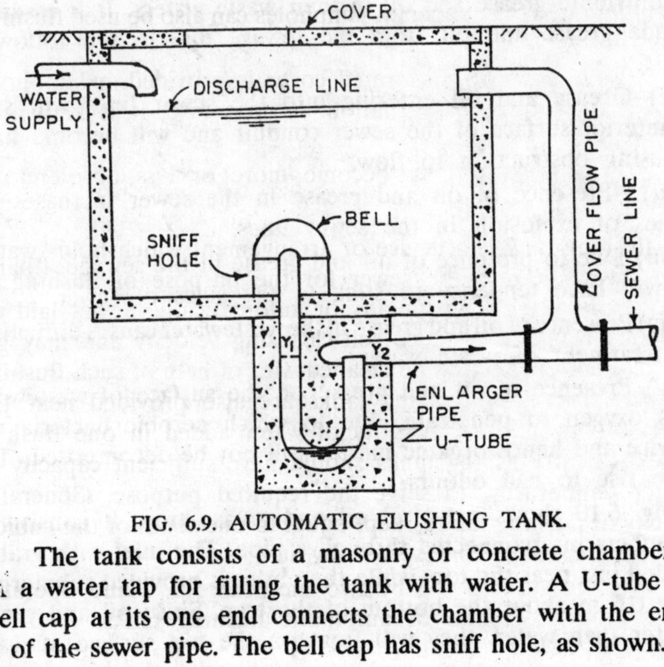

FIG. 6.9. AUTOMATIC FLUSHING TANK.

The tank consists of a masonry or concrete chamber filled with a water tap for filling the tank with water. A U-tube in the a bell cap at its one end connects the chamber with the enlarged end of the sewer pipe. The bell cap has sniff hole, as shown. When the water level in the tank is below the sniff hole, the water level in the U-tube, (consisting of one long arm and other short arm) is at level $Y_1 - Y_2$. As the water level in the tank rises above the sniff hole, further entry of air into the bell is sealed, and the air caught up in the bell is compressed. This compressed air exerts pressure on the water surface at Y_1, due to which water level in the long arm of the U-tube is depressed. As the water level in the tank increases, the water level in the longer arm of U–tube goes on depressing more and more, till a stage is reached when water level in the tank just reaches the *discharge level.* At this stage some compressed air gets released through the shorter arm and the siphonic action so caused starts releasing water from the tank into the sewer through the enlarger pipe. The siphonic action continues till the water level in the tank just falls to the level of the sniff hole. At tnis stage, air enters the bell portion through the sniff hole, thus breaking the siphonic action. At this stage, water level in the U-tube stands at level $Y_1 - Y_2$.

6.9. GREASE AND OIL TRAPS

Grease and oil traps are specially built chambers on the sewers to exclude grease and oil from sewage before they enter the sewer line. Such traps are located near those sources, such as automobile repair workshops, garages, kitchens of hotels, oil and grease industries, which contribute grease and oil in their waste waters. It is essential to exclude grease and oil from the sewage due to the following reasons :

(*i*) Grease and oil entering into the sewer lines will stick to the interior surface of the sewer conduit and will become hard, thus causing obstruction to flow.

(*ii*) Presence of oil and grease in the sewer increases the possibities of explosion in the sewer lines.

(*iii*) Due to presence of oil and grease in the sewage, floating matter will have tendency to stick to the sides of sewer.

(*iv*) Presence of oil and grease in the wastewater causes difficulties in the treatment of weste water.

(*v*) Presence of oil and grease on the surface of wastewater prevents oxygen to penetrate, due to which aerobic bacteria will not survice and hence organic matter will not be decomposed. This will give rise to bad odours.

Fig. 6.10 shows a typical oil and grease trap. It consists of two chambers interconnected through a pipe. The inlet with grating is provided just near the top, while the outlet is provided submerged, at about 0.6 m above the bottom of the trap. Since oil and grease are lighter than water, they will float on the top surface of waste

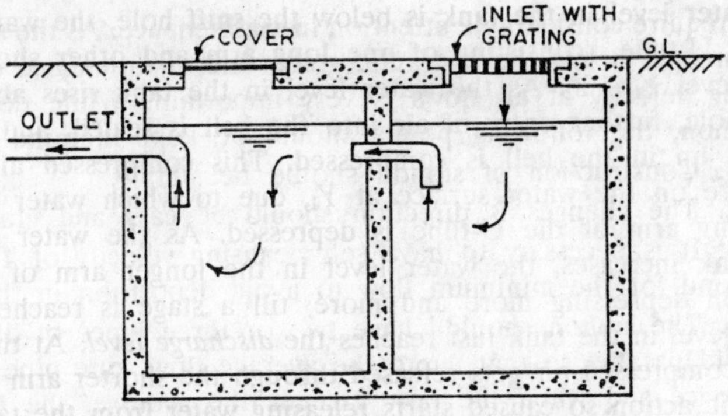

FIG. 6.10. OIL AND GREASE TRAP

water. Since the outlet is much lower, oil and grease will be easily excluded. However, wastewater from garages, particularly from floor drains and wash racks, contains oil, mud and sand. To trap all these, a combined sand, grease and oil trap may be provided as shown in Fig. 6.11.

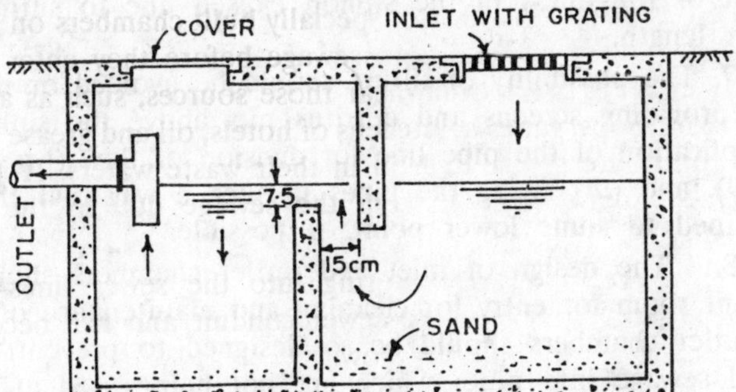

FIG. 6.11. COMBINED SAND, GREASE AND OIL TRAP.

6.10. INVERTED SIPHONS

When a sewer line dips below the hydraulic grade line, it is called an inverted siphon. The purpose is to carry the sewer under the obstruction, such as roadway, railway, stream, river etc., and regain as much elevation as possible. The sewage through such a pipe line (or inverted siphon) will flow under pressure which is greater than atmospheric. However, in the inverted siphon, the hydraulic gradient line is above the flow line, whereas in true siphon the hydraulic line is below the flow line. Hence the term 'inverted siphon' is misnamed as it is not a siphon. It will be more appropriate to call it a *depressed sewer*. Inverted siphons should be resorted to only where other means of passing the obstruction are not feasible, as they require considerable attention in maintenance. As the siphons are depressed below the hydraulic grade line, maintenance of self-cleansing velocity at all flows is very important. While designing the siphon, the following points should be kept in mind :

1. Construction of siphon should be simple.

2. The changes is direction should be easy and gradual.

3. It is necessary to have self-cleansing velocity of 1 metre per second for the minimum flow to avoid deposition in the line.

4. The siphon should have two or three pipes of different sizes, laid parallel so that upto the average flows one pipe comes into service and when the flow exceeds the average, the balance flow is taken up by the second and subsequent pipes.

5. For the selection of proper sizes of pipes, the minimum average and maximum flows in the sewer should be considered.

6. The total length of each pipe consists of straight length, and the lengths of fall, bends and rise. Allowance should be made for loss of head due to bends and due to increased friction on account of roughness in the siphon caused due to silting in the straight length.

7. The possibility of silting should be avoided or minimised by (*i*) providing screens and detritus pits above the siphon inlet, (*ii*) duplication of the pipe line for division of flow (provision for standby), and (*iii*) laying the pipes in such a way that they could be drained to some lower point, if possible.

8. The design of inlet and outlet chambers should allow sufficient room for entry for cleaning and maintenance of siphons. The outlet chambers should be so designed to prevent the back flow of sewage into pipes which are not being used at the time of minimum flow.

9. Proper bypass arrangements should be provided from the inlet chamber to the nearby stream if permitted by the pollution control authority ; otherwise, special arrangements should be made for pumping the sewage to the lower reach of the sewer line.

10. If the length of the siphon is more, hatch boxes at intervals of about 100 m should be provided for facility of rodding. There should be a vent pipe in the hatch box to prevent the formation of air-locks in the slphon.

Fig. 6.12 shows the details of an inverted siphon. It consists of inlet chamber, outlet chamber and three pipes laid parallel to each other. Fig. 6.12(*c*) shows the section of the inlet chamber. In order to achieve the self-cleansing velocity even at minimum flows, it is essential that the pipes come into action successively as the

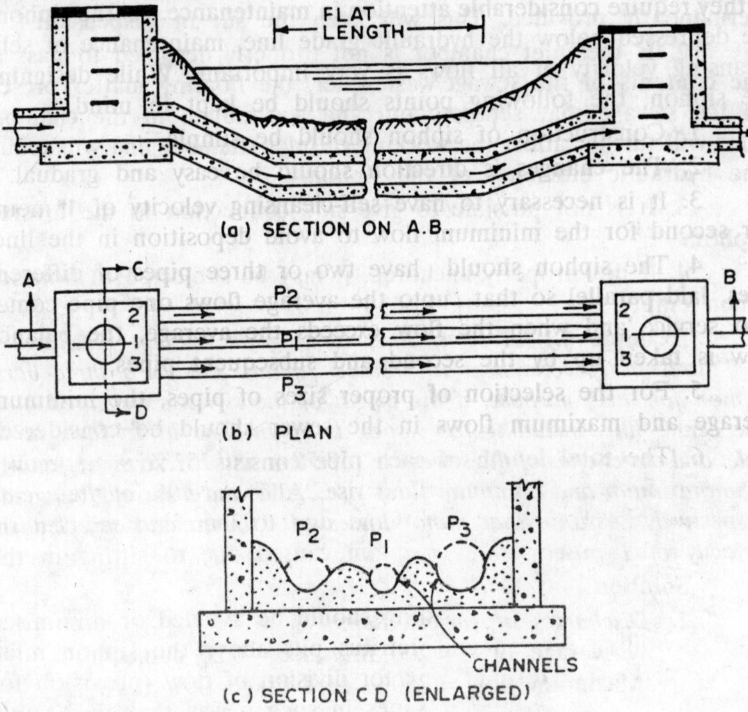

FIG. 6.12 INVERTED SIPHON

flow increases. This is achieved by providing *lateral weirs* with heights kept in accordance with depth of flow at which one or more siphon pipes function. In Fig. 6.12(*c*), *three channels* are shown, one for each pipe. Channel 1 comes into operation first. When channel 1 overflows, sewage enters into channel No. 2 and thus pipe No. 2 comes into operation. Lastly, when channel 2 overflows, sewage enters channel No. 3 and pipe No. 3 comes into operation. For a separate system, consisting of sanitary sewers, only two sections are provided —one for the minimum D.W.F. and other for the max. D.W.F. The design of an inverted siphon has been illustrated in example 6.1.

Disadvantages of inverted siphons. As stated earlier, inverted siphons should be resorted to only where other methods of passing the obstruction are not feasible. An inverted siphon possesses the following disadvantages :

1. The down gradient in the siphon pipes is not continuous, and they are likely to get silted if not properly designed to maintain self-cleansing velocity at all flows. When once silted, it is very difficult to remove the silt. In order to stir up the deposited silt, a chain

extending from inlet chamber to the outlet chamber is some time permanently provided. This will keep the silt in suspension.

2. If the inlet chamber is not properly designed to pass all the contents of the sewage/wastewater, the floating matter present in the sewage may separate out and accumulate in the chamber. This will further affect the working of inlet chamber and will reduce the hydraulic efficiency of the siphon.

3. It is not possible to give side connection to the inverted siphon.

4. If not properly anchored, it may be subjected to flotation, when empty.

Example 6.1. *A sewer line, carrying an average discharge of 200 litres per second, has to cross a stream. Design a three barrel siphon for this purpose, if the length of the siphon, measured along the centre line, including the slopes, is 90 m. The invert level at the inlet and outlet ends of the sewer are 152.50 m and 151.78 m respectively. The maximum and minimum flows are 250% and 40% of the average respectively. Assume that minor losses as 0.07 m, and self-cleansing velocity of 1 m/sec.*

Solution

1. ***Discharges at various flows***

Average flow = 200 litres/sec. = 0.20 cumecs.

Maximum flow = 250% of average flow

= 2.5 × 0.2 = 0.5 cumecs.

Minimum flow = 40% of the average flow

= 0.4 × 0.2 = 0.08 cumecs.

Max. available head = 152.50 − 151.78

= 0.72 m.

2. ***Design for flow at minimum discharge***

$$Q_{min.} = 0.08 \text{ cumecs } = A \times V$$

$$\therefore \quad 0.08 = \frac{\pi}{4} d^2 (1.0)$$

From which $d = 0.319$ m.

Use 300 mm dia. cast iron pipe.

Actual $$V = \frac{0.08}{\frac{\pi}{4}(0.30)^2} = 1.132 \text{ m/sec.}$$

For computation of head loss, let us use Manning's formula:

$$V = \frac{1}{N} R^{2/3} S^{1/2}$$

where $R = d/4 = 0.30/4 = 0.075$

Assume $N = 0.013$

$V = 1.132$ m/sec (computed above)

$\therefore \quad 1.132 = \dfrac{1}{0.013}(0.075)^{2/3} S^{1/2}$

From which, we get $S \approx 0.00685$

Now frictional head loss, $h_f = S \times L$

$= 0.00685 \times 90 = 0.616$ m.

$\therefore$ Total losses $h_L = h_f +$ minor losses.

$= 0.616 + 0.07 = 0.686$

This is less than the available head of 0.72 m.

Hence the pipe dia. is satisfactory.

3. Design for flow at average discharge

Average discharge = 0.2 cumecs.

Min. discharge = 0.08 cumecs.

At the average discharge, the excess discharge passes through the second pipe.

$\therefore$ Discharge for the second pipe = 0.2 − 0.08 = 0.12 cumecs.

Hence assuming a self cleansing velocity of 1 m/sec.

$$d = \sqrt{\frac{0.12 \times 4}{\pi \times 1}} = 0.39 \text{ m.}$$

Use 380 mm dia. cast iron pipe.

Actual $V = \dfrac{0.12}{\frac{\pi}{4}(0.38)^2} = 1.058$ m/sec.

Now $V = \dfrac{1}{N} R^2 S^{1/2}$

$\therefore \quad 1.058 = \dfrac{1}{0.013}\left(\dfrac{0.38}{4}\right)^{2/3} S^{1/2}$

From which $S = 0.00436$

$\therefore \quad h_f = S \times L = 0.00436 \times 90 = 0.392$ m.

and $h_L = 0.392 + 0.07 = 0.463 \text{ m} < 0.72$ m.

Hence OK.

4. Design for flow at max. discharge

$Q_{max} = 0.5$ cumecs.

Hence flow in the third pipe, during flow at max. discharge.

$= 0.5 - 0.2 = 0.3$ cumecs

Let us provide a higher velocity, say 1.5 m/sec. in this case.

$$\therefore \quad d = \sqrt{\frac{0.3 \times 4}{\pi \times 1.5}} = 0.5046 \text{ m.}$$

Provide 500 mm dia. pipe.

Actual $V = \dfrac{0.3}{\frac{\pi}{4}(0.50)^2} = 1.528$ m/sec.

Now $$V = \frac{1}{N} R^{2/3} S^{1/2}$$

$\therefore$ $$1.528 = \frac{1}{0.013}\left(\frac{0.50}{4}\right)^{2/3} S^{1/2}$$

From which $S = 0.00631$

$\therefore$ $h_f = 0.00631 \times 90 = 0.568$ m.

and $h_L = 0.568 + 0.07 = 0.638 < 0.72$ m.

Hence the sizes of the three barrels are :

Barrel 1 : **300 mm** dia. pipe

Barrel 2 : **380 mm** dia. pipe

Barrel 3 : **500 mm** dia. pipe

6.11. STORM REGULATORS

Storm regulators are the appurtenances which are constructed for combined sewerage system, so as to permit the diversion of storm water when the discharge exceeds a certain value. The excess water is diverted to a natural stream or river. In a combined system, the D.W.F. may be very small and the quantity of storm water is large. Generally, a combined sewer is designed for a discharge equal to 3 to 6 times the dry weather flow. Sometimes, when the intensity of rain is more, the quantity of storm water becomes enormous. This excess quantity is therefore diverted through the use of storm regulators. Thus the storm water regulators, also some times called as *overflow devices,* prevent overloading of sewers, pumping stations, treatment plants and other disposal arrangements. The excess waste water will be mainly composed on storm water and its will therefore be not foul in nature.

Following are three types of storm regulators :

1. Over flow weirs (or side flow weirs). 2. Leaping weirs. and 3. Siphon spillways.

1. Overflow weirs (or side flow weirs). This a very common device and consists of a masonry chamber with an overflow weir built in it at a place where a sewer approaches a natural watercourse. The object is to keep the discharge to the storm water down to a predetermined maximum and allow the surplus to overflow to the water course. The *overflow weirs* (side weirs) may be *single acting* (or single side) or *double acting* (or double sides).

Fig. 6.13 shows the plan and sections of a single overflow weir. A curved masonry channel with a weir on the *outer side* is constructed inside the chamber. The top of the weir is set at such a level that the predetermined flow (usually 6 times D.W.F.) flows through it and is deflected by the weir wall into the sewer of reduced size. When the flow increases, the level rises and the excess flow

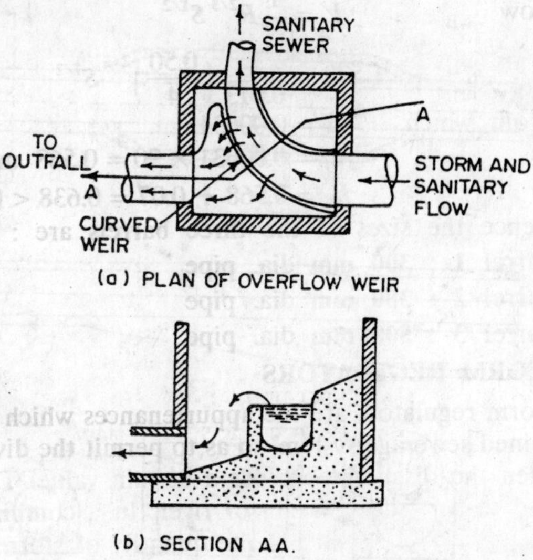

FIG. 6.13. SINGLE OVERFLOW WEIR

drops over the weir and is carried to the watercourse. The weir length should be sufficiently long for effective regulation.

The length of the side flow weir is given by the following expression devised by Babbitt :

$$L = 7.6 \times 10^{-3} VD \log_e \frac{h_1}{h_2} \qquad \text{...(6.2)}$$

where L = required length (m)

V = velocity of approach (m/s)

h_1 and h_2 = heads above the crest of weir upstream and down stream.

The above formula is limited to conditions in which the weir is placed in the side of a circular pipe at a distance above the bottom greater than $d/4$ and less than $d/2$, where d is the diameter of the pipe, and the edge of the weir is sharp and parallel to the invert of the channel. Its usefulness is limited in that it was devised for pipes between 450 and 600 mm in dia. and in that the depth of flow above the weir should not exceed $3d/4$.

The above type of overflow requires a considerable length even in ordinary floods. In order to reduce the length, *double acting overflow* with weirs on both sides may be adopted, as shown in Fig. 6.14.

In Fig. 6.14, the whole structure is compact under a concrete cover. But very often, it is enclosed in a masonry chamber, with one or two manhole covers for inspection. Sometimes the upper main channel is formed by thick steel plates known as *dip plates*

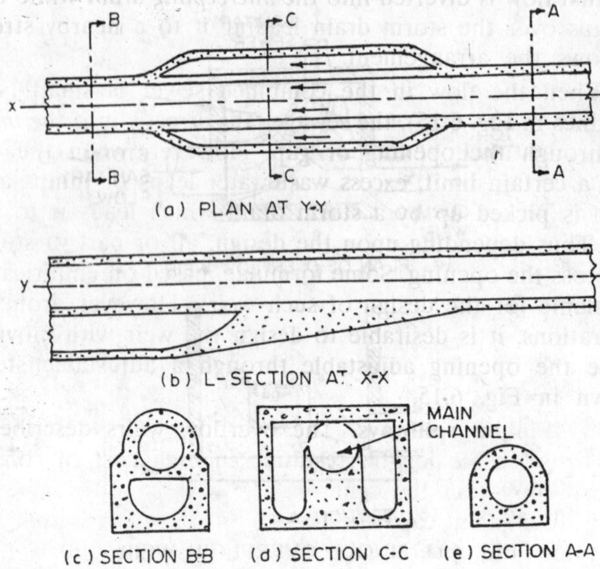

FIG. 16.14. BOTH SIDE OVERFLOW WEIRS

butted to vertical angle irons embeded in concrete. However, a better arrangement is to set the overflow at 0.82 dia. of the pipe sewer by cutting its top at that level in length required for calculations, so as to form an open channel out of the pipe.

2. Leaping weir. A *leaping weir or a jumping weir* is a special device by way of which excess wastewater (during storm) is allowed to escape to a watercourse, while the designed flow is permitted to flow to the intercepting sewer. The arrangement consists of an opening in the invert of the combined sewer, and through this opening

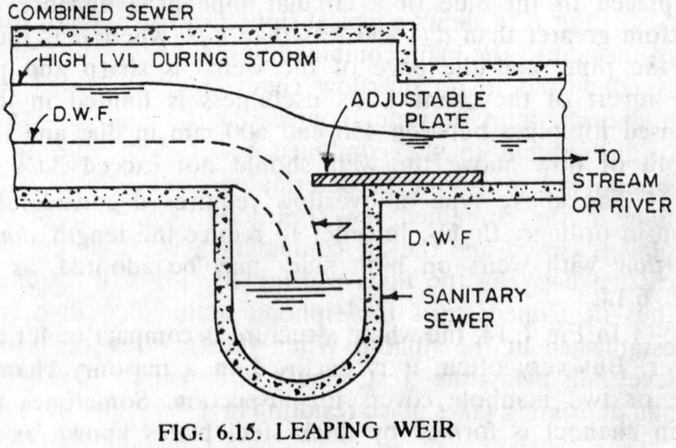

FIG. 6.15. LEAPING WEIR

the normal flow is diverted into the *intercepting drain* while the excess flow leaps over the storm drain leading it to a nearby stream. Fig. 6.15 shows the arrangement.

When the flow in the combined sewer is small (shown by dotted lines in Fig. 6.15), the sewage falls directly into the *intercepting sewer* through the opening or gap. However, when the discharge exceeds a certain limit, excess wastewater leaps or jumps across the gap and is picked up by a storm drain which leads it to a natural stream. Thus depending upon the design, all or part of storm water leaps across the opening. Some formulae, based on empirical findings are available for the design of such weirs. However, from practical considerations, it is desirable to design the weir with moving crests to make the opening adjustable through a adjustable steel plate, as shown in Fig. 6.15.

3. Siphon spillways. The overflow weirs described earlier require considerable length, resulting in high cost of construction. Siphon spillways, on the other hand, are very effective since they have high dischanging capacity. The spillway is very sensitive, automatic in operation and diverts more flow in lesser time.

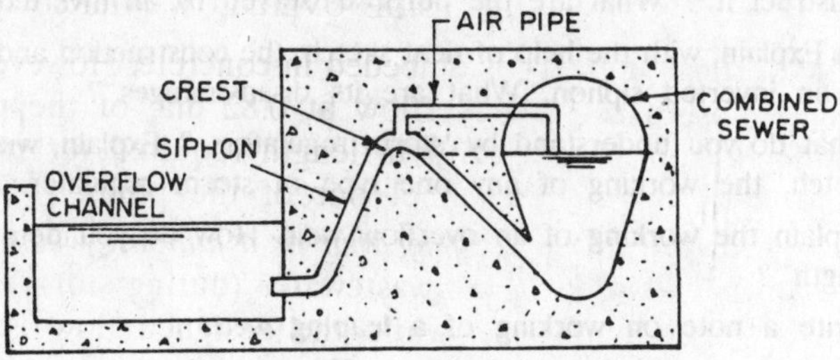

FIG. 6.16. SIPHON SPILLWAY.

Fig. 6.16 shows a typical siphon spillway. It consists of a siphon connected to the main combined sewer. The lower end of the siphon is connected to an overflow channel which leads excess water to a natural drain or water course. An air pipe connects the throat of the siphon to the combined sewer. The level of the crest of siphon is kept at the predetermined desired level. As soon as the flow in the combined sewer exceeds this level, the mouth of air pipe is closed and the air contained in the siphon pipe is suddenly carried away by the flow, causing the priming action. The suction thus developed starts the siphonic action and flow is immediately established in the siphon. When the flow reduces and when water level falls below the crest, mouth of air pipe gets exposed, air enters and depriming takes place resulting in breaking of the siphonic action.

PROBLEMS

1. What do you understand by 'sewer appurtenances' ? Why are they necessary ? Enumerate various appurtenances commonly used.
2. Explain the necessity of providing a manhole in a sewer line. Describe, with the help of a neat sketch the components of a manhole.
3. (*a*) What do you understand by inlets ? Where are these located? Differentiate clearly between a curb inlet and a gutter inlet.

   (*b*) Explain, with neat sketches, various types of combination inlets.
4. Differentiate clearly between an 'inlet' and a catch basin.
5. Define a drop manhole. What are its objects ? How is it constructed ?
6. What do you understand by a 'lamphole' ? Explain the purposes for which it is constructed.
7. What is a flushing tank ? Explain, with the help of a neat diagram, the working of an automatic flushing tank.
8. Explain the principle of working of a grease and oil trap. Mention the reasons for excluding grease and oil from waste water
9. (*a*) What do you understand by an inverted siphon ? When do you construct it ? What are the purposes served by an inverted siphon?

   (*b*) Explain, with the help of neat sketch, the construction and working of an inverted siphon. What are its disadvantages ?
10. What do you understand by 'storm regulators ? Explain, with a neat sketch, the working of any one type of storm regulater.
11. Explain the working of an overflow weir. How do you determine its length ?
12. Write a note on working of a leaping weir.
13. Differentiate between :
    (*i*) Catch basin and inlet
    (*ii*) Vertical inlet and horizontal inlet
    (*iii*) Lamphole and clean out
    (*iv*) Normal manhole and drop manhole
    (*v*) Overflow weir and leaping weir
    (*vi*) Access shaft and working chamber.

7

Sewage Pumping

7.1. INTRODUCTION : NECESSITY

The necessity of lifting wastewater or sewage arises under the following circumstances :

1. When same area of a town is so low-lying that it cannot be drained by gravity to discharge into a submain or main unless the entire sewerage system in the other parts is installed correspondingly at low level. In such circumstances, it is more economical to pump the sewage from the low lying area into the upper branch or main.

2. Pumping is resorted to, at intervals, for a sewage system in a flat country, since laying of sewers at the designed grade continuously all along will mean expensive excavation.

3. Pumping is resorted to when outfall sewer is at a lower level than the body of water into which it is to be discharged, or when the outfall is lower than the entrance to the treatment works.

4. When a sewer has to go across a high ridge, it will be more economical to pump it into sewers laid across the slope of the ridge at reasonable depth, rather than driving a tunnel.

5. Pumping is required to take out sewage from the cellers or sub-basements of buildings, when the level of celler is much lower than the invert level of sewer to which drainage connection is to be made.

7.2. PROBLEMS IN SEWAGE PUMPING

The pumping of sewage is not as simple as pumping of water, since the following *special problems* are to be faced in sewage pumping.

1. Sewage has foul characteristics.

2. Sewage contains a lot of suspended and floating materials. These may make the running of pumps difficult and may cause frequent clogging of the pumps.

3. Sewage contains organic and inorganic wastes which may cause corrosion and erosion of parts of the pumps and reduce the life of the pumps.

4. Sewage contains disease producing bacteria and organisms, which may cause health hazards to the persons working at pumping station.

5. The rate of flow of sewage varies continuously and hence pumping operations are to be adjusted accordingly.

6. The size of sump is limited since large sized sumps will result in the settlements of silt and organic matter at its bottom. The provision of sump or wet well is made to give only a little storage space.

7. The pumps should be of high order reliability since failure of pumps will lead to flooding which may cause unbearable nuisance.

7.3. PUMPING STATIONS

1. Location of pumping stations. Proper location of pumping station requires a comprehensive study of the area to be served to ensure that the entire area can be adequately drained. If a very large quantity of sewage is to be pumped, the site should be near a stream, or a nallah or a storm water drain, into which the sewage could be discharged during emergencies, such as break-down of the pumping plant, failure of power etc. If this precaution is not taken, the station may be flooded with possible damage to machinery specially to the electrical equipment. The site should be aesthetically satisfactory. The pumping station should be located and constructed in such a manner that it will not be flooded at any time. The storm water pumping stations should be located such that water may be impounded without creating undue amount of flood damage if the inflow exceeds the pumping station capacity. The station be so located that it is easily accessible under all weather conditions.

2. Elements of pumping stations. Apart from the *structure* of the pump house, a sewage pumping station consists of the following elements :

(1) Grit channel or etritus pit. (2) Coarse and fine screens. (3) Sumps or wet well. (4) Pump room or dry well. (5) Pumps with driving engine or motor. (6) Miscellancous accessories such as pipes, valves, float-switch arrangements, flow recorders, emergency overflows, ventilation arrangements such as extraction fans etc. etc. Fig. 7.1 shows typical sewage pumping stations.

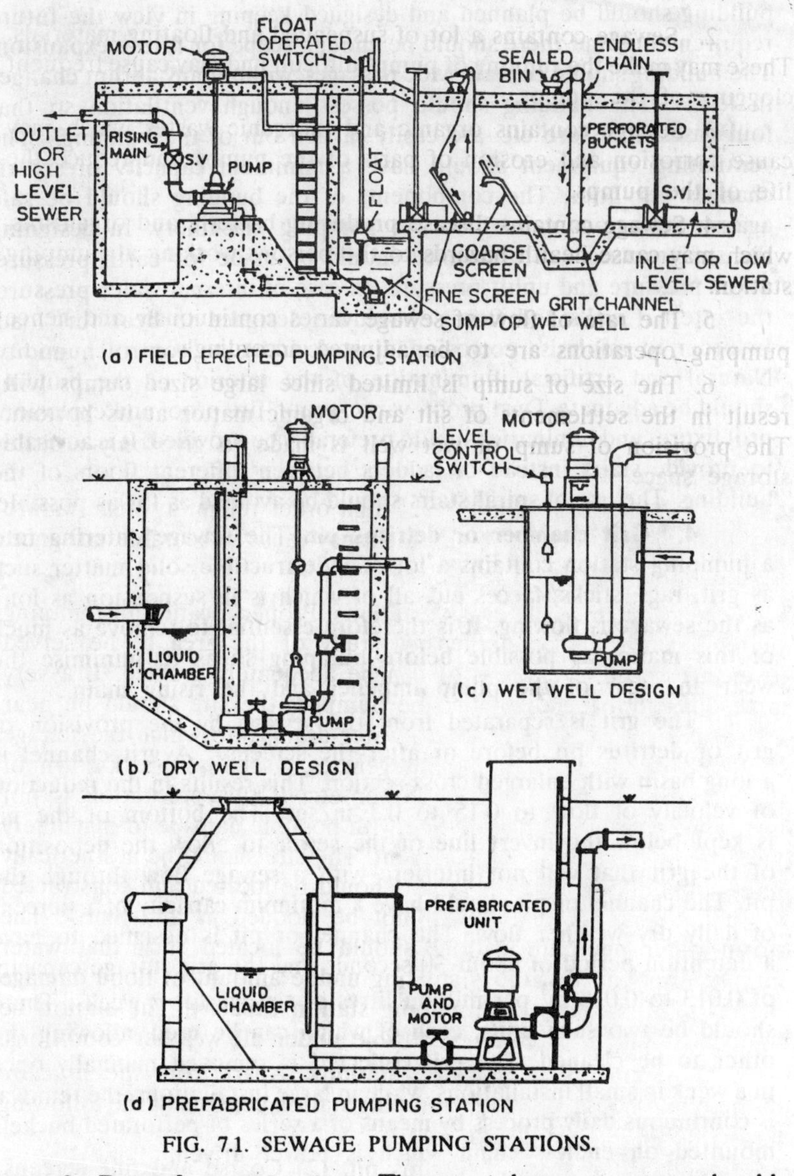

FIG. 7.1. SEWAGE PUMPING STATIONS.

3. Pump house structure. The pump house structure should be designed to withstand flotation forces to which it may be subjected. The substructure of the pumping station may be of mass concrete or R.C. C. while the superstructure may be constructed of any material. The internal walls and floors should be structurally designed to take the weight of machines along with a live load of 5 kN/m^2. The building should be planned and designed keeping in view the future requirements, and there should be enough scope for future expansion. The building should contain wide passages, without any abrupt changes in levels. The building should possess enough ventilation so that foul gases, moisture etc. are easily carried out of the building. The ventilating equipment should have a minimum capacity of six air changes per hour. The components of the building should be safe against vibrations caused due to pumping machinery. In designing the substructure due allowance should be made for earth pressure, water pressure and uplift pressure. In case of internal water pressure, the internal walls and floors should be designed to bear this load. In order to minimise corrosion, it is desirable to control humidity. Natural and artificial illumination of the interior of the building should be adequate. Dust proof, vapor-proof, fire proof and explosion-prof fixures and luminaries should preferably be provided. It is advisable to provide stairs instead of ladders between different floors of the building. The use of spiral stairs should be avoided as for as possible.

4. Grit chamber or detritus pit. The sewage entering into a pumping station contains a lot of indestructible solid matter such as grit, rags, sticks, faeces etc. all of which is in suspension as long as the sewage is flowing. It is therefore essential to remove as much of this matter as possible before pumping so as to minimise the wear and tear of the pump impeller and the rising main.

The grit is separated from the sewage by the provision of grit or detritus pit before or after the screens. A grit channel is a long basin with enlarged cross-section. This results in the reduction of velocity of flow to 0.15 to 0.3 m/sec. The bottom of the pit is kept below the invert line of the sewer to allow the deposition of the grit that will not interfere with a sewage flow through the pit. The channel or pit should have a minimum capacity of 1 percent of daily dry weather flow. The channel or pit is designed to have a detention period of about 30 seconds and the grit storage capacity of 0.013 to 0.028 m^3 per million litres of sewage for 2 weeks. There should be two such units, each of which can be used, allowing the other to be cleaned. The grit collected is removed manually once in a week in small installations, while in large installations, the removal is continuous daily process by means of a series of perforated buckets mounted on endless chain which is power driven.

4. Screens. After the removal of grit from the sewage, it is made to pass through screens to trap the floating matter such as rags, sticks, papers, etc. It is necessary to remove these, otherwise they will choke and damage the pumps. **Screens.** are of two types : Coarse and fine. In large installations, it is usual to provide both, the coarse screen being the first to intercept the flow. Coarse screen is made up of wrought iron bars kept parallel to each other and having a clear spacing of 50 to 100 mm in between them. The material trapped by coarse screen can be removed by hand raking. Fine screens have opening of less than 25 mm and intercept all except very fine particles of sewage. The fine screen should not trap any organic matter as far as possible. The screenings trapped by fine screens are removed by mechanical rakes.

5. Sump well (wet well and dry well). Pumping stations are provided with two separate wells : wet well for receiving the incoming sewage and *dry well* for housing the pumps. The functions performed by a sump or wet well are to act as a *suction pit* from which pumps draw sewage and as an equalising basin to minimise the load-fluctuations on the pumps. The sump (or wet well) is provided either below the floor of a pump house or by the side of dry well, as shown in Fig. 7.2 depending upon the depth of main sewer below the ground level. In the later case, the wet and dry wells may be of the following types : (*i*) Rectangular, with dry and wet wells adjacent to each other. (*ii*) Circular with central dry well and peripheral wet well, and (*iii*) Circular with a dividing wall to separate the wet and dry wells.

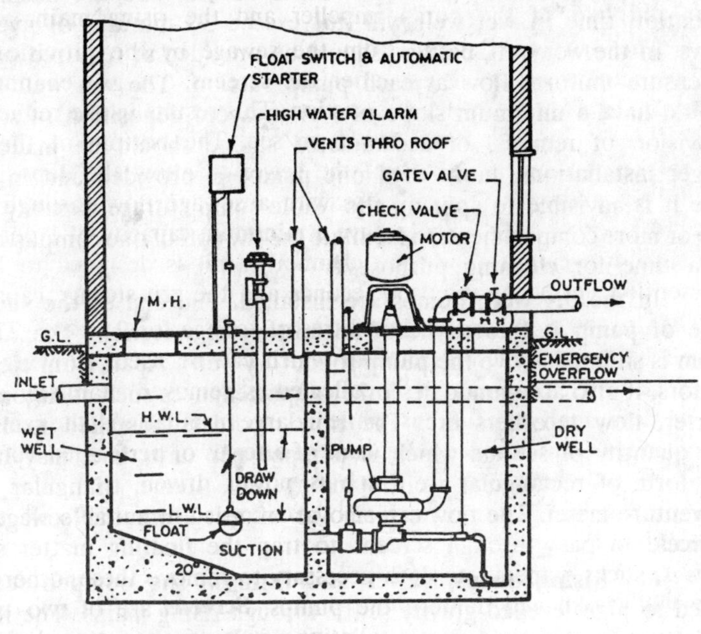

FIG. 7.2 WET AND DRY WELLS

The rectangular type may require thicker walls to withstand the pressure of the lateral soil, subsoil water and sewage and are not recommended except for smaller installations or where the availability of space is a deciding factor.

The sump is so designed that the sewage can collect in it for a time till the rise in the level operates a *float-switch* which starts the pumping unit and the sewage is pumped out. The operation becomes expensive if the power unit is operated at very frequent intervals due to smaller capacity of sump. Usually, a detention period of 30 minutes of average flow is adopted for design. The shape of wet well and the detention time provided for sewage stations shall be such that deposition of solids is avoided and sewage does not turn septic. The *capacity* of wet well is reckoned between the level at which air affects the suction line of the pump of minimum duty installed in the pump house and the designed sewage level in the incoming sewer, i.e. the portion of the well below the upper most *starting point* and the lower most stopping point. It is governed by the pump sets installed to deal with the varying flows. The principle recommended is that any pump should be work for at least 5 minutes before it is stopped. The size of wet well is to be kept such that with any combination of inflow and pumping, the cycle of operation for each pump will not be less than 5 minutes and the maximum detention time in wet well will not exceed 30 minutes of average flows. In the wet well, baffless should be provided at required places to ensure uniform flow at each pump suction. The wet well floor should have a minimum slope of 1 :1 to avoid deposition of solids. Provision of removal of accumulated sludge should be made for bigger installations, more than one pumps is provided and in that case it is advisible to split up the whole sump storage capacity into two or more compartment, to facilitate the shut off of one compartment at a time for cleaning purposes.

In the dry well, pumps are installed. The end of the suction pipe of pump is located near the bottom of wet well. The motor room is situated above the pump room (dry well). Apart from electric motors, it also accommodates other appurtenances such as automatic starter, flow recorders etc. The flow recorders, installed to know the quantity of sewage which is pumped per unit time, may be in the form of rectangular weir, standing wave flume, triangular weir or venture meter. The flow can also be recorded by suitable electrical devices.

6. Rising mains, valves and fittings. The pumped sewage is led to high levelled gravity sewer through rising mains. The rising mains are made up of steel, cast iron, spun iron or asbestos-cement pressure pipes. Generally, cast iron pipes with flanged joints are provided. The flanged joints provide easiness in dismantling and repair of the pumping station equipment. The length of the discharge pipe (*i.e.* rising main) should be kept as small as possible because long detention of sewage in closed pipes under pressure causes their anaerobic deterioration. As far as possible, the rising main should rise steadily from the pumping station to the point of discharge. The velocity of flow in the rising main should not be less than 0.75 m per second. At the same time velocities higher than 2 m/sec should be avoided. For economic design, the velocity of 0.85 m/sec. for normal rate of pumping is desirable.

A reflux or non-return valve is fitted on the rising main, just next to the pump, to prevent a backflow through the pump when the pump is stopped. Check valve should be provided in the sewer line discharging the sewage, to prevent the back flow of sewage during floods in the river or discharge area. When the pump is to be dismantled for repairs etc., the rising main may be isolated by providing a hand operated sluice valve next to the reflux valve. Gate valves should be provided on the sewer line just before the wet well and on the suction and discharge pipe to close the flow of sewage during maintenance, inspection and repair of the pumps. For cleaning out the rising main, chambers with hatch-boxes should be provided at intervals so that drain cleaning apparatus could be worked through them. Pressure gauges should be installed at appropriate position to record the suction and delivery pressure.

7.4. TYPES OF PUMPS

Following are the types of pumps commonly used for sewage and storm water pumping : (*i*) Centrifugal pumps. (*ii*) Reciprocating pumps. (*iii*) Propeller or axial flow pumps. and (*iv*) Air pressure pumps or ejectors.

1. **Centrifugal pumps**. Centrifugal pumps are most widely used for pumping sewage and storm water, as these can be easily installed in pits and sumps, and can easily transport the suspended matter present in the sewage without getting clogged too often. These pumps work on the principle of centrifugal force. They essentially consist of two main parts : (*i*) the casing and (*ii*) the impeller. The impeller rotates with high speed inside the casing. The commonly used horizontal axial flow types pumps are fitted with either open or closed three-vane type

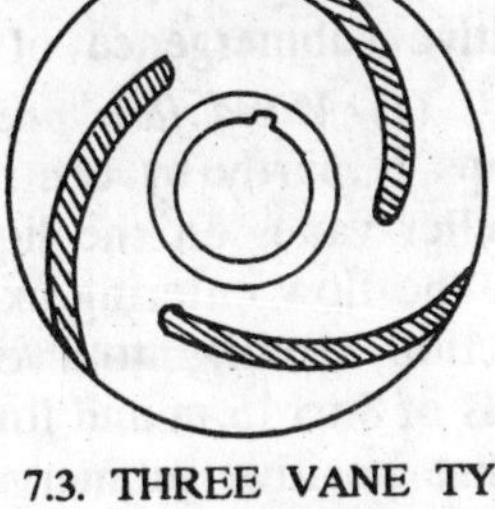

FIG. 7.3. THREE VANE TYPE IMPELLER

impellers (Fig. 7.3). The clearance between the vanes is kept large enough to allow any solid matter entering the pump to pass out with the liquid, thus preventing the *clogging*. Such pumps are therefore, called non-clog pumps.

There are three classes of centrifugal pumps : (*a*) disintegrator pumps, (*b*) full way pumps and (*c*) free way pumps. *The disintegrator type pumps* are provided with conical shaped impeller having sharp edged grooves, along with fixed knives set against the impeller. Due to this the solids present in the sewage are cut up and disintegrated during pumping and hence the sewage can be directly disposed of into the sea. The *full way pumps* can pass down solids upto 1500 mm diameter through the pump casing into the rising main without getting choked. *Freeway pumps* are used when sewage containing solids of more than 150 mm dia. are required to be passed through the pump casing. Centrifugal pumps can be classified under :

(*a*) Axial flow pumps. (*b*) Mixed flow pumps. (*c*) Radial flow pumps (commonly referred to as centrifugal pumps)

The classification is usually based as the specific speed (N_s) at the point of maximum efficiency. The specific speed of an impeller is defined as the speed, in r.p.m., at which a geometrically similar impeller would run, if it were of size to deliver 1 m^3/min. against 1 m head. The specific speed is given by the expression

$$N_s = \frac{3.65\, n}{Q\, H^{0.75}} \qquad ...(7.1)$$

where Q = flow in m^3/min.

H = head in m

and n = speed in r.p.m.

(*a*) ***Axial flow pumps*** : Axial flow pumps develop most of their head by propelling action of the impeller vanes on the liquid. They are characterised by a single inlet impeller with the flow entering axially and generally used for large installations with capacities greater than 2000 m^3/hr and head less than 9 m. The pumps are generally of vertical type. The axial flow pumps have relatively high specific speed ranging from 8000 to 16000. The vertical units should have positive submergence of propeller for proper operation.

(*b*) ***Mixed flow pumps*** : The head developed by mixed flow pumps is partly by centrifugal action and partly by the lift of the impeller vanes on the liquid. The pump has a single inlet impeller with the flow entering axially and discharging in an axial and radial direction, usually into a volume type casing. They are used for medium heads of 8 to 15 m and for medium to large capacities. They generally require positive submergence but may be used for limited suction lift.

(c) ***Radial flow pumps (or centrifugal pumps)*** : The head developed in these types of pumps is principally by the action of centrifugal force. Pumps of this type can be obtained with either single suction or double suction inlet impellers, the flow leaving the impeller radilly and normal to the shaft axis. As almost any range of head and capacity can be obtained, majority of pumps are of this type. These pumps are characterised by relatively low specific speeds, with single inlet impeller having specific speeds upto 4200 and double suction units having specific speeds less than 6000. Single suction pumps are generally used for sewage and storm water pumping as they are less susceptible to clogging. The characteristics of the three type of centrifugal pumps are summarised in Table 7.1.

TABLE 7.1 CHARACTERISTICS OF CENTRIFUGAL PUMPS

	Characteristics	*Axial flow*	*Mixed flow*	*Radial flow*
1	Usual capacity range	Greater than 2000 m^3/hr.	Greater than 100 m^3/hr.	All flows
2	Head range	0–9 m	8–15 m	All heads
3	Shut off head above rated head at max. efficiency point	About 200%	165%	120 to 140%
4	Kilowatts characteristics	Decrease with capacity	Flat	Increase with capacity
5	Suction lift	Usually requires submergence	Usually requre submergence; short suction lift is permitted.	Usually not over 4.5 m
6	Specific speed	8000 to 16000	4200 to 9000	Below 4200 for single suction and below 6000 for double suction
7	Service	Used where space and cost are considerations and load factor is low	Used where load factor is high and where trash or other solid matter is encountered	Used where load factor is high and high efficiency & ease of maintenance are required.

2. Reciprocating pumps : Reciprocating pumps are more or less obsolete in modern sewage pumping station since they are liable to be clogged by solids or fibrous material, even though sewage may have passed through coarse screens. Also, their initial cost is higher and efficiency is lower than those of a centrifugal pump. However, in cases where it is required to deal with difficult sludges and where large quantity is to be pumped against low heads, reciprocating pumps may be used after passing the sewage through screen with 20 mm spacing. Reciprocating pumps used for pumping sewage are generally of two types : (*i*) Ram type and (*ii*) Propeller type.

In the *ram type* reciprocating pump, a piston or plunger moves through glands displacing liquid in a vessel. The piston is not arranged to fit closely in the cylinder. The *diaphragm pump* is an example of the ram type reciprocating pump. In the *impeller type* reciprocating pump, a multiple blade screw rotor or impeller moves vertically into a pump-casing, causing the sewage to lift up. It draws water through inlet guide vanes and discharges through outlet guide vanes – and this action is some what similar to that of ships propeller. *Axial flow screw pump* is an example of this type of pump.

Diaphragm pump. Diaphragm pump is a ram type reciprocating pump. In this pump (Fig. 7.4), a piston or plunger is attached to the centre of a circular rubber diaphragm. The outer edge the rubber diaphragm is bolted to a flange on the pump. The flexibility of the diaphragm permits the up and down motion of the plunger, thus increasing or decreasing the capacity of the pump, through the suction valve, on the upward movement of the plunger. The delivery valve is located on the plunger itself and downward motion of the plunger closes the suction valve and force the liquid through the delivery valve.

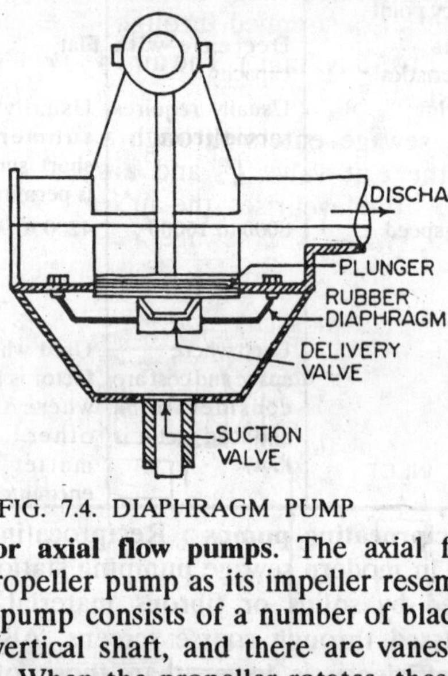

FIG. 7.4. DIAPHRAGM PUMP

3. Propeller or axial flow pumps. The axial flow pump is sometimes called a propeller pump as its impeller resembles the propeller of a ship. The pump consists of a number of blades fixed like screw threads on a vertical shaft, and there are vanes both on the inlet and outlet sides. When the propeller rotates, these vanes guide the flow *axially* along the shaft. The efficiency of axial flow pump is very low, to the extent of about 25% or so. These pumps are suilable to pump large volumes of sewage against low head.

4. Air pressure pumps or pneumatic ejectors. Pneumatic ejectors work on the action of compressed air. These are used for the following conditions : (*i*) where the small quantities of sewage is to be lifted from basements of building, to a high level sewer, (*ii*) where the quantity of waste water from a low-lying area does not justify the construction of a pumping station, and (*iii*) where centrifugal pumps of smaller capacity are likely to be clogged.

The ejectors possess the following *advantages* : (*a*) No sewer gases can escape except through the vent shafts as sewage is completely enclosed. (*b*) Operation is fully automatic and the ejector comes into operation only when needed. (*c*) Only a few parts are in contact with sewages thus necessitating little attention or lubrication. (*d*) Ejectors less susceptible to clogging. (*e*) Screening is not required as check valves and connecting lines will pass all the solids that enter the ejector compartment. The only disadvantage of the ejectors is that they possess very low efficiency to the extent of about 15%. Fig. 7.5 shows Shone's pneumatic ejector.

The ejector consists of a cast iron chamber with a spindle having a bell at its upper end and a cup at its lower end. Two check valves (or reflux valves) V_1 and V_2 are provided at entrance and exit ends respectively. The compressed air inlet valve V_3 is operated by a lever arrangement having a counter weight at its end. Compressed air is supplied through this valve at a pressure of about 1.5 kg/cm^2 (0.15 N/mm^2). The air in the chamber can escape through the exhaust.

The sewage enters through valve V_1 and rises slowly in the chamber, the exit valve V_2 and air inlet valve V_3 being closed at this stage. As the level rises, the air from the chamber escapes through the exhaust. When the sewage reaches the bottom of upper cup

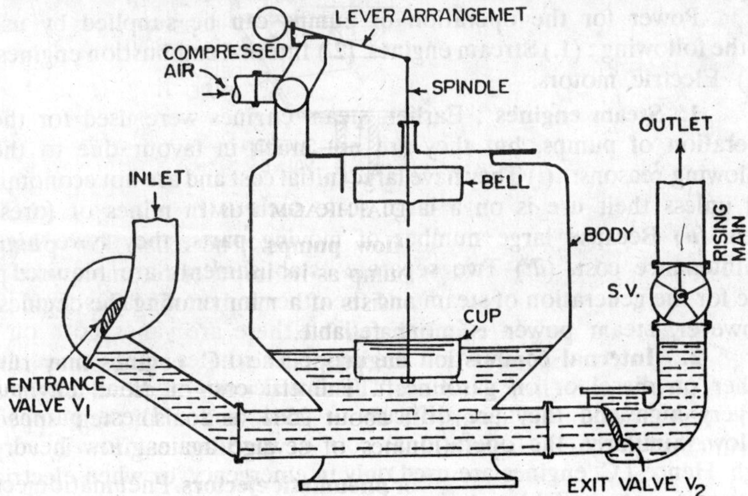

FIG. 7.5. SHONE'S PNEUMATIC EJECTOR.

(or bell) the air inside the upper cup is entrapped. Further rise in level of sewage makes the entrapped air to exert vertical pressure (marked by arrows) on the spindle. Due to this, the spindle is lifted up, the lever is operated, the exhaust is closed and air entrance valve V_3 is opened. The air entering under pressure forces the sewage inside the chamber to rise in the outlet pipe through the exit valve V_2 which also opens at this stage while entrance valve V_1 closes. The sewage in the chamber is discharged. When the sewage level in the chamber falls below the cup, the weight of sewage in the cup causes the spindle to be dragged down. This cuts off the compressed air and opens the ejector to the atmosphere by opening the exhaust. At this stage, valve V_2 is also closed due to back pressure of the sewage in the rising main. The sewage now enters through the valve V_1 and the process is repeated. The pneumatic ejectors are generally used in pairs so that when one is empty, the other is full.

The quantity of air required to operate the ejector is given by the following expression :

$$V = \frac{Q\,(H + 10.3)}{12.2} \quad \text{...(7.2)}$$

where V = volume of free air required in m^3/min.

H = total head in m.

and Q = rate of sewage discharge in m^3/min.

The volume of air storage tank and the characteristics of the compressor should be adequate to provide the necessary volume of air at a pressure at least 40% higher than that required to raise all the sewage to the maximum computed lift.

7.5. POWER FOR PUMPS

Power for the operation of pumps can be supplied by use of the following : (1.) Stream engines. (2.) Internal combustion engines. (3.) Electric motors.

1. Steam engines : Earlier, steam engines were used for the operation of pumps, but they are not much in favour due to the following reasons: (*i*) They have large initial cost and are not economical unless their use is on a large scale such as in mines or forest area. (*ii*) Because large number of moving parts, they have high maintenance cost. (*iii*) Two separate establishments are required : one for the generation of steam and the other for running the engines. However, steam power is more reliable.

2. Internal combustion engines : The I.C. engines may run either on diesel or on gasoline. The initial cost of diesel engines is very high and they are difficult to start, and also their speed is low. Similarly, the operational cost of gasoline engines is very high. Hence I.C. engines are used only in emergency, or when electric power is not available.

3. Electric motors : Pumps are operated through electric motors, in the majority of the cases. They are very convenient and free from most of the objections mentioned above. They are compact, silent in operation, automatic in action and free from nuisance of smoke. They occupy less building space, and can be installed immediately above the pumps. The constant speed squirrel case induction motor is genrally used. However, a diesel standby unit should always be provided in all pumping stations to meet the requirements in case of power failure. A full voltage type of starting equipment is always preferred from the point of view of cost, floor space and reliability. Where motors are started more than three times a day, the use of power circuit breakers as motor starting equipment may not prove economical in the long run. The electric cable should be heavy duty, PVC sheathed, and be of adequate capacity. Cables inside the pump house should be laid in specially made trenches covered with lean concrete.

Example 7.1. *Design a Shone pneumatic ejector to lift the sewage from a low level area having a population of 7500 persons, supplied with a water supply of 180 litres per day. Assume the velocity of sewage in sewage main as 0.9 m/sec and velocity of compressed air in the air main as 6 m/sec.*

Solution. Let us assume that the entire water supplied appears as sewage. $\therefore$ $\text{D. W. F.} = \dfrac{7500 \times 180}{24 \times 60 \times 60} \times 10^{-3} = 0.015625$ cumecs.

Assume peak flow to be 3 time D.W.F.,

$$Q_{max} = 3 \times 0.015625 = 0.0469 \text{ cumecs.}$$

(a) Design of ejector body : Let us assume that the ejector works 10 times per hour. Hence the capacity (C) of the ejector at maximum rate of flow will be

$$C = \frac{0.0469 \times 60 \times 60}{10} = 16.884 \text{ m}^3$$

Let the effective height of ejector body be 2.5 m. Hence its diameter D is given by

$$D = \sqrt{\frac{16.884 \times 4}{\pi \times 2.5}} = 2.93 \text{ m.}$$

Hence provide the ejector of 3 m dia. and 2.5 m height. Provide one more ejector as stand-by

(b) Design of air pipe : Peak flow= 0.0469 cumecs.
Velocity of air= 6 m/sec.

$\therefore$ Area of air pipe= $0.0469/6 = 7.8167 \times 10^{-3} \text{ m}^2$

and dia. of air pipe= $\sqrt{\dfrac{7.8167 \times 10^{-3} \times 4}{\pi}} \approx$ **0.1 m.**

Hence provide 10 cm dia. air pipe.

(c) ***Design of sewer main***

Area of cross-section = 0.0469/0.9 = 0.05211

$$\therefore \text{ Dia. of sewer pipe} = \sqrt{\frac{0.05211 \times 4}{\pi}} = 0.2576 \text{ m.}$$

$$\approx \mathbf{26 \text{ cm}}$$

Example 7.2. *A town having a population of 60000 is supplied with a per capita water supply of 180 litres per day. A separate sewer from this town enters a pumping station through a low level sewer at R.L. 120.00 m. This sewage is to be pumped to a high level sewer at R.L. 129.00 m. Assuming that 80% of water reaches the sewer, determine (a) size of sump well, (b) B.H.P. of the pump motor required, and (c) size of the rising main, if the length is 120 m. Assume suitable data wherever required.*

Solution.

1. ***Assumption of data.*** Let us assume the following data :

(*i*) Peak flow= 3 times the average flow

(*ii*) Velocity of flow in rising mains= 1 m/sec.

(*iii*) Min. time of pumps running continuously= 15 min.

(*vi*) Head losses in bends etc.= 0.4 m.

(*v*) Efficiency of pump = 65%

(*vi*) Efficiency of driving units = 75%

2. ***Peak sewage flow*** : Average quantity of sewage produced

$$Q_{av} = 0.8\left[\frac{60000 \times 180}{24 \times 60 \times 60} \times 10^{-3}\right] = 0.1 \text{ cumec.}$$

$$\therefore \quad Q_{max} = 3\,Q_{av} = 3 \times 0.1 = 0.3 \text{ cumecs.}$$

3. Dia. of rising main : Velocity of flow in rising main = 1 m/sec.

Area of cross-section = 0.3/1 = 0.3 m^2

$$\therefore \text{ Dia. } D = \sqrt{\frac{0.3 \times 4}{\pi}} = 0.618 \text{ m.}$$

Provide rising main of 62 cm dia.

$$\text{Actual velocity of flow} = \frac{0.3}{\frac{\pi}{4}(0.62)^2} = 0.994 \text{ m/sec.}$$

4. Design of sump well : Sump will be designed for 15 minutes low. Peak flow rate = 0.3 cumecs.

∴ Quantity of sewage collected in 15 minutes

$$= 0.3 \times 15 \times 60 = 270 \text{ m}^3$$

Quantity of sewage in rising main

$$= \frac{\pi}{4}(0.62)^2 \times 120 = 36.23 \text{ m}^3$$

$\therefore$ Total capacity of the sump well

$$= 270 + 36.23 = 306.23 \text{ m}^3$$

Let us provide 3 sump wells, two for storing the above sewage and third as a stand by. Let the depth of each unit= 3 m.

$$\therefore \text{ Surface area of each unit} = \frac{306.23}{2 \times 3} = 51.04 \text{ m}^2$$

$$\therefore \text{ Dia. of sump well} = \sqrt{\frac{51.04 \times 4}{\pi}} \approx \mathbf{8.1 \text{ m.}}$$

Hence provide three units of sumps wells, each of 8.1 m dia. and 3 m depth.

*5. **Design of pumps*** : Each pump has to lift a sewage of 306.23/2= 153.115 m^3 in 15 minutes.

$$\therefore \text{ Capacity of each pump} = \frac{153.115}{15 \times 60} = 0.170 \text{ cumecs.}$$

Let us assume Darcy's friction factor = 0.04

$$\therefore \qquad h_f = \frac{f L V^2}{2 gd} = \frac{0.04 \times 120 (0.994)^2}{2 \times 9.81 (0.62)} \approx 0.39 \text{ m.}$$

Losses in bends etc. = 0.4 m.

$$\therefore \text{ Total losses } H_L = 0.39 + 0.4 = 0.89 \text{ m.}$$

$$\text{Static lift} = 129.00 - 120.00 = 9.0 \text{ m.}$$

$$\therefore \qquad \text{Total lift} = H + H_L = 9.0 + 0.89 = 9.89 \text{ m.}$$

$$\text{H.P. of pump motor} = \frac{Q w H}{75 \eta_p \eta_d} = \frac{0.17 \times 1000 \times 9.89}{75 \times 0.65 \times 0.75}$$

$$\approx \mathbf{46}$$

PROBLEMS

1. (*a*) When does it become necessary to pump the sewage ?

   (*b*) What are the requirements of a sewage pump ?

2. Discuss various points that should be kept in mind while selecting the site for a sewage pumping station.

3. Write a note on selection of type and capacities of sewage pump.

4. Explain how do you determine the following.

   (*i*) size of sump well (*ii*) size of rising main.

   (*iii*) B.H.P. of pump motor.

5. Describe with the help of a neat sketch, working of a pneumatic ejector.

6. Draw a neat sketch of a typical sewage pumping station, showing all necessary components, and describe in brief the functions of each.

7. Explain, with the help of a neat sketch, the working of a diaphragm type pump.

8

Wastewater Characteristics

8.1. INTRODUCTION : CONSTITUENTS OF SEWAGE

Sewage or wastewater is a dilute mixture of various wastes from residential, commercial, industrial and other public places. Before we can decide about the line of its treatment and disposal, it is essential to know its composition, quality and characteristics. Though the characteristics of sewage or wastewater depends upto the source of its discharge, sewage in general contains organic matter, inorganic matter and living organisms. The organic and inorganic matter may be in dissolved, suspended and colloidal state. The inorganic or mineral matter consists of ash, cinder, sand, grit, mud and other mineral salts. The organic matter may be either nitrogenous or nitrogen-free. The chief sources of nitrogenous matter are urea and protein, while the nitrogen free compounds include carbohydrates fats and soaps. The *living organisms* may be divided into *plant life* (such as algae, fungi etc.) and *animal life* consisting of various types of micro-organisms such as protozoa, bacteria, virus etc. The bacteria convert the complex organic constituents of sewage into simpler, more stable, organic and mineral compounds.

As mentioned above, the composition or constituent of sewage/waste water largely depends upon the source from which it is found in domestic *wastewater* which may be classified as strong, medium and weak, depending upon the concentration of these constituents. It should be noted that sewage contains only a very small percentage of solids in relation to huge amount of water. Liquid content of sewage is 99.9% while total amount of solids (both suspended as well as dissolved) is only 0.1 percent.

TABLE 8.1 TYPICAL COMPOSITION OF DOMESTIC WASTEWATER.

Constituent		Concentration		
		Strong	Medium	Weak
1. Solids : Total	(mg/l)	1200	720	350
Dissolved, total	(mg/l)	850	500	250
Fixed	(mg/l)	525	300	145
Volatile	(mg/l)	325	200	105
Suspended, total	(mg/l)	350	220	100
Fixed	(mg/l)	75	55	20
Volatile	(mg/l)	275	165	80
2. Settleable solids	(mg/l)	20	10	5
3. Biochemical oxygen damand (BOD_5, 20°C)	(mg/l)	400	220	110
4. Total organic carbon (TOC)	(mg/l)	290	160	80
5. Chemical oxygen demand (COD)	(mg/l)	1000	500	250
6. Nitrogen (total as N)	(mg/l)	85	40	20
Organic	(mg/l)	35	15	8
Free ammonia	(mg/l)	50	25	12
Nitrites	(mg/l)	0	0	0
Nitrates	(mg/l)	0	0	0
7. Phosphorus (Total as P)	(mg/l)	15	8	4
Organic		5	3	1
Inorganic		10	5	3
8. Chlorides*	(mg/l)	100	50	30
9. Alkalinity (as $CaCO_3$)*		200	100	50
10. Grease	(mg/l)	150	100	50

*Depends upon its amount in domestic water supply.

8.2. CHARACTERISTICS OF WASTEWATER

The characteristics or properties of wastewater can be classified under the following three heads : (*a*) Physical characteristics (*b*) Chemical characteristics (*c*) Biological characteristics.

The most important *physical characteristic* of water is its total solids content, consisting of floating matter, matter in suspension, colloidal matter and matter in solution. Other physical characteristics are : (*i*) smell or odour (*ii*) colour and (*iii*) temperature.

Sewage contains complex organic matters derived from urine, faeces etc. and inorganic chemicals. Fresh domestic sewage is slightly alkaline but tends towards acidic as it becomes stale. Important chemical characteristics of sewage are : (*i*) pH value, (*ii*) chloride content (*iii*) nitrogen content (*iv*) fat, grease and oil content (*v*) sulphides, sulphates and H_2S gas, (*vi*) dissolved oxygen, (*vii*) chemical oxygen demand, and (*viii*) biochemical oxygen demand.

Biological characteristics relate to various micro-organisms found in wastewater, some of which may be pathogenic. However, all bacteria present in wastewater are not harmful; some of these help to treat the wastewater and reduce the cost of treatment plants.

Table 8.2 gives a summary of physical, chemical and biological characteristics of wastewater and their sources.

TABLE 8.2 CHARACTERISTICS OF WASTEWATER AND THEIR SOURCES.

		Characteristic	Sources
1.		***Physical characteristics***	
	(*i*)	Colur	Domestic and industrial wastes ; natural decay of organic materials.
	(*ii*)	Odour	Decomposing wastewater ; industrial wastes
	(*iii*)	Solids	Domestic water supply ; domestic and industrial wastes; soil erosions; inflow-infiltration
	(*iv*)	Temperature	Domestic and industrial wastes.
2.		***Chemical characteristics***	
	(*a*	*Organic*	
	(*i*)	Carbohydrates	Domestic, commercial and industrial wastes
	(*ii*)	Fats, oils and greases	Domestic, commercial and industrial wastes
	(*iii*)	Pesticides	Agricultural wastes.
	(*iv*)	Phenols	Industrial wastes
	(*v*)	Proteins	Domestic and commercial wastes
	(*vi*)	Surfactants	Domestic and industral wastes
	(*vii*)	Others	Natural decay of organic materials.
	(*b*)	*Inorganic*	
	(*i*)	Alkalinity	Domestic wastes, domestic water supply, ground water infiltration.
	(*ii*)	Chlorides	Domestic water supply, domestic wastes, groundwater infiltratin, water softeners.
	(*iii*)	Heavy metals	Industrial wastes.
	(*iv*)	Nitrogen	Domestic and agricultural wastes.
	(*v*)	pH	Industrial wastes
	(*vi*)	Phosphours	Domestic and industrial wastes, natural runoff
	(*vii*)	Sulfur	Domestic water supply, domestic and industrial wastes.
	(*viii*)	Toxic compounds	Industrial wastes.
	(*c*)	*Gases*	
	(*i*)	*Hydrogen sulfide*	Decomposition of domestic wastes
	(*ii*)	Methane	Decomposition of domestic wastes
	(*iii*)	Oxygen	Domestic water supply, surface water infiltration
3.		***Biological characteristics***	
	(*i*)	Animals	Open water courses and treatment plants
	(*ii*)	Plants	Open water courses and treament plants
	(*iii*)	Protista	Domestic wastes ; treatment plants.
	(*iv*)	Viruses	Domestic wastes

The organic matter in sewage consists of urea from urine, proteins, carbohydrates, fats and soaps. These undergo continuous decomposition and in that process, pass through several stages, resolving into simpler elements such as nitrogen, hydrogen, carbon and oxygen with small quantities of sulphur and phosphorous. These elements ultimately combine by means of *chemical* and *biological actions* to form inorganic substance. Hence before we take up physical, chemical and biological characteristics, let us first study the *cycles of decomposition* through which the organic matter pass with passage of time.

8.3. CYCLES OF DECOMPOSITION

Most of the organic matter present is sewage decomposes readily through combined biological and chemical processes. These processes, which are *bio-chemic* in nature, decompose highly complex organic compounds in the sewage into materials or constituents to a much simpler chemical structure. The organic matter which can be decomposed by bacteria, under biological action, is called *biodegradable organic matter*. The decomposition of such organic matter takes place through the agency of different types of bacteria (*i.e.* aerobic bacteria, anaerobic bacteria or facultative bacteria) and the nitrogenous and carbonaceous materials in the sewage serve as food for them. The *biological decomposition* can be of two types :

(*i*) Aerobic decomposition (also called *aerobic oxidation*) and (*ii*) Anaerobic decomposition (also called *putrefaction*).

Aerobic decomposition may be caused by both aerobic bacteria as well facultative bacteria operating aerobically, in presence of air or oxygen which is available in the wastewater in the dissolved form. During this process, organic matter is broken up and *oxidised* to form stable and non objectionable end products like carbon dioxide, nitrates, sulphates etc. Treatment units which work on aerobic decomposition alone are aeration tanks, trickling filters, contact beds, oxidation ponds etc.

Anaerobic decomposition or *putrefaction* is caused by anaerobic bacteria, as well as facultative bacteria, operating anaerobically. The end products of putrefaction include black residue, nitrogen, hydrogen etc. The anaerobic bacteria survive by extracting and consuming the *bounded molecular oxygen* present is compounds like nitrates (NO_3) and sulphates (SO_4). Treatment units which work on putrefaction alone are septic tanks, Imhoff tanks, and sludge digestion tanks.

The matter of the universe remains constant, but its form changes because of bio-chemical reactions. The complex organic compounds of biodegradable nature are broken up by biochemical reaction into simple compounds which are in turn consumed by animal and plant life for their growth. This cycle goes on. From sewage treatments point of view, the cycles of decomposition of the following five elements are of importance : (1) Nitrogen cycle. (2) Carbon cycle. (3) Sulphur cycle. (4) Calcium cycle. (5) Phosphrous cycle.

The first three cycles relate to the decomposition of nitrogeneous organic matter, carbonaceous organic matter and sulphurous organic matter respectively.

1. The Nitrogen cycle. The nitrogen cycle, corresponding to the biochemical degradation or decomposition of nitrogeneous organic matter, is in fact an endless chain connecting the processes of life and decay of both animal and plant worlds. This is illustrated in Fig. 8.1. The nitrogen cycle consists of following steps :

(a) Main steps :

1. The *nitrogenous organic matter*, in the form of waste animal and plant matter, consisting of urea, proteins and hydrocarbons, undergoes decomposition either by way of oxidation by aerobic bacteria or by way of putrefaction by anaerobic bacteria, resulting in the formation of ammonia (NH_3) and other gases.

2. By process of nitrification, ammonia is first converted into nitrites (NO_2) by partial oxidation and then finally to nitrates (NO_3) by the action of aerobic bacteria.

3. The products formed in step 2 are consumed by plant life as food, through photosynthesis. The plant life grows, due to which plant tissues, *plant protein* (seeds) and free oxygen are produced.

4. The plant protein is consumed by animals, resulting in the production of *animal proteins* (meat, milk etc.)

5. The *animal wastes* in the form of urine and other excretions, as well as the dead bodies of the animals result in the formation of nitrogenous organic matter.

The cycle thus continues. However, there may be some short circuits in the above cycle, as shown by dotted line in Fig. 8.1, and as described below.

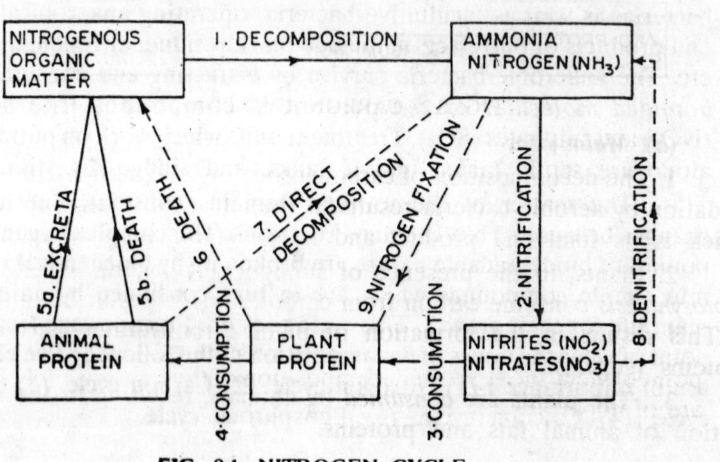

FIG. 8.1. NITROGEN CYCLE.

(b) Short circuit steps

6. The death or decay of plant life (as it happens in forests where leaves, flowers etc. of plants lie in layers on ground surface to decay) may directly result in the formation of nitrogenous organic matter.

7. Waste products such as urea and excretions of animals may sometimes decompose directly and form ammonia nitrogen.

8. The nitrates formed after step 2 may be converted or reduced into free nitrogen (and some times to ammonia) by anaerobic bacteria, by a process known as *denitrification*

9. However, free nitrogen, produced in step 8, may directly be converted into plant proteins by certain type of bacteria present in the roots of plants, through a process known as *nitrogen fixation.*

2. Carbon cycle

The carbon cycle corresponding to biochemical degradation or decomposition of *carbonaceous organic matter* is also an endless chain connecting the processes of life and decay of both animal and plant worlds. This is illustrated in Fig. 8.2.

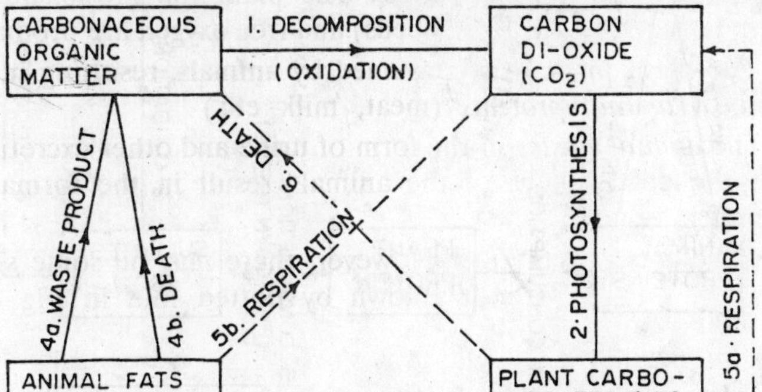

FIG. 8.2. CARBON CYCLE.

(a) Main steps

1. The decomposition of carbonaceous organic matter, through oxidation by aerobic bacteria results in the release of carbon dioxide which is its final end product and is the stable form of carbon.

2. Plants, in the presence of sunlight and by the process of *photosynthesis* consume carbon from carbon dioxide produced in step 1. This results in the formation of plant carbohydrates, fats and proteins (sugars).

3. The plants are *consumed* by animals, resulting in the formation of animal fats and proteins.

4. Waste products or death of animals results in the formation of carbonaceous organic matter..

The carbon cycle is thus completed. This cycle continues in nature. However, there may be some short circuits in the above cycle, as shown by dotted lines in Fig. 8.2, and as described below.

(*b*) ***Short circuit steps***

5. (*a*) Plant life gives off carbon dioxide at night, through the process of respiration. (*b*) Similarly, animal life gives off CO_2 during respiration.

6. Carbonaceous organic matter may also be formed directly by the death/decay of plant life.

3. **Sulphur cycle.** The sulphur cycle corresponds to biochemical degradation or decomposition of *sulphurous organic matter*, and is also an endless chain connecting the processes of life and decay of both animal and plant worlds. This is illustrated in Fig. 8.3.

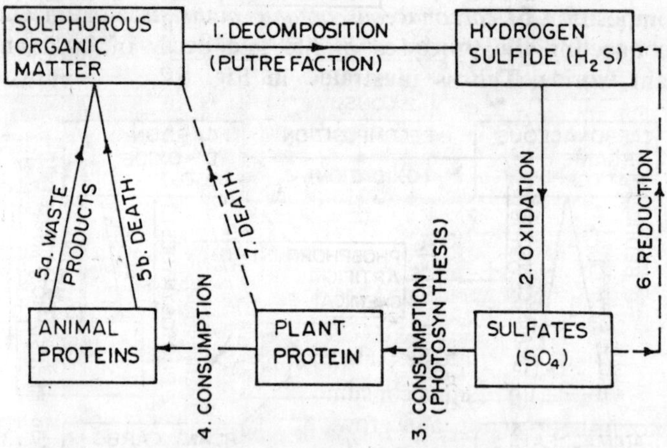

FIG. 8.3. SULPHUR CYCLE.

Main steps

1. The decomposition of sulphurous organic matter, through the action of anaerobic bacteria, in absence of oxygen, results in the formation of hydrogen sulphide (H_2S).

2. By process of oxidation, H_2S is first converted into sulphur and finally to sulphates.

3. Sulphates, when consumed by plants, through photosynthesis, changes into plant proteins.

4. Animals consume the plant proteins. This results in the formation of animal proteins.

5. Wastes produced by animals and their dead bodies results in the formation of sulphurous organic matter.

The sulphur cycle is thus completed. However, there may be some short circuits in the above cycle, as shown by dotted lines in Fig. 8.3, and as described below :

(*b*) ***Short circuit steps***

6. Sulphates, in the absence of oxygen, are converted into H_2S, by the process of *reduction.*

7. Organic sulphurous matter may directly be produced by the death or decay of plants.

4. Phosphorus cycle. Phosphorous cycle relates to the maintenance of level of phosphorus in the soil. The phosphorous cycle is shown in Fig. 8.4.

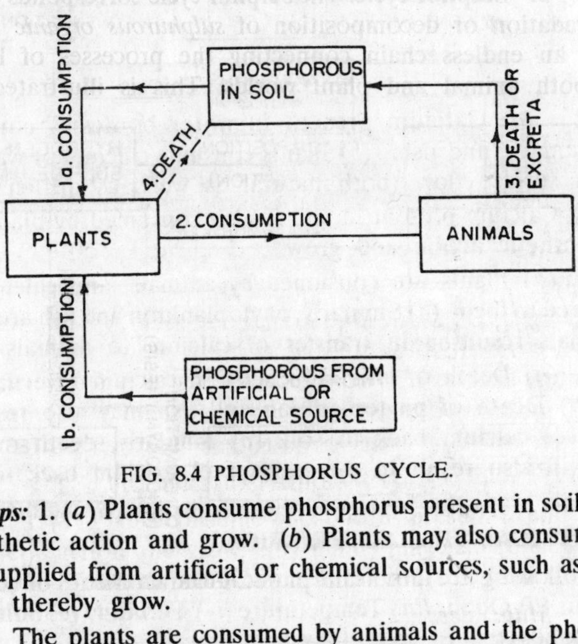

FIG. 8.4 PHOSPHORUS CYCLE.

Steps: 1. (*a*) Plants consume phosphorus present in soil through photosynthetic action and grow. (*b*) Plants may also consume phosphorus supplied from artificial or chemical sources, such as manure etc., and thereby grow.

2. The plants are consumed by animals and the phosphorus elements is thus transferred to animals.

3. The wastes produced by animals and their dead bodies results in the transfer of phosphorus back to soil.

4. The death or decay of plant life also results in the transfer of phosphorus back to soil. Thus, the phosphorus cycle is completed.

5. Calcium cycle. Calcium cycle also relates to the maintenance of level of calcium in the soil. Calcium cycle is shown in Fig. 8.5.

Steps

1. Calcium present in soil is carried away to water bodies (such as rivers, lakes sea etc.) through surface runoff.

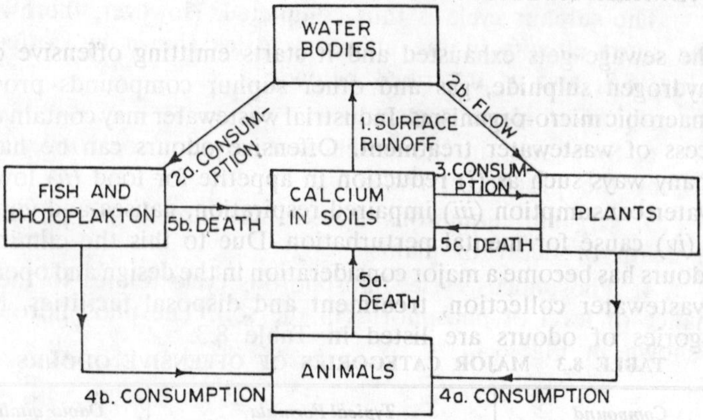

FIG. 8.5. CALCIUM CYCLE.

2. (*a*) Calcium present in water bodies is consumed by phytoplankton and fish. (*b*) Plants consume calcium of water bodies, through surface flow (both natural as wells as artificial).

3. Calcium present in soil is also consumed by plants through photosynthetic action and grow.

4. (*a*) Plants are consumed by animals, and calcium is thus transferred to them. (*b*) Similarly, phytoplankton and fish are consumed by animals resulting in transfer of calcium to animals.

5. (*a*) Death of animals results in calcium reverting back to soil. (*b*) Death of phytoplankton and fish may also result in the transfer of calcium back to soil. (*c*) Similarly, death or decay of plants will also result in the transfer of calcium back to the soil.

This way, the level of calcium in soil is maintained.

8.4. PHYSICAL CHARACTERISTICS

Following are important physical characteristics of waste water: (*i*) Colour (*ii*) Odour (*iii*) Temperature (*iv*) Turbidity (*v*) Solid contents (Total solids)

1. Colour : Fresh domestic sewage is grey, somewhat resembling a weak solution of soap. With the passage of time, as putrefaction starts, it begins to get black. The colour of *septic sewage* is more or less black or dark in colour. The colour of industrial wastewater depends upon the chemical process used in the industries. Industrial waste water, when mixed with domestic sewage, may also add colour to it.

2. Odour : Normal fresh sewage has a *musty odour* which is normally not offensive, but as it starts to get stale, it begins to give offensive odour. Within 3 or 4 hours, all the oxygen present in the sewage gets exhausted and it starts emitting offensive odour of hydrogen sulphide, gas and other suphur compounds produced by anaerobic micro-organisms. Industrial wastewater may contain either process of wastewater treatment. Offensive odours can be harmful in many ways such as (*i*) reduction in appetite for food (*ii*) lowering in water consumption (*iii*) impaired respiration, nausea and vomiting and (*iv*) cause for mental perturbation. Due to this the elimination of odours has become a major consideration in the design and operation of wastewater collection, treatment and disposal facilities. Major categories of odours are listed in Table 8.3.

TABLE 8.3 MAJOR CATEGORIES OF OFFENSIVE ODOURS

Compound	*Typical Formula*	*Odour quality*
1. Amines	CH_3NH_2, $(CH_3)_3N$	Fishy
2. Ammonia	NH_3	Ammoniacal
3. Diamines	$NH_2(CH_2)_4NH_2$, $NH_2(CH_2)_5NH_2$	Decayed flesh
4. Hydrogen sulphide	H_2S	Rotten eggs
5. Mercaptans	CH_3SH, $CH_3(CH_2)_3SH$	Skunk
6. Organic sulphides	$(CH_3)_2S$, CH_3SSCH_3	Rotten cabbage
7. Skatole	$C_8H_5NHCH_3$	Faecal

3. Temperature : Generally, the temperature of wastewater is higher than that of the water supply, due to addition of warm water from the house holds and from industries. When the wastewater flows in closed circuits, its temperature rises further. This results in the increase in the viscosity of water and also increase in its bacterial activity. Average temperature of wastewater in India is around 20°C, which is quite close to the ideal temperature for the biological activities. The change in the temperature affects the wastewater in the following ways : (*i*) As the temperature rises, its viscosity increases with a corresponding increase in its tendency to precipitalte. Extremely low temperatures affects adversely the efficiency of sedimentation. (*ii*) The bacterial activity increases with increase in temperature, upto about 60°C. After this, if again falls. This characteristic of temperature thus effects the design in wastewater units and their efficiency. (*iii*) The solubility of gases in wastewater decreases with the increase in temperature. This results in the separation of dissolved oxygen and other gases from the wastewater, and consequent reduction in the self-purification of streams and increase in bacteriological growth. The increase in the rate of biochemical reactions that accompanies an increase in temperature, combined with the decrease in the quantity of oxygen present in surface waters, can cause serious depletion in dissolved oxygen concentrations-specially when large quantities of relatively warm wastewater are discharged into the receiving water (*iv*) The increase in temperature of wastewater, when discharged into receiving waters, affects the aquatic life (*v*) Abnormally high temperatures can foster the growth of undesirable water plants and wastewater fungus in the receiving waters.

4. Turbidity : The turbidity of wastewater depends on the quantity of solid matters present in the suspension state. Turbidity is a measure of light-emitting properties of wastewater, and *turbidity test* is used to indicate the quality of waste discharges with respect to colloidal matter. The turbidity depends upon the strength of sewage or waste water. The stronger or more concentrated the sewage, the higher is its turbidity. Turbidity can be determined either by turbidity rod or by Jackson's turbidimeter.

5. Total solids : Sewage normally contains 99.9 percent of water and 0.1 percent of solids. Analytically, the *total solids* content (S_T) of a wastewater is defined as all the matter that remains as residue upon evaporation to 103 to 105°C. Total solids in wastewater exist in three different forms : (*i*) suspended solids (*ii*) colloidal solids and (*iii*) dissolved solids. *Suspended solids* (S_S) are those which can be *filtered out* on an asbestos mat or filter paper *i.e.* suspended solids are non-filterable solids.

The *suspended solids* may be further subdivided into (*a*) settleable solids and (*b*) non-settleable solids. *Settleable solids* are those that will settle to the bottom of a cone-shaped container, called *Imhoff cone*, in 2–hour, period. The *Imhoff cone* (Fig. 8.6) consists of a conical glass vessel, having a capacity of 1 litre and graduated upto about 50 ml. Sewage is allowed to stand in the cone for a period of two hours and the quantity of *settleable solids* (S_{SS}) in the cone is then read. However, to obtain the exact proportion of settleable solids, the liquid from the cone is decanted off and the settleable solids collected in the bottom of the cone is weighed. *Non-settleable* solids do not settle down by mere detention, but may be arrested by special laboratory filters.

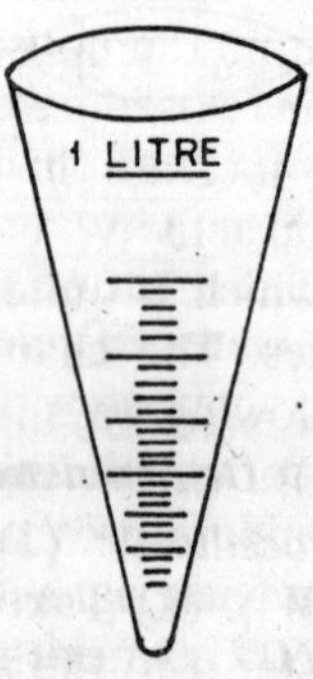

FIG. 8.6 IMHOFF CONE.

The total solids in wastewater are further classified as (*a*) fixed solids (S_F) and (*a*) volatile solids (S_V). The fixed solids are generally classed as inorganic while the volatile solids represent the *organic matter*. In order to determine the quantity of *volatile solids* (S_V), the dry residue (*i.e.* total solids), obtained by evaporating the wastewater at 105° C, is further heated and *ignited*. The loss of weight due to this operation represents the volatile solids (S_V) or the *strength* or nuisance capacity of the wastewater.

When sewage in passed through laboratory filter, the filtrate is not clear but is turbid, because of presence of *filterable solids*. The filterable-solids fraction consists of *colloidal solids* as well as *dissolved solids* (S_D). The *colloidal fraction* consists of the particulate matter of diameters ranging from 1 millimicron ($m\mu$) to 1 micron (μ). The colloidal matter consists of finely divided particles of gels, emulson and foam, and it cannot be removed by gravitational setting; they are generally removed by biological oxidation or coagulation, followed by sedimentation. The dissolved solids consists of both organic and inorgation molecules and ions that are present in true solution in water. *Turbidity* test in used to indicate the quálity of wastewaters with respect to colloidal matter. Colloidal matter scatters or absorbs light and thus prevents its transmission.

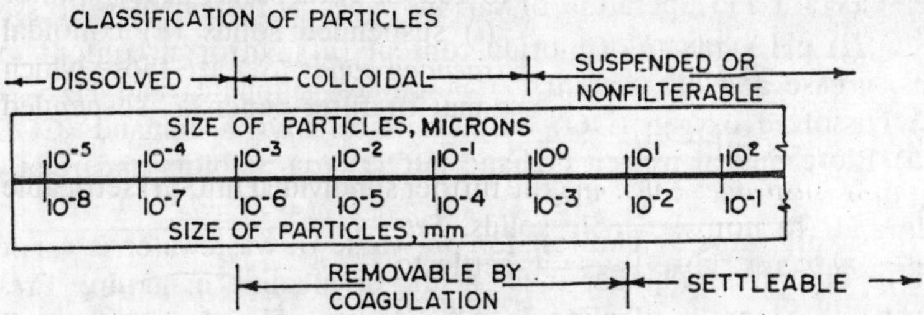

FIG. 8.7. CLASSIFICATION AND SIZE RANGE OF SOLIDS IN WASTEWATER

Fig. 8.7 shows the classification and size range of solids found in waste water. Table 8.1 gives the typical values of solids found in domestic wastewater.

(a) Determination of total solids and volatile solids.

Procedure : (1) Take a tarred dish and ignite it to constant weight (W_1). (2) Take a known volume (V) of a well mixed sample and transfer it to the above dish. (3) Evaporate the sample to dryness at 103°C for 24 hours, in constant temperature oven. (4) Cool the dish in a desiccator and determine its weight (W_2). (5) Ignite the dish at 600° C in a muffle furnace, for 30 miniatures. (6) Cool the dish in a dissicator and determine its weight (W_3).

Then, total solids $(S_T) = \dfrac{(W_2 - W_1)\,100}{V}$ (mg/l) ...(8.1)

Total volatile solids $(S_V) = \dfrac{(W_2 - W_3)\,100}{V}$ (mg/l) ...(8.2)

(b) Determination of non-filterable solids (suspended solids)and volatile non-filterable solids. *Procedure* : (1) Filter a suitable aliquote of sample (vol. V) through a tarred gooch crucible (ignited to constant weight W_1), applying suction. (2) Dry the crucible along with the retained matter in a constant temperature oven maintained at 103°C, for 24 hours. (3) Cool it in a desiccator and find its weight (W_2). (4) Ignite the crucible at 600°C in a muffle furnace, for 30 minutes. (5) Cool in a desiccator and record the weight (W_3).

$$\text{Then, Non-filterable solids} = \frac{(W_2 - W_1)\,100}{V} \text{ mg/l} \quad ...(8.3)$$

$$\text{and} \quad \text{Non-fiterable volatile solids} = \frac{(W_2 - W_3)\,100}{V} \text{ mg/l} \quad ...(8.4)$$

8.5. CHEMICAL CHARACTERISTICS AND THEIR DETERMINATION

A detailed list of various chemical characteristics along with their sources is given in Table 8.2. However, we shall discuss the following important chemical characteristics of wastewater which affect the selection and operation of various types of treatment processes:

(*i*) pH value (*ii*) Chloride content (*iii*) Nitrogen content (*iv*) Fats, grease and oil content (*v*) Sulphides, sulphates and H_2S gas (*vi*) Dissolved oxygen (DO) (*vii*) Chemical oxygen demand (COD) (*viii*) Bio-chemical oxygen demand (BOD) (*ix*) Stability and relative stability.

1. pH value : The test for pH value of wastewater is carried out to determine whether it is acidic or alkaline in nature. Fresh sewage is generally alkaline in nature, (its pH value between 7.3 to 7.5). However, as the time passes, pH value tends to fall due to production of acids by bacterial action, and the sewage tends to become acidic. However, after oxidation when it is relatively stable, it becomes alkaline again. Properly oxidised effluent should have a pH value of about 7.3 or so. A high concentration of either an acid (pH $\ll$ 7) or alkali (pH $\gg$ 7) in wastewater is indicative of industrial wastes. The determination of pH value of sewage is important since certain treatment methods depend on proper pH value of the wastewater for their efficient workings. Sometimes, lime is added for creating alkaline condition.

2. Chlorides content : Chlorides are mineral salts and, therefore, are not affected by biological action of sewage. Chlorides in natural water result from the leaching of chloride-containing rocks and soils with which the water comes in contact. Chlorides found in domestic sewage is derived from kitchen wastes, human faeces and urinary discharges etc. Human excreta, for example, contains about 6 g of chlorides per person per day. Water softeners also add large quantities of chlorides. Large amounts of chlorides may also enter in wastewaters from industries like ice cream plants, meat salting etc. Infiltration of ground water into sewers adjacent to saltwater is also a potential source of high chlorides. The chloride content of wastewater can be measured by titrating the sample of wastewater with standard silver nitrate solution, using potassium chromate as indicator.

3. Nitrogen contents : The presence of nitrogen in wastewater indicates the presence of organic matter in it. Nitrogen is essential to the growth of protista and plants and as such is known as nutrient or biostimulant. Since nitrogen is an essential building element in the synthesis of protein, nitrogen data is required to evaluate the treatability of wastewater by biological processes. Nitrogen appears in the following five different forms in wastewater :

(*i*) Ammonia nitrogen or free ammonia (*ii*) Organic nitrogen (*iii*) Albuminoid nitrogen (*iv*) Nitrites nitrogen and (*v*) Nitrates nitrogen

The various forms of nitrogen that are present in nature and the pathways by which the forms are changed as depicted through the nitrogen cycle in Fig. 8.1.

(i) Ammonia nitrogen or free ammonia : Ammonia nitrogen exists in aqueous solution as either the ammonium ion or ammonia, depending upon the pH of the solution, in accordance with the following equilibrium reaction :

$$NH_3 + H_2O \Leftrightarrow NH_4^+ + OH^- \qquad ...(8.5)$$

When pH > 7, the equilibrium is displaced to the left while for pH < 7, the ammonium ion is prediminant. For measuring the free ammonia which is given off when sewage decomposes, it is distilled and collected in the distillate. The age of wastewater is indicated by the relative amount of ammonia that is present. The presence of considerable amount of free ammonia indicates stale or old sewage.

(ii) Organic nitrogen. Organic nitrogen present in wastewater is determined by *Kjeldahl method.* In this method, ammonia is first driven off from the aqueous sample by boiling it. The sample is then *digested,* during which the organic nitrogen is converted to ammonia, and from the latter the quantity of nitrogen is calculated. *Total Kjeldahl nitrogen* (or the total organic nitrogen), which is the sum of organic and ammonia nitrogen, is determined in the same manner as the organic nitrogen, except that the ammonia is not driven off before the digestion step.

(iii) Albuminoid nitrogen : Just as *free ammonia* indicates the very first stage of decomposition of organic matter, *albuminoid nitrogen* indicates quantity of nitrogen present in wastewater before the decomposition of organic matter is started. In other words, albuminoid nitrogen (or albuminoid ammonia) indicates the amount of undecomposed nitrogenous material in the wastewater. However, unlike free ammonia, this does not really exist in the wastewater. It is produced by first freeing from the sample all free ammonia and then treating the sample with the alkaline solution of potassium permanganate. The ammonia given out as a result of the chemical reaction is called *albuminoid ammonia* which is taken as a measure of the easily decomposable organic nitrogen present in the waste water.

(*iv*) ***Nitrites nitrogen*** : Nitrites indicate the presence of partly decomposed (not fully oxidised) organic matter. The presence of nitrites in effluent indicates that oxidation is in progress. Nitrite nitrogen is relatively unimportant in wastewater or water pollution studies because it is unstable and is easily oxidised to the nitrate form. The presence of nitrites show that the treatment given to wastewater is still incomplete, and the sewage is still stale. Hence it is an indicator of past pollution in the process of stabilisation. Its quantity seldom exceeds 1 mg/l in wastewater. Nitrite nitrogen is determined by *colourimetric methods*. The colour is developed by adding sulphonilic acid and naphthamine, which is then compared with standard colours of known concentration.

(*v*) ***Nitrate nitrogen*** : Nitrates indicate the presence of fully oxidised organic matter. They indicate the most stable form of nitrogenous matter contained in wastewater, thus indicating the well oxidised and treated wastewater. Increase in proportion of nitrates during the process of the sewage treatment serves as a guide for measuring the progress achieved in the wastewater treatment. Nitrates may vary in concentration from 0 to 20 mg/l as nitrogen in wastewater effluents. The nitrate concentration is also usually determined by *colourimetric methods.* The colour is developed by adding phenol-di-sulphonic acid and potassium hydroxide, which is then compared with standard colours of known concentrations.

4. Fats, grease and oils : Fats and oil are mainly contributed from kitchen wastes, because they are major components of food stuffs such as butter, lard, margarine, and vegetable oils and fats. Fats are also commonly found in meats, seeds, nuts and some fruits. Grease and oils are also discharged from industries like garages, workshops, factories etc. Fats and oils are compounds (esters) of alcohol or glycerol (glycerin) with fatty acids. Such matters float on the top of sedimentation tanks, often choke pipes in the winter, and clog filters. They thus interfere with functioning of normal treatment plants. The particles interfere with biological action and cause maintenance problems. Fats are among the more stable of organic compounds and are not easily decomposed by bacteria. It is therefore necessary to detect and remove these from wastewater.

Fats, grease, mineral oils and lubriation oils are soluble in hexane or ether. In order to determine their content, the sample of wastewater is first treated with dilute hydrochloric acid which results in liberation of fatty acids. The later are then evaporated. The residue is then mixed with ether. When the ether is driven off, the ether-soluble matter, which remains behind, represent the fats, grease and oil content.

5. Surfactants : Surfactants come primarily from synthetic detergents. These are discharged from bathrooms, kitchens, washing machines etc. Surfactants (or surface-active agents) are large organic molecules which cause foaming in wastewater treatment. Due to this, aeration of wastewater is hindered. Alkyl-benzene-sulfonate (ABS), a type of surfactant commonly used in synthetic detergents, is more troublesome since it is not biodegradable. Hence some countries have banned the use of ABS in detergents, and hence ABS has been replaced in detergents by linear-alkyl-sulfonate (LAS) which is biodegradable. The content of surfactants in wastewater is determined by measuring the colour change in standard solution of methylene blue dye.

6. Phenols, pesticides and agricultural chemicals : Phenols are mostly found in industrial wastewater. If such wastewaters are directly discharged into receiving streams, they cause serious taste problems in drinking water, specially when water is disinfected by chlorination. However, phenols can be biologically oxidized if the concentrations are upto 500 mg/l. Pesticides, herbicides and other agricultural chemicals primarily result from surface runoff from agricultural, vacant and park lands, specially in a combined sewerage system. The concentration of these trace contaminants is measured by the carbon-chloroform extract method.

7. Toxic compounds : Copper, lead, silver, chrominum, arsenic and boron are some of the cations which are toxic to micro-organisms resulting in the malfunctioning of the biological treatment plants. These results from industrial wastewaters. Some toxic anions, including cyanides and chromates, present in some industrial wastes also hinder the wastewater treatment facilities. Hence their presence should be taken into consideration in the design of biological treatment plants.

8. Sulphates, sulphides and H_2S gas : Sulphates and sulphides are formed due to decomposition of various sulphur containing substances present in wastewater. The sulphate ions (SO_4) occur naturally in most water supplies and hence they are also present in wastewater. Sulphur, required in the synthesis of proteins is released in their degradation. Anaerobic bacteria chemically reduce sulphates to sulphides and to hydrogen sulphide, as indicated by the following equations:

$$SO_4^{-} + \text{organic matter} \xrightarrow{\text{bacteria}} S^{--} + H_2O + CO_2 \quad ...(8.6\ a)$$

$$S^{--} + 2H^{+} \longrightarrow H_2S \quad ...(8.6\ b)$$

The hydrogen-sulphide gas so produced cause bad smells and odours. Besides this, H_2S gas gets oxidised biologically to sulphuric acid resulting in corrosion to sewer pipes. The biological process of sludge digesters is severely hindered when sulphates are reduced to sulphides, specially when their concentration exceeds 200 mg/l. Also, H_2S gas, which is evolved and mixed with wastewater gas ($CH_4 + CO_2$), is corrosive to the gas piping.

9. Other gases : Following are the gases that are commonly found in untreated wastewater : (*i*) Nitrogen (N_2) (*ii*) Oxygen (O_2) (*iii*) Carbon-dioxide (CO_2) (*iv*) Hydrogen sulphide (H_2S) (*v*) Ammonia (NH_3) (*iv*) Methane (CH_4). While the first three are the gases of the atmosphere which are found in all waters (including wastewaters) exposed to air, the later three are as a result of decomposition of organic matter present in the waster water. *Methane gas* is the principal by-product of the anaerobic decomposition of the organic matter in wastewater. This gas is colourless, and odourless and is highly combustible. Since its explosion hazard is high, manholes, sewer junctions, junction chambers etc. should be kept well ventilated. *Oxygen* in a sample of wastewater is reported in the following three ways : (*a*) Oxygen consumed (*b*) Dissolved oxygen and (*c*) Oxygen demand

10. Oxygen consumed. *'Oxygen consumed'* is the oxygen required for the oxidation of carbonaceous matter. This quantity of oxygen is determined by adding standard amount of potassium permanganate with dilute sulphuric acid to a sample of wastewater. The reaction is allowed to take place for periods of 15 minutes and 4 hours, at a constant temperature of 18°C. The potassium permanganate liberates oxygen which is consumed by the wastewater. This test is made to determine the relative strength of sewage (i.e. whether strong, medium or weak), instead of BOD test. However, this test does not give the total oxygen needed for the biological oxidation of all or the bulk of the organic matter.

11. Dissolved oxygen : *Dissolved oxygen* (DO) is the amount of oxygen in the dissolved state in the wastewater. Though wastewater generally does not have DO, its presence in untreated wastewater indicates that the wastewater is fresh. Similarly, its presence in treated wastewater/effluent indicates that considerable oxidation has been accomplished during the treatment stages. While discharging the treated wastewater into receiving waters, it is essential to ensure that atleast 4 p.p.m. of DO is present in it. If DO is less, the aquatic animals like fish etc. are likely to be killed near the vicinity of disposal. The presence of DO in wastewater is desirable because it prevents the formation of noxious odours.

It is essential to determine the DO content of wastewater before it is subjected to treatment, so as to select proper treatment method. The small DO content of fresh wastewater is soon depleted due to aerobic decomposition. The actual quantity of DO is governed by (*i*) solubility of oxygen (*ii*) partial pressure of oxygen in atmosphere (*iii*) the temperature, and (*iv*) purity (salinity, suspended solids etc.) of the water. DO of the wastewater decreases as the temperature increases. The solubility of DO in wastewater is only 95% of that in distilled water. DO content of wastewater may be determined by *Winkler's method.*

8.6. OXYGEN DEMAND

As stated earlier, the presence of oxygen is essential for the livelihood of organisms. The aeroblc action continues only till the oxygen is present in wastewater, and after that anaerobic action begins resulting in putrefaction. Thus, oxygen is *demanded* in wastewater for the oxidation of both inorganic as well as organic matter. Thus demand of oxygen may be expressed in the following ways :

(*i*) Biochemical oxygen demand (BOD) (*ii*) Chemical oxygen demand (COD) (*iii*) Total oxygen demand (TOD) (*iv*) Theoretical oxygen demand (Th.OD)

In addition to these, the amount of organic matter present may also be determined by the *total organic carbon* (TOC) *test*.

8.7. BIOCHEMICAL OXYGEN DEMAND (BOD)

The biochemical oxygen demand (BOD) is a measure of the oxygen required to oxidize the organic matter present in a sample, through the action of micro-organisms contained in a sample of wastewater. It is the most widely used parameter of organic pollution applied to both wastewater as well as surface water. *The BOD may be defined as the oxygen required for the micro-organisms to carry out biological decomposition of dissolved solids or organic matter in the waterwater under aerobic conditions at standard temperature.* The BOD test results are used for the following purposes :

(*i*) Determination of approximate quantity of oxygen required for the biological stabilization of organic matter present in the wastewater

(*ii*) Determination of size of wastewater treatment facilities.

(*iii*) Measurement of efficiency of some treatment processes.

(*iv*) Determination of strength of sewage

(*v*) Determination of amount of clear water required for the efficient disposal of wastewater by dilution.

The organic matter present in wastewater may belong to two groups : (*i*) Carbonaceous matter and (*ii*) Nitrogenous matter.

The *ultimate carbonaceous* BOD of a liquid waste is the amount of oxygen necessary for the micro-organisms in the sample to decompose the carbonaceous materials that are subject of microbial decomposition. This is the first stage of oxidation and the corresponding BOD is also sometimes called the *first stage demand.* In the second stage, the nitrogenous matter is oxidised, and the corresponding BOD is known as *second stage* BOD or *nitrification demand.* In fact, pollution waters will continue to absorb oxygen for a long time. Biochemical oxidation is a slow process and theoretically takes an infinite time to go to completion, though the ultimate first stage BOD of a given wastewater is equal to the initial oxygen equivalent of the organic matter present. Generally, a 5 day period is chosen for standard BOD test, during which oxidation is about 60 to 70 percent complete, while within 20 days period, the oxidation is about 95 to 99 percent complete. A constant temperation of 20°C is maintained during the incubation. The BOD value of 5-day incubation period is commonly written as BOD_5 or 5-day BOD.

1. Test for BOD : BOD test can be performed by two methods: (*i*) Direct method and (*ii*) Dilution method.

(i) Direct method : The test consists of keeping the sample of wastewater in contact with a definite air or oxygen, in a specially prepared vessel. The BOD is then measured menometrically.

(ii) Dilution method : This is the commonly used method. In this method the sample is suitably diluted with a specially prepared dilution water. The dilution water is carefully manufactured so as to include a mixture of salts providing all trace nutrients necessary for biological activity, along with phosphate buffer to maintain a neutral pH. The water is aerated to saturate it with oxygen before mixing it with the sewage sample. The initial DO of this diluted sample is measured. The diluted sample is then incubated for 5 days at 20°C in an air tight glass vessel. The DO of the sample is then again measured. The loss of oxygen or the oxygen consumed during incubation is then found out by calculating the difference between the initial content of DO and the final content of DO. BOD is then computed from the relation.

$$BOD_5 = [\text{oxygen consumed}] \times \text{dilution ratio} \quad ...(8.7)$$

$$\text{where dilution ratio} = \frac{\text{Vol. of diluted sample}}{\text{Vol. of undiluted sewage sample}} \quad ...(8.7\,a)$$

For example, if 1 ml of wastewater is diluted to make 100 ml of test sample the *dilution ratio* or *dilution factor* will be 100/1 = 100. Generally, BOD tests are performed in 300 ml size BOD-bottles. If 5 ml of wastewater is placed in the bottle and mixed with the specially prepared dilution water (or pure aerated water) to make 300 ml diluted sample, the dilution ratio will be equal to 300/5 = 60. The dilution ratio is selected on the basis of expected range of BOD_5 so that adequate nutrients and oxygen will be available during the incubation period. Normally, several dilutions are prepared to cover the complete range of possible values. Table 8.4 gives the range of BOD that can be measured with various dilutions expressed either as percentage mixtures or as ml directly pipetted into 300-ml BOD bottles. BOD is expressed either in ppm or in mg per litre.

TABLE 8.4. BOD_5 MEASURABLE WITH VARIOUS DILUTIONS OF SAMPLES

By using percent mixture		*By direct pipetting into 300-ml BOD-bottles*	
% mixture	*Range of BOD*	*ml*	*Range of BOD*
0.01	20000–70000	0.02	30000–105000
0.02	10000–35000	0.05	12000–42000
0.05	4000–14000	0.10	6000–21000
0.10	2000–7000	0.20	3000–10500
0.20	1000–3500	0.50	1200–4200
0.50	400–1400	1.0	600–2100
1.0	200–700	2.0	300–1050
2.0	100–350	5.0	120–420
5.0	40–140	10.0	60–210
10.0	20–70	20.0	30–105
20.0	10–35	50.0	12–42
50.0	4–14	100.0	6–21
100.0	0–7	300.0	0–7

Table 8.5 gives typical value of 5 day BOD of raw and treated sewage.

TABLE 8.5. VALUES OF BOD FOR RAW AND TREATED SEWAGE

	Nature of sewage	*5 day BOD at 20°C (ppm or mg/l)*
1.	Strong sewage	450 to 550
2.	Average sewage	350
3.	Weak sewage	250
4.	Standard filter sewage effluent	20
5.	Very good filter sawage effluent	5 to 10

2. First stage BOD formulation. At a given temperature, the rate at which BOD is satisfied at any time (*i.e.* rate of *deoxygenation*) may be assumed to be directly proportional to the amount of organic matter present in sewage. In other works, the exertion of BOD is considered to be first order reaction defined by

$$\frac{dL_t}{dt} = -K' \cdot L_t \qquad ...(8.8)$$

where L_t = amount of first stage BOD remaining in the sample at any time t (or oxygen equivalent of carbonaceous oxidisable organic matter present at any time t), expressed as mg/l.

K' = rate constant signifying the rate of oxidation of organic matter, having a unit $(\text{day})^{-1}$. Its value depends upon the nature of organic matter present and the temperature during the reaction.

t = time in days.

Integrating Eq. 8.8 between time $t = 0$ (as which $L_t = L_0$, say) to $t = t$, we get

$$\int \frac{dL_t}{L_t} = -K' \int dt$$

or
$$\log_e \frac{L_t}{L_0} = -K' t \qquad ...(8.9\ a)$$

or
$$\frac{L_t}{L_0} = e^{-K't} = 10^{-Kt} \qquad ...(8.9\ b)$$

where the rate constant $K = \dfrac{K'}{2.303}$ ...(8.9 *c*)

In the above equation, L_0 is the oxygen equivalent of organic matter present in sewage at the beginning. Also K is known as base 10 rate constant (or deoxigenation constant) while K' is known base e rate constant.

The amount of BOD remaining at any time t is

$$L_t = L_0 (10^{-Kt}) \qquad ...(8.9)$$

Hence y_t, the amount of BOD that has been *exerted* at any time t is given by :

$$y_t = (L_0 - L_t) = L_0 (1 - 10^{-Kt}) \qquad ...(8.10)$$

In the above equation, y_t is the BOD of t days (*i.e.* $y_t = \text{BOD}_t$).

From Eq. 8.10, 5-day BOD is evidently given by

$$\text{BOD}_5 = y_5 = L_0 - L_5 = L_0 (1 - 10^{-5K}) \qquad ...(8.10\ a)$$

The relationships expressed by Eqs. 8.9 and 8.10 are represented by the corresponding curves shown in Fig. 8.8.

The *ultimate first stage* BOD (*i.e.* y_u) will be obtained by substituting $t = \infty$ in Eq. 8.10. Thus,

$$y_u = L_0\,(1 - 10^{-K\infty}) = L_0 \qquad \text{...(8.11)}$$

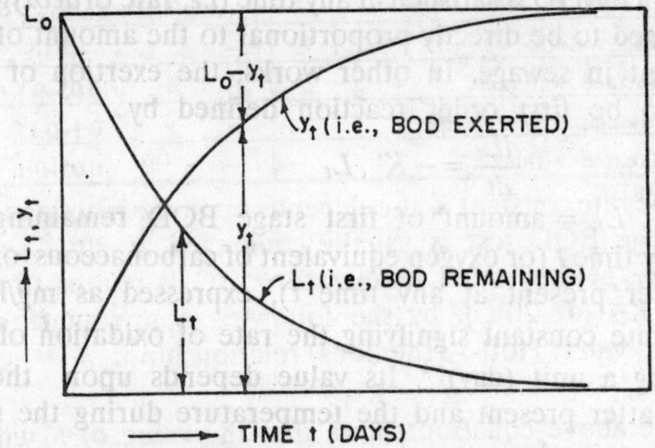

FIG. 8.8. FIRST STAGE BOD CURVE

Thus we conclude that the ultimate first stage BOD of a given wastewater is equal to the initial oxygen equivalent of the organic matter present in the sample of wastewater. y_u is thus a fixed quantity for the given specimen, and does not depend on the temperature during the reaction.

The value of K (*i.e.* rate constant to the base, 10, or deoxygenation constant) determines the speed of BOD reaction, without influencing the ultimate BOD (*i.e.* y_u). The value of K depends upon (*i*) type of wastewater, and (*ii*) temperature during the reaction. The range of K at 20° C reaction temperature, may typically vary between 0.05 to 0.3 day^{-1}, depending upon the type of waste. For the same ultimate BOD, the oxygen uptake will vary with the value of K, as illustrated in Fig. 8.9. Higher the value of K, faster will be the reaction. Table 8.6 gives typical value of K at 20°C, for various types of waters and wastewaters.

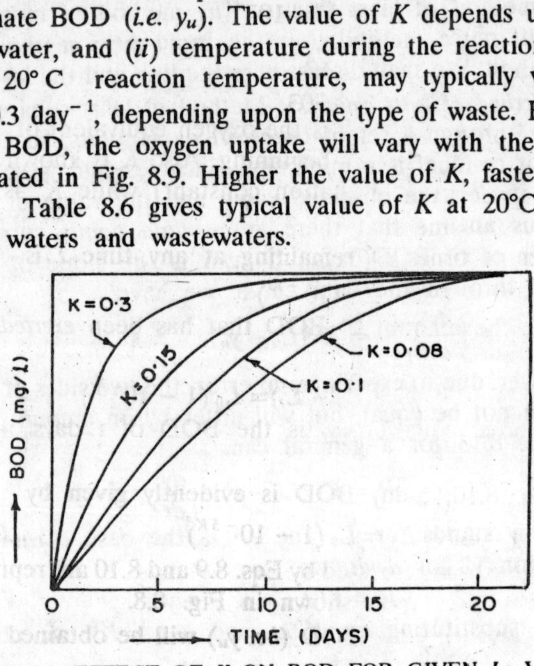

FIG. 8.9. EFFECT OF K ON BOD FOR GIVEN L_0 VALUE

TABLE 8.6. TYPICAL VALUES OF *K* AT 20°C

	Type of water/wastewater	*K* (day^{-1})
1.	Tap waters	<0.05
2.	Surface waters	0.05–0.1
3.	Municipal wastewater	0.1–0.15
4.	Treated effluents	0.05–0.1

The rate constant K also depends upon the reaction temperature. Though the BOD test is usually conducted at 20° C, the reaction contant K_T, at any other temperature T°C, is related to the reaction constant K_{20} (at 20°C), by the following approximate relationship, known as van't Hoft-Arrhenious relationship :

$$K_T = K_{20}\, \theta^{(T-20°)} \qquad ...(8.12)$$

The above equation shows that the value of K will be higher at higher temperatures. The value of θ may be taken as 1.056 for temperature range of 20° to 30° C, and as 1.135 for temperature range of 4° to 20°C. Sometimes, a constant value of $\theta = 1.047$ is adopted, though this value does not apply in cold temperatures.

3. Determination of *K* and L_0. The values of ultimate BOD L_0 (which is approximately equal to 20-day BOD) and rate constant K can be computed form a series of BOD measurements at various times t (days). The usual methods used for computations of K and L_0 are (*i*) method of least squares, (*ii*) method of moments, (*iii*) the daily difference method, (*iv*) the rapid-ratio method, and (*v*) Thomas method. We shall discuss here the first and the last methods.

(i) Method of least squares : Method of least squares involves curve fitting through a set of observed data points, such that sum of squares of residual is a minimum. Each residual is equal to the difference between the observed value and the value of the fitted curve. Let us assume that there are *n data points* corresponding to n number of times t (days). If dy/dt is the slope of the curve to be fitted through the data curve, we have.

$$\left(\frac{dy}{dt}\right)_{t=n} = K'(L_0 - y_n) \qquad ...(8.13)$$

However, due to experimental error, the two sides of the above equation will not be equal, but will differ by an amount R. Hence rewriting Eq. 8.13 for a general can. :

$$R = K'(L_0 - y) - \frac{dy}{dt} \qquad ...(8.14\ a)$$

where y stands for y_t and K' is the base *e rate constant.* Using notation y' for dy/dt, we have

$$R = K' L_0 - K' y - y' \qquad ...(8.14\ b)$$

Again substituting $a = K' L_0$ and $-b = K'$, we obtain

$$R = a + by - y' \quad ...(8.14)$$

In the above equation, the unknowns are a and b. Hence if the sum of the squares of the residual R is to be a minimum, the following equation must hold good :

$$\frac{\partial}{\partial a}(\Sigma R^2) = \Sigma\, 2R\frac{\partial R}{\partial a} = 0 \quad ...(8.15\ a)$$

and $$\frac{\partial}{\partial b}(\Sigma R^2) = \Sigma\, 2R\frac{\partial R}{\partial b} = 0 \quad ...(8.15\ b)$$

If the operations of Eq. 8.15 are carried out using the residual R defined by Eq. 8.14, for each set of data, we get the following equations :

$$na + b\,\Sigma y - \Sigma y' = 0 \quad ...(8.16\ a)$$

and $$a\Sigma y + b\,\Sigma y^2 - \Sigma yy' = 0 \quad ...(8.16\ b)$$

The simultaneous solution of the above two equations yield:

$$K' = -b \quad ...(8.17\ a)$$

and $$L_0 = -a/b \quad ...(8.17\ b)$$

and $$K = K'/2.303 = -0.434\,b \quad ...(8.17\ c)$$

In the above procedure, the slope (dy/dt) is computed by the relation :

$$\frac{dy}{dt} = y' = \frac{y_{n+1} - y_{n-1}}{2\,\Delta t} \quad ...(8.18)$$

See Example 8.19 for illustration.

(ii) ***Thomas method.*** Thomas method is a graphical procedure based on similarity between series expansions for $(1 - e^{K't})$ and $K'\,t\,[1 + (K'\,t/6]^{-3}$. He developed the following approximate formula:

$$\text{BOD}_t = y_t = L_0 K'\,t\left[1 + \frac{K'\,t}{6}\right]^{-3} \quad ...(8.19)$$

which can be linearlised as

$$\left(\frac{t}{y_t}\right)^{1/3} = (K'\,L_0)^{-1/3} - \frac{K'^{2/3}}{6\,L_0^{1/3}}\,t \quad ...(8.20)$$

The above equation has the form :

$$Z = a - bt$$

where $z = (t/y_t)^{1/3}$

$a = (K'\,L_0)^{-1/3}$ ...(8.21 *a*)

$b = K'^{2/3}/6\,L_0^{1/3}$ ...(8.21 *b*)

Hence by plotting $(t/y_t)^{1/3}$ *v/s t*, a straight line will be obtained, having intercept $a = (K'\,L_0)^{-1/3}$ at $t = 0$ and slope $b = K'^{2/3}/6L^{1/3}$, as shown in Fig. 8.10.

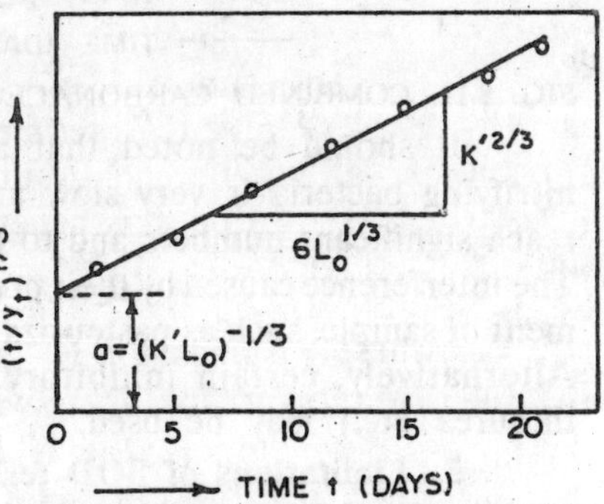

FIG. 8.10. THOMAS METHOD

knowing the values of a and b, we get

$$K' = 6\frac{b}{a} \qquad ...(8.22\ a)$$

and
$$L_0 = \frac{1}{K' a^3} \qquad ...(8.22\ b)$$

Hence
$$K = \frac{K'}{2.303} = 2.61\frac{b}{a} \qquad ...(8.22\ c)$$

It is observed that not all experimental data will fit the theoretical curve. The data which is best is generally that which falls between $t = 2$ to $t = 10$ days. See example 8.18 for illustration.

4. Second stage BOD. As stated earlier, carbonaceous matter is oxidised in the first stage of biochemical reaction, while nitrogenous matter is oxidised in the second stage. Some of the *autotrophic bacteria* are capable of using oxygen to oxidize noncarbonaceous matter, such as ammonia to nitrites and nitrates. This second stage reaction is called *nitrification. Thus, the nitrogenous oxygen demand caused by the autotrophic bacteria is called the second stage BOD.* Fig. 8.11 shows the influence of nitrification on BOD curve.

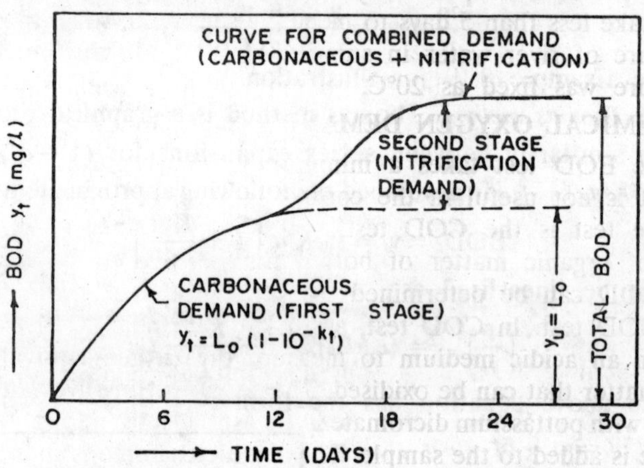

FIG. 8.11. COMBINED CARBONACEOUS AND NITRIFICATION DEMAND.

It should be noted that at 20°C, the reproduction rate of nitrifying bacteria is very slow and it takes about 6 to 10 days to reach significant numbers and to exert a measurable oxygen demand. The interference caused by their presence can be eliminated by pretreatment of sample, such as pasteurization, chlorination or acid treatment. Alternatively, certain inhibitory agents (such as methylene blue, thiourea etc.) may be used.

5. Limitations of BOD test. Despite widespread use of BOD test, it has following limitations :

(*i*) BOD test measures only biodegradable organics.

(*ii*) Pretreatment is needed if the sample contains toxic wastes.

(*iii*) The effects of nitrifying bacteria should be reduced or eliminated by pretreatment or by use of inhabitory agents before the test is applied.

(*iv*) It is essential to have high concentration of active bacteria present in the sample wastewater. If relatively clean industrial wastes are to be analysed, a high concentration of active, acclimated *seed bacteria* is required to be applied.

(*v*) The test looses its stoichiometric validity after the soluble organic matter present in the solution has been used or exhausted.

(*vi*) The test uses an arbitrary long period of time, to obtain results.

(*vii*) For the BOD_5 test, the 5-day period may or may not correspond to the point where soluble organics matter present has been used.

It is interesting to note that the 5-day BOD test was originally devised by the British Royal Commission of Sewage Disposal, and the 5-day period was specifically selected because all the rivers in England take less than 5 days to meet the sea. Also, since the mean temperature of river water in summer was 18.3°C, the incubation temperature was fixed as 20°C.

8.8. CHEMICAL OXYGEN DEMAND (COD)

The BOD test takes a minimum of 5 days time, and due to this, it is not useful in the control of treatment processes. An alternative test is the COD test, which can be used to measure content of organic matter of both wastewater as well as natural waters. COD can be determined only is 3 hours in contrast to 5 days of BOD test. In COD test, a strong chemical oxidising agent is used in an acidic medium to measure the oxygen equivalent of organic matter that can be oxidised. The COD test involves an acidic oxidation with pottassium dicromate. A measured amount of potassium dicromate is added to the sample. The acidified sample is then boiled for 2 hours, cooled and the amount of dicromate remaining is measured by titration with ferrous ammonium sulphate. To accelerate the oxidation of certain types of organic compounds, a catalyst, usually silver sulphate, is required to aid the oxidation. When dicromate is used as oxidizing agent, the principal reaction may be represented in a general way by the following unbalanced equation.

$$\underset{\text{organic matter}}{(C_aH_bO_c)} + Cr_2O_7^{--} + H^+ \xrightarrow[\text{Heat}]{\text{Catalyst}} Cr^{3+} + CO_2 + H_2O \quad ...(8.23)$$

The COD test is specifically more suitable to measure organic matter present in industrial wastes having compounds that are toxic to biological life. However, COD results are generally higher than BOD values since the test will oxidize materials such as fats and lignins which are only slowly biodegradable. No clear correlation exists between BOD and COD in general, but at specific treatment plants, a correlation is possible. When once a correlation has been established, the COD measurements, which can be concluded within 3 hours, can be used to good advantage for the control and operation of those treatment plants. For typical untreated domestic wastes, the ratio COD/BOD_5 is found to vary from 1.25 to 2.5. A higher value of the ratio indicates that the wastewater is difficult to biodegrade. For non-biodegradable wastewater, the ratio exceeds 10. The limiting value of COD of wastewater, generally specified by the authorities is 250 mg/l.

8.9. TOTAL ORGANIC CARBON (TOC)

One approach used to evaluate the amount of organic mater present in the wastewater is to determine the amount of *organic carbon* in the wastewater. The *total organic carbon test* (TOC test) is specially applicable to small concentrations of organic matter. The TOC test consists of acidification of the wastewater sample to convert inorganic carbon to CO_2 which is then stripped. The sample is then injected into a high temperature furnace where it is oxidised in the presence of a catalyst. The CO_2 that is produced is quantitatively measured by means of an infrared analyser, and converted instrumentally to original organic carbon content. The error due to the presence of inorganic carbon can be eliminated by acidification and aeration of the sample prior to the analysis.

The *Carbonaceous Analyzer*, patented by Dow Chemical Company, provides a measure of TOC of an aqueous sample in approximately two minutes. The test is rapid, accurate, and correlates moderately well with BOD. However, certain organic compounds may not be oxidized, and the measured TOC will be slightly less than the actual amount present in the sample. The major obstacle to widespread use of TOC is the cost of the equipment and the skill necessary in its operation. For typical untreated domestic wastewater, the BOD_5/TOC ratio varies from 1.0 to 1.6. Table 8.1 gives some typical TOC values for domestic wastewater.

8.10. TOTAL OXYGEN DEMAND (TOD)

This is yet another instrumental method to measure the organic content of wastewater. The TOD method is based on the quantitative measurement of the amount of oxygen used to burn the organic substances and to a minor extent, inorganic substances. It is thus a direct measure of the oxygen demand of the sample The test is conducted in a platinum-catalyzed consbustion chamber. The oxidizable components into a liquid sample introduced into the combustion chamber are converted to their stable oxides by a reaction that disturbs the oxygen equilibrium in the nitrogen *carrier gas* stream. The momentary depletion in the oxygen concentration in the carrier gas is detected by an oxygen recorder. The TOD for the sample is obtained by comparing this peak height to the peak height of standard TOD calibration solutions. This test can be carried out rapidly and the results have been correlated with the COD.

8.11. THEORETICAL OXYGEN DEMAND (ThOD)

This a theoretical method of computing the oxygen demand of various constituents of the organic matter present in wastewater. The organic matter present in the wastewater may be of animal or vegetable origin, consisting of principal groups such as carbohydrates, protein, fats and products of their decomposition. Each one of these is a typical combination of carbon, hydrogen, oxygen and nitrogen, based on its chemical formula. Hence if the chemical formulae of the constituents of the organic matter are known, ThOD can be easily computed. For example, glycine, commonly present in wastewater has a chemical formula [CH_2 (NH_2) COOH]. Its Th OD can be analytically determined by assuming the following steps in the reactions:

(*i*) In the first step, carbon is converted to CO_2 and nitrogen is converted to ammonia.

(*ii*) In the second step, ammonia is oxidized to nitrite.

(*iii*) In the third step, nitrite is oxidized to nitrate.

The ThOD will then be the sum of the oxygen required for all the three steps of reactions, mentioned above.

8.12. RELATIVE STABILITY

Relative Stability of wastewater is defined as the ratio of *available* oxygen to the *required* oxygen satisfying first stage BOD. The available oxygen will include dissolved oxygen (DO) as well as oxygen present as nitrite or nitrate. It is generally expressed as percentage of total oxygen required. The *test for relative stability* is carried out in the following steps :

1. The wastewater sample is filled in a glass-stoppered bottle and a small quantity of methylene blue is added to it.

2. The mixture is then incubated either at a temperature of 20°C or at 37°C. In countries like India, a temperature of 37°C is preferred.

3. During the incubation period, the anaerobic bacteria start their function, the available DO is consumed and H_2S is produced

which decolourises the mixture. The time t (in days) required for bleaching the blue colour is noted.

The relative stability (S_R) is worked out from the following expressions :

$$S_R = 100\,(1 - 0.794^{t_{20}}) \quad ...(8.24\ a)$$

or

$$S_R = 100\,(1 - 0.605^{t_{37}}) \quad ...(8.24\ b)$$

where t_{20} and t_{37} are the number of days of incubation at 20° C and 37° C respectively.

The relative stability test is not suitable for raw sewage beause the colour is precipitated out due to the presence of some dissolved and colloidal solids. However,the test is quite suitable for studying the quality of effluents coming out of the wastewater treatment plants. If the value of t comes out to be less, it will indicate that availability of oxygen is less and therefore the wastewater is less stable. If the decolourisation during the test takes place in a time lesser than 4 days, the effluent may be taken as relatively unstable (with its relativity less than about 60%) and the effluent cannot be safely discharged into the receiving waters (*i.e.* streams etc.). However, if $t \geq 5$, the effluent is relatively stable and hence it can be safely discharged into the receiving water. Table 8.7 gives the valves of relative stability at 20°C and 37°C required for decolourization.

TABLE 8.7. VALVES OF S_R

Description	*t (days)*									
	1	*2*	*3*	*4*	*5*	*6*	*7*	*8*	*9*	*10*
1. Values of S_R at 20°C	20.6	37	49.9	60.3	68.4	74.9	80.1	84.2	87.5	90
2. Value of S_R at 37°C	39.5	63.4	77.9	86.6	91.9	95.1	97	98.2	98.9	99.3

8.13. POPULATION EQUIVALENT

The wastewater carried by a sewer consists mainly of domestic sewage and the industrial wastewater. Since the contribution of solids to sewage should be nearly constant on a per capita basis, the BOD contribution (expressed in grams/person per day) should also be constant. Generally, BOD contribution per capita per day may be taken as 80 g/day (or 0.08 kg/day). Industrial wastewaters are generally compared with per capita domestic sewage, through the concept of population equivalent (P_E) using per capita BOD value as the basis. Thus we have

$$P_E = \frac{\text{Total BOD}_5 \text{ of the industrial wastewater (kg/day)}}{\text{BOD}_5 \text{ value per capita/day}} \quad ...(8.25)$$

For example, if the total BOD_5 of an industrial wastewater is 800 kg/day and BOD_5 value is 0.08 kg/capita/day,

$$\text{Population equavatent,} \quad P_E = \frac{800}{0.08} = 10{,}000.$$

Similarly, if at the point of measurement, we have a combined wastewater, consisting of domestic sewage as well as industrial wastewater, the *population equivalent* of the combined wastewater can be found by dividing BOD_5 of the combined wastewater by the per capita/day BOD value. Generally, the *population equivalent* is used to indicate the strength of industrial wastewater required for (*i*) estimating the treatment required at the common treatment plant (municipal treatment plant) and (*ii*) charging appropriate levy on the industries to meet the proportionate cost of treating waste in the municipal treatment plant.

Example 8.1. *The domestic sewage of a town was tested for total solids and following results were obtained :*

Weight of sample of sewage = 1000 gm.

Weight of solids after evaporation of liquid = 0.952 gm.

Weight of dry residue after ignition = 0.516 gm.

Determine (i) total solids (ii) fixed solids and (iii) volatile solids.

Solution

$$\text{Total solids,} \quad S_T = \frac{0.952}{1000} \times 10^6 = \mathbf{952\ ppm}$$

$$\text{Fixed solids,} \quad S_F = \frac{0.516}{1000} \times 10^6 = \mathbf{516\ ppm}$$

$$\therefore \text{ Volatile solids,} \quad S_V = S_T - S_F = 952 - 516 = \mathbf{436\ ppm}$$

Example 8.2. *In order to conduct a 5-day BOD test, the sample of wastewater was diluted with specially prepared dilution water, with a dilution factor of 150. The contents of dissolved oxygen in the beginning and end of test were found to be 11 ppm and 7 ppm respectively. Compute the 5-day BOD. What is the nature of the wastewater?*

Solution : From Eq. 8.7,

$$BOD_5 = [\text{oxygen consumed}] \times \text{dilution factor}$$

$$\therefore \quad BOD_5 = (11 - 7) \times 150 = \mathbf{600\ ppm}$$

Hence the wastewater is very strong and requires heavy treatment before disposal.

Example 8.3. *5 ml of raw sewage was diluted by specially prepared water, in a 300 ml capacity BOD bottle. The DO concentration of the diluted sample at the beginning of the test was 9 mg/l and 6 mg/l after 5-day incubation at 20°C. Find the BOD of raw sewage.*

Solution

$$\text{Dilution ratio} = \frac{\text{Vol. of diluted sample}}{\text{Vol. of undiluted sample}} = \frac{300}{5} = 60$$

$$\text{Loss of DO} = 9 - 6 = 3 \text{ mg/l}$$

$$\therefore \quad BOD_5 = 3 \times 60 = \mathbf{180\ mg/l}$$

Example 8.4. *A 1% solution of sewage sample is incubated for 5 days at 20°C. The depletion of oxygen was found to be 3 ppm. Determine the BOD of raw sewage.*

Solution

$$\text{Dilution ratio} = \frac{100}{\text{per cent of solution}} = \frac{100}{1} = 100.$$

$$\therefore \quad BOD_5 = 3 \times 100 = \mathbf{300\ ppm}$$

Example 8.5. *Determine ultimate BOD for a sewage having 5-day BOD at 20°C as 160 ppm. Assume the deoxygenation constant as 0.2 per day.*

Solution. From Eq. 8.1 (*a*) we have,

$$BOD_5 = y_5 = L_0 - L_5 = L_0 (1 - 10^{-5K})$$

Here $BOD_5 = 160$ ppm ; $K = 0.2$ per day.

$$\therefore \quad 160 = L_0 (1 - 10^{-5 \times 0.12}) = L_0 (0.7488)$$

$$\therefore \quad L_0 = \mathbf{213.7\ ppm} \text{ (or mg/l)}$$

Example 8.6. *Determine 2 day BOD for the sewage of example 8.5.*

Solution. From Eq. 8.10, we have

$$y_t = L_0 (1 - 10^{K_t})$$

$$\therefore \quad y_2 = 213.7 (1 - 10^{-0.12 \times 2}) = \mathbf{90.7\ ppm}$$

Example 8.7. *A sample of wastewater has 4-day 20°C BOD value of 75% of the final. Find the rate constant (to the base 10) per day.*

Solution : From Eq. 8.10, we have

$$y_t = L_0 (1 - 10^{-Kt})$$

Here, it is given that $y_4 = 75\%$ of L_0

$$\therefore \quad y_4 = 0.75\, L_0.$$

Hence $\quad 0.75\, L_0 = L_0 (1 - 10^{-K \times 4})$

or $\quad 10^{-4K} = 0.25$

or $\quad -4K = \log_{10} 0.25 = -0.602$

From which $\quad K \approx \mathbf{0.151}$ ***per day***

Example 8.8. *Find the rate constant (to the base 10) at a temperature of 30°C, if its value at 20°C is 0.12 per day.*

Solution. From Eq. 8.12,

$$K_T = K_{20}\, \theta^{T - 20°}$$

Since the temperature range is 20° to 30°, take value of θ as 1.056.

$$\therefore \quad K_{30} = K_{20} (1.056)^{T - 20°}.$$

or $\quad K_{30} = 0.12 (1.056)^{30 - 20} = \mathbf{0.207}$ ***per day***

Example 8.9. *The* BOD_5 *of a waste has been measured as 500 mg/l. If the rate constant K' = 0.26/day (base e), what is the ultimate*

BOD of the waste ? What proportion of BOD_u would remain unoxidised after 20 days ?

Solution Given : $K' = 0.26$/day (base e)

From Eq. 8.9 (c), $K = \frac{K'}{2.303} = \frac{0.26}{2.303} \approx 0.113$/day.

Now $y_t = L_0 (1 - 10^{-K.t})$ ...(8.10)

$\therefore$ $BOD_5 = y_5 = L_0 (1 - 10^{-0.113 \times 5})$

or $500 = L_0 (1 - 0.2723)$

From which $L_0 \approx$ **687** mg/l = BOD_u

Again, $y_{20} = L_0 (1 - 10^{-0.113 \times 20}) = 0.9945\, L_0$

This means that 99.45% of BOD_u is utilized in 20 days and only **0.55%** of ultimate BOD would be left unutilised after 20 days.

Example 8.10. *The BOD of a sewage incubated for one day at 30°C has been found to be 100 mg/l. What will be the 5-day 20°C BOD. Assume K = 0.12 (Base 10) at 20°C.*

Solution Given y_1 (*or* $BOD_1)_{30°}$ = 100 mg/l.

$K_{20} = 0.12$

Let us first find K_{30}, using Eq. 8.12,

$K_T = K_{20}\, \theta^{T - 20°}$

Since the temperature range is 20° to 30°, take $\theta = 1.056$.

$\therefore$ $K_{30} = 0.12\, (1.056)^{30 - 20} \approx 0.207$/day.

Again, applying Eq. 8.10, for reaction at 30° C,

$y_{t(30)} = L_0 (1 - 10^{-K_{30}.t})$

$\therefore$ $100 = L_0 (1 - 10^{-0.207 \times 1})$

From which $L_0 = 263.8$ mg/l.

This is thus the ultimate BOD, the value of which does not depend on temperature of incubation. Hence applying Eq. 8.10 again, for reaction at 20°C, we have

$y_{t\,(20°)} = L_0 (1 - 10^{-K_{20}t})$

or $y_{5\,(20°)} = 263.8\,(1 - 10^{-0.12 \times 5}) = 197.5$

Hence BOD_5 at 20° C = **197.5 mg/l.**

Example 8.11. *For a wastewater sample, 5-day BOD at 20°C is 200 mg/l and is 67% of the ultimate. What will be 4-day BOD at 30°C.*

Solution. From Eq. 8.10 (a) we get

$BOD_5 = y_5 = L_0 (1 - 10^{-5K})$

Here, $y_5 = 200$ mg/l ; Also, y_5 = 67% of L_0

$\therefore$ $L_0 = \frac{y_5}{0.67} = \frac{200}{0.67} = 298.5$ mg/l.

$\therefore$ $200 = 298.5\,(1 - 10^{-5K})$

or $$10^{-5K} = 1 - \frac{200}{298.5} = 1 - 0.67 = 0.33$$

or $$-5K = \log_{10} 0.33 = -0.481$$

From which $K = K_{20} = 0.0963/\text{day}$

Again, $K_{30} = K_{20}(1.056)^{30-20} = 0.0963\,(1.056)^{10}$
$= 0.166/\text{day}$

Now, from Eq. 8.10, at 30°C,

$$y_4 = L_0(1 - 10^{-K_{30}t})$$
$$= 298.5\,(1 - 10^{-0.166 \times 4}) = \mathbf{233.8\ mg/l}$$

Example 8.12. *During BOD test conducted on a 5% dilution of waste, the following observations were taken.*

(*i*) *DO of aerated water used for dilution = 3.6 mg/l*
(*ii*) *DO of original sample = 0.8 mg/l.*
(*iii*) *DO of diluted sample after 5 day incubation = 0.7 mg/l.*

Compute : (a) 5 day BOD and (b) Ultimate BOD.

Assume deoxygenation constant at test temperature as 0.12.

Solution : The diluted sample contain 5% wastewater and 95% aerated water.

DO of test specimen= [DO of wastewater × its content] + [DO of aerated water × its content]
$= (0.8 \times 0.05) + (3.6 \times 0.95) = 3.46$ mg/l

DO of incubated sample after 5 day = 0.7 mg/l.

∴ DO consumed = 3.46 − 0.7 = 2.76 mg/l.

$$\text{Dilution factor} = \frac{100}{5} = 20$$

∴ $\text{BOD}_5 = (\text{DO consumed}) \times \text{dilution factor}$
$= 2.76 \times 20 = \mathbf{55.2\ mg/l}$

Also, $$y_t = L_0[1 - 10^{-Kt}]$$

∴ $$y_5 = L_0[1 - 10^{-0.12 \times 5}] = 0.7488\,L_0$$

∴ $$L_0 = \frac{y_5}{0.7488} = \frac{55.2}{0.7488} = \mathbf{73.72\ mg/l}$$

Example 8.13. *In a test for relative stability, the period of incubation comes out to be 8 days. Determine the relative stability if the test temperature is (a) 20°C, (b) 37°C.*

Solution

(*a*) Test temperature = 20°C.

$$S_R = 100\,(1 - 0.794^{t_{20}}) = 100\,(1 - 0.794^8) = \mathbf{84.2\%}$$

(*b*) Test temperature = 37° C.

$$S_R = 100\,(1 - 0.605^{t_{37}}) = 100\,(1 - 0.605^8) = \mathbf{98.2\%}$$

Example 8.14. *Calculate the population equivalent of a city, given (i) the average sewage from the city is 80 × 10⁶ l/day and (ii) the average 5-day BOD is 250 mg/l.*

Solution Average $BOD_5 = 250$ mg/l

Average sewage flow $= 80 \times 10^6$ l/day.

$\therefore$ Total BOD_5 load in daily sewage $= 250 \times 80 \times 10^6$

$= 20000 \times 10^6$ mg/day $= 20000$ kg/day.

Assuming domestic sewage quantity $= 0.08$ kg/capita/day,

Population equivalent $= \dfrac{20000}{0.08} =$ **250000** persons.

Example 8.15. *If the per capita contribution of suspended solids and BOD is 100 gm and 60 gm, find the population equivalents of*

(a) A combined system serving 12000 persons and having 80 gm of per capita daily BOD, and (b) 200 000 litres daily of industrial wastewater containing 1500 mg/l of suspended solids.

Solution

(*a*) In the first case, population equivalent is based on BOD criterion. Total daily BOD load $= 12000 \times 80 = 96 \times 10^4$ g.

Standard per capita BOD $= 60$ g.

$\therefore$ Population equivalent $= \dfrac{96 \times 10^4}{60} =$ **16000**

(*b*) In the second case, population equivalent is based on the criterion of suspended solids.

Total daily load of suspended solids $= 200000 \times 1500 \times 10^{-3}$

$= 300000$ g.

Standard per capita suspended solids $= 100$ g.

$\therefore$ Population equivalent $= \dfrac{300000}{100} =$ **3000**

Example 8.16. *The 4 day 15°C BOD of a sample of sewage is 200 mg/l. Draw a graph of 5 day BOD as a function of temperature in the range of 10° C to 30° C in steps of 5° C.*

Solution. Let us assume the value of K at 20°C as 0.1/day.

Now $K_T = K_{20}\,(\theta)^{(T-20°)}$ ...(8.12)

Taking $\theta = 1.135$ for a temperature range of 4° to 20°C,

$$K_{15} = 0.1\,[1.135]^{15-20} = 0.053$$

Now, applying Eq. 8.10 for 15° C temperature and 4-day BOD

$$y_4 = L_0\,(1 - 10^{-K \times 4})$$

or $$200 = L_0\,(1 - 10^{-0.053 \times 4}) = 0.386\,L_0$$

From which $L_0 = 517.8$ mg/l

This is value of ultimate BOD which is not dependent of temperature. Let us now find the values of K at various temperature ranging from 10° to 30°. For temperature from 10°C to 20°C, θ is taken equal to 1.135 while for temperature range of 20° to 30°C, θ is taken equal to 1.056.

$\therefore \quad K_{10} = 0.1\,(1.135)^{10-20} = 0.028/\text{day}$

$K_{15} = 0.056$ (as computed above)

$K_{20} = 0.1$ (assumed earlier)

$K_{25} = 0.1\,(1.056)^{25-20} = 0.1313/\text{day}$

$K_{30} = 0.1\,(1.056)^{30-20} = 0.1724/\text{day}$

Now applying Eq. 8.10 (*a*) at various temperatures, we have

$y_{5\,(10°)} = 517.8\,(1 - 10^{-5 \times 0.028}) \approx 143$ mg/l.

$y_{5\,(15°)} = 517.8\,(1 - 10^{-5 \times 0.056}) \approx 246$ mg/l

$y_{5\,(20°)} = 517.8\,(1 - 10^{-5 \times 0.1}) \approx 354$ mg/l

$y_{5\,(25°)} = 517.8\,(1 - 10^{-5 \times 0.1313}) \approx 404$ mg/l.

$y_{5\,(30°)} = 517.8\,(1 - 10^{-5 \times 0.1724}) = 447$ mg/l.

Fig 8.12 shows the required graph between BOD_5 and temperature. From the plot we observe that variation of BOD with temperature is *practically* linear.

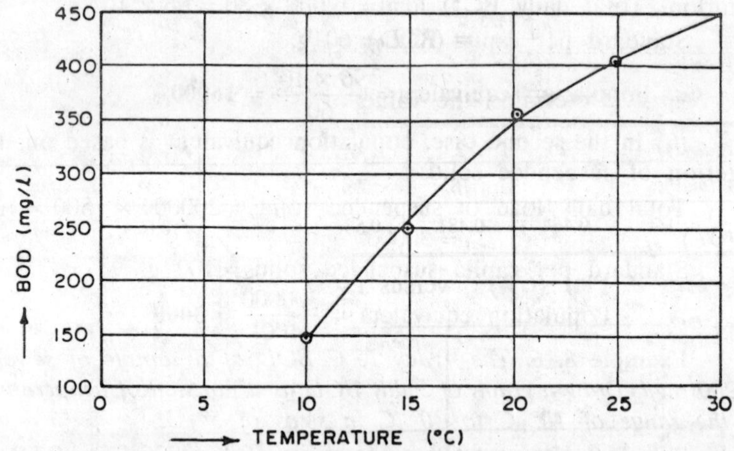

FIG. 8.12.

Example 8.17. *Change in concentration of organic matter L, with time t, is given by* $\frac{dL}{dt} = -KL$. *Calculate the organic matter remaining after 4 days if the initial concentration was 300 mg/l and K = 0.3 per day.*

Solution. $\frac{dL}{dt} = -KL.$

$\therefore \quad \int \frac{dL}{L} = -K \int dt$

$\therefore \quad \log_e L = -Kt + C$

or $\quad 2.303 \log_{10} L = -Kt + C \quad \text{...}(a)$

When $t = 0$, $L = 300$ mg/l.

Hence $C = 2.303 \log_{10}(300) \approx 5.7$

Substituting this value of C in Eq. (a), we get

$$2.303 \log_{10} L = -Kt + 5.7 \quad \text{...}(b)$$

(where $K = 0.3$ per day)

Hence the value of L after 4 days is given by

$$2.303 \log_{10} L_4 = -0.3 \times 4 + 5.7 = 4.5$$

or $$\log_{10} L_4 = 1.954$$

From which $L_4 = 10^{1.954} = 89.9$ mg/l.

$\therefore$ Organic matter left after 4 days= **89.9 mg/l.**

Example 8.18. *From the following test data, determine the value of ultimate BOD and rate constant, using Thomas method.*

Time (days	1	2	4	6	8	10
BOD (mg/l)	10	18	26	31	33	34

Solution. Thomas equation is

$$\left(\frac{t}{y_t}\right)^{1/3} = (K' L_0)^{-1/3} - \frac{K'^{2/3}}{6 L_0^{1/3}} + \quad \text{...(8.20)}$$

Step 1. Determine the value of $(t/y_t)^{1/3}$ for the given data

t (days)	1	2	4	6	8	10
y_t *(mg/l)*	10	18	26	31	33	34
$(t/y_t)^{1/3}$	0.464	0.481	0.536	0.578	0.624	0.665

Step 2. Plot $(t/y_t)^{1/3}$ versus t, as shown in Fig. 8.13.

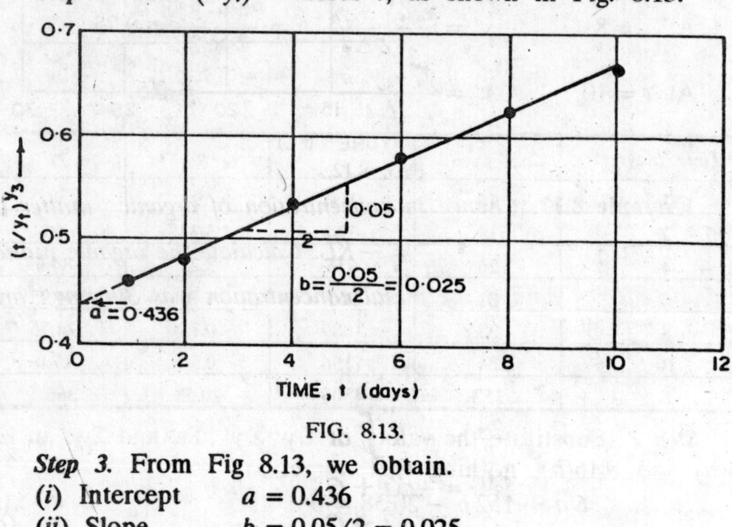

FIG. 8.13.

Step 3. From Fig 8.13, we obtain.

(*i*) Intercept $a = 0.436$

(*ii*) Slope $b = 0.05/2 = 0.025$

Step 4. Compute K' and L_0 from Eqs. 8.22 (***b***) and (***c***)

$$K' = 6\frac{b}{a} = \frac{6 \times 0.025}{0.436} = \mathbf{0.344/day}$$

or $$K = \frac{K'}{2.303} = 2.6\frac{b}{a} = \frac{2.6 \times 0.025}{0.436} = \mathbf{0.149/day}$$

$$L_0 = \frac{1}{K' a^3} = \frac{1}{0.344\,(0.436)^3} \approx \mathbf{35.1\ mg/l}$$

Example 8.19. *Calculate BOD constants K' and L_0 using method of least squares, for the data of Example 8.18.*

Solution

Step1 : Compute $y^2, y'\left(=\frac{dy}{dt}\right)$ and $y y'$ in the tabular form as shown below, in which the slope y' is computed by Eq. 8.18:

$$y' = \frac{dy}{dt} = \frac{y_{n+1} - y_{n-1}}{2\,\Delta t}$$

For example, at $t = 1$, $y' = \frac{y_{n+1} - y_{n-1}}{2\,\Delta t} = \frac{18 - 0}{(1+1)} = 9$

(Note that $y_{n-1} = y_0 = 0$ and $2\,\Delta t = \Delta t_1 + \Delta t_2 = 1 + 1 = 2$)

At $y = 2$, $y' = \frac{y_{n+1} - y_{n-1}}{2\,\Delta t} = \frac{26 - 10}{(1+2)} = 5.33$

At $t = 4$, $y' = \frac{y_{n+1} - y_{n-1}}{2\,\Delta t} = \frac{31 - 18}{2 \times 2} = 3.25$

At $t = 6$, $y' = \frac{y_{n+1} - y_{n-1}}{2\,\Delta t} = \frac{33 - 26}{2 \times 2} = 1.75$

At $t = 8$, $y' = \frac{y_{n+1} - y_{n-1}}{2\,\Delta t} = \frac{34 - 31}{2 \times 2} = 0.75$

At $t = 10$, $y' = \frac{y_n - y_{n-1}}{\Delta t} = \frac{34 - 33}{2} = 0.5$

Time (days)	*y*	y^2	y'	yy'
1	10	100	9	90
2	18	324	5.33	96
4	26	676	3.25	84.5
6	31	961	1.75	54.25
8	33	1089	0.75	24.75
10	34	1156	0.50	17.0
Σ	152	4306	20.58	366.5

Step 2. Substitute the values of Σy, $\Sigma y'$, Σy^2 and $\Sigma yy'$ in Eq. 8.16(*a*) and 8.16(*b*), nothing that $n = 6$.

$$\therefore \quad 6a + 152b - 20.58 = 0 \qquad ...(a)$$

$$152a + 4306b - 366.5 = 0 \qquad ...(b)$$

Solving Eqs. (*a*) and (*b*), we get

$$a \approx 12.03 \quad \text{and} \quad b = -0.34$$

Hence $K' = -b = \mathbf{0.34 / day}$

and $$L_0 = -\frac{a}{b} = \frac{12.03}{0.34} = \mathbf{35.39\ mg/l}$$

It is to be noted that the values of K' and L_0 computed by the method of least squares are quite close to the ones computed by Thomas method.

8.14. BIOLOGICAL CHARACTERISTICS

Domestic sewage, by its nature, contains enormous quantities of micro-organisms. The biological characteristics of sewage are related to the presence of these micro-organisms. The sanitary engineer must have considerable knowledge of (*i*) principal groups of micro-organisms found in water and wastewater (*ii*) pathogenic organisms in wastewater, and (*iii*) organisms used as indicators of pollution. Excremental matter contains myriads of micro-organisms – as many as 320 billions in the sewage per head per day. Most of them are not only harmless, but are friendly to mankind, since they help in the biological treatment of sewage to render it harmless.

A micro-organism that requires living tissues to grow is called a *pathogen* and is a parasite which must depend upon host organism as a proper environment for growth and reproduction. Pathogenic micro-organisms are harmful to man. Micro-organisms can be either plants or animals. When the assemblages of aquatic organisms drift more or less passively with waves and current, they constitute the *plankton*. If however, they change their position or location due to their own effort, they are called *nekton*. If they maintain their stability due to surface tension forces, they are referred to as *neuston*.

The various micro-organisms found in water or wastewater may be broadly classified under three categories : 1. Aquatic plants 2. Aquatic animals 3. Aquatic molds, bacteria and viruses.

1. Aquatic plants : Under this category, the following are included :

(*i*) *Spermophyta* : Water weeds
(*ii*) *Bryophyta* : Mosses and lever words
(*iii*) *Pteridophyta*: Ferns and horsetails
(*iv*) *Thallophyta* : Algae.

Out of these, *water weeds* and *algae* are of practical importance to sanitary engineers. If favourable conditions of light, temperature and nutrition exist, *water weeds* grow in dense form. Their decay during autumn add organic matter to water. *Algae* are considered as simple, photo-synthetic plants with unicellular organs of reproduction. They are, thus the organisms that are self nourishing by deriving energy from simple inorganic substances with the aid of sunlight.

One of the most important problems facing the sanitary engineers is how to treat wastes of various origins so that their effluents discharged into receiving waters do not encourage growth of algae and other aquatic plants. This solution may lie in the removal of carbon, various forms of nitrogen and phosphorus.

2. Aquatic Animals : They include the following

(*i*) *Vertebrate* : Fish and amphibians

(*ii*) *Mollusca* : Mussels, snails, slugs, limplets, cocklets.

(*iii*) *Arthopoda* : Crustacea, insects, spiders, mites.

(*iv*) *Worms* : Aquatic earthworms, thread worms, rotifera

(*v*) *Metazoa* : Hydra, polyzoa

(*vi*) *Protozoa* : Endameba histolytica etc.

Protozoa : Protozoa of importance to sanitary engineers include amoebas, flagellates and free swimming and stalked ciliates. Protozoa are essentially unicellular (single celled) animals that reproduce by binary fission. *They are the lowest and simplest forms of animal life.* One form of protozoa–*Endemeba histolytica* causes amoebic dysentery. If forms cysts which are excreted in the bowel discharges of infected persons and which will live for long periods in water. The protozoa are bacteria eaters and survive in dilute organic wastes by eating bacteria and thus destroy the pathogens. When the concentration of organic waste is sufficiently great, the protozoa can ulitize the soluble organic compounds for their food. They are essential in the operation of biological treatment processes and in the purification of streams because they maintain a natural balance among the different groups of micro-organisms.

3. Aquatic molds, bacteria and viruses : Strictly speaking, molds (or fungi), bacteria and viruses come under the category of aquatic plant, but because of their special importance, they are generally kept in a separate category.

Fungi : Fungi are unicellular, non-photosynthetic plants capable of growing in low temperature and low pH environments. They flourish over a wide range of pH (4 to 10) and themselves modify the pH by producing organic acids and ammonia as well. The reproductive stage of fungi is a spore. Spores are transmitted long distances in the air by wind currents. Absence of suitable environment prevents most fungi spores from germinating.

Viruses : Viruses are infectious agents of both plant and animal cells. They are ultramicroscopic, obligate introcellular parasites that manifest their presence by destruction or impairment of host cells. They can pass through an ultra-microscopic filter and they fall in the size range of 10 to 500 milli-microns. Because of their small size, viruses lack the biochemical systems needed for normal metabolic cell functions and are essentially units organised solely for self-replications. A typical virus particle consists an outer protein coat enclosing a core of nuclei acid. One group of viruses, the bacteriophages, are infectious agents of bacteria and are parasitic to bacteria. It initiates infection by attaching itself by its tail to the wall of a bacterial cell. Out of several other forms of viruses, *adenoviruses* are associated with upper respiratory infections in children. *Entero-viruses* are found in gastro-intestinal tract and faeces of man and many lower animal. *Enteric viruses* include *coxsackie viruses*, infectious *hepatitis viruses, poliviruses*, reoviruses etc.

Bacteria : Bacteria are single-celled microorganisms with rigid cell walls that take in soluble food and convert it to new cells. Bacteria may be classified according to *type* as (*i*) *saprophytic* bacteria, (*ii*) *parasitic* bacteria (*iii*) *pathogenic* bacteria and (*iv*) *non-pathogenic* bacteria. Saprophytic bacteria are beneficial to man, and obtain organic matter in solution from dead and decaying tissue in plants and animals. Parasitic bacteria live and multiply on or within the body of a living organism of a higher type. *Pathogenic bacteria* and pathogenic organisms are capable of causing disease within the living organisms on which they subsist. According to *oxygen need*, bacteria can be classified as (*i*) *aerobic bacteria* which need oxygen to live, (*ii*) *anaerobic bacteria* which survive in the absence of oxygen and (*iii*) *faculative* bacteria which can live and multiply with or without oxygen. According to temperature of which they flourish, bacteria can be (*i*) *psychrophilic bacteria* which can survive between 10° to 20°C (*ii*) *mesophilic bacteria* which can servive between 20° to 40° and (*iii*) *thermophilic bacteria* which can survive between 40° to 65° C. Bacteria are less sensitive to cold than heat. At low temperatures, they are essentially dormant and can survice for long periods of time. Some bacteria tolerate low pH while others tolerate high pH, the optimum being between 6.5 and 7.5.

The bacteria of *excretal origin* may be three types : (*a*) The coli-aerogenes group bacteria (*b*) Clostridium Welchii (*c*) Faecal streptococci.

Coli-aerogeneous group : The intestines of men and warm blooded animals contain certain harmless bacteria which are excreted with faeces. These are numerous in sewage. Formerly, these organisms were known as *colon bacilli* but at present they are known as *coliforms*. The number of coliform organisms in human faeces is estimated to lie between 10^{11} and 10^{13} per capita daily. The *coliform group* includes all of the aerobic and facultative anaerobic nonsporeforming, Gram-negative rod shaped bacteria that ferment lactose (milk sugar) with production at 35° C of gas within 48 hours. The coliform group is comprised of two important species. *Escherichi coli* (E-coli) and *Aerobacteria aerogenes*. They are of advantage as indicators of water contamination.

Clostridium Welchi is a stout Gram-positive rod having elliptical sub-terminal spores. The organism is found in cultivated soils, sewage and polluted water. Intestines are its natural habitat where it causes non harm but assists digestion.

Faecal streptococci are found in human intestines. It is a Gram-positive coccus occurring in pairs or short chains. It grows in presence of bile salt and is capable of growing at temperatures as high as 45°C. However, they are not as numberous as E-coli in all normal cases. Hence test for faecal streptococci offers no advantage over the *E-coli test* except in case of doubt.

The coliforms and the E-coli are normally used as indicator organisms since they are present in large numbers than pathogens. They also lend themselves to numerical evaluation as well as qualitative distinction. They are also identifiable by relatively simple analytical procedures, providing information quickly and economically. The usual procedure for determining the presence of coliforms consists of (*i*) *presumptive* (*ii*) the *confirmed* tests, and (*iii*) completed test, and the detaited procedures for these tests can be seen in Author's book 'Water Supply Engineering'. There are two accepted methods for obtaining the number of coliform organisms in the specimen : (*i*) the *coliform index* and (*ii*) the *most probable number* (M.P.N). The *most probable number* (M.P.N) of coliform is defined as that bacterial density, which if it had been actually present in the sample under examination, would more frequently than any other, have given the observed analytical, result. Thus, MPN is not the absolute concentration of organisms that are present but only a statistical estimate of that concentration.

Anaerobic processes : The work done by anaerobic bacteria, viz. decomposition of organic matter is called *putrefaction* and the result is called *liquefaction*, as the solid organic matter is dissolved by enzymes. Anaerobic bacteria oxidize organic matter utilizing *electron acceptors* other than oxygen. In carrying out their metabolic processes, they produce CO_2, H_2O, H_2S, CH_4, NH_3, N_2, reduced organics and more bacteria. A large part of the available energy appears in the form of end products, and hence cell production is low. Treatment units which work on putrefaction alone are septic tanks, Imhoff tanks, and sludge digestion tanks. The end products of an anaerobic fermentation are likely to be odorous. The production of a stable effluent is unlikely since wastes dot not usually contain sufficient electron acceptors to permit complete oxidation.

Aerobic processes : The work of the aerobic bacteria, i.e. combination with oxygen is called *oxidation*. Aerobic bacteria utilize free oxygen as an electron acceptor. The end products of aerobic activity are CO_2, H_2O, SO_4, NO_3, NH_3 and more bacteria. The bulk of the available energy finds its way into cell mass or heat, yielding a stable effluent which will not undergo further decomposition.

Though each of the above two *processes* work in the opposite direction–the former by splitting up and the latter by, building up, there is co-ordination between the two. In the first stage, the anaerobic bacteria decompose complex organic matter into simple organic compounds while in the second stage, the aerobic bacteria oxidise them to form stable compounds.

PROBLEMS

1. Give a brief list of various constituents of wastewater.

2. Write a note on various physical properties of wastewater.

3. (*a*) Differentiate clearly between BOD, COD and TOD.

(*b*) A 2% solution of a sewage sample is incubated for 5 days at 20°C. The depletion of oxygen was found to be 5 mg/l. Determine the BOD of the sewage. **Ans.** [250 mg/l]

4. (*a*) Differentiate between 'oxygen demand' and 'biochemical oxygen demand' ? How do you determine BOD ? What are the limitations of BOD test ?

(*b*) The 5-day BOD (20° C) of a sewage sample was found to be 100 mg/l. Calculate 2 day BOD at 30° for the same sample. Assume $K_{20} = 0.1$/day. **Ans.** [80 mg/l]

5. Write short notes on the following :

(*i*) Anaerobic process. (*ii*) Aerobic process. (*iii*) Nitrogen cycle. (*iv*) Carbon cycle.

6. Differentiate between the following

(*i*) Organic solids and inorganic solids. (*ii*) Volatile solids and fixed solids. (*iii*) COD and ThOD (*iv*) Free ammonia and albuminoid ammonia. (*v*) Denitrifications and nitrogen fixation.

7. Define BOD. Deduce expression for first stage BOD.

8. The BOD_5 of a waste has been measured as 600 mg/l. If $K = 0.23$/day (base e), what the ultimate BOD_u of waste ? What proportion of BOD_u would remain unoxidised after 20 days ? (UPSC, CES, 1978) **Ans.** [1 %]

9. Data from an unseeded domestic wastewater BOD test are : 5 ml of waste in 300 ml bottle, initial DO of 7.8 mg/l and 5 days DO equal to 4.3 mg/l. Compute : (*a*) The BOD ; (*b*) the ultimate BOD, assuming a K-rate of 0.1 per day. (UPSC civil service, 1979) **Ans.** [(*a*) 210 mg/l; (*b*) 307.1 mg/l.]

10. (*a*) Differentaite between first stage BOD and second stage BOD.
(*b*) What are the factors on which K-value depends.
(*c*) Explain with the help of curve, how the value of *K* determines the speed of BOD reaction without influencing the ultimate BOD.

11. A dairy processing about 1,33,000 kg of milk daily produces an average of 246 cubic metre per day of wastewater with a BOD of 1400 mg/l. The principal operations are bottling of milk and making ice cream, with limited production of cheese. Compute the wastewater flow and BOD per 1000 kg of milk received, and the equivalent population of the daily waste discharge. (UPSC, Civil services, 1979)

Ans. [1.85 m^3 ; 2.59 kg ; 4305]

12. What do you understand by population equivalent ? How do you determine it ?

13. (*a*) State and describe four important tests that may be carried out to know the characteristics of, sanitary sewage. (*b*) Distinguish between aerobic, anaeobic and facultative micro-organisms and their role in the decomposition of sewage.

14. The 3 day 15° C BOD of sample of sewage is 150 mg/l. Draw a graph of 5-day BOD as a function of temperature in the range 10° C to 30° C in steps of 5°C. (UPSC Engg. Services, 1989)

Ans. [L_0 = 488.6 mg/l]

15. If the period of incubation is 10 days at 20° C in the relative stability test on sewage, calculate the percentage of relative stability. (UPSC, CES, 1988) Ans [90.04%]

16. Explain how do you determine the ultimate BOD and the rate constant, using BOD test data, by (*a*) Thomas method (*b*) method of least squares.

17. Calculate the population equivalent of a city, given (*i*) the average sewage from the city is 95 × 10^6 l/day and (*ii*) the average 5 day BOD is 300 mg/l. (UPSC, CES, 1987) **Ans.** [356250]

18. The following BOD results were obtained on a sample of untreated wastewater at 20° C.

t (days)	0	1	2	3	4	5
y (mg/l)	0	65	109	138	158	172

Compute the reaction constant *K* and ultimate first-stage BOD using Thomas method.

19. For the data problem 18, determine *K* and BOD_u, using the method of least squares.

20. Change in concentration of organic matter *L*, with time *t* is given by $\frac{dL}{dt} = -KL$. Calculate the organic matter remaining after 3 days if the initial concentration was 200 mg/l and $K = 0.4$ per day. (UPSC, Engg. Services, 1974) **Ans.** [60.3 mg/l]

9

Natural Methods of Wastewater Disposal

9.1. INTRODUCTION

After conveying the wastewater through sewers, the next step is its *disposal*, either after treatment or even before treatment. The methods of disposal of wastewater may be classified under the following categories.

(*i*) Natural methods : (*a*) By dilution, and
(*b*) By land treatment

(*ii*) Artificial methods : (*a*) Primary treatment, and
(*b*) Secondary treatment.

(*iii*) Combined methods.: (*a*) Primary treatment, and
(*b*) Effluent disposal by natural methods

Dilution, a prominent method of natural disposal, consists of discharging the wastewater into receiving water body (such as river, sea, lake etc). This is done on the assumption that sufficient dissolved oxygen is available in the water body so that biochemical oxygen demand is satisfied. If however, the diluting water is not sufficient to supply the biological (or biochemical) oxygen demand to oxidise the entire matter present, there will be nuisance of foul odour and unsightly islands of half-digested floating, putrefying matter at the surface. In addition to this problem, the depletion of oxygen would kill the aquatic life, and if this dilution water is used at the downstream side for drinking water puspose, it will cause danger to public health. *Hence modern practice is to give atleast primary treatment to the waste-water before the effluent is disposed off by natural methods.* In other words, we normally follow the combined method of primary treatment of wastewater and then the disposal of the effluent by natural methods. In some cases, it may be even necessary to have both primary treatment as well as secondary treatment before the effluent is disposed off by natural methods. In this chapter we shall discuses only the *natural methods* of effluent disposal. The methods of treatment of wastewater are discussed is subsequent chapters.

9.2. DISPOSAL BY DILUTION

Disposal by *dilution* is the process whereby the treated wastewater or effluent from treatment plants is discharged either in large static water bodies (such as lake or sea) or in moving water bodies such as rivers or streams. The discharged wastewater or effluent is purified, in due course of time, by the so called *self purification process* of natural waters. The limit of effluent discharge and the degree of treatment of wastewater depend upon the *self purification capacity* of natural waters as well as the intended use of the water body at the downstream side.

Conditions favouring dilution without treatment. Wastewater, can be directly discharged into receiving waters under the following conditions favourable for dilution (*i*) where the wastewater is quite fresh, *i.e.* it is discharged within 2 to 3 hours of its collection (*ii*) where the floating matter and settleable solids have been removed (*iii*) where water body has large volume in comparision to the volume of wastewater (*iv*) where the diluting water has high content of DO, so that not only the BOD is satisified, but suffient DO remains available for the aquatic life (*v*) where it is possible to thoroughly mix or diffuse the wastewaters through the water body (*vi*) where swift forward currents are available, so that there is no deposition of sewage at the site (*vii*) where the wastewater does not contain industrial wastewater having toxic substances, and (*viii*) where the receiving water is not a source of drinking water collection immediately to the downstream side.

Conditions essential for treatment before dilution. (*i*) where the wastewater discharge is detrimental to aquatic life (*ii*) where the wastewater contains industrial wastes containing toxic substance or where industrial effluent is quite warm (*iii*) where the volume of diluting water is insufficient (*iv*) where the receiving waters are used for inland navigation (*v*) where the receiving water is a source for drinking water (*vi*) where wastewater is not fresh but is stale, and (vii) where the effluent is not likely to be dispersed easily due to tides, winds, cross-currents etc.

Standards of dilution. The ratio of quantity of receiving water to that of wastewater or effluent discharge is called the *dilution* *factor*. The Royal Commssion Report on sewage disposal has laid down certain standards indicated in Table 9.1.

TABLE 9.1. STANDARDS OF DILUTION

Dilution Factor	*Standards of purification requird*
1. Above 500	No treatment required. The raw sewag or wastewater can be discharged directly in the receiving water.
2. Between 300 to 500	Primary treatment consisting of plain sedimentation is required so that the effluent does not contain more than 150 ppm of suspended solids
3. Between 150 to 300	Treatment such as sedimentation, screening and chemical precipitation are required, so that the effluent does not contain more than 50 ppm of suspended solids.
4. Less than 150	The wastewater/sewage should be treated thoroughly so that the effluent does not contain more than 30 ppm of suspended solids and its 5 day BOD at 18.3°C does not exceed 20 ppm.

The standards given above have been operative in England since 1912. However, the requirements laid down in IS 4764-1973 should be adhered to while disposing sewage or sewage-effluent in a body of water. Also, where the dilution is low, these standards may have to be made stringent to ensure that the receiving water conforms to the standards IS : 2296-1974.

9.3. TYPES OF RECEIVING WATERS FOR DILUTION

The following is the list of receiving waters (or natural waters) into which wastewater or effluent can be discharged for dilution: (*i*) Perennial rivers and streams (*ii*) Lakes. (*iii*) Ocean or sea (*iv*) Esturies (*v*) Creeks.

Perennial rivers or streams are probably the best type of receiving waters, since the water is in continuous motion. Also in natural streams there is balance between plant and animal life, with considerable interaction among the various life forms. However the discharge flowing during summer and during winter varies. During summer, there may be minimum flow in the stream, with the result that dilution factor may be low, and also high temperature of water may result in low solubility of oxygen, necessitating proper treatment before dilution.

Sometimes, specially when perennial streams are not available, lakes may be used for dilution. Various charateristics of lakes, such as its size, shape, volume of fresh water flowing into it etc., should be critically examined before deciding its self-purifying capacity. Ocean has abundant water and the dilution factor is unlimited. However, sea water has about 20% less DO than river or stream, water is turbid due to dissolved impurities and penetration of sun's rays is less. Due to this, dilution by sea results usually into *anaerobic conditions* leading to formation of sludge banks and emission of foul odour. A *creek* is in the form of an inlet on sea coast, which may not have dry weather flow during some part of the year. Due to this, great care should be taken in disposal of effluent into it. *Estuary* is wide lower tidal part of a river. Hence the dilution in a estuary is affected both by ocean water as well as river water. However, the process of dilution is generally satisfactory in estuaries.

9.4. SELF PURIFICATION OF NATURAL STREAMS

When the wastewater or the effluent is discharged into a natural stream, the organic matter is broken down by bacteria to ammonia, nitrates, sulphates, carbon dioxide etc. In this process of oxidation, the dissolved oxygen content of natural water is utilised. Due to this, deficiency of dissolved oxygen is created. As the excess oganic matter is *stabilized*, the normal cycle will be restabilished in a process known as *self-purification* wherein the oxgen in replenished by its reaeration by wind. Also, the stable byproducts of oxidation mentioned above are utiltized by plants and algae to produce carbohydrates and oxygen. Water quality standards are often based upon maintenance of some minimum dissolved oxygen concentration which will protect the natural cycle in the stream while taking advantage of its natural assimilative capacity. Actions involved in self purifications are physical, chemical and biological in nature, as explained below.

Actions involved in self purification. The following are various factors which affect the process of self purification of stream.

1. Dilution : When wastewater is discharched into the receiving water, dilution takes place due to which the concentration of organic matter is reduced and the potential nuisance of sewage is also reduced. If C_S and C_R are the concentrations of any impurity such as organic content, BOD, suspended solids in the sewage and river having discharge rates Q_S and Q_R respectively, the resulting concentration C of the mixture is given by

$$C_S Q_S + C_R Q_R = C\,(Q_S + Q_R)$$

or

$$C = \frac{C_S Q_S + C_R Q_R}{Q_S + Q_R} \qquad \text{...(9.1)}$$

When the dilution ratio is quite high, large quantities of DO are always available which will reduce the chances of putrefaction and pollutional effects. Aerobic conditions will always exist because of dilution. This will however, not be there if dilution ratio is small, i.e. when large quantities of effluent is discharged into a small stream.

2. Dispersion due to Currents : Self-purification of stream largely depends upon currents which will readily disperse the wastewater in the stream, preventing locally high concentration of pollutants. High velocity impoves reaeration which reduces the concentration of pollutants. High velocity improves rearation which reduces the time of recovery, though length of stream affected by the wastewater is increased.

3. Sedimentation : If the stream velocity is lesser than the scour velocity of particles, sedimendation will take place, which will have two effects : (i) the suspended solids, which contribute largely to oxygen demand will be removed by settling and hence water quality to the downstream will be increased, (*ii*) due to settled solids, anaerobic decomposition may take place.

4. Oxidation : The organic matter, present in the wastewater is oxidised by aerobic bacteria utilising dissolved oxygen of the natural water. This process prevails till complete oxidation of organic matter takes place. The stream which is capable of absorbing more oxygen rapidly through reaeration etc. can purify heavily polluted water in a short time.

5. Reduction : The reduction occurs in the streams due to hydrolysis of the organic matter biologically or chemically. Anaerobic bacteria will split the organic matter into liquids and gases, thus paving way for their ultimate stabilization by oxidation.

6. Temperature : At low temperature, the activities of bacteria is low and hence rate of decomposition will also be slow, though DO will be more because of increased solubility of oxygen in water. At higher temperatures, however, the self-purification takes lesser time, though the quantity of DO will be less.

7. Sunlight : Sunlight helps certain micro-organisms to absorb carbon dioxide and give out oxygen, thus assisting in self-purification. Sunlight acts as a disinfectant and stimulates the growth of algae which produce oxygen during daylight but ultilise oxygen at night. Hence wherever there is algal growth, the water may be supersaturated with DO during day light hours, though anearobic conditions exist in night.

Zones of pollution in the stream. The self purification process of stream polluted by the wastewater or effluent discharged into it can be divided into the following four *zones* : (1) Zone of degradation. (2) Zone of active decomposition. (3) Zone of recovery. (4) Clear water zone.

1. Zone of degradation : This zone is situated just below the outfall sewer when discharging its contents into the stream. In this zone, water is dark and turbid, having the formation of sludge deposits at the bottom. The DO is reduced to 40% of the saturation values. There is an increase in CO_2 content, and reaeration is much slower than deoxygenation. Though conditions are unfavourable for aquatic life, fungi at higher points and bacteria at lower points breed small worms which 'work over' and stabilize the sewage sludge. The decomposition of solid matter takes place in this zone and anarobic decomposition prevails.

2. Zone of active decomposition : This zone is just after the degradation zone and is marked by heavy pollution. Water in this zone becomes greyish and darker than the previous zone. The DO concentration in this zone falls down to zero. Active anaerobic organic decomposition takes place, with the evolution of methane (CH_4), hydrogen sulphide (H_2S), carbon dioxide (CO_2) and nitrogen (N_2), bubbling to the surface with masses of sludge forming black scum. Fish life is absent in this zone but bacteria flora will flourish with the presence of anaerobic bacteria at upper end and aerobic bacteria at the lower end. Protozoa and fungi will first disappear and then reappear, while algae will mostly be absent. However, near the end of this zone, as the decomposition slackens, reaeration sets in and DO again rises to its original level of 40% (Fig. 9.1).

3. Zone of recovery : In this zone, the process of recovery starts, from its degraded condition to its former condition. The stabization of organic matter takes place in this zone. Due to this, most of the stabilised organic matter settles as sludge, BOD falls and DO content rises above the 40% value. Mineralisation is active, with the resulting formation of products like nitrates (NO_4), sulphates (SO_4), carbonates (CO_3). Near the end of the zone, microscopic aquatic life reappears, fungi decrease and algae reappears.

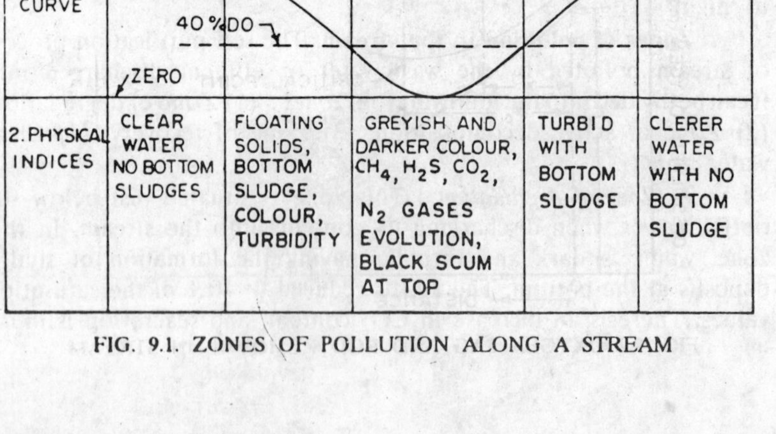

FIG. 9.1. ZONES OF POLLUTION ALONG A STREAM

4. Clearer water zone : In this zone, the natural condition of stream is restored with the result that (*i*) water becomes clearer and attractive in appearance, (*ii*) DO rises to the saturation level, and is much higher than BOD, and (*iii*) oxygen balance is attained. Thus recovery is said to be complete in this zone, though some pathogenic organisms may be present in this zone.

9.5. OXYGEN SAG ANALYSIS

The oxygen sag or *oxygen deficit* in the stream, at any point of time during the self purification process is the difference between the *saturation* DO *content* and the *actual* DO *content* at that time:

$$\text{Oxygen deficit,} \quad D = \text{Saturation DO} - \text{Actual DO} \qquad ...(9.2)$$

The normal saturation DO value for fresh water depends upon the temperature, and its value varies from 14.62 mg/l at 0°C to 7.63 mg/l at 30°C respectively.

At the point where wastewater is discharged into the stream, it is not necessary that the DO content of the stream will be equal to the saturation, DO; there may be *intial oxygen deficit* D_0 (say). At this stage, when the wastewater effluent, with an initial BOD load L_0, is discharged into the stream, the DO content of the stream starts depleting and the *oxygen deficit* (*D*) increases. The variation of oxgyen deficit (*D*) with the distance along the stream (and hence time of flow from the point of pollution) is depicted by the *oxygen sag curve*, as shown in Fig. 9.2 along with BOD removal curve.

The major point of interest in the oxygen sag analysis is the point of minimum DO, or point of maximum deficit. The maximum or *critical deficit*, lebelled as D_C occurs at the infleetion point of the oxygen sag curve.

Deoxygenation and reoxygenation curves. When pollution load is discharged into the stream, the DO content of the stream goes

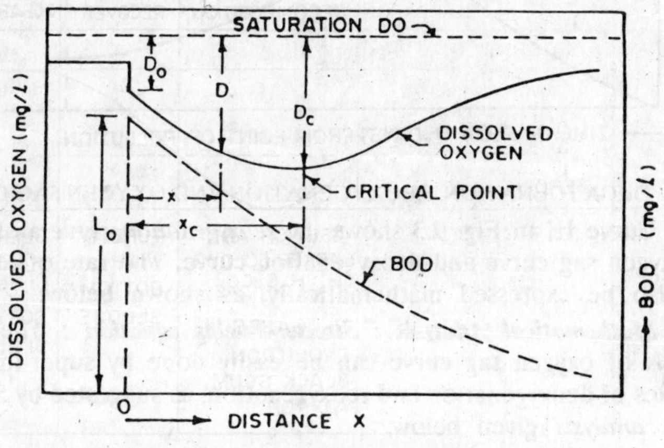

FIG. 9.2 OXYGEN SAG AND BOD REMOVAL IN STREAM

on depleting. This depletion of DO content is known as *deoxygenation*. The rate of deoxygenation depends upon the amount of organic matter remaining (L_t) to be oxidised at any time t, as well as temperature (T) of the reaction. The variation or depletion of DO content of the stream, with time is depicted by the *deoxygneation curve*, shown by curve II (Fig. 9.3), in absence of aeration. The ordinates below the deoxygenation curve indicate the oxygen remaining in the natural stream after satisfying the biochemical demand of oxygen. It should be noted that the *deoxygenation curve* is similar to the *first stage BOD curve* and it can be expressed mathematically as per Eq. 8.9.

Though the DO content of the stream is gradually consumed due to BOD load, atmosphere supplies oxygen continuously to the water through the process of *rearation* or *reoxygenation*. In other words, along with deoxygenation, reoxygenation or reaeration also continuonly takes place. The *rate of reoxygenation* depends upon: (*i*) Depth of water in the stream (more for shallow depth) ; (*ii*) velocity of flow in the stream (less for stagnant water) ; (*iii*) oxygen deficit below satuation DO ; and (*iv*) temperature of water.

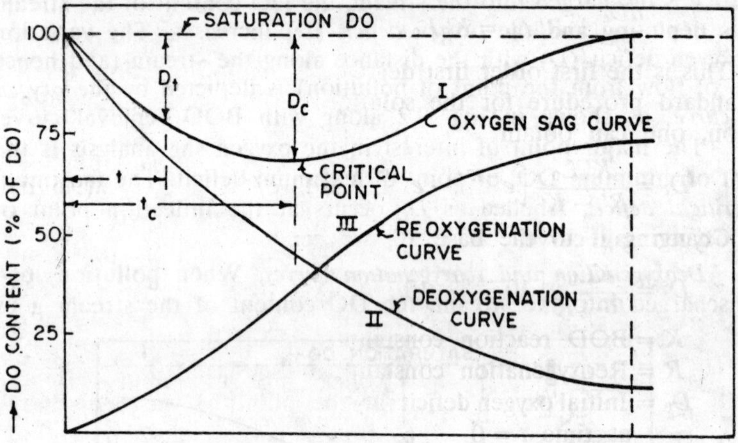

FIG. 9.3. DEOXYGENATION, REOXYGENATION AND OXYGEN SAG CURVES.

Curve III in Fig. 9.3 shows the *reoxygenation curve* along with the oxygen sag curve and deoxygenation curve. The rate of reoxygen can also be expressed mathematically, as shown below.

Mathematical Analysis : Streeter-Phelps equation : The entire analysis of oxygen sag curve can be easily done by super-imposing the rates of deoxygenation and reoxygenation, as suggested by *Streeter-Phelps analysis* given below.

$$\frac{dD_t}{dt} = f\,(\text{deoxygenation and reoxygenation})$$

or
$$\frac{dD_t}{dt} = K' L_t - R' D_t \qquad ...(9.3)$$

where D_t = DO deficit at any time t.

L_t = amount of first stage BOD remaining in the sample at any time t.

K' = BOD reaction rate constant or deoxygenation constant to the base e.

R' = reoxygenation constant to the base e

t = time (in days)

$\frac{dD_t}{dt}$ = rate of change of DO deficit.

But $L_t = L_0 . e^{-K't}$ (From Eq. 8.9 b)

where L_0 = BOD remaining at time $t = 0$.

Hence
$$\frac{dD_t}{dt} = K' L_0 e^{-K't} - R' D_t$$

or
$$\frac{dD_t}{dt} + R' D_t = K' L_0 e^{-K't} \qquad ...(9.4)$$

This is the first order first degree differential equation. Using the standard procedure for the solution of the above differential equation, one can obtain

$$D_t = \frac{K' L_0}{R' - K'}\left[e^{-K't} - e^{-R't}\right] + D_0 e^{-R't} \qquad ...(9.5\ a)$$

Changing it to the base 10, we get

$$D_t = \frac{KL_0}{R - K}\left[10^{-Kt} - 10^{-Rt}\right] + D_0 10^{-Rt} \qquad ...(9.5)$$

where K = BOD reaction constant to the base 10.

R = Reoxygenation constant to the base 10.

D_0 = Initial oxygen deficit at the point of waste discharge at time $t = 0$

Eq. 9.5 is the classic Streeter-Phelps oxygen sag equation which is most commonly used in river analysis. The graphical representation of Eq. 9.5 in shown in Fig. 9.4.

Active biological decomposition starts immediately after the waste discharge, and such decomposition utilises oxygen. This increases the oxygen deficit. However, *atmospheric reaeration* also similtaneously takes place. Since the atmospheric reaeration (or reoxygenation) is proportional to DO deficit, its rate will increase with increasing deficit, till a *critical point* (distant x_c) is reached where the rate of oxygen utilised for waste decomposition is equal to the rate of

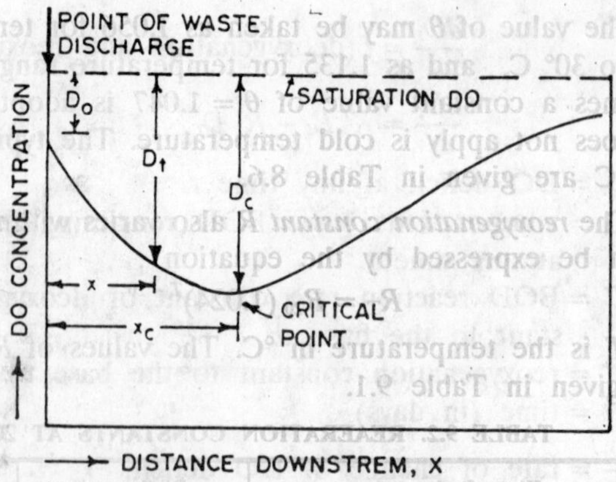

FIG. 9.4. CHARACTERISTIC OXYGEN SAG CURVE OBTAINED USING STREETER-PHELPS EQUATION

atmospheric reaeration. Downstream from this critical point, the rate of reaeration is greater than the rate of utilisation and hence DO begins to increase. Eventually, the stream will show no effects due to waste discharge, due to the above phenomenon of *self-purification*.

Determination of critical DO deficit (D_c) at the point x_c is of engineering significance. The value of D_c can be obtained by putting $\frac{dD_t}{dt} = 0$ in Eq. 9.4. Thus, we obtain.

$$D_c = \frac{K'}{R'} L_0 e^{-K' t_c} \qquad \text{...(9.6 } a)$$

or

$$D_c = \frac{K}{R} L_0 (10)^{-K t_c} \qquad \text{...(9.6)}$$

where t_c = time required to reach the critical point.

The value of t_c can be obtained by differentiating Eq.9.5*a* (or Eq. 9.5) with respect to t and setting $\frac{dD_t}{dt}$ equal to zero. Thus,

$$t_c = \frac{1}{R' - K'} \log_e \frac{R'}{K'} \left[1 - \frac{D_0 (R' - K')}{K' L_0} \right] \qquad \text{...(9.7 } a)$$

or

$$t_c = \frac{1}{R - K} \log_{10} \frac{R}{K} \left[1 - \frac{D_0 (R - K)}{K L_0} \right] \qquad \text{...(9.7)}$$

The distance x_c is given by

$$x_c = t_c \cdot v \qquad \text{...(9.8)}$$

where v = velocity of flow in the stream

The *deoxygenation constant* K, which can be determined either by laboratory or field tests, varies with temperature as indicated by Eq. 8.12 :

$$K_T = K_{20}\, \theta^{(T - 20°)} \qquad \text{...(8.12)}$$

where the value of θ may be taken as 1.056 for temperature range of 20° to 30° C, and as 1.135 for temperature range of 4° to 20°C. Sometimes a constant value of $\theta = 1.047$ is adopted, though this value does not apply is cold temperature. The typical values of K at 20° C are given in Table 8.6.

The *reoxygenation constant* R also varies with the temperature and can be expressed by the equation :

$$R_T = R_{20}\,(1.024)^{T-20} \qquad ...(9.9)$$

where T is the temperature in °C. The values of R' and R at 20° C are given in Table 9.1.

TABLE 9.2. REAERATION CONSTANTS AT 20° C

Water Body	*R' (base e)*	*R (base 10)*
1. Small ponds and backwaters	0.10 – 0.23	0.04 – 0.10
2. Sluggish streams and large lakes	0.23 – 0.35	0.1 – 0.15
3. Large stream of low velocity	0.35 – 0.46	0.15 – 0.20
4. Large stream of normal velocity	0.46 – 0.69	0.20 – 0.30
5. Swift streams	0.69 – 1.15	0.30 – 0.50
6. Rapids and water falls	> 1.15	> 0.50

Note that $R = R'/2.303$.

The ratio of constants R/K (or R'/K') is called the *self-purification constant* f_s :

$$f_s = \frac{R'}{K'} = \frac{R}{K} \qquad ...(9.10)$$

The values of self-purification constant (f_s) are in Table 9.3.

TABLE 9.3 VALUES OF SELF-PURIFICATION CONSTANT

	Type of water body	*Range of f_s*
1	Small ponds and back waters	0.5 to 1.0
2	Sluggish streams, large lakes and impounding reservoirs	1.0 to 1.5
3	Large streams of low velocity	1.5 to 2.0
4	Large streams of normal velocity	2.0 to 3.0
5	Swift streams	3.0 to 5.0
6	Rapids and water falls	5.0 and above

Substituting $R/K = f_s$ in Eq. 9.7, we get

$$t_c = \frac{1}{K(f_s - 1)} \log_{10}\left[f_s\left\{1 - (f_s - 1)\frac{D_0}{L_0}\right\}\right] \qquad ...(9.7\ b)$$

Also, from Eq. 9.6, we get

$$D_c = \frac{L_0}{f_s}(10)^{-Kt_c} \qquad ...(9.6\ b)$$

Taking log of both sides, we get

$$\log_{10} D_c = \log_{10} \frac{L_0}{f_s} - Kt_c \qquad ...(9.11)$$

Substituting the value of t_c from Eq. 9.7(*b*), we get.

$$\log_{10} D_c = \log_{10} \frac{L_0}{f_s} - \frac{1}{f_s-1} \log_{10} \left[f_s \left\{ 1 - (f_s-1) \frac{D_0}{L_0} \right\} \right]$$

$$(f_s-1)\left[\log_{10} \frac{L_0}{f_s} - \log_{10} D_c\right] = \log_{10}\left[f_s\left\{1 - (f_s -) \frac{D_0}{L_0}\right\}\right]$$

or $$\log_{10}\left[\frac{L_0}{f_s D_c}\right]^{f_s-1} = \log_{10}\left[f_s\left\{1 - (f_s-1)\frac{D_0}{L_0}\right\}\right]$$

or $$\left(\frac{L_0}{f_s D_c}\right)^{f_s-1} = f_s\left\{1 - (f_s-1)\frac{D_0}{L_0}\right\} \qquad ...(9.12)$$

This is important equation, in which L_0 is the first stage BOD of mixture of wastewater and stream. The self-purification constant f_s corresponds to the temperature of mixure of wasterwater and stream at the outfall. The solubility of oxygen in water which affects the reoxygenation, is a function of temperature. Table 9.4 gives the values of dissolved oxygen in fresh water at various temperatures.

TABLE 9.4. SOLUBILITY OF OXYGEN IN FRESH WATER

Temp. °C	*Dissolved oxygen (mg/litre)*	*Temp.* °C	*Dissolved oxygen (mg/l)*
0	14.62	16	9.95
1	14.23	17	9.74
2	13.84	18	9.54
3	13.48	19	9.35
4	13.13	20	9.17
5	12.80	21	8.99
6	12.48	22	8.83
7	12.17	23	8.68
8	11.87	24	8.53
9	11.59	25	8.38
10	11.33	26	8.22
11	11.08	27	8.07
12	10.83	28	7.92
13	10.60	29	7.77
14	10.37	30	7.63
15	10.15		

Example 9.1. *The domestic sewage of a town is to be discharged into a stream after treatment. Determine the maximum permissible effluent*

BOD and the percentage purification required in the treatment plant, given the following particulars :

Population of town : 50000.
D.W.F. of sewage : 150 litres per capita per day
BOD contribution per capita : 0.075 kg per day.
Minimum flow of stream : 0.20 m^3/s
BOD of stream : 3 mg/l
Max. BOD of stream on downstream : 5 mg/l.

Solution

D.W.F. of sewage = 150 × 50000 litres per day.

$$= \frac{150 \times 50000}{24 \times 60 \times 60 \times 1000} = 0.0868 \text{ m}^3/\text{s}.$$

Stream discharge = 0.20 m^3/s.

The BOD (y_m) of the mixture at the downstream is

$$y_m = \frac{y_1 Q + y_2 Q_2}{Q_1 + Q_2}.$$

where y_m = 5 mg/l (given), y_1 is the BOD of the effluent having discharge Q_1 and y_2 is the BOD of the stream having discharge Q_2.

$$\therefore \quad 5 = \frac{(y_1 \times 0.0868) + (3 \times 0.2)}{0.0868 + 0.2}$$

From which y_1 = BOD of effluent = 9.608 mg/l.

Now BOD per captia per day = 0.075 × 1000 × 1000
= 75000 mg/day.

Sewage DWF = 150 litres/day.

$$\therefore \text{ Actual BOD of effluent } = \frac{75000}{150} = 500 \text{ mg/l.}$$

$$\text{Hence percentage purification required} = \frac{500 - 9.608}{500} \times 100$$

$$\approx \mathbf{98.1\%}$$

Example 9.2. *A stream, saturated with DO, has a flow of 1.2* m^3/s*, BOD of 4 mg/l and rate constant of 0.3 per day. It receives an effluent discharge of 0.25* m^3/s *having BOD 20 mg/l, DO 5 mg/l and rate constant 0.13 per day. The average velocity of flow of the stream is 0.18 m/s. Calculate the DO deficit at point 20 km and 40 km downstream. Assume that the temperature is 20° C throughout and BOD is measured at 5 days. Take saturation DO at 20°C as 9.17 mg/l.*

Solution

Let y_5 = BOD of the mix.
Q_s = stream flow = 1.2 m^3/s
y_s = BOD of stream = 4 mg/l.
Q_e = Effluent discharge = 0.25 m^3/s

$y_e =$ BOD of effluent $= 20$ mg/l.

Then $$y_5 = \frac{Q_s y_s + Q_e y_e}{Q_s + Q_e} = \frac{1.2 \times 4 + 0.25 \times 20}{1.2 + 0.25}$$
$$= 6.759 \text{ mg/l.}$$

Now $$y_5 = L_0 (1 - 10^{-Kt})$$

or $$6.759 = L_0 (1 - 10^{-0.13 \times 5})$$

From which $L_0 \approx 8.71$ mg/l ...(*i*)

Again, $(DO)_s =$ Saturation DO of stream at 20°C

$(DO)_e =$ DO of effluent $= 5$ mg/l.

$$\therefore \quad (DO)_{mix} = \frac{(DO)_s \times Q_s + (DO)_e \times Q_e}{Q_s + Q_e}$$
$$= \frac{(9.17 \times 1.2) + (5 \times 0.25)}{1.2 + 0.25} = 8.45 \text{ mg/l.}$$

$\therefore$ Initial DO deficit $= D_0 = 9.17 - 8.45 = 0.72$ mg/l. ...(*ii*)

(a) DO deficit at a point 20 km downstream

$$t = \frac{\text{distance}}{\text{velocity}} = \frac{20 \times 1000}{0.18 \times 60 \times 60 \times 24} = 1.286 \text{ days.}$$

Using Streeter-Phelps Equation (Eq. 9.5)

$$D_t = \frac{KL_0}{R - K}\left[10^{-Kt} - 10^{-R.t}\right] + D_0\, 10^{-R.t}$$
$$= \frac{0.13 \times 8.71}{0.3 - 0.13}[10^{-0.13 \times 1.286} - 10^{-0.3 \times 1.286}] + 0.72 \times 10^{-0.3 \times 1.286}$$
$$\approx \mathbf{2.089 \text{ mg/l}}$$

(b) DO deficit at a point 40 km downstream

$$t = \frac{40 \times 1000}{0.18 \times 60 \times 60 \times 24} = 2.572 \text{ days.}$$

$$D_t = \frac{0.13 \times 8.71}{0.3 - 0.13}[10^{-0.13 \times 2.572} - 10^{-0.3 \times 2.572}] + 0.72 \times 10^{-0.3 \times 2.572}$$
$$= \mathbf{2.079 \text{ mg/l}}$$

Example 9.3. *A city discharges sewage at the rate of 1200 litres per second, into a stream whose minimum flow is 5000 litres per second, the temperature of both being 20°C. The 5 day BOD at 20° C for sewage is 160 mg/l and that of river water is 2 mg/l. The DO content of sewage is zero while that of stream is 90% of the saturation DO. Find out the degree of treatment required if the minimum DO to be maintained in the stream is 4 mg/l. Assume deoxygenation coefficient as 0.1 and reoxygenation coefficient as 0.3. Given saturation DO at 20°C as 9.17 mg/l.*

Solution

DO of stream $= 90\%$ of saturation DO

$= 0.9 \times 9.17 = 8.253$ mg/l.

$$\text{DO of the mix} = \frac{8.253 \times 5000 + 0 \times 1200}{5000 + 1200} = 6.656 \text{ mg/l.}$$

$\therefore$ Initial DO deficit $= D_0 = 9.17 - 6.656 = 2.514$ mg/l.

Min. DO to be maintained in the stream $= 4$ mg/l.

$\therefore$ Max. permissible saturation deficit is

$$D_c = 9.17 - 4 = 5.17 \text{ mg/l.}$$

Now, the first stage BOD of mixture of sewage and stream is

$$\left(\frac{L_0}{f_s D_c}\right)^{f_s - 1} = f_s\left\{1 - (f_s - 1)\frac{D_0}{L_0}\right\} \quad ...(9.12)$$

where $f_s = R/K = 0.3/0.1 = 3$

$D_0 = 2.514$ mg/l and $D_c = 5.17$ mg/l.

$$\therefore \quad \left(\frac{L_0}{3 \times 5.17}\right)^{3-1} = 3\left\{1 - (3-1)\frac{2.514}{L_0}\right\}$$

or

$$\frac{L_0^2}{240.56} = 3\left[1 - \frac{5.028}{L_0}\right]$$

Solving this by trial and error, we get $L_0 = 23.86$ mg/l.

Now

$$y_5 = L_0(1 - 10^{-K.t})$$
$$= 23.86(1 - 10^{-0.1 \times 5}) = 16.31 \text{ mg/l.}$$

This is permissible BOD of the mix.

$$\therefore \quad 16.31 = \frac{(2 \times 5000) + y_e \times 1200}{5000 + 1200}$$

From which $y_e = 75.94$ mg/l.

Hence *permissible* BOD of effluent $= y_e = 75.94$ mg/l.

Actual BOD of effluent $= 160$ mg/l.

$$\therefore \text{ Degree of treatment required } = \frac{160 - 75.94}{160} \times 100$$
$$= \mathbf{52.5\%}$$

Example 9.4. *A town discharges 80 cumecs of sewage into a stream having a rate of flow of 1200 cumecs during lean days, at a 5-day BOD of sewage at the given temperature is 250 mg/l. Find the amount of critical DO deficit and its location in the downstream portion of the stream. Assume deoxygenation coefficient K as 0.1 and coefficient of self purification (f_s) as 3.5. Assume saturation DO at given temperature as 9.2 mg/l.*

Solution : $(DO)_{stream} = 9.2$ mg/l ; $(DO)_{effluent} = 0$

$$\therefore \quad (DO)_{mix} = \frac{(9.2 \times 1200) + (0 \times 80)}{1200 + 80} = 8.625 \text{ mg/l.}$$

$\therefore$ Initial DO deficit $= D_0 = 9.2 - 8.625 = 0.575$ mg/l.

Also, 5-day BOD of the mix is given by

$$y_5 = \frac{(0 \times 1200) + (250 \times 80)}{1200 + 80} = 15.625 \text{ mg/l.}$$

But $\quad y_5 = L_0\,[1 - (10)^{-K \times 5}]$

$\therefore \quad 15.626 = L_0\,[1 - (10)^{-0.1 \times 5}]$

From which $\quad L_0 = 22.85$ mg/l.

$$\therefore \quad t_c = \frac{1}{K\,(f_s - 1)} \log_{10}\left[f_s \left\{ 1 - (f_s - 1)\,\frac{D_0}{L_0} \right\} \right] \qquad \text{...(9.7 a)}$$

$$= \frac{1}{0.1\,(3.5 - 1)} \log_{10}\left[3.5 \left\{ 1 - (3.5 - 1) \times \frac{0.575}{22.85} \right\} \right]$$

$= \mathbf{2.063\ days}$

$$\text{Also,} \quad D_c = \frac{L_0}{f_s}\,(10)^{-K.t_c} \qquad \text{...(9.6 a)}$$

$$= \frac{22.85}{3.5}\,(10)^{-0.1 \times 2.063} = \mathbf{4.06\ mg/l}$$

Hence $\quad x_c = v \times t_c = 0.12\,(2.063 \times 24 \times 60 \times 60) \times 10^{-3}$

$= \mathbf{21.39\ km}$

Example 9.5. *The population of a town is 60000 and the domestic sewage is 160 litres/capita/day, having per capita BOD of 70 g/day. The dairy waste of the town is 2.5 × 10⁶ litres/day with a BOD of 1600 mg/l and the waste from other industries is 1.7 × 10⁶ litres/day with BOD of 2000 mg/l. An overall expansion factor of 15% is to be provided. The wastewater is discharged in the natural stream having a minimum discharge of 7500 litres/sec and a saturation dissolved oxygen content of 9.0 mg/l. It is necessary to maintain a dissolved oxygen content of 4 mg/l in the stream. Determine the degree of treatment required to the wastewater prior to its discharge into the stream. Assume suitable values of coefficients of de-oxygenation and reoxygenation.*

Solution. Domestic sewage $= 60000 \times 160 = 9.6 \times 10^6$ litres/day.

Per capita BOD of domestic waste $= 70 \times 1000$ mg/day

Per capita sewage $= 160$ litres/day.

$$\therefore \text{BOD of domestic waste} = \frac{70 \times 1000}{160} = 437.5 \text{ mg/l.}$$

$$\text{Average BOD of effluent} = \frac{437.5 \times 9.6 + 1600 \times 2.5 + 2000 \times 1.7}{9.6 + 2.5 + 1.7}$$

$= 840.6$ mg/l.

Total effluent $= (9.6 + 2.5 + 1.7)\,10^6 = 13.8 \times 10^6$ litres/day.

$= 159.7$ litres/sec.

Using an expansion factor of 15%,

Design effluent $= 1.15 \times 159.7 \approx 184$ litres/sec.

Initial DO of stream $= 9.0$ mg/l.

$$\therefore \quad (DO)_{mix} = \frac{9.0 \times 7500 + 0 \times 184}{7500 + 184} = 8.78 \text{ mg/l.}$$

$\therefore$ Initial DO deficit $= D_0 = 9 - 8.78 = 0.22$ mg/l.

Critical DO deficit, $= D_c = 9 - 4 = 5$ mg/l.

Now, from Eq. 9.12.

$$\left(\frac{L_0}{f_s D_c}\right)^{f_s - 1} = f_s\left\{1 - (f_s - 1)\frac{D_0}{L_0}\right\}$$

where $D_c = 5$ mg/l and $D_0 = 0.22$ mg/l.

Assuming $K = 0.1$, $R = 0.3$ and hence $f_s = 3$, we get

$$\left(\frac{L_0}{3 \times 5}\right)^{3-1} = 3\left\{1 - (3-1)\frac{0.22}{L_0}\right\}$$

or
$$\frac{L_0^2}{675} = 1 - \frac{0.44}{L_0}$$

Solving this by trial and error, we get $L_0 = 25.76$ mg/l.

Hence $y_5 = L_0\,[1 - (10)^{-0.1 \times 5}]$

$= 25.76\,[1 - (10)^{-0.5}] = 17.61$ mg/l.

Hence 5-day BOD of the mixture $= 17.61$ mg/l.

$$\therefore \quad 17.61 = \frac{y_e \times 184 + 0 \times 7500}{184 + 7500}$$

or $y_e = 736.4$ mg/l

$$\therefore \text{ Degree of treatment required} = \frac{840.6 - 736.4}{840.6} \times 100$$

$= \mathbf{12.5\ \%}$

Example 9.6. *If in the previous example, no treatment is provided, determine the stream discharge (and hence the dilution ratio).*

Solution. If no treatment is provided, $y_e = 840.6$ mg/l.

$$\therefore \quad 17.61 = \frac{840.6 \times 184 + 0 \times Q_s}{184 + Q_s}$$

From which $Q_s = 8599$ litres/sec.

The above calculations are based on the assumption that $(DO)_{max}$ does not change appreciably and hence D_0 remains the same. Actually, as Q_s increases, $(DO)_{mix}$ will *increase* and hence D_0 will decrease. This will require repeat of calculation to obtain precise results. For example, for $Q_s = 8599$ litres/sec.

$$(DO)_{mix} = \frac{9.0 \times 8599 + 0 \times 184}{8599 + 184} = 8.812 \text{ mg/l.}$$

$D_0 = 9 - 8.812 = 0.188$ mg/l

$D_c = 5$ mg/l (as before)

$$\therefore \quad \left(\frac{L_0}{3 \times 5}\right)^{3-1} = 3\left\{1 - (3-1)\frac{0.188}{L_0}\right\}$$

Solving this by trial and error, we get $L_0 = 25.79$ mg/l.

$\therefore \quad y_5 = 25.79\,[1 - (10)^{-0.5}] = 17.634$ mg/l.

$$\therefore \quad 17.634 = \frac{840.6 \times 184 + 0 \times Q_s}{184 + Q_s}$$

$$\therefore \quad D_0 = 9 - 8.812 = 0.88 \text{ mg/l}$$

From which, $Q_s = 8587$ litres/sec.

Thus, we find that result does not change appreciably and no repeat calculations are required.

Example 9.7. *Using the following data, find out DO at the end of 1 and 2 days :*

	River	*wastewater*
Flow (m^3/s)	25	2
DO (mg/l)	*9.1*	*0*
5-day BOD (mg/l)	2	200

Take deoxygenation constant as 0.1 per day and reoxygenation constant as 0.3 per day. Take saturation DO as 9.10 mg/l.

Solution. $$(DO)_{mix} = \frac{9.1 \times 25 + 0 \times 2}{25 + 2} = 8.426 \text{ mg/l}:$$

$$(y_{mix})_5 = \frac{(2 \times 25) + (200 \times 2)}{25 + 2} = 16.667 \text{ mg/l}.$$

$\therefore$ Initial DO deficit $= D_0 = 9.10 - 8.426 = 0.674$ mg/l.

Also, $$(y_{mix})_5 = L_0 (1 - 10^{-K.t})$$

or $$16.667 = L_0 (1 - 10^{-0.1 \times 5})$$

From which, $L_0 = 24.375$ mg/l.

(*a*) ***At the end of 1 day***

$$D_1 = \frac{K L_0}{R - K} [10^{-K.t} - 10^{-R.t}] + D_0 (10)^{-R.t}$$

$$= \frac{0.1 \times 24.375}{0.3 - 0.1} [10^{-0.1 \times 1} - 10^{-0.3 \times 1}] + 0.674 (10)^{-0.3 \times 1}$$

$$= 3.91 \text{ mg/l}.$$

$\therefore$ $(DO)_1 = 9.1 - 3.91 =$ **5.19 mg/l**

(*b*) ***At the end of 2 days***

$$D_2 = \frac{0.1 \times 24.375}{0.3 - 0.1} [10^{-0.1 \times 2} - 10^{-0.3 \times 2}] + 0.674 (10)^{-0.3 \times 2}$$

$$= 4.80 \text{ mg/l}.$$

$\therefore$ $(DO)_2 = 9.1 - 4.80 =$ **4.3 mg/l.**

Example 9.8. *A wastewater effluent of 600 l/s with a BOD = 60 mg/l, DO = 2.5 mg/l and temperature of 25°C enters a river where the flow is 30 m^3/sec and BOD = 3 mg/l, DO = 8.5 mg/l and temperature of 16°C. Deoxygenation constant for the waste is 0.10 per day at 20°C. The velocity of water in the river downstream is 0.15 m/s and depth of flow is 1.5 m. Determine the following after mixing of wastewater with the river water : (i) combined discharge. (ii) BOD (iii) DO, and (iv) Temperature*

Solution : Wastewater discharge $= 600\ l/s = 0.6\ m^3/s$.

(*i*) Combined discharge $= Q_R + Q_E = 30 + 0.6 =$ **30.6 m³/s**

(*ii*) $(BOD)_{mix} = \dfrac{(30 \times 3) + (0.6 \times 60)}{30 + 0.6} =$ **4.118 mg/l**

(*iii*) $(DO)_{mix} = \dfrac{(30 \times 8.5) + (0.6 \times 2.5)}{30 + 0.6} =$ **8.382 mg/l**

(*iv*) $(Temp.)_{mix} = \dfrac{(30 \times 16) + (0.6 \times 2.5)}{30 + 0.6} =$ **16.18°C**

Example 9.9. *100 cumece of sewage of a city is discharged in a perennial river which is fully saturated with oxygen and flows at a minimum rate of 1250 cumecs with a minimum velocity of 0.15 m/sec. If the 5-day BOD of the sewage is 260 mg/l, find out where the critical DO will occur in the river. Assume :*

(*i*) *The coefficient of purification of river as 4.0*

(*ii*) *Coefficient of DO as 0.11*

and (*iii*) *The ultimate BOD as 125% of the 5 day BOD of the mixture of sewage and river water.*

Solution. Let us assume a temperature of 20° C, for which saturation DO is equal to 9.17 mg/l. Also, assume that DO of effluent is zero.

$$\therefore \quad (DO)_{mix} = \frac{(1250 \times 9.17) + (100 \times 0)}{1250 + 100} = 8.49\ \text{mg/l}.$$

Hence initial DO deficit $= D_0 = 9.17 - 8.49 = 0.68$ mg/l.

Also, 5-day BOD of sewage = 260 mg/l.

Let 5-day BOD of river = 0

$$\therefore \quad (y_5)_{mix} = \frac{(1250 \times 0) + (100 \times 260)}{1250 + 100} = 19.26\ \text{mg/l}.$$

It is given that ultimate BOD is equal to 125% of the 5-day BOD of the mixture of sewage and river water.

$$\therefore \quad L_0 = 1.25\,(y_5)_{mix} = 1.25 \times 19.26 = 24.07\ \text{mg/l}.$$

Thus, both D_0 and L_0 are known. Now, from Eq. 9.7 (*b*),

$$t_c = \frac{1}{K(f_s - 1)} \log_{10}\left[f_s\left\{1 - (f_s - 1)\frac{D_0}{L_0}\right\}\right]$$

Here, $K = 0.11$ (given) and $f_s = 4.0$ (given)

$$\therefore \quad t_c = \frac{1}{0.11\,(4 - 1)} \log_{10}\left[4\left\{1 - (4 - 1)\frac{0.68}{24.07}\right\}\right] = 1.7078$$

$$\therefore \quad x_c = \text{velocity} \times \text{time} = 0.15\,(1.7078 \times 24 \times 3600) \times 10^{-3}$$

$= \mathbf{22.13\ km}$

Hence critical deficit will occur at 22.13 km downstream of the sewage disposal point.

Example 9.10. *A river having a flow of 5 m^3/s, 5-day BOD 1 mg/l and saturated with oxygen receives at a point A, a sewage effluent discharge of 2 m^3/s with 5-day BOD 20 mg/l and DO 2 mg/l.*

A tributary meets the stream at a point B, 20 km on the downstream, having a flow of 2 m^3/s, 5-day BOD 2 mg/l and DO 9 mg/l.

As the river further advances by a distance of 25 km, at point C on downstream, another effluent having flow of 2 m^3/s, 5-day BOD 16 mg/l and DO 3 mg/l joins the main stream.

Compute the DO deficiency at point P 20 km downstream of C. Assume saturation DO as 9.17 mg/l for a constant temperature of 20°C, K = 0.12 per day and R = 0.36 per day. Assume uniform velocity of flow equal to 0.3 m/s.

Solution

(a) At junction point A

$$(y_A)_5 = \frac{(5 \times 1) + (2 \times 20)}{5 + 2} \approx 6.43 \text{ mg/l}$$

But $$(y_A)_5 = L_0 (1 - 10^{-K.t})$$

or $$6.43 = L_0 (1 - 10^{-0.12 \times 5})$$

From which, $$L_0 = 8.59 \text{ mg/l}$$

$$(DO)_A = \frac{(5 \times 9.17) + (2 \times 2)}{5 + 2} = 7.12 \text{ mg/l.}$$

$$\therefore \quad (D_0)_A = 9.17 - 7.12 = 2.05 \text{ mg/l.}$$

(b) At a point just upstream of B

$$v = 0.3 \text{ m/s}; \; x_B = 20 \text{ km} = 20 \times 10^3 \text{ m.}$$

$$\therefore \quad t_B = \frac{20 \times 10^3}{0.3\,(24 \times 3600)} = 0.772 \text{ days.}$$

DO deficit prior to point B is given by Streater-Phelps Equation:

$$(D)_{0.772} = \frac{KL_0}{R - K}\left[10^{-K.t} - 10^{-R.t}\right] + D_0\, 10^{-R.t}$$

$$= \frac{0.12 \times 8.59}{0.36 - 0.12}\left[10^{-0.12 \times 0.772} - 10^{-0.36 \times 0.772}\right] + 2.05\,(10)^{-0.36 \times 0.772}$$

$$= 2.29 \text{ mg/l.}$$

$\therefore$ DO of river $= 9.17 - 2.29 = 6.88$ mg/l.

(c) Just after junction point B,

$$(DO)_B = \frac{6.88\,(5 + 2) + (9 \times 2)}{(5 + 2) + 2} = 7.35 \text{ mg/l.}$$

$$\therefore \quad (D_0)_B = 9.17 - 7.35 = 1.82 \text{ mg/l.}$$

Also, $$(y_5)_B = \frac{6.43\,(5 + 2) + (2 \times 2)}{(5 + 2) + 2} = 5.45 \text{ mg/l.}$$

But $(y_5)_B = L_0\,(1 - 10^{-K.t})$

or $5.45 = L_0\,(1 - 10^{-0.12 \times 5})$

$\therefore$ $(L_0)_B = 7.28$ mg/l.

(d) At a point just before junction point C

Distance $x_{BC} = 25$ km ; $v = 0.3$ m/s

$$t_{BC} = \frac{25 \times 10^3}{0.3\,(24 \times 3600)} = 0.965 \text{ days.}$$

$$\therefore \quad (D)_{0.965} = \frac{0.12 \times 7.28}{0.36 - 0.12}\left[10^{-0.12 \times 0.965} - 10^{-0.36 \times 0.965}\right] + 1.82(10)^{-0.36 \times 0.965}$$

$$= 1.97 \text{ mg/l.}$$

$\therefore$ DO of river = 9.17 – 1.97 = 7.2 mg/*l*.

(e) Just after junction point C

$$(\text{DO})_C = \frac{7.2\,(5 + 2 + 2) + 3 \times 2}{(5 + 2 + 2) + 2} = 6.44 \text{ mg/l.}$$

$\therefore$ $(D_0)_C = 9.17 - 6.44 = 2.73$ mg/l.

Also, $$(y_5)_C = \frac{5.45\,(5 + 2 + 2) + 16 \times 2}{(5 + 2 + 2) + 2} = 7.37 \text{ mg/l.}$$

But $(y_5)_C = L_0\,(1 - 10^{-K.t})$

$\therefore$ $7.37 = L_0\,(1 - 10^{-0.12 \times 5})$

$\therefore$ $(L_0)_C = 9.84$ mg/l.

Thus, D_0 and L_0, just after the junction point *C* are known.

(f) At point P, 20 km from C

Distance $x_{CP} = 20$ km ; $v = 0.3$ m/s

$$\therefore \quad t_{CP} = \frac{20 \times 10^3}{0.3\,(24 \times 3600)} = 0.772 \text{ days.}$$

DO deficit at point *P* is given by Streeter Phelps equation:

$$(D)_{0.772} = \frac{0.12 \times 9.84}{0.36 - 0.12}\left[10^{-0.12 \times 0.772} - 10^{-0.36 \times 0.772}\right] + 2.73\,(10)^{-0.36 \times 0.772}$$

$$= \mathbf{2.82 \text{ mg/l}}$$

Example 9.11. *A wastewater tratment plant disposes off its effluents into a stream at a point A. Characteristics of the stream at a location fairly upstream of A and of the effluent are as below:*

Item	*units*	*effluent*	*stream*
Flow	*m^3/s*	*0.16*	*0.4*
Dissolved oxygen	*mg/l*	*1.60*	*8.20*
Temperature	*°C*	*25*	*22*
BOD_5 at 20°C	*mg/l*	*32*	*2.0*

Assume that the deoxygenation constant K', at 20°C (base e) = 0.20 d^{-1} and re-aeration constant R' at 20°C (base e) = 0.40 d^{-1} for *the mixture. Equilibrium concentration of dissolved oxygen C_s for the fresh water is as follows.*

Temp.°C	*18*	*20*	*22*	*23*	*24*	*25*	*26*
C_c(mg/l)	*9.54*	*9.17*	*8.99*	*8.83*	*8.53*	*8.38*	*8.22*

The velocity of the stream downstream of the point A is 0.16 m/s. Determine the critical oxygen deficit and its location. Use temperature coefficients of 1.04 for K' and 1.02 for R'.

Solution

1. ***DO of the mixture***

$$(\mathrm{DO})_{mix} = \frac{(0.4 \times 8.20) + (0.16 \times 1.60)}{0.4 + 0.16} = 6.31 \text{ mg/l.}$$

2. ***Temperature of the mixture***

$$(T)_{mix} = \frac{(0.4 \times 22) + (0.16 \times 25)}{0.4 + 0.16} = 22.86°\text{C.}$$

3. ***BOD₅ of mixture at 20°C***

$$(BOD_5)_{mix} = \frac{(0.4 \times 2.0) + (0.16 \times 32)}{0.4 + 0.16} = 10.57 \text{ mg/l.}$$

4. ***Values of K' and R' at T = 22.86°C***

$$K_T' = K_{20}'(\theta)^{T-20°} = 0.2\,(1.04)^{22.86-20} = 0.224\,d^{-1}$$

$$R_T' = 0.4\,(1.02)^{22.86-20} = 0.423\,d^{-1}$$

5. ***Ultimate BOD of mixture***

$$(y_5)_{mix} = L_0\,(1 - e^{-K't})$$

Since L_0 does not depend on the temperature of the reaction, we can find it on the basis of BOD of mix at 20° C.

$$\therefore \quad 10.57 = L_0\,(1 - e^{-0.2 \times 5})$$

From which $L_0 = 16.72$ mg/l. ...(*i*)

6. ***Initial DO deficit of mixture***

$$(DO)_{mix} = 6.31 \text{ mg/l.}$$

Saturation DO at mix. temp. of 22.86°C

= 8.85 mg/l (found by interpolation)

∴ DO deficit of mix $= D_0 = 8.85 - 6.31 = 2.54$ mg/l ...(*ii*)

7. ***Critical time t_c and location x_c***

$$t_c = \frac{1}{0.423 - 0.224} \log_e \frac{0.423}{0.224}\left[1 - \frac{2.54\,(0.423 - 0.224)}{0.224 \times 16.72}\right]$$

= **2.466 days.**

$$\therefore \quad x_c = v \,.\, t_c = 0.16\,(2.466 \times 24 \times 3600)\,10^{-3} = \mathbf{34.09 \text{ km}}$$

8. ***Critical deficit D_c***

$$D_c = \frac{K'}{R'} L_0\, e^{-K' t_c} \qquad \text{...(9.6 } a\text{)}$$

$$= \frac{0.224}{0.423} \times 16.72\,(e)^{-0.224 \times 2.466}$$

$$= \mathbf{5.1\ mg/l}$$

Hence critical DO deficit, equal to 5.1 mg/l, occurs at 34.09 km downstream of point *A*, after 2.466 days.

Example 9.12. *A city discharges 1.25 m^3/s of wastewater into a stream whose minimum rate of flow is 8 m^3/s. The velocity of stream is 0.8 m/s. The temperature of the wastewater is 20° C and that of the stream is 15°C. The 20°C BOD_5 of the wastewater is 250 mg/l, and that of the stream is 2.0 mg/l. The wastewater contains no dissolved oxygen but the stream is 90 percent saturated upstream of the discharge. At 20°C, K' is estimated to be 0.3 per day and R' is 0.9 per day. Determine the critical-oxygen deficit and its location. Also, estimate the 20°C BOD_5 of a sample taken at the critical point. Use temperature coefficients of 1.135 for K' and 1.024 for R'. Also, plot the dissolved-oxygen-sag curve. Given : saturation concentration at 15°C = 10.15 mg/l.*

Solution

1. ***DO of the stream.*** Saturation concentration = 10.15 mg/l

∴ DO of stream = 0.9 × 10.15 = 9.14 mg/l.

2. ***Characteristics of the mixture***

(*i*) $$(T)_{mix} = \frac{(1.25 \times 20) + (8 \times 15)}{1.25 + 8} = 15.68°C$$

(*ii*) $$(BOD_5)_{mix} = \frac{(1.25 \times 250) + (8 \times 2)}{1.25 + 8} = 35.51 \text{ mg/l.}$$

(*iii*) $$(L_0)_{mix} = \frac{35.51}{1 - e^{-0.3 \times 5}} = 45.71 \text{ mg/l.}$$

(*iv*) $$(DO)_{mix} = \frac{(1.25 \times 0) + (8 \times 9.14)}{12.5 + 8} = 7.90 \text{ mg/l.}$$

3. ***Initial dissolved oxygen deficit at T = 15.68°C***

From Table 9.4, DO saturation concentration at 15.68°*C* ≈ 10.01 mg/l.

∴ Initial deficit D_0 = (10.01 − 7.90) = 2.11 mg/l

4. ***Rate constants at 15.68°C***

$$K' = 0.3\,(1.135)^{15.68-20} = 0.174/d$$

$$R' = 0.9\,(1.024)^{15.68-20} = 0.812/d$$

5. ***Determination of t_c and x_c***

$$t_c = \frac{1}{R' - K'} \log_e \left[\frac{R'}{K'} \left\{ 1 - \frac{D_0\,(R' - K')}{K'\,L_0} \right\} \right]$$

$$= \frac{1}{0.812 - 0.174} \log_e \left[\frac{0.812}{0.174} \left\{ 1 - \frac{2.11\,(0.812 - 0.174)}{0.174 \times 45.71} \right\} \right]$$

$= \mathbf{2.124\ days}$

$\therefore \quad x_c = v \cdot t_c = 0.8\,(2.124 \times 24 \times 3600)\,10^{-3} = \mathbf{146.8\ Km}$

6. *Determination of D_c and DO at x_c*

$$D_c = \frac{K'}{R'} L_0\, e^{-K' t_c}$$

$$= \frac{0.174}{0.812} \times 45.71\,(e)^{-0.174 \times 2.124}$$

$$= \mathbf{6.77\ mg/l}$$

DO at $x_c = 10.01 - 6.77 = 3.24$ mg/l.

7. *Determination of BOD_5 of a sample taken at x_c*

$$L_t = L_0\, e^{-K' t_c} = 45.71\,(e)^{-0.174 \times 2.124}$$

$$= 31.59 \text{ mg/l.}$$

$$\therefore \quad 20°C\ BOD_5 = 31.59\,[1 - e^{-0.3(5)}] = \mathbf{24.54\ mg/l}$$

8. *Computation of additional points on the oxygen sag curve*

$$D_t = \frac{K' L_0}{R' - K'}\left[e^{K' t} - e^{-R' t}\right] + D_0\,(e)^{-R' t}$$

$$= \frac{0.174 \times 45.71}{0.812 - 0.174}\left[e^{-0.174 t} - e^{-0.812 t}\right] + 2.11\,(e)^{-0.812 t}$$

$$= 12.466\left[e^{-0.174 t} - e^{-0.812 t}\right] + 2.11\,(e)^{-0.812 t}$$

$$(DO)_t = \text{sat. concentration of DO} - D_t$$

$$= 10.01 - D_t \qquad \text{...(b)}$$

Distance x (km)	0	50	100	146.8	200	250
Time t (day)	0	0.723	1.447	2.124	2.894	3.617
DO (mg/l)	7.9 (step 2)	4.71	3.52	3.24 (step 6)	3.46	3.92

The oxygen-sag curve is shown in Fig. 9.5.

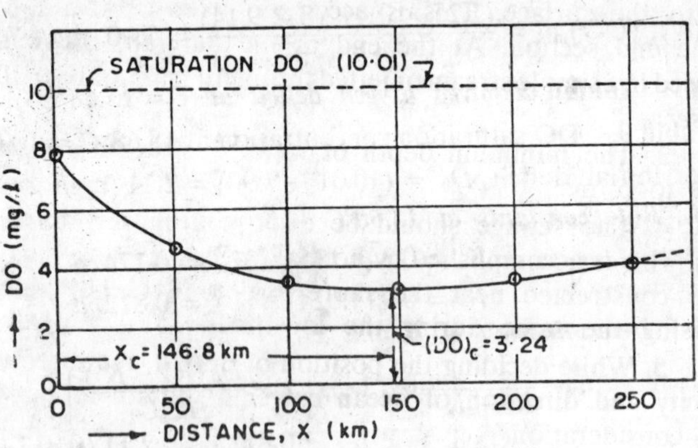

FIG. 9.5. OXYGEN-SAG CURVE.

9.6. DILUTION INTO SEA

The saturation concentration of dissolved oxygen in water decreases with increasing salt content. Due to this reason, the saturation concentration in sea water is approximately 80% of that in water. In addition to this deficiency, the temperature of sea water is lower than the sewage temperature, whereas the specific gravity is higher. Due to these reasons, when sewage is discharged into the sea water, the lighter and warmer sewage will rise up to the surface, resulting in the spreading of the sewage at the top surface of sea in a thin film or *'sleek'*. Moreover, sea water contains a large amount of dissolved matter which chemically react with the sewage solids, resulting in the precipitation of some of the sewage solids, giving a *milky appearance* to the sea water and resulting in formation of *sludge banks*. These sludge banks and thin milky layer formed at the top of sea water produce offensive hydrogen sulphide gas by reacting with the sulphate rich water of the sea. The various chemical actions and the prevailing dissolved matter in the sea water reduce its capacity to absorb more quantity of sewage. However, since the sea contains large volume of water, most of these deficiencies can be overcome if the sewage is discharged deep into the sea, much away from the coast line, with extreme care. The following points should be kept in mind.

1. The sewage should be discharged deep into the sea, preferably 1 to 1.5 km away from the shore. This is accomplished by submarine outfalls that consist of a long section of pipe to transport the wastewater some distance from the shore. Such outfalls are placed on a firm rocky foundation. The cast iron pipe used for this purpose is encased in thick stone masonry or special type of concrete.

2. The out fall should be so designed that proper dilution of waste with seawater is a accomplished before the waste tries to rise to the surface. This is accomplished by providing a diffuser at the end section. At the end of the out fall, the wastewater is released in a simple stream or jetted through a mainfold or *multiple-point diffuser*.

3. The minimum depth of water at the outfall point should be 3 to 5 m.

4. The sewage should be disposed off only during the low tides. This is accomplished by holding the wastewater into large sized tanks constructed near the shore during the high tides and then releasing the same during the low tides.

5. While deciding the position of outfall, the direction of wind velocity and direction of ocean currents should be carefully taken into consideration.

Table 9.5 gives the comparison of sewage dilution by sea and river or stream.

TABLE 9.5
COMPARISON BETWEEN SEA WATER AND STREAM WATER DILUTION.

	Item	*Sea water*	*Stream water*
1.	Specific gravity	High	Low
2.	Quantity of solids in suspension	Large	Small
3.	Penetration of sun's rays	Less penetration due to turbid water	More penetration due to clear water
4.	Dissolved oxygen	About 20% less than stream water	More DO
5.	Maximum sewage load	No limit	Depends on stream discharge
6.	Condition of sewage after discharge	Usually anaerobic, formation of sludge banks and giving foul odour	Mostly remain aerobic
7.	Suitability	Suitable for towns situated on sea shores	Suitable for towns situated on banks of streams

9.7. DISPOSAL BY LAND TREATMENT

When the wastewater, either raw or partly treated, is applied or spread on the surface of land, the method is called *disposal by land treatment*. Some part of the wastewater evaporates while other part percolates in the ground leaving behind suspended solids which are partly acted upon by the bacteria and partly oxidised by exposure to atmospheric actions of air, heat and light. The sewage adds to the fertilising value of the land, and crops can be profitably raised on such land. Due to this, the disposal by land treatment is also sometime called as *sewage farming.* The three principal processes of land treatment of wastewater are : (1) Broad irrigation or sewage farming. (2) Rapid infiltration. and (3) Overland runoff

These are shown diagrammatically in Fig. 9.6.

The first two processes depend upon moving or percolating the water downward through the soil and thus are limited by infiltration and percolating capacity of the land. While the percolating capacity is a function of soil characteristics, the infiltration depends upon the degree of clogging at the soil surface. If the waste is sufficiently pretreated, clogging will be minimised and percolation will limit the rate at which liquid can be applied. For percolation rate of 6 to 25 mm/min. rapid infiltration is practicable ; for 2 to 6 mm/min., irrigation is suitable and below 2 mm/min. overland runnoff should be adopted.

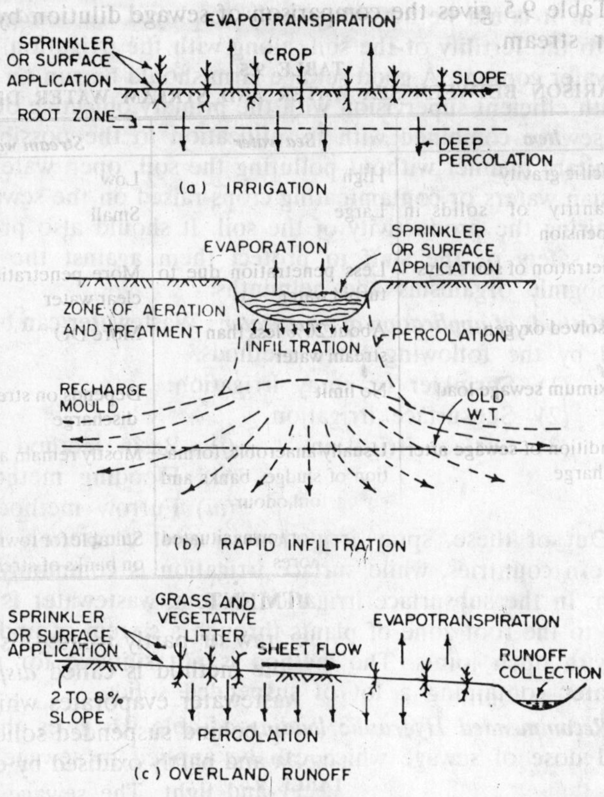

FIG. 9.6. THREE PRINCIPAL PROCESSES OF LAND TREATMENT.

Rapid infiltration may be used for waste disposal, ground water recharge or both. For this process, wastewater is discharged into large basin underlined by sand and soils of high permeability. Bottom of the basin may be covered with grass like bermuda which can persist in wet or dry condition. The grass assists in nitrogen removal and helps maintain the infiltration capacity of the surface.

The technique of *overland runoff* is applied when soils have poor permeability. It is not a true disposal system since wastewater must be collected after passage over the soil, as shown in Fig. 9.6(*c*). Plant or tree cover is essential to minimise and assist in nutrient removal. Hay grasses are usually employed. The process by which the purification of sewage takes place in first two methods described above, is the same, *i.e.* *mechanical straining* or *filtration* of organic matter present in the sewage. The complex compounds in the sewage are thus converted into harmless mineral salts which serve as valuable fertilizer. The nutrients in sewage like nitrogen, phosphorous and potassium along with the micro-nutrients as well as organic matter present in it could be advantageously employed for *sewage farming* to add to the fertility of the soil, along with the irrigation potential of the water content. A good sewage farm should be run on scientific lines with efficient supervision with the primary objective of disposal of the sewage combined with its utilization to the possible extent in a sanitary manner without polluting the soil, open water courses or artesian waters or contaminating crops raised on the sewage farm; or impairing the productivity of the soil. It should also provide for hygienic safety of the staff to protect them against the infection by pathogmic organisms and helminths.

Methods of application of wastewater : Wastewater can be applied to land by the following three methods.

(1) Sprinkler or spray irrigation.
(2) Subsurface irrigation
(3) Surface irrigation
  (*i*) Basin method
  (*ii*) Flooding method
  (*iii*) Furrow method.

Out of these, spray irrigation is most commonly employed in western countries, while surface irrigation is commonly adopted in India. In the subsurface irrigation system, wastewater is supplied directly to the root zone of plants through a system of underground pipes with open joints. The method is not suitable for untreated wastewater containing a lot of suspended solids.

Recommended Hydraulic loading : Table 9.6 gives the recommended dose of sewage which can be applied in sewage farming.

TABLE 9.6.

RECOMMENDED DOSES OF SEWAGE FOR SEWAGE FARMING.

Type of soil	*Hydraulic loading of setted sewage (m^3/hectare/day)*
1. Sandy soil	220–250
2. Sandy loam	150–200
3. Loam	100–150
4. Clayey loam	50–100
5. Clayey soil	30–50

In case of farms using settled sewage, the sludge from the sedimentaion tanks should be mixed with manure or peat and composted for which adequate land should be provided. The duration of composting is so chosen as to ensure complete dehelminthisation of the sludge. The composted material is applied on the fallow fields or in the farm itself.

Water quality considerations : The quality of water for irrigation is determined by the effects of its constituents both on the crop and the soil. The deleterious effects of the constituents of the irrigant on plant grown can result from (*i*) direct osmotic effects of salts in preventing water uptake by plant (*ii*) direct chemical effects upon the metabolic reactions in the plants (toxic effect), and (*iii*) any indirect effect through changes in soil structures, permeability and aeration. The suitability of an irrigant is judged on the basis of the soil properties, quality of irrigtion water and the salt tolerance behaviour of the crop grown in a particular climate.

Degree of pretreatment : Though sewage after primary treatment can be applied to the farms, the temptation of providing only primary treatment and eliminating secondary treatment merely on cost considerations should be resisted. Under no conditions, application of raw sewage on sewage farms should be permitted.

IS : 3307–1965 lays down standards of sewage effluents for sewage farming, as given in Table 9.7.

TABLE 9.7
INDIAN STANDARDS OF SEWAGE EFFLUENTS FOR SEWAGE FARMING

S.N.	*Type of load present*	*Prescribed limit*
1.	Total concentration of solids (inorganic)	≤ 2100 ppm
2.	Percentage of sodium with respect to total content of sodium, calcium, magnesium and potassium	≤ 60%
3.	Boron	≤ 2 ppm
4.	pH value of sewage	5.5–9.0
5.	BOD	≤ 500

Conditions favourable for land treatment. Disposal of sewage by land treatments is favoured under the following conditions. (1) When natural rivers or streams are not located in the vicinity, land treatment is the only alternative. (2) When rivers run dry or have a very small flow during summer, discharging sewage into them is out of question. (3) When plentiful land with sandy, loamy or alluvial soil overlying soft murram, sand or gravel is available, land treatment is favoured. Such soils are easily aerated and it is easy to maintain aerobic conditions in them. (4) When climate is arid, land treatment is favoured. (5) Land treatment is favoured when subsoil water table is low even during the wet season. (6) Land treatment is favoured when rainfall is low and there is an acute demand for irrigation water. (7) When large open areas in the surround locality are available, broad irrigation by sewage can be easily practiced. (8) Cash crops can be easily grown on sewage farms.

Broad Irrigation. *Broad irrigation* is similar to *sewage farming* though the purposes of the two are different. In sewage farming, the prime consideration is the successful growing of the crops while in broad irrigation (also known as effluent irrigation), the prime consideration is the successful disposal of sewage. However, in both the processes, the ultimate result is that crop is raised and at the same time sewage is disposed by land application. In *sewage farming*, raw sewage is not used while in broad irriation, raw or settled sewage is applied on vacant land having a provision of properly laid under drains to collect the small quantity of percolating effluent. In broad irrigation, applied on relatively more pervious soil having higher percentage of voids, the sewage is stabilised by aerobic action.

Sewage sickness. After continuous application of sewage on land, the pores of the soil get clogged, preventing oxidation and causing noxious smells. The land is unable to take any further load of sewage. This phemenon of soil is known as *sewage sickness* of land. Sewage sicknees can be *prevented* by adopting the following measures.

1. Pretreatment of sewage : By giving primary treatment to the sewage, the suspended solids are removed. Due to this measure, the pores of the soil will not get clogged quickly. Also, BOD load will be reduced by about 30%.

2. Provision of extra land : Extra land, as reserve or stand by should be available so that the land with sewage sickneess can be given the desired rest. During the rest period, the sick land should be properly ploughed so that it is broken up and aerated.

3. Under Drainage of soil : Subsoil drains should be provided to collect the percolating effluent. This will minimise the possibility of sewage sickness.

4. Proper choice of land : The land chosen for this purpose should be sandy or loamy, having higher permeability. Clayey soil should be avoided.

5. Rotation of crops: Rotation of crops minimise the chances of sewage sickness.

6. Shallow depth application : Sewage should be applied in shallow depths. If sewage is applied in greater depths, chances of sewage sickness are increases.

Example 9.13. *A town, having population of 60000 and water supply at the rate of 180 l/day, disposes its sewage through land application. The land used for such a purpose can absorb 75 cubic metre of sewage per hectare per day. If 80 percent of water supply is available as sewage, determine the land area required.*

Solution. Quantity of sewage $= 0.8 \times 60000 \times 180$

$= 8640 \times 10^3$ litres $= 8640\ m^3$ per day

$\therefore$ Area of land required $= \frac{8640}{75} = 115.2$ hectares.

Providing 50% extra for rest and rotation,

Actual area $= 1.5 \times 115.2 =$ **172.8 hectares.**

Example 9.14. *A town, having a population of 50000 and the rate of water supply as 160 l/day, disposes off its sewage successfully by land treatment. The area of land available is 180 hectares. If 80% of the water supplied is converted into sewage, find out the consuming capacity of soil.*

Solution. Quantity of sewage$= 0.8 \times 50000 \times 160$

$= 6400 \times 10^3$ litres $= 6400\ m^3/day$

Total area of land $= 180$ hectares

Providing 50% exta area as reserve for rest and rotation,

Actual available area for land application $= \frac{180}{1.5} = 120$ hectares.

$\therefore$ Consuming capacity of soil $= \frac{6400}{120}$

$=$ **53.33** cubic metres/hectare/day.

9.8. COMPARISION OF DISPOSAL METHODS

1. For disposal by dilution, large volumes of natural waters, either in the form of stream or lake or sea, are required, while for disposal by land application, large area of pervious land is required.

2. In urban areas, the cost of land is quite high and hence disposal by land application may be costly. If such areas are situated in close vicinity of natural waters, disposal by dilution may be preferred.

3. In rural areas, the cost of land is relatively less. At the same time, the rate of water supply is generally less, resulting in concentrated sewage. For such cases, therefore, disposal through land application may be preferred.

4. Disposal by land application, in general, requires efficient land management, Dilution method, however, does not require such efficient management.

5. In disposal by land application, sewage can be applied either raw or after primary treatment (in case of sewage farming), while in disposal by dilution, efficient pretreatment, to meet the effluent quality requirements, is must.

6. In disposal by land application, natural gradients may not be available, necessitating high head pumping, while in dilution methods, such pumping may not be required because rivers/streams flow through lowest contours of the area.

7. In hot climates, DO content of natural waters is low and the quantity of water flow is also less, resulting in pollution hazards

if high degree of pretreatment is not applied to the effluents before discharging them. In such cases, disposal by land application may be preferred.

8. If the receiving waters are used as source for water supply at the downstream side, disposal by dilution can not be adopted unless high degree of pretreatment is applied to the effluents.

7. In disposal by dilution, the recreational use of natural waters will vanish completely.

8. Disposal by land application specially through sewage farming will increase the yield of crops etc., but it should be practiced with caution since it may cause serious health hazards. Disposal by land application may also pollute ground water.

PROBLEMS

1. Mention various methods of wastewater disposal. Discuss their merits and demerits. Explain the conditions favourable for their adoption.
2. Discuss fully the process of self purification of natural water.
3. What is meant by disposal of wastewater by dilution ? Mention the conditions favourable for it.
4. What is land treatment ? Discuss the conditions under which it is suitable.
5. Write short notes on the following :
   (*i*) Dilution factor.
   (*ii*) Oxygen sag curve.
   (*iii*) Sewage farming.
   (*iv*) Zones of pollution in a stream.
   (*v*) Sewage sickness.
6. What do you understand by self-purification property of a stream ? Explain the factors affecting this property.
7. Explain the difference between dilution process if the wastewater effluents are disposed of in stream water and sea water.
8. Write a note on disposal of sewage in sea water.
9. What do you understand by oxygen-sag curve. Derive the classic Streeter-Phelps oxygen sag equation.
10. A waster water effluent of 560 l/s with a BOD = 50 mg/l, DO = 3.0 mg/l and temperature of 23°C enters a river where the flow is 28 m^3/sec, and BOD = 4.0 mg/l, DO = 8.2 mg/l and temperature of 17°C. K_1 for waste is 0.10 per day at 20°C. The velocity of water in the river downstream is 0.18 m/s and depth of 1.2 m. Determine the following after mixing of waster water with the river water :

(*i*) Combined discharge
(*ii*) BOD
(*iii*) DO, and
(*iv*) Temperature. (UPSC, Civil Services, 1981)

Ans [(*i*) 28.56 m^3/sec ; (*ii*) 4.90 mg/l
(*iii*) 8.098 mg/l (*iv*) 17.12°C.]

11. 125 cumecs of sewage of a city discharged in a perennial river which is fully saturated with oxygen and flows at a minimum rate of 1600 cumecs with a minimum velocity of 0.12 m/sec. If the 5-day BOD of sewage is 300 mg/l, find out where the critical DO will occur in the river. Assume.

(*i*) The coefficient of purification of the river as 4.0
(*ii*) The coefficient of DO as 0.11, and
(*iii*) The ultimate BOD as 125% of the 5-day BOD of the mixture of sewage and river water

(UPSC, Engg. Service, 1987)

Ans. [17.88 km]

12. A wastewater treatment plant disposes off its effluents into a stream at a point *A*. Characteristics of the stream at a location fairly upstream of *A* and of the effluent are as follows :

Item	Units	Effluent	Stream.
Flow	m^3/s	0.20	0.50
Dissolved oxygen	mg/l	2.0	8.0
Temperature	°C	26	22
BOD_5 at 20°C	mg/l	40	3

Assume that the deoxygenation constant K_1 at 20°C (base *e*) $= 0.20\, d^{-1}$ and the reaeration constant K_2 at 20°C (base *e*) $= 0.4\, d^{-1}$ for the mixture equilibrium concentration of dissolved oxygen C_s for the fresh water is as follows .

Temperature °C	18	20	22	23	24	25	26
C_s (mg/l)	9.54	9.17	8.99	8.83	8.53	8.38	8.22

The velocity of stream downstream of the point *A* is 0.2 m/s. Determine the critical oxygen deficit and its location.

[Use temperature coefficients of 1.04 for K_1 and 1.02 for K_2)

(UPSC, Civil Services, 1989)

Ans. [47.37 km; 6.29 mg/l]

10

Unit Operations for Wastewater Treatment

10.1. UNIT OPERATIONS AND PROCESSES

The wastewater treatment is a broad term that applies to any operation/process or combinations of operations and processes, that can reduce the objectionable properties of water-carried waste and render it less dangerous and repulsive to man. Wastewater treatment is a combination of physical, chemical and biological processes. Methods of treatment in which the application of physical forces predominate are known as *unit operations* while methods of treatment is which chemical or biological activities are involved are known as *unit processes.* There are three types of unit operations and processes : 1. Physical unit operations, 2. Chemical unit processes, and 3. Biological unit processes.

The term *'unit operations'* first became popular in chemical engineering in which any chemical process may be resolved into a coordinated series of the so called 'unit actions'. In chemical engineering, the term 'unit operation' is applied to procedures in which changes are essentially physical. However, in water and wastewater treatment, unit operations are those procedures in which changes produced are physical, chemical or biological. The unit operation approach in water and wastewater treatment has the following advantages :

1. It gives better understanding of the processes inherent in the treatment, and of the capabilities of these processes in attaining the objectives.

2. It helps in the development of mathematical and physical models of treatment mechanisms and the consequent design of treatment plants.

3. It helps in the coordination of effective treatment procedure to attain desired plant performance and effluent quality.

The output of wastewater treatment works is an *acceptable effluent* through the available unit operations. Most wastewater treatment processes bring about changes in concentration of a specific substance by moving the substance either into or out of the wastewater itself. This is called *phase transfer*. The principal phase transfers are : (*i*) gas transfer, (*ii*) ion transfer (*iii*) solute stabilisation (*iv*) solids transfer, (*v*) nutrient or molecular transfer etc. Following are the important *operations* which constitute the physical, chemical and biological unit operations/processes, employed in water and wastewater treatment.

1. Gas transfer : Aeration

2. Ion transfer

(*a*) Chemical coagulation (*b*) Chemical precipitation
(*c*) Ion exchange (*d*) Adsorption

3. Solute stabilization

(*a*) Chlorination (*b*) Liming
(*c*) Recarbonation (*d*) Super-chlorination

4. Solids transfer

(*a*) Straining (*b*) Sedimentation
(*c*) Floatation (*d*) Filtration

5. Nutrient or molecular transfer

6. Interfacial contact

7. Miscellaneas operations

(*a*) Disinfection (*b*) Copper sulfating
(*c*) Fluoridation (*d*) Thermal desalination

8. Solid concentration and stabilization

(*a*) Thickening (*b*) Centrifuging
(*c*) Chemical conditioning (*d*) Elutriation
(*e*) Biological floatation (*f*) Vaccum filtration
(*g*) Air drying (*h*) Heat drying
(*i*) Sludge digestion (*j*) Incineration
(*k*) Wet combustion

10.2. PHYSICAL UNIT OPERATIONS

The physical unit operations in which application of physical forces predominate, consist of the following :

(*i*) Screening (*ii*) Comminution
(*iii*) Flow equalization (*iv*) Mixing

(*v*) Flocculation (*vi*) Sedimentation
(*vii*) Floatation (*viii*) Elutriation
(*ix*) Vacuum filtration (*x*) Microscreening
and (*xi*) Air drying.

Because the physical unit operations were derived originally from observations of the physical world, they were the first treatment methods to be used. They form the basis of most process flowsheets. Table 10.1 gives the applications of physical unit operations in wastewater treatment.

TABLE 10.1. APPLICATIONS OF PHYSICAL UNIT OPERATIONS IN WASTEWATER TREATMENT

	Operation	*Application*
1.	Screening	Removal of coarse and settleable solids by surface straining
2.	Comminution	Griding of coarse solids
3.	Flow equalization	Equalization of flow and mass loadings of BOD and suspended solids
4.	Mixing	Mixing of chemicals and gases with wastewater, and maintaining solids in suspension
5.	Flocculation	Promotion of the aggregation of small particles into larges particles
6.	Sedimentation	Removal of settleable solids and thicknening of sludges
7.	Floatation	Removal of finely divided suspended solids and particles with densities close to that of water. Also thickens biological sludges.
8.	Filtration	Removal of fine residual suspended solids remaining after biological or chemical treatment.
9,	Microscreening	Same as filtration. Also removal of algae from stabilization-pond effluents.

10.3. CHEMICAL UNIT PROCESSES

Chemical unit processes are those in which removal of contaminants are brought about by chemical activity. In the field of wastewater treatment, chemical unit operations are usually used in conjunction with physical unit operations and biological unit processes. The following chemical unit processes are commonly used for wastewater treatment :

(*i*) Chemical precipitation (*ii*) Gas transfer
(*iii*) Adsorption (*iv*) Disinfection
(*v*) Combustion (*vi*) Ion exchange, and
(*vii*) Electrodialysis

It should be clearly noted that chemical unit processes are ***additive processes*** whereas physical unit operations and biological unit processes are *substractive processes*. This is an inherent disadvantage of chemical unit operations because there is usually a net increase in the dissolved constituents in the wastewater because of the chemical unit processes. In most of the cases, we add something to the wastewater to achieve the removal of something else. Another disadvantage of chemical unit processes is that they are all *intensive* in local *operating costs.* Table 10.2 gives the summary of the applications of chemical unit processes in wastewater treatment.

TABLE 10.2. APPLICATIONS OF CHEMICAL UNIT PROCESSES IN WASTEWATER TREATMENT

Process	*Application*
1. Chemical precipitation	Removal of phosphorous and enhancement of suspended solids removal in primary sedimentation
2. Gas transfer	Addition and removal of gases
3. Adsorption	Removal of organics
4. Disinfection	Disinfection of disease-causing organisms
5. Dechlorination	Removal of total combined chlorine residuals
6. Miscellaneous	Achievement of specific objectives in waste-water treatment.

10.4. BIOLOGICAL UNIT PROCESSES

Biological unit processes are those in which removal of contaminants are brought about by biological activity. In biological treatment of wastewater, the objectives are to coagulate and remove the non-settleable colloidal solids and to stabilise the organic matter. The wastewater may generally be from three sources (*i*) domestic wastewater (*i.e* sewage), (*ii*) agricultural return wastewater, and (*iii*) industrial wastewater. For domestic wastewater, the objective is to remove various nutrients, specifically nitrogen and phosphorous which are otherwise capable of stimulating growth of aquatic plants. For industrial wastewater, the objective is to remove/reduce the concentration of organic and inorganic compounds which may otherwise be toxic.

Biological processes are classified by the oxygen dependence of the primary micro-organisms responsible for waste treatment. Biological unit processes may, therefore, be classified in the following heads.

(a) Aerobic processes : Aerobic processes are biological treatment processes that occur in the presence of dissolved oxygen. The bacteria that can survive only in the presence of dissolved oxygen are known as obligate aerobes. The aerobic processes include the following : (*i*) Activated sludge process (*ii*) Trickling filters (*iii*) Aerobic stabilisation ponds and (*iv*) Aerated lagoons.

(*b*) ***Anaerobic processes*** : Anaerobic waste treatment involves the decomposition of organic and/or inorganic matter in absence of molecular oxygen. Anaerobic processes consist of the following:

(*i*) Anaerobic sludge digestion.
(*ii*) Anaerobic contact processes.
(*iii*) Anaerobic filters.
(*iv*) Anaerobic lagoons or ponds.

(*c*) ***Aerobic-anaerobic processes*** : In this method, stabilization of waste is brought about by a combination of aerobic, anaerobic and facultative bacteria.

10.5. METHODS OF TREATMENT OF WASTEWATER

The wastewater treatment or sewage treatment is a broad term that applies to any process/operation or combination of processes and operations that can reduce the objectionable properties of water carried waste and render it less dangerous with the following :

(*i*) Removal of suspended and floatable material
(*ii*) Treatment of biodegradable organics
and (*iii*) Elimination of pathogenic organisms.

Later, major effort was undertaken to achieve more effective and wide spread treatment of wastewater. As a consequence, the required degree of treatment has significantly increased and additional objectives and goals have been added. The recently added treatment objectives are the removal of nitrogen, phosphorous and toxic organic compounds. Thus, the treatment methods may be divided into the following heads :

(*a*) ***Conventional treatment methods***
1. Preliminary processes
2. Primary treatment
and 3. Secondary (or biological) treatment

(*b*) ***Advanced wastewater treatment***
4. Tertiary treatment.

1. Preliminary processes

Preliminary processes consist of pumping, screening and grit removal. Wastewater contains varying quantities of floating and suspended solids, some of considerable size. The purpose of the preliminary processes (some times also known as *preliminary treatment*) is to remove *floating materials* (like dead bodies of animals, pieces of wood, papers, rags, metal containers, plastic or rubber containers, grease etc.) and also heavy settleable inorganic solids (like grit, fragments of masonry etc.), including flow measuring devices. The preliminary processes remove the constituents the presence of which would otherwise interfere with subsequent treatment processes or mechanical equipment. The preliminary processes/treatment reduces the BOD load of the wastewater by about 5 to 10 percent. Various units involved in preliminary treatment are :

(*i*) *Screens* for removal of floating matter. (*ii*) *Grit chambers* or Detritus tanks for removal of sand and grit. (*iii*) *Comminutors* for grinding or chopping large size suspended solids. (*iv*) *Floatation units, skimming tanks* for removal of oils and grease. (*v*) *Flow measuring units* (*vi*) *Pumping* (*vii*) *Pre-aeration.*

Sometimes, flocculation with or without chemical additions are practiced on high-strength municipal wastes to provide increased *primary removal* and prevent excess loads on the secondary treatment processes. Similarly, chorination of raw wastewater is sometimes used for odor control and to improve settling characteristics of the waste. Table 10.3 gives the processes or units employed in preliminary treatment, along with their functions and efficiencies.

2. Primary treatment

Primary treatment includes those physical unit operations and chemical unit processes by way of which larger suspended solids are removed from the wastewater. These suspended solids will be *organic* if the wastewater has been subjected to preliminary treatment prior to primary treatment. If however, no preliminary treatment is given, then the primary treatment removes both the inorganic as well as organic suspended solids. When the wastewater has already been subjected to preliminary treatment, the primary treatment consists of *sedimentation* and *chemical coagulation. Sedimentation* of raw wastewater is practiced in all large municipal plants and must precede conventional trickling filtration. Chemicals are sometimes added in *primary clarifiers* to assist in removal of finely divided and colloidal solids, or to precipitate phosphorus. *Primary clarifiers* are usually designed to remove particles with settling rates of 0.3 to 0.7 mm/s. Organic solids, settled in primary clarifiers are often *stabilised* by anaerobic decomposition in a *digestion tank* or *incinerated. Septic tanks* or *Imhoff tanks* incorporate organic solids separation and digestion in a *single unit*. Hence septic and Imhoff tanks also fall under the category of *other primary processes*. Table 10.4 gives the processes or units employed in the primary treatment, along with their functions and efficiencies.

3. Secondary Treatment

Secondary treatment systems remove the soluble and colloidal organic matter which remains after primary treatment. Since removal of this material is effected by physico-chemical means, secondary treatment is usually understood to *imply biological treatment process.* Secondary treatment, as defined by U.S. Environmental Protection Agency, directed principally towards the removal of ***biodegradable***

TABLE 10.3
PRELIMINARY TREATMENT UNITS, THEIR FUNCTIONS AND EFFICIENCIES.

Purification Effected	*Process or unit employed*	*BOD removal (% of original)*	*Removal of suspended and dissolved solids (% of original)*	*Removal of bacterial load (% of original).*	*Disposal of Residuals*
1. Removal of *floating materials* such as dead animals, wood pieces, tree branches, rags, papers, metal containers, plastic containers & other large sized floating materials	Coarse and fine screens of different designs	5–10	2–10	10–20	Screenings disposed of by burials or burnings
2. Removal of *heavy settleable inorganic solids*	Grit chambers or Detritus tanks.	10–20	20–40	10–20	Grits disposed of by burials or burnings or raising low lying area
3. Removal of *fats and greases*	Floatation units and skimming tanks	20–30	20–40	10–20	Skimmings being unstable volatile materials, are disposed of by first stabilising them in digestion tanks by anaerobic process

TABLE 10.4

PRIMARY TREATMENT UNITS, THEIR FUNCTIONS & EFFICIENCIES

Purification Effected	*Process or unit employed*	*BOD removal (% of original)*	*Removal of suspended solids (% of original)*	*Removal of bacterial load (% of original)*	*Disposal of Residuals.*
Removal of large *suspended organic solids*	(*i*) Plain sedimentation tanks or	30–35	60–65	25–75	Sludge, after stabilisation in digestion tanks, is used as a manure or soil builder.
	(*ii*) Septic tanks or (*iii*) Imhoff Tanks	20–30	40–60	25–75	These units combine sedimentation and sludge digestion. Sludge used as manure while effluent is either treated for secondary treatment or disposed of by land treatment.

organics and suspended solids. Biological treatment systems are designed to maintain a large active mass of bacteria within the system confines.

The secondary treatment include all treatment methods enumerated in § 10.4 under the *biological unit processes.* While the basic principles remain the same in all the secondary treatment processes, the techniques used in their application may vary widely, but may be classified under the following heads :

(*i*) *Attached (film) growth process*
(*ii*) *Suspended growth process*
(*iii*) *Combined processes*

Attached-growth processes (or *fixed film processes*) utilize a solid medium (such as rocks, slag or specially designed ceramic or plastic materials) upon which bacterial solids are accumulated in order to maintain a high population. They convert the organic matter or other constituents in the wastewater to gases and cell tissue. Such *surface growth processes* include the following :

1. Intermittent sand filters.
2. Trickling filters.
3. Rotating biological contactors
4. Packed-bed reactors
5. Anaerobic lagoons (Ponds)
6. Fixed film denitrification.

Suspended growth processes maintain adequate biological mass in suspension within the liquid in the reactor, by employing either natural or mechanical mixing. The organic matter and other constituents of the wastewater are converted into gases and cell tissues. In most processes, the required volume in reduced by returning bacteria from the secondary clarifier in order to maintain a high solid concentration. Suspended growth processes include the following :

1. Activated sludge process
   (*i*) Conventional (plug flow)
   (*ii*) Continuous-flow stirred-tank
   (*iii*) Step aeration
   (*iv*) High purity oxygen system
   (*v*) Modified aeration
   (*vi*) Contact stabilization
   (*vii*) Extended aeration
   (*viii*) Oxidation ponds.
2. Suspended growth nitrification and suspended growth denitrification
3. Aerated lagoons
4. Sludge digestion systems.

Combined processes consist of both surface growth processes as well as suspended growth processes. They include the following in sequence :

1. Tricking filter, activated sludge 2. Activated sludge, trickling filter. 3. Facultative lagoons.

Almost all the units (aerobic as well as anaerobic) use primary settled sewage, and hence they are called *secondary units.* In contrast to this, *septic tanks* and *Imhoff tanks* (which are also anaerobic biological units) make use of raw sewage and hence they are not classified as secondary units ; they fall under primary treatment units.

In secondary or biological treatment processes, *secondary settling tanks* (also called *final settling tanks*) are essentially provided. The organic solids separated out in secondary settling tanks are disposed of by stabilising them under aerobic process in a *sludge digestion tank.* The effluent from the secondary biological plant will usually contain a little BOD (usually 5 to 10% of the initial). The effluent may even contain some DO. Table 10.5 gives the characteristics of the effluent after secondary treatment, as promulgated by the U.S. Environmental Protection Agency. Table 10.6 gives the processes or units employed in the secondary treatment, along with their functions and efficiencies.

TABLE 10.5. DEFINITION OF SECONDARY TREATMENT, AS PROMULGATED BY U.S. ENVIRONMENTAL PROTECTION AGENCY.

Characteristic of discharge	*Unit of measurement*	*Average monthly concentration*	*Average weekly concentration.*
1. BOD_5	mg/l	30 [a, b]	45 [b]
2. Suspended solids	mg/l	30 [a, b]	45 [b]
3. Hydrogen-ion concentration	pH units		6–9 [d]

[a] Or, in no case more than 15 percent of influent value.

[b] Arithmetic mean

[c] Treatment plants with stabilisation ponds and flows <7570 m^3/d are exempt.

[d] Continuous, only enforced if caused by industrial wastewater or in-plant treatment.

Fig. 10.1 shows a typical flow sheet for the treatment of wastewater to meet secondary-treatment standards, as defined by U.S. Environmental protection Agency (Table 10.5).

4. Advanced or tertiary treatment

Tertiary treatment or advanced treatment includes all operations and processes used to remove the *pollutants* not removed in preliminary, primary and secondary treatment. These pollutants may include soluble inorganic compounds such as phosphorus or nitrogen which may support algal growth in receiving waters; organic materials contributing biochemical oxygen demand (BOD), chemical oxygen demand (COD), colour, taste and odour ; bacteria; viruses ; colloidal solids contributing turbidity; or soluble minerals which may interfere with subsequent re-use of the wastewater (Culp, 1978). Tertiary treatment may be aimed at the *re-use* of wastewater. The need for direct and deliberate reuse of reclaimed sewage effluents is increasing in many parts of the World. The city of Windhoek, South West Africa, is successfully

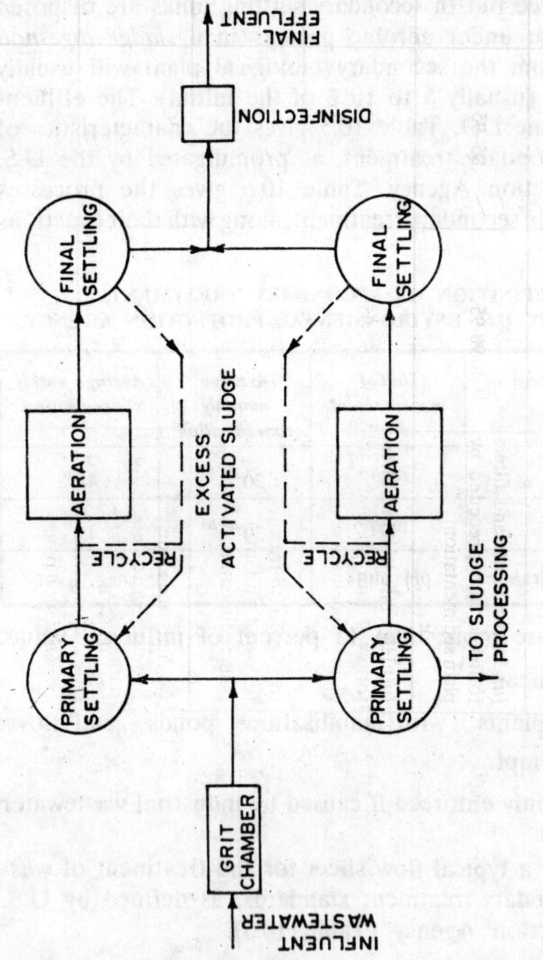

FIG. 10.1. PROCESS FLOW SHEET FOR TREATMENT PLANT DESIGNED TO MEET SECONDARY TREATMENT STANDARDS OF THE U.S. ENVIRONMENTAL PROTECTION AGENCY.

TABLE 10.6. SECONDARY TREATMENT UNITS, THEIR FUNCTIONS AND EFFICIENCIES.

Purification effected	*Process or unit employed*	*BOD removal (% of original)*	*Removal of suspended solids (% of original)*	*Removal of bacterial load (% of original)*	*Disposal of Residuals*
Removal of fine suspended and dissolved organic matter	(*i*) Chemical flocculation & sedimentation or	50–85	70-90	40-80	Sludge containing organic matter has to be stabilised in digestion tanks before use as manure or soil builder
	(*ii*) Intermittent sand filters followed and preceded by plain sedimentation or	90-95	85–95	95–98	Same as above
	(*iii*) High rate trickling filters followed and preceded by plain sedimentation or	60-95	65–92	80–85	Same as above
	(*iv*) Low rate trickling filters followed and preceded by plain sedimentation or	90-95	70–92	90–95	Same as above
	(*v*) Activated slude treatment process and secondary settling tanks or	75-95	85–90	90–98	Same as above
	(*vi*) Oxidation ponds	85-90	85–90	90–98	Effluents are generally disposed of by using them for irrigation

recycling advanced wastewater treatment plant effluent directly to the inlet of a water treatment plant.

The common processes that are used in the tertiary treatment are as follows :

(*i*) Chemical clarification
(*ii*) Recarbonation
(*iii*) Filtration
(*iv*) Activated carbon adsorption
(*v*) Disinfection
(*vi*) Nitrogen removal
(*vii*) Phosphorous removal, and
(*viii*) Demineralisation.

Wastewater treatment units configurations

In normal practice wastewater treatment is confined up to secondary treatment only. Various types of physical, chemical and biological processes are available for wastewater treatment and the choice of the treatment methods/units depends upon several factors such as wastewater characteristics, degree of purification required, funds available and disposal facilities available. It should be noted that the distinction between preliminary, primary, secondary and tertiary treatment is arbitrary, and some of modern treatment units/plant incorporate physical, chemical and biological processes in the same operation. Figs. 10.2, 10.3 and 10.4 give some alternative flow diagrams of wastewater treatment configurations.

Table 10.7 gives the list of units operations/ processes and treatment systems used to remove major contaminants found in wastewater.

10.6. METHODS OF SLUDGE PROCESSING AND DISPOSAL

Various processes and methods of treating the liquid part of wastewater have been enumerated and highlighted in the previous article. Of equal importance is the processing and disposal of *sludge* so obtained. The bulk of the suspended soilds which enter a wast-treatment plant, and the waste solids generated in biological treatment must be handled as sludge at some point of treatment process. Disposal of the accumulated waste sludge is a major economic factor in wastewater treatment, since the construction cost of a sludge processing facility is approximately one third that of a treatment plant.

Fig. 10.5 shows the flow schemes for withdrawal, holding, thickening, processing and disposal of waste sludge. Table 10.8 gives a list of various methods of processing and disposal of wastewater sludges. Since the processing and treatment of sludge has become very much specialised, a separate Chapter (chapter 16) has been devoted entirely to this topic.

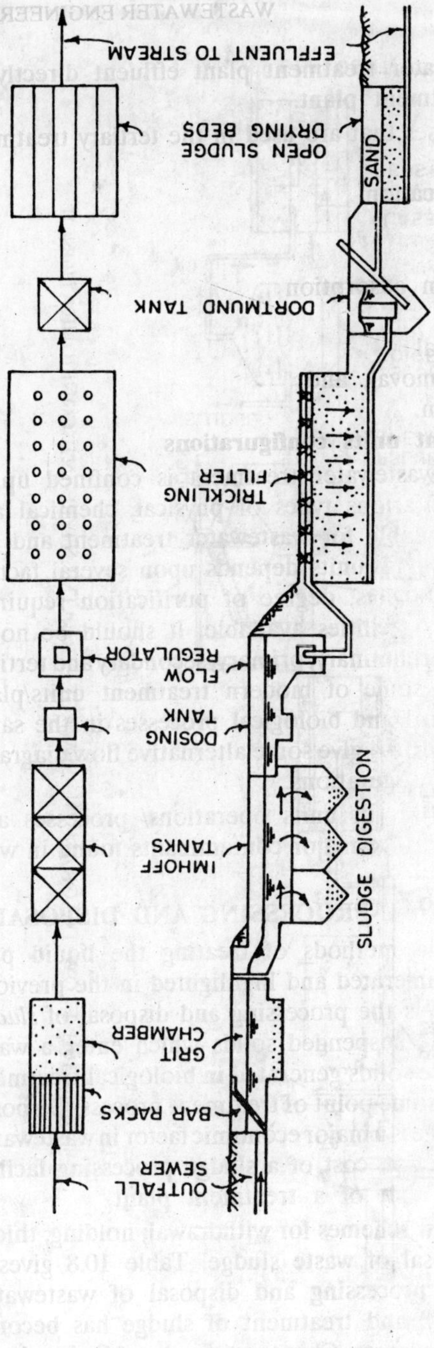

FIG. 10.2 FLOW DIAGRAM OF SEWAGE TREATMENT PLANT CONFIGURATION I

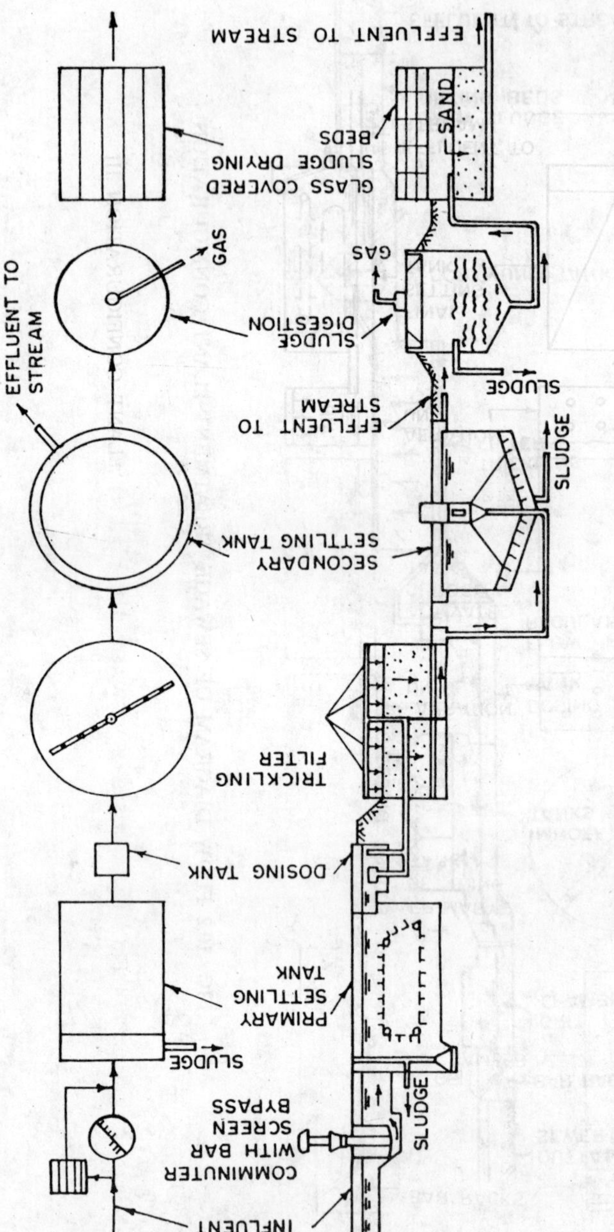

FIG. 10.3 FLOW DIAGRAM OF SEWAGE TREATMENT PLANT CONFIGURATION II

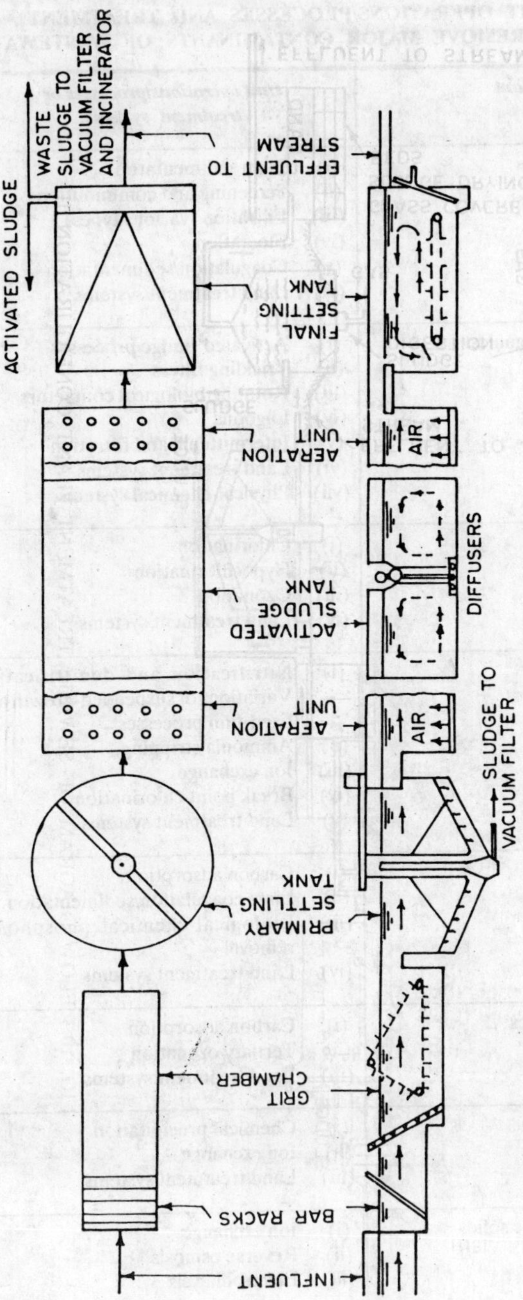

FIG. 10.4 FLOW DIAGRAM OF SEWAGE TREATMENT PLANT CONFIGURATION III

TABLE 10.7. UNIT OPERATIONS/PROCESSES AND TREATMENT SYSTEMS USED TO REMOVE MAJOR CONTAMINANTS OF WASTEWATER

Contaminant	*Unit operations/processes or treatment systems.*
1. Suspended solids	(i) Plain sedimentaton (ii) Screening and comminution (iii) Filtration (various types) (iv) Floatation (v) Coagulation/sedimentation (vi) Land treatment systems
2. Biodegradable organics	(i) Activated sludge processes (ii) Trickling filters (iii) Rotating biological contactors (iv) Lagoons (v) Intermittent sand filtration (vi) Land treatment systems (vii) Physical-chemical systems.
3. Pathogens	(i) Chlorination (ii) Hypochlorination (iii) Ozonation (iv) Land treatment systems
4. Nitrogen	(i) Nitrification and denitrification Variations of suspended-growth and fixed-film processes (ii) Ammonia stripping (iii) Ion exchange (iv) Break point chlorination (v) Land treatment systems.
5. Phosphorous	(i) Carbon adsorption (ii) Lime coagulation/sedimentation (iii) Biological-chemical phosphorus removal (iv) Land treatment systems
6. Refractory organics	(i) Carbon adsorption (ii) Tertiary ozonation (iii) Land treatment systems
7. Heavy metals	(i) Chemical precipitation (ii) Ion exchange (iii) Land treatment systems
8. Dissolved inorganic solids	(i) Ion exchange (ii) Reverse osmosis (iii) Electrodialysis

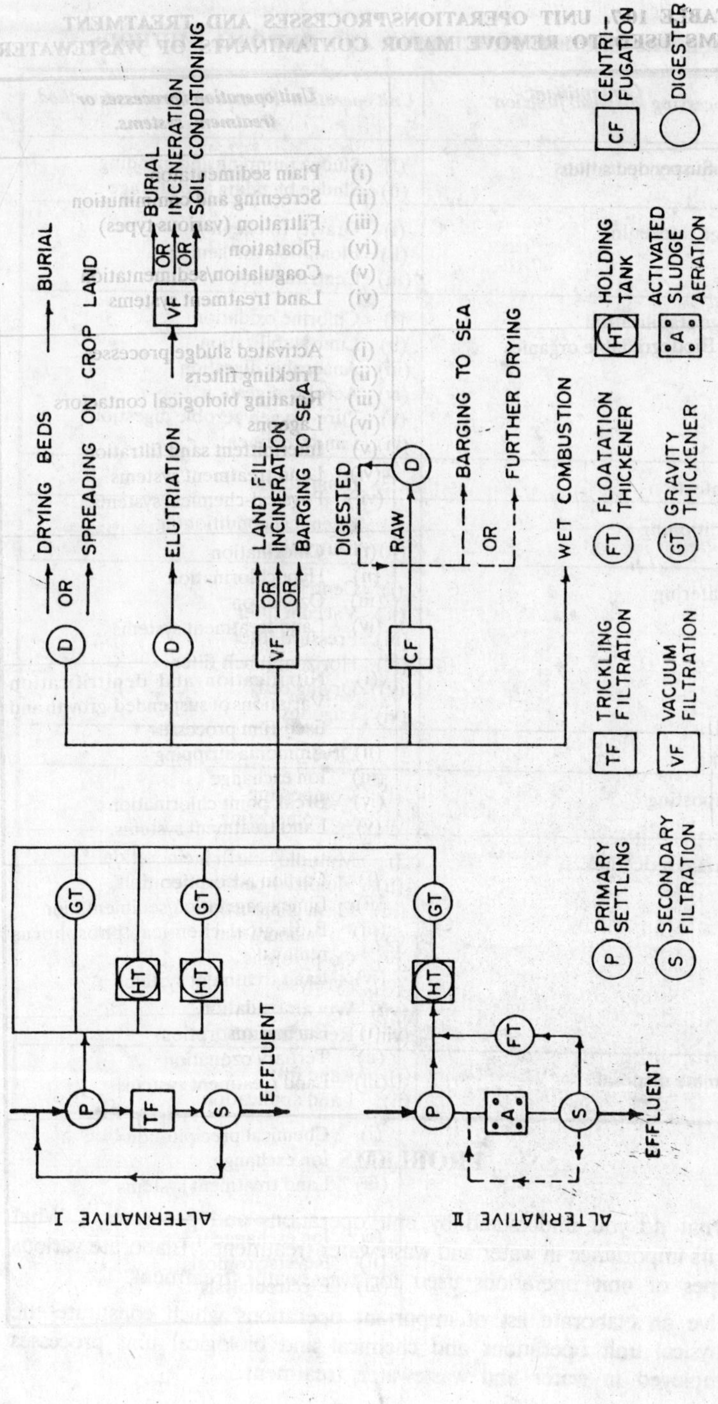

FIG. 10.5 FLOW SCHEMES FOR PROCESSING AND DISPOSAL OF SLUDGES.

TABLE 10.8 SLUDGE PROCESSING AND DISPOSAL METHODS.

Processing /disposal function	*Unit operation/ Process or treatment method.*
1. Preliminary operations	(i) Sludge pumping and grinding (ii) Sludge blending and storage
2. Sludge thickening	(i) Gravity thickening (ii) Floatation thickening (iii) Centrifugation
3. Sludge stabilization	(i) Chlorine oxidation (ii) Lime stabilization (iii) Anaerobic digestion (iv) Aerobic digestion (v) Pure-oxygen aerobic digestion (vi) Land treatment
4. Disinfection	Disinfection
5. Conditioning	(i) Chemical conditioning (ii) Elutriation
6. Dewatering	(i) Centrifuge (ii) Vaccum filter (iii) Pressure filter (iv) Horizontal belt filter (v) Drying bed (vi) Lagoons
7. Drying	Dryer
8. Composting	(i) Composting (ii) Co-composting
9. Thermal reduction	(i) Multiple hearth incineration (ii) Fluidized-bed incineration (iii) Flash combustion (iv) Co-incineration (v) Co-pyrolysis (vi) Pyrolysis (vii) Wet air oxidation (viii) Recalcination
10. Ultimate disposal	(i) Land fill (ii) Land application

PROBLEMS

1. What do you understand by unit operations and processes ? What is its importance in water and wastewater treatment ? Elaborate various types of unit operations used for wastewater treatment.

2. Give an elaborate list of important operations which constitute the physical unit operations and chemical and biological unit processes employed in water and wastewater treatment.

3. What do you understand by physical unit operations ? Write a note on applications of various physical unit operations employed in wastewater treatment.
4. What is meant by chemical unit processes. Enlist the applications of various chemical unit processes employed in wastewater treatment.
5. Write a detailed note on biological unit processes, employed in wastewater treatment.
6. What are various methods of treatment of wastewater ? Distinguish clearly between them.
7. Write a note on preliminary and primary treatment of wastewater. To what extent various impurities/pollutants are removed in these operations ?
8. Explain, with the help of flow sheets, possible arrangements of preliminary treatment units in wastewater processes.
9. What do you understand by secondary treatment of wastewater ? Enumerate various processes employed for this purpose.
10. Write a detailed note on advanced or tertiary treatment of wastewater.
11. Explain, with the help of flow diagrams, various operations/ processes employed in conventional wastewater treatment.
12. Enumerate various unit operations/processes and treatment systems used to remove major contaminants of wastewater.
13. Write a note on 'sludge processing and disposal' using a flow diagram.

11

Preliminary Treatment

11.1. INTRODUCTION

The preliminary treatment consists of the processes which remove those constituents of the wastewater the presence of which would otherwise interfere with subsequent treatment processes or mechanical equipment. Wastewater contains varying quantities of floating and suspended solids, some of considerable size. Preliminary treatment processes remove floating materials (like dead bodies of animals, pieces of wood, papers, rags, metal containers, plastic or rubber containers, grease etc.) and also heavy settleable inorganic solids like grit, and fragments of masonry etc. The process units also include pumping as well as flow measurement systems.

Various units involved in preliminary treatment are :

(*i*) *Screenings*, for removal of floating matter.
(*ii*) *Grit chamber* or *detritus tank*, for removal of sand and grit
(*iii*) *Comminuters* for grinding or chopping large size suspended solids.
(*iv*) *Floatation units* and *skimming tanks* for the removal of soils and grease.
(*v*) *Flow measuring units*, such Parshall flume etc.
(*vi*) *Pumping*
and (*vii*) *Pre-aeration*

Sometimes, *flocculation* with or without chemical additions are practiced on high-strength municipal wastes to provide increased *primary removal* of solids with the aim of preventing excess loads on the secondary treatment processes (*i.e.* biological processes). Similarly,

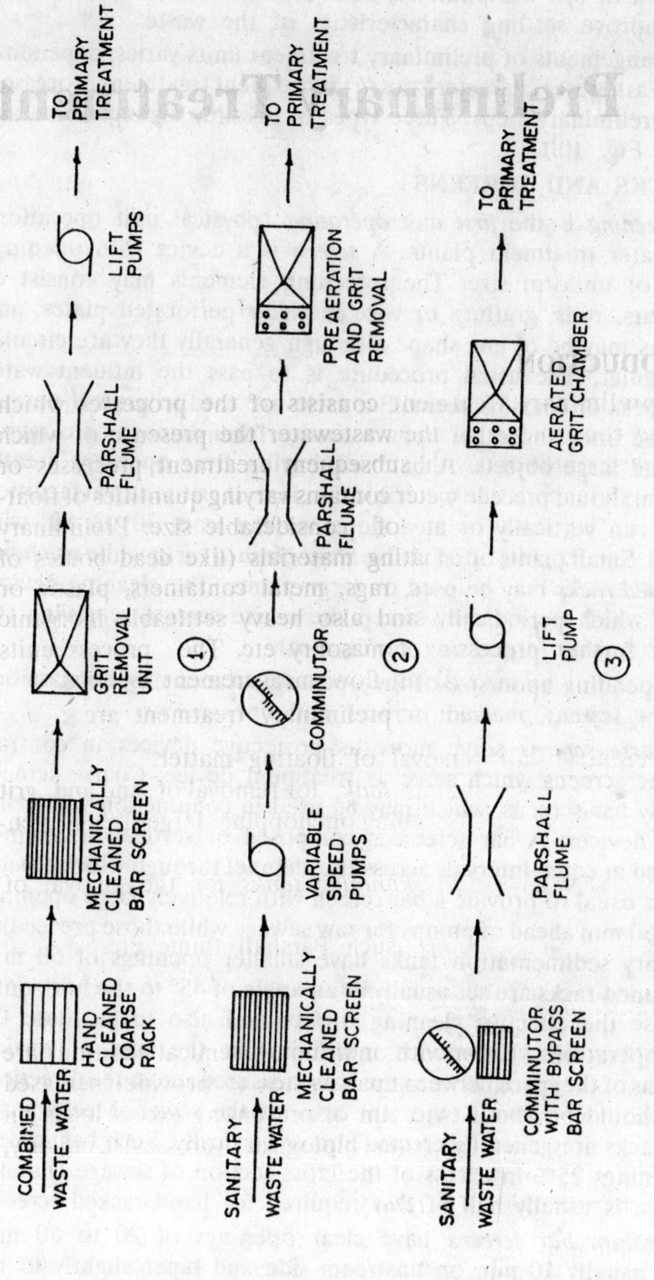

FIG. 11.1 POSSIBLE ARRANGEMENTS OF PRELIMINARY TREATMENT UNITS IN WASTEWATER PROCESSES.

chlorination of raw wastewater is sometimes used for odour control and to improve settling characteristics of the waste.

Arrangements of preliminary treatment units varies, depending upon (*i*) Wastewater characteristics (*ii*) subsequent treatment processes, and (*iii*) preliminary steps. Three types of possible arrangements are shown in Fig. 10.1.

11.2. RACKS AND SCREENS

Screening is the *first unit operation* (physical unit operation) in wastewater treatment plants. A *screen* is a device with openings generally of uniform size. The screening elements may consist of parallel bars, rods, gratings or wire mesh or perforated plates, and the screens may be of any shape although generally they are circular or rectangular. The usual procedure is to pass the influent water through racks or coarse screens. Bar racks are used to protect pumps, valves, pipe lines and other appurtenances from damage or clogging by rags and large objects. A bar rack having clear openings greater than 40 mm should precede mechanically cleaned grit-removal facilities. The bars run vertically or at a slope varying from 30° to 80° with horizontal. Small plants often utilise *hand-cleaned racks* while *mechanically cleaned racks* may be used on large plants. The cleaning devices are rakes which periodically sweep the entire screen, removing the solids for further processing or disposal.

Depending upon size of the openings, screen may be classified as *coarse screens*, *medium screens* and *fine screens.*

Coarse screens serve more as protective devices in contrast to the fine screens which serve as treatment device. Coarse screens are usually bar screens, which may be used in conjunction with comminuting devices. A bar screen is composed of vertical or inclined bars spaced at equal intervals across the channel through which sewage flows. It is usual to provide a bar screen with relatively large openings of 75 to 150 mm ahead of pumps for raw sewage while those preceeding the primary sedimentation tanks have smaller openings of 50 mm. Hand cleaned racks are set usually at an angle of 45° to the horizontal to increase the effective cleaning surface and also to facilitate the racking operations. Experience indicates that the area of vertical projections of the space between the bars measured across the directions of flow should be about twice the area of the sewer. Mechanically cleaned racks are generally erected almost vertically. Such bar screens have openings 25% in excess of the cross-section of sewage channel. Their area is usually half of that required for hand racked screens.

Medium bar screens have clear openings of 20 to 50 mm. Bars are usually 10 mm on upstream side and taper slightly to the down stream side. These mechanically raked units are used before all pumps or treatment units. The bars used for screens are rectangular in cross-section usually about 10 mm × 50 mm and are placed with large dimension parallel to the flow.

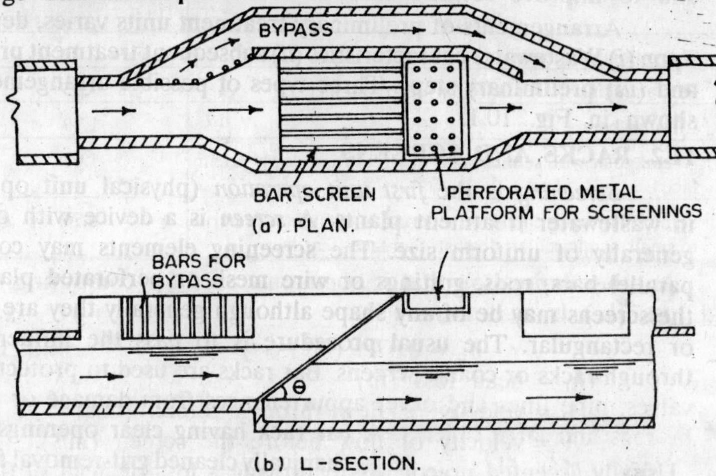

FIG. 11.2. HAND-CLEANED BAR SCREEN WITH OVERFLOW BYPASS.

The design velocity through the screens should be such as to permit 100% removal of material of certain size without undue depositions. Velocities of 0.6 to 1.2 m/s through the open area for the peak flows have been found satisfactory. When considerable amounts of storm water are to be handled, *approach velocity* of about 0.8 m/s is desirable, to avoid grit deposition at the bottom of the screen, though a lower value of 0.6 m/s is used in current practice. Further, the velocity at low flows in the approach channel should be less than 0.3 m/s to avoid deposition of solids.

The *loss of head* through bar racks is a function of (*i*) bar shape and (*ii*) velocity of flow between the bars. Kirschmer has proposed the following expression for the head loss :

$$h_L = \beta \left(\frac{w}{b}\right)^{4/3} h_v \sin\theta \quad ...(11.1)$$

where h_L = head loss (m)

β = a bar-shape factor (Table 11.1)

w = max. cross-sectional width of the bars facing the direction of the flow (m)

b = minimum clear spacing of the bars (m)

h_v = velocity head of flow ($V^2/2g$) approaching the rack (m)

θ = angle of the rack with the horizontal.

The head loss (h_L) calculated from Eq. 11.1 applies only to clean bar racks. Head increases with the degree of clogging.

TABLE 11.1 KIRSCHMER FACTOR (β)

	Bar type	β
1.	Sharp-edged rectangular	2.42
2.	Rectangular with semicircular upstream face	1.83
3.	Circular bars	1.79
4.	Rectangular with semicircular upstream and down stream face	1.67
5.	Semicirular upstream face tapering in symmetrical curve to small, semicircular, downstream face (teardrop)	0.76

Another expression that is commonly used for computing the head loss (h_L) is as follows :

$$h_L = 0.0729\,(V^2 - v^2) \qquad ...(11.2)$$

where h_L = head loss (m)

V = velocity of flow through screen (m/s)

v = velocity of flow before the screen (m/s).

Usually accepted practice is to provide a loss of head of 0.15 m but the maximum loss with clogged hand cleaned screen should not exceed 0.3 m. For the mechanically cleaned screen, the head loss is specified by the manufacturers. Table 11.2 gives typical design information for hand cleaned and mechanically cleaned bar racks.

TABLE 11.2 DESIGN INFORMATION FOR BAR RACKS.

	Item		*Hend Cleaned*	*Mechanically cleaned*
1.	Bar size :			
	Width	(mm)	5-15	5-15
	Depth	(mm)	27-75	25-75
2.	Clean spacing between bars	(mm)	25-50	15-75
3.	Slope from vertical	(deg.)	30-45	0-30
4.	Approach velocity	(m/s)	0.3-0.6	0.6-1.0
5.	Allowable head loss	(mm)	150	150

The submerged area of surface of the screen, including bars and openings should be about 200% of the cross-sectional area of the entrant sewer for separate sewers and 300% for combined sewers. The net submerged area of the opening of the screen should be about 500 cm^2 per mLd of flow for separate sewerage and 7500 cm^2 per mLd of flow for combined sewage system.

Fine screens. Fine screens are *mechanically cleaned* devices using perforated plate, woven wire cloth or very closely spaced bars with clear openings of less than 20 mm. They are used for pretreatment of industrial wastes to remove materials which tend to produce excessive scum or foam on the top of digestion tank contents. Fine screens are normally not suitable for sewages because of the clogging possibilities. Fine screens may be of drum or disc type, mechanically cleaned and continuously operated. Fig. 11.3 shows a disc type fine screen which is cleaned by a cone brush.

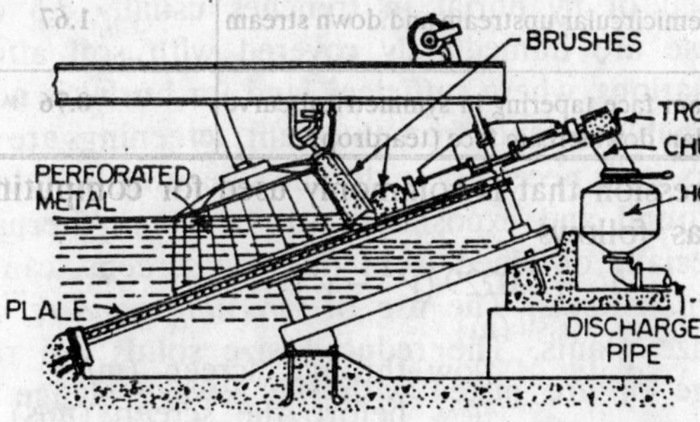

FIG. 11.3. DISC TYPE FINE SCREEN.

Brass or bronze plates are used for constructing fine screen so as to provide resistance to rust and corrosion. They have perforations of 1.5 mm to 3 mm in size. Fine screens have generally a net submerged open area of not less than 0.05 m^2 for the every 1000 m^3 of average daily flow of wastewater from a separate system, the corresponding figure being 0.075 m^2 for combined systems. Table 11.3 gives the usual bar sizes and openings commonly comployed in bar racks.

TABLE 11.3. USUAL BAR SIZES AND OPENINGS.

Dimension of bar facing the flow (mm)	*Clear spacing between bars (mm)*	$\frac{\textit{Area of the opening}}{\textit{Gross surface area of screen}} \times 100$
6	18	75
6	24	80
6	30	83.3
6	36	85.6
9	18	66.7
9	24	72.8
9	30	77
9	36	80
12	18	60
12	24	66.7
12	30	71.5
12	36	75

Screenings and their disposal. The quantity to screenings varies with the size of the screen used and on the nature of sewage. Generally, it has been found that the screenings from sanitary sewage vary from 0.0015 m^3/mL with screen size of 10 cm to 0.015 m^3/mL in case of 2.5 cm size. Screenings are disposed of by the following methods : (*i*) Burial (*ii*) Incineration (*iii*) Digestion and (*iv*) Grinding.

Medium sized screenings, particularly in small installations, can be disposed of by burial in trenches usually 7.5 cm to 10 cm deep. These are immediately covered with soil after burying. At large installations, where sufficient land for burial is not available without a reasonable distance from the plant, screenings are *incinerated.* Before doing so, its is necessary to dry the screening first, by spreading it over the ground and exposing it to the sun. Alternatively, the screenings, specially obtained from the fine screens, can be placed in *sludge digestion tanks*. The use of *screening grinders* is desirable for medium size plants. The reduced size solids are returned to the raw sewage or are mixed with the sewage sludge.

Example 11.1. *Design a bar screen for a peak average flow of 40 million litres per day.*

Solution. Assuming manual cleaning, let us keep the bars at an inclination of 45° with vertical. Let us use bars of size 9 mm ×50 mm, with 9 mm dimension facing the flow. Let us keep a clear spacing of 36 mm between the bars. Let us assume a desired velocity through the screen as 0.8 m/s at peak flow.

$$\text{Max. rate of flow} = \frac{40\times10^6\times10^{-3}}{24\times60\times60} = 0.4630 \text{ m}^3/\text{sec.}$$

$$\therefore \quad \text{Net area of screen} = \frac{0.4630}{0.8} = 0.5787 \text{ m}^2$$

$$\therefore \quad \text{Gross area of screen} = 0.5787 \times \frac{45}{36} = 0.7234 \text{ m}^2$$

Since the screen is inclined at 45° with the horizontal, gross area of screen needed $= \dfrac{0.7234}{\sin\theta} = \dfrac{0.7234}{\sin 45°} = \mathbf{1.023\ m^2}$

Also, velocity of flow above screen $= 0.8 \times \dfrac{36}{45} = 0.64$ m/s

Thus, we have $V = 0.8$ m/s and $v = 0.64$ m/s

$$\therefore \quad h_L = 0.0729\,(V^2 - v^2) \qquad ...(11.2)$$

$$= 0.0729\,[(0.8)^2 - (0.64)^2] = \mathbf{0.017\ m = 1.7\ cm}$$

This will be the head loss when the screen is clean. If, however, the screen is half clogged,

$$V = 2 \times 0.8 = 1.6 \text{ m/s}$$

$$\therefore \quad h_L = 0.0729\,[(1.6)^2) - (0.64)^2] = \mathbf{0.157\ m = 15.7\ cm}$$

Thus we find that when the screen is clean (in the initial stage of operation), the head loss is only 1.7 cm, while it shoots up to 15.7 cm (nearly *nine times*) when the screen is half clogged. Hence the bar screen should be frequently cleaned, in order to keep the head loss within a desirable limit.

11.3. COMMINUTORS AND BARMINUTORS

A comminuting device is a mechanically cleaned screen which incorporates a cutting mechanism that cuts the retained material enabling it to pass along the sewage. Comminuting devices may be preceded by grit chambers to prolong the life of the equipment. Frequently, they are installed in the wet well of pumping stations to protect the pumps against clogging by rags and large objects. However, provision must be made to *bypass* comminutors in case flows exceed the capacity of the comminutor or in case there is a power or mechanical failure. The use of comminutors tends to reduce odours, flies and unsightliness.

A comminutor consists of a vertical revolving drum-screen with 6 mm to 10 mm slots. The coarse material is cut by cutting teeth and the shear bars on the revolving drum as solids are carried past a stationary comb (Fig. 11.4). The small sheared particles then pass through the slots of the drum out and of a bottom opening through an inverted siphon. The head loss across comminutors depends upon screen details and flow, the normal values being on the order of 50 to 100 mm. Another form of comminutor (known as *stationary screen type comminutor*) consists of a stationary semicircular screen and a rotating circular *cutting disk*. The grid intercepts the large solid particles where as smaller solids pass through the space between the grid and cutting discs.

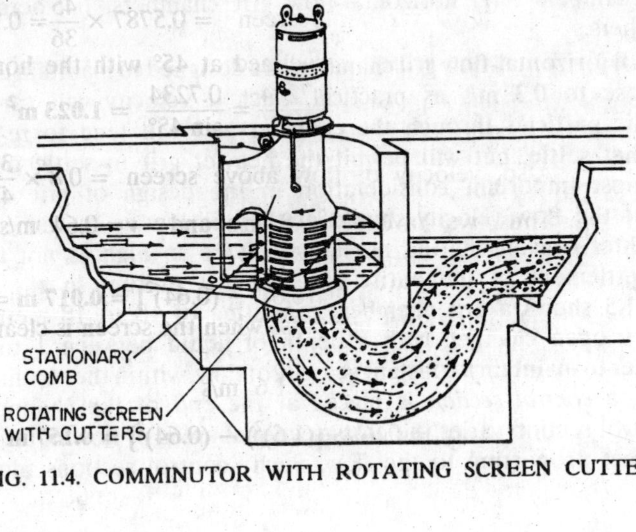

FIG. 11.4. COMMINUTOR WITH ROTATING SCREEN CUTTER.

A *barminutor* is another comminuting unit which is used for flows exceeding 50 mLd. It is a combined screening and cutting machine consisting of a bar screen of special design with small openings, along with rotating cutters comprising a comminuting unit which travels up and down the screen cutting the retained solids.

11.4. GRIT CHAMBERS

Grit is the heavy mineral material found in raw sewage, and it may contain sand, gravel, silt, cinders, broken glass, small fragments of metal and other small inorganic solids, that have subsiding velocities or specific gravities substantially greater than those of organic putrescible solids in wastewater. Grit also includes eggshells, bone chips, coffee grounds, seeds and large organic particles, such as food wastes. Grit is generally nonputrescible. Grit settles more rapidly then does the organic or putrescible material in the wastewater, thus allowing a reasonably clean separation from the waste stream. Grit has a specific gravity ranging from 2.4 to 2.65.

Grit chambers are provided to protect moving mechanical equipment from abrasion and accompanying abnormal wear. They reduce the formation of heavy deposits in pipe lines, channels and conduits. They also reduce the frequency of digester cleaning that may be required as a result of excessive accumulations of grit in such units. High speed equipment such as centrifuges require that practically all grit be eliminated to prevent rapid wear and reduce maintenance. The removal of grit is also essential ahead of heat-exchanger and high pressure diaphragm pumps.

Grit channels, grit chambers or *grit basins* are intended to remove the grit present in the wastewater. There are two general types of grit chambers : (*i*) horizontal-flow grit chambers. (*ii*) aerated grit chambers.

Horizontal-flow grit chambers are designed to maintain a velocity as close to 0.3 m/s as practical. Such a velocity will carry most organic particles through the chamber and will tend to re-suspend any that settle, but will permit the heavier grit to settle out. Thus the most important consideration in the design of grit chambers is that the flow velocity should neither be low as to cause settling of lighter organic matter, nor should it be so high as not to cause the settlement of the entire silt and grit present in the sewage. Fig 11.5 shows a grit chamber, consisting of 10 to 18 metres long narrow open channel, with a depth of liquid between 1 to 1.3 m. In order to maintain a constant velocity of flow, within the recommended range, a *control section* is placed at the end of the channel. Such a control section varies the cross-sectional areas of flow in the section in direct proportion to the flow. Such control sections include (*i*) proportional flow weirs, (*ii*) Sutor weirs, (*iii*) Parshall flumes, (*iv*) Palmer-Bowlus flumes etc.

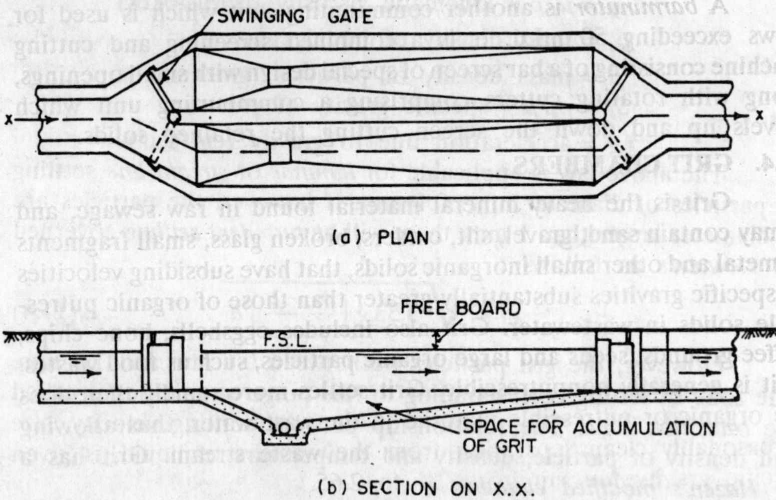

FIG. 11.5. GRIT CHAMBER

Grit chamber may be either *mechanically cleaned* or manually cleaned. The choice depends upons (*i*) quality and quantity of grit to be handled, (*ii*) head loss requirements, (*iii*) space requirements, (*iv*) topography and (*v*) economic considerations with respect to capital costs as well as maintenance costs. For all sewage plants where flows exceed 10 mLd, mechanised grit removal units are preferred.

11.5. DESIGN OF GRIT CHAMBERS

1. Settling velocity. Grit chamber may be designed on a rational basis, by considering it as a sedimentation basin having *discrete settling* (Chapter 12). *The settling velocity* for such a case is governed by the size and specific gravity of grit particles to be separated and the viscosity of the sewage. The size of separation based on the minimum size of grit to be removed is 0.2 mm although 0.15 mm is preferred for conditions where considerable amount of ash is likely to be carried by the sewage. The specific gravity may be as low as 2.4 but for design purposes, a value of 2.65 may be used.

The *settling velocity* is given by Stoke's law, applicable for particles of diameter less than 0.1 mm :

$$v_s = \frac{g}{18}\left(\frac{\rho_s - \rho}{\mu}\right) d^2 \qquad \text{...(11.3 a)}$$

or

$$v_s = \frac{g}{18} \frac{S_s - 1}{\nu} d^2 \qquad \text{...(11.3)}$$

where v_s = settling velocity (cm/sec)

d = size of particles (cm)

μ = dynamic (or absolute) viscosity of liquid (centipoise)
$\nu = \mu/\rho$ = kinematic viscosity (centistokes)
ρ_s = mass density of liquid (gm/cm^3
ρ = mass density of particles (gm/cm^3)
$S_s = \rho_s/\rho$ = specific gravity of particles.
g = acceleration due to gravity (cm/sec^2)

The above law is applicable for *laminar* or *stream line* settling of particles of diameter upto 0.1 mm. If however, the particles are of diameter larger than 1 mm, there will be *turbulent settling* governed by Newten's as follows :

$$v_s = \sqrt{3.33\, g \left(\frac{\rho_s - \rho}{\rho}\right) d} \qquad \text{...(11.4)}$$

However, the grit particles lie between 0.1 mm and 1.0 mm. The zone of setting corresponding to this range of particles is called the *transition zone*. The relationship between settling velocity, size and density of particle, density and temperature of liquid, is given by *Hazen's modified equation* :

$$v_s = 60.6\,(S_s - 1)\, d \left(\frac{3\,t + 70}{100}\right) \qquad \text{...(11.5)}$$

where v_s = settling velocity in cm/sec.
t = temperature of liquid (°C)

For grit having specific gravity equal to 2.65, we get from Eq. 11.5 :

$$v_s = 60.6\,(2.65 - 1)\, d\, \frac{(3t + 70)}{100}$$

or $$v_s = d\,(3\,t + 70) \qquad \text{...(11.6 a)}$$

For organic solids having specific gravity of 1.20 (assumed), we get, from Eq. 11.5.

$$v_s = 60.6\,(1.20 - 1)\, d \left(\frac{3t + 70}{100}\right)$$

or $$v_s = 0.12\, d\,(3t + 70) \qquad \text{...(11.6 b)}$$

The settling velocity (*i.e.* the rate of subsidence) of a particle in liquid medium is also denoted as *Hydraulic Subsidence Value* (HSV).

2. Over flow rate : The efficiency of grit removal can be expressed as the percentage removal of grit above a specified particle size. A grit chamber designed to remove 100% of grit particles of smallest size would also remove all grit particles larger than this. To obtain a 100% removal of the smallest size particles, it would be theoretically necessary for the detention time in the tank to equal the time required for the minimum sized particles to reach the tank bottom. It can be shown that the settling velocity v_s of the minimum sized particle is equal to the surface loading rate (Q/A), or overflow rate in order to obtain a theoretical 100% removal of the particles.

Table 11.4 gives settling velocity of different size particles of specific gravity 2.65 (mineral matter) and 1.20 (organic matter) and corresponding overflow rates for 100% removal of these particles based on Hazen's modified equation (Eq. 11.5).

TABLE 11.4 SETTLING VELOCITIES AND OVERFLOW RATES FOR GRIT CHAMBERS (TEMP. = 10°C)

Diameter of particles (mm)	*Settling velocity (cm/sec)*		*Overflow rate in an ideal grit chamber $m^3/d/m^2$*	
	$S_s = 2.65$	$S_s = 1.20$	$S_s = 2.65$	$S_s = 1.20$
0.20	2.0	0.24	1700	210
0.18	1.8	0.22	1600	190
0.15	1.5	0.18	1300	160

The values given in Table 11.4 should be corrected for any other temperature by the factor $\frac{3t + 70}{100}$. However, in practice, there are turbulence and short circuiting effects due to several factors such as wind currents, and density currents. Hence the overflow rate should be diminished to account for the *basin performance*. Generally, value of two thirds to one half are used in the design, depending upon the type of grit chamber. These values are much higher than those needed for organic solids of specific gravity 1.2. The surface area of grit chamber should be worked out based on the reduced loading rates.

3. Detention period : The detention period for grit chambers may vary from 45 to 90 seconds. A detention period of 60 seconds is usually adopted.

4. Bottom scour and flow through velocity : The grit chamber efficiency is very much affected by the bottom scour. The scouring process itself determines the *optimum velocity of flow* through the grit chamber. The flow velocity should be enough to scour out the settled organic matter and reintroduce it into the flow stream. The critical velocity for scour may be calculated from the *modified Schield's formula*

$$v_c = 3 \text{ to } 4.5 \sqrt{g\,(S_s - 1)\,d} \qquad ...(11.7)$$

For a grit particle size of 0.2 mm, the formula gives critical velocity values of 17.1 to 25.6 cm/sec. In actual practice, a horizontal velocity of flow of 15 to 30 cm/sec is used at peak flows.

5. Velocity control devices : The horizontal velocity of flow should be maintained constant at other flow rates also, to ensure that only organic solids and not the grit are scoured from the bottom.

Numerous devices have been designed in an attempt to maintain a constant horizontal velocity of flow through grit chambers in the recommended range of 15 of 30 cm/sec, with a limit of variance of 5-10% above and below the desired value. A satisfactory method of controlling velocity of flow through the grit channels is by using a *control section* which, when placed at the end of the channel, varies the cross sectional area of flow in the section in direct proportion to the flow. Such control sections may consist of the following : (*i*) Proportional flow weirs. (Fig. 11.6 *a*) (*ii*) Sutro weirs (Fig. 11.6 *b*) (*iii*) Parshall flumes (Fig. 11.6 *d*). (*iv*) Palmer Bowlus flumes, etc.

We shall discuss here the first three devices.

It should be clearly noted that when a *proportional weir* (Fig. 11.6 *a*) is used as a velocity control device, then a *rectangular cross-section* is required for the grit channel, while if a Parshall flume is used as a control device, then a *parabolic cross-section* is required for the grit channel in order to keep the flow velocity constant.

6. Number of units. In case of manually cleaned grit chambers, at least two units should be provided. All mechanically cleaned units should be provided with a manually cleaned unit to act as a bypass.

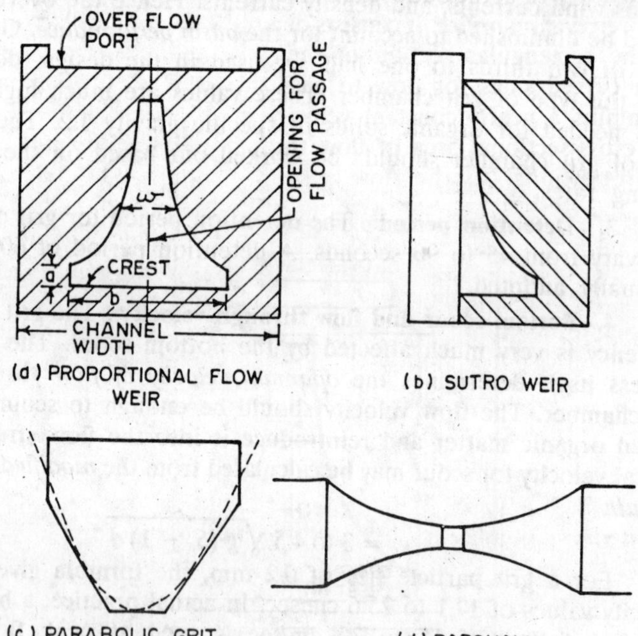

FIG. 11.6. FLOW CONTROL DEVICES.

7. Dimensions of each unit : (*i*) The surface area of each unit is determined on the basis of overflow rate. (*ii*) The breadth of section is fixed with reference to the control device adopted. (*iii*) The length of the unit is then worked out on the basis of selected overflow rate. (*iv*) For mechanised units, the horizontal dimensions selected above are readjusted to suit the standard sizes of mechanical equipment. (*v*) The depth of flow is determined on the basis of horizontal velocity and the peak flow. (*vi*) Additional depth is provided for storage of grit between the intervals of cleaning. (*vii*) A free board of 150 to 300 mm is provided. (*viii*) Bottom slopes are based on the type of scraper mechanism used.

8. Loss of head : Loss of head in a grit chamber may vary from 0.06 m to 0.6 m depending upon the device adopted for velocity control.

11.6. PROPORTIONAL FLOW WEIR

For proper functioning of the grit chamber, the velocity through the grit chamber should not be allowed to change inspite of change in flow. One of the most satisfactory types of automatic control for rectangular section grit chamber is achieved by providing a *proportional flow weir* at the outlet end. *A proportional flow weir is a combination of a weir and an orifice.* It consists of a rectangular plate (corresponding to the rectangular section of the grit chamber), with an opening with curved sides for flow to pass through. Such an arrangement maintains a nearly constant velocity in the grit channels by varying the cross-sectional area of flow through the weir so that the depth is proportional to the flow.

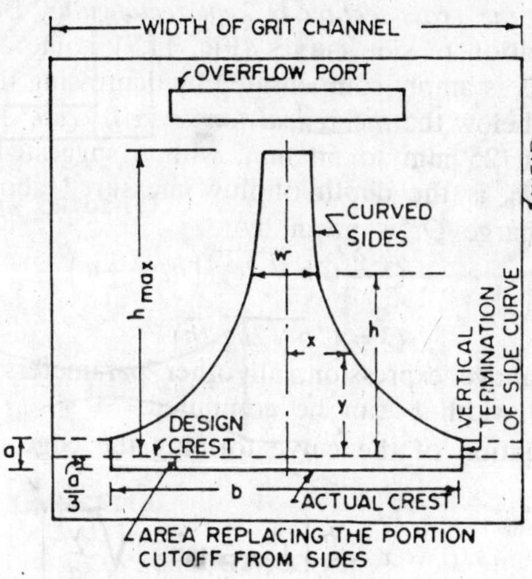

FIG. 11.7. GEOMETRY OF PROPORTIONAL FLOW WEIR

The shape of the opening between the plates of the proportional flow weir is made in such a way that the chamber depth will vary directly as the discharge, as a result of which the chamber velocity will remain constant for all flow conditions. The sides are so curved that *the area decreases as the three half power of the increasing depth (h) of the flow over the weir.* Hence the rate of flow over the weir will vary directly as the head over the weir. The discharge Q (in litres per second) over the weir is given by :

$$Q = 1570\, C \sqrt{2g}\, (wh^{1/2})\, h \qquad ...(11.8)$$

where Q = discharge (litres per second)
h = depth of flow measured above the design crest (m)
w = width of the opening, at height h (m)
g = acceleration due to gravity (m/sec^2)
C = discharge coefficient.

Taking $C = 0.6$ for symmetrial sharp edged weir, the above expression becomes

$$Q = 4170\, (wh^{1/2})\, h \qquad ...(11.9)$$

Hence if $wh^{1/2}$ is made constant, then the depth h will vary directly with the discharge. For different values of h, the corresponding values of the width w can be determined and hence the parabolic curvature of the sides of weir could be worked out. The simple form of the curve would result in the width (b) of the weir approaching infinity as the head (h) approaches zero. Since this is not practical, the base is limited to a convenient width b and for a height a over the crest, *the cross-section is kept rectangular.* Because of the vertical termination of side curves (Fig. 11.7), some area reduction takes place; this is amply compensated by depressing the *actual crest* by a value $a/3$ below the *theoretical (or desired) crest.* The dimension a may be from 25 mm to 50 mm, with a suggested value of 35 mm. Thus, if h_a is the depth of flow measured above the *actual crest,* the discharge Q, is given by

$$Q = C\, b \sqrt{2ag}\left(h_a - \tfrac{1}{3} a\right) \qquad ...(11.10)$$

or

$$Q = C\, b \sqrt{2ag}\, (h) \qquad ...(11.10\,a)$$

In the above expression, all other parameters except b are known. Hence width b can be computed.

The equation of the curve forming the edge of the weir is given by

$$x = \frac{b}{2}\left(1 - \frac{2}{\pi} \tan^{-1}\sqrt{\frac{y}{a}}\right) \qquad ...(11.11)$$

where depth y is measured above the theoretical crest.

The weir should be set 100 to 300 mm above the bottom of grit chamber to provide for the grit storage or for operation of mechanical grit scraper. The weir should also be set at an elevation as to provide a free fall into the outlet channel as such weir cannot function under submerged conditions. Each grit chamber should be provided with a separate control weir.

Sutro weir. Sutro weir is a half proportional flow weir cut symmetrically and centrally along the vertical axis, as shown in Fig. 11.6 (*b*). The orifice has a straight horizontal bottom forming the weir.

11.7. PARSHALL FLUME WITH PARABOLIC SECTION OF GRIT CHAMBER

Instead of rectangular cross-section of grit chamber with proportional flow weir, one can use a parabolic cross-section of the grit chamber (Fig. 11.6 *c*) with *fixed width control section* in the form of a Parshall flume. The flume, placed at the downstream end of the parabolic-section grit chamber, can control the velocity with moderate limits and with a moderate range of flow. A Parshall flume is an open constricted channel which can be also be used as a measuring device. The flume has distinct advantage over the proportional, flow weir, as it involves negligible head loss and can work under submerged flow conditions up to certain limits. The limits of submergence are 50% in case of 150 mm throat width and 70% for wider throat widths upto 1 m. Another advantage is that one control section can be installed for 2 to 3 grit chambers. The flume is also self cleansing and there is no problem of clogging. As Parshall flume

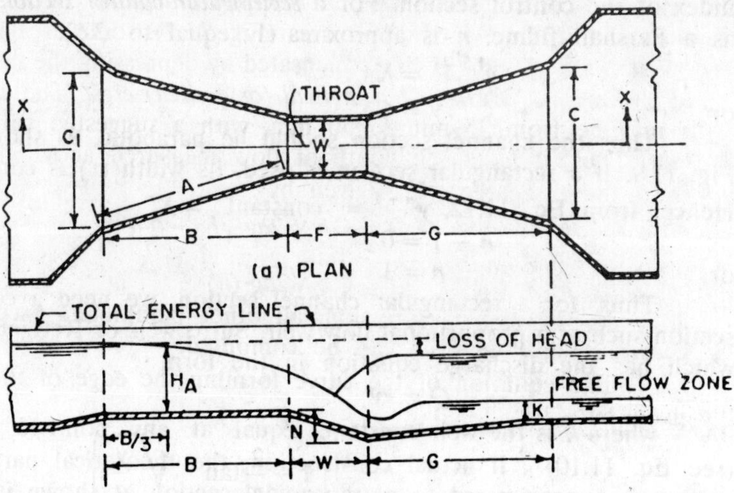

FIG. 11.8. PARSHALL FLUME

is a rectangular control section, the grit chamber above it must be designed to approach a parabolic section.

Parabolic section of grit channel. As stated earlier, it is essential to maintain a constant horizontal flow velocity v_h in the grit channel, at all the discharges. For that, the cross-sectional area of the channel should change in direct proproportion to the change in the discharge.

Hence we have

$$v_h = \frac{Q}{\int_0^y x\,dy} = \text{constant} = C_1 \quad \text{...}(i)$$

Also, in a control section, in general, we have

$$Q = C_2 y^n \quad \text{...}(ii)$$

Equating (*i*) and (*ii*), we get

$$C_1 \int_0^y x\,dy = C_2 y^n$$

$$\therefore \quad C_1 x y = C_2 y^n$$

or $$y^{n-1} = \frac{C_1}{C_2} x = Kx \quad \text{...(11.10)}$$

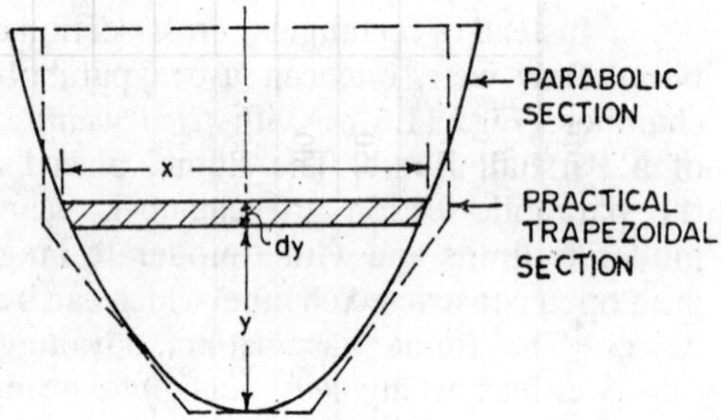

FIG. 11.9.

Hence the condition of constant velocity is maintained in the basin if its width varies so that $y^{n-1} = Kx$, where n is the discharge index of the control section. For a *rectangular control section* such as a Parshall flume, n is approximately equal to 3/2.

$$y^{1.5-1} = Kx$$

or $$y = Cx^2 \quad \text{...(11.13)}$$

Thus the channel section should be parabolic, as shown in Fig. 11.9. If a rectangular section is used, its width (x) is constant. Hence, from Eq. 11.12, $y^{n-1} =$ constant $= K'$

$$\therefore \quad n - 1 = 0$$

or $$n = 1$$

Thus, for a rectangular channel section, we need a control section such as a proportional flow weir, Sutro weir or Rettger weir, which has the discharge equation in the form

$$Q = ky$$

where k is the weir constant, equal at any point to $w\sqrt{y}$ (see Eq. 11.10). It actual construction, the theoretical parabolic section is approximated to a trapezoidal section as shown in Fig. 11.9.

TABLE 11.5
DIMENSIONS OF PARSHALL FLUME (mm)

Flow range $Q_{max.}$ (mLd)	W	A	B *	C	C_1 **	F	G	K	N
up to 5	75	460	450	175	255	150	300	25	56
5–30	150	610	600	315	391	300	600	75	113
30–45	225	865	850	375	566	300	750	75	113
45–170	300	1350	1322	600	831	600	900	75	225
170–250	450	1425	1397	750	1010	600	900	75	225
250–350	600	1500	1472	900	1188	600	900	75	225
350–500	900	1650	1619	1200	1547	600	900	75	225
500–700	1200	1800	1766	1500	1906	600	900	75	225
700–850	1500	2100	2060	2100	2625	600	900	75	225
850–1400	2400	2400	2353	2700	3344	600	900	75	225

* The value should be equal to $1.5\,(Q_{max})^{1/3}$, but not less than those shown in the Table.

** For higher values of B (than shown in the Table), the values of C_1 also to be increased to keep C_1/B ratio same as shown the Table.

Rectangular section grit chamber with trapezoidal bottom. As a parshall flume is a rectangular control section, the grit chamber above it must be designed to approach a parabolic section. However, a rectangular section with trapezoidal bottom may be used with a parshall flume. In such a case, the variation in velocity at maximum and minimum flow conditions from the designed velocity of flow should be within permissible limits as given by the following equations:

$$Q = 2264\,W\,(H_A)^{3/2} \qquad \text{...(11.14)}$$

$$D + Z = 1.1\,H_A \qquad \text{...(11.15)}$$

$$\frac{D_{min}}{D_{max}} = \frac{1.1\,(Q_{min.}/2264\,W)^{2/3} - Z}{1.1\,(Q_{max}/2264\,W)^{2/3} - Z} \qquad \text{...(11.16)}$$

$$D = 1.1\left[\frac{Q}{2264\,W}\right]^{2/3} - Z \qquad \text{...(11.16 a)}$$

$$b = \frac{Q_{max}}{1000\,D_{max}\,v_{max}} - \frac{Q_{min}}{1000\,D_{min}\,v_{min}} \qquad \text{...(11.17)}$$

$$v = \frac{Q}{1000\,b\,D} \qquad \text{...(11.18)}$$

where Q = rate of flow in lps.

$Q_{min.}$ = min. rate of flow in lps.

$Q_{max.}$ = max. rate of flow in lps.

W = throat width, in m.

H_A = depth of flow in upstream leg of the flume at one third point, in m (Fig. 11.8)

Z = a constant, in m

D = depth of flow in grit chamber in m

and v = velocity of flow in mps at a particular depth of flow.

Recommended throat width for different ranges of flow along with the dimensions of the various *elements* of the flume (Fig. 11.8) for the different throat width are given in Table 11.5, which should be strictly adhered to. Smaller Parshall flumes are manufactured often of fiberglass, and are supplied with rating curves. The depth of flow can be measured by a float in a stilling well. or by an electronic pressure sensor.

11.8. DISPOSAL OF GRIT

Clean grit is characterised by the lack of odour. Washed grit may resemble particles of sand and gravel, interspersed with particles of egg shell, and other similar relatively inert materials from the house holds. Grit washing mechanism has to be included whenever the detention time is more and flow through velocity is less. Unless washed, it may contain considerable amount of organic matter. This becomes an attraction to rodents and insects and is also unsightly and odorous. The grit may be disposed of by dumping or burying or by sanitary land fill. The ultimate method used, however, depends upon the quantity and characteristics of the grit, availability of land for dumping, filling or burial. In general, unless grit is washed, provision for burial should be made. However, unwashed grit, when mixed with soil, is valuable as soil conditioner and will give good yield of garden crops like cucumbers, squashes, tomatoes etc.

11.9. AERATED GRIT CHAMBERS

The discovery of grit accumulation in spiral-flow aeration tanks preceded by grit chambers led to the development of the aerated grit chambers. An aerated grit chamber is a special form of grit chamber consisting of a standard spiral flow aeration tank provided with air-diffusion tubes placed on one side of the tank, 0.6 to 1 m from the bottom. Aerated grit chambers control the separation of inorganic and organic solids by producing a *rolling flow pattern.* The grit particles tend to settle down to the bottom of the tank at the rates dependent upon the particle size and their specific gravity, and the bottom velocity of roll of the spiral flow. The heavier grit particles with their higher settling velocities drop down to the floor whereas lighter organic particles are carried with the roll of spiral motion and eventually out of the tank. Air flow is regulated to give the separation desired.

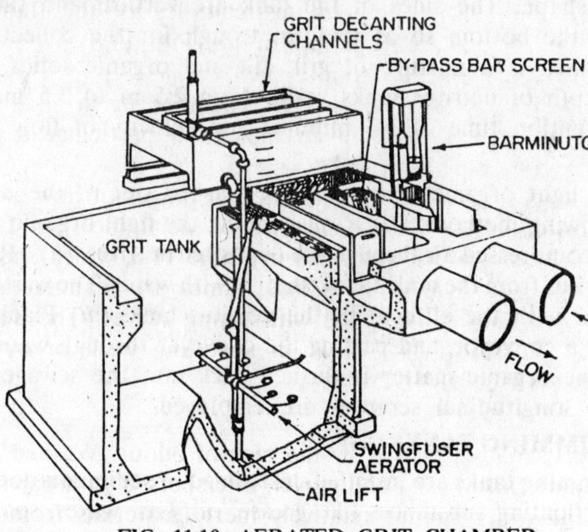

FIG. 11.10. AERATED GRIT CHAMBER.

The velocity of roll, however, should not exceed critical velocity of scour of grit particles. Table 11.6 gives typical design information for aerated grit chambers. Head loss through aerated grit chambers is negligible.

TABLE 11.6. AERATED GRIT CHAMERS.

Item		*Value*	
		Range	*Typical*
1. Dimensions			
(i) Depth	(m)	2–5	
(ii) Length	(m)	7.5–20	
(iii) Width	(m)	2.5–7.0	
(iv) Width-depth ratio		1:1 to 5:1	2:1
2. Detention time at peak flow	(min.)	2-5	3
3. Air supply m^3/min. per in of length		0.15-0.45	0.3

11.10. DETRITUS TANKS

A detritus tank is a grit-removal unit which also removes silt as well as some organic matter along with grit. This is because the flow through velocity is less and detention time is more in a detritus tank. The main idea of installing a detritus tank is therefore, to remove finer particles than those removed by a grit chamber.

A detritus tank is a continuous flow settling tank of rectangular, or square shape. The sides of the tank are vertical and they are tapered at the bottom so as to form trough for the collection of detritus, which is a mixture of grit, silt and organic solids. The over all depth of detritus tanks varies from 2.5 m to 3.5 m. They have a detention time of 3-4 minutes and velocity of flow of 0.2 to 0.3 m/sec.

The light organic matter can be washed out of the detritus by the following methods : (*i*) Resuspending the light organic matter by passing compressed air through the deposited detritus. (*ii*) Removing the detritus from the tank and washing it with water. The wastewater is then sent with the effluent of the detritus tank. (*iii*) Placing the detritus on a conveyor, and passing the conveyer through water such that the fine organic matter is flushed back into the sewage. Both rotary and longitudinal scrapers are employed.

11.11. SKIMMING TANKS

Skimming tanks are installed, just ahead of sedimentation tanks to remove floating substances like grease, oil, fats, waxes, soap, free fatty acids, pieces of cork and wood, fruit skins and vegetable debris etc., Much of these oily, greasy substances enter the sewers from kitchens of restaurants and houses, motor garrages, oil refineries, soap and candle factories etc. If these are not removed, they seriously affect the working of various treatment units. Skimming tank is a chamber so arranged that floating matter rises and remains on the surface of wastewater until removed, while the liquid flows out continuouly through outlet located at depth, or under partitions, curtain walls, deep scum board etc. Most skimming tanks are rectangular or circular, having a detention time of 3 to 5 minutes.

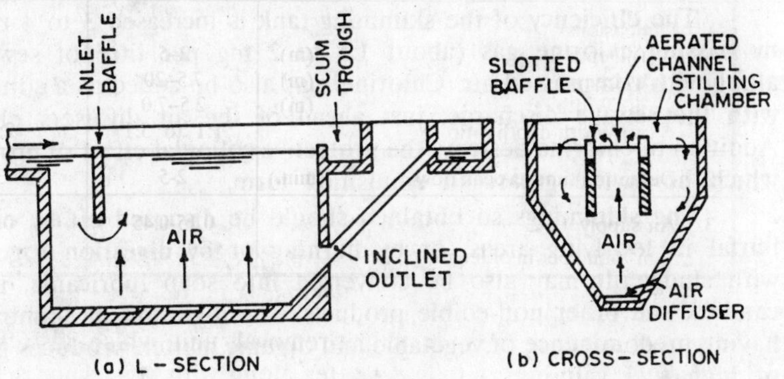

FIG. 11.11. SKIMMING TANK.

Fig. 11.11 shows a long trough-shaped skimming tank, divided into two or three lateral compartments by vertical baffles having slots for a short distance below the wastewater surface, permitting oil and grease to escape into stilling compartments. For efficient working of skimming tank, air diffusers are provided at the bottom of the tank. The rising air bubbles congeal the greasy and oily material and push it to the side compartment (stilling chambers). The amount of air required is about 0.2 m^3/m^3 of sewage. The floating matter can be removed either manually or with the help of some mechanical equipment. The volume of skimmings may vary from 0.625 to 37.5 m^3 per million cubic metre of sewage.

The surface area required for the tank can be found from the following formula :

$$A = 0.00622 \frac{q}{V_r} \quad ...(11.17)$$

where A = surface area of tank (m^2)

q = rate of flow of sewage (m^3/day)

V_r = minimum rising velocity of greasy material to be removed in m/min. (= 0.25 m/min. in most cases)

The sewage enters the tank from one end, flows longitudinally and leaves out through a narrow inclined duct. This is so narrow that the suspended heavier particles are swept up the slope and out of the tank.

The aeration of sewage in skimming tanks has following advantages :

1. It skims the oil and greasy matter out of the sewage and raises it to the surface.
2. It freshens the sewage, supplying some dissolved oxygen.
3. Objectionable gases likes H_2S etc. are driven out.
4. It causes flocculation of colloidal matter.

The efficiency of the skimming tank is increased 3 to 4 times by passing chlorine gas (about 1.5 to 2 mg per litre of sewage) along with compressed air. Chlorine may also be added as a solution with the sewage discharge, just ahead of the air diffusers plates. Addition of chlorine destroys the protective colloidal effect of protein, which holds the grease in emulsified form.

The skimmings so obtained should be disposed off by either burial in low-lying areas, or by burning or by digestion together with sludge. It may also be converted into soap lubricants, pitch candles, and other non-edible products. The digestion of skimmings having predominence of vegetable and organic matter, produces gases of high fuel value.

11.12. VACUUM FLOATATION

Grease, oils etc. can also be removed by subjecting the aerated sewage to vacuum pressure of about 0–25 cm of mercury for 10 to 15 minutes in a *vacuater*. The process is known as vacuum floatation. The vacuum pressure cause the air bubbles, present in the aerated sewage, to expand and move upwards. The rising bubbles lift the grease and other lighter matter to the surface.

The vaccum floatation unit consists of a covered cylindrical tank in which a partial vaccum is maintained. The tank is provided with scum- and sludge-removal mechanism. The floating material is continuously swept to the tank periphery, automatically discharged into scum trough, and removed from the unit to a pump also under partial vacuum. However, prior to the application of partial vaccum, the wastewater is saturated with air, either (*i*) directly in an aeration tank, or (*ii*) by permitting air to enter on the suction side of the wastewater pump. Hence in the vacuum floatation method, auxiliary equipment includes an aeration tank for saturating the wastewater with air, a short-period detention tank for removal of large air bubbles, vacuum pumps, and vacuum and sludge pumps.

Example 11.2. *Design a grit chamber for a maximum wastewater flow of 8000 m^3/day, to remove particles upto of 0.2 mm dia. having specific gravity of 2.65. The settling velocities of these particles is found to range from 0.018 to 0.022 m/sec. Maintain a constant flow through velocity of 0.3 m/sec through the provision of a proportional flow weir.*

Solution. Since the velocity control device is in the form of a proportional flow weir, let us provide a rectangular section for the grit chamber.

Now $v_h = 0.3$ m/sec. and $Q = \dfrac{8000}{24 \times 60 \times 60} = 0.0926 \text{ m}^3/\text{sec}$

$$\therefore \quad A = \frac{Q}{v_h} = \frac{0.0926}{0.3} = 0.3086 \text{ m}^2$$

Providing a depth of 1 m, the width of grit chamber is,

$$B = \frac{0.3086}{1} = 0.3086 \text{ m}$$

Provide a width of 0.35 m.

The settling velocity of the particles to be removed in the grit chamber varies from 0.018 to 0.022 m/sec. Hence let us assume a settling velocity $v_s = 0.02$ m/sec.

$$\therefore \text{ Detention time} = \frac{\text{Depth of chamber}}{\text{settling velocity}} = \frac{1}{0.02} = 50 \text{ sec.}$$

$$\text{Also, length of tank} = v_h \times \text{detention time}$$
$$= 0.3 \times 50 = 15 \text{ m}$$

Hence provide a grit chamber of length 15 m, width 0.35 m and depth 1 m.

Example 11.3. *Design a detritus tank for a D.W.F. of 350 lps in a separate sewage system. Make suitable assumptions wherever required.*

Solution. Let us assume the following :

Detention time : 3 m ; Flow velocity : 0.2 m/sec.

Max. flow : 3 times DWF.

Hence $Q_{max} = 3 \times 350$ litres/second.

Let us provide 3 tanks attached and running parallel to each other. Hence design discharge for each tank is

$$Q = 350 \text{ } l/s = 0.35 \text{ m}^3/\text{s}.$$

$$\therefore \text{ Cross-section area required} = \frac{Q}{v} = \frac{0.35}{0.2} = 1.75 \text{ m}^2$$

Let us provide a water depth of 1.2 m, in the rectangular portion.

$$\therefore \text{ Width of tank} = \frac{\text{Area}}{\text{depth}} = \frac{1.75}{1.2} = 1.458 \text{ m.}$$

Provide a width of 1.5 m.

$$\text{Also, length of tank} = \text{velocity} \times \text{detention time}$$
$$= 0.2\,(3 \times 60) = 36 \text{ m.}$$

Making a provision of 6 metres for inlet and outlet arrangement, the total length of tank = 42 m. Thus, each unit of the detritus tank will be of 1.5 m width and 42 m length. Provide a free board of 0.3 m. Also, provide a bottom depth of 0.5 m for the accumulation

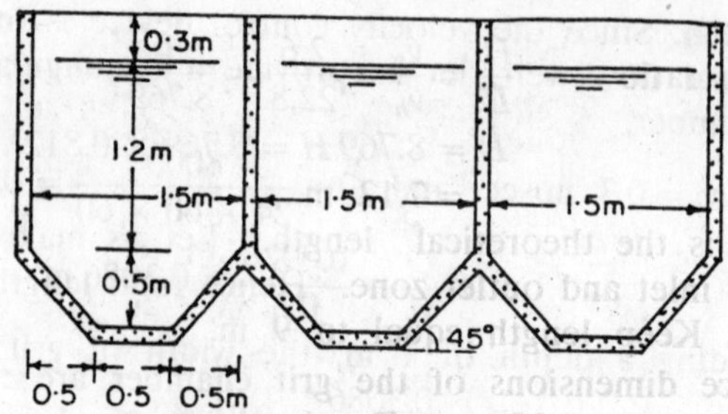

FIG. 11.12. DETRITUS TANK.

of detritus and this depth be tapered at an angle of 45°. Fig 11.12 shows the section of the detritus tank.

Example 11.4. *Design a grit chamber having rectangular cross-section and a proportional flow weir as the velocity control device, for the following data.*

Maximum flow : 20 mLd

Dia. of the smallest grit particles to be removed : 0.2 mm.

Average temperature : 20°C

Specific gravity of grit particles : 2.65

Solution. For grit particles, the settling will be in the transition zone, for which settling velocity is given by Hazen's modified equation:

$$v_s = 60.6\,(S_s - 1)\,\frac{3t + 70}{100} \qquad \text{...(11.3)}$$

$$= 60.6\,(2.65 - 1)\,0.02\,\frac{3 \times 20 + 70}{100}$$

$$\approx 2.6 \text{ cm/sec.}$$

Again, critical velocity is given by the modified Shield's formula. Taking the multiplying factor as 4 in Eq. 11.5, we have

$$v_c = 4\sqrt{g\,(S_s - 1)d} = 4\sqrt{981\,(2.65 - 1)\,0.02}$$

$$\approx 22.8 \text{ cm/sec.} = 0.228 \text{ m/sec.}$$

$$\therefore \quad v_h = v_c = 0.228 \text{ m/sec.}$$

Now $\quad Q = 20 \text{ mLd} = \dfrac{20 \times 10^6 \times 10^{-3}}{24 \times 3600}$

$$= 0.231 \text{ m}^3\text{/sec.}$$

$\therefore$ Cross-sectional area, $\quad A = \dfrac{0.231}{0.228} = 1.0153 \text{ m}^2$

Providing a width of 1.25 m, liquid depth (H) required = 0.812 m. Let us provide a free board of 0.3 m and a space of 0.25 m for sludge accumulation.

$\therefore$ Total depth = 0.812 + 0.3 + 0.25 = 1.362 m.

Hence keep depth = 1.4 m.

Now, ratio $\frac{H}{L} = \frac{v_s}{v_h} = \frac{2.6}{22.8} = \frac{1}{8.769}$

or $L = 8.769\,H = 8.769 \times 0.812$

$= 7.12$ m.

This is the theoretical length. Let us make an allowance of 25% for inlet and outlet zone. Hence total length= 1.25 × 7.12 = 8.901 m. Keep length equal to 9 m.

Hence dimensions of the grit chamber are :

Width = 1.25 m ; Depth = 1.4 m ; Length = 9 m.

For the proportional flow weir as a control section to be used with the rectangular section of the above grit chamber, let us keep $a = 0.035$ m. Then, from Eq. 11.10, taking $C = 0.6$, we get.

$$Q = 0.6\, b \sqrt{2ag}\left(h_a - \frac{1}{3}a\right) \qquad ...(11.10)$$

$$\therefore \quad b = \frac{0.231}{0.6\sqrt{2 \times 0.035 \times 9.81}\left(0.812 - \frac{0.035}{3}\right)} = 0.58 \text{ m.}$$

The co-ordinates of the curve forming the edge of the weir is given by Eq. 11.11 :

$$x = \frac{b}{2}\left(1 - \frac{2}{\pi}\tan^{-1}\sqrt{y/a}\right)$$

or

$$x = \frac{0.58}{2}\left(1 - \frac{2}{\pi}\tan^{-1}\sqrt{y/0.035}\right)$$

The values of x for various values of y will be as under.

***y* (m)**	***x* (m)**	***y* (m)**	***x* (m)**
0.01	0.200	0.40	0.053
0.02	0.171	0.50	0.048
0.03	0.152	0.60	0.044
0.04	0.139	0.70	0.041
0.05	0.128	0.80	0.038
0.10	0.099	0.90	0.036
0.20	0.073	1.00	0.034
0.30	0.061		

Example 11.5. ***Design of parabolic grit chamber :*** *Design a parabolic section of a grit channel for the following data :*

(i) Maximum flow : 60,000 m^3/day

(ii) Minimum flow : 14000 m^3/day

(iii) Average flow : 37000 m^3/day

(iv) Horizontal velocity : 0.25 m/sec.

Solution.

1. Design discharges. Let us provide 2 grit channels, with each having the following discharges :

(*i*) Max. flow $= 60{,}000/2 = 30000 \text{ m}^3/\text{day}$

$$= \frac{30000}{24 \times 60 \times 60} = 0.3472 \text{ m}^3/\text{sec.}$$

(*ii*) Min. flow $= \dfrac{14000/2}{24 \times 3600} = 0.0810 \text{ m}^3/\text{sec.}$

(*iii*) Average flow $= \dfrac{37000/2}{24 \times 3600} = 0.2141 \text{ m}^3/\text{sec.}$

2. Areas of flow. Let B = top width and D = depth of flow, for each condition of flow. Hence, $v_h = 0.25$ m/sec.

$$A_{max.} = \frac{0.3472}{0.25} = 1.3888 \text{ m}^2$$

$$A_{min.} = \frac{0.0810}{0.25} = 0.324 \text{ m}^2$$

$$A_{av.} = \frac{0.2141}{0.25} = 0.8564 \text{ m}^2$$

3. Depth of flow. Let us limit the max. width of channel to 1.8 m at the max. discharge $Q_{max.}$

$$\therefore \quad A_{max.} = \frac{2}{3} B_{max.} D_{max} \quad \text{(for parabolic section)}$$

$$\therefore \quad 1.3888 = \frac{2}{3} \times 1.8\, D_{max.}$$

From which $D_{max.} = 1.157$ m.

4. Determination of critical velocity at max. discharge

Critical velocity (v_c) will take place at the point of jump formation in the throat. The *total energy* E_1 at the upstream of control section is

$$E_1 = D + \frac{v_1^2}{2g} \approx D.$$

(Neglecting the velocity head at approach)

Again, total energy at the critical point is

$$E_c = y_c + \frac{v_c^2}{2g} + h_L$$

Taking *head loss* (h_L) equal to 10% of the velocity head at the control section, and further noting that $y_c = v_c^2/g$, we have.

$$E_c = \frac{v_c^2}{g} + \frac{v_c^2}{2g} + \frac{0.1\, v_c^2}{2g}$$

or
$$E_c = 3.1 \frac{v_c^2}{2g} \qquad \text{...(11.20 } a\text{)}$$

However, since the total energy remains the same, we get

$$E_c = E_1$$

or $$3.1\frac{v_c^2}{2g} = D$$

or $$1.55\frac{v_c^2}{g} = D \qquad ...(11.20)$$

or $$1.55\frac{v_c^2}{g} = 1.157$$

From which $$v_c = \sqrt{\frac{1.157 \times 9.81}{1.55}} \approx 2.71 \text{ m/sec.}$$

*5. **Determination of critical depth of flow and width of throat***

At max. discharge

$$y_c = \frac{v_c^2}{g} = \frac{(2.71)^2}{9.81} \approx 0.75 \text{ m.}$$

Also, flow area at the throat $= W . y_c$.

$$\therefore \quad Q = (W . y_c)\, v_c$$

$$\therefore \quad W = \frac{0.3472}{0.75 \times 2.71} = 0.1708 \text{ m.}$$

Hence let us provide $W = 0.17$ m.

*6. **Parameters at other flow conditions.*** Having determined B, y_c and D at max. discharge, let us find these parameters at other discharges.

$$y_c = \frac{v_c^2}{g} \text{ or } v_c = (g\, y_c)^{1/2}$$

But $$(W . y_c)\, v_c = Q$$

$$\therefore \quad W . y_c\, (g\, y_c)^{1/2} = Q$$

or $$g\, W^2 . y_c^3 = Q^2$$

$$\therefore \quad y_c = \left(\frac{Q^2}{gW^2}\right)^{1/3} \qquad ...(11.21)$$

Since W has already been fixed (= 0.17 m), y_c can be calculated corresponding to each discharge Q.

Also, $$D = 1.55\frac{v_c^2}{g} \qquad ...(11.20)$$

$$\therefore \quad D = 1.55\, y_c \qquad ...(11.22)$$

Thus, D can be determined at various discharges.

Again $$\frac{2}{3}BD = A$$

$$\therefore \quad B = \frac{3A}{2D} \qquad ...(11.23)$$

Since area (A) of flow at various charges are already known in step 2, the width of flow can be computed. The results are tabulated below.

Condition of flow	Q (m^3/sec.)	A (m^2)	$y_c = \left(\frac{Q^2}{gW^2}\right)^{1/3}$	$D = 1.55\, y_c$	$B = \frac{3A}{2D}$
Max.	0.3472	1.3888	0.752	1.165	1.787*
Min.	0.0810	0.324	0.285	0.442	1.100
Av.	0.2141	0.8564	0.545	0.844	1.521

*B fixed as 1.80 m.

From the values of D and B computed above, the parabolic section can be drawn. However, in actual construction, the parabolic section is approximated by the trapezoidal section as shown in Fig. 11.13.

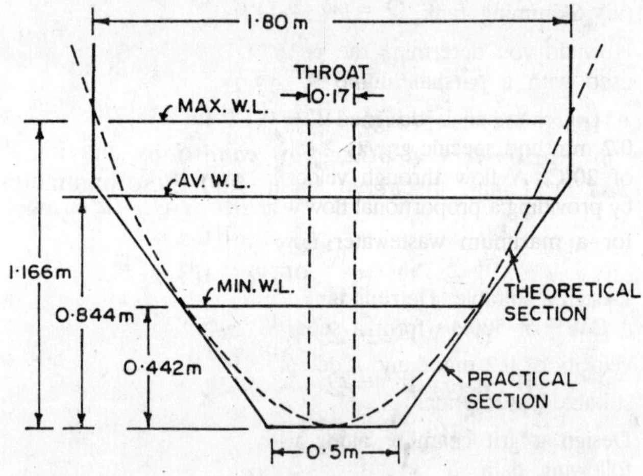

FIG. 11.13. PARABOLIC SECTION OF GRIT CHAMBER.

PROBLEMS

1. What do you understand by preliminary treatment of wastewater ? Enumerate various unit operations falling under this. Also, draw flow diagrams for the possible arrangements of various units falling under preliminary treatment.
2. (*a*) Write a note on racks and screens.
   (*b*) How do you determine head loss through bar screens ?
3. (*a*) Give, in a tabular form, design information for bar racks.
   (*b*) Write a short note on disposal of screenings.

4. Write a note on comminutor and barminutors.

5. What do you understand by grit chambers ? Why it is necessary to provide a grit chamber ? Explain the configuration of a grit chamber with the help of neat sketches.

6. Write a detailed note on various design parameters, such as settling velocity, overflow rate, detention time etc.

7. What do you understand by velocity control device. What are the different velocity control devices used with a grit chamber ?

8. What is a proportional flow weir ? How do you determine the geometry of such a weir ?

9. Write short notes on the following :

(*i*) Sutro weir.
(*ii*) Parshall flume.
(*iii*) Aerated grit chamber.
(*iv*) Vaccum floatation.
(*v*) Detritus tank
(*vi*) Skimming tank.

10. How do you determine the parabolic section of grit chamber to be used with a parshall flume ?

11. A grit chamber is designed to remove particles with a diameter of 0.2 mm and specific gravity 2.65, for an average working temperature of 20°C. A flow through velocity of 0.25 m/sec will be maintained by providing a proportional flow weir. Determine the channel dimension for a maximum wastewater flow of 12000 m^3/day.

Ans. [$D = 1$ m ; $B = 0.6$ m ; $L = 9.6$ m]

12. Design a suitable Detritus tank for a sewage treatment plant receiving a DWF of 500 l/s from a separate sewerage system. Provide a flow velocity of 0.2 m/sec and a detention time of $2\frac{1}{2}$ minutes. Make other suitable assumptions.

13. Design a grit chamber along with proportional flow weir for the following data.

Maximum flow : 12 mLd
Size of grit to be removed : 0.16 mm
Average temperature : 25°C
Specific gravity of grit particles : 2.65

14. Design the parabolic section of a grit chamber for the following data:

Max. flow : 50000 m^3/day.
Min. flow : 12000 m^3/day
Average flow : 30000 m^3/day
Horizontal velocity : 0.25 m/sec.

15. Design a suitable bar screen for a plant treating a peak flow of 50 million litres per day of sewage. Also, compute the head loss through such a screen.

12

Sedimentation and Chemical Clarification

12.1. INTRODUCTION

As stated earlier, *primary treatment* includes those *physical unit operations* and *chemical unit processes* by way of which larger *suspended solids* present in wastewater are removed through sedimentation/ clarification. *Sedimentation* is the separation from water, by gravitational settling of suspended particles that are heavier than water. These suspended solids will be *organic* if the wastewater has been subjected to *preliminary treatment* prior to primary treatment. If, however, no preliminary treatment is given, then primary treatment removes both inorganic as well as organic solids. In general, *sedimentation* is used for grit removal in grit chamber, particulate matter removal in the primary settling basin, biological-floc removal in the activated-sludge settling basin (secondary settling basin) and chemical-floc removal when the chemical coagulation process (*i.e.* chemical clarification) is used. Sedimentation is also used for solids concentration in sludge thickeners.

Sedimentation of raw wastewater is practiced in all large municipal plants and must precede conventional trickling filtration. Chemicals are sometimes added in *primary clarifiers* to assist in removal of finely divided and colloidal solids. *Primary clarifiers* are usually designed to remove particles with settling rates of 0.3 to 0.7 mm/s. Organic solids, settled in primary clarifiers are often stabilized by anaerobic decomposition in a *digestion tank* or incinerated. *Septic tanks* or *Imhoff tanks* incorporate organic solids separation and digestion.

Plain sedimentation : When the impurities are separated from suspending fluid by action of natural forces alone, *i.e.* by gravitation and natural aggregation of settling particles, the operation is called *plain sedimentation*.

Sedimentation with coagulation (Clarification) : When chemicals or other substances are added to induce or hasten aggregation and settling of finely divided suspended matter, colloidal substances, and large molecules, the operation is called sedimentation with coagulation or simply *clarification*.

Chemical precipitation : When chemicals are added to throw dissolved impurities *out of solution*, the operation is called *chemical precipitation*.

Discrete particles : A particle that does not alter its shape, size and weight while settling or rising in water, is known as discrete particle.

12.2. TYPES OF SETTLINGS

Particles may settle out of a suspension in the following four ways, depending upon the concentration of the suspension and the flocculating properties of the particles (Fig. 12.1) :

1. Types I Sedimentation (Discrete settling)
2. Type II Sedimentation (Flocculant settling)
3. Type III Sedimentation (Hindered or zone settling)
4. Type IV Sedimentation (Compression settling)

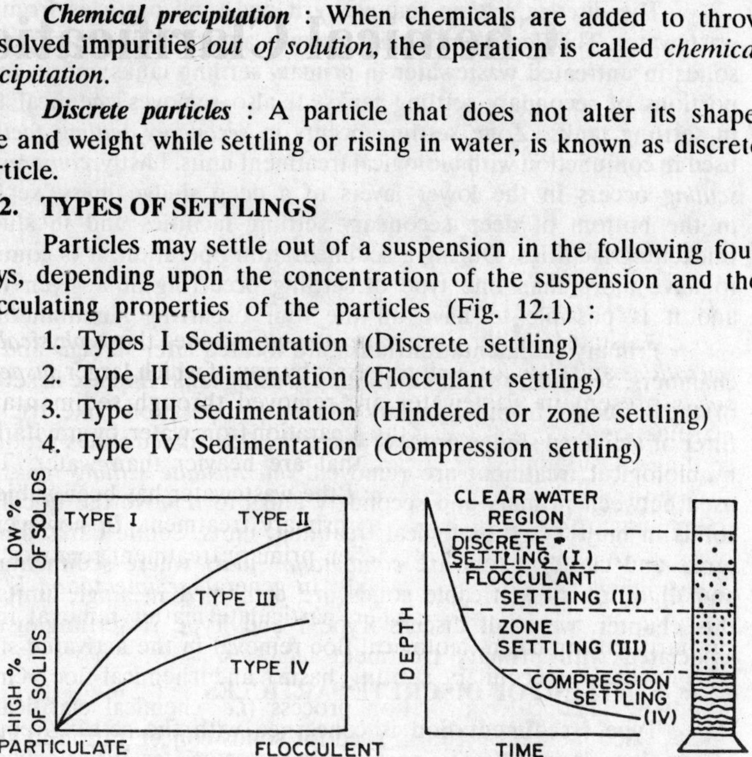

FIG. 12.1. TYPES OF SETTLINGS

Type I : Discrete settling : This corresponds to the sedimentation of discrete particles in a suspension of low solids concentration. This is also known as *free settling* since the particles have little tendency to flocculate or coalesce upon contact with each other.

Type II : Flocculant settling : This type of settling refers to rather dilute *suspension* of particles that coalesce or flocculate during sedimentation process. Due to flocculation, particles increase in mass and settle at a faster rate.

Type III : Hindered or zone settling : This type of settling refers to flocculent suspension of *intermediate concentration*. Inter -particle forces hold the particles together and hence the mass of the particles subside as a whole.

Type IV : Compression settling : This refers to flocculent suspension of so *high concentration* that particles actually come in contact with each other, resulting in the formation of a *structure*. Further settling can occur only by compression of structure brought about due to weight of particles which are constantly being added to the structure.

The *discrete settling* removes grit and sand particles from the wastewater. The *flocculant settling* removes a portion of the suspended solids in untreated wastewater in primary settling tanks and in upper portions of secondary settling tanks. It also removes chemical flocs in settling tanks. Zone settling occurs in *secondary settling facilities* used in conjunction with biological treatment units. Lastly, *compression settling* occurs in the lower layers of a deep sludge mass, such as in the bottom of deep secondary settling facilities and in sludge-thickening facilities. During a sedimentation operation, it is common to have more than one type of settling occurring at a given time, and it is possible to have all the four occurring simultaneously.

Primary sedimentation tanks are located after screens and grit chambers. Secondary sedimentation or settling tanks find use in settling of the effluents from secondary treatment operations like trickling filter or activated sludge proeess where the flocculated solids produced by biological treatment are removed. *Intermediate settling tanks* are used between primary and secondary tanks to remove the floculated solids in multi-stage biological treatment units. Septic tanks, Imhoff tanks and claridigestors are *combination units* where sedimentation and digestion of settleable solids are combined in single units. In this chapter, we shall discuss Type I and Type II settling usually associated with primary treatment facilities.

12.3. SETTLING OF DISCRETE PARTICLES

Type I sedimentation is concerned with the settling/removal of non flocculating, discrete particles from wastewater. When a discrete particle is placed in a quiescent fluid, it will accelerate until the *frictional resistance* (called drag force F_D) of the fluid equals the impelling force F_I (or driving force) acting on the particle. At this stage, the particle attains a uniform or *terminal velocity* and settles down with this constant velocity known as *settling velocity*.

The *impelling force* (F_I) is evidently equal to the *effective weight* of the particle :

$$F_I = (\rho_s - \rho)\, g\, V \qquad \text{...(12.1)}$$

where ρ_s = mass density of particle

ρ = mass density of fluid

V = volume of particle $= \frac{\pi}{6} d^3$, where d is the diameter of spherical particle.

(*a*) **Newton's Law : Turbulent settling.** The *drag force* (f_D) depends upon (*i*) dynamic viscosity (μ) of fluid, (*ii*) mass density (ρ) of the fluid, and (*iii*) shape and size of the particle. The dry force is given by *Newton's law* for frictional drag in the following form :

$$F_D = C_D \, . \, A \frac{\rho v_s^2}{2} \quad ...(12.2)$$

where C_D = drag coefficient = projected area of the particle $= \frac{\pi}{4} d^2$.

Equating the two (*i.e.* Eqs. 12.1 and 12.2) we get an equation for the settling velocity in the form :

$$v_s = \sqrt{\frac{4}{3} \frac{g}{C_D} \frac{(\rho_s - \rho)\, d}{\rho}} \quad ...(12.3\ a)$$

or closely,

$$v_s \approx \sqrt{\frac{4}{3} \frac{g}{C_D} (S_s - 1)\, d} \quad ....(12.3\ b)$$

where S_s = specific gravity of the particles.

Use of Eq. 12.3(*a*) requires the determination of drag coefficient C_D which is related to Reynolds number R by the following observational relationships :

(*i*) For R between 0.5 to 10^4

$$C_D = \frac{24}{R} + \frac{3}{\sqrt{R}} + 0.34 \quad ...(12.4)$$

(*ii*) For high Reynolds number ($R > 10^3$ *to* 10^4), $C_D \approx 0.4$

(*iii*) For low Reynolds number ($R < 0.5$)

$$C_D = \frac{24}{R}$$

where R = Reynold's number $= \frac{\rho\, v_s \, . \, d}{\mu}$

The variation of C_D with Reynolds number R is shown in Fig. 12.2.

For Reynold's number $R > 10^3$ to 10^4, taking $C_D = 0.4$, Eqs. 12.3 (*a*) and (*b*) takes the following form :

$$v_s = \sqrt{3.33\, g \left(\frac{\rho_s - \rho}{\rho}\right) d} \approx \sqrt{3.33\, g\, (S_s - 1)\, d} \quad ...(12.3)$$

This equation is applicable for particles greater than 1 mm and the corresponding nature of settling is called *turbulent settling.*

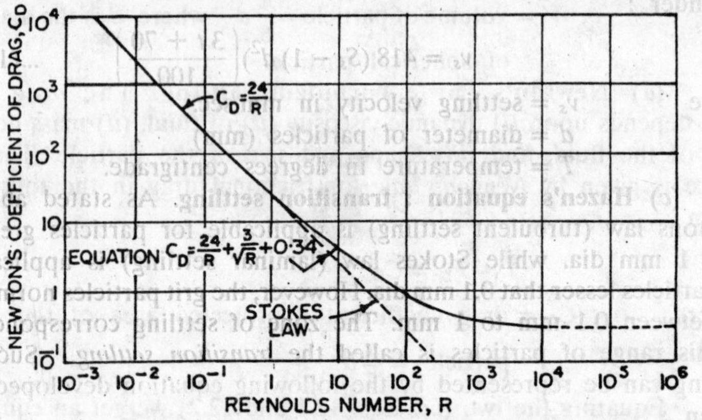

FIG. 12.2 VARIATION OF NEWTON'S C_D WITH R.

(*b*) **Stoke's law : stream line settling.** Stoke's law for the drag of small settling spheres in a viscous fluid, neglecting the inertia force is given by

$$F_D = 3\,\pi\,\mu\,v_s\,d \qquad ...(12.5)$$

Equating this to Eq. 12.2 indicates that the drag coefficient in Newton's law (Eq. 12.2) is given by

$$C_D = \frac{24}{R} \qquad ...(12.4\ a)$$

Substituting this value of C_D in Eq. 12.3 (*a*), we get

$$v_s = \frac{1}{18}\frac{g}{\mu}(\rho_s - \rho)\,d^2 \qquad ...(12.6)$$

or

$$v_s = \frac{g}{18}\frac{S_s - 1}{\nu}\,d^2 \qquad ...(12.7)$$

where v_s = settling velocity (cm/sec)
d = diameter of particles (cm)
μ = dynamic (or absolute) viscosity of liquid (centipoise)
$\nu = \frac{\mu}{\rho}$ = kinematic viscosity of liquid (centipoise)
g = acceleration due to gravity (cm/sec^2)

The above law is applicable for *laminar* or *stream line settling* of particles of diameter up to 0.1 mm, involving the value of R less than 1.

Taking g = 981 cm/sec^2 = 9810 mm/sec^2 and ν = 1.31 centistokes at 10°C, the factor $\frac{1}{18}\frac{g}{\nu}$ comes out to be about 418. Hence Eq. 12.7 reduces to the following form at 10°C :

$$v_s = 418\,(S_s - 1)\,d^2 \qquad ...(12.8)$$

At any other temperature, the above equation may be modified as under :

$$v_s = 418\,(S_s - 1)\,d^2 \left(\frac{3t + 70}{100} \right) \qquad ...(12.9)$$

where v_s = settling velocity in mm/sec.
d = diameter of particles (mm)
T = temperature in degrees centigrade.

(*c*) **Hazen's equation : transition settling.** As stated above, Newtons law (turbulent settling) is applicable for particles greater than 1 mm dia. while Stokes law (laminar settling) is applicable for particles lesser that 0.1 mm dia. However, the grit particles normally lie between 0.1 mm to 1 mm. The zone of settling corresponding to this range of particles is called the *transition settling.* Such a settling can be represented by the following equation developed by Hazen :

$$v_s = 418\,(S_s - 1) \left(\frac{3t + 70}{100} \right) \qquad ...(12.10)$$

The above formula represents theoretical velocities of discrete *spherical* particles. However, the suspended matter is seldom spherical. They may be rod-like or disc-like spheroidal particles which possess greater surface area per unit volume than do spheres. Due to this, they settle more slowly than do spheres of equivalent volume. Also, frictional drag changes with the orientation of the particles relative to the direction of settling. The *actual settling velocity* is generally expressed by *Hazen's modified equation*, applicable for the transition zone :

$$v_s = 60.6\,(S_s - 1)\,d\,\frac{3t + 70}{100} \qquad ...(12.11)$$

where v_s = settling velocity (cm/sec)
d = diameter of particles (cm)
t = temperature of liquid (°C)

12.4. TYPES OF SETTLING TANKS

Settling tanks can be classified in the following three ways.

(*a*) ***Classification based on nature of working***

(*i*) Fill and draw type
(*ii*) Continuous flow type

(*b*) ***Classification based on its location***

(*i*) Preliminary settling tank (grit chambers)
(*ii*) Primary settling tank
(*iii*) Secondary settling tank.

(*c*) ***Classification based on shape***

(*i*) Rectangular settling tank

(*ii*) Circular settling tank

(*iii*) Hopper bottom settling tank.

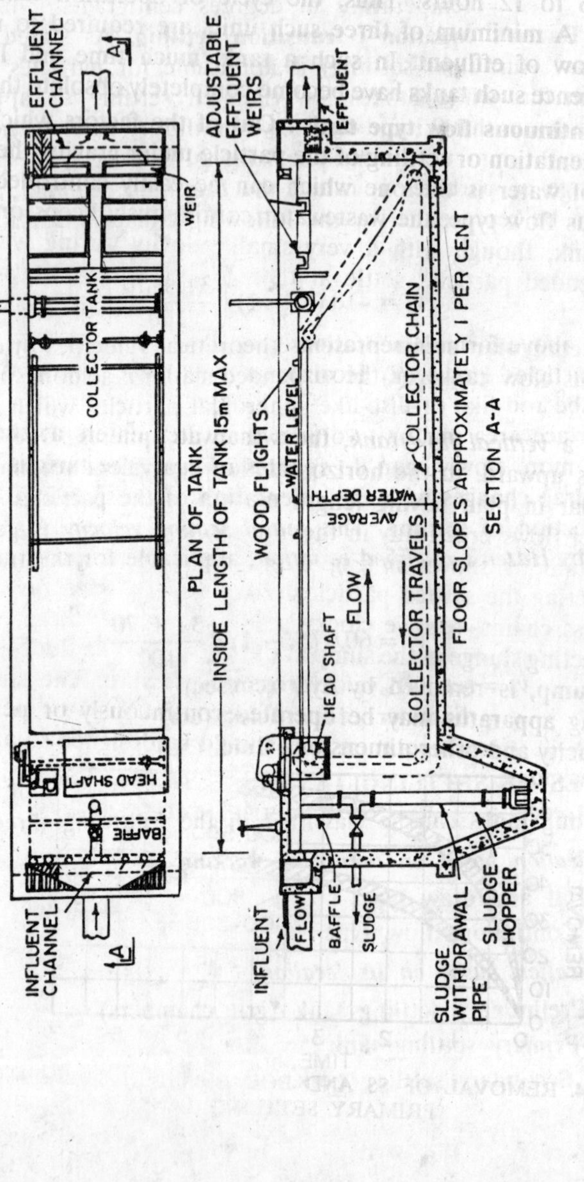

FIG. 12.3. RECTANGULAR SETTLING TANK.

Fill and draw type (or intermittent) settling tank : In this type, the sedimentation tank is first filled with incoming wastewater and is allowed to rest for a certain time. During this rest period, the suspended particles settle down at the bottom of the tank. Generally, a detention period of 24 hours is allowed. At the end of the period, the effluent is drawn off through the outlet weir. The tank is then cleaned of settled particles and filled again. The cleaning may take another 6 to 12 hours. Thus, the cycle of operation takes 30 to 36 hours. A minimum of three such units are required to maintain contant flow of effluent. In such a tank, much time and labour is wasted. Hence such tanks have become completely obsolete these days.

Continuous flow type tank : Out of the factors which hinder the sedimentation or settling of the particle under gravity, the velocity of flow of water is the one which can be easily controlled. In the continuous flow type, the wastewater continuously keeps on moving in the tank, though with a very small velocity during which time the suspended particles settle at the bottom before they reach the outlet.

Rectangular sedimentation tanks : There are two types of continuous flow tanks. (*i*) Horizontal flow tank. (*ii*) Vertical flow tank.

In a *vertical flow tank*, the wastewater enters at the bottom and flows upward. In the horizontal flow type, the tank is generally rectangular in plan having length equal to atleast twice the width. The water flows practically in the horizontal direction. A rectangular settling tank is shown in Fig. 12.3, with mechanical arrangement for collecting the settled particles. The scrapers, which are mounted on endless chains, revolve clockwise, forcing scum to the outlet end and collecting sludge in the sump at the inlet end. The sludge, collected in the sump, is removed by hydrostatic pressure. The sludge concentrating apparatus may be operated continuously or periodically. The capacity and other dimensions should be designed so at to effect

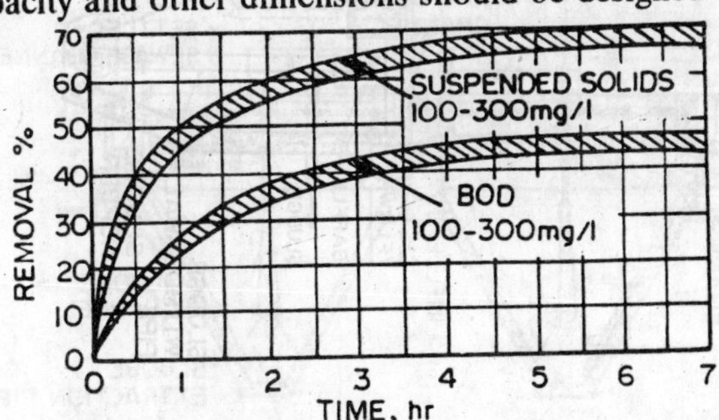

FIG. 12.4. REMOVAL OF SS AND BOD BY PLAIN SEDIMENTATION IN PRIMARY SETTLING TANKS.

a high percentage of removal of suspended organic material. A plain sedimentation may remove about 60% of the solids and 35% of BOD from the sewage. Capacities larger than normal are provided when BOD removal as well as suspended-solids removal is to be high. Fig 12.4 shows the effect of time on percentage removal of suspended solids (SS) and BOD, by plain sedimentation in primary tanks.

Circular settling tanks : In a circular settling tank, the wastewater enters at the center and flows radially outwards to the periphery. Fig. 12.5 shows such a tank. In such a tank, the scrapper shoes are mounted on a rotating frame and the sludge is collected in a sludge sumps located at the centre.

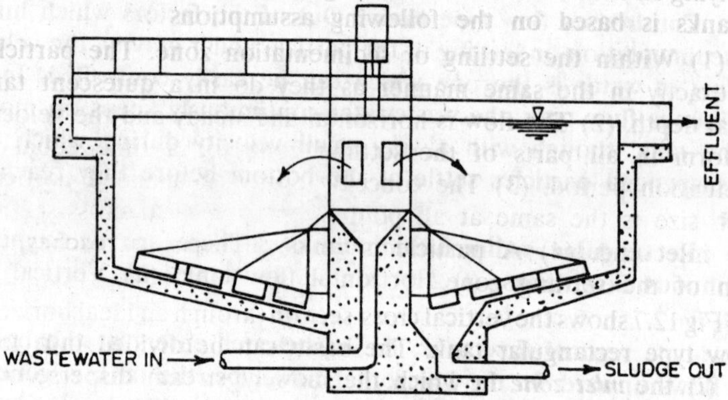

FIG. 12.5. CIRCULAR SETTLING TANK

Vertical flow settling tanks : Vertical flow settling tanks are usually in the form of small diameter circular tanks with deep conical hoppers as shown in Fig. 12.6. The diameter of the tank may vary from 7 to 9 m while the total height may vary from 7.5 to 9 m. They are constructed partly below the ground and partly above the ground. Wastewater, after passing though screens, enters the tank

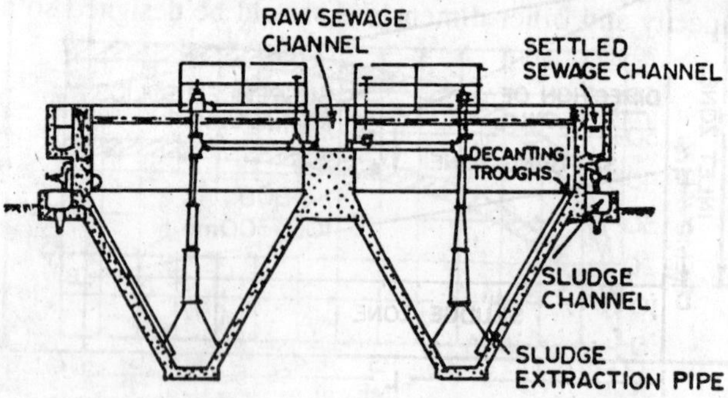

FIG. 12.6. VERTICAL FLOW SETTLING TANKS.

from near the bottom, rises upwards through the sludge blanket, gets clarified and then escapes through an effluent trough located along the top periphery. The tank is so designed that the upward velocity of flow is less than the settling velocity of smallest particle to be removed. Thus, the wastewater can be passed with a relatively higher velocity. Also, shorter detention times are required resulting in small volumes.

12.5. HORIZONTAL FLOW SEDIMENTATION TANK

In the design of a horizontal flow tank, the aim is to achieve as nearly as possible the ideal conditions of equal velocity at all points lying on each vertical in the settling zone, the design of horizontal flow tanks is based on the following assumptions :

(1) Within the settling or sedimentation zone. The particles settle exactly in the same manner as they do in a quiescent tank of equal depth. (2) The flow is horizontal and steady and the velocity is uniform in all parts of the settling zone for a time equal to the detention period. (3) The concentration of suspended particles of each size is the same at all points of the vertical cross-section at the inlet end. (4) A particle is removed when it reaches the bottom of the settling zone.

Fig 12.7 shows the vertical cross-section through an ideal horizontal flow type rectangular tank. The basin can be divided into four zones. (*i*) the *inlet zone* in which the influent stream disperse over the cross-section at right angles to flow, (*ii*) the *settling zone*, (*iii*) the bottom zone or sludge zone., and (*iv*) the *outlet zone,* in which the water and remaining suspended particles assemble and are carried to the effluent conduit.

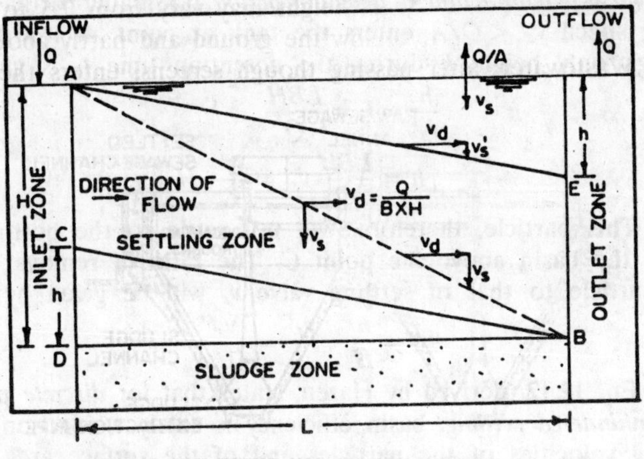

FIG. 12.7. VERTICAL SECTION THROUGH RECTANGULAR TANK.

Let L and H be the length and depth, respectively of the settling zone. Let B be the width of the tank and Q be the discharge rate. The horizontal discharge velocity v_d is given by :

$$v_d = \frac{Q}{H \times B} \quad ...(i) \quad ....(12.11\ a)$$

The time of horizontal flow

$$t_0 = \frac{L}{v_d} = \frac{LBH}{Q} \quad ...(ii) \quad ...(12.11\ b)$$

Consider a particle entering the tank to have a vertical falling velocity v_s. The time for falling through distance (height) H will be H/v_s. For the particle to reach the bottom before the water leaves the tank, the time of fall should be equal to the time of horizontal flow.

Hence, $$\frac{H}{v_s} = \frac{LBH}{Q}$$

From which, $$v_s = \frac{Q}{LB} = \frac{Q}{A} \quad ...(12.11)$$

Eq. 12.11 defines the *overflow rate*, *overflow velocity* or *surface loading*, which is numerically equal to flow divided by plan area of the basin.

Also, from Eqs. 12.11 (a) and 12.11, $$\frac{v_d}{v_s} = \frac{L}{H} \quad ...(12.11\ c)$$

and from Eqs. 12.11 (b) and 12.11, $$t_0 = \frac{H}{v_s} \quad ...(12.11\ d)$$

Eq. 12.11. suggests that all those particles which have falling speed greater than Q/A will reach the bottom before the outlet end of the tank. The time t_0 of the horizontal flow, given by Eq. 12.11 b is known as *detention time* or *detention period*. If a smaller particle having falling speed $v_s' < Q/A$ enters the tank at point A, it will settle through only height h during the detention time t_0 is given by

$$\frac{h}{v_s'} = t_0 = \frac{LBH}{Q}$$

or $$h = \frac{LBH}{Q} v_s' = \frac{v_s'}{v_s} . H$$

This particle, therefore, will not settle to the bottom if it enters the basin above the point C. The ratio of removal of this size particle to that of settling valve v_s will be given by

$$x_r = \frac{h}{H} = \frac{v_s'}{v_s} = \frac{v_s'}{Q/A} \quad ...(12.12)$$

Eq. 12.12, derived by Hazen, states that for *discrete* particles and *unhindered settling*, basin efficiency is solely a function of the settling velocities of the particles and of the surface area of the basin relative to the flow rate. *The efficiency is otherwise independent of the basin depth and the detention period.*

Flowing through period (t_d) and displacement efficiency (η_d) :

Flowing through period (t_d) is the average time required for a batch of wastewater to pass though the settling tank. It is always less than the detention period due to short circuit effects. How close this will approximate the detention period will depend upon the design of the tank.

The displacement efficiency (η_d) is the ratio of flowing through period to the detention period : $\eta_d = t_d/t_0$.

Removal efficiency : The efficiency of a sedimentation basin indicates the overall percentage removal of suspended matter at a given *overflow rate* (v_s). Eq. 12.11 physically represents a settling velocity of the slowest settling particle which is 100% removed. These particles which settle at velocity equal to or greater than overflow rate (v_s) will be entirely removed while those which settle at lower velocities will be removed in direct proportion to the ratio of their settling velocity v_s' to v_s, assuming that they are uniformly distributed upon entering the basin.

For actual suspensions of particles with considerable variety of sizes and densities, prediction of efficiency of basin requires a settling column analysis from which the *cumulative frequency distribution curve*, similar to that of Fig. 12.8, may be obtained.

According to Eq. 12.12, the weight fraction removal of particles having velocity $v_s' < v_s$ will be $\int_0^{x_s} \frac{v_s'}{v_s} dx$

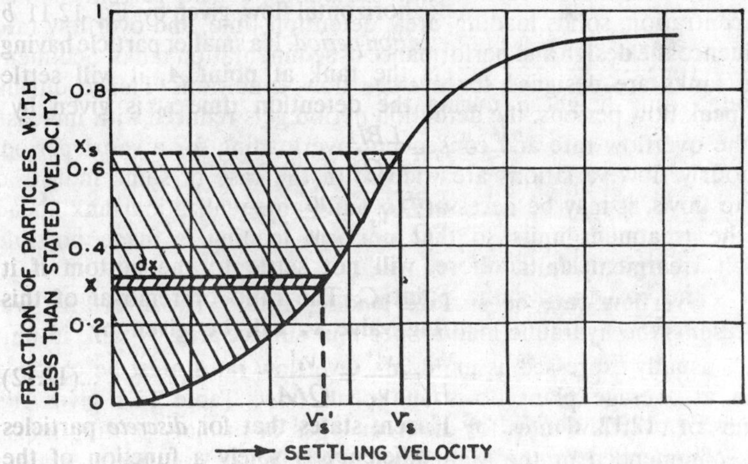

FIG. 12.8 CUMULATIVE DISTRIBUTION OF PARTICLE SETTLING VELOCITY.

Hence the overall removal (R) from the clarified liquid will be

$$R = (1 - x_s) + \frac{1}{v_s} \int_0^{x_s} v_s' \, dx \qquad ...(12.13)$$

in which $(1 - x_s)$ is the fraction with $v_s' \geq v_s$ and the integral is the fraction of particles with $v_s' < v_s$ which are removed in the basin.

The second term in the above equation can be determined by graphical integration of the curve, shown shaded in Fig. 12.8. The actual calculations are performed with a finite number of points on the distribution curve, so that, in practice, Eq. 12.13. is approximated by

$$R = (1 - x_s) + \frac{1}{v_s} \Sigma \, v_s' \, \Delta x \qquad ...(12.13\ a)$$

12.6. DESIGN OF SEDIMENTATION TANKS

The settleable solids to be removed in sedimentaion tanks are mainly organic in nature, dispersed or flocculated. The specific gravity of these suspended solids may vary from 1.01 to 1.20. Generally, raw sewage is a dilute heterogeneous suspension of low specific gravity solids ranging from fully dispersed to completely flocculated ones. In primary sedimentation, the bulk of solids reaching the tank is in a completely flocculated stage or susceptible to flocculation due to fluid motion within the tank. Hence law of sedimentation governing discrete and non-flocculant particles cannot be readily applied for purposes of design. Since the particles are subject to flocculation, the sedimentation tank cannot be designed on the basis of surface area alone and we will have to take into consideration detention time also.

Several factors such as flow variations, density currents, solids concentration, solids loading, area, detention time, and overflow rate influence the design and performance of sedimentation tanks. Sedimention tanks are designed for *average* flow conditions. Hence during the peak flow periods, the detention period gets reduced with increase in the overflow rate and consequent overloading for a short period. If hourly flow variations are wide as in the case of some industrial waste flows, it may be necessary to build an equalization tank ahead of the treatment units so that uniform loading is made possible in all treatment units.

Overflow rate or surface loading rate : The overflow rate represents the hydraulic loading per unit surface area of tank in unit time, usually expressed as $m^3/d/m^2$. Overflow rates must be checked both at average plant flows and peak flow. Table 12.1 gives the values of overflow rates for different types of primary settling tanks, as recommended by the Manual on sewerage and sewage treatment prepared by the Ministry of Urban Development (MUD), New Delhi.

TABLE 12.1. DESIGN PARAMETERS FOR SETTLING TANKS.

Type	Overflow rate ($m^3/d/m^2$)		Depth (m)
	Average	Peak	
1. Primary settling only	25–30	50–60	3.0–3.5
2. Primary settling followed by secondary treatment	35–50	80–125	3.0–3.5
3. Primary settling with activated sludge return	25–35	50–60	3.5–4.5
4. Secondary settling for trickling filter	10–25	40–50	3.0–3.5
5. Secondary settling for activated sludge (excluding extended aeration)	15–35	40–50	3.5–4.5
6. Secondary settling for extended aeration	8–15	35	3.5–4.5

In case of secondary clarifiers following trickling fitters, the design overflow rate must include recirculated flows when clarified effluent is used for recirculation. In the case of activated sludge plants, however, the design overflow rate is to be applied only to the sewage flow and not to the mixed liquor flow to the settling tank. Secondary settling tank of activated sludge plants must be designed not only for hydraulic overflow rates but also for solids loading rates so that adequate sludge thickening may take place and there may be concentrated sludge return. Solids loading rates (based on MLSS, *i.e.* mixed liquor flow to the settling tanks) must be in the range of 100-150 $kg/d/m^2$ for conditions of average flow and should not exceed 250 $kg/d/m^2$ for peak flow conditions. The surface area for activated sludge settling tanks is to be designed for larger value adopted.

Detention period : The rate of removal of BOD and SS is maximum during the first 2 to $2\frac{1}{2}$ hours of settling and thereafter decrease appreciably. Hence increase in detention time beyond 2 to $2\frac{1}{2}$ hours will not increase the percentage removal of BOD or SS proportionately. Longer detention beyond 4 hours may affect the tank performance adversely due to settling in septic conditions particularly in tropical climates. Experience has shown that a detention time of 2 to $2\frac{1}{2}$ hours for primary settling tanks and $1\frac{1}{2}$ to 2 hours for secondary settling tanks will produce optimum results. Longer detention may be necessary in case of industrial wastes with or without chemical precipitation.

Weir loading : The outlet weirs draw off wastewater and take it out without disturbing the quiescent conditions in the tank. Weir loading influence the removal of solids in sedimentation tank, particularly in secondary settling tanks where flocculated solids are settled. There is no positive evidence that weir loading has any significant effect on removal of solids in primary settling tanks. However, certain loading rates based on practice are recommended both for primary as well as secondary tanks. For all primary, intermediate and secondary settling tanks, in all cases, except in the case of secondary tanks for activated sludge process, weir loading not greater than 100 $m^3/d/m$, for average flows is recommended. For secondary settling tanks in activated sludge or its modifications, weir loading should not exceed 150 $m^3/d/m$. The loading should however ensure uniform withdrawal over the entire periphery of the tank to avoid short circuiting or dead pockets. Performance of existing sedimentation tanks can be improved by merely increasing their weir length.

This restriction in weir overflow rate requires *special outlet weir design* including a total weir length several times the tank width (see Fig. 12.3) for rectangular tanks and often two weirs with an outlet channel between them for circular tanks. Very long weirs cannot be maintained truly level over their full length, except perhaps at considerable expense ; and satisfactory distribution of flow is more readily obtained by forming indentations at regular intervals, such as shallow-V-notches say 50 mm deep spaced 0.15 to 0.3 m apart. In addition to the head above the V-notches, reasonable free fall of 0.05 to 0.15 m should be allowed for maximum flows depending in part, on the total head available.

Depth : The depth sets the detention time in the settling tank and also influences sludge thickening in secondary settling tanks of activated sludge plants. The depth recommended for horizontal flow tanks are given in Table 12.1. In vertical flow tanks, depth may be 2.0 m excluding hoppers.

Inlet and outlet devices : Performance of sedimentation tanks is very much influenced by inlet devices which are intended to distribute and draw the flow evenly across the basin. All inlets must be designed to keep down the entrance velocity to prevent formation of eddy or inertial currents in the tank to avoid short circuiting.

(*a*) ***Rectangular Tanks*** : In horizontal flow rectangular tanks, inlet and outlets are placed opposite each other separated by the length of the tank with inlet perpendicular to the direction of flow. Following methods are used to distribute the flow uniformly across the tank : (*i*) Multiple pipe inlets with baffle boards of depth 0.45 to 0.6 m in front of the inlets, 0.6 to 0.9 m away from it and with the top of baffle being 25 mm below water surface for the scum to pass over ; (*ii*) Channel inlet with perforated baffle-side wall between the tank and the channel (Fig. 12.3) ; and (*iii*) Inlet channel with submerged weirs discharging into tank followed by a baffle board inside the tank. A stilling chamber is necessary ahead of inlets if the sewage is received under pressure from pumping mains.

Outlet is generally an overflow weir located near the effluent end, preferably adjustable for maintaining the weir at a constant level. V-notches are provided on the weir to provide for uniform distribution of flow at low heads of discharge over the weir. Weir lengths could be increased by placing outlet channel inside the tank with weirs on both sides (Fig. 12.3). Scum baffles are provided ahead of outlet devices to prevent the escape of scum with the effluent.

(*b*) ***Circular tanks*** : In radial flow circular tanks, the usual practice is to provide a central inlet and a peripheral outlet. The central inlet pipe may be either a submerged horizontal pipe from wall to centre or an inverted siphon laid beneath the tank floor. An inlet baffle is placed concentric to the pipe mouth generally with a diameter of 10 – 20% of the tank diameter and extending 1 to 2 m below water surface. Where the inlet pipe discharges into central hollow pillar, the top of the pillar is flared to provide adequate number of inlet diffusion ports through which sewage enters the tank with an entry velocity of 0.1 to 0.25 m/s through the ports. The entry ports are submerged 0.3 to 0.6 m below water surface.

Outlet is generally a peripheral weir discharging freely into a peripheral channel. The crest of the weir is provided with V-noteches for uniform drawoff at low flows. In all primary settling tanks, a peripheral scum baffle extending 0.2 to 0.3 m below water surface is provided ahead of the effluent weir. If the length of the peripheral weir is not adequate, a weir trough mounted on wall brackets near the periphery with adjustable overflow weir on both sides is provided to increase the length of weir.

Scum removal : One distinct feature of primary settling tank is the skimming device. The skimming device could be operated by the same scraper mechanism used for sludge scraping at the bottom of the tank. It generally consists of a skimmer arm to which a scraper blade is attached and moved, partly submerged and partly projecting above the water surface, from the outlet end towards the inlet end in case of rectangular tanks or in a circular path in the case of circular tanks. The floating scum is thus collected at the forward end of the scraper blade and moved till it is tipped manually or automatically into a scum trough which discharges the scum to a sump outside the tank from where it is removed for burial, burning or feeding to the digester. A scum baffle at least 0.15 m above and extending to at least 0.30 m below water level is provided along the periphery, ahead of outlet device, to prevent the escape of scum with effluent.

Tank dimensions : Circular tanks are more common than rectangular or square tanks. However, rectangular tanks need less space than circular tanks and could be more economically designed where multiple units are to be constructed in a large plant.

Rectangular tanks : For rectangular tanks maximum length and widths of 90 m and 30 m respectively with length to width ratios of 1.5 to 7.5 and length to depth ratios of 5 to 25 are recommended. A minimum depth of 2 m in case of primary settling tanks and 2.5 m in case of secondary settling tanks for activated sludge should be provided. Bottom slopes of 1% are normally adopted. Peak velocities greater than 1.5 mph should be avoided.

Circular tanks : Diameter of circular tanks vary widely from 3 to 60 m although most common range is 12 to 30 m. Diameter and depths could be chosen at the discretion of the designer in conformity with the manufactured sizes of scraper mechanism. The water depth is kept as 2 m. Floors are sloped from periphery to centre at a rate of 7.5 to 10%. The inlet to the tank is generally at the centre and the outlet is a peripheral weir, the flow being radial and horizontal from centre to the periphery of the tank.

Performance : Primary sedimentation of domestic sewage may be expected to accomplish 45 to 60% removal of SS and 30 to 45% removal of BOD, depending on concentration and characteristics of solids in suspension. Secondary settling tanks if considered independently, remove a very high percentage of flocculated solids, even more than 99%, particularly following an activated sludge unit where a high mixed liquor suspended solids is maintained in the aeration chamber. However, the efficiency of secondary biological treatment process is always defined in terms of combined efficiency of the treatment units and its secondary settling tank with reference to the characteristics of the incoming sewage.

12.7. CHEMICAL CLARIFICATION

Chemical clarification is a method of increasing and hastening the settling of suspended and colloidal solids in wastewater by the addition of floc forming or coagulating chemicals. Chemical precipitation of wastewater is analogous to coagulation in water purfication. Coagulation means gathering together of a number of smaller particles into bigger and heavier masses. Certain salts of heavy metals like iron and alumimum when added to wastewater containing alkaline substances develop heavy precipitates which brings down with them colloidal substances. The coagulation and rapid settling is due to two reasons : (*i*) formation of gelatinous insoluble floc which enmeshes or absorbs to its sticky surface the suspended and colloidal solids in wastewater. (*ii*) dissolution of the chemical into high-valent ions which neutralize the electrical charges on the colloidal particles thus causing them to coagulate and settle.

The method produces intermediate results between plain sedimentation and secondary treatment. With proper dose of chemicals, this treatment method may be expected to remove 60 to 80% of SS and 45 to 65% of BOD when it is not preceded by plain sedimentation. Chemical precipitation will not, however, remove dissolved solids. On the contrary, they may add to the dissolved solids content of sewage due to addition of chemicals. The main advantage of this method is its suitability for seasonal and batch operations. However large volumes of sludge (2 to 3 times) produced pose serious disposal problems.

Chemicals used : Table 12.2 gives the most commonly used chemicals for chemical clarification of wastewater.

TABLE 12.2 CHEMICALS USED IN WASTEWATER TREATMENT

Chemical	*Formula*	*Molecular weight*
1. Alum	$Al_2 (SO_4)_3 18 H_2O$	666.7
2. Ferrous sulphate (copperas)	$Fe SO_4 . 7H_2O$	278.0
3. Lime	$Ca (OH)_2$	56 as CaO
4. Ferric chloride	$Fe Cl_3$	162.1
5. Ferric sulphate	$Fe_2 (SO_4)_3$	400

Aluminium salts : Aluminium chloride and sulphate of alumina (filter alum) are the commonly used aluminium salts. Where alum is used, the sludge produced is greater in volume and also bulky than with iron salts making it less easily settleable.

Iron salts : Ferric salts are better coagulants than ferrous salts because of their higher valency and their efficiency over a wider pH range. Ferric salts are effective at all pH values above 3, the efficiency increasing with increase in pH, while the useful pH range of ferrous salts is above 10. But when wastewaters are highly alkaline due to presence of trade wastes, it may be cheaper to use larger dosage of ferrous salts. Chlorinated copperas, which is an equi-molecular mixture of ferric sulphate and ferric choride formed by the addition of chlorine to ferrous sulphate is also used in place of ferric salts.

Lime and sodium carbonate : These are used for pH adjustments to favourable ranges of coagulants especially when sewage is highly

TABLE 12.3.
EFFECTIVENESS OF COAGULANTS FOR SEWAGE TREATMENT

Coagulant	*Dosage (mg/litre)*	*BOD removal as percentage of total present*	*Suspended solids (SS) removal, as percentage of total present*	*pH range for proper functioning*	*Remarks*
1. Alum	40–90	60	80	6.0–8.5	Generally not used for sewage treatmet because of spongy floc resulting in large volume of sludge
2. Sulphate (chlorinated copperas)	35–80	70–80	80–90	5.5–7.0 & 9.0–9.5	Effective for producing sludge for activated sludge process
3. Ferric chloride	25–35	80–90	90–95	5.5–7.0	Forms heavy, dense floc which settles rapidly. Hence used widely for a sewage treatment
4. Ferric sulphate	35–40	60	80	8.0–8.5	Found to be more effective than chlorinated copperas if used in conjunction with lime.

acidic. Lime is sometimes used independently as a precipitant, particularly when iron pickling liquors are present in wastewater. The action may be due to formation of calcium carbonate floc or reactions with small amounts of aluminium or iron salts present in wastewater. Lime incidently help in grit separation, oil and grease removal and is perhaps the cheapest chemical used in chemical precipitation/clarification.

Coagulation-Flocculation aids : Other materials, which are used in relatively small concentrations may be classified as follows : (*i*) *Oxidants*, such as chlorine or ozone ; (*ii*) *Weighting agents*, such as bentonite clay ; (*iii*) *Activated silica* ; and (*iv*) *Polyelectrolytes*. These materials are added either to the coagulation and/or flocculation basins. Table 12.3 shows the effectiveness of commonly used coagulants.

Advantages of coagulation process in sewage treatment :

(1) Chemical clarification is found to be more effective than plain sedimentation, because it results in greater reduction of SS, BOD and colour. (2) Coagulated settling tanks requires lesser space than plain settling tanks. (3) The process is simple in operation and can be suitably adjusted with the changing conditions, (4) The coagulation process also removes phosphates and other nutrients from the wastewater, which may help in controlling *eutrophication*.

Disadvantages of coagulation process in sewage treatment

(1) Biological treatment method, presently adopted for sewage treatment are more effective and complete in themselves and therefore coagulation is not of much assistance. Moreover, coagulation rather makes these processes more difficult. (2) Chemicals used for coagulation, which reacting with sewage, kill some of those micro-organisms which are responsible for the digestion of sludge. (3) The process of coagulation adds to the cost of sedimentation. (4) The process of chemical clarification produces large volumes of sludge which require careful processing and disposal. (5) The process of chemical clarification requires skilled supervision and handling of chemicals.

Uses : Due to above disadvantages, chemical clarification is not widely practiced, except for the following purposes : (1) For treatment of wastewater from industries using some specific chemicals in their processes. (2) For the precipitation of some of the nutrients, such as phosphorous, nitrogen etc. from wastewater. (3) For situation where there is large seasonal variations in sewage flow, such as at hill stations. (4) For emergent situations aimed at increasing the settling capacity of an overloaded plain sedimentation tank.

Unit operations in chemical clarification : The term chemical clarification/precipitation is a treatment process made up to three distinct unit operations : (1) coagulation, (2) flocculation and (3) sedimentation.

1. ***Coagulation*** : It is the process where by chemicals are added to a wastewater resulting in a reduction of the forces tending to keep suspended particles apart. This process physically occurs in a rapid mix or flash mix basin. Mixing is accomplished in a rapid or flash mixing unit designed to create velocity gradients of 300 s^{-1} or more with detention period of 15 to 50 sec. The paddles or propellers of mixing unit are mounted on a vertical shaft and driven by a constant speed motor through reduction gears. The size and speed of the propeller is so selected as to give a propeller capacity of twice the maximum flow through the tank. The shaft speed is generally of the order of 100-200 rpm and power requirement is about 0.1 kW/mLd.

2. ***Flocculation*** : The purpose of flocculation is to increase the collisions of coagulated solids so that they agglomerate to form settleable or filterable solids. The principle of flocculation is accomplished by prolonged agitation of coagulated particles (known as *floccules*) in order to promote an increase in size and/or density. This is achieved either by air flocculation or by mechanical flocculation. Both diffused air and mechanical vertical draft tube are used for air flocculation. Revolving paddle type is the most common of the mechanical flocculator.

Flocculation can be carried out in a separate basin or in an integral part of the clarifier structure. Flocculation tanks are usually in duplicate, with a detention period of 30-90 minutes. The peripheral speed of the paddles is kept in the range in the range of 0.3 to 0.4 m/s. The flow through velocity through the flocculator should be in the range of 0.15 to 0.24 m/s to prevent sedimentation.

Camp and Stein studied the establishment and effect of velocity gradients and developed the following equation for use in the design and operation of flocculation system*.

$$P = \frac{1}{2} C_D \cdot \rho A_p \cdot v_r^3 \quad \text{...(12.14)}$$

$$G = \sqrt{P/\mu V} \quad \text{...(12.15)}$$

$$G t_d = \frac{V}{Q} \sqrt{P/\mu V} = \frac{\sqrt{P V/\mu}}{Q} \quad \text{...(12.16)}$$

where P = power requirement (watts)

C_D = coefficient of drag of paddles (= 1.8 for paddle with flat blade)

---

*See Author's Environmental Engineering Vol. 1 : 'Water Supply Engineering' for details.

v_r = relative velocity of paddles in fluid (usually 0.7 to 0.8 of paddle-tip speed.

A_p = area of paddles

G = mean velocity gradient (s^{-1})

V = flocculator volume (m^3)

μ = absolute viscosity ($N \cdot s/m^2$)

t_d = displacement time ($= V/Q$) in seconds

The following *design criteria* are commonly adopted in a flocculator.

(*i*)	Depth of tank	: 3 to 4.5 m.
(*ii*)	Detention time	: 10 to 40 min. ; normal, 30 min.
(*iii*)	Velocity of flow	: 0.2 to 0.8 m/s ; normal 0.4 m/s.
(*iv*)	Total area of paddles	: 10 to 25% of the cross-section area of the tank.
(*v*)	Peripheral velocity of blades	: 0.2 to 0.6 m/s ; normal, 0.3 to 0.4 m/s
(*vi*)	Velocity gradient (G)	: 10 to 75 s^{-1}
(*vii*)	Factor Gt_d	: 10^4 to 10^5
(*viii*)	Power consumption	: 10.0 to 36.0 kW/mLd
(*ix*)	Outlet flow velocity	: 0.15 to 0.25 m/s.

3. Clarification/sedimentation : The flocculated sewage solids are settled out in a subsequent sedimentation tank. Usually, detention period of 2 hours and an overflow rates of not more than 50 $m^3/d/m^2$ for average flows is adopted in the design of these sedimentation tanks. Table 12.4 gives the recommended surface-loading rates of various chemical suspensions.

TABLE 12.4. RECOMMENDED SURFACE LOADING RATES FOR VARIOUS CHEMICAL SUSPENSIONS.

Suspension	*Loading rate* $m^3/m^2/d$	
	Range	*Peak flow*
1.. Alum floc*	25–50	50
2. Iron floc*	25–50	50
3. Lime floc*	30–60	60
4. Untreated wastewater	25–50	50

*Mixed with the settleable solids in the untreated wastewater and colloidal or other suspended solids swept out by the floc.

12.8. TYPE 2 SETTLING : FLOCCULENT SETTLING

Particles in relatively dilute solution sometimes will not act as discrete particles but will coalesce during sedimentation. As coalescence or flocculation occurs, the mass of the particles increase and settles faster. The density of the composite particle will decrease due to inclusion of water. No mathematical relationship exists that can readily be used to analyse flocculent settling. To develop data, such as those plotted in Fig. 12.9, it is necessary to perform a *settling column test* in which a suspension is placed in a long column (glass cylinder) and is allowed to settle under quiescent conditions. Sample are withdrawn from this cylinder, at several depths at several time intervals and the percentage removal is determined. The numerical value of percentage removal is located at the coordinates identifying the time and depth of the sample and iso-concentration lines connecting points of equal removal are drawn, as shown in Fig. 12.10. Based on these data, for example, a detention time of 2 hours can be expected to remove about 60% of suspended solids. To account for the less than optimum conditions encountered in the field, the design settling velocity or overflow rate obtained from column studies often is multiplied by a factor of 0.65 to 0.85 and the detention times are multiplied by a factor of 1.25 to 1.5.

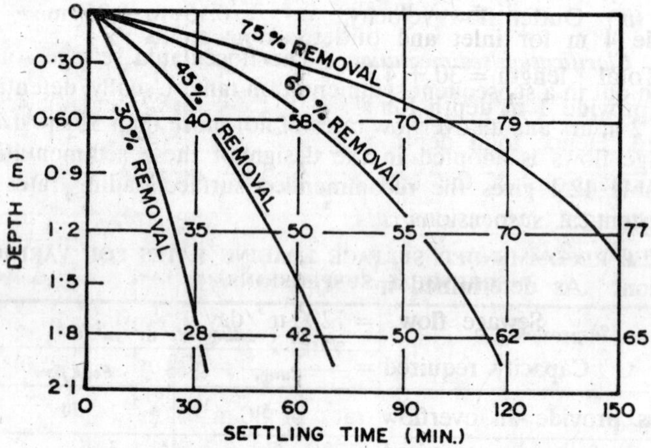

FIG. 12.9. RESULTS OF BATCH SETTLING TEST

In discrete settling, *overflow rate* is an important parameter. But in flocculant settling, in addition to overflow rate, other necessary *design parameters* are depth, detention time and effluent weir flowrate. If any two of the three parameters of depth, detention time or overflow rates are set, the third is established automatically. The relationship is obvious if the overflow rate is considered as a velocity. The effluent weir flow rate is usually limited by standards or design manuals, although there is little evidence that these standards are realistic.

Example 12.1. *Design a primary settling tank of rectangular shape for a town having a population of 50,000, with a water supply of 180 litres per capita per day.*

Solution : Assuming that 80% of water supplied to the city is converted into sewage, total sewage flow= $0.8 \times 50000 \times 180$ $= 7200 \times 10^3$ litres/day. Let us assume a detention period of 2 hours.

$\therefore$ Capacity required= $\dfrac{7200}{24} \times 2 = 600 \text{ m}^3$ ...(1)

Again, let us assume an overflow rate of 30 $\text{m}^3/\text{d}/\text{m}^2$ for average flow (Table 12.1).

Surface area $= 7200/30 = 240 \text{ m}^2$ ...(2)

$\therefore$ Effective depth $= 600/240 = 2.5$ m. ...(3)

Again, $B \times L = 240 \text{ m}^2$

Taking $L = 4\ B$,

$B\,(4\,B) = 240$

From which, $B = 7.46$ m.

Keep $B = 7.5$ m and $L = 30$ m.

Provide 4 m for inlet and outlet arrangements.

$\therefore$ Total length $= 30 + 4 = 34$ m.

Also, provide 1 m depth for sludge accumulation and 0.5 m as free board. Hence total depth $= 2.5 + 1 + 0.5 = 4$ m.

Hence the dimensions of the tank will be 34 m $\times$ 7.5 m $\times$ 4 m.

Example 12.2. *Design a circular primary settling tank for the data of example 12.1.*

Solution. As determined in Example 12.1,

Sewage flow $= 7200 \text{ m}^3/\text{day}$

$\therefore$ Capacity required $= \dfrac{7200 \times 2}{24} = 600 \text{ m}^3$ ...(1)

Let us provide an overflow rate of $30 \text{ m}^3/\text{d}/\text{m}^2$.

$\therefore$ Surface area required $= 7200/30 = 240 \text{ m}^2$

Hence dia. of tank $= \sqrt{\dfrac{240 \times 4}{\pi}} \approx 17.5$ m.

$\therefore$ Effective depth of tank $= \dfrac{\text{Capacity}}{\text{surface area}} = \dfrac{600}{240} = 2.5$ m.

Provide 1 m extra depth of sludge accumulation and 0.5 m depth as free board.

PROBLEMS

1. What do you under stand by sedimentation of wastewater ? Describe in brief various types of settlings.
2. Discuss in brief various types of settling ʻanks.
3. (*a*) Write a note on removal of suspended solids and BOD by plain sedimentation in primary settling tank.

   (*b*) Prove that the area and the overflow rates, rather than detention period govern the design of an ideal settling tank.
4. Why coagulants are used in the sewage treatment ? List various coagulants used along with their effectiveness in sedimentation of sewage.
5. Discuss in brief various design parameters used for settling tanks.
6. Write notes on the following ·

   (*a*) Various types of settlings.

   (*b*) Hazen's modified equation for settling velocity

   (*c*) Inlet and outlet devices for settling tanks.

   (*d*) Outlet weir design and weir loading.

   (*e*) Uses of chemical clarification.
7. Describe in brief various unit operations for chemical clarification along with the design recommendations.

13

Biological Treatment I : Sewage Filtration

3.1. INTRODUCTION

The objectives of the biological treatment of wastewater are to coagulate and remove the nonsettleable colloidal solids and to stabilise the organic matter. *Biological treatment systems are living systems which rely on mixed biological cultures to break down waste organics and remove organic matter from the solution.* Biological treatment systems are designed to maintain a large active mass of bacteria within the system confines. A treatment unit provides a controlled environment for the desired biological process. Domestic wastewater (*i.e.* sewage) supplies the biological food, growth nutrients, and inoculum.

Biological processes are classified by oxygen dependence of primary micro-organisms responsible for waste treatment. Based on this, biological processes may be (*a*) aerobic, (*b*) anaerobic, and (*c*) aerobic-anaerobic. *Aerobic processes* are those which occur in the presence of dissolved oxygen. The aerobic processes include the following : (*i*) Trickling filters ; (*ii*) Activated sludge processes ; (*iii*) Aerobic stabilization ponds ; (*iv*) Aerated lagoons.

Anaerobic waste treatment involves the decomposition of organic and/or inorganic matter in absence of molecular oxygen. Anaerobic processes consist of the followings : (*i*) Anaerobic sludge digestion; (*ii*) Anaerobic contact processes ; (*iii*) Anaerobic filters, and (*iv*) Anaerobic lagoons and ponds.

Aerobic-anaerobic processes are those in which stabilization of waste is brought about by a combination of aerobic, anaerobic and facultative bacteria. Most of the biological treatment processes are *preferred* to work on *aerobic bacterial decomposition* because such decomposition does not produce bad smells and gases as produced by anaerobic decomposition, and also because aerobic bacteria are about three times more active than anaerobic bacteria at 30°C.

13.2. BIOLOGICAL TREATMENT TECHNIQUES

The *biological treatment techniques* used may be classified under the following three heads : (*i*) Attached growth processes (or fixed film processes), (*ii*) Suspended growth processes, and (*iii*) Combined processes.

(*a*) *Attached growth processes (or fixed film processes)* : These are the biological treatment processes in which the microorganisms responsible for the conversion of the organic matter or other constituents in the wastewater to gases and cell tissue are attached to some inert medium, such as rock, slag or specially designed ceramic or plastic materials. Such processes include the followings :

(*i*) Intermittent sand filters; (*ii*) Trickling filters ; (*iii*) Rotating biological contactors ; (*iv*) Packed bed reactors ; (*v*) Anaerobic lagoons (ponds) ; (*vi*) Fixed film denitrification.

(*b*) *Suspended growth processes* : These are the biological treatment processes in which the micro-organisms responsible for the conversion of the organic matter or other constituents in the wastewater to gases and cell tissue are maintained in suspension within the liquid in the reactor by employing either natural or mechanical mixing. In most processes, the required volume is reduced by returning bacteria from the secondary clarifier in order to maintain a high solids concentration. The suspended growth processes include the following:

(*i*) Activated sludge processes; (*ii*) Aerated lagoons; (*iii*) Sludge digestion systems; (*iv*) Suspended growth nitrification and suspended growth denitrification.

(*c*) *Combined processes* : These consist of both attached growth processes as well as suspended growth processes. They include the following in sequence :

(*i*) Trickling filter, activated sludge ; (*ii*) Activated sludge, trickling filter ; (*iii*) Facultative lagoons.

It should be noted that various types of biological processes only help in changing the unstable organic matter into stable forms which are then removed in the secondary settling tanks. *Thus, in the biological treatment processes, secondary settling tanks are essential.* Out of various biological treatment processes enumerated above, *sewage filtration* and *activated sludge processes* are the one which are most commonly used. In this chapter, we shall discuss sewage filtration. Activated sludge processes have been discussed in the next chapter, while other miscellancous processes have been briefly discussed in chapter 15.

13.3. TYPES OF SEWAGE FILTERS

The action involved in sewage filtration consists of building up a few new compounds, which are stable, by combining with oxygen. Filtration satisfies the BOD of organic waste with the help of aerobic bacteria. The *filter units* used for biological treatment consist of *open beds* of coarse aggregates over which effluent from the primary clarifier is applied or sprinkled intermittently. In sewage filtration, mechanical action of filtration is very little only, since only the coarsest particles are arrested by the filtering media which oxidises and nitrifies the organic matter. The necessary *contact surface*, essential for the growth of aerobic bacteria, is provided by the aggregates in the bed and the aeration is provided by nature. The finer the filtering material, the larger is the total superficial area giving *lodgment* to the aerobic bacteria, but the possibility of clogging and reduction in the voids which supply the air are the factors which limit the size of the filtering media. The effluent from the filter units, containing stabilised new compounds, is settled out in *secondary clarifiers.*

Common filters, used for sewage filters, are of the following types : (*i*) Intermittent sand filters ; (*ii*) Contact beds ; (*iii*) Trickling filters.

13.4. INTERMITTENT SAND FILTERS

Intermittent sand filter, falling under the early developments of biological sewage treatment, consists of a layer of sand with an effective size of 0.2 to 0.5 mm, uniformity coefficient of 2 to 5 and of depth 75 to 100 cm. Sewage effluent from primary clarifiers is applied by means of a dosing tank and siphon. The effluent then escapes from the side openings of trough and ultimately, flows on the surface of sand bed. In order to facilitate the drainage of filtered effluent, a layer of about 15 cm to 30 cm depth of gravel is provided at the bottom of sand layer. To carry off the effluent, open jointed drainage pipes are laid in the gravel layer. The filters are generally rectangular in plan, with length to width ratio between 3 to 4, and area of each unit varying from 0.2 to 0.4 hectares. Usually 3 to 4 beds are provided adjacent to each other, so that they can work in rotation.

Operation : On each bed, sewage effluent from primary settling tank is applied intermittently, through a dosing tank containing a siphon. The flooding is done from 5 to 10 cm in depth after an interval of 24 hours. The sewage effluent then percolates through the sand bed, and in this process, the suspended organic matter gets trapped in the voids of top portion of sand, through *straining action*. During the *rest period*, the trapped organic matter is acted upon by aerobic bacteria present in the filter layer. These aerobic bacteria flourish well in the presence of free oxygen available from atmosphere during the rest period when the sewage dose has percolated down.

The effluent from the intermittent sand filter is quite clear, containing suspended solids less than 10 ppm which is well nitrified and stable. Also the BOD of the effluent is less than 5 ppm. However, in order to maintain the efficiency of the filter, the topmost layer of depth of about 25 mm should be raked at regular intervals to break up the materials caught up in the top part of the filter. The sand of the filter should also be renewed from time to time.

The rate of filtration depends upon the size of sand. Finer sand will result in better quality effluent but the rate of filtration is less. Coarser sand will permit higher filtration rate, but in that case the penetration of solids will be too deep. Generally, top 15 cm layer is kept of finer variety, than the rest, which is kept of uniform size.

Advantages : (1) The effluent from intermittent sand filter is of better quality. It is more clean and more stable, and hence does not need further treatment before disposal. (2) The filter works under aerobic conditions, and hence there is no trouble of odour, flies

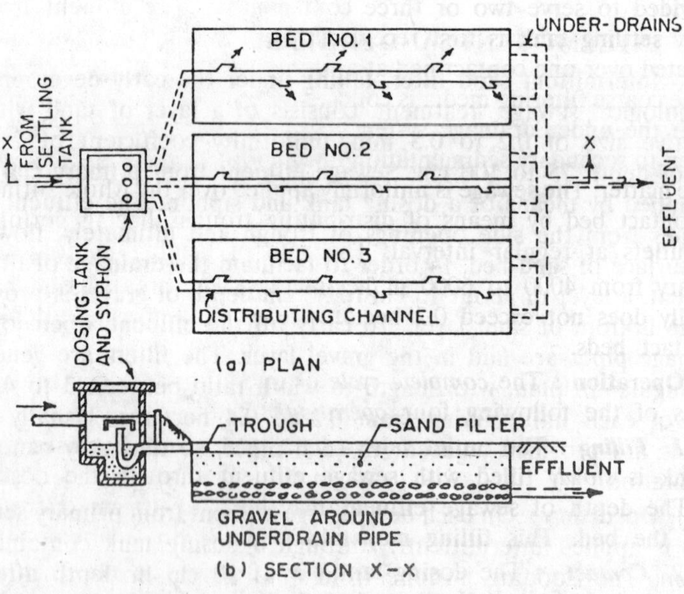

FIG. 13.1. INTERMITTENT SAND FILTER.

and insects. (3) The operation is very simple, requiring no mechanical equipment except for dosing. (4) Smaller head is required for applying the sewage on the surface of the filter. (5) There is no secondary sludge which is to be disposed of, except for the occasional sand scraping.

Disadvantages : (1) The rate of filtration, and hence that of loading is very small, per unit surface area of the filter. Hence they cannot be employed for medium size or bigger plants. (2) They requires large area and large quantity of sand, due to which their construction is very costly.

Due to the above defects, intermittent sand filtration is nowadays not employed except for very small plants such as those for hotels, hospitals, or other social centres placed in remote areas where large area of land is available.

13.5. CONTACT BEDS

Contact beds, also called contact filters, are similar to intermittent sand filters in construction, except that the filtering media is very coarse, consisting of broken stone, called *ballast* of 20 to 50 mm gauge. A contact bed is a water tight tank of masonry walls and of rectangular shape. The depth of filtering media is kept between 1 to 1.8 m (common depth being 1.2 m). The tank is generally dug below ground level and are lined with water-tight cement plaster, or of concrete surfaces instead of masonry. A siphonic dosing tank is provided to serve two or three contact beds. The effluent from primary settling tank is first received by the dosing tank and then distributed over one contact bed at a time. The effluent, after passing over the coarse filtering media is collected at the bottom and conveyed through the under drainage system to the effluent pipe which may be taken to secondary sedimentation tank for settling out the oxidised organic matter. The sewage is uniformly applied over the whole surface of a contact bed by means of distributing troughs having perforations/outlets at regular interval. The rate of loading is slow and may vary from 4000 to 6000 m^3/hectare/day. The area of one bed generally does not exceed 0.2 hectares. Fig. 13.2 shows the details of contact beds.

Operation : The *complete cycle* of operation of a contact bed consists of the following four *operations*

*1. **Filling*** : The outlet valve of the underdrain is closed and the tank is *slowly* filled with sewage effluent through the dosing tank. The depth of sewage effluent may be 5 to 10 cm over the top of the bed. This filling may take about 1 to 2 hours.

*2. **Contact*** : The dosing tank outlet is then closed, and the sewage admitted over the contact bed is allowed to stand for about

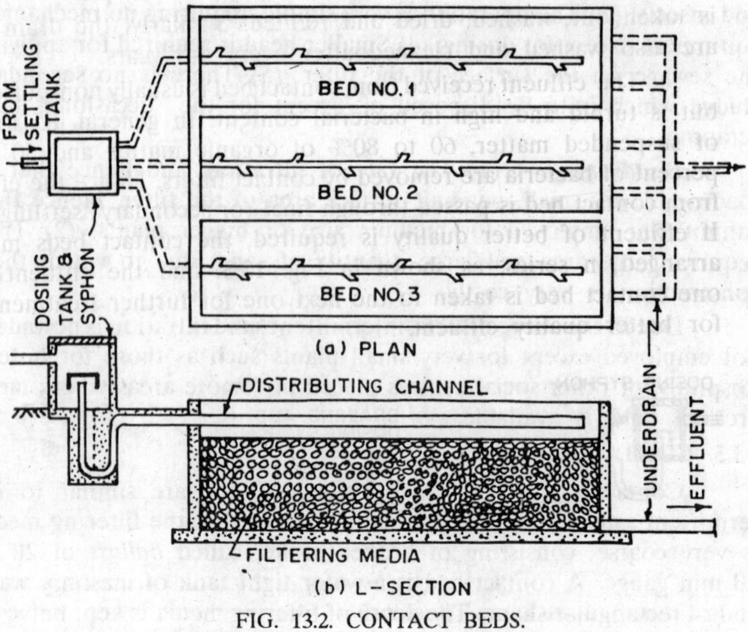

FIG. 13.2. CONTACT BEDS.

2 hours. During this period, the colloidal and dissolved matter gets transferred to the filter media, and comes in contact with the bacterial film covering the filter media.

3. Emptying : The outlet valve of the underdrain is then opened and the sewage present in the contact bed is withdrawn slowly, without disturbing the organic film. This operation may take about 1 to 2 hours.

4. Oxidation : The contact bed is then allowed to stand empty for about 4 to 6 hours. During this period of rest, atmospheric air enters into the void space of the contact media, thus supplying oxygen to the aerobic bacteria, resulting in the oxidation of the organic matter present in the film.

This completes the cycle of operations, which may be repeated after every 8 to 12 hours. Thus each contact bed is given two to three loading per day. The stabilised organic matter from the voids is finally washed down the contact material in the next emptying operation of the tank. However, since the flow of sewage from the primary settling tank is continuous while the bed is operated intermittently, a number of units are constructed side by side, so that sewage can be applied in turn to different units. The voids inside the filtering media go on reducing continuously due to accumulation of solids in them. Hence after every 4-5 years, the filtering media is taken out, washed, dried and refilled. Similarly, the drain pipes are also washed and cleaned after every 4-5 years.

The effluent received from contact bed is usually non-putresible, but is turbid and high in bacterial content. In general, 85 to 90% of suspended matter, 60 to 80% of organic matter and 50 to 75 percent of bacteria are removed by contact filters. Hence the effluent from contact bed is passed through final (or secondary) settling tank. If effluent of better quality is required, the contact beds may be arranged in series, as shown in Fig. 13.3, and the effluent from one contact bed is taken to the next one for further treatment and for better quality effluent.

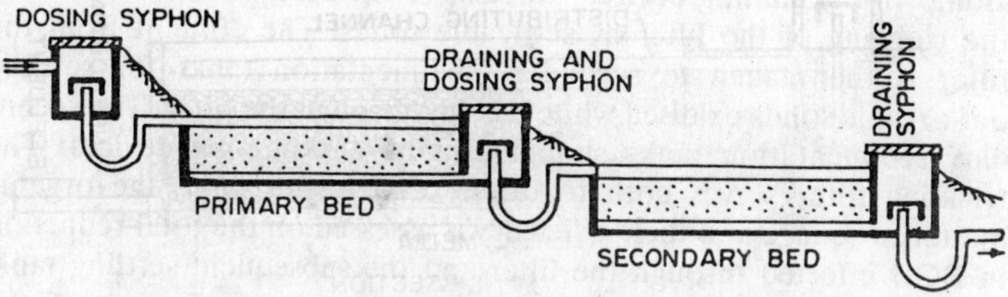

FIG. 13.3. TWO CONTACT BEDS IN SERIES

Advantages of contact of beds : (1) Contact beds can work under small heads. (2) Contact beds can be operated without exposing the sewage effluent to view. (3) There is no nuisance of filter flies. (4) The problem of odour is much less as compared to trickling filters.

Disadvantages : (1) Rate of loading is much less in comparison to trickling filters. (2) Large areas of land is required for their installation. (3) Intermittent operation requires continuous attendance. (4) The cost of contact beds is much more as compared to trickling filters.

Because of the above disadvantages, the contact bed filtration is now only of historical interest, and is not commonly used except for very small installations. These drawbacks led to the evolution of trickling filters where the operation is continuous and also at a much higher rate.

13.6. TRICKLING FILTERS

Trickling filters, also known as *percolating filters* or *sprinkling filters* are similar to contact beds in construction, but their operation is continuous and they allow constant aeration. In this system, sewage is allowed to sprinkle or trickle over a bed of coarse, rough, hard filter media, and it is then collected through the underdrainage system. Spray nozzles or rotary distributors are used for this purpose. The biological purification is brought about mainly by aerobic bacteria which form a bacterial film, known as *bio film*, around the particles of the filtering media. The colour of this film is blackish, greenish and yellowish, and apart from bacteria, it may consist of fungi, algae, lichens, protozoa etc. For the existence of this film, sufficient oxygen is supplied by providing suitable ventilation facilities in the body of the filter and also to some extent by the intermittent functioning of the filter. The straining due to mechanical action of filter bed is much less. Organic removal occurs by biosorption from the rapidly moving part of the flow, and by progressive removal of soluble constituents from the more slowly moving portion.

The trickling filter is always preceded by primary sedimentation along with skimming devices to remove the scum. This will prevent the clogging of the filter by settleable solids. The effluent from the filter is then taken to secondary sedimentation tanks for settling out organic solids oxidised while passing through the filter. The secondary sedimentation tanks should also have skimming devices. The trickling filter serves both to *oxidise* and *bioflocculate* the organic matter in sewage and their efficiency is assessed on the total reduction in BOD effected through the filter and the subsequent settling tank, since the effluent quality is reckoned after the settlement of the bioflocculated solids.

Biological process in a trickling filter : Though, trickling filter is classified as an aerobic treatment device, the *microbial film* (or the *biofilm*) or the *slime layer* formed on the filter medium is aerobic to a depth of only 0.1 to 0.2 mm (Fig. 13.4), and the remaining part of the film is anaerobic. As the wastewater flows over the microbial film, the soluble organic material in the sewage is rapidly metabolised while the colloidal organics are adsorbed onto the surface. In the outer portions of the biological film (or slim layer), the organic matter is degraded by the aerobic micro-organisms. Since food concentration is higher at the outer layer, the micro-organisms near the outer surface are in a rapid growth phase. As the micro-organisms at the outer surface grow, the thickness of the slime layer increases and the diffused oxygen is consumed before it can penetrate the full depth of the slime layer. Hence the lower zone of the film is in a state of starvation, due to which anaerobic environment is established near the surface of the media. As a result of having no external organic source available for cell carbon, the micro-organisms near the media surface enter into an endogenous phase of growth and lose their ability to cling to the media surface. Eventually, there is scouring of the slime layer due to flowing liquid and a fresh slime layer begins to grow on the media. This phenomenon of scouring of the slime is called *sloughing* or unloading of the filter.

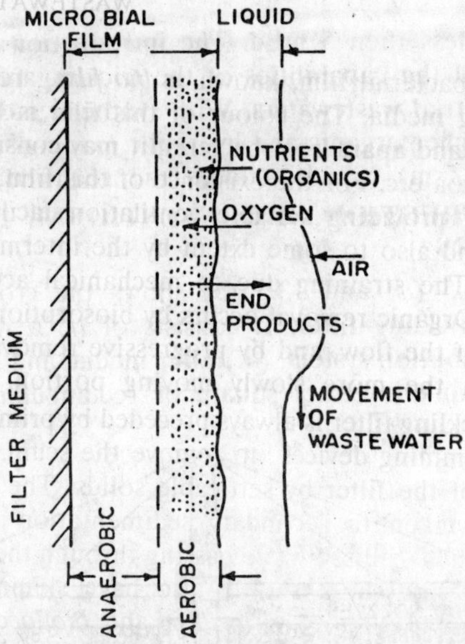

FIG. 13.4. SCHEMATIC DIAGRAM OF ATTACHED GROWTH PROCESS IN A TRICKLING FILTER.

Classification of trickling filters : On the basis of hydraulic and organic loading rates, filters are usually divided into two classes: (*i*) Low rate filters and (*ii*) High rate filters

Although there is no well defined practice, the ranges of loading usual for low rate and high rate filters are given in Table 13.1.

TABLE 13.1. CLASSIFICATION OF FILTERS

	Hydraulic loading $m^3/d/m^2$	*Organic loading* $g/d/m^{-3}$
1. Low rate filters	1–4	80 to 320
2. High rate filters	10 to 30 (including recirculation)	500 to 1000 (excluding recirculation)

The hydraulic loading rate is the total flow including recirculation (see § 13.11) applied on unit area of the filter in a day while the organic loading rate is the 5-day 20 °C BOD, excluding the BOD of the recirculant, applied per unit volume in a day.

Use of trickling filter : Trickling filters are used for biological treatment of domestic sewage and industrial wastes which are amenable to aerobic biological processes. They find use for complete treatment of moderately strong wastes and as roughing filter for very strong wastes prior to activated sludge units. Trickling filters possess a unique capacity to handle shock loads and provide dependable performance with a minimum of supervision. They are particularly suited for plants of capacities less then 5 mLd. The introduction of synthetic media has broadened the capabilities of the trickling filters in successfully treating industrial wastewaters. With synthetic media, trickling filters can handle higher organic and hydraulic loadings and can use deeper beds (upto 12 m), which reduces the required land area.

13.7. CONSTRUCTION OF CONVENTIONAL TRICKLING FILTERS

Fig. 13.5 (*a*) and 13.5 (*b*) shows the cross-section through a slow rate trickling filter. It consists of (*i*) a water tight holding tank, (*ii*) distribution system, (*iii*) filter media, and (*iv*) under drainage system. The tank is either square or rectangular in shape if fixed

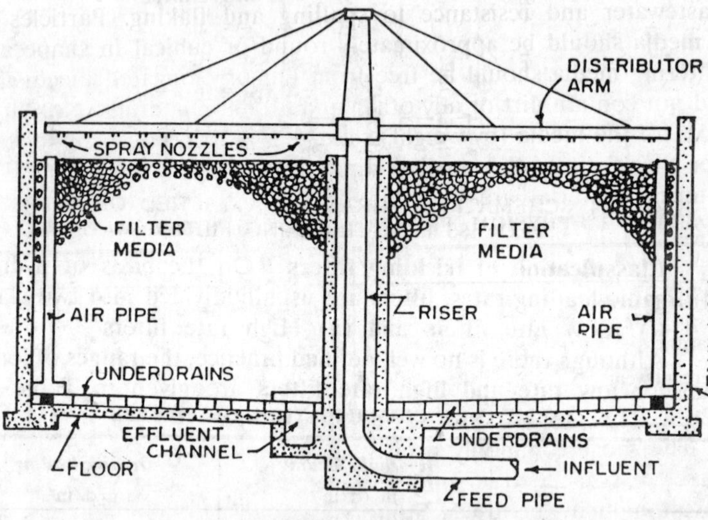

(*a*) TYPICAL CROSS-SECTION

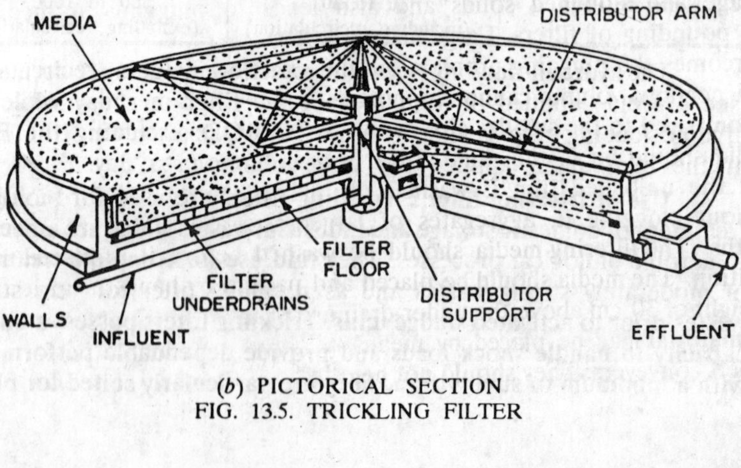

(*b*) PICTORICAL SECTION

FIG. 13.5. TRICKLING FILTER

nozzles are used and circular if rotatry distributors are used. As rotary distributors are more reliable and easy to maintain and operate, circular shape is most commonly used. The tank walls of either masonry or concrete, are made water tight. The distribution system spreads primary sedimentation tank effluent over a bed of filter media supported by a tile under-drain system which also provides adequate ventilation. The walls of the tank are designed to withstand the pressure of sewage from inside. The under-drain system is supported by a floor which slopes to a collection channel.

1. Filter media : The filter media used for trickling filters should have high specific surface area, high percent void space, resistance to abrasion or disintegration during placement, insolubility in sewage or wastewater and resistance to spalling and flaking. Particles of filter media should be approximately round or cubical in shape and the filtering media should be free from flat or elongated pieces and should not contain dirt or any other undesirable materials. Not more than 5% of the media (by weight) should have the longest dimension greater then 3 times the smallest dimension. The physical properties of filtering material are given in Table 13.2.

TABLE 13.2. PHYSICAL PROPERTIES OF FILTERING MATERIAL

Property	*Desired value*
1. Crusting strength	Not less than 100 N/mm^2
2. Hardness	Not less than 12
3. Percent wear	Not more than 4
4. Specific gravity	Not less than 2.6

The most commonly used filter media is broken stone, slag or gravel of size 25 to 75 mm. For stone media, trap rock, granite or limestone may be used. Stones less than 25 mm dia. do not provide sufficient pore space between them to permit free flow of sewage and sloughed solids and also leads to plugging of media and pounding of filters. Large size stones greater than 100 mm dia. overcomes the plugging problem but due to relatively smaller surface area per unit volume cannot support as large a microbial population as compared to the smaller size media. As far as possible, the filtering media should be of the same size from top to bottom. But in order that the unloaded film is washed through the pore space of filter without clogging it, aggregates of larger size are placed at lower depths. The filtering media should be washed before it is placed in position. The media should be placed and packed by hand for at least a height of 30 cm above the under-drainage system. The remainder of the material may be placed by means of wheel barrows or boxes or by belt conveyers. They should not be dumped or tripped from lorries.

Filter depth : It has been observed that the concentration of remaining impurities in sewage decreases as it passes downward through the filtering media. In other words, the effectiveness of filter decreases as depth increases. Hence the modern practice is to restrict the depth of filter and *recirculate* the effluent through the filter. For low rate operation, the filter depth usually varies from 1.8 m to 3.0 m.

Under-drainage System : The purpose of under-drainage system is two fold : (*i*) to carry away the liquid effluent and sloughed biological solids, and (*ii*) to distribute air through the bed. The under-drains cover the entire floor of the filter to form a false bottom and consist of drains with semicircular or equivalent inverts. They are formed of precast vitrified clay or concrete blocks, complete with *perforated cover*, or else they may be formed *in-situ* with concrete or brick and covered with perforated precast concrete slabs. Fig. 13.6 shows a variety of commercially available underdrains.

The slope of the under-drains should be the same as that of the floor, sloping towards a common collection point or channel. The drains shall be so sized that the flow occupies less than 50% of the cross-sectional area with velocity not less than 0.75 m/s at peak instantaneous hydraulic loading. At design flow (including recirculation), the velocity may be between 0.6 to 0.9 m/s. The cover over the drains shall be perforated to provide a total area of inlet openings into the drains not less than 15% of the surface area of the filter. Underdrains may be open at both ends so that they may be inspected easily and flushed out if they become clogged.

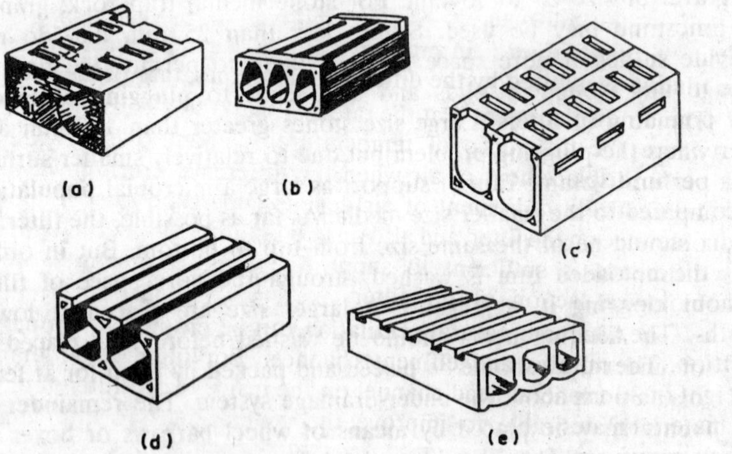

FIG. 13.6. COMMERCIALLY AVAILABLE BLOCKS FOR UNDER-DRAINAGE SYSTEM.

Main collecting channel : The main collecting channel is provided to carry away the flow from the underdrains and to admit air to the filter. In a circular filter, the main channel may be located along a diameter, suitably curved around the central feed well or parallel to the diameter with a slight offset from the centre (Fig. 13.4). Alternatively, the channel may be provided along the periphery of the filter. In the former case, the channel shall be provided with perforated covers to enable drainage and also the ventilation of the filter media above the channel. The channel should be extended outside the filter, both at the upper and lower ends with vented manholes to facilitate ventilation and access for cleaning.

The main collecting channel shall have semicircular or other rounded inverts. The velocity in the channels shall not be less than 0.6 m/s for the average instantaneous hydraulic loading and the flow shall be only half-depth particularly where recirculation is low. At the peak instantaneous hydraulic loading, the water level in the channel should not rise above the inverts of the under-drains at their junctions with the channel.

Filter floor : The filter floor should be strong enough to support the under-drainage system along with fully loaded superimposed filter media. The present practice is to provide a nominally reinforced cement concrete slab, 10 to 15 cm thick, over a proper levelling course. The slope of the floor towards the main collecting channel may be between 0.5 to 5%, usual value being 1%. Flatter slopes are provided in larger filters.

Filter walls : Filter walls may be either of fully plastered stone or brick masonry, or of reinforced concrete. For flooding operation, reinforced concrete is preferred. The walls of the filter are made *honey combed* or otherwise provided with openings for circulation of air, all through.

Ventilation : In properly designed beds, adequate *natural* ventilation is induced by the difference in temperature of air and sewage. The air displacement is at the rate of about 0.3 $m^3/sec/m^2$ of filter bed where the difference of temperature is about 6°C, but there will be no displacement of air when the difference drops down to about 1.9°C. The requirement of air depends upon the rate of application of sewage on the filter, and also on its strength. Usually, 6°C temperature difference is sufficient to supply the required quantity of air for biological action of the filter.

Adequate natural ventilation can be ensured by proper design of under-drains and effluent channels. For filter larger than 30 m dia., a peripheral head channel on the inside of the filter with vertical vents is desirable to improve the ventilation (see Fig. 13.4). One square metre of open grating in the ventlating mannoles and vent stacks should be provided for 250 m^2 of filter area. The vertical vents can also be used in flushing the under drains. In case of deep filters, *forced ventilation* may be provided with advantage if it is properly designed, installed and operated. It consists of forcing the air vertically upwards through the filter by the use of fans or other suitable mechanical equipment. Such a design of forced ventilation should provide for an air flow of 1 $m^3/min/m^2$ of filter area in either direction. It may be necessary during periods of extreme low air temperature to *restrict* the flow of air through the filter to keep it from freezing. However, a minimum of 0.1 $m^3/min/m^2$ of filter area should be provided.

Flow distribution : The function of a *distributor* is to spread or spray the influent evely on the surface of the trickling filter. This can be done from either *fixed sprays* or *moving sprays*. In the fixed sprays, used for rectangular tanks, the sewage is sprinkled through *stationary nozzles* connected to a net work of pipes. These nozzles spray the incoming sewage into the air which then falls over the bed of the filter, under gravity. However, fixed nozzle distributors are no longer used because of the elaborate piping requirements and the necessity of dosing tanks, siphons or motor operated valves to obtain variable dosing rates.

The moving sprays system may be either of longitudinal travelling type or rotary distributor type. Among the moving types, the longitudinal travelling distributors (used for longitudinal or square tanks) are not popular in the country because of long resting period associated with their time of travel from one end of the bed to the other and the need for reversing the gear at each end of the bed to change the direction of motion. The present practice is to provide circular tanks and to use only reaction type *rotary distributors* (Fig. 13.5). Such distributors are available in our country upto 60 m dia. The pipe to the distributor is generally taken below the filter floor and in rare cases through the filter media just above the under-drains. The pipes should be designed for a peak velocity of not greater than 2.0 m/s and an average velocity not less than 1 m/s.

The reaction type rotary distributor consists of a feed column at the centre of the filter, a turn table assembly at the top and two or more hollow radial distributor arms with orifices. When multiple arms (greater than two) are provided, low flows are distributed through two arms only and as flow increases, it is distributed through additional pair of arms. This is achieved by overflows from weirs, incorporated in the central column diverting the higher flows into the additional arms. The peak velocities in the distributor arms should not exceed

1.2 m/s. The orifices in the distributor arms should have spreader plates at the bottom to spread out the discharge. The clearance between the distributor pipe and the top of filter media should be between 15 to 25 cm. The distributors should ensure that the entire surface of the filter is wetted and no area is left dry. Under average flow conditions, the rate of dosing per unit area at any one point in a filter should be within ±10% of the calculated average dosing rate per unit area for the whole filter.

Reaction type rotary distributors require adequate hydraulic head for operation. The head required is generally 1 to 1.5 m measured from the centre line of the distribution arm to the lower water level in the distribution well or in the siphon dosing tank preceding the filter. When the hydraulic head is inadequate to provide the reaction drive, the rotary distributor may be driven by an electric motor. The speed of rotation of the distributors shall ensure that the intervals of successive dosing is between 15 and 20 seconds. The rate of rotation may vary from 2 rpm for small distributors to less than 1/3 to 1/2 for large distributors.

13.8. LOADING, EFFICIENCY AND PERFORMANCE OF CONVENTIONAL TRICKLING FILTERS

1. Loading : The permissible loading on standard rate trickling filters can be expressed in the following units.

(a) Hydraulic loading :

1. As quantity of sewage per hectare of surface area per day

According to this unit the loading varies from 25 to 40 million litres per hectare of surface area per day.

2. As quantity of sewage per unit volume of filter bed

According to this unit, the loading varies from 7.5 to 22.5 million litres per hectare-metre of filter volume per day.

(b) Organic loading

3. As kg of 5-day of BOD per unit volume of filter bed.

According to this unit the loading varies from 1000 to 2200 kg of 5-day BOD per hectare metre of filter volume per day, or as 15 to 30 kg of 5-day BOD per 1000 m^3 of filter material.

(c) Population based loading

4. As number of persons served per hectare of filter surface area.

5. As number of persons served per unit volume of filter media.

The population based loading units are not normally followed.

2. Efficiency : The efficiency of conventional filter can be expressed by the following equation evolved by National Research Council (NRC) of USA :

$$E(\%) = \frac{100}{1 + 0.0044\sqrt{u}} \qquad \text{... (13.1 a)}$$

or

$$E = \frac{100}{1 + 0.44\sqrt{U}} \qquad \text{...(13.1)}$$

where E = Efficiency of the filter and its secondary clarifier, in terms of % of applied BOD removed.

$$= \frac{C_i - C_e}{C} \times 100$$

C_i = influent BOD concentration

C_e = effluent BOD concentration

u = organic loading, in kg/ha-m/day, applied to the filter (*called unit hydralic loading*)

U = unit hydralic loading in $kg/m^3/day$

Eq. 13.1 can also be expressed in the following form

$$\frac{C_i - C_e}{C_i} = \frac{1}{1 + 0.44\sqrt{U}} \qquad \text{... (13.1 b)}$$

Eq. 13.1 indicates that the efficiency of filter decreases with increase in organic loading. The above equation holds good when there is no recirculation.

3. Performance : The normal rate trickling filter has high efficiency in the removal of BOD and other organic matter. The effluent obtained from a conventional trickling filter plant is highly nitrified and stabilized and the slough formed is easily digestible. The suspended solids and BOD are easily reduced by 90% each. The BOD left in the effluent is generally less than 20 ppm. The sludge from the secondary clarifier is thick (moisture content 92%), heavy and easily digestible.

13.9. MERITS AND DEMERITS OF CONVENTIONAL TRICKLING FILTERS

Merits : (1) The effluent obtained from trickling filters is highly nitrified and stabilised. The effluent can, therefore, be disposed of in smaller quantity of dilution water. (2) It has good dependability to produce good effluent under very widely varying weather and other conditions. It possess unique capacity to handle shock loads. Even if they are over loaded, they can recoupe after rest. (3) They can remove about 80% of suspended solids and about 75 to 80% of BOD. (4) The rate of filter loading is relatively higher, in comparision to contact beds or intermittent sand filters. Hence it requires lesser land space. (5) The working of trickling filter is simple and cheap and does not require any skilled supervision. (6) They are self-cleansing. (7) As it contains less mechanical equipment, mechanical wear and tear is small. (8) Operation of trickling filters requires less electrical power to run the mechanical equipment. (9) The moisture content of sludge obtained from trickling filter system is as high as 99%.

Demerits : 1. The loss of head through the filter system is high, thus making the automatic dosing through siphonic dosing tanks necessary. 2. The cost of construction of the filter is high. 3. They require large area in comparison to other biological treatment processes. 4. They require preliminary treatment and therefore cannot treat raw sewage as such. 5. Final settlement in *humus tank* is necessary. 6. The process may develop odour and fly nuisance due to Psychoda which may be carried away into human habitation, proving serious nuisance and health problems.

Example 13.1. *The sewage flows from a primary settling tank to a standard rate trickling fiter at a rate of 5 million litres per day having a 5-day BOD of 150 mg/l. Determine the depth and the volume of the filter, adopting a surface loading of 2500 $l/m^2/day$ and an organic loading of 165 $g/m^3/day$. Also, determine the efficiency of the filter unit, using NRC formula.*

Solution

Total 5-day BOD per day $= (5 \times 10^6 \times 150)\, 10^{-3}$

$= 750000$ g per day

$$\therefore \text{ Volume of filter media required} = \frac{\text{Total BOD}}{\text{Organic loading}}$$

$$= \frac{750000}{165} = 4545.5 \text{ m}^3$$

$$\text{Also, required surface area} = \frac{\text{Total flow}}{\text{Hydraulic loading}}$$

$$= \frac{5 \times 10^6}{2500} = 2000 \text{ m}^2$$

$$\therefore \text{ Depth of bed} = \frac{4545.5}{2000} = 2.273 \text{ m.}$$

Hence provide a filter depth of **2.3 m**

$$\text{Filter diameter} = \sqrt{\frac{2000 \times 4}{\pi}} = 50.5 \text{ m}$$

However, provide the filter of 50 m dia.

$$\text{Hence actual surface area} = \frac{\pi}{4}(50)^2 = 1963.5 \text{ m}^2$$

$$\therefore \text{ Actual volume} = 1963.5 \times 2.3 = \mathbf{4516\ m^3}$$

$$\text{Actual organic loading, } U = \frac{750000}{4516} = 166.1 \text{ g/m}^3\text{/day}$$

$$= 0.1661 \text{ kg/m}^3\text{/day}$$

$$\text{Efficiency} \quad E = \frac{100}{1 + 0.44\sqrt{0.1661}} = \mathbf{84.8\ \%}$$

Examp!e 13.2. ***Design of standard rate trickling filter.*** *Design a circular trickling filter unit for treating 4 million litres of sewage per day, having a 5 day BOD of 160 mg/l. Also design the underdrainage system as well as rotary system for the filter. Assume suitable design data wherever required.*

Solution

(*a*) Design of filter dimensions : While designing the plant, the filter is designed for average flow only, while the distribution arms and under-drainge system etc are designed for peak flow and checked for average flow. The hydralic loading of standard rate filter varies between 1 to 4 $m^3/d/m^2$. Let us assume a hydralic loading of 2 $m^3/d/m^2$

$$\therefore \text{ Surface area required} = \frac{(4 \times 10^6)\ 10^{-3}}{2} = 2000\ m^2$$

Again, the organic loading of standard rate filter varies from 80 to 320 $g/d/m^3$. Let us assume an organic loading of 150 $g/d/m^3$.

Now total BOD present $= (160 \times 10)^{-3} \times (4 \times 10^6) = 640000$ g/day

$$\therefore \text{ Volume of filter media required} = \frac{640000}{150} = 4266.7\ m^3$$

$$\therefore \quad \text{Depth of filter} = 4266.7/2000 = 2.134\ \text{m.}$$

$$\text{Also, dia. of filter} = \sqrt{\frac{2000 \times 4}{\pi}} = 50.46\ \text{m}$$

Let us provide 50 m dia. filter unit.

$$\text{Actual surface area} = \frac{\pi}{4}(50)^2 = 1963.5\ m^2$$

$$\text{Required Depth} = 4266.7/1963.5 = 2.173\ \text{m}$$

Provide filter of 2.2 m depth.

$$\text{Hence, actual organic loading} = \frac{640000}{1963.5 \times 2.2}$$

$$= 148.16\ g/d/m^3 \ \text{(satisfactory)}$$

and $\quad$ actual hydraulic loading $= \dfrac{4 \times 10^6 \times 10^{-3}}{1963.5}$

$$\approx 2.04\ m^3/d/m^2 \ \text{(satisfactory)}$$

(*b*) Design of rotary distributors

1. Design of central column : As stated earlier, the pipe of rotary distributor is designed for a peak velocity of not greater than 2.0 m/s and for average velocity not less than 1 m/s. Let us take peak flow factor as 2.25.

$$\text{Peak flow} = \frac{2.25 \times 4 \times 10^6 \times 10^{-3}}{24 \times 60 \times 60} = 0.1042 \text{ m}^3/\text{sec}$$

$\therefore$ Flow area of central column $= 0.1042/2 = 0.0521 \text{ m}^2$

$$\therefore \text{ Dia. of central column} = \sqrt{\frac{0.0521 \times 4}{\pi}} = 0.2575 \text{ m}$$

$$\text{Average flow} = 0.1042/2.25 = 0.0463 \text{ m}^3/\text{sec}$$

$\therefore$ Velocity at average flow $= 0.0463/0.0521 = 0.89$ m/sec.

This is less than the permissible value of 1 m/sec. To bring it to the permissible value, we have to *reduce* the dia. of central column. However, reduction of dia. of central column will result in the increase in the velocity at the peak flow, which has to be restricted to a value of 2 m/sec. Thus, we find that both the conditions cannot be satisfied simultaneously, specially in absence of recirculation (see example 13.5). Hence provide 25 cm dia. central column.

2. ***Design of arms*** : Let us provide 4 arms for the rotary reaction spray type distributor.

$\therefore$ Peak discharge per arm $= 0.1042/4 = 0.02605 \text{ m}^3/\text{sec}$

$$\text{Length of arm} \approx \frac{50 - 0.25}{2} \approx 24.875 \text{ m}$$

Let us provide 24.88 m long arms with its size reducing from the centre to the end. For this purpose, let us provide 3 sections of arm, with first two sections of 8 m length and the third (or end) section of 8.875 m length. The flow in tnese sections of each arm has to be adjusted in proportion to the filter area covered by these lengths of the arm. Let A_1, A_2 and A_3 be the circular filter areas covered by each length of the arm. Let us provide fcr 0.34 m dia. in the centre for the central column etc.

$$\therefore \quad A_1 = \pi\,[(8.17)^2 - (0.17)^2] = 209.61 \text{ m}^2$$

$$A_2 = \pi\,[(16.17)^2 - (8.17)^2] = 611.73 \text{ m}^2$$

$$A_3 = \pi\,[(25)^2 - (16.17)^2] = 1142.07 \text{ m}^2$$

$$\text{Sum} = 1963.41 \text{ m}^2$$

Check : Total area, $A = \pi\,[(25)^2 - (0.17)^2] = 1963.4 \text{ m}^2$

Hence proportionate areas served by each section of arm are:

$$p_{a1} = \frac{A_1}{A} \times 100 = \frac{209.61}{1963.41} \times 100 = 10.67\%$$

$$p_{a2} = \frac{A_2}{A} \times 100 = \frac{611.73}{1963.41} \times 100 = 31.16\%$$

$$p_{a3} = \frac{A_3}{A} \times 100 = \frac{1142.07}{1963.41} \times 100 = 58.17\%$$

$$\text{Total} = 100\%$$

Discharge through each arm = 0.02605 m^3/sec. The flow through velocity in the arm, at peak flow, should be less than 1.2 m/s.

Design of first section of the arm

$$\text{Discharge} = 0.02605 \ m^3/sec$$

$$\text{Design velocity} = 1.2 \ m/s$$

$$\therefore \text{ Area required} = \frac{0.02605}{1.2} = 0.02171 \ m^2$$

$$\therefore \text{ Dia. required} = \sqrt{\frac{0.02171 \times 4}{\pi}} = 0.166 \ m$$

Hence provide 170 mm dia.

Design of second section

$$\text{Discharge} = \frac{100 - 10.67}{100} \times 0.02605 = 0.02327 \ m^3/sec$$

$$\therefore \text{ Area required} = \frac{0.02327}{1.2} = 0.01939 \ m^2$$

$$\therefore \text{ Dia. required} = \sqrt{\frac{0.01939 \times 4}{\pi}} = 0.1571 \ m$$

Hence provide 160 mm dia.

Design of third section

$$\text{Discharge} = \frac{[100-(10.67+31.16)]}{100} \times 0.02605 = 0.01515 \ m^3/sec$$

$$\therefore \text{ Area required} = 0.01515/1.2 = 0.012625 \ m^2$$

and $$\text{Dia. required} = \sqrt{\frac{0.012625 \times 4}{\pi}} = 0.1268 \ m$$

Hence provide 130 mm dia.

Thus, each arm is made up of three sections : the first section of 8 m length and 170 mm dia, the second section of length 8 m and of 160 mm dia, and the last section of 8.875 m length and of 130 mm dia.

3. Design of orifices : Let us provide 12 mm dia. orifices with a coefficient of discharge (C_d) equal to 0.6 and head causing flow equal to 1.5 m.

$$\therefore \text{ Discharge through each orifice} = C_d \, . \, a \sqrt{2gh}$$

$$= 0.6 \times \frac{\pi}{4} (0.012)^2 \sqrt{2 \times 9.81 \times 1.5}$$

$$= 3.6813 \times 10^{-4} \ m^3/sec.$$

$$\therefore \text{ No. of orifices required in each arm} = \frac{0.02605}{3.6813 \times 10^{-4}} \approx 71$$

No. of orifices (n) in each section of the arm will be as under:

First section, $n_1 = \frac{10.67}{100} \times 71 = 7.58$, say 8

Second section, $n_2 = \frac{31.16}{100} \times 71 = 22.12$, say 22

Third section, $n_3 = \frac{58.17}{100} \times 71 = 41.3$, say 41

Total $= 71$

The spacing (s) of orifices in each section will be as under:

First section, $s_1 = \frac{800}{8} = 100$ cm c/c

Second section, $s_2 = \frac{800}{22} = 36.36$ cm c/c

Third section, $s_3 = \frac{887.5}{41} = 21.65$ cm c/c.

(c) ***Design of underdrainage system***

Peak flow $= 0.1042\ m^3/sec$.

Let us provide central channel of rectangular section, fed by radial laterals of semi-circular section discharging into the central channel. The radial laterals, laid at a slope of 1 in 40, will be in the form of under-drain block lengths containing semi-elliptical openings.

(i) *Design of rectangular effluent channel*

The velocity of flow should not be less than 0.75 m/s at peak instantaneous hydraulic loading or not less than 0.6 m/s at average instantaneous hydraulic loading.

Let us provide a flow velocity of 1 m/s at peak flow.

Peak flow $= 0.1042\ m^3/sec$

$\therefore$ Area of channel $= 0.1042/1 = 0.1042\ m^2$

Let us provide a width of 0.25 m

$$\therefore \quad \text{Depth} = \frac{0.1042}{0.25} = 0.4168$$

Let us provide a width of 0.25 m and a depth of 0.40 m so that :

Area $A = 0.25 \times 0.4 = 0.10\ m^2$

Actual velocity $= 0.1042/0.1 = 1.042$ m/sec.

$$R = \frac{A}{P} = \frac{0.10}{(0.25 + 2 \times 0.4)} = 0.0952$$

The bed slope of the channel is determined by Mannings formula:

$$Q = \frac{1}{N} A R^{2/3} S^{1/2}$$

Taking $N = 0.018$ (say), we have

$$0.1042 = \frac{1}{0.018} \times 0.10\,(0.0952)^{2/3} \,.\, S^{1/2}$$

From which $S = \frac{1}{123.6}$, say 1 in 120

Hence, provide the central effluent channel of width 25 cm, and depth 40 cm below the level of lateral, and lay the channel at a slope of 1 in 120.

(*ii*) *Design of radial laterals*

Let us lay radial under-drains at a slope of 1 in 40. The under-drain block length can be placed in rows, discharging into the effluent channel. In order to ensure proper ventilation, the laterals are designed to run approximately half full,

i.e. $a/A = 0.25$

Corresponding to this, we get, from Table 4.8, by extrapolation

$$d/D = 0.298,\ q/Q = 0.194 \text{ and } r/R \approx 0.680$$

Now permissible velocity at peak flow ≥ 0.75 m/s

$$\therefore \quad q/a = 0.75 \qquad ...(i)$$

$$\text{But} \quad Q = \frac{1}{N} A\, R^{2/3}\, S^{1/2} \quad \text{and} \quad q = \frac{1}{N} a\, r^{2/3}\, S^{1/2}$$

$$\therefore \quad \frac{q}{Q} = \frac{a}{A}\left(\frac{r}{R}\right)^{2/3}$$

$$\text{or} \quad \frac{q}{a} = \frac{Q}{A}\left(\frac{r}{R}\right)^{2/3} \qquad ...(ii)$$

Equating (*i*) and (*ii*) and noting that $\frac{r}{R} = 0.68$, we get

$$0.75 = \frac{Q}{A}\left(\frac{r}{R}\right)^{2/3} = \frac{Q}{A}\,(0.68)^{2/3}$$

From which $\frac{Q}{A} = 0.9699$

$$\text{or} \quad \frac{1}{N} R^{2/3}\, S^{1/2} = 0.9699$$

Taking $N = 0.015$ and noting that $R = D/4$ and $S = 1/40$ we get

$$\frac{1}{0.015}\left(\frac{D}{4}\right)^{2/3} \frac{1}{\sqrt{40}} = 0.9699$$

From which $D = 0.112$ m.

Hence provide 12 cm dia. semi-circular drain blocks.

$\therefore$ Discharge Q through a circular sewer of $D = 0.12$ m is given by

$$Q = \frac{1}{0.015}\,\frac{\pi}{4}\,(0.12)^2\left(\frac{0.12}{4}\right)^{2/3} \frac{1}{\sqrt{40}}$$

$$= 0.01151 \text{ m}^3/\text{sec}$$

Hence $q = 0.194\,Q = 0.194 \times 0.01151 = 0.00223 \text{ m}^3/\text{sec}.$

No. of laterals $= \dfrac{0.1042}{0.00223} = 46.7 \approx 47$ laterals.

Check for velocity at peak flow :

$$a = 0.25A = 0.25 \times \frac{\pi}{4}(0.12)^2 = 2.827 \times 10^{-3} \text{ m}^2$$

$$\therefore \quad v = \frac{q}{a} = \frac{0.00223}{2.827 \times 10^{-3}} = 0.789 \text{ m/sec.}$$

($>$ 0.75 required)

Check for velocity at average flow :

Average discharge $= \dfrac{0.1042}{2.25} = 0.0463 \text{ m}^3/\text{sec}$

Average discharge per lateral, $q = \dfrac{0.0463}{47} = 0.000985 \text{ m}^3/\text{sec}$

$$\therefore \quad \frac{q}{Q} = \frac{0.000985}{0.01151} = 0.0856$$

Corresponding to this value of q/Q, we get, from Table 4.8, by extrapolation,

$$\frac{d}{D} \approx 0.1984 \text{ and } \frac{a}{A} \approx 0.1415$$

Hence $$a = 0.1415 \times \frac{\pi}{4}(0.12)^2 = 1.6 \times 10^{-3} \text{ m}^2$$

$$\therefore \quad v_{av} = \frac{q_{av}}{a_{av}} = \frac{0.000985}{1.6 \times 10^{-3}}$$

$$= 0.615 \text{ m/s } (> 0.6 \text{ required})$$

Hence provide 47 radial laterals of semi-circular section, of 12 cm dia., laid at a slope of 1 in 40, each discharging into the rectangular effluent channel of width 25 cm and depth 40 cm.

13.10. HIGH RATE FILTRATION

The basic difference between high rate trickling filter and slow rate or conventional trickling filter is that the rate of filter loading (both hydraulic as well as organic) of the former is several times more than that of the latter. The main defect of the conventional trickling filter is that it has high initial cost, it requires large area of construction and it requires large quantity of filtering media. Experiments conducted on trickling filters with increasing rate of sewage flow revealed the following :

1. As the sewage flow is increased, the thickness of the gelatinous biofilm is reduced and the organic materials deposited on the contact surface is continuously washed away with the effluent.

2. Thinner biofilm is more efficient and supplies more continuous nutrients to the aerobic bacteria.

3. The precipitation and biological coagulation of the dissolved and colloidal matter is more or less of the same degree as in normal rate filters.

4. However, there is lesser oxidation of organic matter because of reduction in the contact period.

5. Since large quantity of unloaded putrescible organic material reachs the secondary settling tank, the load on the secondary settling tank is increased.

6. The sludge produced as a result of high rate filtration is not easily digestible.

7. The cost of construction and land etc. decreases with the increase in rate of filtration.

Due to the above favourable observations, high rate trickling filters have become more popular, and they have replaced conventional trickling filters in many countries.

To achieve high rate filtration, the following modifications are made to the conventional or slow rate trickling filters.

1. Better quality filtering material is used, so as to give higher specific surface. The recent trend is towards the use of larger size stone media, or to use plastic synthetic media.

2. The depth of filter media is reduced to about 1.5 to 2.0 m, so as to obtain better aeration in order to obtain high rate of biological activity.

3. The size of under-drains is increased and their slope is also made steeper so that the filter effluent can be collected and conveyed to the secondary settling tank quickly.

4. The speed of radiation of the rotating arm is increased to 2 rpm for increased hydraulic loading.

5. The size of secondary settling tank is also correspondingly increased to cope with the increased quantity of flow and bio-flocculent solids coming out with the trickling filter effluent.

Filter media : The size of filter media is of considerable significance as the specific surface decreases with increase in media size but the percent void space increases, as indicated in Table 15.3.

Stone media. The recent trend is towards the use of larger size media specially for high rate filters. The current specification for *stone media* is that when mechanically graded over vibratory screens :

100% should pass through 110 mm square mesh
95-100% should be retained on 75 mm square mesh
0.2% alone should pass through 50 mm square mesh
0.1 % alone should pass through 25 mm square mesh.

The above percentages are by weight.

TABLE 13.3. PHYSICAL PROPERTIES OF TRICKLING FILTER MEDIA.

Media	Size (mm)	Unit weight (kg/m^3)	Specific surface (m^2/m^3)	Void space (%)
1. Granite	25–75	1440	62	46
2. Granite	100	1440	47	60
3. Slag	50–75	1090	67	49
4. Plastic sheet	600×600×1200	32–96	82–115	94–97
5. Red wood slats	1200×1200×500	165	46	76

Plastic media : Plastic media have high specific surface area, high voids space and low unit weight. Synthetic filter media have of late been used successfully in *super-rate filters* for the treatment of strong industrial wastes or sewage mixed with strong industrial wastes having hydraulic loading rates in the range of 30-90 $m^3/d/m^2$ and organic loading rates of 1000 to 2000 $g/d/m^3$. The media consists of interlocking sheets of plastics which are arranged in a honey comb fashion to produce a porous and nonclog filter media. The sheets are corrugated so that a strong, light weight media is obtained. In trickling filters employing synthetic media, aerobic organism (heterotrophic) are present throughout the depth of the filter, primarily because of the higher oxygen transfer capability even with the higher hydraulic loadings. Therefore, in the trickling filters using synthetic media, organic removal occurs throughout the depth of the filter. Filters as deep as 10 m have been used with this type of synthetic media.

Wood media : Wood media have been used as a substitute for rock or plastic, and they are used in *super-rate filters* or *roughing filter*. Redwood slats are fabricated into flats which are stacked upon one another to fill the basin.

Recirculation : Recirculation consists in returning a portion of the treated or partly treated sewage to the treatment process. Recirculation is necessary to provide uniform hydraulic loading as well as to dilute the high strength wastewaters. In contrast to the low rate filters, in high rate filters a part of settled or filter effluent is recycled through the filter. Further details on recirculation are given in the next article.

Single and two stage plants : A single stage plant is the one in which the sewage is passed through single filter; if there are two or more filters, they are operated *parallel*. Sewage may be, and usually is, recirculated to single stage filters. Two stage filters consist of two filters in series with primary settling tank, an intermediate settling tank which may be omitted in certain cases, and a final settling tank. Recirculation facilities are provided for each stage. The effluent from the first stage filter is applied on the second stage filter either after settlement (in the intermediate settling tank) or without settlement. Two stage filtration will provide a higher degree of treatment than the single stage for the same total volume of media. Two stage units are used for strong sewage when the effluent BOD has to be less than 30 mg/l. The same volume of medium is used for single stage and two stage plants. With two stages, however, the volume of filter medium is divided, usually equally, between the first-stage and second-stage units. After primary settling, the entire organic load is applied to the first-stage (or *primary*) filter, due to which the first-stage filter is a so-called *roughing filter*, the principal function of which is to remove BOD and prepare the sewage for treatment in the second-stage filter.

TABLE 13.4. COMPARISON BETWEEN CONVENTIONAL AND HIGH RATE TRICKLING FILTERS

Characteristics	*Conventional or low rate filter*	*High rate filter*
1. Depth of media	1.8 to 2.4 m	1.2 to 1.8 m
2. Hydraulic loading ($m^3/d/m^2$)	1 to 4	10 to 30 (including recirculation)
3. Organic loading as 5-day BOD in $g/d/m^3$	80 to 320	500 to 1000 (excluding recirculation)
4. Recirculation system	Usually not provided, but can be provided if the hydraulic load does not exceed the limit	Always provided. Recirculation ratio 1:1 to 4:1
5. Volume of bed	5 times	1
6. Interval of dosing	≯ 5 minutes. The sewage is applied at intervals.	≯ 15 seconds. Sewage is thus applied continuously.
7. Sloughing	Intermittent	Continuous
8. Cost of operation	More	Less
9. Land required	More	Less
10. Characteristics of final effluent	Contains BOD ≤20 mg/l; it is highly nitrified into nitrate stage.	Contains BOD ≥30 mg/l; it is not fully nitrified.
11. Secondary sludge	Highly oxidized, Black colour, having light fine particles	Not fully oxidized; brownish black colour containing fine particles

Advantages of high rate filters :

(1) Since smaller filter volume is required, initial cost is less.

(2) Operating costs are also low.

(3) The trouble of bad smell or odour is much less.

(4) There is absence of trickling filter flies.

(5) The working is flexible. Hence the efficiency of filter is not seriously affected due to variations in the strength and character of sewage.

Disadvantages : (1) The effluent is not highly nitrified. Hence the effluent from the system requires more volume of dilution water. (2) Raw sewage cannot be treated and the process requires primary treatment of sewage.

Comparison with standard rate filter : Table 13.4 gives the comparison of standard rate and high rate filters with respect to various essential features.

13.11. RECIRCULATION AND TRICKLING-FILTER FLOW SHEETS

Recirculation is the return of a portion of treated or partly treated sewage to the treatment process. Usually, the return is from the secondary settling tank to the primary settling tank or to the dosing tank of the filter. Recirculation of sewage is an important feature of high rate filter. It is expressed in terms of *recirculation ratio* (R).

Recirculation ratio : The ratio of recirculated flow to the flow of raw sewage is called recirculation ratio (R). This ratio determines the required capacity of recirculating pumps and the hydraulic load placed upon the filters. Thus, we have

Capacity of recirculating pump $= R \times$ (influent sewage flow)

Hydraulic load of filter $= (1 + R) \times$ (influent sewage flow).

Recirculation factor : The number of *effective* passes through the filter is known as *recirculation factor* (F) and is given by the equation :

$$F = \frac{1 + R}{[1 + (1 - f)R]^2} = \frac{1 + R}{(1 + 0.1R)^2} \qquad ...(13.2)$$

where F = treatability factor (= 0.9 for sewage)

Hydraulic recirculation factor : The number of hydraulic passes of recirculated sewage through the filter is called hydraulic recirculation factor (F_h) and is given by

$$F_h = \frac{\text{Inflow + Recirculation}}{\text{Inflow}} = 1 + R \qquad ...(13.2\ a)$$

Advantages of recirculation : (1) Recirculation keeps the self propelled distributors running at the time of reduced flows (*i.e.* during night hours). (2) The thickness of biological film on contact media is reduced by forced film sloughing. (3) The filter influent is freshened due to which foul odour is prevented. (4) The filter influent is diluted and weakened so that filter works at a constant efficiency and the quality of filter effluent is improved. (5) The applied sewage is seeded with active organisms and enzymes of effluent, due to which the efficiency of filter is increased. (6) Recirculation loads deeper portions of the filters more effectively.

Recirculation ratio usually range from 0.5 to 3 and values exceeding 3 are considered to be uneconomical in the case of domestic sewage, but ratio of 8 and above have been used in industrial wastes. If the sewage has high concentration of 5-day BOD, it can be effectively treated by passing it a number times through the trickling filters and settling tanks. However, recirculation of sewage for more than 3 times is not beneficial because there is less response of the organic material (to the removed) to removal and this response goes on reducing in each successive recirculation cycle.

Table 13.5 gives common recommended recirculation ratios for high rate trickling filter units.

TABLE 13.5 RECOMMENDED RECIRCULATION RATIOS FOR HIGH RATE TRICKLING FILTERS

BOD of Raw sewage (ppm)	*Recirculation Ratio*			*Hydraulic recirculation factor for designing capacity of secondary clarifiers*	
	Single stage		*Two stage*		
	Intermediate treatment	*Full treatment*	*Each stage*	*Single stage*	*Two stage*
upto 150	1	0.75	0.50	1.75	1.50
150 –300	2	1.50	1.00	2.50	2.00
300–450	3	2.25	1.50	3.25	2.50
450–600	4	3.00	2.00	4.00	3.00
600–750	5	3.75	2.50	4.75	3.50
750–900	6	4.60	3.00	5.60	4.00

Flow diagrams of high rate trickling filter plants

(a) Single Stage Filter : High rate filters may be single stage or two stage. Media depths of 0.9 to 2.5 m have been used for high rate filters with an optimum range of 1.5 to 2.0 m for the first stage and 1 to 2 m for the second stage filters. *Single stage* units consist of a primary settling tank, the filter, secondary settling tank and facilities for recirculation of the effluent. Settling before recirculation is carried out either in the primary or the secondary clarifier as shown in Fig. 13.7 (*ai*), (*aii*) and (*aiii*). Recent studies have shown that for municipal sewage, direct recirculation without settling is as effective as recirculation of settled effluent.

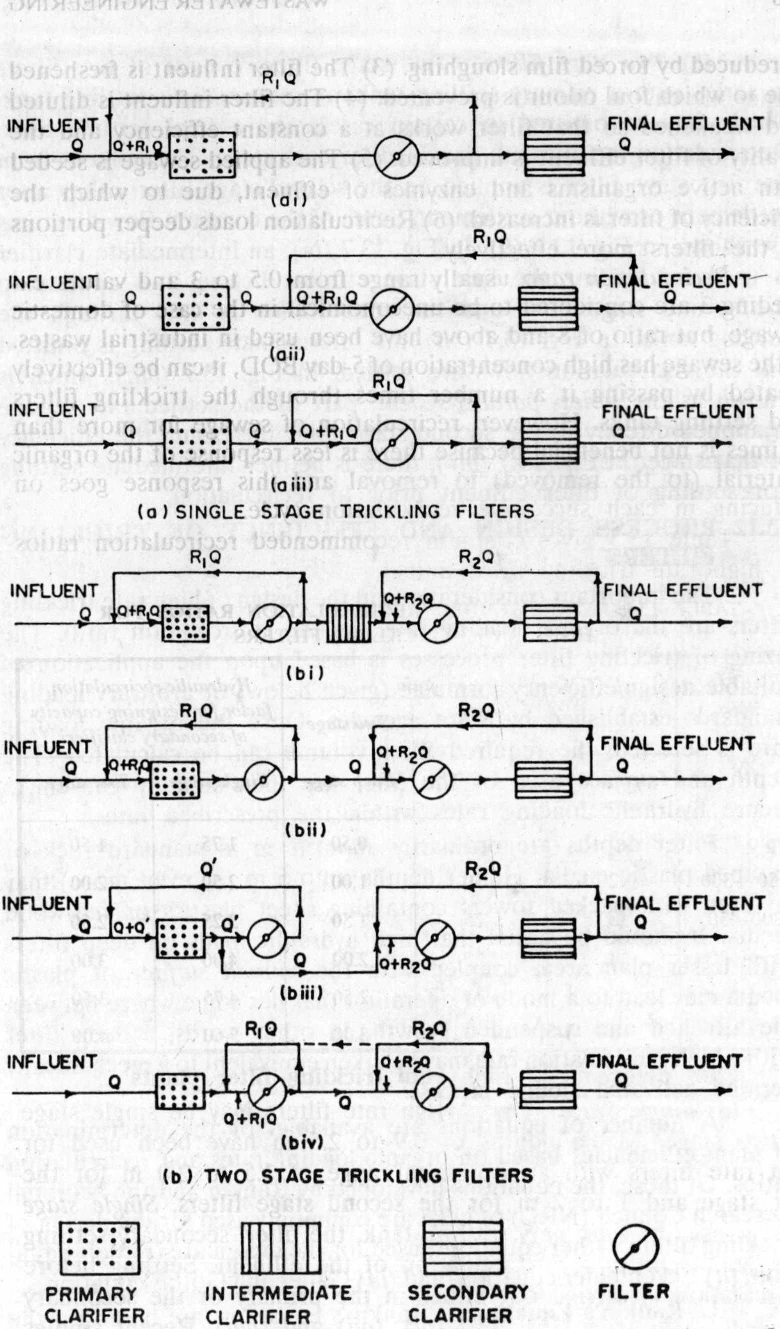

FIG. 13.7. FLOW DIAGRAMS FOR HIGH RATE TRICKLING FILTERS.

(*b*) ***Two stage filters*** : Two stage filters consist of two filters in series with a primary settling tank, an intermediate settling tank, which may be omitted in certain cases, and a final settling tank. Recirculation facilities are provided for each stage. The effluent from the first stage filter is applied on the second stage filter either after settlement or without settlement. Some of the common flow diagrams are shown in Fig. 13.7 (*b*). In Fig. 13.7 (*bi*), an intermediate clarifier is used for settling the first stage effluent before it is applied to the second stage filter and the recirculation is only through the settling tanks. In Fig. 13.7 (*bii*), the intermediate settling is omitted but the recirculation flows are settled. In Fig. 13.7 (*biii*), which is known as the *series- parallel-system*, part of the settled raw sewage is applied directly to the second stage filter increasing the efficiency of that stage. In Fig. 13.7 (*biv*), there is neither intermediate settling nor settling of filter effluent prior to recirculation.

13.12. PROCESS DESIGN AND EFFICIENCY OF TRICKLING FILTERS

The important considerations in the design of high rate trickling filters are the organic loading rate and the recirculation ratio. The sizing of trickling filter processes is based upon the application of suitable design/efficiency formulae (given below) or arbitrary loading standards established by State agencies. Once the organic loading rate is selected, the required filter volume can be calculated. The depth and surface area of the filter are then suitably chosen to secure hydraulic loading rates within the prescribed rates.

Filter depths are ordinarily about 2 m in standard rock or dumped plastic media. Higher depths upto 5 m (or even more) may be used with packed towers containing sheet plastics or red-wood media. It should be noted that high hydraulic loads on deep filters with lesser plan area, coupled with the *smooth surface* of plastic media may lead to a mode of operation that lies some where between the attached and suspended growth. In other words, a deep filter with a high recirculation rate may be taken equivalent to a mechanically aerated activated sludge process.

A number of equations are available for the determination of plant efficiencies based on organic loading rates and recirculation ratios. Of these, the equations developed by Rankin and the National Research Council (NRC) of USA are commonly used for the design of trickling filters. Other equations, used for the design are (*i*) Velz equation, (*ii*) Eckenfelder equation, and (*iii*) Galler and Gottas equation.

1. Rankin's Equations : Rankin's equations are based on the requirements of the *Ten State Standards*. There equations are applicable for various flow diagrams of high rate trickling filters.

(a) For single stage filters : For *single stage filters*, the Ten State Standards states that the BOD of the influent to the filter (recirculation included) shall not exceed three times the BOD of the required settled effluent. Rankins equation for single stage high rate filter of flow diagrams of Fig. 13.7 (*ai*) and 13.7 (*aii*) express the above relation as under:

$$C_i + R\,C_e = 3\,(1 + R)\,C_e$$

or $$C_e = \frac{C_i}{3 + 2R} \qquad \text{...(13.3)}$$

and $$e = \frac{C_i - C_e}{C_i} = \frac{1 + R}{1.5 + R} \qquad \text{...(13.4)}$$

where C_e = BOD of settled filter effluent (mg/l)
C_i = BOD of influent sewage after settling (mg/l)
R = Recirculation ratio
e = Efficiency of the filter, expressed as ratio.

The above equations are applicable only when the organic loading rate on the filter, including recirculation, is less than 1800 $g/d/m^3$ and the hydraulic loading rate, including recirculation, is maintained between 10 to 30 $m^3/d/m^2$.

When the organic loading ranges between 1800 to 2800 $g/d/m^3$, the following equation is used :

$$C_e = \frac{C_i}{2.78 + 1.78\,R} \qquad \text{...(13.5)}$$

For all loadings in excess of 2800 $g/d/m^3$, the BOD removal is assumed to be 1800 $g/d/m^3$ only.

The above equations are applicable to flow diagram of Fig. 13.7 (*aiii*) and the first stage filter of Fig. 13.7 (*bi*).

(a) First stage of two-stage filter : When the effluent of the first stage filter is applied to second stage filter without settling, such as shown in the flow diagram of Fig. 13.7 (*bii*) and 13.7 (*biv*), following equations are applied to the first stage plant effluent :

$$C_e = \frac{C_i}{2 + R} \qquad \text{...(13.6)}$$

and $$e = \frac{1 + R}{2 + R} \qquad \text{...(13.7)}$$

For the above equations, the loading limitations are the same as for Eq. 13.3.

If the first stage effluent consists partly of settled sewage, which does not pass through the first stage filter, such as shown in the flow diagram of Fig. 14.7 (*biii*), the following equation is used :

$$C_e = \frac{1.5 + C_i}{2.5 + r} \qquad ...(13.8)$$

where r is the *assumed recirculation* given by

$$r = \frac{Q_1 - Q}{Q} \qquad ...(13.8\ a)$$

Q being the raw sewage flow to the plant and Q_1 is the total flow through the first stage filter.

The BOD loading on the filter is given by the following equation:

$$L_{a1} = Q\,(C_i + R_1\,C_e) \qquad ...(13.9)$$

where L_{a1} = BOD_5 loading in kg/day
and Q = Sewage flow in mLd

(*b*) *Second stage of two stage filter*

The efficiency of removal of BOD for the second stage of two stage filter is less than that of the first stage because the amenability to treatment of the applied BOD is afteted by previous treatment. Hence for the second stage filters, the Ten State Standards stipulate that the BOD of the sewage applied to the second stage filter, including recirculation, shall not exceed two times the BOD expected in the settled effluent.

For the flow diagrams of Fig. 13.7 *b* (*i*), (*ii*) and (*iv*), Rankin's equations are as under :

$$C_e + R'\,C_e' = 2\,(1 + R')\,C_e'$$

or

$$C_e' = \frac{C_e}{2 + R'} \qquad ...(13.10)$$

and

$$e' = \frac{1 + R'}{2 + R'} \qquad ...(13.11)$$

where C_e' = BOD of settled effluent from second stage (mg/l)

R' = Recirculation ratio in the second stage filter

and e' = Efficiency of second stage filter (expressed as ratio)

For the flow diagram of Fig. 13.7 *b* (*iii*) where a part of the influent is fresh settled sewage which allows for greater efficiency, the following equation used :

$$C_3' = \frac{C_e}{3 + R'} \qquad ...(3.12)$$

2. NRC Equations : The NRC equation for trickling filter performance are *empirical* expressions developed from a study of operation results of trickling filters serving military installations in USA. These equations are applicable both for low rate as well as high rate filters. The *efficiency* (*E*) of single stage filter or first stage of two stage filters is given by

$$E = \frac{100}{1 + 0.44\sqrt{W/VF}} \quad ...(13.13)$$

or
$$E = \frac{100}{1 + 0.44\sqrt{U}} \quad ...(13.13\ a)$$

For the second stage of the two stage filters, efficiency (E') is given by

$$E' = \frac{100}{1 + \frac{0.44}{1-e}\sqrt{W'/V'F'}} \quad ...(13.14)$$

or
$$E' = \frac{100}{1 + \frac{0.44}{1-e}\sqrt{U'}} \quad ...(13.14\ a)$$

where E = Percentage efficiency in BOD removal of the single stage or first stage of the two stage filter

$e = E/100$

E' = Percentage efficiency of second stage filter

W = BOD loading of settled raw sewage in the single stage or first stage of the two stage filter (kg/day)

V = Volume of first stage filter (m^3)

F = Recirculation factor, or number of effective passes for the first stage filter

$$= \frac{1 + R}{[1 + (1-f)R]^2}$$

$$= \frac{1 + R}{(1 + 0.1\,R)^2}, \text{ taking } f = 0.9$$

$\frac{W}{VF} = U$ = first stage unit organic loading (kg/m^3/day)

R = recirculation ratio for the first stage filter

f = treatability factor (= 0.9 for sewage)

$W' = W(1-e)$ = BOD loading on second stage filter (kg/day)

V' = volume of second stage filter (m^3)

F' = recirculation factor or number of effective passes for second stage filter

$$= \frac{1 + R'}{[1 + (1-f)\,R']^2}$$

$\frac{W'}{V'F'} = U'$ = second stage unit organic loading (kg/m^3/day)

R' = recirculation ratio for the second stage filter.

3. Velz Equation : In contrast to previous equations, which are based on the data analysis, Velz equation is based on a fundamental law relating to BOD remaining at depth D as follows :

$$\frac{L_D}{L} = 10^{-kD} = e^{-KD} \quad ...(13.15\ a)$$

where
L = total removable BOD (mg/l)
L_D = removable BOD at depth D (mg/l)
D = filter depth (m)
k, K = experimental constants

The removable BOD in the above equation is the maximum fraction of applied BOD removed at a specified hydraulic loading range. When recirculation is practiced, the total applied BOD, L_a, after dilution by recirculation, R, may be determined as follows :

$$L_a = \frac{L_i + R\,L_e}{1 + R} \quad ...(13.15\ b)$$

where
L_i = Total influent BOD, not including recirculation (mg/l)
L_e = Total effluent BOD (mg/l)

Hence Velz formula takes the following form

$$L_e = \left(\frac{L_i + R\,L_e}{1 + R}\right) e^{-KD} \quad ...(13.16)$$

where the experimental constant K ranges from 0.49 for high rate filters to 0.57 for low-rate systems.

4. Eckenfelder equation : Eckenfelder improved on the Velz equation to include (*i*) the effect of changes in filter depth on the BOD removal per unit of depth, and (*ii*) the hydraulic loading rate. He suggested the following equation.

$$\frac{L_e}{L_a} = \frac{1}{1 + \dfrac{18.6\,D^{0.67}}{\sqrt{Q/A}}} \quad ...(13.17)$$

where
L_a = influent BOD (including recirculation (mg/l)
L_e = BOD of unsettled filter effluent (mg/l)
D = depth of filter (m)
Q = flow (mLd)
A = area in hectares.

5. Galler and Gotaas equation : Galler and Gotaas, based on a multiple regression analysis of data from existing plants and providing for the effects of recirculation, hydraulic loading, filter depth and temperature of sewagc, developed the following equation:

$$L_e = \frac{0.182\,K\,(i\,L_i + r\,L_e)^{1.19}}{(1 + K)^{0.78}\,(D + 0.305)^{0.67}\,(a)^{0.25}} \quad ...(13.18)$$

where $K = \frac{9.731}{(i)^{0.28}\,(T)^{0.15}}$ ...(13.18 *a*)

L_e = unsettled filter effluent BOD (mg/l)

L_i = filter effluent BOD (mg/l)

D = filter depth (m)

i = influent flow (mLd)

r = recirculation flow (mLd)

a = filter radius (m)

T = sewage temperature (° C).

Galler and Gotaas suggested that recirculation improves the performance of the filter but established that a ratio of 4 : 1 was the practical upper limit of recirculation.

Applicability of different equations : The NRC and Eckenfelder equations show fairly close agreement for filter volumes when no recirculation is adopted while the Galler and Gotaas equation give volumes significantly different. With recirculation, there is good agreement in filter volumes between the Eckenfelder and Galler and Gotaas equations but the NRC equation gives very conservative design volumes. In general, the NRC equation would seem to apply when recirculation is not considered, when seasonal variations in temperature are not large and when the sewage load is highly variable and of high strength.

13.13. PROPRIETORY TYPES OF HIGH RATE FILTERS

There are following three principal proprietory types of high rate filters on the market :

1. Bio-filter (Dorr CO.)
2. Accelo-filters (Infilco Inc.)
3. Aero-filters (Yeomans Bros.)

The patents for two of them cover the method of recirculation while the patent for the third covers the method of application of the sewage to the filter. Through cross-licensing and expiry of patents, these features no longer control the use of high-rate trickling filters.

1. Bio-Filters : This is a proprietory design of Dorr CO. It is a relatively shallow filter, generally 1.2 to 1.5 m in depth, which utilizes *recirculation* of a portion of the filter effluent to the primary settling tank for a second passage through the filter. Fig. 13.8 shows flow diagrams of biofilter plants giving the following types of treatment:

(*a*) Single stage intermediate treatment

(*b*) Single stage complete treatment

(*c*) Two stage complete treatment

In case of industrial or trade wastes which are strong, two stage bio filtration may be used. In two stage bio-filtration, recirculation is used for each stage. The total filter volume of two stages is the

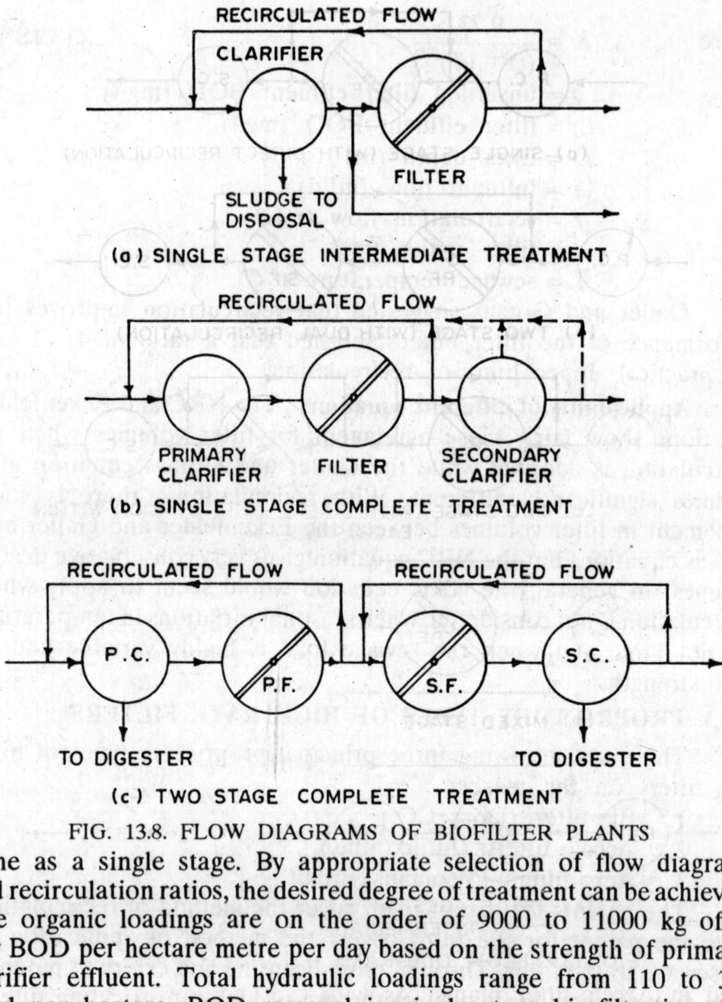

FIG. 13.8. FLOW DIAGRAMS OF BIOFILTER PLANTS

same as a single stage. By appropriate selection of flow diagrams and recirculation ratios, the desired degree of treatment can be achieved. The organic loadings are on the order of 9000 to 11000 kg of 5-day BOD per hectare metre per day based on the strength of primary-clarifier effluent. Total hydraulic loadings range from 110 to 330 mLd per hectare. BOD removal in single stage bio-filtration may be from 80 to 90 %, while two stage bio-filtration may produce 95% BOD removal.

The efficiency of BOD removal in bio-filters can be obtained by Rankin's equations given earlier.

2. Accelo-filters : Accelo-filter is a proprietory design of Infilco Inc. The filter is relatively deeper, normal depth varying between 1.8 m to 2.4 m. It utilises *direct recirculation* of unsettled filter effluent from the bottom of the filter directly back to the top of the filter through a two-arm distributor. The filter uses 100 mm size stone filter media and hydraulic loadings are maintained at less than 340 mLd per hectare of filter area. Fig. 13.9 shows some of the arrangements

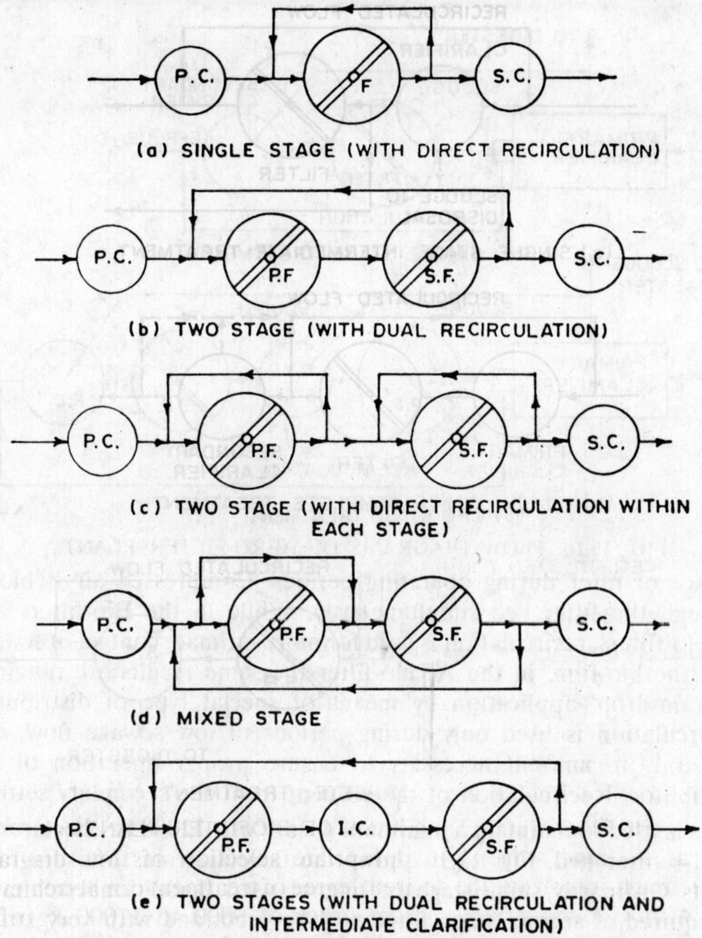

FIG. 13.9. FLOW DIAGRAMS OF ACCELO-FILTERS

used in Accelo-filter plants. As with the Biofilter, a large variety of flow diagrams is possible, including use of a primary high rate filter and a secondary low rate filter. The system of recirculating the effluent from the bottom of the filter directly back to the top of the filter provides a means of reducing odours, increasing the concentration of dissolved oxygen and continuously reseeding the influent to the filter with active organisms. The organic loading in single stage operation may be of the order of about 9000 kg/hec-m/day, with recirculative rates of 100 to 200 per cent of raw sewage flow.

3. Aero-filters : Aero-filters are the proprietory designs of Yeomans Bros. The process is based upon the principle of obtaining a rain like distribution, more or less continuously over the entire

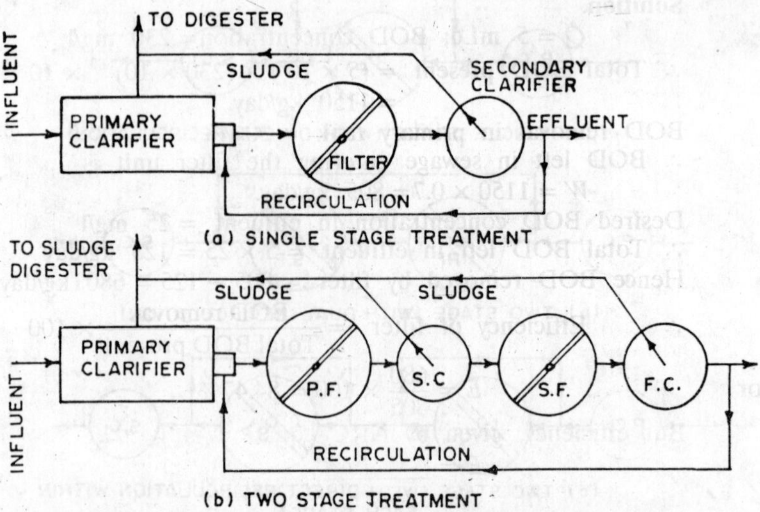

FIG. 13.10. FLOW DIAGRAMS OF AERO-FILTER PLANTS

surface of filter during operating periods. Compressed air is blown through the filter bed simultaneously. While in the Bio-filters and Accelo-filters, recirculation is used for more intimate contact of sewage with the bio-film, in the Accelo-filter the same results are obtained by 'rain-drop' application by means of special type of distributor. Recirculation is used only during periods of low sewage flow, and that only in amount necessary to ensure proper operation of the distributor. Recirculation of the effluent from the secondary settling tanks is used to maintain a minimum dosage of 1,12000 m^3/hectare/day on the filter bed. Fig. 13.10 shows the flow diagrams of Aero-filter plants. Ordinarily, single stage treatment is used. If additional treatment is required, a second stage filter may be used, and with very strong sewages, intermediate clarification may be provided.

Aero-filter beds are more than 1.8 m deep. The recommended organic loading may vary from 11000 to 12000 kg of 5-day BOD per hectare metre perday. Two stage Aero-filters produce better BOD removal than single stage, specially when 60 to 65% of the filter media is in the first stage.

Example 13.3. *Determine the size of a high rate trickling filer for the following data :*

(i) Sewage flow = 5 mLd
(ii) Recirculation ratio = 1.5
(iii) BOD of raw sewage = 230 mg/l
(iv) BOD removal in primary clarifier = 30 %
(v) Final effluent BOD desired = 25 mg/l

Solution.

$Q = 5$ mLd; BOD concentration $= 230$ mg/l

$\therefore$ Total BOD present $= (5 \times 10^6) \times (230 \times 10)^{-3} \times 10^{-3}$
$= 1150$ kg/day.

BOD removal in primary tank $= 30\ \%$

$\therefore$ BOD left in sewage entering the filter unit is

$W = 1150 \times 0.7 = 805$ kg/day.

Desired BOD concentration in effluent $= 25$ mg/l

$\therefore$ Total BOD left in effluent $= 5 \times 25 = 125$ kg/day

Hence BOD removed by filter $= 805 - 125 = 680$ kg/day

$$\therefore \quad \text{Efficiency of filter} = \frac{\text{BOD removal}}{\text{Total BOD present}} \times 100$$

or $$E = \frac{680}{805} \times 100 = 84.47\%$$

But efficiency, given by NRC equation is

$$E = \frac{100}{1 + 0.44\sqrt{W/VF}} \qquad ...(13.13)$$

where W = total BOD applied to filter $= 805$ kg/day

V = Volume of filter (m^3) in the first stage filters.

F = recirculation factor

$$= \frac{1+R}{(1+0.1\,R)^2} = \frac{1+1.5}{(1+0.1 \times 1.5)^2} = 1.89$$

$$\therefore \quad 84.47 = \frac{100}{1 + 0.44\sqrt{805/(V \times 1.89)}}$$

From which $V = 2439.5\ m^3$

Let us assume the depth filter as 1.8 m.

$$\therefore \text{ Surface area of filter} = \frac{2439.5}{1.8} = 1355.28\ m^2$$

$$\text{Hence dia. of filter} = \sqrt{\frac{1355.28 \times 4}{\pi}} = 41.5 \text{ m.}$$

Hence provide a single stage high rate filter of 41.5 m dia, and 1.8 m deep filter media, with a recirculation ratio of 1.5.

Example 13.4. *Design an equivalent standard rate filter for the data of example 13.3, without recirculation.*

Solution

For standard rate filter having no recirculation, $F = 1$

$$\therefore \quad 84.47 = \frac{100}{1 + 0.44\sqrt{805/V}}$$

$$1 + 0.44\sqrt{805/V} = 1.1838$$

From which $V = 4610.7\ m^3$

Providing a depth of filter media equal to 2 m,

Surface area $= 4610.7/2 = 2305.3$

$$\therefore \quad \text{Dia. of filter} = \sqrt{\frac{2305 \times 4}{\pi}} = 54.18 \text{ m.}$$

Hence provide a standard rate filter of 54.2 m dia and 2 m deep filter media.

Example 13.5. *Design a high rate single stage trickling filter for treating the following wastewater of a town having a population of 40000 persons :*

(a) Domestic sewage @ 150 lpcd having 200 mg/l of BOD

(b) Industrial wastewater @ 0.25 million litres per day having 600 mg/l of BOD.

Assume the following :

(i) BOD removal in primary clarifier = 35%

(ii) Permissible organic loading of filter = 8000 kg/hec-m/day (excluding recirculated sewage)

(iii) Recirculation ratio = 1.0

(iv) Permissible surface loading = 160 mL/hec/day (including recirculated sewage)

Also, determine the efficiency of the filter and BOD of the effluent.

Solution. Domestic sewage $= (150 \times 40000)\ 10^{-3} = 6000 \text{ m}^3/\text{day}$

Industrial wastewater $= 0.25 \times 10^6$ litres/day $= 250 \text{ m}^3/\text{day}$

Total BOD of domestic sewage $= (150 \times 40000 \times 200)\ 10^{-6}$
$= 1200$ kg/day

Total BOD of industrial wastewater $= (0.25 \times 10^6 \times 600)\ 10^{-6}$
$= 150$ kg/day

$\therefore$ Total BOD of combined flow $= 1200 + 150 = 1350$ kg/day.

Hence BOD applied to the filter is

$$W = 1350 \times 0.65 = 877.5 \text{ kg/day.}$$

$$\therefore \text{ Filter volume } V = \frac{\text{Total BOD removed}}{\text{permissible organic loading}}$$

$$= \frac{877.5}{8000} \times 10^4 = 1096.9 \text{ m}^3 \qquad ...(i)$$

Also, total volume of wastewater influent, $Q = 6000 + 250$
$= 6250 \text{ m}^3/\text{day}$

Now total inflow, including recirculation $= (1 + R)\,Q$
$= (1 + 1)\,Q = 2 \times 6250 = 12500 \text{ m}^3/\text{day}$

$$\therefore \quad \text{Filter area} = \frac{\text{Total volume of flow}}{\text{Permissible surface loading}}$$

Here, permissible surface loading per day = 160 mL/hectare

$$= \frac{(160 \times 10^6) \times 10^{-3}}{10^4} = 16 \text{ m}^3/\text{m}^2$$

$\therefore$ Filter area $= \dfrac{12500}{16} = 781.25 \text{ m}^2$ ...(2)

Hence, dia. of filter $= \sqrt{\dfrac{781.25 \times 4}{\pi}} = 31.54$ m

However, provide a tank of **32 m** dia.

Actual surface area $= \dfrac{\pi}{4}(32)^2 = 804.24 \text{ m}^2$

Hence filter depth $= 1096.9/804.25 = 1.364$ m.

Provide the depth of media equal to **1.4 m.**

Efficiency, $E = \dfrac{100}{1 + 0.44\sqrt{W/VF}}$

Here $W = 877.5$ kg/day

$V =$ Volume of filter $= 804.25 \times 1.4 = 1125.95 \text{ m}^3$

$F =$ Recirculation factor

$$= \frac{1+R}{(1+0.1R)^2} = \frac{1+1}{(1+0.1\times 1)^2} = 1.653$$

$$\therefore \quad E = \frac{100}{1 + 0.44\sqrt{877.5/1125.95 \times 1.653}}$$

$$\approx 76.8\%$$

$\therefore$ BOD of effluent $= \left(\dfrac{100 - 76.8}{100}\right) \times 877.5$

$= 203.6$ kg/day

Total volume of effluent $= 6250 \text{ m}^3/\text{day}$

$\therefore$ BOD concentration $= \dfrac{203.6 \times 10^6 \text{ (mg)}}{(6250 \times 10^3)\ (l)} =$ **32.58 mg/l**

Example 13.6. *A single stage filter is designed for an organic loading of 10000 kg of BOD in raw sewage per hectare metre per day with a recirculation ratio of 1.2. This filter treats a flow of 4 mLd of raw sewage with a BOD of 220 mg/l. Using NRC formula, determine the strength of the effluent.*

Solution.

Total BOD of raw sewage $= (4 \times 10^6) \times (220 \times 10^{-6})$

$= 880$ kg/day

Required filter vol. $= \dfrac{\text{Total BOD of raw sewage}}{\text{Permissible BOD loading}}$

$= \dfrac{880}{10000} \times 10^4 = 880 \text{ m}^3$

Let us assume that primary clarifier removes 35% of BOD.

$\therefore$ BOD of influent applied to filter $= 0.65 \times 880 = 572$ kg.

The efficiency of filter is given by

$$E = \frac{100}{1 + 0.44\sqrt{W/VF}}$$

Here $W = 572$ kg/day.

V = Volume of filter $= 880 \text{ m}^3$

$$F = \text{Recirculation factor} = \frac{1+R}{(1+0.1R)^2}$$

$$= \frac{1+1.2}{(1+0.1\times 1.2)^2} = 1.754$$

$\therefore$ Total BOD of effluent $= (1 - 0.789) \times 572$
$= 120.85$ kg.

BOD concentration of effluent

$$= \frac{\text{Total BOD}}{\text{sewage volume}}$$

$$= \frac{120.85 \times 10^6}{4 \times 10^6} = \mathbf{30.21 \ mg/l}$$

Example 13.7. *Instead of single stage plant of Example 13.6, it is proposed to use a two stage plant. The total volume of the filter remains the same as that of single-stage plant, with filter volume of each stage equal to half of the total volume. The recirculation ratio for each filter is 1.2. What will be the BOD of the plant effluent?*

Solution

For each filter, $R = 1.2$ and hence $F = 1.754$

For the first stage, efficiency is given by Eq. 13.13.

$$E = \frac{100}{1 + 0.44\sqrt{W/VF}}$$

where W = total BOD applied to first stage filter
$= 572$ kg/day

V = volume of first stage filter $= 880/2 = 440 \text{ m}^3$

$F = 1.754$

$$\therefore \quad E = \frac{100}{1 + 0.44\sqrt{\dfrac{572}{440 \times 1.754}}} = 72.53\%$$

$\therefore$ Amount of BOD left in the effluent from the first stage filter $= 572\,(1 - 0.7253) = 157.15$ kg/day.

For the second stage filter, efficiency is given by Eq. 13.14

$$E' = \frac{100}{1 + \dfrac{0.44}{1-e}\sqrt{\dfrac{W'}{V'F'}}}$$

where $W' = 157.15$ kg/day and $V' = 880/2 = 440 \text{ m}^3$

$$F' = 1.754$$
$$e = E/100 = 0.7253$$

$$\therefore \quad E' = \frac{100}{1 + \frac{0.44}{1 - 0.7253}\sqrt{157.15/(440 \times 1.754)}} = 58.05\%$$

$\therefore$ Total BOD of effluent from the plant $= 157.15\,(1 - 0.5805)$

$= 65.95$ kg/day

$\therefore$ BOD concentration of effluent $= \dfrac{65.95 \times 10^6}{4 \times 10^6} = \mathbf{16.48\ mg/l}$

Example 13.8. *Design a single stage Bio-filter to produce an effluent having max. BOD concentration of 50 mg/l, for a flow of sewage of 4 million litres per day containing BOD of raw sewage of 320 mg/l. Make suitable assumptions.*

Solution

BOD of raw sewage $= (4 \times 10^6) \times (320 \times 10^{-6}) = 1280$ kg/day

Let us assume that the primary clarifier removes 35% of BOD of raw sewage. Total BOD of filter influent $= 0.65 \times 1280 = 832$ kg

Permissible total BOD of effluent $= (4 \times 10^6) \times (50 \times 10^{-6})$

$= 200$ kg

$\therefore$ Total BOD to be removed by the filter $= 832 - 200 = 632$ kg

$$\therefore \quad \text{Efficiency of filter, } E = \frac{632}{832} \times 100 = 75.96\%$$

Again, let us assume an organic loading of 0.8 kg/day/m^3 (*i.e.* between 500 to 1000 g/day/m^3)

$\therefore$ Volume of filter required $= 632/0.8 = \mathbf{790\ m^3}$

$$\text{Again,} \quad E = \frac{100}{1 + 0.44\sqrt{W/VF}}$$

where $W = 832$ kg/day
$V = 790$ m^3
$E = 75.96$

$$\therefore \quad 75.96 = \frac{100}{1 + 0.44\sqrt{832/792\,F}}$$

From which $F = 2.0356$

$$\therefore \quad 2.0356 = \frac{1 + R}{(1 + 0.1\,R)^2}$$

Solving this by trial, we get $R \approx \mathbf{1.85}$

Example 13.9. *Calculate the effluent* BOD_5 *of a two stage trickling filter with the following flows,* BOD_5 *and dimensions, using (a) NRC formula (b) Velz formula (c) Rankin's formulae.*

$$Q = 5000\ m^3/day$$
$$BOD_5 = 280\ \ mg/l$$

Volume of primary filter $= 1000\ m^3$

Volume of secondary filter $= 800\ m^3$

Filter depth $= 2\ m.$

Recirculation ratio for primary filter $= 1.5$

Recirculation ratio for secondary filter $= 1.25$

Solution

(*a*) *NRC formula* : For the *first stage*

$$e = \frac{L_i - L_e}{L_i} = \frac{1}{1 + 0.44\sqrt{W/VF}}$$

Here $W = 5000 \times 10^3 \times (280 \times 10^{-6}) = 1400$ kg/day

$$F = \frac{1 + R}{(1 + 0.1\,R)^2} = \frac{1 + 1.5}{(1 + 0.1 \times 1.5)^2} = 1.89$$

$$V = 1000\ \text{m}^3\ ; \qquad L_i = 280\ \ \text{mg/l}$$

$$\therefore \qquad e = \frac{280 - L_e}{280} = \frac{1}{1 + 0.44\sqrt{\dfrac{1400}{1000 \times 1.89}}} = 0.7253$$

From which $L_e = 76.92$ mg/l

For the second stage

$$\frac{L_e - L_e'}{L_e} = \frac{1}{1 + \dfrac{0.44}{1 - e}\sqrt{W'/V'\,F'}}$$

where $W' = (5000 \times 10^3) \times (76.92 \times 10^{-6}) = 384.6$ day/day

$$V' = 800\ \text{m}^3$$

$$F' = \frac{1 + R'}{(1 + 0.1\,R')^2} = \frac{1 + 1.25}{(1 + 0.1 \times 1.25)^2} = 1.778$$

$$L_e = 76.92\ ; \qquad e = 0.7253$$

$$\therefore \qquad \frac{76.92 - L_e'}{76.92} = \frac{1}{1 + \dfrac{0.44}{1 - 0.7253}\sqrt{\dfrac{384.6}{800 \times 1.778}}} = 0.5456$$

From which $L_e' = 34.95$ mg/l

(*b*) *Velz formula*

$$L_e = \left(\frac{L_i + R\,L_e}{1 + R}\right) e^{-KD} \qquad \text{...(13.16)}$$

Here $D = 2$ m.

Let us take $K = 0.49$.

In the first stage $L_e = \left(\dfrac{280 + 1.5\, L_e}{1 + 1.5} \right) e^{-0.49 \times 2}$

or $L_e = (280 + 1.5\, L_e)\, 0.15$

From which $L_e = 54.24$ mg/l

In the second stage

$$L_e' = \left(\frac{54.24 + 1.25 \times L_e'}{1 + 1.25} \right) e^{-0.49 \times 2}$$

or $L_e' = (54.24 + 1.25\, L_e')\, 0.1668$

From which $L_e' =$ **11.43 mg/l**

The Velz equation is valid for BOD removals of 90% or less.

(c) Rankin's formulae . Assume that the organic loading, including recirculation, is less than 1.8 kg/d/m^3, in which case, Eq. 13.3 will be applicable. Let us assume that effluent BOD_5 from the first stage is about 100 mg/l.

∴ Total organic loading, including recirculation

$= (280 \times 10^{-6})(5000 \times 10^3) + (100 \times 10^{-6})(5000 \times 10^3)(1.5)$

$= 1400 + 750 = 2150$ kg/day

Filter volume = 1000 m^3

∴ Unit organic loading $= \dfrac{2150}{1000} = 2.15$ kg/d/m^3

Hence Eq. 13.3 in not applicable, but instead, Eq. 13.5 is applicable.

$$\therefore \quad L_e = \frac{L_i}{2.78 + 1.78\,R} = \frac{280}{2.78 + 1.78 \times 1.5} = 51.38 \text{ mg/l}$$

Check : Organic loading

$= (280 \times 10^{-6})(5000 \times 10^3) + (51.38 \times 10^{-6})(5000 \times 10^3)(1.5)$

$= 1400 + 385.4 = 1785.4$ kg/day

∴ Loading rate = 1785.4/1000 = 1.785 kg/d/m^3

This is just less than the limiting rate of 1.8 kg/d/m^3

Hence from Eq. 13.3,

$$L_e = \frac{L_i}{3 + 2R} = \frac{280}{3 + 2 \times 1.5} = 46.67 \text{ mg/l}$$

Hence $L_e = 46.67$ mg/l

For the second stage filter, Eq. 13.10 is applicable :

$$L_e' = \frac{L_e}{2 + R'} = \frac{46.67}{2 + 1.25} = \mathbf{14.36 \text{ mg/l}}$$

Example 13.10. *Design a single stage trickling filter to yield an effluent BOD_5 of 25 mg/l. The influent BOD_5 following primary clarification is 170 mg/l and the flow is 8000 m^3/day. Maintain a hydraulic loading rate of 20 m/day and a filter depth of 2 m.*

Solution : From NRC equation, we have

$$e = \frac{L_i - L_e}{L_i} = \frac{1}{1 + 0.44\sqrt{W/VF}}$$

where $W = (8000 \times 10^3)(170) \times 10^{-6} = 1360$ kg/day

$$\therefore \quad \frac{170 - 25}{170} = \frac{1}{1 + 0.44\sqrt{1360/VF}}$$

From which $\sqrt{\dfrac{1360}{VF}} = 0.3918$

$$\therefore \quad VF = 8857.3 \qquad \text{...(a)}$$

Now for a depth of 2 m, $V = 2A$ ...(*i*)

Also, for a hydraulic loading of 20 m/day (m³/day/m²),

$$A = \frac{8000\,(1 + R)}{20} = 400\,(1 + R) \qquad \text{...(ii)}$$

Also, $$F = \frac{1 + R}{(1 + 0.1\,R)^2} \qquad \text{...(iii)}$$

Substituting the values in (*a*), we get

$$2 \times 400(1 + R) \times \frac{1 + R}{(1 + 0.1\,R)^2} = 8857.3$$

or $$\frac{(1 + R)^2}{(1 + 0.1)^2} = 11.072$$

From which $R = \mathbf{3.49}$

Hence $A = 400\,(1 + 3.49) = 1796 \text{ m}^2$

$$\therefore \quad \text{Filter dia.} = \sqrt{\frac{1796 \times 4}{\pi}} = 47.82 \text{ m}$$

Alternatively, providing two filter units,

$$\text{Dia. of each filter} = \sqrt{\left(\frac{1796}{2}\right) \times \frac{4}{\pi}} = 33.8 \text{ m}$$

Hence provide 2 filter units each of dia. 34 m., and of constant depth 2 m.

Check from Velz formula:

$$L_e = \left(\frac{L_i + R\,L_e}{1 + R}\right) e^{-KD} \qquad \text{...(13.16)}$$

Taking $K = 0.49$ for high rate fiters, we get

$$L_e = \left(\frac{170 + 3.49\,L_e}{1 + 3.49}\right) e^{-0.49 \times 2}$$

Which gives $L_e = 20.1$ mg/l

Example 13.11. *Design a high rate trickling filter plant to treat settled domestic sewage with a BOD_5 of 190 mg/l for an average flow* *of 25 mLd to satisfy an effluent BOD_5 of 12 mg/l. Take a peak factor as 2.25.*

Solution

(*a*) Design of filter units

1. Filter dimensions for first stage filter

Since the effluent BOD_5 required is less than 30 mg/l, a *two-stage filtration plant* has to be adopted. The filter units are to be designed for average flow while the distributing arms, underdrainage system, piping etc are to be designed for peak flow and checked for average flow. We will adopt flow diagram of Fig. 13.7 *b* (*i*).

The organic loading for high rate filters varies from 500 to 1000 g/d/m^3 (excluding recirculation). Let us adopt an organic loading value of 800 g/d/m^2.

$\therefore$ Vol. of first stage filter

$$= \left(25 \times 10^6 \times \frac{190}{10^3}\right) \times \frac{1}{800} = 5937.5 \text{ m}^3$$

Adopting a filter depth of 1.5 m,

Filter area, $$A = \frac{5937.5}{1.5} = 3958.3 \text{ m}^2$$

$$\therefore \quad \text{Filter dia.} = \sqrt{\frac{3958.3 \times 4}{\pi}} \approx 71 \text{ m.}$$

Since rotary distributors are available indigenously only upto 60 m, it is desirable to have at least two units.

$$\therefore \quad \text{Dia. of each filter} = \sqrt{\frac{3958.3}{2} \times \frac{4}{\pi}} = 50.2 \text{ m.}$$

However, provide two units of filters, each of 50 m, for first stage filtration. This will increase organic loading slightly.

2. Recirculation ratios for first stage and second stage filters

The efficiency of first stage filter is given by Rankin's equation:

$$E = \frac{1 + R}{1.57 + R} \times 100 \qquad \text{...(13.4)}$$

Let us apply the above equation for various values of $R = 0.5$, 0.75, 1.0, 1.5, 2.0, 2.5 and 3.0. This gives the corresponding values of E as 75, 77.78, 80, 80.33, 85.77, 87.50 and 88.88% respectively.

These values have been entered in columns 2 and 3 of Table 13.6. Also, the value of first stage effluent BOD_5 (L_e) is given by Eq. 13.3

$$L_e = \frac{L_i}{3 + 2R} \qquad \text{...(13.3)}$$

These values of L_e for various values of R have been entered in column 4 of Table 13.6.

Similarly, the efficiency E' of the second stage filter is given by Eq. 13.7 :

$$E' = \frac{1 + R'}{2 + R'} \quad ...(13.11)$$

where R' is the recirculation ratio for the second stage filter.

The values of E', for various value of R' (= 0.5, 0.75, 1.0, 1.5, 2.0, 2.5 and 3.0) are computed from Eq. 13.11 and entered in column 6 of Table 13.6.

TABLE 13.6

S. No.	*First stage filter*			*Second stage*		
	R	E (%)	L_e (mg/l)	R'	E' %	*R' for various values of R to give 95% combined efficiency (E_c)*
1	2	3	4	5	6	7
1.	0.50	75.00	47.50	0.50	60.00	3.00
2.	0.75	77.78	42.22	0.75	63.64	2.42
3.	1.00	80.00	38.00	1.00	66.67	2.00
4.	1.50	83.33	31.67	1.50	71.43	1.34
5.	2.00	85.71	27.14	2.00	75.00	0.88
6.	2.50	87.50	23.75	2.50	77.78	0.5
7.	3.00	88.89	21.11	3.00	80.00	–

Let us adopt the combined efficiency (E_c) of the filters, required to given an effluent BOD_5 of 12.5 as 95%.

Now E_c (for two stages) $= E + E' \, (1 - E/100)$

or $$95 = E + E' \, (1 - E/100)$$

or $$E' = \frac{95 - E}{1 - E/100} \quad ...(a)$$

The values of $E'$are computed for various values of R (and hence of E), and for each value of E' so calculated, the value of R' is selected from column 5 and entered in column 7.

For example, for $R = 0.5$, $E = 75\%$, and hence

$$E' = \frac{95 - 75}{1 - 75/100} = 80 \text{ \%}.$$

From column 5, we get $R' = 3.00$ for $E' = 80\%$. Hence this value of $R' = 3$ is entered in column 7 against $R = 0.5$.

Hydraulic loading (including recirculation) for a given value of R is given by

$$\text{Hyd. loading} = \left(\frac{25}{2} \times 10^6 \times 10^{-3} \right) \times \frac{4}{\pi \, (50)^2} (1 + R)$$

or Hyd. loading = 6.3662 (1 + R) ...(b)

Hydraulic loading values for various values of R are calculated with the above equation and tabulated as under :

R	*Hyd. loading*	*R*	*Hyd. loading*
0.50	9.55	2.00	19.10
0.75	11.14	2.50	22.28
1.00	12.73	3.00	25.46
1.50	15.92		

The permissible hydraulic loading rates (including recirculation) varies from 10 to 30 $m^3/d/m^2$.

Hence choose $R = 2$, for first stage recirculation.

Corresponding $R' = 0.86$, for second stage filter.

However, choose $R' = 1$ for the second stage.

For $R = 2$, the corresponding value of $L_e = 27.14$ mg/l (from col. 4). For the first stage filter units (*i.e.* two filters each of 50 m dia. and 1.5 m depth), organic loading (including recirculation) is

$$= 25 \times 10^6 (190 + 2 \times 27.14) \times 10^{-3} \times \frac{4}{2 \times \pi (50)^2 \times 1.5}$$

$$= 1036.8 \text{ g/d/m}^3$$

This is less than 1800 $g/d/m^3$ and hence Rankin's equations used above are applicable.

3. ***Filter dimensions for second stage filter***

Let us choose an organic loading rate of 500 $g/d/m^3$.

$$\text{Vol. of second stage filter} = \frac{(25 \times 10^6) \times (27.14 \times 10^{-3})}{500}$$

$$= 1357 \text{ m}^3$$

Let us adopt a depth of 1 m.

$\therefore$ Area of filter = 1357/1 = 1357 m^2

Let us check this for hydraulic loading on second stage filter.

$$\text{Hyd. loading} = \frac{Q(1 + R')}{A'} = \frac{25 \times 10^3 (1 + 1)}{1357} = 36.84 \text{ kLd/m}^2$$

which is much more than permissible value of 30 kLd/m^2.

For a permissible value of 30 kLd/m^2, corresponding filter area is

$$A' = \frac{25 \times 10^3 (1 + 1)}{30} = 1666.7 \text{ m}^2$$

$$\therefore \quad \text{Filter dia.} = \sqrt{\frac{1666.7 \times 4}{\pi}} \approx 46 \text{ m.}$$

Hence adopt one unit of 46 m dia. and 1 m depth for 2nd stage filter.

(b) Design of rotary distributor and under-drainage system

The design of the rotary distributor and under-drainage system of each filter (first stage as well as second stage) is done exactly in the same manner as illustrated in Example 13.2.

Typical details of trickling filter are shown in Fig. 13.11.

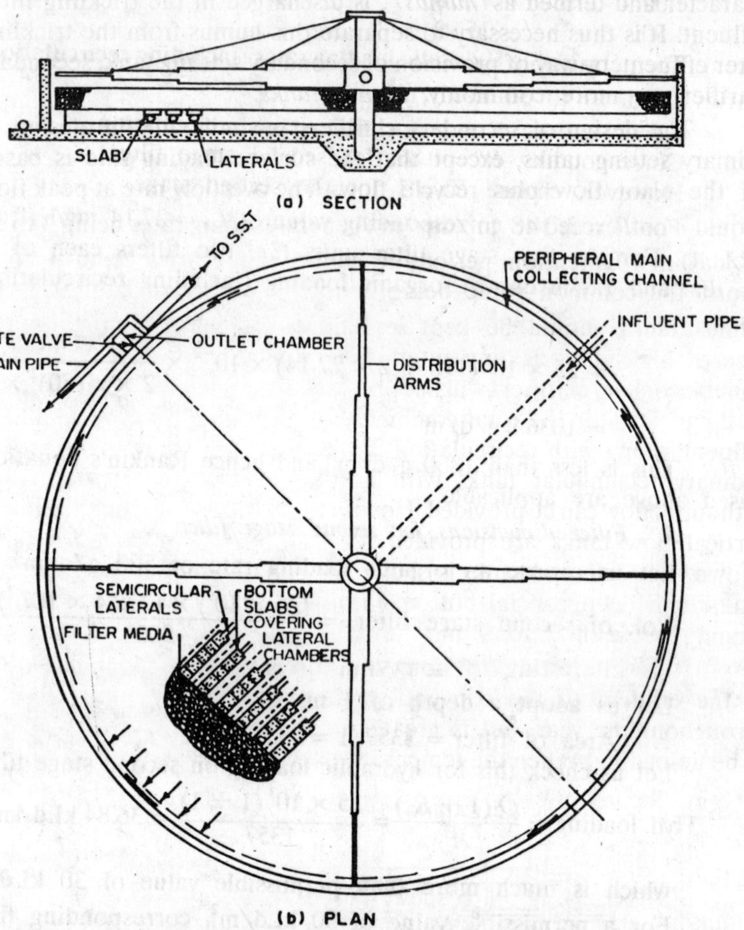

FIG. 13.11. TYPICAL DETAILS OF TRICKLING FILTER

13.14. SECONDARY CLARIFIERS (HUMUS TANK)

The function of settling tank that follows trickling filter is to produce clarified effluent. As pointed out earlier, the trickling filter removes only a relatively small percent of solids present in the primary tank effluent applied to it, through mechanical straining. In fact, the trickling filter nearly changes the character of colloidal (non- settleable) and dissolved organic solids due to biological oxidation and nitrification taking place in the filter resulting in formation of coagulated suspended mass which is hearier and bulkier. Such coagulated suspended mass, having dark brown colour and flocculent in character and termed as *'humus'* , is discharged in the trickling filter effluent. It is thus necessary to separate this humus from the trickling filter effluent, by way of provision of secondary settling tank, secondary clarifier, or more commonly, *humus tanks*.

The design of secondary clarifiers is similar to the design of primary setting tanks, except that the surface loading rate is based on the plant flow plus recycle flow. The overflow rate at peak flow should not exceed 48 $m^3/day/m^2$, general loading rates being 3 to 50 $m^3/day/m^2$. Detention period in humus tank varies from 1 to 3 hours, the common value being 2 hours. The weir loading in the humus tanks should be kept less than 225 m^3/m length. In the case of high-rate trickling filters, where recirculation of effluent is done, secondary clarifiers will have higher capacities to provide adequate settling time for the total combined flow of sewagc and recirculated effluent, along with recirculation devices. For secondary clarification, ordinary rectangular tanks with horizontal flow, or circular tanks with spiral flow can be provided. However, some times, hopper bottomed vertical flow tanks are provided for this purpose. One such tank, known as Dortmund tank, is shown in Fig. 13.12. These are circular tanks with deep concial bottom, having diameter as well as height ranging between 7 to 9 m.

It is interesting to note that the amount of humus solids in the effluent of the conventional trickling filter is not constant throughout the year, but is greater in spring when the filter is said to be unloading. When the temperature rises during spring, the biofilm

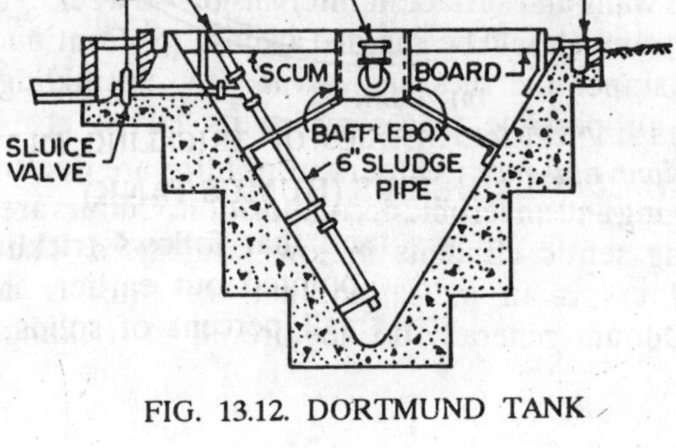

FIG. 13.12. DORTMUND TANK

gets loosened and sloughs off. Hence during the unloading period, the amount of suspended solids in the filter effluent is considerably greater than the amount in the primary effluent applied to the filter. However, only half of such suspended matter will get removed in the final clarification process.

The specific gravity of humus is about 1.02, and its moisture content varies from 90 to 95 percent. The quantity of humus collected in the humus tank may be generally taken as $\frac{1}{5}$ to $\frac{1}{6}$ of the quantity of sludge deposited in the plain sedimentation process. The humus or sludge settled at the bottom of the humus tanks should be continuously and rapidly removed and sent to the digestion tanks either directly or through primary clarifiers. Humus is generally removed weekly and hence storage space should be provided accordingly.

13.15. TRICKLING FILTER TROUBLES AND REMEDIES

Following are common troubles occuring at the site and operation of trickling filter filtes.

(*a*) ***Fly nuisance*** : Slow rate trickling filters often become infested with small moth like, deceptively fragile flies called 'Psychoda.' These flies do not bite, but may get into the eyes, nostrils and ears of men and animal. Its size is so small that it can even enter through the window mesh or screens. The life cycle of these flies varies from 22 days at 15.5°C temperature to 7 days at 29°C. Its radius of flight is usually only a few hundred feet, but it may be carried long distances due to wind. The flies lay their eggs in the filter and the wormlike larvae feed upon the filter by constantly consuming the dead and dying zoological masses. Heavy infestation is associated with thick films and high temperatures. Following are the *remedial measures* : (1) Flooding the filter for about 24 hours, at weekly or biweekly intervals. (2) Jetting down the inside walls of the filter with a high pressure hose. (3) Chlorinating the filter influent (o.5 to 1.0 mg/l) for several hours at one to two week intervals. (4) Application of larvicides such as D.D.T., chordane, malathion etc., when the filter is not in operation. These should be applied to the filter side walls and surface at intervals of 4-6 weeks. Development or resistant strains should be guarded against. (5) Continuous hydraulic loading to hamper the flies and larvae. (6) Sprinkling lime at the site and at all possible places where its larvee are breeding.

(*b*) ***Odour nuisance*** : Odours from filter are due to undesirable growth, sludging and anaerobic decomposition. Odours are most serious when treating septic effluents in low rate filters. The presence of excessive odours is an indication that the filter is not operating efficiently. Odours generally do not prevail in trickling filters using rotary distributors, but they are more common when fixed nozzles are used. Following are *remedial measures*: (1) Maintaining a well ventilated filter, either by natural ventitation or by forced ventilation. (2) Recirculation of filter effluent or secondary clarifier effluent, which will wash fine solids through the filter bed, dilute the influent sewage and add dissolved oxygen. (3) Aeration or chlorination of sewage before primary settling of sewage.

(*c*) ***Ponding nuisance*** : This nuisance is caused when all the voids of the trickling filters are filled up due to choking by heavy fungus or other suspended matters, due to which the sewage cannot pass through the filter and accumulate at the surface in the form of pond. Ponding decreases filter ventilation, reduces the effective volume of the filter and reduces filter efficiency. Ponding or clogging is due to excessive organic loading, inadequate hydraulic loading and inadequate size of media. Following are the *remedial measures*: (1) Opening the clogged section by flushing with a fire hose and simultaneously loosening the aggregates by a steel bar. (2) Reducing the strength of filter influent by re-circulation. (3) Flooding the filter once in a day and allowing of it to stand for 24 hours. (4) Chlorinating the influent with a dose not exceeding 5 kg/100 m^2 of filter area, once in every 4 to 7 days. This may be done at night when sewage flow is low. (5) Stopping he distributor over the ponded area. (6) Allowing the zoogleal masses to dry by keeping the filter out of operation for 12 to 48 hours.

In addition to the above mentioned filter troubles, there may be additional trouble of formation of ice cover on the surface of the bed, in conditions of extreme cold weather. Reduction of the recirculation flow, adjustment of nozzle or construction of wind breakers are methods used to reducing *icing problems*.

13.16. OTHER TYPES OF FILTERS

In addition to intermittent sand filters, contact beds, conventional trickling filters, high rate trickling filters, super-rate filters, other types of miscellanceous filters are as under :

(*i*) Dunbar filters (*ii*) Magnetite filters

(*iii*) Rapid sand filters.

Dunbar filter : This filter is so named because it was first developed and used at Dunbar, a town in Germany. The filter, rectangular in shape with length to breadth ratio of 3 to 4, is similar in construction to intermittent sand filter. However, the filter media is graded from top to bottom from 1 mm to 150 mm size. The depth of filter is kept from 1 to 1.2 m. The walls may be of masonry or concrete, and the floor of concrete, sloping cross-ways on either side towards a central longitudinal drain.

The filter is loaded intermittently. At each dose, the filter is flooded by forming a pond of 75 to 10 cm at the top, and is then allowed to rest. In a few hours, the sewage sinks down, leaving a film of gelatinous material at the top. This film is allowed to dry in a day or two, after which it is gently brushed over and removed in the form of powder. This operation makes the filter ready for use again. Hence several such units are needed (at least three) to be used alternatively. The rate of filter loading may be about 25 million litres per hectare of surface area per day. The BOD removal is upto about 85%. The effluent is so stable that it requires no further settling treatment and there is no sludge produced. However, such filters are useful only for very small populations such as of tubercular or other sanatoria.

Magnetite filter : The filter consists of a bed of magnetite sand, about 75 to 80 mm thick, resting on a fine screen of non-corrosive phosphor-bronze wire. The magnetite sand is of about 0.85 mm grain size. When an effluent from a plain or chemical precipitation tank is filtered through the magnetite sand layer, the suspended solids are retained in the voids, purely by mechanical straining action. Fig. 13.13 shows the section of such a filter.

The rate of loading may vary from 120 to 300 litres per m^2 per minute. The maximum permissible head loss is between 7.5 to 10 cm. Filter is cleaned automatically when the head loss exceeds the above value, through the operation of a float switch. The cleaning operation is done by means of a *travelling electromagnet* along the total length of the tank. The energized magnet lifts the magnetite sand clear of the supporting screen, and the partially treated sewage rushes through the bed. On demagnetization, the sand falls back on the wire screen. Alternate action of magnetization and demag-

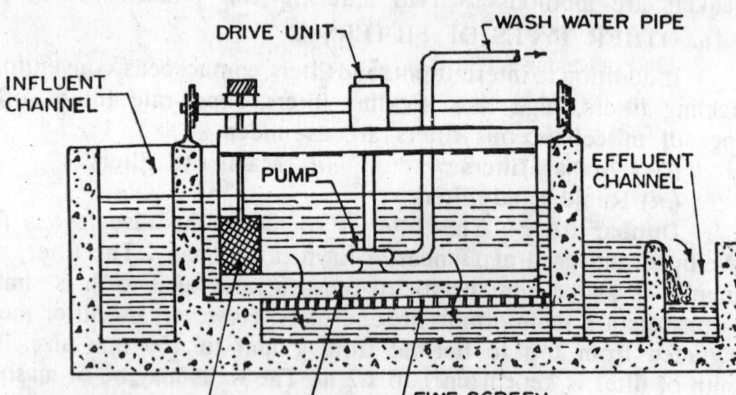

FIG. 13.13. MAGNETITE FILTER.

netization cleans the magnetite sand particles. The dirty water is removed through a small centrifiugal pump provided in the filter tank. These filters are not in common use.

3. Rapid sand filters : Such filters are similar to the one used for treating raw water for public water supplies. These are used for treating the sewage, with the sole purpose of reclaiming the used water ; the reclaimed water can subsequently be used for gardening etc. These filters get clogged very soon, and are therefore not generally used.

PROBLEMS

1. What do you under stand by secondary treatment (or biological treatment) of wastewater ? Enumerate various treatment techniques used for biological treatment.
2. Differentiate clearly between attached growth processes and suspended growth processes. List various treatment techniques falling under each such process.
3. Explain the construction and working of intermittent sand filters. What are its advantages and disadvantages.
4. What do you understand by a 'contact bed'. Explain various operations of its working. List various advantages and disadvantages of contaet beds.
5. What do you understand by a 'trickling filter' ? Explain, with the help of a neat sketch, the biological process involved in working of a trickling filter.
6. Compare, in a tabular form, low rate and high rate trickling filters.
7. Explain, with the help of neat sketch, the construction and working of a conventional trickling filter. What are its loading rates ?
8. What do you understand by the efficiency of a trickling filter ? How do you determine the efficiency, using (*i*) NRC formulae and (*ii*) Velz equation ?
9. State advantages and disadvantages of a conventional trickling filter.
10. What do you understand by (*i*) recirculation, (*ii*) two stage high rate trickling filters.
11. Give various flow diagrams used for single stage and two stage trickling filters. How do you determine the efficiency of these, using Rankin's equations.
12. The effluent from a primary settling tank is applied to a standard rate filter at the rate of 3 million litres perday, having a BOD_5 of 175 mg/l. Determine the depth and volume of filter, adopting a surface loading of 2000 l/m^2/day and an organic loading of 150 g/m^3/day. Also determine the efficiency of such filter unit, using NRC formula.

Ans [3500 m^3; 2.33 m; 85.4%]

13. Determine the size of a high rate trickling filter for the following data :

(i) Sewage flow = 4 mLd

(ii) Recirculation rato = 1.4

(iii) BOD of raw sewage = 260 mg/l

(iv) BOD removal in primary clarifier = 30%

(v) Final effluent desired = 40 mg/l.

Ans. [V= 962 m^3; Dia. = 30 m; Depth = 1.36 m]

14. A single stage, high rate trickling filter is to treat a flow of 3 mLd loading of 12000 kg of BOD in raw sewage per hectare metre, and the recirculation ratio is to be 1. What will be the strength of the effluent ? Use NRC formula. Assume that 35% of BOD is removed in the primary clarifier.

Ans. [37.72 mg/l]

15. Calculate the effluent BOD_5 of a two-stage trickling filter with the following data :

Q= 4500 m^3/d
BOD_5= 290 mg/l
Volume of filter No. 1 = 830 m^3
Volume of filter No. 2 = 830 m^3
Filter depth = 2 m.
Recirculation (filter No. 1) = 125 percent Q
Recirculation (filter No. 2) = 100 percent Q.
Use NRC formula.

Ans [37.5 mg/l]

16. Solve problem 15, using Rankin's formulae.

Ans [19.3 mg/l]

17 Design a single-stage trickling filter to yield an effluent BOD_5 of 30 mg/l. The influent BOD_5 following primary clarification is 160 mg/l and the flow is 10^4 m^3/day. Maintain a hydraulic loading rate of 20 m/day and a filter depth of 2 m.

Ans. [A = 1500 m^2; Two units each of dia. 31 m;]

18. Write short notes on the following :
 (i) Humus tank
 (ii) Contact bed
 (iii) Dunbar filter
 (iv) Bio-filter
 (v) Magnetite filter.

19. Explain various filter troubles, giving remedial measures for each.

20. Explain, with the help of flow diagrams, the following :
 (a) Aero-filters
 (b) Accelo-filters

14

Biological Treatment II : Activated Sludge Process

14.1. INTRODUCTION

As indicated earlier, the objectives of the biological treatment of wastewater are to coagulate and remove the nonsettleable colloidal solids and to stabilise the organic matter with the help of living systems which rely on mixed biological cultures. The *biological treatment techniques* used may be classified as (*i*) attached film processes, (*ii*) suspended growth processes, and (*iii*) combined processes.

In the suspended growth processes, the micro-organisms responsible for the conversion of organic matter or other constituents in the wastewater to gases and cell tissues are maintained in *suspension* within the liquid in the reactor by employing either natural or mechanical mixing. The suspended growth processes include (*i*) activated sludge process (*ii*) aerated lagoons, and (*iii*) sludge digestion systems. This chapter deals with various aspects of activated sludge processes, while aerated lagoons and sludge digestion systems have been discussed in subsequent chapters.

It should be noted that various types of biological processes only help in changing the unstable organic matter into stable forms which are then removed easily in the secondary settling tanks. Thus, in the biological treatment processes, secondary settling tanks are essential.

14.2. ACTIVATED SLUDGE PROCESS MECHANISM

The *activated-sludge process* was developed in England in 1914 by Ardern and Lockett and was so named because it involved the production of an activated mass of micro-organisms capable of aerobically stabilizing a waste. The *activated sludge* is the sludge which is obtained by settling sewage in presence of abundant oxygen so as to be supercharged with favourable aerobic micro-organisms. The activated sludge process of sewage treatment is based on providing *intimate contact* between the sewage and *biological active sludge*. Such sludge is developed initially by prolonged aeration under conditions which favour the growth of organisms with special ability to oxidise organic matter.

The effluent from the primary settling tank is mixed with a dose of activated sludge and is aerated in an *aeration tank* for a period of some hours. During the aeration, the micro-organisms in the sewage multiply by assimilating part of the influent organic matter. In this process, part of organic matter is synthesised into new cells and part is oxidised to derive energy. The synthesis reaction, followed by subsequent separation of the resulting biological mass and the oxidation reaction are the main mechanisms of BOD removal in the activated sludge process. The *biological mass* generated in the aeration tank consists of zoogleal bacteria, protozoa, rotifers etc. The biomass is generally *flocculant* and *quick settling*. It is separted from the aerated sewage in a secondary settling tank and is recycled continuously to the aeration tank as an essential feature of the process. The mixture of recycled sludge and sewage in the aeration tank is referred to as *'mixed liquor'*. The *recycling* of sludge helps in the initial build-up of a high concentration of active micro-organisms in the mixed liquor which accelerates the BOD removal. Once the required concentration of micro-organisms in the mixed liquor has been reached so as to maintain proper food/microorganism ratio (F/M) for optimum operation, its further increase is prevented by regulating the quantity of sludge recycled and wasting the excess from the system.

The BOD removal is evaluated based on the BOD_5 of the aeration tank influent and the BOD_5 of the final effluent after sludge separation. The effluent obtained from properly operating activated sludge plant is clear and sparkling, containing very low amount of organic matter. The effluent is of high quality, usually having a lower BOD than that of a trickling filter. The BOD removal is upto 80-95 percent and the degree of bacteria removal is upto 90-95 percent.

The activated sludge process is truly an aerobic process since the biological floc are suspended in a liquid media containing dissolved oxygen. Aeration conditions must be maintained in the tank. However, in the final clarifier, the dissovled oxygen concentration can become extremely low. Dissovled oxygen extracted from the mixed liquor is replenished by air supplied to the aeration tank. Fig. 14.1 shows the generalised biological process reactions in the activated sludge process. The microorganisms require food, oxygen and nutrients and yield carbon dioxide, water and new micro-organisms, as indicated by the following equation :

$$\text{Organic material} + O_2 + \text{Nutrients} \xrightarrow{\text{micro-organisms}}$$
$$CO_2 + H_2O + \text{Engery} + \text{Micro-organisms} \quad ...(14.1)$$

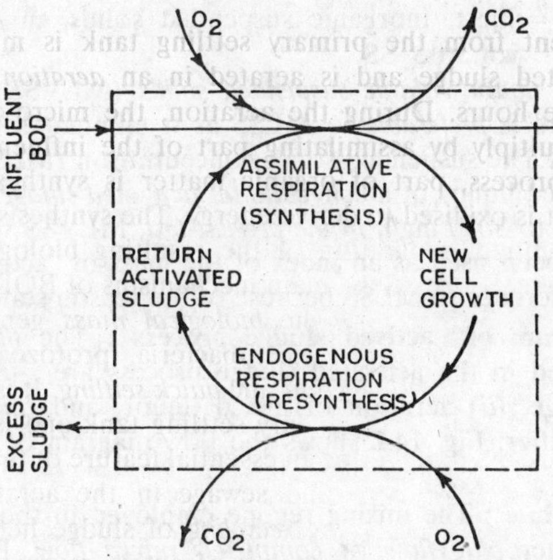

FIG. 14.1. GENERALISED BIOLOGICAL PROCESS REACTION IN THE ACTIVATED SLUDGE PROCESS

The total oxidation of organics in the sewage by microorganisms takes place in two phases. The first phase involves the conversion of organic matter to CO_2, water or new microoganisms. The second phase involves endogenous respiration during which process the microorganisms consume their own cell material for energy. The endogenous respiration rate is relatively slow, and at the end of this phase a nonbiodegradable residue remains. At this point the organic compounds are considered to be oxidised and are no longer pollutants. Carbonaceous compounds are oxidised first and on continuation of the process, the nitrogenous materials are converted into nitrites and nitrates.

Index of the mass of active micro-organisms : The *mixed liquor suspended solids* (MLSS) content generally taken as an index of the mass of micro-organisms in the aeration tank. However, the MLSS will contain not only active microorganisms but also their dead cells as well as inert organic and inorganic matter derived from the influent sewage. Composition of MLSS in the aeration basin is determined by the following equation :

$$x_t = x + x_{ii} \quad ...(14.2)$$

where x_t = MLSS, (mg/l)

x = MLVSS, (mg/l) $= x_a + x_e + x_i$ ...(14.3)

x_a = active microbial mass, mg/l of VSS

x_e = endogenous respiration mass, mg/l of VSS

x_i = inert, non-biodegradable organic suspended solids, mg/l of VSS

x_{ii} = inert, inorganic suspended solids, mg/l of non voltatile SS.

The *mixed liquor volatile suspended solids* (MLVSS) value is generally used and is *preferable* to MLSS as it eliminates the effect of organic matter. The suspended-solids concentration (MLSS) maintained in the mixed liquor of a *conventional activated sludge proccess.* ranges from 1500 to 3000 mg/l. *In the present text,* MLVSS *symbolised by* x(mg/l), has been used as an index of the mass of active micro-organisms, in preference to MLSS because of the reason stated above.

Flow diagram of actived sludge process : The *three basic operations* involved in the activated sludge process are : (*a*) mixing of activated sludge, (*ii*) aeration of mixed liquor, and (*iii*) settling in secondary clarifier. Fig. 14.2 shows the flow diagram of activated sludge process.

Mixing regime : The mixing regime employed in the aeration tank may be either *plug flow* or *completely mixed flow*. Plug flow

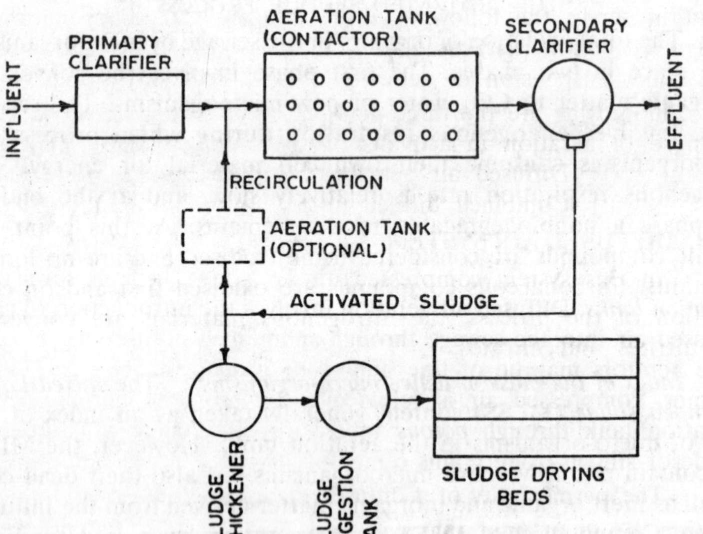

FIG. 14.2. FLOW DIAGRAM OF ACTIVATED SLUDGE PROCESS.

implies that the sewage moves down progressively along the aeration tank, essentially *unmixed* with the rest of the tank contents. *Complete mix flow* involves the rapid dispersal of the incoming sewage throughout the tank. In the plug flow system, the F/M and the oxygen demand will be highest at the inlet end of the aeration tank and will then progressively decrease. In the complete mix system, the F/M and oxygen demand will be uniform throughout the tank.

Flow Scheme : The flow scheme involves the pattern of sewage addition and sludge return to the aeration tank and also the pattern of aeration. *Sewage addition* may be at a single point at the inlet end of the tank or it may be at several points along the aeration tank. The *sludge return* may be directly from the settling tank to the aeration tank or through a sludge aeration tank. Air may be applied uniformly along the whole length of the tank or it may be tapered from the head of the aeration tank to its end.

14.3. AERATION UNITS

Aeration is the most important operation in the activated sludge process, so as to provide *oxygenation* and *mixing*. The aeration facilities are designed to meet the calculated oxygen demand of the process while maintaining in the aeration tank a minimum DO of about 1 - 2 mg/l which is necessary for proper development of biological sludge. In addition to supplying dissolved oxygen, the aeration devices have also to provide adequage mixing and agitation so that the mixed liquor suspended solids do not settle down. This way, aeration increases the contact opportunity between the floc and sewage. To summarise, aeration serves the following *three* functions : (*i*) Oxygenation of the mixed liquor. (*ii*) Flocculaion of the colloids in sewage influent and (*iii*) Suspension of activated sludge floc.

Following are the three methods which are employed for the purpose of aeration in activated sludge process :

(*i*) Diffused air aeration (*ii*) Mechanical aeration

(*iii*) Combined diffued air and mechanical aeration.

14.4. DIFFUSED AIR SYSTEM

In this system compressed air is blown through sewage in *aeration tanks*. Diffused air aeration involves the introduction of compressed air into the sewage through submerged diffusers or nozzles. The aerators may be of fine bubble or coarse bubble type. In the former, compressed air is released at or near the bottom of the aeration tank through *porous* tubes or plates made of aluminium oxide or silicon oxide grains cemented together in a ceramic matrix.

The permeability of a diffuser plate is defined as the volume of free air in m^3/min that will pass through 1 m^2 of diffuser at

50 mm differential pressure under dry conditions at temperature of 21°C. Recommended permeabilities lie between 12 and 24, larger rating tending to give uneven distribution.

Standard ceramic plate diffusers have dimensions 0.3 m×0.3 m ×25 mm with pores of about 0.3 mm dia. Such plates, under water, pass 1.2 m^3 of air/min/m^2 with pressure losses between about 100 mm and 200 mm of water. For the best uniformity of distribution and prevention of clogging problems, a minimum of 0.6 m^3 of air/min/m^2 of plate surface under water is advisable and should not exceed 2.5 m^3 because of pressuere losses, poor air economy and more rapid clogging due to corrosion.

The air supply should be at least 0.25 m^3 per min. per metre length of channel. Spacing of the diffusers should be 0.6 m and preferably 1 m apart between centres to avoid rising streams of bubbles.

Other types of diffusers using porous material having the form of long tubes or mushroom shaped domes installed at a level of 0.3 to 0.6 m above the floor of the aeration tank are preferred sometimes to the diffuser plates which are placed at the floor level, because of the ease of cleaning and replacement with the unit under operation. Tube diffusers are generally 0.6 m long with ineternal dia. as 75 mm and thickness of wall equal to 15 mm. The 180 mm dome diffusers pass about 0.85 – 1.4 m^3 of air per hour through each dome. Porous diffusers are liable to clogging from the inside by the dust carried in the air and also clogging from the outside by the suspended solids in the sewage. The air supplied must therefore be free of dust, by providing air filters.

Coarse buble aerators consist of proprietory devices such as Monosparj, Deflectofuser, Discfuser etc. They have slightly lower aeration efficiency than fine bubles aerators, but are cheaper in first cost and are less liable to clogging and donot require filtration of air. Air diffusers are generally placed along one side of the aeration tank, helping to set up a *spiral flow in the tank (Fig. 14.4) which improves mixing and prevents the solids from settling. They are located 0.3 to 0.6 m above tank floor to aid in tank cleaning and reduce clogging during shutdown.*

The quantity of oxygen to be delivered through the diffuser system depends on the oxygen demand of the sewage and the efficiency of oxygen transfer of the diffusers with the latter being controlled by the size of the air bubble produced and the depth of submersion of the diffusers. The oxygen transfer efficiency at 1 to 2 mg/l of DO in aeration tank varies from 5 to 15 percent for most diffusers with 8 percent being common for fine buble diffusers and 6 percent for coarse bubble diffusers. The quantity of air to be delivered through the diffuser system can be worked out from the quantity of oxygen to be delivered assuming 23.2 percent oxygen in air and air density of 1.43 kg/m^3 under standard conditions. The air requirements for different types of activated sludge system are given in Table 14.3.

The air delivery systems are designed to deliver 1.5 times the normal air requirements and compressors are installed in multiple units do enable increase or decrease of the air supply. The pressure developed by the air compressors should equal the depth of sewage to the diffuser units plus losses in diffuser plus about 25% extra for losses in transmission or about 0.4 to 0.65 kgf/cm^2 total. Air-pipings are designed for velocities of 6 to 30 m/s for pipe diameters of 25 mm to 1500 mm. Air header pipes should be located above the tank water level to avoid back siphonage when the compressers trip.

The advantages of flexibility in the aeration arragements are given particular attention for large plants. Greater flexibility requires costlier air supply piping systems–additional fittings, valves, air flow meters, perhaps less efficient compressor operation, etc–but the added costs may be justified by improved purification efficiencies. Automatic dissolved oxygen or redox potential measurments may also be provided.

Types of Aeration tanks : Following are *two* types of aeration tanks generally used in the diffused air aeration ; (*i*) Ridge and furrow type tank ; (*ii*) Spiral flow type tank

1. Ridge and Furrow type tank : Fig. 14.3 shows the ridge and furrow type aeration tank, used with fine buble aeration through diffuser plates placed in the furrows or depressions. These tanks are in the form of narrow rectangular channels, 30 to 120 m long, 4.5 to 9 m wide and 3 to 4.5 m deep, laid parallel to each other. The diffuser tiles are *fixed* in the furrow portion by cement or bituminous

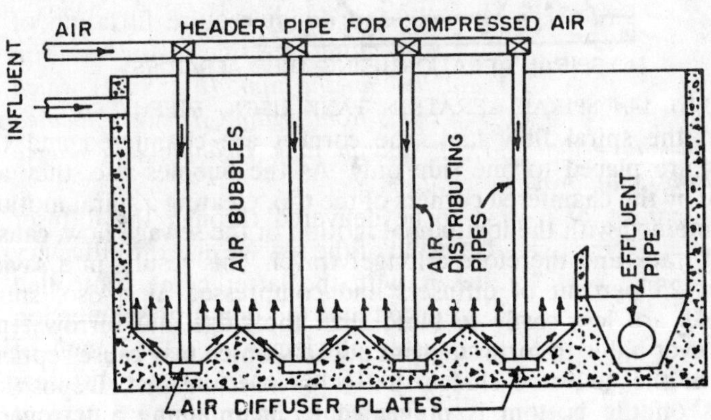

FIG. 14.3. RIDGE AND FURROW TYPE AERATION TANK

compounds and are made air-tight by rubber rings. Air is supplied to the diffuser by header pipe through air distributing pipes. The diffuser plates are provided at right angles to the direction of flow. A free board of 60 cm is provided at the top.

2. ***Spiral flow type tank*** : In this type of tank, diffused air may be supplied either through *plate diffusers* placed at the bottom along one side of the tank, or through *tube diffusers* kept suspended from the top, though tube diffusers are more commonly used. Fig. 14.4 (*a*) and (*b*) show the details of spiral flow type tanks.

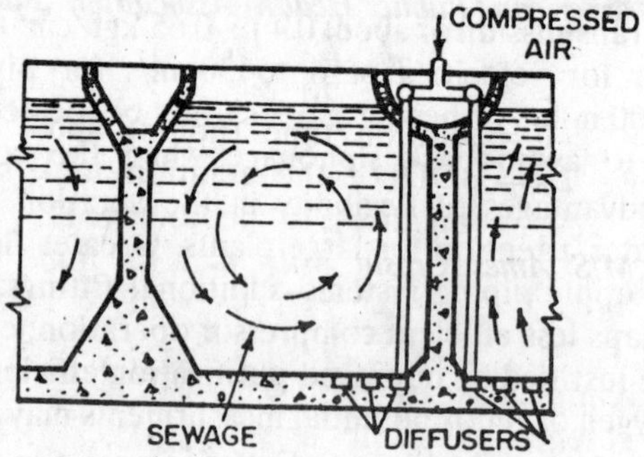

(a) SPIRAL AERATION USING PLATE DIFFUSERS

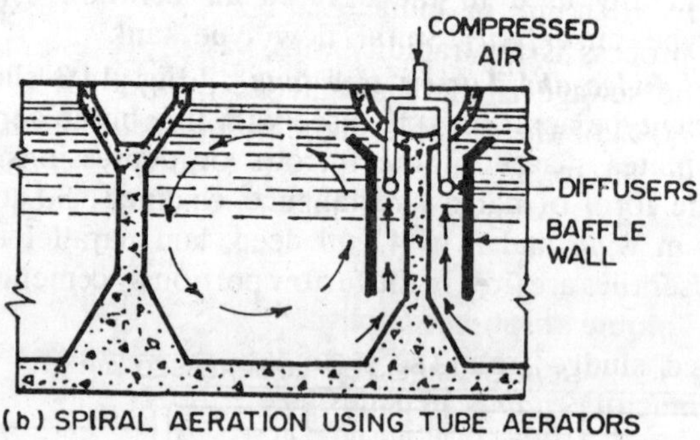

(b) SPIRAL AERATION USING TUBE AERATORS

FIG. 14.4 SPIRAL AERATION TANK USING DIFFUSERS

In the spiral flow tank, the corners are chamfered and the diffusers are placed to one side only. As the bubbles rise, they are deflected by the chamfered corners of the top, creating a spiral motion. This, combined with the logitudinal motion of the sewage flow, causes a helical track and therefore a longer travel. This results in a saving of about 25 percent of diffusers and compressed air. Also, spiral flow tanks are less costly to build than the ridge and furrow type. It is however liable to form stagnant pockets which may cause septicily in part of the activated sludge; unless guarded against. Deposition of solids on the bottom is prevented by maintaining a transverse velocity across the bottom of about 0.4 to 0.5 m/s.

Aeration period : The aeration period is the detention time of the raw-wastewater flow in the aeration tank, expressed in hours. Period of aeration depends upon the following : (*i*) strength of sewage and MLSS concentration ; (*ii*) desired degree of purification in term of BOD removal ; (*iii*) rate of aeration and (*iv*) proportion of returned activated sludge.

A number of empirical formulae and charts are available to determine the aeration period. The following are two available *empirical* formuale :

(a) American Public Health Association Formula

$$T = \frac{L_a}{20} - 1 \qquad ...(14.4)$$

where T = aeration time (hours) and
L_a = BOD of the aeration tank sewage influent (mg/litre) to be removed

(b) M/S Ames Crosta Mills and Co. Ltd. (England) Formula

$$T = \left(\frac{L_a}{10}\right)^{3/4} \qquad ...(14.5)$$

For complete treatment (such that the effluent is fairly stabilised with the presence of nitrates and some dissolved oxygen) a period of 4 to 6 hours is required in America, 10 to 12 hours in Britain and 6 to 10 hours in India. A noteworthy feature of the activated sludge process is the rapidity with which organic matter is oxidised when the sewage is first brought in contact with the active sludge. About 60% of the organic matter is oxidised during the first hour and only 30 to 35 percent in the next 5 or 6 hours. Generally, aeration time varies from 4 to 8 hours, the common value being $4\frac{1}{2}$ hours.

Volume of returned activated sludge : The volume of returned activated sludge from the secondary clarifier to the aeration tank (or contactor) *mainly* depends upon the extent of BOD desired to be removed. Table 14.1 gives the volume of activated sludge to be added to remove the desired BOD.

TABLE 14.1. VOLUME OF RETURNED ACTIVATED SLUDGE

Desired BOD removal (ppm or mg/l)	*Percentage of returned activated sludge*
150	25
250	30
300	35
400	40
500	48
600	53

Fig. 14.5 gives an emipiral chart to determine the aeration time as well as returned activated sludge, both of which are primarily dependant on the desired BOD removal.

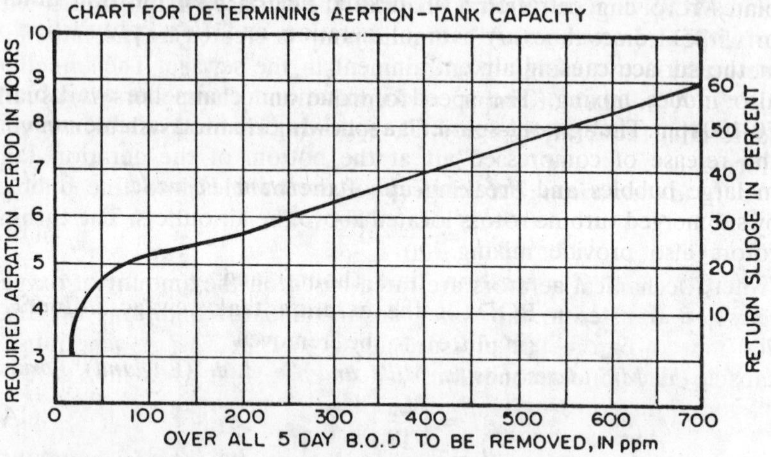

FIG. 14.5. CHART FOR DETERMINATION OF AERATION TANK CAPACITY

Capacity of aeration tank : The capacity (*C*) of aeration tank depends upon the following three factors : (1) Aeration period ; (2) Volume of returned sludge ; (3) Volume of flow of sewage.

Out of these, the first two can be determined from the chart of Fig. 14.5. Thus we have,

$$V = (Q + Q_s)\frac{T}{24} \quad ...(14.6)$$

where V = Capacity of aeration tank (m^3)

Q = Volume of flow of sewage (m^3/day)

Q_s = Volume of returned sludge (m^3/day)

T = aeration period (hours).

14.5. MECHANICAL AERATION SYSTEMS

In the diffused air system, the air which is actually utilised in oxidation is only about 5 percent, while the remainder is required for the purpose of agitating the liquid to meet the mixing requirements. A need was therefore led to the development of some cheaper system to achieve the purpose. This has led to many mechanical devices. With the development of efficient mechanical surface aerators requiring very little operational attention, diffused air aeration is falling out of use. Mechanical aerators were linked to small installations in the past but with recent improvements in their design, they are being increasingly used for large plants in preference to diffused air aeration systems. Some of their advantages are (*i*) higher oxygen transfer capacity (*ii*) absence of air piping and air filter, and (*iii*) simplicity of operation and maintenance.

Mechanical aeraters generally consist of large diameter impeller plates revolving on vertical shaft at the surface of the liquid with or without draft tubes. A hydraulic jump is created by the impellers at the surface causing air entrainment in the sewage. The impellers also induce mixing. The speed of rotation of impellers is usually 70-100 rpm. The agitator-sparjer is a special mechanical system involving the release of compressed air at the bottom of the aeration tank in large bubbles and breaking up of the bubbles into fine bubbles by submerged turbine rotors located above the air outlets. The turbine rotors also provide mixing.

Mechanical aerators are rated based on the amount of oxygen they can transfer to tap water under standard conditions of 20°C, 760 mm kg barometric pressure and zero DO. The oxygen transfer capacity under field conditions can be calculated from the standard oxygen transfer capacity by the following formula :

$$N = N_S \cdot \frac{C_S - C_L}{9.17} \times 1.024^{T-20} \times \alpha \qquad ...(14.7)$$

where N = oxygen transfer under field conditions (kg O_2/h)

N_S = oxygen transfer capacity under standard conditions

C_S = dissolved oxygen saturation value for sewage at operating temperature

C_L = operating DO level in aeration tank (usually 1 to 2 mg/l)

T = temperature °C

α = correction factor for oxygen transfer of sewage, usually 0.8 to 0.85.

The oxgyen transfer capacity of mechanical aerators under standard conditions is about 1.9 kg/kwh compared to 1.3 kg/kwh for sparjer system, 1.5 kg/kwh for diffused air fine buble aeraters and 0.9 kg/kwh for diffused air coarse bubble aerators. Oxygen transfer capacities of surface aeraters should be supported by actual test data for the model and size offered.

Mixing requirements : The aeration equipment has also to provide adequate mixing in the aeration tank to keep the solids in suspension. Mixing considerations require that the power input in mechanical aerators should not be less than 0.015 – 0.026 kw/m^3 of tank volume. The power input of mechanical aerators derived from oxygenation considerations should be checked to satisfy the mixing requirements and increased where required.

Systems of mechanical aeration : Following are some of the patented systems of mechanical aeration :

1. Haworth paddle or Sheffield aeration system.
2. Hartley paddle or Birmingham Bio-flocculation system.
3. Simplex aeration system.
4. Link Belt aeration system.
5. Kessner Brush aeration system.

*1. **Haworth paddle or Sheffield aeration system*** : In the Haworth system, adopted in Sheffield (England), the aeration tank is divided with a number of long and relatively narrow parallel channels by means of thin dividing walls. Two rows of paddles mounted on horizontal shafts, provided about midway between ends of each channel, rotate at a speed of 15 rpm. The rotation of paddles arranged in staggered fashion, cause spiral motion of sewage required for the aeration. The inlet and outlet positions are arranged as shown in Fig. 14.6, and they are cross connected to achieve the recirculation of a part of the flow through the tank. The depth of the tank is kept about 1.2 m and the width of each channel is kept between 1.2 to 2 meters.

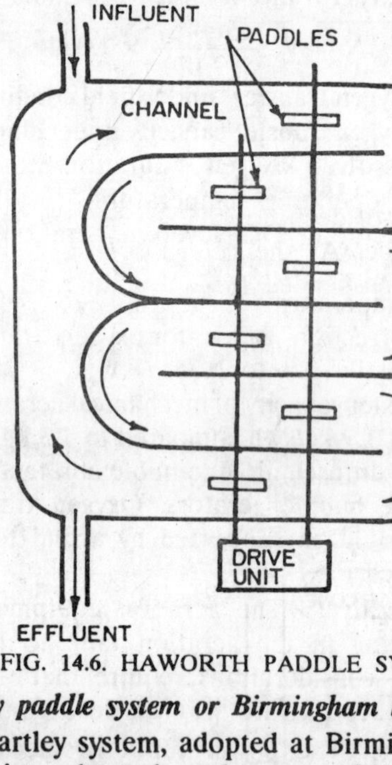

FIG. 14.6. HAWORTH PADDLE SYSTEM

*2. **Hartley paddle system or Birmingham bio-flocculation system:***

In the Hartley system, adopted at Birmingham (England) the tank is divided into channels and at one of the ends of each bend, partially submerged paddles are provided, inclined at a small angle from the vertical, as shown in Fig. 14.7. These paddles give forward movement to the sewage and at the same time set up a wave action which brings the sewage in intimate contact with the atmoshpheric oxygen. Diagonal flat baffle plates are provided in each channel, at regular intervals to reduce the velocity of sewage and to maintain spiral flow. This prevents sedimentation of sludge and also increases the efficiency of aeration.

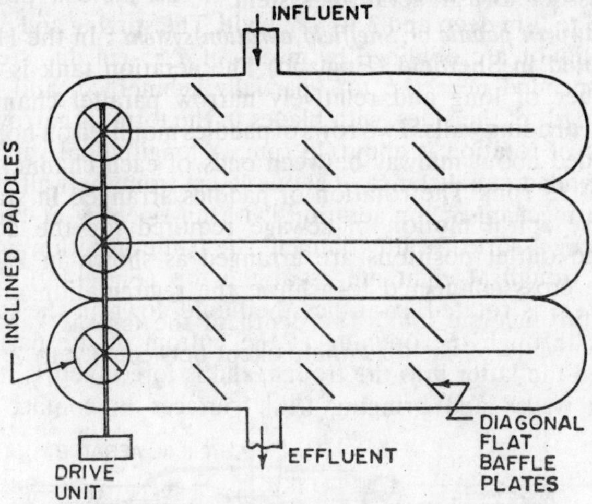

FIG. 14.7. HARTLEY SYSTEM

3. *Symplex system* : The symplex system, also known as bio-aeration system, consists of deep hopper bottom vertical flow tanks, generally of square shape. At the centre of each tank, a hollow *uptake* or *draft tube* is suspended from the top, keeping it about 15 cm above the tank floor. There is a steel cone with vanes, which, by rotating by means of gear at the top end of the tube, draws

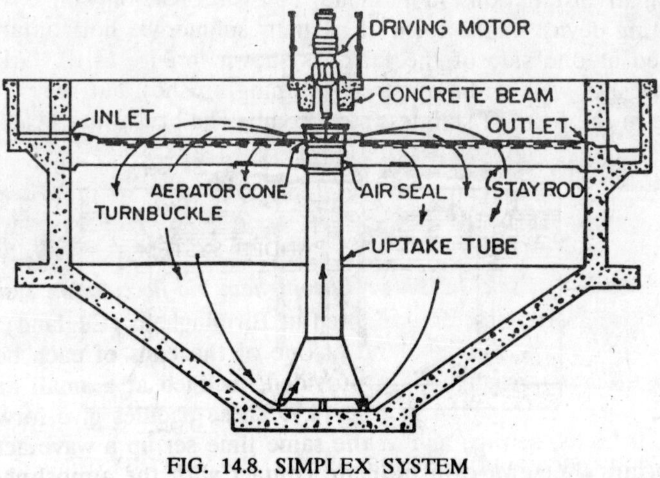

FIG. 14.8. SIMPLEX SYSTEM

the mixed liquor up through the tube, by the suction effect, and sprays it over the surface. The spray absorbs oxygen from the atmosphere and keeps the surface in motion. The speed of rotation is kept about 60 rpm, by means of motor located at the top (Fig. 14.8).

4. ***Link-Belt aeration system*** : In this system, the aeration tank is 2.5 to 3 m deep and 3 to 4 m wide. The stirring and circulation of mixed liquor are obtained by means of a 75 cm dia. steel paddle wheel suspended near the top, partially submerged, and extending the full length of the sides, with blades in the form of narrow ribbons. The speed of rotation is about 48 rpm. A longitudinal vertical baffle wall is provided at a distance of about 45 cm from the wall supporting the above mechanism for aeration, with an opening at the bottom, thus forming a sort of a lift channel. The baffle partition wall carries a narrow trough at right anlges at the top (Fig. 14.9). When the paddle wheel is rotated, it pushes the liquid down in the tank which then rises through the opening at the bottom of the partition and up behined the latter into the trough, and is forced across the surface producing waves and bringing fresh surfaces in contact with air.

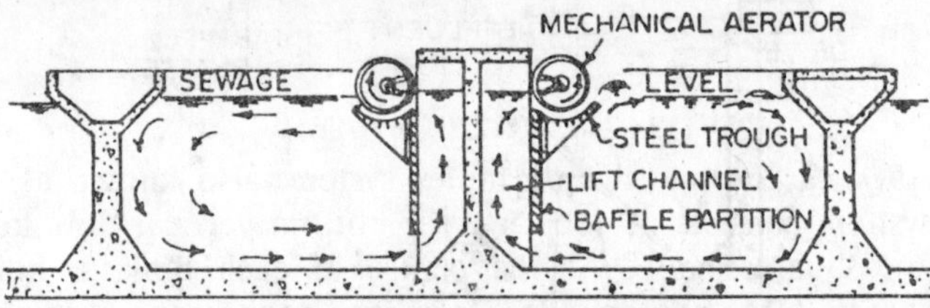

FIG. 14.9. LINK-BELT AERATION SYSTEM

5. ***Kessner Brush aeration system*** : This process, develped for small installations in Holland, consists of a long tank with an agitating device in the form of partially submerged horizontal brush located at one side of the tank, as shown in Fig. 14.10. Originally the brushes were ordinary street cleaning brushes, but they are now made in the form of stainless steel comb. The brush may be partially

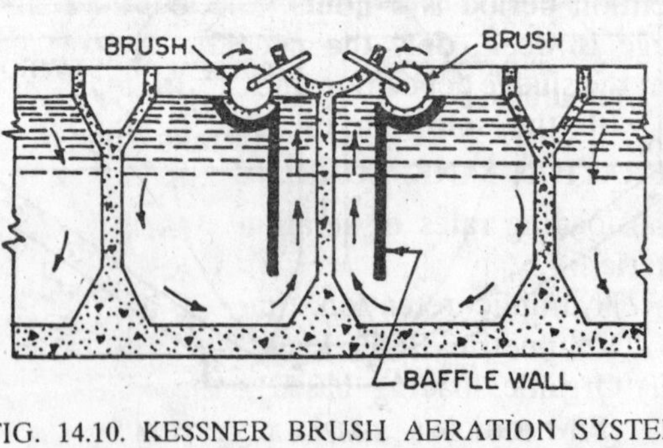

FIG. 14.10. KESSNER BRUSH AERATION SYSTEM

submerged to a depth of about 5 mm to 40 mm, and is rotated at a speed of about 40 to 50 rpm. The rotation of the brush cause wave action in the sewage to bring about the necessary aeration.

14.6. COMBINED MECHANICAL AND DIFFUSED AIR SYSTEM

In this system, diffused air aeration and mechanical aeration are combined in a single unit, so as to achieve both the efficiency of diffused air system and economy of the mechanical aeration system simultaneouly. This reduces the cost of construction and maintenance. In the combined system, aeration of sewage is done by compressed air, using air diffusers placed at the bottom of the tank, and the agitation of sewage is carried out by means of mechanical paddles. An example of this is found in Imhoff type aeration tank patented by Dorr Oliver Co. (America), under the name 'Dorr Aerator', shown in Fig. 14.11.

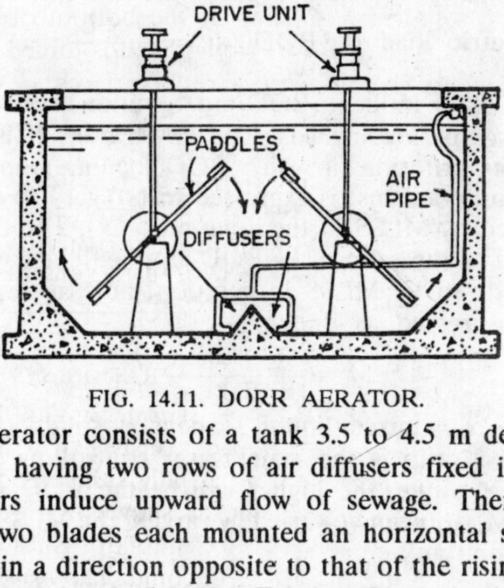

FIG. 14.11. DORR AERATOR.

Dorr aerator consists of a tank 3.5 to 4.5 m deep and about 7.5 m square, having two rows of air diffusers fixed in the bottom. These diffusers induce upward flow of sewage. There are paddle wheels with two blades each mounted an horizontal shafts, making 10 to 12 rpm in a direction opposite to that of the rising air bubbles. The detention period is 5 hours. For low flows at night, and when the sewage is weak, only the paddle wheel may be kept working to prevent the sludge deposition and the diffused air may be completely stopped, with the sole aim of achieving of saving in power.

14.7. AERATION TANK LOADING CRITERIA

The loading rates of aeration tank is based on the following four criteria :

(*i*) Hydraulic retention time (HRT)
(*ii*) Volumetric BOD loading
(*iii*) Organic loading based on food to micro-organisms ratio (F/M ratio)

and (*iv*) Solids Retention Time (SRT) or Mean Cell Residence Time (MCRT) or Sludge age.

(*i*) **Hydraulic retention time (HRT) or aeration period** : The aeration period or loading rate expresses the rate at which sewage is applied in the aeration tank. A loading parameter that has been developed *empirically* over the years is the hydraulic retention time (HRT) which is expressed as follows :

$$\text{HRT(hours)} = \frac{V}{Q \times 1000} \times 24 \qquad ...(14.8)$$

where V = Volume of aeration tank (m^3)

Q = sewage inflow, mLd (excluding sludge recycle)

(*ii*) **Volumetric BOD_5 loading** : Another empirical loading parameter is volumetric loading which is defined as the BOD_5 applied per unit volume of aeration tank, expressed as under.

$$\text{Volumetric load (kg } BOD_5/m^3) = \frac{Q \times L_a}{V} \qquad ...(14.9)$$

where L_a = influent BOD_5 to aeration tank (mg/l)

(*iii*) **Organic loading based on F/M ratio** : It is an important organic loading criterion in which BOD loading (representing food F to the micro-organisms) is expressed with regard to microbial mass M (represented by MLSS in the aeration tank). The organic loading rate is defined as the ratio of kg BOD_5 applied per day (representing microbial food) to kg MLSS in aeration tank (representing micro-organisms), expressed as under

$$F/M = \frac{Q \times L_a}{(V/1000)\, x_t} \qquad ...(14.10)$$

where x_t = mixed liquor suspended solids (MLSS), mg/l

The F/M ratio is the main factor controlling BOD removal. Lower the F/M value, the higher will be the BOD removal in the plant. The F/M ratio can be varied by varying the MLSS concentration in the aeration tank.

(*iv*) **Sludge age : solids retention time (SRT)** : Another parameter that could be used for checking the design of activated sludge systems is the solids retention time (SRT), also known as mean cell residence time (MCRT) or sludge age (θ_c) defined by the expression:

$$\theta_c = \frac{X}{(\Delta X/\Delta t)} \qquad ...(14.11)$$

where X = total microbial mass in a reactor

$\Delta X/\Delta t$ = total quantity of solids *withdrawn* daily, including solids deliberatey wasted and those in the effluent.

Thus, the *mean cell residence time* (θ_c) or sludge age may be defined as average time for which the mass of suspended solids (or the biological solids) remain under aeration. Though the hydraulic retention time may be only few hours, the residence time of biological solids is much greater, and while the sewage (liquid) passes through the aeration tank only once, within the hydraulic retention time (HRT), the resultant biological growth and the extracted organic solids are repeatedly recycled from the secondary settling tank back to the aeration tank thereby increasing the retention time of solids known as *solids retention time* (SRT) or mean cell residence time (MCRT).

Completely mixed-cellular recycle system

Fig. 14.12 shows a completely mixed process of solids recycle applicable to activated sludge systems of any configuration (see § 14.11). The model was specifically developed for *completely mixed systems*, but is conservative in its prediction of effluent quality for plug-flow systems. In the model, it is assumed that there are no micro-organisms in the influent, and that the system contains a unit in which the cells from the reactor are *settled* and then *returned* to the reactor. Because of the pressence of *settling unit*, following two *additional assumptions* are made : (*i*) waste utilization by the microorganisms occurs only in the reactor unit. This assumption leads to a conservative model, since in some systems, there may be some waste utilization in the settling unit. (*ii*) The volume used in calculating the mean cell residence time for the system includes only the volume of the reactor unit (*i.e.* aeration unit).

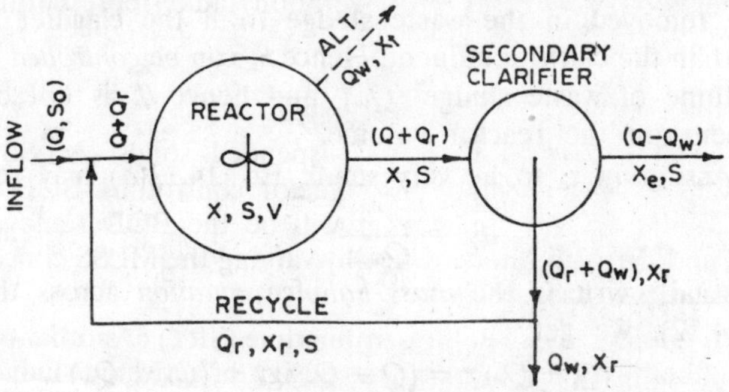

FIG. 14.12. COMPLETELY MIXED PROCESS OF SOLIDS RECYCLE.

Hence the mean cell residence time (θ_c) or the *sludge age* (days) for this system by definition (Eq. 14.11) is given by

$$\theta_c = \frac{\text{mass of suspended solids in the system}}{\text{mass of solids/leaving the system per day}} = \frac{X}{(\Delta X/\Delta t)}$$

Let Q = the waste flow rate (or volume of flow per day)

x = microbial mass concentration = mixed liquor volatile suspended soils concentration, MLVSS or simply, VSS.

V = volume of the reactor (or aerator)
Q_w = volume of wasted sludge, per day
x_r = clarifier underflow suspended solids concentration (*i.e.* concentration of solids in the returned sewage or in the wasted sludge, both being equal).
x_e = concentration of solids in the effluent. Then :

(*i*) Mass, of solids in the reactor $= V.x$

(*ii*) Mass of solids removed with the wasted sludge per day $= Q_w . x_r$

(*iii*) Mass of solids removed with the effluent, per day $= (Q - Q_w) x_e$

$\therefore$ Total mass of solids removed per day

$$= Q_w . x_r + (Q - Q_w) x_e$$

$$\therefore \qquad \theta_c = \frac{V.x}{Q_w x_r + (Q - Q_w) x_e} \qquad \text{...(14.12 } a)$$

If the solids are *wasted* from the reactor, as shown by the dashed line, rather than from the clarifier, Eq. 14.12 (*a*) takes the following form :

$$\theta_c = \frac{V.x}{Q_w . x + (Q - Q_w) x_e} \qquad \text{..(14.12 } b)$$

If the clarifier is functioning efficiently, the bulk of solids will be removed in the waste sludge from the clarifier underflow rather than the clarifier effluent. Hence θ_c can be *controlled* by varying the volume of waste sludge (Q_w), and hence θ_c is not completely dependent on the reactor volume.

Assuming x_e to be very small, Eq. 14.12(*a*) may rewirten us

$$\theta_c \approx \frac{Vx}{Q_w x_r} \qquad \text{...(14.12 } b)$$

Agai.. writing the *mass balance equation* across the reactor (Fig. 14.12), we get

$$(Q + Q_r) x = (Q - Q_w) x_e + (Q_r + Q_w) x_r$$

$$(Q - Q_w) x_e + Q_w . x_r = (Q + Q_r) x - Q_r x_r$$

Substituting the value of L.H.S. of the above expression in Eq. 14.12 (*a*), we get

$$\theta_c = \frac{V.x}{(Q + Q_r) x - Q_r x_r}$$

or
$$\frac{1}{\theta_c} = \frac{1}{V}\left[Q + Q_r - Q_r \frac{x_r}{x} \right]$$

or
$$\frac{1}{\theta_c} = \frac{Q}{V}\left[1 + r - r\frac{x_r}{x} \right] \qquad \text{...(14.13)}$$

or $$r = \frac{Q_r}{Q} = recycle\ ratio$$

x_r/x = ratio of return solids concentration in the clarifier underflow to the mixed liquor concentration.

Thus, from Eq. 14.13, we observe that θ_c is a function of both the *recycle ratio* (r) as well as x_r/x ratio. The value of x_r is dependent upon the efficiency of the clarifier and the settling characteristics of the biological mass.

Also, the rate of return sludge flow (Q_r) depends on the volatile suspended solids concentration (x_r) in the secondary settling tank underflow and the mixed liquor volatile suspended solids (MLVSS), x, and is given by the following equation:

$$(Q + Q_r)\,x = Q_r \,.\, x_r \qquad ...(14.14)$$

Hence from Eq. 14.14, recirculation flow (Q_r) can be directly calculated.

Lawrence and McCarty have shown that the *total microbial mass* in the reactor is given by :

$$x\,V = \frac{Y Q\,(S_o - S)\,\theta_c}{1 + k_d\,\theta_c} \qquad ...(14.15\ a)$$

From which $$x = \frac{\theta_c}{\theta} \cdot \frac{Y\,(S_0 - S)}{1 + k_d\,\theta_c} \qquad ...(14.15)$$

where Y = *growth yield coefficient* relating cell yield to the material metabolised (Table 14.2)

$$= \frac{\text{mass of cells formed}}{\text{mass of substrate consumed}}$$

S_o = influent substrate concentratoin or influent *soluble* BOD

S = effluent substrate concentration or effluent soluble BOD

θ = liquid retention time $= V/Q$

k_d = microorganism decay coefficient, per day (Table 14.2)

TABLE 14.2 TYPICAL KINETIC COEFFICIENTS FOR THE ACTIVATED SLUDGE PROCESS

S.N.	*Coefficient*	*Basis*	*Value at 20° C*	
			Range	*Typical*
1.	k	d^{-1}	2–10	5.0
2.	K_s	mg/l BOD_5	25–100	6.0
3.	Y	mg VSS/mg BOD_5	0.4–0.8	0.6
4.	k_d	d^{-1}	0.04–0.075	0.06

It should be noted that the kinetic growth equations are applicable for *soluble* substrate only. *Hence S_o and S do not represent total BOD of influent and effluent respectively, but represent only the portion of BOD which is soluble.* However, in the activated sludge process, which treats settled wastes only, total BOD may be considered as representing the influent soluble BOD.

$$\therefore \quad S_o \approx L_a \text{ (mg/l)} \quad \text{...(14.16 a)}$$

However, BOD of effluent is attributed to that due to *biological cells* and that due to *soluble organics* escaping the treatment. *Hence effluent soluble* BOD (*S*) *is not equal to total BOD of effluent.* The effluent soluble BOD(*S*) may be estimated after giving a credit to that due to biological cells. The *ultimate BOD* of the waste cells may be taken as 1.42 mg/l. The BOD_5 of effluent solids can be estimated by the following empirical relationship :

$$BOD_5 \text{ of effluent solids} = 0.63(SS) \quad \text{...(14.16 b)}$$

Hence $$S = \text{effluent } BOD_5 - 0.63(SS) \quad \text{... (14.16 c)}$$

Factor of safety : Two particular values of θ_c, significant in the design of biological processes are : (*i*) θ_c^m, and (*ii*) θ_c^d. Here θ_c^m is defined as the lowest value of θ_c at which operation is just possible. At retention times less than θ_c^m, organisms are removed more quickly than they are synthesized, and hence failure will occur. Specific value of θ_c^m is a function of the waste concentration and various biological kinetic parameters. θ_c^d is the design value of θ_c and must be significantly greater then θ_c^m. The ratio θ_c^d/θ_c^m gives the *safety factor* for the system. Required safety factors depend upon the anticipated variability of the waste and should be atleast 4.

Food-micro-organism ratio (*F/M* ratio)

$$\text{Food } (F) = Q \, . \, S_o$$

$$\text{Microbial mass } (M) = V \, . \, x$$

$$\therefore \quad \frac{F}{M} = \frac{Q \, . \, S_o}{V \, . \, x} = \frac{S_o}{\theta \, . \, x} \quad \text{...(14.17)}$$

where $\theta = \frac{Q}{V}$ = liquid retention time.

Process efficiency (*E*)

The process efficiency is defined by the following equations

(*a*) *Based an soluble* BOD_5

$$E_s = \frac{S_o - S}{S_o} \times 100 \quad \text{...(14.18 a)}$$

The treatment process fails or becomes ineffective when the substrate concentration at the effluent becomes equal to that at the inlet, *i.e.* $S = S_o$.

(*b*) *Overall plant efficiency* :

$$E = \frac{L_a - L}{L_a} \times 100 \qquad ...(14.18\ b)$$

Plug flow–cellular recycle system : Fig. 14.13 shows plug flow-cellular recycle system. A plug flow reactor model assumes that no longitudinal mixing occurs between the contents of the reactor. Each element of the mixed liquor simply traverses along the length of the reactor without losing its identity, and behaves just like a moving unit batch reactor. In the reactor, the waste concentration decreases along its length while the microbial mass concentration increases due to the utilisation of the substrate by the microorganisms. Hence the kinetic model of plug-flow system is mathematically difficult. However, in order to develop a useful kinetic model, Lawrence and McCarty *assumed* that the concentration of micro-organisms in the influent to the reactor is approximately the same as that in the effluent from the reactor. This assumption applies if $\theta_c/\theta > 5$. The resulting *average concentration* of micro-organisms (or average concentration of microbial mass) in the reactor is symbolised by $\bar{x}$.

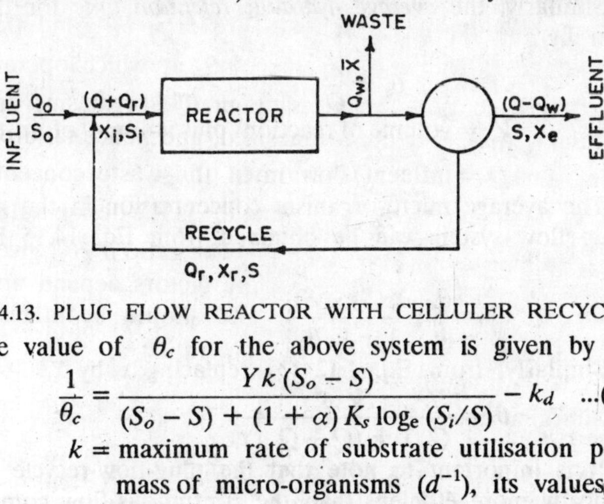

FIG. 14.13. PLUG FLOW REACTOR WITH CELLULER RECYCLE.

The value of θ_c for the above system is given by

$$\frac{1}{\theta_c} = \frac{Y k (S_o - S)}{(S_o - S) + (1 + \alpha) K_s \log_e (S_i/S)} - k_d \qquad ...(14.19)$$

where k = maximum rate of substrate utilisation per unit mass of micro-organisms (d^{-1}), its values varies from 2 to 10 (Table 14.7)

K_s = half velocity constant ; substrate concentration at one-half the maximum growth rate, mass/unit volume (mg/l BOD_5). Its value varies from 25 to 100 (Table 14.2)

S_i = influent substrate concentration to reactor after dilution with recycle flow

$$= \frac{S_o + r S}{1 + r} \qquad ...(14.19\ a)$$

$$r = \text{recycle ratio} = \frac{Q_r}{Q}$$

It is seen from Eq. 14.19 that θ_c is a function of both S_o and S in a plug flow celluler recycle system, while in the complete mix system, θ_c is function of S only.

It should be noted that in Fig. 14.13, the excess micro-organisms are *essentially* wasted from the reactor and not from the recycle line. Hence θ_c for the plug-flow recycle system could also be defined by Eq. 14.12 (*c*) modified as under, with the same assumptions applying:

$$\theta_c \approx \frac{V\bar{x}}{Q_w \cdot x_r} \qquad ...(14.12\ d)$$

The *average hydraulic retention time* of the waste in the *reactor* is given by :

$$\theta = \frac{V}{Q} \qquad ...(4.20\ a)$$

where V is the volume of the reactor.

Similarly, the *average hydraulic retention time* for the *system* is given by

$$\theta_s = \frac{V_s}{Q} \qquad ...(4.20\ b)$$

where V_s = volume of reaction plus volume of settling tank.

Q = influent flow rate

The average micro-organism concentration in the reactor of the plug flow system can be obtained from Eq. 14.15 by noting that $\bar{x}$:

$$\bar{x} = \frac{\theta_c}{\theta} \cdot \frac{Y(S_o - S)}{1 + k_d \theta_c} \qquad ...(14.21)$$

Similarly, from Eq. 4.12 (*b*) replacing x by $\bar{x}$,

$$\theta_c = \frac{V \cdot \bar{x}}{Q_w \bar{x} + (Q - Q_w) x_e} \qquad ...(14.22)$$

It is important to note that the plug-flow-recycle system is theoretically more efficient than the continuous-flow complete mix recycle system. However, complete mix system can easily alsorb shock loads. In actual practice, a true plug-flow regime is difficult to obtain because of *longitudinal dispersion* or back mixing. When a high degree of back mixing occurs in the tank, the condition approaches a complete-mix reactor.

14.8. SLUDGE PRODUCTION AND PROCESS CONTROL

1. Rate of slude production

The rate of sludge (or biological solids) production is obtained from Eq. 14.11 :

$$P_x = \frac{dX}{dt} = \frac{x\,V}{\theta_c} \qquad ...(14.23)$$

where $\frac{dX}{dt} = P_x$ = quantity of microbial mass (or volatile suspended solids, VSS) produced per unit time (kg/day)

The value of xV (*i.e.* total microbial mass in the reactor) is given by Eq. 14.15 (*a*). Substituting the value of xV is Eq. 14.23, we get

$$P_x = \frac{dX}{dt} = \frac{Q\,Y\,(S_o - S)}{1 + k_d\,\theta_c} \qquad ...(14.24)$$

The above expression of microbial mass production represents the *volatile suspended solids* (VSS) production in the tank, from which suspended solids production (*i.e.* SS = VSS/0.8) and hence sludge production can be estimated.

Sludge recycle : Slude Volume Index (SVI) : The purpose of the return of sludge is to maintain a sufficient concentration of activated sludge in the aeration tank so as to obtain the required degree of treatment in the given time interval. As stated earlier, the most essential part of the activated sludge process is the return of activated sludge from the secondary clarifier to the inlet of the aeration tank. Control of the sludge recycle rate is the method of maintaining the desired MLSS (and hence MLVSS) and F/M level in the aeration tank which determine the degree of purification. The rate of sludge recycle can be formulated by the following *two methods*.

Method 1 : In this method, the return sludge pumps are so set that the return flow is approximately equal to the percentage ratio of the volume ocupied by the settleable solids from the aeration tank effluent to the volume of clarified liquid (supernatant) after settling for 30 min. in a 1000 ml graduated cylinder. For example, if the settleable solids occupy a volume of 200 ml in a 1000 ml cylinder, after 30 min of settling, the percentage volume will be equal to 200/(1000−200) = 25%.

Thus, if the plant flow is 4 m^3/s, the return sludge rate should be $0.25 \times 4 = 1\ m^3/s$.

Method 2 : Sludge Volume Index : Another method often used to control the rate of return sludge pumping is based an *empirical measurement* known as *Sludge Volume Index*(SVI). The index is defined as the volume in millilitres occupied by one gm of solids in the activated sludge mixed liquor after settling for 30 min. in a 1000–ml graduated cylinder, and is determined experimentally.

The standard test for the determination of SVI consists of collection of 1 litre of sample of mixed liquor from the discharge end of the aeration tank, in a 1000–ml measuring cylinder. This sample, collected in the cylinder is allowed to settle for 30 minutes. The setted sludge (V_s) is recorded in ml. After remixing the settled solids in the cylinder, the mixed liquor is further tested for MLSS by the available standard methods. Let the concentration of MLSS be x_t (mg/l).

Then, $$\text{SVI} = \frac{V_s\,(\text{ml/l})}{x_t\,(\text{mg/l})}\ \text{ml/ mg} \qquad ...(14.25\ a)$$

or $$\text{SVI} = \frac{1000\,V_s}{x_t}\ \text{ml/g} \qquad ..(14.25)$$

When the SVI value in ml/g is divided by 10^6, it will give SVI value in l/mg.

Hence $$\text{SVI (l/mg)} = \frac{\text{SVI (ml/g)}}{10^6} \qquad ...(14.25\ b)$$

Since the concentration of micro-organisms (MLSS) maintained in the contactor has a direct effect on the oxidation of organic pollutants, the liquid-solids *separation characteristics* of these solids are very important. An indication of the ability of the micro-organisms to be separated from the wastewater after contact is indicated by the sludge volume index (SVI) value. The value of SVI is of operational importance since it reflects changes in the treatment system. Any increase of SVI with no increase in MLSS concentration indicates that the settling characteristics of the solids *are changing* and that a plant upset may occur.

Settleability of activated sludge can be hypothetically related to the quantity and solids-concentration in the return sludge and aeration system, as depicted in Fig. 14.14. Thus, we have

$$\frac{Q_r}{Q + Q_r} = \frac{V_s\,(\text{ml/l})}{1000\,(\text{ml})} \qquad ...(14.26\ a)$$

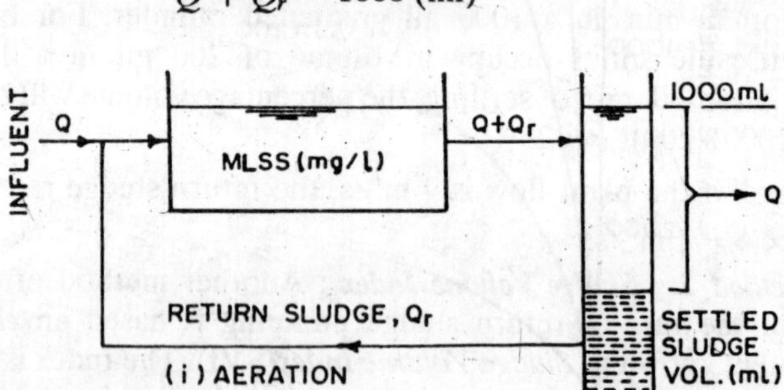

FIG. 14.14. HYPOTHETICAL RELATIONSHIP BETWEEN SETTLED SLUDGE VOLUME AND Q_r IN THE SVI TEST.

or $$\frac{Q+Q_r}{Q_r}=\frac{1000}{V_s}=\frac{1}{\frac{\text{SVI}}{10^6}\cdot x_t} \quad \text{(from Eq. 14.25)}$$

$$\therefore \quad \frac{Q}{Q_r}=\frac{1}{\frac{\text{SVI}}{10^6}\cdot x_t}-1=\frac{1-\frac{\text{SVI}}{10^6}\cdot x_t}{\frac{\text{SVI}}{10^6}\cdot x_t}$$

or $$\frac{Q_r}{Q}=r=\frac{x_t}{\frac{10^6}{\text{SVI}}-x_t} \quad ...(14.26)$$

where x_t = MLSS expressed as mg/l.

SVI = sludge volume index, expressed as ml/g

Q_r = return sludge flow (mLd)

Q = sewage inflow (mLd)

r = return sludge ratio (Q_r/Q)

The relationship between MLSS concentration, SVI and recycle ratio (Q_r/Q) are shown in Fig. 14.15. The amount of recycle flow depends to a great extent on the settling charactereristies of MLSS. For instance, if SVI value is 400 and the desired MLSS concentration is 2000 mg/l, a recycle ratio of 3.5 is required. On the other hand,

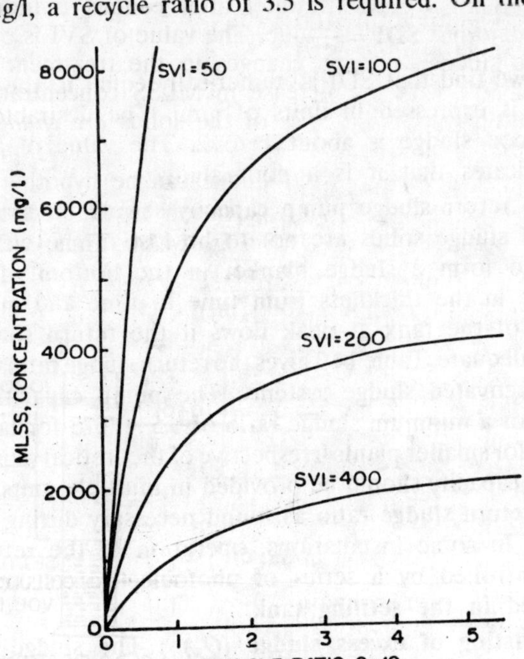

FIG. 14.15. RELATIONSHIP BETWEEN SVI, Q_r/Q AND MLSS CONCENTRATION.

if the SVI is 50, the recycle ratio required would be about 0.2. This demonstrates that the settling characteristics of the biological solids is of primary importance to the successful operation of the activated sludge process.

The concentration of suspended solids in the return sludge can be calculated from the SVI values as follows :

$$\text{SS in return sludge} = \frac{1000\,(\text{mg/g}) \times 1000\,(\text{ml/l})}{\text{SVI}\,(\text{ml/g})}$$

or $$\text{SS in return sludg} = \frac{10^6}{\text{SVI}}\ (\text{mg/l}) \qquad ...(14.26\ b)$$

The above relationship assumes that sludge recirculation is exactly at the required rate.

Sludge density index : Some times, the term *'sludge density index'* (SDI) is used to quantify the solids concentration in the return sludge.

It is given by the following expression

$$\text{SDI} = \frac{\%\text{ of SS by weight}}{\%\text{ of sludge by volume}} \times 100$$

$$= \frac{\text{MLSS}\,(\%) \times 100}{\text{Settled vol. of sludge in 30 min}\,(\%)}$$

$$\therefore \qquad \text{SDI} = \frac{100}{\text{SVI}} \qquad ...(14.25\ c)$$

Thus, we find that SDI is numerically equal to the reciprocal of SVI, and is expressed in units of g/ml. The desirable value of SDI for a good sludge is about 1 to 2. The value of SDI lower than 0.3 indicates that it is a poor sludge.

Ample return-sludge pump capacity should be provided and is essential if sludge solids are not to be lost in the effluent. The solids tend to form a sludge blanket in the bottom of the tank which varies in the thickness from time to time and may fill the entire depth of the tank at peak flows if the return sludge pump capacity is inadequate. Table 14.3 gives the return sludge ratios employed in different activated sludge systems. The pump capacities should be designed for a minmum sludge ratio of 0.5 – 0.75 for large plants and 1.0 – 1.5 for smaller plants irrespective of theoretical requirements. The required capacity should be provided in multiple units to permit variation of return sludge ratio as found necessary during operation of the plant. In some installations, operation of the return-sludge pumps is controlled by a series of photoelectric cells or a sonic system located in the settling tank.

3. Wasting of excess sludge (Q_w) : The sludge generated in the aeration tank has to be wasted to maintain a steady level of MLSS (and hence MLVSS) in the system. The excess sludge quantity will increase with increasing F/M ratio and decrease with increasing temeperature. In case of domestic sewage, the excess sludge to be wasted will be about 0.5 – 0.75 kg per by BOD_5 removed for the conventinal systems and its variations operating with F/M values of 0.2 – 0.5 and about 0.4 – 0.5 by per kg BOD_5 removed in the case of extended aeration plants having no primary settling. The volume of sludge to be wasted will depend on the suspended solids concentration in the waste stream.

Excess sludge may be wasted either from the sludge return line or directly from the aeration tank as mixed liquor. The latter procedure is to be preferred as the concentration of suspended solids will then be fairly steady in the waste stream making control easy. The waste sludge is either discharged with the primary settling tank or thickened in a sludge thickening unit and digested directly. In the extended aeration plants, the excess sludge is taken to sludge drying beds directly and the sludge filtrate discharged into the effluent stream.

Example 14.1. *The MLSS concentration in an aeration tank is 2000 mg/l and the sludge volume after 30 minutes of settling in a 1000 ml graduated cyclinder is 176 ml. Calculate (i) SVI (ii) SDI (iii) required return sludge ratio and (iv) SS concentration in the recirculated sludge.*

Solution

$$\text{SVI} = \frac{1000\,V_s}{x_t}\ (\text{ml/g}) = \frac{1000 \times 176}{2000} \qquad \ldots(14.25)$$

$$= \mathbf{88\ ml/g}$$

$$\frac{Q_r}{Q+Q_r} = \frac{V_s}{1000} = \frac{176}{1000} = 0.176 \qquad \ldots(14.26\ a)$$

or $$\frac{Q}{Q_r} + 1 = 5.68$$

From which $$\frac{Q_r}{Q} = \mathbf{0.214}$$

Alternatively, From Eq. 14.26,

$$\frac{Q_r}{Q} = \frac{x_t}{\dfrac{10^6}{\text{SVI}} - x_t} = \frac{2000}{\dfrac{10^6}{88} - 2000} = 0.214$$

SS concentration in return sludge $= \dfrac{10^6}{\text{SVI}}$ mg/l (*i.e.*, 1.136 %)

$$= \frac{10^6}{88} = \mathbf{11364}\ \text{mg/l}\ \ (i.e.\ 1.136\%)$$

Also $$\text{SDI} = \frac{100}{\text{SVI}} = \frac{100}{88} = \mathbf{1.136}\ \ (\text{g/ml})$$

14.9. OXYGEN REQUIREMENTS AND CONTROL

Oxygen is required in the activated sludge process for the *oxidation* of a *part* of the influent organic matter and also for the *endogenous respiration* of the micro-organisms in the system. The former is a function of BOD removed, while the latter is the function of MLVSS in the aeration tank. However, a part of BOD is removed from the system, without being oxidised, in the form of wastage of excess sludge (synthesized biomass) from the system. Hence the oxygen requirement will be equal to the amount that would be required if all the BOD is to be removed by oxidation only (*i.e.* the BOD) less a credit for the fraction of BOD removed by sludge wasting.

It has been found from Stoichiometry of cell oxidation that 1.42 kg of O_2 is required per kg of cells.

$$O_2 \text{ required, kg/d} = \left(\begin{array}{l}\text{total mass of BOD}_L\\ \text{utilised , kg/d}\end{array}\right) - 1.42\left(\begin{array}{l}\text{mass of organisms}\\ \text{wasted , kg/d}\end{array}\right) \quad ...(14.27)$$

Also, taking a BOD rate constant = 0.23/day at 20°C

$$\frac{BOD_L}{BOD_5} = \frac{1}{1 - e^{-0.23 \times 5}} = 1.47$$

which gives $BOD_L = 1.47\ BOD_5$ ...(14.20 *a*)

or $BOD_5 = 0.68\ BOD_L$ ...(14.28 *b*)

Hence from Eq. 14.27, we obtain

O_2 required/day= 1.47(BOD_5 removed/day) − 1.42 (excess sludge wasted/day)

or O_2 required/day $= 1.47\, Q\,(S_o - S) - 1.42\, V\,(x/\theta_c)$ ...(14.29)

or O_2 required/day $= 1.47\, Q\,(S_0 - S) - 1.42\, P_x$ ...(14.29*a*)

where S_o, S = substrate concentration at the inlet and outlet of the system respectively, BOD_5 in mg/l

V = volume of the reactor.

x = micro-organism concentration in the reactor, which may be taken equal to MLVSS, mg/l.

θ_c = mean cell residence time.

Eckenfelder and O'Connor gave a *similar* expression for the O_2 requirement, as given below :

O_2 required/day $= a'\,Q\,(S_o - S) + b'\,V.x$ ...(14.30)

where a' and b' are constants determined *experimentally*. For municiple sewage, the values of a' and b' have been reported to be 0.4 to 0.65 and 0.1 to 0.3 respectively.

The formula does not allow for nitrification but allows only for carbonaceous BOD removal. The extra oxygen requirements for nitrification is 4.56 kg O_2/per kg NH_3-N oxidised to NO_3-N.

The total oxygen requirements per kg BOD_5 removed for different activated sludge processes are given in Table 14.3. The amount of oxygen required for a particular process will increase within the range shown in the table as the F/M value decreases.

After having known the total O_2 requirement, the actual quantity of air to be supplied is found by considering the fraction of oxygen in air and the *oxygen transfer efficiency* of aerators. The specific weight of air, at mean sea level is 1.2 kg/m^3 at 20°C (and 1.16 kg/m^3 at 30°C), while the fraction of oxygen in air is 23.2 %.

$$\text{Hence theoretical } Q_{air} = \frac{O_2 \text{ demand}}{0.232\,(1.20)} = \frac{O_2 \text{ demand}}{0.278} \quad ...(14.31)$$

For porous tube diffusers, used in conventional activated sludge units, oxygen transfer efficiency is 8%.

$$\therefore \quad \text{Actual } Q_{air} = \frac{O_2 \text{ demand}}{0.278\,(0.08)} \quad ...(14.31\ a)$$

The oxgyen transfer efficiency for coarse bubble diffusers is about 6%.

$$\therefore \quad \text{Actual } Q_{air} = \frac{O_2 \text{ demand}}{0.278\,(0.06)} \quad ...(14.31\ b)$$

The air supply must be adequate to (*i*) satisfy the BOD of the waste, (*ii*) satisfy the *endogenous respiration* by the sludge organisms (*iii*) provide adequate mixing and (*iv*) maintain of minimum dissolved-oxygen concentration of 1 to 2 mg/l throughout the aeration tank.

The following information about air supply is noteworthey.

1. For F/M ratios greater than 0.3, air requirements for conventional process amount to 30 to 55 m^3/kg of BOD_5 removed.

2. At F/M ratios lesser than 0.3, endogenous respiration, nitrification and prolonged aeration periods increase air use to 75 to 115 m^3/kg of BOD_5 removed.

3. For diffused air aeration, the amount of air used has commonly ranged from 3.75 to 15.0 m^3/m^3 at different plants, with 7.5 m^3/m^3 an early rule-of-thumb design factor.

4. The Ten States Standards require the air diffusion system to be capable of delivering 150 percent of normal requirements, which are assumed to be 62 m^3/kg of BOD in the waste water applied to the aeration tanks.

14.10. AERATION TANK DESIGN CONSIDERATIONS

1. **Capacity** : The aeration tank capacity is determined from the F/M and MLSS values selected for the plant. The F/M and MLSS levels generally employed in different types of activated sludge process are given in Table 14.3 along with their corresponding BOD removal efficiences. The lower F/M values are recommended also when winter operating temperatures are low and near freezing point. The tank capacity is determiend on the basis of organic loading

formula. The capacity obtained should be checked against the empirical design criteria for HRT and volumetric loading.

2. Configuration : Except in the case of extended aeration plants and complete mix plants, the aeration tanks are designed as narrow channels. This configuration is achieved by provision of round-the-end baffles in small plants when only one or two tank units are proposed and by construction as long and narrow rectangular tanks with common intermediate walls in large plants when several units are proposed. In extended aeration plants other than oxidation ditches, and in complete mix plants, the tank shape may be circular or sequare when the plant capacity is small or rectangular with several side inlets and equal number of side outlets, when the plant capacity is large.

3. Dimensions : The width and depth of aeration channel depends on the type of aeration equipment employed. The depth controls the aeration efficiency and usually ranges from 3 to 4.5 m, the latter depth being formed to be more economical for installations treating more than 50 mLd. Beyond 70 mLd, duplicate units are preferred. The width controls the mixing and is usually kept between 5 and 10 m. Width-depth ratio should be adjusted to be between 1.2 to 2.2. The length should not be less than 30 m or not ordinarily longer than 100 m in a single section length before doubling back. The horizontal velocity should be around 1.5 m/min. Excessive width may lead to settlement of solids in the tank. Triangular baffles and fillets are used to eliminate dead spots and induce spiral flow in the tanks. Tank *free board* is generally 0.5 m. Due consideration must be given in the design of aeration tanks to the need for emptying them for maintenance and repair of the aeration equipment. Intermediate walls should be designed for empty conditions on either side. The method of dewatering should be considered in the design and provided for during construction.

4. Inlet and outlet arrangements : The inlet and outlet channels of the aeration tank should be designed to maintain a minimum velocity of 0.2 m/s to avoid deposition of solids. The channels or conduits and their appurtenances should be sized to carry the maximum hydraulic load to the remaining aeration tank units when any one unit is out of operation.

The inlet should provide for free fall into the aeration tank when more than one tank unit or more than one inlet is proposed. The free fall will enable positive control of the flows through the different inlets. Outlets usually consist of free-fall weirs. The weir length should be sufficient to maintain a reasonably constant water level in the tank. When multiple inlets or multiple tanks are involved, the inlets should be provided with valves, gates or stop planks to enable regulation of flow through each inlet.

Example 24.2. *An average operating data for conventional activated shadge treatment plant is as follows :*

1. *Wastewater flow* $= 50000\ m^3/d.$
2. *Volume of aeration tank* $= 15500\ m^3$
3. *Influent BOD* $= 200\ mg/l$
4. *Effluent BOD* $= 25\ mg/l$
5. *Mixed liquor suspended solids(MLSS)* $= 3000\ mg/l$
6. *Effluent suspended solids* $= 40\ mg/l$
7. *Waste sludge suspended solids* $= 12000\ mg/l$
8. *Quantity of waste slugde* $= 250\ m^3/d$

Based on the information above, determine:

(*a*) *Aeration period(hours)*
(*b*) *Food to micro-organisms ratio (F/M) (kg BOD per day/kg MLSS)*
(*c*) *Percentage efficiency of BOD removal*
(*d*) *Sludge age (days).*

Solution : In this example, all the data about solids are in terms of suspended solids (SS) and not in terms of volatile suspend solids (VSS).

(*a*) ***Aeration period*** : The aeration period (T) or hydraulic retention time (HRT) is given by Eq. 14.8 :

$$T = \frac{V}{Q} \times 24 \text{ (where } Q \text{ is m}^3\text{/day)}$$

$$= \frac{15500}{50000} \times 24 = \mathbf{7.44 \text{ hours}}$$

(*b*) ***F/M ratio*** : F/M ratio is to be expressed in terms of MLSS, (as desired) Eq. 14.10 :

$$F = Q \,.\, L_a = 50000 \times 200 \text{ g/day}$$

$$= \frac{50000 \times 200}{1000} = 10000 \text{ kg/day}$$

$$M = \text{total mass of MLSS}$$

$$= V \,.\, x_t = 15500 \text{ (m}^3) \times 3000 \text{ (gm/m}^3)$$

$$= \frac{15500 \times 3000}{1000} = 46500 \text{ kg.}$$

$$\therefore \quad \text{F/M ratio} = \frac{10000}{46500}$$

$$= \mathbf{0.215} \text{ kg BOD per day/kg of MLSS.}$$

Note: Actually, F/M ratio should be expressed in terms of MLVSS. In this example, MLVSS = 0.8 × 3000 = 2400 mg/l, taking the ratio MLVSS/MLSS as equal to 0.8.

Then $M = V.x = \frac{15500 \times 2400}{1000} = 37200$ kg

$\therefore$ $\text{F/M ratio} = \frac{10000}{37200}$

$= 0.269$ kg BOD per day/kg of MLVSS.

(*c*) ***Percentage efficiency of BOD removal***

$$\eta = \frac{200 - 25}{200} \times 100 = 87.5\%$$

(*d*) ***Sludge age :*** Sludge age can be found by Eq. 14.12 (*a*), by expressing the equation in terms of suspended solids (SS), rather than in terms of volatile suspended solids (VSS).

$$\therefore \quad \theta_c = \frac{V x_t}{Q_w x_r + (Q - Q_w) x_e}$$

$$= \frac{15500 \times 3000}{(250 \times 12000) + (50000 - 250)\, 40}$$

$$= \mathbf{9.32\ days}$$

Example 14.3. Design of complete-mix activated sludge process

Design a continuous flow completely-mixed activated sludge process to yield an effluent BOD_5 of 20 mg/l and suspended solids of 23 mg/l. The influent BOD_5 following primary clarification is 200 mg/l. The waste flow is 0.2 m^3/sec. Take $Y = 0.65$, $k_d = 0.05$, $\theta_c = 10$ days, MLVSS = 3000 mg/l, return-sludge concentration as 15000 mg/l of suspended solids (SS) and MLVSS ratio = 0.8.

Solution

1. ***Determination of concentration of soluble BOD_5 in the effluent***

BOD_5 of effluent solids ≈ 0.63 (SS)

$= 0.63 \times 23 \approx 14.5$ mg/l

$\therefore$ S = effluent $BOD_5 - 0.63(\text{SS})$

$= 20 - 14.5 = 5.5$ mg/l.

Also, $S_o = L_a = 200$ mg/l

2. ***Determination of treatment efficiency***

(*a*) Efficiency based on soluble BOD_5 is

$$E_s = \frac{S_o - S}{S_o} \times 100 = \frac{200 - 5.5}{200} \times 100 = 97.25\%$$

(*b*) Overall plant efficiency :

$$E = \frac{L_a - L}{L_a} \times 100 = \frac{200 - 20}{200} \times 100 = 90\%$$

3. ***Computation of reactor volume***

$$Q = 0.2\ m^3/\text{sec} = (0.2 \times 10^3)(60 \times 60 \times 24)$$

$$= 17.28 \times 10^6 \text{ litres/day} = 17280\ m^3/\text{day}$$

$$x V = \frac{Y Q \theta_c (S_0 - S)}{1 + k_d \theta_c} \qquad \text{...(14.15 } a)$$

$$= \frac{0.65\,(17.28 \times 10^6)(10)(200 - 5.5)}{1 + 0.05\,(10)}$$

$$= 14.564 \times 10^9 \text{ mg}$$

But $\quad x = \text{MLVSS} = 3000 \text{ mg/l}$

$$\therefore \quad V = \frac{14.564 \times 10^9 \text{ (mg)}}{30000 \text{ (mg/}l)} = 4.855 \times 10^6 \text{ litres}$$

$$= 4855 \text{ m}^3$$

4. ***Computation of sludge to be wasted*** : The rate of slude (biological solids) production is obtained by Eq. 14.23 :

$$P_x = \frac{dX}{dt} = \frac{xV}{\theta_c} = \frac{14.564 \times 10^9 \text{ (mg)}}{10 \text{ (days)}} = 1.456 \times 10^9 \text{ mg/day.}$$

$$\approx 1456 \text{ kg/day}$$

The above value of microbial mass production respresents volatile solids (VSS) production in the tank.

$\therefore$SS production = VSS/0.8 = 1456/0.8= 1820 kg/day.

The underflow solids concentration, = 15000 mg/l (SS)

$$\therefore \quad Q_w = \frac{1820 \times 10^6 \text{ (mg/day)}}{15000 \text{ mg/l)}}$$

$$= 121.3 \times 10^3 \text{ l/day} = 121.3 \text{ m}^3\text{/day}$$

Alternatively, from Eq. 14.12 (*c*), we obtain directly,

$$Q_w \approx \frac{Vx}{\theta_c x_r}$$

where $\quad x_r = (15000 \text{ mg/l SS}) \times 0.8 = 12000 \text{ mg/l VSS}$

$$\therefore \quad Q_w = \frac{14.564 \times 10^9}{10\,(12000)} = 121.4 \times 10^3 \text{ l/day}$$

$$= 121.4 \text{ m}^3\text{/day.}$$

5. ***Determination of recirculation flow*** (Q_r)

$$(Q + Q_r)\,x = Q_r\,x_r \qquad ...(14.14)$$

or

$$Q_r = \frac{Qx}{x_r - x}$$

where $\quad Q = 0.2 \times 60 \times 60 \times 24 = 17280 \text{ m}^3\text{/day.}$

x = aerator VSS concentration = 3000 mg/l

x_r = return VSS concentration= 15000 × 0.8

= 12000 mg/l VSS

$$\therefore \quad Q_r = \frac{17280 \times 3000}{12000 - 3000} = 5760 \text{ m}^3\text{/day}$$

$$\therefore \quad r = \frac{Q_r}{Q} = \frac{5760}{17280} = 0.333$$

6. Determination of hydraulic retention time in the reactor

$$\theta = \frac{V}{Q} = \frac{4855}{17280} = 0.281 \text{ day}$$
$$= 6.74 \text{ hours.}$$

7. Computation of oxygen requirements

$$O_2 \text{ demand } = 1.47\, Q\, (S_o - S) - 1.42\, P_x \qquad ...(14.29\, a)$$

where
$$Q = 17.28 \times 10^6 \text{ litres/day}$$
$$S_0 - S = 200 - 5.5 = 194.5 \text{ mg/l.}$$
$$P_x = \frac{x\, V}{\theta_c} = 1.4564 \times 10^9 \text{ mg/day}$$

$$\therefore O_2 \text{ demand (kg/day)} = \left[1.47 \times 17.26 \times 10^6 \times 194.5 - 1.42 \times 1.4564 \times 10^9\right] \times 10^{-6} \approx 2867 \text{ kg/day.}$$

8. Computation of volume of air at STP

$$\text{Theoretical} \quad Q_{air} = \frac{O_2 \text{ demand}}{0.232\,(1.20)} = \frac{O_2 \text{ demand}}{0.278}$$
$$= 2867/0.278 = 10313 \text{ m}^3/\text{day.}$$

Assuming oxygen transfer efficiency of porous tube diffusers as 8%

$$\text{Actual} \quad Q_{air} = \frac{10313}{0.08} = 128912 \text{ m}^3/\text{day.}$$

$$\therefore \text{Design air requirement} = 2 \times 128912 = 257824 \text{ m/day}$$

Check for air volume on the basis of actual Q_{air}

(*i*) Air requirement per unit volume $= \dfrac{128912}{17280}$

$$= 7.46 \text{ m}^3/\text{m}^3$$

(*ii*) Air requirement per unit of BOD_5 removed :

$$= \frac{128912}{[(200 - 5.5) \times 17.28 \times 10^6]\, 10^{-6}}$$
$$= 38.36 \text{ m}^3/\text{kg of } BOD_5 \text{ removed.}$$

(Normal value ranges from 50 to 75 m^3/kg of BOD_5 removed for complete mix process)

9. Check for F/M ratio :

$$\frac{F}{M} = \frac{S_o}{\theta . x} = \frac{200 \text{ (mg/l)}}{0.281 d \times 3000 \text{ (mg/l)}}$$
$$= 0.237\, d^{-1}$$

(This falls within the prescribed range of 0.2 to 0.6 for complete mix process)

10. Check for volumetric loading

$$\text{Volumetric loading} = Q\, L_a / V \qquad ...(14.9)$$

$$= \frac{17.28 \times 10^6 \text{ (litres/day)} \times 200 \text{ (mg/l)} \times 10^{-6}}{4855 \text{ (m}^3\text{)}}$$

$= 0.712 \text{ kg BOD}_5/\text{m}^3$

(This is less than permissible loading of 0.8 to 2.0 kg BOD_5/m^3 in complete mix process.

Example 14.4. Design of conventional activated sludge plant

Design a conventional activated sludge plant to treat settled domestic sewage with diffused air aeration system, for the following data:

(i) Population	...	*1,20,000*
(ii) Per capita sewage contribution	...	*160 lpcd*
(iii) Settled sewage BOD_5	....	*200 mg/l*
(iv) Effluent BOD_5 required	...	*15 mg/l*

Solution

1. ***Influent flow and process efficiency required***

$$\text{Average flow} = 120000 \times 160 = 19.2 \times 10^6 \text{ lpd}$$
$$= 19.2 \text{ mLd} = 19200 \text{ m}^3/\text{d}$$
$$\text{Efficiency required} = \frac{200 - 15}{200} \times 100 = 92.5\%$$

2. ***Determination of volume of aeration tank***

From Table 14.3, select $F/M = 0.2$ and MLSS $(x_t) = 3000$ mg/l

$$\text{Now} \quad \frac{F}{M} = \frac{Q \cdot L_a}{(V/1000)\, x_t} \quad \text{...(14.10)}$$
$$\therefore \quad 0.2 = \frac{19.2 \times 200 \times 1000}{V \times 3000}$$
$$\text{From which} \quad V = 6400 \text{ m}^3$$

3. ***Check for hydraulic retention time***

$$\text{HRT} = \frac{V}{Q \times 1000} \times 24$$
$$= \frac{6400}{19.2 \times 1000} \times 24 = 8 \text{ hrs.}$$

(This is within the prescribed range of 4 to 8 hours)

4. ***Check for volumetric loading***

$$\text{Volume loading} = \frac{Q \cdot L_a}{V} \quad \text{...(14.9)}$$
$$= \frac{19.2 \times 200}{6400} = 0.6 \text{ kg BOD}_5/\text{m}^3$$

(This is within the prescribed range of 0.3 to 0.7)

5. ***Return sludge ratio***

$$\frac{Q_r}{Q} = \frac{x_t}{(10^6/\text{SVI}) - x_t} \quad \text{...(14.26)}$$

Taking SVI = 100, we get

$$\frac{Q_r}{Q} = \frac{3000}{(10^6/100) - 3000} = 0.429$$

$\therefore \quad r = 42.9\%$

(This is within the prescribed range of 25 to 50%)

6. ***Tank dimensions*** :

Let us adopt a depth of 3 m and width of 4.5 m

$$\therefore \text{Length of aeration tank} = \frac{6400}{3 \times 4.5} = 474 \text{ m}$$

Provide a *continuous channel* with six baffles, so as to get seven sections, the length of each section being 68 m to give a total length of 68 × 7 = 476 m. Let the thickness of each baffle be 0.25 m.

$\therefore$ Total width of tank = (7 × 4.5) + (6 × 0.25) = 33 m

Provide a free board = 0.5 m.

Hence overall inner dimensions of the tank are 68 m × 33 m × 3.5 m.

7. ***Check for horizontal velocity***

$$Q + Q_r = \frac{19200\,(1 + 0.429)}{24 \times 60} = 19.05 \text{ m}^3/\text{min}$$

$$\therefore \quad v = \frac{19.05}{3 \times 4.5} = 1.41 \text{ m/min.}$$

(OK)

8. ***Air requirement and arrangement of diffuser plates***

Air needed = 100 m³/day per kg BOD_5 removed.

$$= \frac{100\,(200 - 15)\,19.2}{24 \times 60}$$

$$= 246.7 \text{ m}^3/\text{min} \approx 250 \text{ m}^3/\text{min. (say)}$$

Choose standard diffuser plates of 0.3 m × 0.3 m × 25 mm, passing 1.2 m³ of air/min/m² with 0.3 mm pores.

$$\therefore \text{No. of plates required} = \frac{250}{0.3 \times 0.3 \times 1.2} = 2315$$

Let us provide 2320 plates.

A plate concentration of 30% extra is provided in the first half of the tank to take care of more frequent clogging of this zone.

$\therefore$ No. of plates needed in the first half = 1.3 [2320/2]

= 1508

Let us provide clear distance between the plates of 0.9 m to avoid interference from the rising streams of bubbles.

Hence c/c distance of rows of plates = 0.9 + 0.3 = 1.2 m

No. of rows in the first half length (of 234 m) = 234/1.2 ≈ 195 rows.

Provide 8 plates in a row of 1.2 m spacing.

Hence plates provided in the first 234 m length = 195 rows × 8 in each row = 1560.

Balance plates to be provided = 2320 − 1560 = 760

This is to be provided in the balance length of 476 − 234 = 242 m.

$$\text{Hence spacing} = \frac{242 \times 8}{760} = 2.55 \text{ m (or say 2.5 m)}$$

9. ***Check for minimum air availability***

In the second half, 760 diffuser plates give air

$$= 760 \times 0.3 \times 0.3 \times 1.2 = 82.08 \text{ m}^3/\text{min}$$

This is in a length of 242 m.

∴ Air available per metre length of channel

$$= 82.08/242 = 0.34 \text{ m}^3/\text{min/m length.}$$

This is more than the prescribed value of 0.25 m^3/min/m length. Hence satisfactory.

14.11. ACTIVATED-SLUDGE-TREATMENT SYSTEMS

The activated sludge process originally was contained in long narrow tank with air for oxygen supply and mixing being supplied through diffusers at the basin bottom. As its use increased, many *modifications* were made both with respect to basin configuration, as well as aeration techniques. The more important processes are:

(*i*) Conventional process
(*ii*) Tapered aeration process
(*iii*) Step aeration process
(*iv*) Contact stabilisation process
(*v*) Completely mixed process
(*vi*) Modified aeration process
(*vii*) Extended aeration process.

1. **Conventional activated sludge process** : The conventional system is always *preceded* by primary settling. The plant consists of an aeration tank, a secondary settling tank, a sludge return line and excess sludge waste line. Both influent settled wastewater and recycled sludge enter the tank at the head end and pass through the reactor via '*plug-flow*' fashion (Fig. 14.16 *b*). Plug flow implies that the wastewater moves down progressively along the aeration tank, essentially unmixed with the rest of the tank contents. Each element of the mixed liquor simply traverses along the length of the reactor without losing its identily. Because of plug flow regime, the oxygen demand at the head of the aeration tank is high and then tapers down as each element of mixed liquor traverses along the length. However, air is provided *uniformly* along the basin length with the help of porous diffusers. This leads to either oxygen deficiency

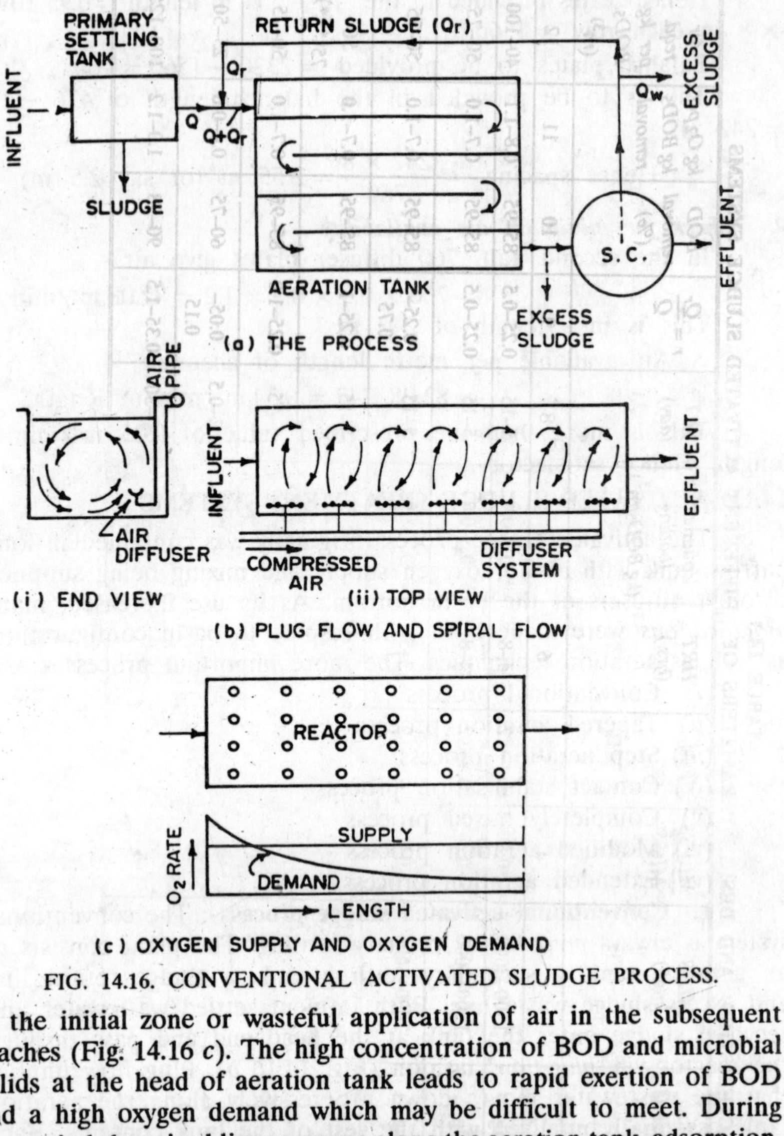

FIG. 14.16. CONVENTIONAL ACTIVATED SLUDGE PROCESS.

in the initial zone or wasteful application of air in the subsequent reaches (Fig. 14.16 *c*). The high concentration of BOD and microbial solids at the head of aeration tank leads to rapid exertion of BOD and a high oxygen demand which may be difficult to meet. During the period the mixed liquor moves along the aeration tank, adsorption, flocculation and the oxidation of the organic matter takes places.

The HRT, volumetric loading and F/M usually adopted for the conventional system are given to Table 14.3. The BOD removal in the process is 85 – 95 percent. The plug flow regime is achieved by a long and narrow configuration of the aeration tank with length equal to 5 to 50 times the width. The sewage and mixed liquor

TABLE 14.3
CHARACTERISTICS AND DESIGN PARAMETERS OF DIFFERENT ACTIVATED SLUDGE SYSTEMS

Process Type	Flow Regime	MLSS (mg/l)	$\frac{MLVSS}{MLSS}$	F/M	HRT (hrs.)	Volumetric loading ($kg\ BOD_5/m^3$)	SRT (day)	$r = \frac{Q_r}{Q}$	BOD removal (%)	kg O_2 per kg BOD_5 removal	Air reqd. per kg BOD_5 (m^3)
1	2	3	4	5	6	7	8	9	10	11	12
1. Conventional	Plug	1500–3000	0.8	0.4–0.2	4–8	0.3–0.7	5–15	0.25–0.5	85–95	0.8–1.1	40–100
2. Tapered aeration	Plug	1500–3000	0.8	0.4–0.2	4–8	0.3–0.8	5–15	0.25–0.5	85–95	0.7–1.0	50–75
3. Step aeration	Plug	2000–3000	0.8	0.4–0.2	3–5	0.7–1.0	5–15	0.25–0.75	85–95	0.7–1.0	50–75
4. Contact Stabilisation	Plug	1000–3000 (*) 3000–6000 (**)	0.8	0.5–0.2	0.5–1.5(*) 3–6(**)	1.0–1.2	5–15	0.25–1	85–95	0.7–1.0	50–75(***)
5. Complete mix.	Complete mix.	3000–6000	0.8	0.6–0.2	3–5	0.8–2.0	5–15	0.25–1.0	85–95	0.7–1.0	50–75
6. Modified aeration	Plug	300–800	0.8	5–1.5	1.5–3	1.2–2.4	0.2–0.5	0.05–0.15	60–75	0.4–0.6	25–50
7. Extended aeration	Complete mix.	3000–8000	0.5–0.6	0.15–0.05	18–36	0.2–0.4	20–30	0.35–1.5	90–98	1.0–1.2	100–125

(*)In contact aeration tank ; (**)In sludge reaeration tank ;

(***)Divided equally between contact aeration tank and sludge reaeration tank

are let in at the head of the tank and withdrawn at its end. The mixed liquor is settled in the activated-sludge settling tank (secondary clarifier), and sludge is returned at a rate of approximate 25 to 50 percent of the influent flow rate.

The main limitation of the conventional system is that the aeration tank volume requirement is high. Futher, there is a lack of operational stability at times of excessive variation in rate of inflow and in influent strength. For histroical reasons, the conventional system is the most widely used type of the activated sludge process. Capacities of plants in use range from 5 mLd to above 1000 mLd.

2. Tapered aeration process : The objective of tapered aeration is to match the quantity of air supplied to the demand exerted by the microorganisms, as the mixed liquor traverses the aeration tank. Thus, tapered aeration involves a minor modification of the conventional system in that the air is supplied at a high rate at the head of the aeration tank to meet the high oxygen demand there and then gradually reduced to match the reduced oxygen demand; this can be achieved by varying the spacing of aerators, as indicated in Fig. 14.17. The main advantages of tapered aeration is the optimal application of air. The second benefit of the method is the avoidance of over aeration which will inhibit the growth of nitrifying organisms, which cause high oxygen demands. The method lends itself readily for the upgrading of conventional plants employing diffused air aeration.

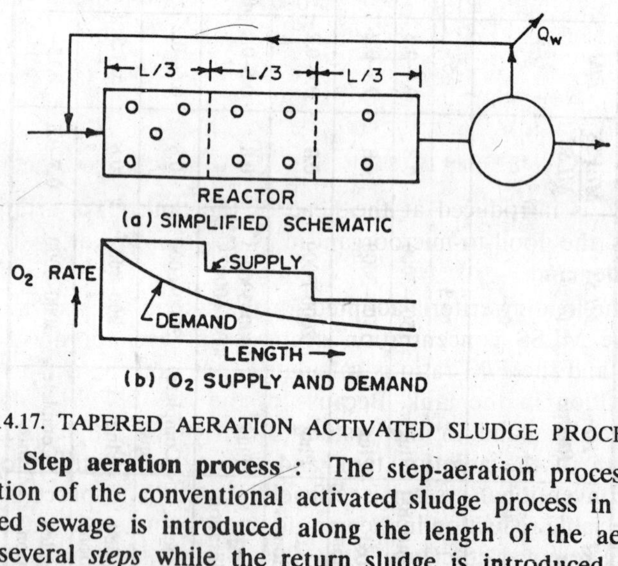

FIG. 14.17. TAPERED AERATION ACTIVATED SLUDGE PROCESS.

3. Step aeration process : The step-aeration process is a modification of the conventional activated sludge process in which the settled sewage is introduced along the length of the aeration tank in several *steps* while the return sludge is introduced at the head. Thus, as shown in Fig. 14.18 (*b*), the influent discharge Q is introduced in four steps Q_1, Q_2, Q_3 and Q_4, while the return

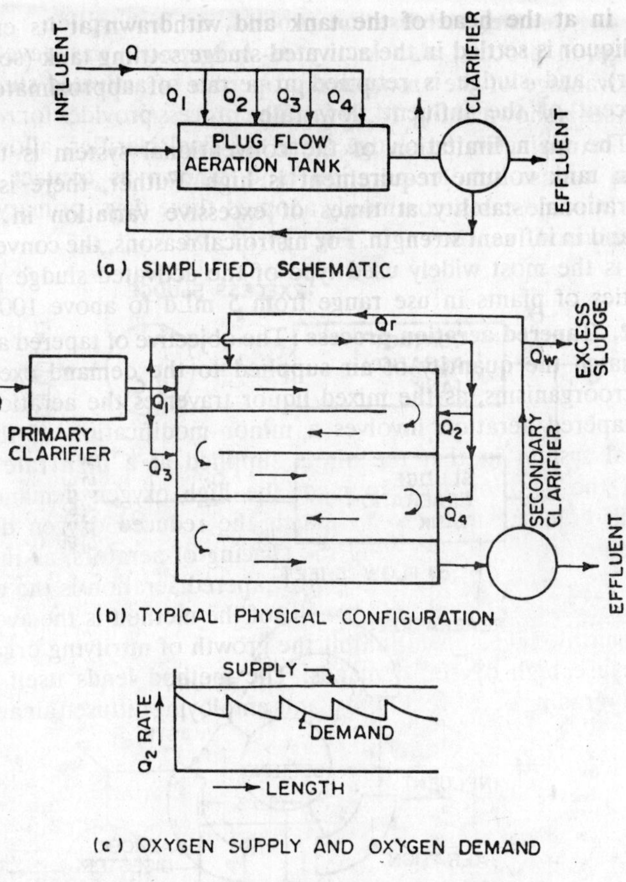

FIG. 14.18. STEP-AERATION PROCESS

sludge Q_r is introduced at the head of the tank. This arrangement equalises the food-to-microorganism ratio, thus lowering the peak oxygen demand.

The loading criteria adopted for the system are shown in Table 14.3. The MLSS concentration decreases at each point of sewage addition and the F/M ratio is calculated based on the average MLSS concentration in the tank. Because of the *step addition* of sewage, the oxygen demand in the system is fairly uniformly spread over the length of the aeration tank and the air supplied uniformly as in the conventional system is efficiently used. The process enables an appreciable reduction in the aeration tank volume without lowering the BOD removal efficiency. Step aeration method has considerable capacity to absorb shock organic loadings. The method has found application for larger plants of capacities upto about 1000 mLd.

4. Contact stabilisation process, or Biosorption : Contact stabilisation process, also known as *biosorption*, was developed to take advantage of the absorptive properties of activated sludge. The flow sheet is shown in Fig. 14.19 (*a*). The process provides for *reaeration* of the return activated sludge from the final clarifier, allowing this method to use smaller aeration tank (known as contact aeration tank). In some cases, commonly adopted these days, primary settling in eliminated, as shown in Fig. 14.19 (*b*).

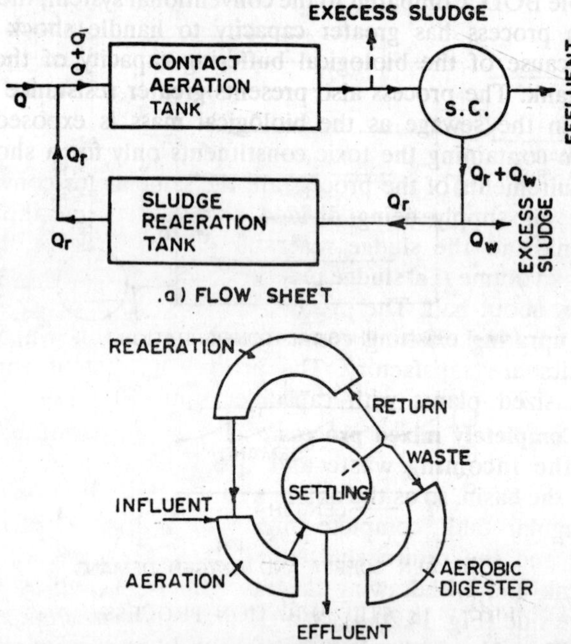

FIG. 14.19. CONTACT STABILISATION PROCESS (BIOSORPTION)

The incoming sewage is aerated in the contact aeration tank, along with return sludge, for a comparatively short period of 30 to 90 minutes. This is known as *absorptive phase,* during which the sludge absorbs the organic matter in the sewage. The mixed liquor is then settled in a secondary settling tank. The return fraction of the sludge withdrawn from the clarifier, with its burden of adsorbed organics, is *reaerated* in a separate sludge reaeration tank for a period of 3 to 6 hours before it is fed back into the contact aeration tank. During the sludge reaeration, adsorbed organics are re-released and stabilised, restoring the adsorptive capacity of the sludge. A portion of the return sludge is wasted prior to recycle, to maintain a constant mixed-liquor VSS concentration in the tanks. The loading criteria for the process are given in Table 14.3. It is to be noted that in determining F/M for the process, both the SS in the contact aeration tank and SS in the sludge reaeration tank are to be taken into account.

The contact stabilisation process is quite effective in the removal of colloidal and suspended organic matter, but it is not very effective in removing soluble organics. The method is well suited for the treatment of fresh domestic sewage containing only a low percent age of soluble BOD. Compared to the conventional system, the contact stabilisation process has greater capacity to handle shock organic loadings because of the biological buffering capacity of the sludge reaeration tank. The process also presents greater resistance to toxic substances in the sewage as the biological mass is exposed to the main stream containing the toxic constituents only for a short time. The air requirements of the process are the same as for conventional system, the air supply being divided equally between the contact aeration tank and the sludge reaeration tank. However, the total aeration tank volume (*i.e.* sludge reaeration tank plus contact aeration tank) is only about half. The process, therefore presents an effective method of uprating exisiting conventional type plants when sewage characteristics are satisfactory. The process has found application in medium sized plants with capacities upto 40 mLd.

5. Completely mixed process : The *completely mixed* process disperses the incoming waste and the return sludge uniformly throughout the basin, so as the obtain a completely mixed flow regime. In a rectangular tank, complete mixing is achieved by distributing the sewage and the return sludge uniformly *along one side* of the aeration tank and withdrawing the aerated sewage uniformly along the opposite side (Fig. 14.20 *a*). In a circular or square tank, complete mixing is achieved by mechanical aerator with adequate mixing capacity installed at the centre of the tank. In a complete-mix-process (also known as *continuous-flow stirred-tank process*), the organic load on the aeration tank and the oxygen demand are uniform from one end to the other (Fig. 14.20 *b*). As the mixed liquor passes across the aeration tank from the influent ports to the effluent channel, it is complelety mixed by diffused or mechanical aeration.

Following are the *advantages* of this process :

(1) There is less variation in the organic loading of the contactor resulting in a more uniform effluent quality and in a more uniform oxygen demand. (2) There is dilution of the wastewater into the entire basin volume, resulting in a reduced effect of shock loadings, either organic or toxic in nature. (3) There is maximum efficiency of the entire contactor contents at all times due to complete mixing

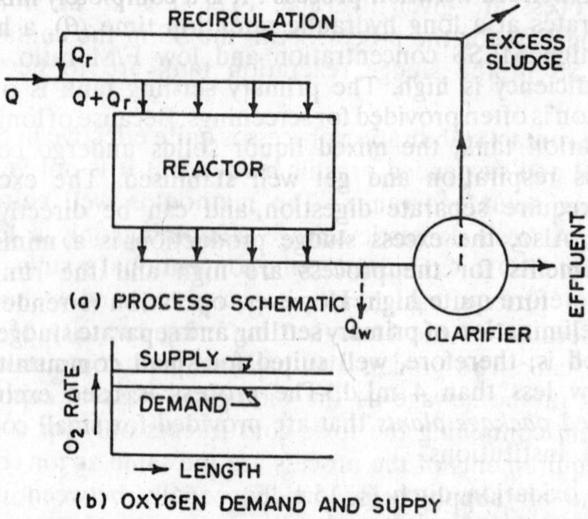

FIG. 14.20. COMPLETELY-MIXED PROCESS

of the wastewater. (4) There is stabilisation of the operation by the leveling out of organic loading. (5) The plant has the capacity to hold a high MLSS level in the aeration tank, enabling the aeration tank volume to be reduced.

Due to the above advantages, the completely mixed activated sludge process represents the best activated sludge system. The loading parameters of the process are shown in Table 14.3. The complete mix process has been used mainly for small plants with capacity less than 25 mLd.

6. Modified aeration process : The modifed aeration process, also termed as *minimal solids aeration,* uses the flow scheme similar to the conventional process or tapered aeration process. However, primary settling is dispensed with frequently. The process is *distingushed* by : (*i*) Short aration period (1.5 to 3h) (*ii*) high volumetric loading. (*iii*) high F/M ratio (*iv*) low percentage of sludge return. (*v*) low concentration of MLSS (*vi*) High organic loading. (*vii*) Lesser air requirements.

Modified aeration develops dispersed biological growth which does not flocculate and quickly settle. Alum and polyelectrolytes are sometimes employed to improve secondary settling. The BOD removal in the process is only 60–75%. Hence the method is suitable when only intermediate quality effluent is desired as for example, when the effluent is to be used for sewage farming. The process is attended by substantial savings in construction and aeration costs. The process has been employed mainly in large plants with capacities above 200 mLd.

7. Extended aeration process : It is a completely mixed process which operates at a long hydraulic retention time (θ), a high sludge age (θ_c), high MLSS concentration and low F/M ratio. The BOD removal efficiency is high. The primary settling tank is omitted but communition is often provided for screenings. Because of long detention in the aeration tank, the mixed liquor solids undergo considerable endogenous respiration and get well stabilised. The excess sludge does not require separate digestion and can be directly dried on sand beds. Also, the excess sludge production is a minimum. The air requirements for the process are high and the running costs are also therefore quite high. However, operation is rendered simple due to the elimination of primary settling and separate sludge digestion. The method is, therefore, well suited for small communities having sewage flow less than 4 mLd. The process is used exclusively for prefabricated *package plants* that are provided for small communities or isolated institutions.

The oxidation ditch (§ 15.1, Fig. 15.1) also conforms to the extended aeration principle and relies on aerating the mixed liquor in an endless ditch.

14.12. SECONDARY SETTLING

Secondary settling assumes considerable importance in the activated sludge process as the effluent separation of the biological sludge is necessary not only for ensuring final effluent quality but also for return of adequate sludge to maintain the MLSS level in the aeration tank. The secondary settling tank of the activated sludge process is particularly sensitive to fluctuations in flow rate and on this account it is recommended that the units be designed not only for average overflow rate but also for peak overflow rates. The high concentration of suspended solids in the effluent requires that the solids loading rates should also be considered.

The settling characteristics of a concentrated suspension, like the mixed liquor of the activated sludge process, are very much different from those of dilute suspensions. For such suspension, both *hindered or zone settling* (type 3) and *compression settling* (type 4) usually occur in addition to *discrete* (free) and *flocculant settling*. Fig 14.21 shows the schematic of settling regions of the mixed liquor of activated sludge process.

Operationally, secondary settling tank must perform two functions : (*i*) separation of the mixed liquor suspended solids from the treated wastewater, which results in the clarified effluent, and (*ii*) thickening of treated wastewater to obtain desired concentration of solids (x_r). Both these functions must be taken into consideration very carefully, if secondary clarifiers are to be designed properly. Adequate consideration must be given to the selection of depth that will provide necessary volume for both the functions. The conventional

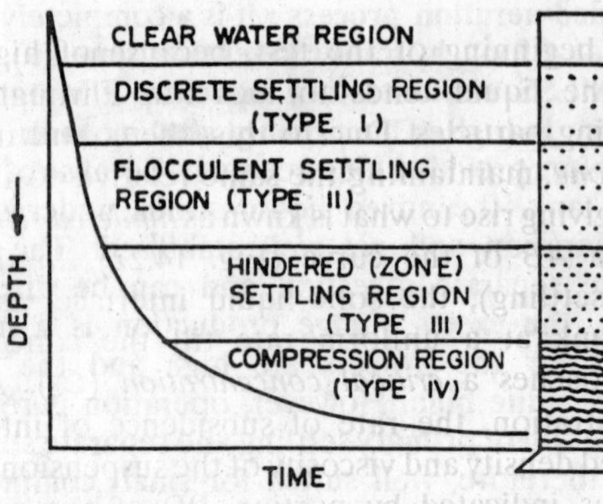

FIG. 14.21. SCHEMATIC OF SETTLING REGIONS OF MIXED LIQUOR FOR ACTIVATED SLUDGE PROCESS.

design usually overlooks the requirements for the thicknening but is clear from the following expression (Fig. 14.14) that the MLSS concentration is entirely dependent on the concentration of the solids (x_r) in the underflow of the settling tank :

$$x = x_r \frac{Q_r}{Q + Q_r} \qquad ...(14.14)$$

Hence any failure on the part of the settling tank in the production of the desired underflow concentration will lead to an unsatisfactory performance of the entire activated sludge process.

Settling characteristies of concentrated suspension : The behaviour of a particular concentrated-suspension in both clarification and thickening is determined by conducting settling tests in a graduated cylinder. The test is usually known as *column settling test.* The test is started with a uniformly mixed suspension of height H_o and concentration C_o. Time required by solid-liquid interface to cross each mark are recorded, and are plotted in a height-time graph, as shown in Fig. 14.22.

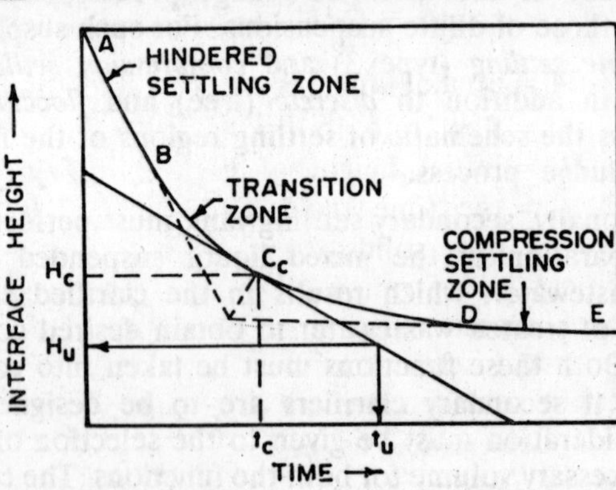

FIG. 14.22. LABORATORY BATCH SETTLING TEST

At the beginning of the test, because of high concentration of particles, the liquid tends to move up through the interstices of the contacting particles. Due to this, the contacting particies tend to settle as a *zone*, maintaining the same relative position with respect to each other, giving rise to what is knwn as *hindered settling*, represented by the portion *AB* of the curve (Fig. 14.22). In the zone settling (or hindered settling), the solid-liquid interface visibly subsides in the settling tank at a uniform rate till the concentration at the interface approaches a *critical concentration* (C_c). At this stage of critical concentration, the rate of subsidence of interface decreases due to increased density and viscosity of the suspension, and a transition zone occurs, as indicated by portion *BD* of the curve. After that, as the time passes, a *compression zone* occurs, when the solid particles come into physical contact with each other, and get compacted due to the weight of the sludge in the upper layers. Compression is represented by portion *DE* of the curve.

Area requirement based on single-batch test results : For the purposes of design, the *final overflow rate* selected should be based on the following three factors : (*i*) Area required for clarification (*ii*) Area required for thickening. and (*iii*) Rate of sludge withdrawal. The area required for thickening is usually greater than the area required for clarification.

The area required for thickening is determined according to a method developed by Talmadge and Fitch (1955). A column of height H_o is filled with a suspension of solids of uniform concentration (C_o). The position of the interface as the time lapses and the suspension settles is given in Fig. 14.22. The rate of subsidence of the interface at any time is equal to the slope of the curve. The critical area is given by

$$A = \frac{Q}{H_o} t_u \qquad ...(14.32)$$

where A = area of tank, required for sludge thickening. (m^2)
Q = flow rate in the tank (m^3/s)
H_o = initial height of interface in column (m)
t_u = the time required to reach the desired underflow solids concentration.

The critical concentration (C_c) is determined by extending tangents from the hindered and compressive settling lines to their point of intersection and bisecting the angle so formed. The bisector will then intersect the subsidence curve at C_c, the critical concentration. The time t_u is determined as under :

1. Draw a tangent to the subsidence curve at point C_c.

2. Determine the height H_u that corresponds to the depth at which the solids are at the desired underflow concentration (C_u), using the following expression :

$$H_u C_u = H_0 C_0$$

or
$$H_u = \frac{C_o H_o}{C_u} \quad ...(14.33)$$

3. Draw a horizontal line at height H_u. The intersection of this line with the tangent at C_c will yield the required time t_u.

Knowing time t_u required to reach the desired underflow concentration (C_u), the area required for sludge thickening is found using Eq. 14.32. The area required for *clarification* is then determined from the slope of subsidence curve in the hindered settling zone (*i.e.* line *AB*). The slope of the line *AB* gives the surface oveflow rate necessary for hindered settling.

Surface and solids loading rates : The design of secondary clarifiers for the activated-sludge processes, is based both on *solids loading* as well as *surface overflow rate* (SOR).

Solids loading : Since hindered settling is likely to occur and settling velocity of the discrete particles may not govern the basis of design, *solids loading* is important in the secondary clarifiers. The solids loading rate on an activated-sludge settling tank may be computed by dividing the total solids applied by the surface area of the tank. The *solids loading rate* is expressed in kg SS/m^2 per hour and is related to the surface overflow rate (SOR) by the following expression.

$$\text{kg/m}^2 \text{ per hour} = \frac{\text{MLSS (mg/l)} \times \text{SOR (m/day)}}{24 \times 10^3} \quad ...(14.34)$$

Typical solids loading rates vary from 2.5 to 6.2 kg/m^2 per hour at average and peak loading conditions respectively. In effect, the solids loading rate represents a characteristic value for the suspension under consideration. It has been observed that in a settling tank of fixed surface area, the effluent quality will deteriorate if the solids loading is increased beyond the characteristic value for the suspension.

Table 14.4 gives the surface overflow rates (SOR) at peak flow, as recommended by Metcaff and Eddy. The SOR is expressed either as m^3/day/m^2, or simply as m/day. These values are based on maximum solids loading of 6.2 kg/m^2 per hour. These values may be adopted in absence of any actual test data.

Table 14.5 gives the values of overflow rates and solids loading rates for secondary settling tanks of activated sludge, as recommended by the Manual on sewage and sewage treatment (MUD). These design overflow rates should be applied to the sewage flow and *not* to

TABLE 14.4 RECOMMEDED SOR ($m^3/day/m^2$) AT VARIOUS MLSS CONCENTRATION AND RECYLCE RATES

MLSS (mg/l)	*Recycle percent*	
	25 (%)	*50 (%)*
500	57	57
1000	57	57
1500	49	49
2000	49	49
2500	47	39
3000	39	33
3500	33	28

the mixed liquor flow to the settling tank. The solids loading rates must be in the range of 100 – 150 kg/d/m^2 for condition of average flow and should not exceed 250 kg/d/m^2 for peak flow conditions The surface area for activated sludge settling tank is to be designed for both SOR, and solids loading rate and the larger value should be adopted. Sludge thickening is a special function of the secondary settling tank of a activated sludge plant as thicknened sludge is an advantage in sludge recycling. Sludge thickening is effected by providing adequate tank depth as indicated in Table 14.5.

TABLE 14.5.
DESIGN PARAMETERS FOR SECONDARY SETTLING TANKS

Process	*Overflow rate* (m/d/m^2)		*Depth of* (m^3)
	Average	*Peak*	
1. Secondary settling for activated sluge (excluding extended aeration)	15–35	40–50	3.5–4.5
2. Secondary settling for extended aretion	8–15	35	3.5–4.5

Secondary settling tank types : The settling tanks for activated sludge process may be either circular or rectangular. The diameter of circular tanks may vary from 3 m to 60 m; however the tank radius should preferably not exceed five times the sidewater depth. Circular tanks may be either of *central feed type* or *rim-feed type*, each using a revolving mechanism to transport and remove the sludge from the bottom of the clarifier. In rectangular tanks, the maximum length, where possible, should not exceed ten times the depth, though lengths upto 90 m have been used successfully in large plants. Width of tank is generally limited to 6 m, but where the width of the tank exceeds 6 m, multiple sludge collection mechanisms may be used to permit tank widths upto 24 m.

Flow through velocity : In order to avoid the troubles due to density currents and scouring of the deposited sludge, the horizontal flow through velocities based on maximum mixed liquor flow should be limited to 30.5 m/h in a rectangular tank. In circular tanks, of center-feed type, the inlet baffle should have a diameter of 15 to 20% of tank diameter and should not extend more than 1 m below the surface to avoid the scouring of the deposited sludge.

Placement and loading of effluent weir : Circular settling tanks are manufactured with overflow weirs located near both the center and the perimeter of the tank. The minimum water depth below effluent weirs so located should be 3.1 m to prevent overflow of density currents. When weirs are located at tank perimeter in circular tanks, or at end walls in rectangular tanks, the minimum depth should be 3.7 m. Optimum position of the circular weir troughs to intercept well clarified effluent is at a distance of 2/3 to 3/4 of the radial distance from the centre. The *weir loading rates* should be kept below 375 m^3/m per day for weirs located away from the tank periphery, and below 250 m^3/m per day if located either against the wall where density currents might turn upward or in basins having depth less than 3.7 m. These loading rates are based on peak flow rate. In small tanks, the weir loading rate should not exceed 150 m^3/m per day at maximum flow or 125 m^3/m per day at average flow. The upflow velocity in the vicinity of the weir should not exceed 3 to 5 m/hour.

Example 14.5. *The subsidence curve shown in Fig. 14.23 was obtained for a sludge with an initial concentration of 3200 mg/l, using a settling cyclinder 400 mm high. Determine the surface overflow rate required to obtain a thickened sludge with a solids concentration of 2.2 percent by weight.*

Solution

For solids concentration of 2.2% by weight, solids concentration

$$\text{value} = \frac{2.2}{100} \times 10^6 = 22000 \text{ mg/l}$$

Hence
$$H_u = \frac{C_o H_o}{C_u} = \frac{3200 \times 400}{22000} = 58.2 \text{ mm}$$

In Fig. 14.23, from the point of intersection of the two tangents *AB* and *DE*, draw a bisector of the angle, thus getting point C_c. At this point, draw a tangent to the subsidence curve. Also, draw a horizontal line corresponding to H_u= 58.2 mm. The point of intersection of this line with the tangent to the curve gives values of t_u = 34.5 min.

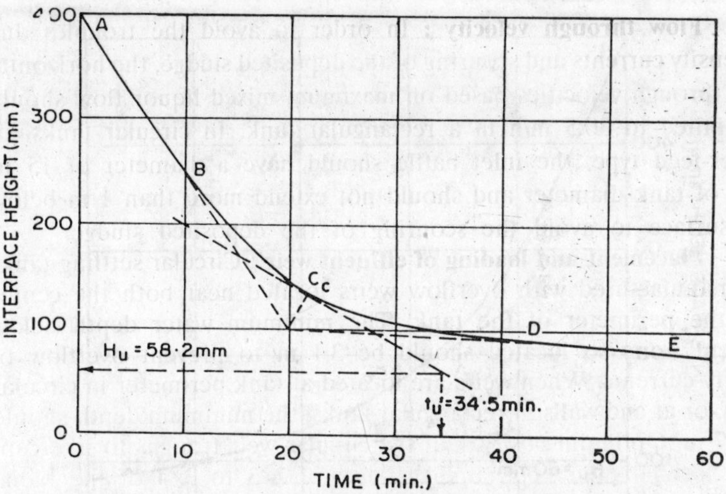

FIG. 14.23.

(i) Area from sludge thicknening point of view

$$A = \frac{Q}{H_o} \cdot t_u = \frac{Q\,(34.5)}{400} = 0.08625\,Q \qquad ...(1)$$

$$\text{SOR} = Q/A = \frac{Q}{0.08625} = 11.6 \text{ mm/min.}$$

$$= \frac{11.6 \times 60 \times 24}{1000} = 16.7 \text{ m}^3/\text{d}/\text{m}^2$$

(ii) Area from clarification criterion : From the slope of the subsidence curve in the hindered settling zone (*i.e.* from slope of line *AB*), we get, velocity of settling

$$v = 150 \text{ mm/10 min} = 15 \text{ mm/min} = (15 \times 60 \times 24)10^{-3}$$
$$= 21.6 \text{ m/day}$$

Since the over flow rate is proportional to the liquid volume above the sludge zone, over flow rate

$$Q' = Q \times \left(\frac{400 - 58.2}{400}\right) = 0.8545\,Q$$

$$\therefore \text{ Area required } = \frac{Q'}{v} = \frac{0.8545\,Q}{21.6} = 0.03956\,Q \qquad ...(2)$$

Hence the sludge thickening governs the SOR, which will be equal to 16.7 $m^3/d/m^2$. However, actual hydraulic loading rate

$$= \frac{\text{over flow rate } (Q')}{A} = \frac{0.8545\,Q}{0.08625\,Q} = 9.91 \text{ m}^3/\text{d}/\text{m}^2$$

Example 14.6. *In a settling cylinder, 0.5 m high, the settling curve shown in Fig. 14.24 was obtained for an activated sludge with an initial solids concentration of 3200 mg/l. Determine the area to yield a thickened sludge concentration of 20000 mg/L with an inflow of 350 m^3/d. Also, determine the solids loading and the over flow rate.*

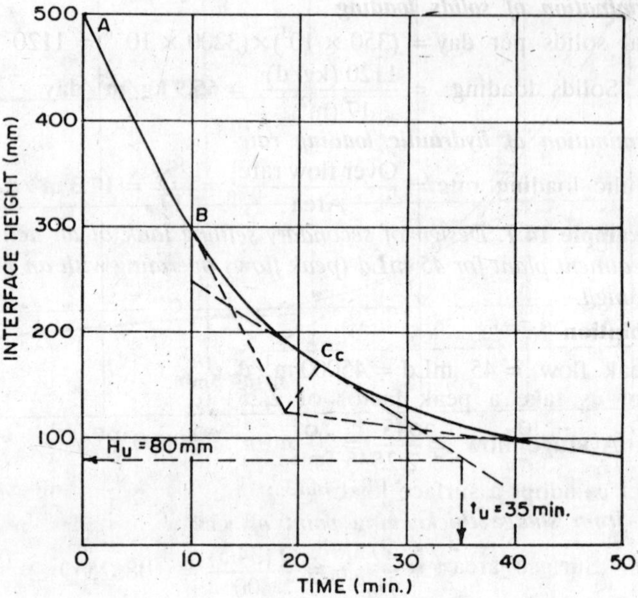

FIG. 14.24.

Solution :

1. Determination of H_u and t_u

$$H_u = \frac{C_o H_o}{C_u} = \frac{3200 \times 500}{20000} = 80 \text{ mm}$$

Corresponding to this value of H_u = 80 mm, we get t_u = 35 min from Fig. 14.24.

2. Determination of area on the basis of sludge thickening

$$A = \frac{Q\,t_u}{H_o} = \left(\frac{350}{24 \times 60}\right) \times \frac{35}{0.5} \approx 17 \text{ m}^2$$

3. Determination of area on the basis of clarification

The slope of the lile *AB* in the hindered settling portion of the curve gives settling velocity :

$$v = \frac{500 - 300}{10} = 20 \text{ mm/min.}$$

$$= \frac{20 \times 60}{10000} = 1.2 \text{ m/h} = 28.8 \text{ m/day}$$

The *over flow rate* (Q') is proportional to the liquid volume above the critical sludge zone, and is may be computed as under:

$$Q' = 350 \times \frac{500 - 80}{500} = 350 \times 0.84 = 294 \text{ m}^3\text{/d}$$

∴ Area required $= Q'/v = 294/28.8 = 10.2 \text{ m}^2$

Hence the controlling requirement is the thicknening area of 17 m².

4. Determination of solids loading

Total solids per day = $(350 \times 10^3) \times (3200 \times 10^{-6}) = 1120$ kg/d

$\therefore$ Solids loading $= \dfrac{1120\ (\text{kg/d})}{17\ (\text{m}^2)} = \mathbf{65.9\ kg/m^2/day}$

5. Determination of hydraulic loading rate

Hydraulic loading rate $= \dfrac{\text{Over flow rate}}{\text{Area}} = \dfrac{294}{17} = \mathbf{17.3\ m^3/d/m^2}$

Example 14.7. *Design of secondary settling tank of an activated sludge treatment plant for 45 mLd (peak flow) operating with an MLSS of 3000 mg/l.*

Solution

Peak flow = 45 mLd = 45000 m³/d

Let us take a peak factor of 2.25

$\therefore$ Average flow $= \dfrac{45}{2.25} = 20$ mLd $= 20000\ \text{m}^3/\text{d}$

Let us adopt a surface loading rate of 20 $\text{m}^3/\text{d}/\text{m}^2$ at average flow.

$\therefore$ Surface area $= \dfrac{20000}{20} = 1000\ \text{m}^2$ ...(*i*)

Check surface loading at peak flow :

Surface loading $= \dfrac{45000}{1000} = 45\ \text{m}^3/\text{d}/\text{m}^2$

This is within the prescribed range of 40 to 50 (Table 14.5).

Again, on the basis of solids loading of 125 kg/day/m² at average flow,

Area required $= \dfrac{(20 \times 10^6) \times (3000 \times 10^{-6})}{125} = 480\ \text{m}^2$ ...(*ii*)

Also, on the basis of solids loading of 250 kg/day/m² at peak flow,

Area required $= \dfrac{(45 \times 10^6) \times (3000 \times 10^{-6})}{250} = 540\ \text{m}^2$ ...(*iii*)

Hence adopt a surface area of 1000 m² which is highest of the three values. Adopting a circular tank,

Dia., $d = \sqrt{\dfrac{1000 \times 4}{\pi}} = 35.7$ m

Adopt a dia. of 36 m. Actual area = 1017.9 m²

Actual solids loading at average flow

$$= \frac{(20 \times 10^6) \times (30000 \times 10^{-6})}{1017.9}$$

$$\approx 59\ \text{kg/day/m}^2$$

Let us provide effluent weir at the outer periphery.

Length of weir $= \pi \times 36 = 113$ m

$$\therefore \text{ Weir loading at av. flow} = \frac{20000}{113} \approx 177 \text{ m}^3/\text{d}/\text{m}$$

This is more than recommended value 125 $m^3/m/d$ for small tanks. Hence provide a trough instead of a single weir, at the outer periphery thus getting double edged effluent channel, for with available over flow length will be 2×113 and weir loading will reduce to $177/2 = 88.5$ $m^3/d/m$ length. Keep the depth of tank equal to 4 m. Typical details of the tank are shown in Fig. 14.25.

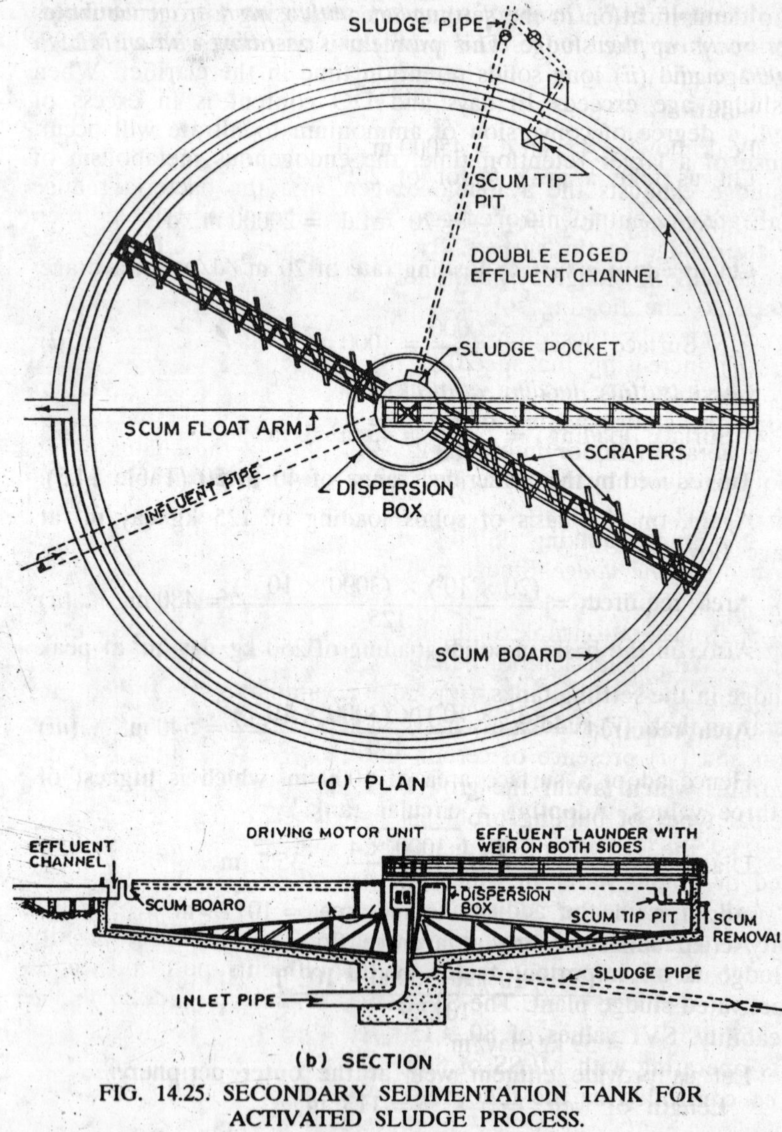

FIG. 14.25. SECONDARY SEDIMENTATION TANK FOR ACTIVATED SLUDGE PROCESS.

14.13. OPERATIONAL DIFFICULTIES

There are two common difficulties encountered in the operation of activated-sludge plants : (*i*) Rising sludge, and (*ii*) Bulking sludge

These problems hinge upon the inability to maintain the desired sludge age. Because of these troubles, sludge may be lost from a clarifier even when the clarifier is properly designed from the stand point of solids loading.

1. Rising sludge or floating sludge : Rising sludge may be due to denitrification in the settling tank releasing nitrogen bubbles which buoys up the sludge. This problem is associated with (*i*) high sludge age and (*ii*) long solids retention time in the clarifier. When the sludge age exceeds 10 days and DO content is in excess of 2 mg/l, a degree of conversion of ammonium to nitrate will occur. Because of a larger retention time, the endogenous metabolism of the sludge exhausts the available oxygen, and the bacteria reduce any nitrate present to nitrogen gas. The resulting gas bubbles then buoy the sludge to the surface. Rising sludge can be differentiated from the bulking sludge by noting the presence of small gas bubbles attached to the floating solids. The problem of sludge-rising can be over come by the following : (*i*) increasing the return sludge age ; (*ii*) increasing the speed of the sludge scraper mechanism, where possible, (*iii*) decreasing the mean cell residence time by increasing the sludge wasing rate ; and (*vi*) decreasing the rate of flow of aeration liquor into the offending tank if the sludge depth cannot be reduced by increasing the return activated sludge withdrawal rate.

2. Sludge bulking : Sludge with poor settling characteristics is termed *bulking sludge*. Sludge bulking results in poor influent due to the presence of excessive suspended solids and also in rapid loss of MLSS from aeration tank. Sludge bulking occurs due to (*i*) inadequate air supply (*i.e.*, under-aeration), (*ii*) low pH, (*iii*) prolonged detention of sludge in the settling tanks, (*iv*)sludge accumulation on the bottom of aeration tank, (*v*) sudden discharge of septic solids from the sewage system and (*vi*) presence of certain industrial waste containing carbohydrates which favour the growth of filamentous fungi. The two principal types of bulking problems have been indentified. One is caused by the growth of filamentous organisms and the other is caused by bound water in which the bacterial cells composing the *floc swell* through the addition of water to the extent that their density is reduced and they do not settle. The quick settleability of sludge is an important factor in the effluents performance of the activated sludge plant. The SVI serves also as an index of sludge settleability. SVI values of 80 – 150 are considered satisfactory in plants operating with MLSS of 800 – 3500 mg/l. Bulking of sludge can be controlled by the following : (*i*) Reducing the sewage flow to the aeration tanks for a short period. (*ii*) Reducing the suspended solids in the mixed liquor. (*iii*) Prolonged aeration and reaeration of return sludge. (*iv*) Application of chlorine either to the sewage or to the return sludge, so as to control filamentous growth. Chlorine requirements are 0.2 to 1.0 percent of dry solids weight in the return sludge. (*v*) Application of hydrogen peroxide, to control the growth of filamentous organisms. (*vi*) Addition of hydrated lime to raise pH to 8 or more.

14.14. ADVANTAGES AND DISADVANTAGES

Advantages of activated sludge process : (1) Clear sparkling and non-putrescible effluent possible. (2) No odours during the process as compared to other biological processes. (3) No fly nuisance. (4) Highly efficient. Removal of SS, BOD and bacteria are around 90% each. (5) Degree of stabilisation or nitrification is controllable between limits, so as to match with the quantity and character of receiving waters. The treatment may be partial or full as desired/required. (6) Relatively low cost of installation, as compared to the total cost of trickling filter installations, cost of land, cost of huge filtering material, cost of distribution mechanism etc. (7) Smaller area required, as compared to trickling filters. (8) The excess sludge has higher fertilizing values as compared to sludge obtained from other treatment methods. The excess activated sludge may contain 6.3% Nitrogen, 1.44% phosphorous, and 75% volatile matter, on the basis of dry weight. (9) Amount of hydraulic head consumed by the process is less.

Disadvantages : (1) Very sensitive to variations in the quality of sewage, particularly in respect of industrial wastes which may cause sludge bulking. Trickling filters are the best in this repect. (2) Hight cost of operation. (3) Necessity of constant skilled attendance. (4) Uncertainly of the expected results under all conditions. (5) Large quantity of sludge is produced which is difficult to dewater, digest and dispose of.

14.15. HIGH-PURITY-OXYGEN ACTIVATED SLUDGE

The use of pure oxygen in the activated sludge process attracted renewed attention in 1977. With advances in oxygen production technology, the potential of pure oxygen usuage in the activated sludge process has improved. The primary distinguishing feature of this process is the application of high purity oxygen instead of air for aerobic treatment by activated sludge. Because one of the limitations of the activated sludge process is the capability of the aeration device to supply the oxygen needed, it is important that with the use of pure oxygen, the oxygen transfer capability increased upto five times. High purity oxygen gas can be efficiently generated by cryogenic air separation or pressure-swing adsorption (PSA) processes. The high-purity-oxygen activated sludge process consists of the following major components: (*i*) Oxygen gas generator (cryogenic air separation system or PSA process) (*ii*) Aeration tank having three or four compartments separated by means of baffles to simulate plug flow, and covered with air tight enclosure. (*iii*) Final clarifier. (*vi*) Pumps for recirculating activated sludge and (*v*) Sludge disposal facilities.

Fig. 14.26 shows the schematic diagram. The preliminary treatment facilities may consist of (*i*) screening device (*ii*) degritting device, and (*iii*) primary settling, which is optional.

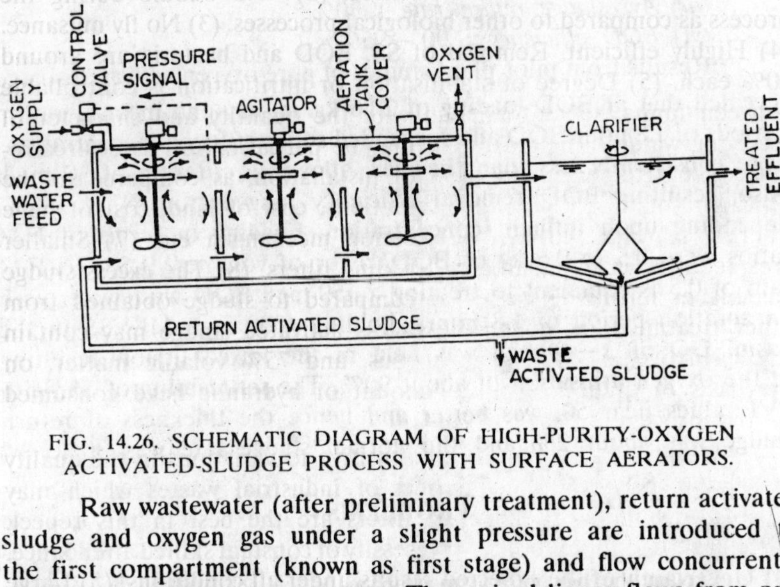

FIG. 14.26. SCHEMATIC DIAGRAM OF HIGH-PURITY-OXYGEN ACTIVATED-SLUDGE PROCESS WITH SURFACE AERATORS.

Raw wastewater (after preliminary treatment), return activated sludge and oxygen gas under a slight pressure are introduced to the first compartment (known as first stage) and flow concurrently through succeeding comparments. The mixed liquor flows from one section to the other through submerged ports. Oxygen can be mixed with the tank contents by injection through a hollow shaft to a rotating sparger device. Head gases pass freely from one stage to the other from the top space, with only a slight pressure. The exhausted waste gas, consisting of CO_2, N and about 10 to 20% of applied O_2, come out through the oxygen vent provided in the last stage. The effluent mixed liquor is then conveyed to the final clarifier for clarification. The activated sludge is returned to the aeration tank, while the treated effluent is carried away for final disposal.

The advantages of the process include (*i*) capability to meet high oxygen demands, (*ii*) ability to carry a reserve of oxygen in solution to meet variations in demand (*iii*) superior settling characteristics of MLSS allowing a high concentration of solids (2 to

3%) in the return sludge. (*iv*) reduced aeration period (*vi*) high efficiency even at increased BOD loads (*vii*) reduced sludge production than that generated in air activated sludge systems (*viii*) ability to withstand shock organic loads (*ix*) lesser emission of foul odours because of the highly aerobic environment, (*x*) efficient operation even during cold weather because the tank is covered.

Following are the data about general loading and operational parameters for high-purity-oxygen activated-sludge process.

(*i*) *BOD loading* : (*a*) 1.92 kg of BOD/day/m^3

(*b*) 0.6 – 1.5 kg of BOD per day/kg of MLSS.

(*ii*) *Aeration period* : 1.0 – 3.0 hours

(*iii*) *Av. sludge return rate* : 30%

(*vi*) *BOD efficiency* : 90 – 95%

Results from pilot-plant studies of pure oxygen activated sludge revealed that at BOD loading of 2.56 kg/m^3-day and mean aeration periods of 1.8 hour BOD and suspended solids effluent concentrations were consistently less than the max. allowable criterion of 30 mg/l. Also, resulting BOD removal efficiency ranged from 87 to 97%, depending upon influent concentration. Loading in terms of F/M ratios were 0.5 to 0.6 kg of BOD/day/kg of volatile MLSS. A F/M rato of 0.5 is equialent to treating a 150-mg/l BOD wastewater with an aeration period of 1.8 hours, holding 4000 mg/l MLVSS in the basin. DO of 3 – 9 mg/l was held in the mixed liquor, resulting in the oxygen utilisation of about 90%. The settleability of SS, with SVI values near 50, was better and hence the thickness of return sludge was about 2% and the normal sludge rate was 30%.,

PROBLEMS

1. Explain the biochemical mechanism of the activated sludge process.
2. (*a*) What do you understand by activated sludge ? What are its properties?

   (*b*) What do you understand by (*i*) mixed liquor (*ii*) mixed liqur solids concentration, (*iii*) mixed liquor volatile concentration ? How are (*ii*) and (*iii*) interrelated ?
3. Explain, with the help of a flow diagram, the essentials of activated sludge process.
4. Explain in brief various methods used for 'aeration' in the activated sludge process.
5. Write a note on 'diffused air system'.
6. Explain in brief various systems of mechanical aeration.
7. Explain various loading criteria used for aeration tank of activated sludge process.

8. What do you undersand by 'sludge age'? How do you determine its value for a completely-mixed-cellular recycle system?

9. Write notes of the following :
   (*i*) SVI
   (*ii*) F/M ratio
   (*iii*) Activated sludge process efficiency
   (*iv*) Safety factor in activated sludge process
   (*v*) Plug flow
   (*vi*) Completely mixed flow.

10. What is meant by sludge recycle ? How do you determine the sludge recycle rate if SVI and MLSS concentration in the reactor are known?

11. The MLSS concentration in an aeration tank of activated sludge process is 2500 mg/l and the sludge volume after 30 minutes of settling in a 1000 ml graduated cydinder is 180 ml. Determine (*i*) SVI (*ii*) return sludge ratio required and (*iii*) SS concentration in the return sludge.
Ans. [90 ml/g; 0.29; 11111 mg/l]

12. Explain how do you determine the oxygen requirements in an activated sludge process.

13. An average operating data for conventional activated sludge treatment plant is as follows :
   (1) Wastewater flow = 35000 m^3/d
   (2) Volume of aeration tank = 10900 m^3
   (3) Influent BOD = 250 mg/l
   (4) Effluent BOD = 20 mg/l
   (5) Mixed liquor suspended solids (MLSS) = 2500 mg/l
   (6) Effluent suspended solids = 9700 mg/l
   (7) Quantity of waste sludge = 220 m^3/d

   Based on the information above, determine :
   (*a*) Aeration period (hrs)
   (*b*) Food to micro-organisms ratio (F/M) (kg BOD per day/kg MLSS)
   (*c*) Percentage efficiency of BOD removal
   (*d*) Sludge age (days)

(GATE, 1993)

Ans. (*a*) 7.47 hours (*b*) 0.321 kg/BOD day/kg of MLSS or 0.401 kg BOD/day/kg of MLVSS (*c*) 92% (*d*) 8.576 days.

14. What is meant by activated sludge ? Describe with sketches the treatment of sewage by activated sludge process. Mention the advantages and disadvantages of this system.

15. Design an activated sludge process to yield an effluent BOD_5 of 20 mg/l and suspended solids of 25 mg/l. The influent BOD_5 following primary settling tank is 160 mg/l. The waste inflow is 10 m^3/min. Take $Y = 0.65$, $k_d = 0.05$ and $\theta_c = 10$ days.

Ans. [$V = 3888\ m^3$; Sludge $= 1215$ kg/day, $r = 0.2$, $T = 6.48$ h; Air $= 68.3\ m^3$/min.]

16. Write short notes on the following.

(*i*) Tapered aeration process

(*ii*) Step aeration process

(*iii*) Modified aeration process

(*iv*) Extended aeration process.

17. Explain, with the help of a neat sketch the contact stabilisation process of activated-sludge.

18. Distingnish clearly between the conventional and completely-mixed activated sludge process.

19. Explain clearly how do you determine the area of the secondary settling tank used for activated sludge process. How do you decide the solids loading rate for such a tank ?

20. Explain, in brief, various operational problems commonly encountered in activated-sludge plants.

21. Design a secondary clarifier for an activated sludge process with a recycle rate of 30%, a MLSS concentration of 3000 mg/l and an anticipated flow rate of 9000 m^3/day.

15

Biological Treatment III : Miscellaneous Methods

15.1. OXIDATION DITCH

The oxidation ditch, which is essentially an extended-aeration activated sludge process, was developed for small towns is Netherlands. Fig. 15.1 (*a*) shows two typical oxidation ditches. An oxidation ditch consists of an endless ditch for the aeration tank and a rotor for aeration mechansim. The ditch consists of a long continuous channel, usually oval in plan. The channel may be earthen with *lined sloping sides* and *lined floor* or it may be built in concrete or brick with vertical walls. There is normally no primary tank used in the oxidation ditch process. Raw sewage passes directly through a bar screen to the ditch.

The sewage is aerated by a *surface rotor* (generally, a cage rotor) placed across the channel. The rotor entrains the necessary oxygen into the liquid and keeps the contents of the ditch mixed and moving. They are designed to impart a velocity of 0.3 to 0.4 m/s to the mixed liquor, preventing the biological sludge from settling out. Cage rotors usually have a dia. of 70 cm (27.5 in.) and a speed of 75 rpm. Rotors are manufactured in 30 cm intervals upto 4.5 m length. A rotor assembly can be of multiple lengths but it must be supported by intermediate bearings. The selection of rotor assembly length dictates the approximate cross-section of the ditch. The width of the ditch divided by the rotor length should give a ratio between 1.5 and 2.8. The larger ratios are normally used for short length of 0.9 to 1.2 m.

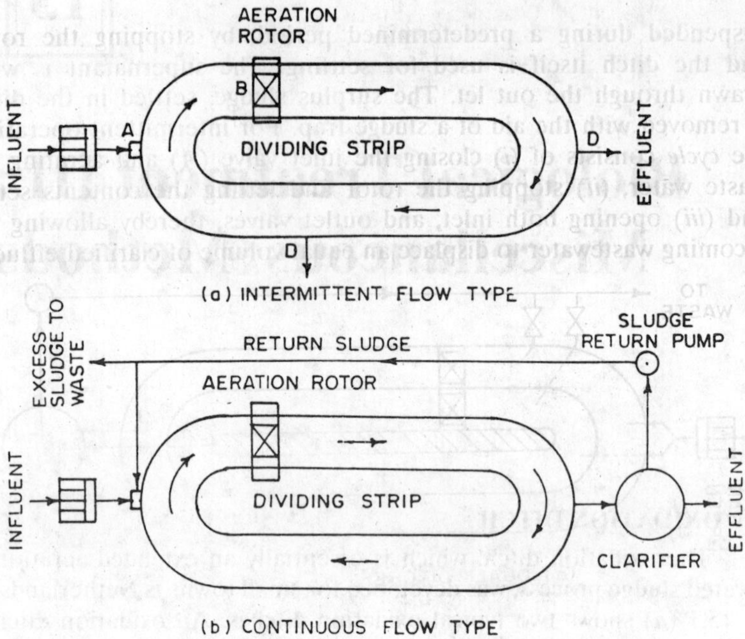

FIG. 15.1. OXIDATION DITCHES

The standard oxygen transfer capacity of rotors is 2.8 kg O_2 per m length at 16 cm depth of immersion. This rotor has been found to impart adequate circulation for 120 – 150 m^3 of ditch volume per meter length of rotor. Power requirements per metre length is about 1.35 kW at the rpm and immersion depth stipulated. The ends of the ditch are rounded to prevent eddying and dead areas. The ditch volume is designed based on the criteria given for extended aeration systems in Table 14.3. The width of ditch is designed to accommodate the length of the aerator required to meet the oxygen demand either as a single rotor or as multiple rotors. The depth of ditch is kept as 1.0 to 1.2 m and the length of the ditch is designed to give the required aeration tank volume.

The raw sewage and return sludge are discharged into the ditch upstream of the rotors. The outlet of the ditch should be located geometrically opposite to the inlet. The outlet weir should be baffled and should be of adequate length so that the ditch water level does not rise excessively and overload the rotors at periods of peak flow.

Oxidation ditches are constructed in two types : (*i*) continuous flow type, and (*ii*) intermittent flow type.

In the *intermittent flow type* oxidation ditch, shown in Fig. 15.1(*a*), the no separate settling tank is used. The flow in the ditch remains suspended during a predetermined period, by stopping the rotor, and the ditch itself is used for settling. The supernatant is withdrawn through the out let. The surplus sludge, settled in the ditch, is removed with the aid of a sludge trap. For intermittent operation, the *cycle* consists of (*i*) closing the inlet valve (*A*) and aerating the waste water, (*ii*) stopping the rotor and letting the contents settle, and (*iii*) opening both inlet, and outlet valves, thereby allowing the incoming wastewater to displace an equal volume of clarified effluent.

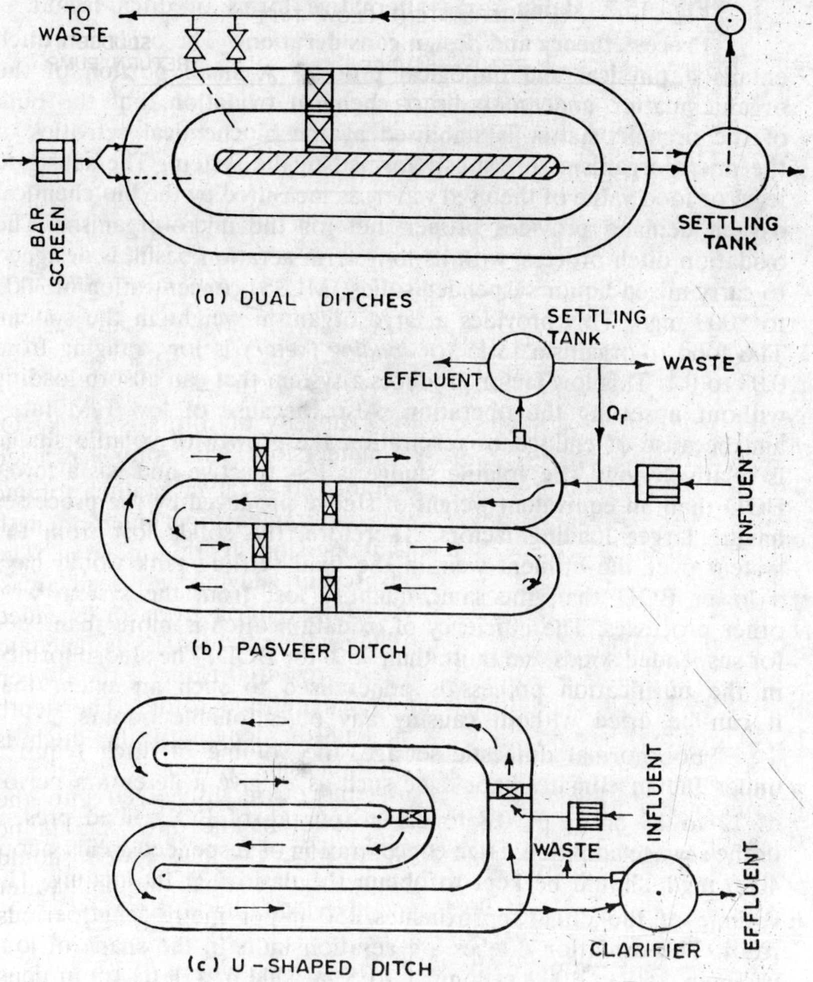

FIG. 15.2. ALTERNATE FORMS OF DITCH LAYOUT

In the *continous flow type* oxidation ditch, shown in Fig. 15.1 (*b*), the operation is kept continous by allowing the mixed liquor to settle in a separate settling tank. Quiescent conditions in the

clarified liquid passes over the effluent weir for final disposal. The settled sludge is removed from the bottom of the clarifier by an air lift or pump and is returned to the ditch. The oxidation ditch is operated as a closed system, and the net growth of the volatile suspended solids will require periodic removal of some sludge from the process. Wasting of the sludge, through a connection to the return sludge line, lowers the concentration in the ditch and keeps the metabolism more active.

Fig. 15.2. shows some alternative forms of ditch layout.

Process theory and design considerations. The oxidation ditch entails a physical and biological process. A small portion of the organic matter undergoes direct chemical oxidation, but the bulk of the organic matter is stabilised by the biochemical activities of the micro-organisms previously formed in the system. The substrate level or food value of the used water, as measured by the bio-chemical oxygen demand provides proper diet for the micro-organism. The oxidation ditch process, with its long term aeration basin, is designed to carry mixed liquor suspended solids (MLSS) concentration of 3000 to 8000 mg/l. This provides a large organism weight in the system. The food-to-organism ratio (or *loading factor*) is low, ranging from 0.03 to 0.1. This low factor produces a system that can absorb loading without upsetting the operation. Also, because of low F/M ratio, and because of endogenos respiration the growth of volatile sludge is relatively low. The volatile sludge is less reactive and has a lower BOD than an equivalent weight of sludge produced by the processes having larger loading factors. Therefore, the solids lost from the system over the effluent weir in the final settling tank would have a lower BOD than the same quantity lost from the system with other processes. The efficiency of oxidation ditch is more than 95% for suspended solids and more than 98% for BOD. The sludge formed in the purification process is mineralised to such an extent that it can be dried without causing any objectionable odours.

For normal domestic sewage, the volume of ditch required under Indian climates should be such as to give a detention period of 12 to 15 hours or 0.8 to 2.5 m^3 per kg of BOD_5 load present in the sewage admitted. High concentration of suspended solids, about 4000 mg/l, should be kept to obtain the desired BOD loading. The volume of the ditch approximates 150 m^3per metre length of the rotor. The oxidation ditches are aeration units in the shape of long channels 150 to 1000 m long, 1 to 5 m wide and 1 to 1.5 m deep.

Kinetic equations. The kinetic equations given in the previous chapter (activated sludge process) are also applicable for the oxidation-ditch-process, but with necessary modifications to the available data. In any system treating soluble waste, the MLVSS are taken as the mass of the active-micro-organisms in the reactor. However, for the present case, unsettled domestic sewage, which is directly applied to the oxidation ditch, consists of a large portion of *inactive* VSS in it. Hence influent soluble BOD S_o is equal to L_a minus BOD of SS contained in the influent. For Indian conditions, the BOD of SS may be taken @ 0.25 kg/kg of SS. Similarly, the BOD of SS escaping with the effluent may be taken @ 0.45 kg/kg of SS. The microbial mass concentration may be taken equal to 60% of the MLSS, to take care of *inactive* VSS in raw sewage.

Thus, we have the following equations applicable for the process:

$$\text{F/M} = \frac{Q\,L_a}{(V/1000)\,x_t} \qquad \text{...(15.1)}$$

(where Q is in mLd).

$$\text{sludge return ratio : } \frac{Q_r}{Q} = \frac{x_t}{\dfrac{10^6}{\text{SVI}} - x_t} \qquad \text{...(15.2)}$$

$$x = \frac{Q}{V}\,\frac{\theta_c\,Y\,(S_o - S)}{1 + k_d\,.\,\theta_c} \qquad \text{...(15.3)}$$

$$\text{Microbial mass in excess sludge, } P_x = \frac{x\,V}{\theta_c} \qquad \text{...(15.4)}$$

$$\text{Wasting of excess sludge} = \frac{V}{\theta_c} \qquad \text{...(15.4 } a\text{)}$$

Active micro-organisms in return sludge is given by

$$x\,(Q + Q_r) = x_r\,Q_r \qquad \text{...(15.5)}$$

$$\text{Hydraulic retention time} = \frac{V}{Q} \qquad \text{...(15.6)}$$

$$\text{Process loading} = \frac{Q\,(L_a - L)}{x_t\,V} \text{ kg total BOD/kg of MLSS} \qquad \text{...(15.7)}$$

where x = mixed liquor volatile suspended solids (MLVSS), mg/l.

Q = sewage inflow.

V = total volume of aeration tank (oxidation ditch)

θ_c = mean cell residence time (days).

k_d = micro-organisms decay coefficient/day.

Y = growth yield coefficient

S_o = influent soluble BOD (mg/l)

S = effluent soluble BOD (mg/l)

P_x = quantity of microbial mass produced per unit time (kg/day).

Q_r = return sludge flow.

SVI = sludge volume index.

Example 15.1. *Design an oxidation ditch for a community with the following data :*

(i) Population of the community : 6000 persons.

(ii) Organic load of sewage : 40 g BOD per capita per day.

(iii) Sewage flow : 160 litres/capita/day.

(iv) Permissible BOD of effluent : 20 mg/l.

Solution

(i) Inflow rate and influent BOD

Total sewage flow = 6000 × 160 = 960 × 10^3 l/day (0.96 mLd)

= 960 m^3/day

Total BOD applied/day = (6000 × 40) 10^{-3} = 240 kg/day

$$\therefore \quad \text{Influent BOD} = \frac{240 \times 10^6}{960 \times 10^3} = 250 \text{ mg/l}$$

$$\therefore \quad L_a = 250 \text{ mg/l}$$

$$\text{BOD removal required} = \frac{250 - 20}{250} \times 100 = 92\%$$

(ii) Volume of ditch

Volume of the ditch can be found from the formula :

$$\text{F/M} = \frac{Q\,L_a}{(V/1000)\,x_t}, \quad (\text{where } Q \text{ is in mLd})$$

Choosing F/M = 0.1 and keeping x_t = 3000 mg/l, we get

$$V = \frac{1000\,(0.96)(250)}{0.1 \times 3000} \approx 800 \text{ m}^3$$

(iii) Volumetric loading rate

$$\text{Volumetric loading} = \frac{Q\,L_a}{V} = \frac{0.96 \times 250}{800}$$

$$= 0.3 \text{ kg BOD}_5/\text{m}^3$$

(which is within the prescribed range of 0.2 to 0.4)

(iv) Hydraulic retention time

$$\text{HRT} = \frac{V}{Q} = \frac{800}{960} \times 24 = 20 \text{ hours.}$$

(which is within the prescribed range of 20 to 30 hours)

(v) Return sludge ratio

$$\frac{Q_r}{Q} = \frac{x_t}{\dfrac{10^6}{\text{SVI}} - x_t}$$

Adopting value of SVI = 100,

$$\frac{Q_r}{Q} = r = \frac{3000}{\dfrac{10^6}{100} - 3000} \approx 0.43$$

(which is within the prescribed range of 0.35 to 1.5)

(*vi*) ***Oxygen requirement and no. of rotors***

O_2 required = 1.2 kg/kg of BOD removed

= 1.2 (250 − 20) 0.96 ≈ 265 kg/day

= 11.04 kg/hour.

Oxygenation capacity of cage rotor of dia. 70 cm, @ 75 rpm, at immersion depth of 16 cm = 2.8 kg of O_2/hr/m length.

∴ Length of rotor needed on the basis of oxygenation capacity = 11.04/2.8 = 3.94 m (or say 4 m)

Length or rotor neeeded from velocity consideration of 0.3 m/s (assuming 150 m^3 of ditch volume per m length of rotor for adequate circulation) = 800/150 = 5.33 m. Hence adopt 2 rotors of 2.7 m length each.

Giving a clearance of 0.25 m on either side, width of ditch = 2.7 + 2 × 0.25 = 3.2 m. Let us keep a depth of 1.5 m.

$$\therefore \text{ Surface area required } = \frac{800}{1.5} = 533.3 \text{ m}^2$$

Adopting 2 ditches, surface area of each ditch = 266.7 m^2

$$\therefore \text{ Length of ditch } = \frac{266.7}{3.2} = 83.3 \text{ m or say 85 m.}$$

(*vii*) ***Power required***

Power required for each rotor @ 1.35 kW/m length

= 1.35 × 2.7 ≈ 3.7 kW.

Total power required for both ditches = 2 × 3.7 = 7.4 kW.

(*viii*) ***Design of settling tank***

Let us provide SOR of 20 m^3/m^2/day and a detention time of 2 hours.

$$\therefore \text{ Surface area } A = \frac{Q}{\text{SOR}} = \frac{960}{20} = 48 \text{ m}^2$$

$$\text{Also, volume } V = \frac{960}{24} \times 2 = 80 \text{ m}^2$$

$$\therefore \text{ Depth of tank} = 80/48 = 1.67 \text{ m.}$$

$$\text{Dia. of tank} = \sqrt{\frac{48 \times 4}{\pi}} = 7.8 \text{ m.}$$

Hence provide 8 m dia. tank of 1.7 m depth, with hopper bottom.

$$\text{Actual area } = \frac{\pi}{4}(8)^2 = 50.27 \text{ m}^2$$

$$\text{Actual SOR } = \frac{960}{50.27} = 19.1 \text{ m}^3/\text{m/day.}$$

The details of the oxidation ditch (one unit) are shown in Fig. 15.3.

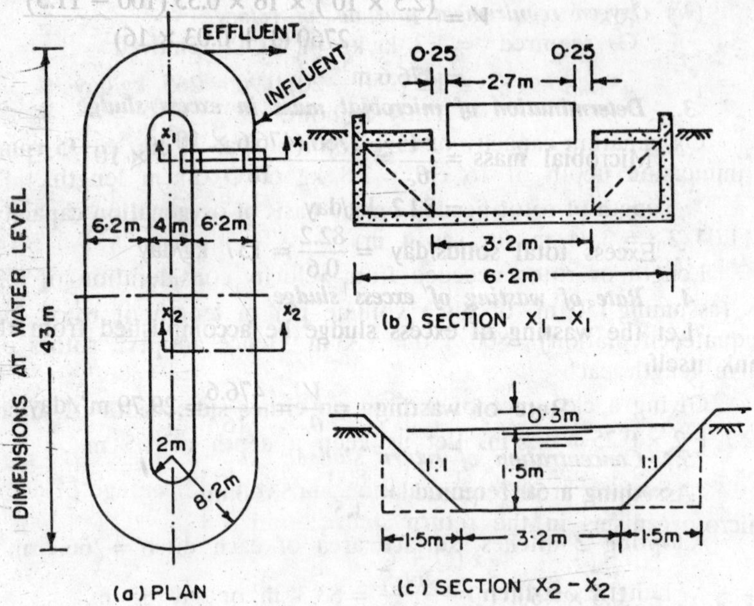

FIG. 15.3. DETAILS OF OXIDATION DITCH

Example 15.2. *Design a continuous flow type oxidation ditch to treat a domestic sewage flow of 2.5 mLd. Given the following data:*

1. *BOD_5 of raw sewage* = *160 mg/l*
2. *Desired BOD_5 of effluent* = *25 mg/l*
3. *SS in raw sewage* = *240 mg/l*
4. *Desired SS in effluent* = *30 mg/l*
5. *Mixed liquor suspended solids* = *4600 mg/l*
6. *Growth yield coefficient* = *0.55*
7. *Micro-organism decay coefficient* = *0.03*
8. *Mean cell residence time* = *16 days.*

Solution

1. ***Determination of influent and effluent soluble BOD***

BOD of SS in influent = 0.25 × 240 = 60

∴ Influent soluble BOD_5, S_o = 160 − 60 = 100 mg/l.

BOD of SS in effluent = 0.45 × 30 = 13.5

∴ Effluent soluble BOD_5, S = 25 − 13.5 = 11.5 mg/l.

2. ***Determination of volume of reactor*** **(ditch)**

$$x = \frac{Q}{V}\,\frac{\theta_c \cdot y\,(S_o - S)}{1 + k_d\,\theta_c}$$

Assuming microbial mass concentration to be 60% of the MLSS (to take care of inactive VSS in the raw sewage), $x = 0.6 \times 4600 = 2760$ mg/l.

$$V = \frac{(2.5 \times 10^3) \times 16 \times 0.55\,(100 - 11.5)}{2760\,(1 + 0.03 \times 16)}$$

$$= 476.6 \text{ m}^3$$

3. ***Determination of microbial mass in excess sludge***

$$\text{Microbial mass} = \frac{xV}{\theta_c} = \frac{2760\,(476.6 \times 10^3)}{16} \times 10^{-6}$$

$$= 82.2 \text{ kg/day}$$

$$\therefore \text{ Excess total solids/day} = \frac{82.2}{0.6} = 137 \text{ kg/day}$$

4. ***Rate of wasting of excess sludge***

Let the wasting of excess sludge be accomplished from the tank itself.

$$\therefore \quad \text{Rate of wasting} = \frac{V}{\theta_c} = \frac{476.6}{16} = 29.79 \text{ m}^3/\text{day.}$$

5. ***Concentration of return sludge***

Assuming a 50% recirculation and same percentage of active micro-organisms in the return sludge we have.

$$x\,(Q + Q_r) = x_r\,(Q_r)$$

or $$0.6 \times 4600\,(Q + 0.5\,Q) = 0.6\,x_r\,(0.5\,Q)$$

or $$4600\,(1.5\,Q) = x_r\,(0.5\,Q)$$

From which $$x_r = 4600 \times \frac{1.5}{0.5} = 13800 \text{ mg/l.}$$

6. ***Oxgygen requirement***

O_2 required (kg/day) @ 1.2 kg of O_2 per kg BOD_5 removal.

$$= 1.2\,(160 - 25)\,2.5 = 405 \text{ kg/day}$$

7. ***Checks***

(*i*) Hydraulic retention time provided $= V/Q$

$$= \frac{476.6 \times 10^3}{2.5 \times 10^6} = 0.191 \text{ days} = 4.58 \text{ hours.}$$

(ii) Process loading

$$\text{Process loading} = \frac{Q\,(L_a - L)}{x_t\,V}$$

$$= \frac{(2.5 \times 10^6)(160 - 25)\,10^{-6}}{(4600 \times 10^{-6})\,(476.6 \times 10^3)}$$

$$= 0.154 \text{ kg total BOD/kg of MLSS}$$

8. ***Rotor length and power requirement.*** Oxygenation capacity of cage rotor of dia. 70 cm @ 75 rpm and at immersion depth of 16 cm = 2.8 kg of O_2/hr/m length.

∴ Length of rotor needed on the basis of oxygenation capacity

$$= \frac{405}{24 \times 2.8} = 6.03 \text{ m}$$

Length of rotor needed from velocity consideration of 0.3 m/s (assuming 150 m^3 of ditch volume per m length of rotor for adequate circulation) = 476.6/150 = 3.18 m. Hence rotor length = 6.03 m.

Let us provide 2 rotors of 3 m length each.

Total power required = 2 × 3 × 1.35 = 8.1 kW.

9. Ditch dimensions

Length of each rotor = 3 m.

Giving a clearance of 0.25 m on either side, width of ditch = 3 + 2 × 0.25 = 3.5 m. Let us keep a depth 1.2 m.

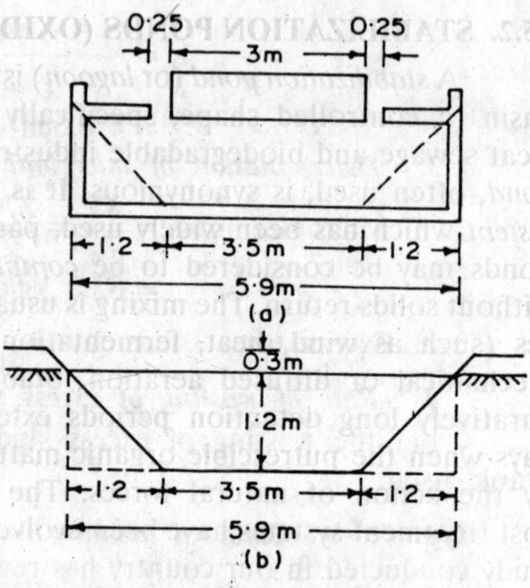

FIG. 15.4. SECTIONS OF THE DITCH

$\therefore$ Surface area required $= \dfrac{476.6}{1.2} = 397\ m^2$

$\therefore$ Providing 2 ditches, surface area of each ditch = 397/2 ≈ 200 m^2.

$\therefore$ Length of each ditch = 200/3.5 ≈ 58 m.

The details of the ditch are shown in Fig. 15.4.

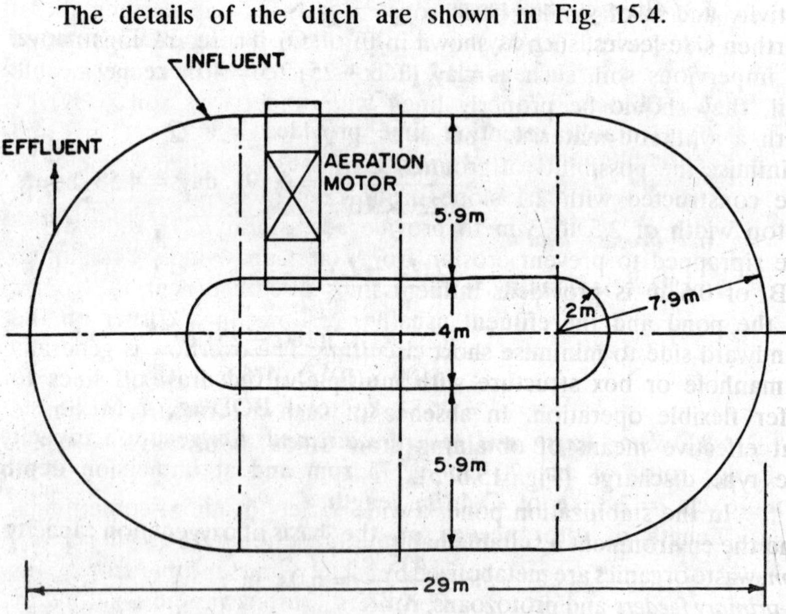

FIG. 15.5 PLAN OF ONE DITCH

15.2. STABILIZATION PONDS (OXIDATION PONDS)

A *stabilization pond* (or *lagoon*) is an open, flow-through earthen basin of controlled shape, specifically designed and constructed to treat sewage and biodegradable industrial wastes. The term *oxidation pond*, often used, is synonymous. It is a relatively *low-cost treatment system* which has been widely used, particularly in rural areas. These ponds may be considered to be *completely mixed biological reactors* without solids return. The mixing is usually provided by natural processes (such as wind, heat, fermentation), but may be augmented by mechanical or diffused aeration. Stabilization ponds provide comparatively long detention periods extending from a few to several days when the putrescible organic matter in the waste gets stabilized by the action of natural forces. The design criteria for such low cost treatment systems have been evolved in our country. An extensive study conducted in our country has revealed that, under certain conditions, the degree of treatment that can be achieved is as good as that of the conventional system, if these low cost systems are adopted.

Classification of stabilization ponds. Stabilization ponds are usually classified according to the nature of biological activity that is taking place as :

(*i*) aerobic (*ii*) anaerobic or (*iii*) facultative (aerobic-anaerobic)

Each sub-type mentioned above utilizes a different biological activity and is of similar basic construction, with an earthen pit of earthen side leeves, such as shown in Fig. 15.6. These are constructed in impervious soil, such as clay. If constructed in more permeable soil, they should be properly lined with impervious soil (clay) or with a synthetic material. This is essential to prevent seepage and minimise the possibility of ground water contamination. Levee walls are constructed with 2:1 slope inside and 3:1 slope outside, with a top width of 2.5 to 3 m to provide accessibility. The side slopes are riprapped to prevent erosion from water and wind. A minimum F.B. of 0.6 m is provided. Influent lines discharge near the centre of the pond and the effluent usually overflows in a corner on the windward side to minimise short circuiting. The overflow is generally a manhole or box structure with multiple-valved draw off lines to offer flexible operation. In absence of such a structure, a simple but effective means of obtaining draw off is to install a sideways tee type discharge (Fig. 15.6 *b*).

In the stabilization pond, a wide variety of microscopic plants find the environment a suitable habitat. In additon to the *algal population*, waste organics are metabolized by bacteria and saprobic protozoans as *primary feeders* and protozoans, rotifers, crustaceans etc as *secondary*

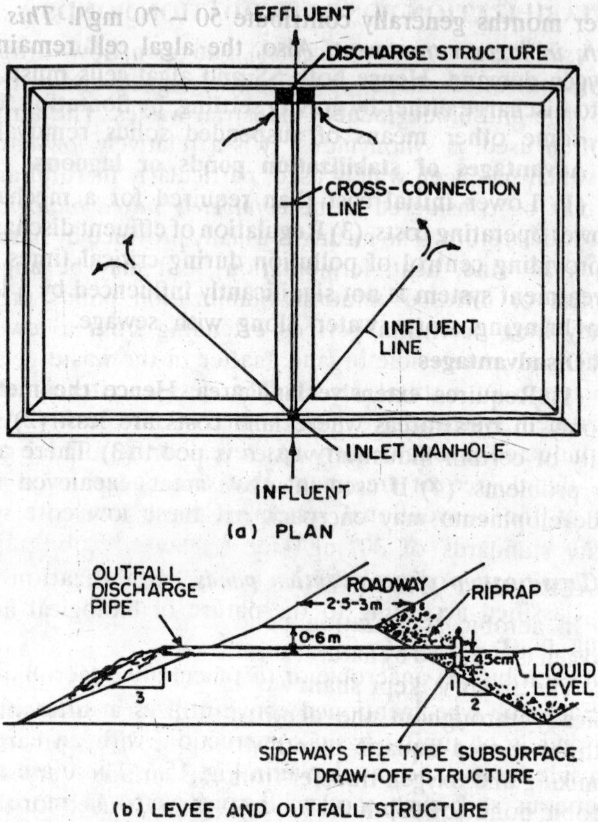

FIG. 15.6. STABILIZATION POND.

feeders. The daily flow of sewage containing organic material provides necessary food to the microbial population which stabilize the putrescible matter by oxidising it to form nitrates and CO_2. The nutrients released by microbial population are used by the *algal population* which in turn release oxygen through the process of photosynthesis. This process, usually known as 'bacterial-algal-symbiosis' results in complete stabilization of organic wastes. When the pond bottom is anaerobic (due to provision of greater depth or due to lack of proper mixing and aeration), biological activity results in digestion of the settled solids.

The pond water becomes super saturated with dissolved oxygen during afternoon, because of release of oxygen by the algae in presence of sun light. The suspended solids and BOD in the pond effluent are primarily from the algae. Though the BOD reduction exceeds 95% (specially in summer), the effluent does not meet the standard of 30 mg/l of suspended solids. Algae suspended in water during summer months generally contribute 50 – 70 mg/l. *This is a serious problem in lagoon treatment.* Also, the algal cell remains represent an oxygen demand. Hence both SS and algal cells must be removed prior to discharge either by gravity settling, by floatation, by filtration, or by some other means of suspended solids removal.

Advantages of stabilization ponds or lagoons

(1) Lower initial cost than required for a mechanical plant. (2) Lower operating costs. (3) Regulation of effluent discharge possible, thus providing control of pollution during critical times of the year. (4) Treatment system is not significantly influenced by a leaky sewage system bringing storm water along with sewage.

Disadvantages

(1) Requires extensive land area. Hence the method can be used only in rural areas where land costs are less. (2) Assimilative capacity of certain industrial wastes is poor. (3) There are potential odour problems. (4) If used in urban areas, expansion of town and new developments may encroach on the lagoon site. (5) Effluent quantity standards of 30 mg/l for suspended solids are not met.

15.3. AEROBIC PONDS (ALGAE PONDS)

In aerobic stabilization ponds (also known as *algae ponds*), the oxygen is supplied by natural surface aeration and by algal photosynthesis. The pond is kept shallow (0.5 to 1.2 m), so that it functions aerobically throughout the depth. Shallower levels will encourage growth of rooted aquatic plants while greater depth may interfere with mixing and oxygen transfer from the surface. Very shallow depth of aerobic pond (of depth 0.15 m to 0.45 m) is used for the treatment of irrigation return water or any other industrial waste where the aim is the removal of nitrogen by algal growth. However for the treatment of domestic waste, the depth is kept between 1 to 1.2 m. The length to width ratio of the pond depends on the geometry of the land but should be maximized to approach but not be exceed 3:1. This tends to prevent short circuiting. The influent and effluent structures are so located that entire lagoon volume is utilised. The contents of the tank are stirred occassionally to prevent anaerobic conditions in the settled sludge. Except for the algal population, the microbiological population present in the ponds are similar to that in activated sludge system. The daily flow of sewage containing organic material provides necessary food to the *aerobic population* which stabilizes the putrescible matter by oxidising it to form nitrates and CO_2. The *algal population* use these products for their growth to produce more algal cells and, during daylight, oxygen which is again used by the aerobic population to decompose the original waste. The action taking place in these ponds is known as *'bac-*

terial-algal-symbiosis'. Thus the symbiotic relation between bacteria and algae leads to the stabilization of the incoming waste. The algael-bacterial interplay in an aerobic lagoon is shown in Fig. 15.7.

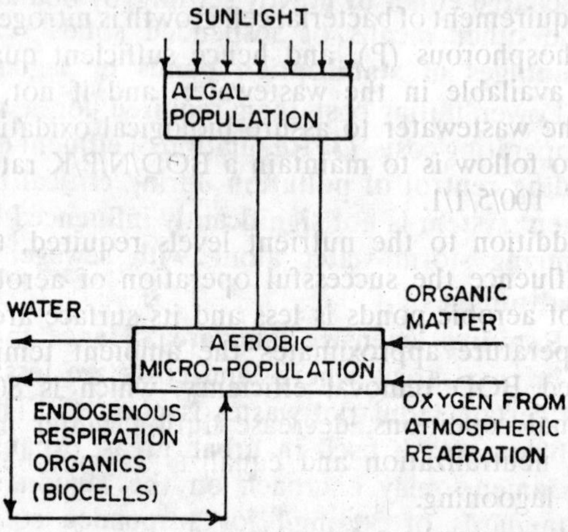

FIG. 15.7. ALGAL-BACTERIAL INTERPLAY IN AN AEROBIC LAGOON

Oxygen transfer in the lagoon depends on the following factors.

1. ***Ratio of Lagoon surface area to volume*** : Larger the ratio, better will be the oxygen diffusion into the lagoon.

2. ***Turbulence*** : Generally provided by the wave action.

3. ***Temperature of lagoon*** : Greater solubility of oxygen in water and hence greater diffusion rate at lower temperature.

4. ***Bacterial oxygen-uptake rate*** : The faster the micropopulation consumes the dissolved oxygen, the greater will be the rate at which oxygen is replenished.

The aerobic decomposition results in the production of carbon dioxide, water and ammonia, as indicated by the following equation:

$$C_{11}H_{29}O_7N + 14\,O_2 + H^+ \longrightarrow 11\,CO_2 + 13\,H_2O + NH_4^+ \quad ...(15.8)$$

Several steps in the oxidation process, including the formation of new bacterial cells, in turn oxidize more organic waste.

The oxygen is produced by algae in the presence of sunlight, and the liquid gets supersaturated with oxygen to an extent of about 20 mg/l. This is consumed by the bacterial community in the night, and the nocturnal concentration falls to zero. Oxygen released by photosynthesis in algae is represented by the following relationship:

$$NH_4^+ + 7.6\,CO_2 + 17\,H_2O \xrightarrow{\text{Sunlight}} 15.2\,H_2O + H^+$$

$$-\,866\ \text{Kcal. per gram} + \underset{\text{algae}}{C_7H_{8.1}O_{2.5}N} + 7.6\,O_2 \quad ...(15.9)$$

The composition of biological cell mass is approximated by the formula $C_{120} H_{180} O_{45} N_{15} K P$. According to this empirical formula, the basic requirement of bacterial cell growth is nitrogen (N), potassium (K) and phosphorous (P), and hence sufficient quantities of these should be available in the wastewater, and if not, they should be added to the wastewater to assure biological oxidation. A good rule of thumb to follow is to maintain a BOD/N/P/K ratio in the lagoon influent of 100/5/1/1.

In addition to the nutrient levels required, temperature and pH also influence the successful operation of aerobic ponds. Since the depth of aerobic ponds is less and its surface area is quite large, liquid temperature approximates the ambient temperature. Hence the expected BOD removal efficiency, which is 80 to 95 percent under normal conditions, decrease during winter months. If pH is a problem, neutralization and equalization may be necessary prior to aerobic lagooning.

Loading may be taken as 150 to 200 kg BOD_5 per hectare per day during cold weather, which may be increased considerably during summer. The detention period should be about 7 days for proper development of algae.

15.4. ANAEROBIC PONDS

In anaerobic pond, the entire depth is in anaerobic condition except an extremely shallow top layer. The anaerobic micro-organisms do not require the presence of dissolved oxygen in the water in order to function. Their requirement is met from the oxygen chemically contained in the organic materials. The *anaerobic decomposition* takes place in two separate but interrelated steps :

Step 1 : Decomposition of dissolved organic waste, by *acid-producing bacteria*, to organic acids (such as acetic, propionic and butyric acid) and,

Step 2 : Futher decomposition of these acids to the end products of methane, carbon dioxide and water, by the *methane-producig-bacteria*.

This is depicted in Fig. 15.8.

Effective operation requires a balance between step 1 and step 2, because the methane producers are sensitive to the concentration of volatile acids. A certain portion of the waste material is used by the anaerobic biosystem as source of energy and in the synthesis of new bacterial cells. Sludge or solids buildup is therefore much less in the anaerobic system. The contents of the anaerobic lagoon are black in colour, which is an effective indication that the lagoon is functioning properly.

The process is some what attended by septic odours and the effluent will be only partially purified. The obnoxious odours are

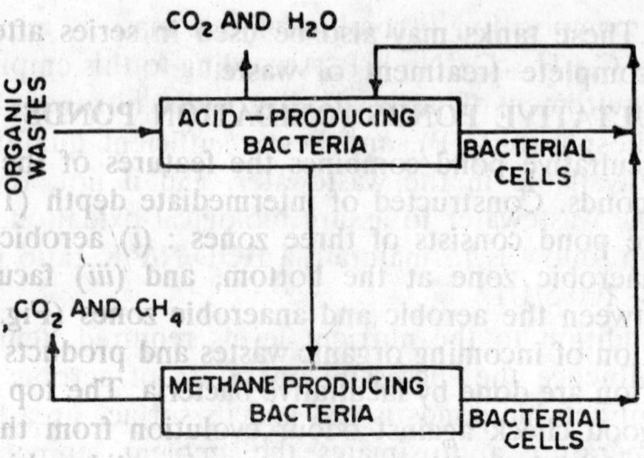

FIG. 15.8. ANAEROBIC DEGRADATION PROCESS

the results of the reduction of sulphate compounds to hydrogen sulphide (H_2S), by the acid-producing bacteria. At high concentration, H_2S attacks painted surface and is deleterious if inhaled for an extended period. The long term solution of this problem is to limit the concentration of sulphate in the influent.

Anaerobic pond requires much less surface area than the aerobic pond. However, the pond can be as deep as practical. Pond depths usually range from 2.5 m to 5 m. The relative greater depth of anaerobic lagoon allows for improved heat retention. The efficient length to width ratio is 2:1. The levee of anaerobic pond is similar to that of aerobic/facultative pond, both in geometry as well as in construction. The BOD loading in summer may be taken at 1000 to 2000 kg per hectare per day with detentions of 2 to 5 days ; the BOD_5 reduction may be expected at 65 to 80%. In winter, BOD loading may be taken at 500 to 1000 kg per hectare per day, with removal of 45 to 65%. For anaerobic ponds, less surface area is required because the organic loading is about 10 times more than for aerobic ponds. The depth can be taken normally upto 3 m instead of 1m, and this will reduce the area requred to about 1/30. Deposition and digestion of sewage solids is the chief function of anaerobic ponds. With suspended solids loadings of about 1000 kg/ha/day, sludge may be allowed to accumulate for 10 years or more before removal is necessary. The whole design aspect of anaerobic lagoon is controlled by the detention time whereas that of aerobic lagoon by surface loading rate. The ideal pH range is 6.6 to 7.6. The anaerobic process functions optimally over two temperature ranges : the mesophylic range of 85° to 100°F and the thermophilic range of 120° to 135°F. The greater depth provided in anaerobic pond thus helps in maximizing heat retention. This type of pond finds use mainly in the treatment of strong industrial wastes and has limited application for the treatment of sewage. These tanks may also be used in series after facultative pond for complete treatment of waste.

15.5. FACULTATIVE PONDS (OXIDATION PONDS)

A facultative pond combines the features of the aerobic and anaerobic ponds. Constructed of intermediate depth (1.0 to 1.5 m), a facultative pond consists of three zones : (*i*) aerobic zone at the top (*ii*) anaerobic zone at the bottom, and (*iii*) facultative zone, situated between the aerobic and anaerobic zones (Fig. 15.9) where decomposition of incoming organic wastes and products of anaerobic decomposition are done by facultative bacteria. The top aerobic layer acts as a good check against odour evolution from the pond. The pond depth inhibits mixing, hence organic solids which settle will remain on the bottom and will be subjected to anerobic decomposition. The action in the aerobic zone is similar to the one found in the aerobic ponds, giving rise to *bacterial-algal-symbiosis*. The treatment effected by this type of pond is comparable to that of conventional secondary treatment processes. Hence the facultative pond is best suited and most commonly used for treatment of sewage.

Most of the stabilization ponds in India are facultative in nature, and hence will be discussed in detail here.

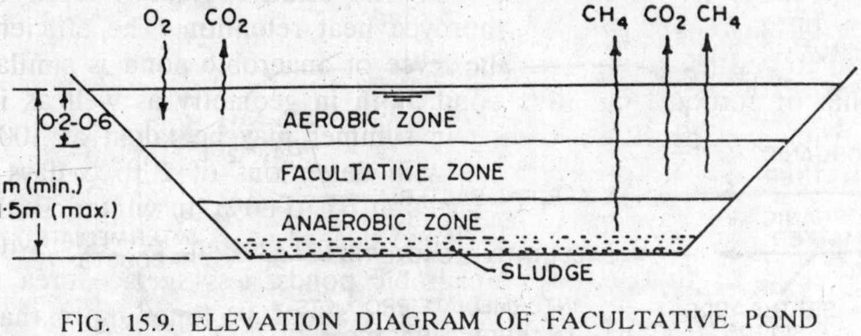

FIG. 15.9. ELEVATION DIAGRAM OF FACULTATIVE POND STRATA AND OPERATION

Mechanism of purification.

In a facultative pond, the influent organic matter is stabilized by methane fermentation in the bottom layers and partly by the bacterial oxidation in the top layers. In the liquid above the bottom sludge layer, there is a zone (known as facultative zone), in which facultative bacteria oxidize the incoming organics as well as the products of anaerobic decomposition of the bottom anaerobic zone. When the sewage enters the pond, the suspended organic matter in the influent as well as the bioflocculated colloidal organic matter settle to the bottom of the pond. In the absence of dissovled oxygen at the pond bottom, the settled sludge undergoes anaerobic fermentation with the liberation of methane which represents a BOD removal from the system—0.25 g of methane being literated for every gramme of ultimate BOD utilized. In the liquid layers of the pond, algae begins to grow under favourable conditins. The algae utilizes the carbon dioxide in the sewage for photosynthesis during day light hours, liberating oxygen which maintains aerobic conditions in the upper layers of the pond. The aerobic conditions promote the oxidation of organic waste matter by aerobic bacteria. Thus it is seen that there is an interdependance between algae and bacteria with the algae supplying oxygen required by the bacteria and bacteria making available the carbon dioxide required by the algae. This type of association between organisms is referred to as *symbiosis*, a relationship where two or more species live together for mutual benefits such that the association stimulates more vigorous growth of each species than if growth were separate. This is illustrated in Fig. 15.10. During periods when the dissolved oxygen is less than saturation level, the surface water level is aerated through wind action. During the wind, both bacterial metabolism and algal synthesis are slowed by cold tempteraures.

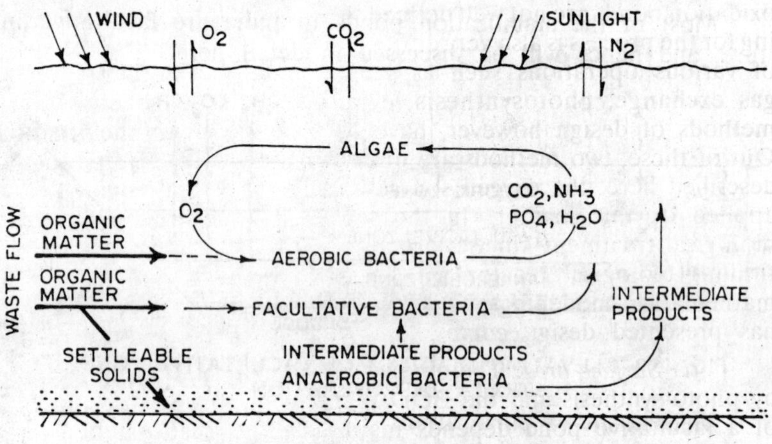

FIG. 15.10. OPERATIONAL LAGOON AS A FACULTATIVE SYSTEM.

Diurnal variation. Both the dissolved oxygen and pH of the pond are subject to diurnal variation due to photosynthetic activity of algae which is dependent on incident solar radiation. A high DO concentration upto about 4 times the saturation value may be observed in the afternoon hours. Simultaneously, the pH value may reach a maximum of 9.0 or more due to the conversion of CO_2 to O_2. Towards the evening or in the night, when photosynthetic activity decreases or stops, there is a gradual decrease in both DO and pH. In properly designed ponds, the DO does not completely disappear from the top layers and the pH does not fall below influent pH.

Odour control. In a facultative pond, the nuisance associated with anaerobic reactions is eliminated due to the presence of oxygen in the top layers. The foul smelling end products of anaerobic degradation which permeate to the top layers are oxidised in an aerobic environment. Furthermore, due to a high pH in top layers, compounds such as organic acids and H_2S, which would otherwise volatilise from the surface of the pond and cause odour problems are ionised and held back in solution.

Algal population. In stabilization ponds, the signficant algae are *green algae* and *bluegreen algae* of various species, which possess relatively high oxygen donation capacity per unit weight. However it is not practical to promote the growth of any particular type of algae in a pond which will depend on such factors as temperature, characteristics of the waste and intensity of sun light. Concentration of algae in a stabilization pond is quite high and gives the pond effluent a typical dark green colour. Floating blue-green algal mats may develop in ponds during summer months. They are undersirable since they lead to insect breeding and cutout sunlight.

Process design considerations. The design parameters for the oxidation ponds are not well defined. The precise mathematical modelling for the process is also very difficult due to simultaneous involvement of various operations such as sedimentation, oxidation, digestion, gas exchange, photosynthesis, evaporation, seepage etc. Various methods of design however, have been proposed in the literature. Out of these, two methods are quite prominent. In the first method, described here, the oxygen sources on the pond are equated to the applied organic loading. In the second method, the principles of biological treatment kinetics are applied, treating the pond as a continuous biological reactor without recycle. Based on this type of mathematical modelling and resulting equations, Thirumurthi (1969) has presented design curves.

*(i) **Surface area.*** The amount of oxygen that can be produced by photo-synthesis and the BOD that can be satisfied per unit area of a facultative pond depends mainly on the quantum of sunlight falling on the pond surface which in turn depends on the latitude of the pond site, its elevation above MSL, time of the year and sky clearance. Based on studies conducted by the National Environmental Engineering Research Institute (NEERI), Nagpur, Table 15.1 gives the values of *yield* of photo synthetic oxygen in different latitudes. IS : 5611 recommends the BOD loading equal to the yield of photosynthetic oxygen, as indicated by column 3 of Table 15.1.

The recommended BOD loadings are for municipal sewage and are inclusive of BOD of the settleable solids in the wastes. The values are applicable to towns at sea levels and where the sky is clear for nearly 75% of the days in a year.

TABLE 15.1
YIELD OF PHOTOSYNTHETIC O_2 AND RECOMMENDED BOD LOADING

Latitude (°*N*)	*Yield of photosynthetic* O_2 (*kg/ha/d*)	*Areal* BOD_5 *loading* (*kg/ha/d*)
36	150	150
32	175	175
28	200	200
24	225	225
20	250	250
16	275	275
12	300	300
8	325	325

The values may be modified for elevation above MSL by dividing by a factor, (1 + 0.003 EL), where EL is the elevation of the pond site above MSL in hundred metres. A further correction is the pond volume has to be made when the sky is clear for less than 75% of the days at the rate of 3% for a fall of every 10%.

At the recommended BOD loading, sufficient photosynthetic O_2 will be produced to stabilize about 90% of the influent BOD. Also, aerobic conditions will be maintained in the top layers of the pond at most of the time and there will generally be no evolution of septic odours from the pond.

When the ponds are intended to serve small communities such as institutions, or when they are to be located close to residence, it will be prudent to adopt lesser BOD loadings than recommended in Table 15.1 so as to fully ensure the absence of septic odours.

In any system, the individual pond area should not exceed 0.5 ha. However, it is desirable to have more than one pond in any system requring 0.5 ha or more pond area.

(ii) Detention period. The detention period of a pond should be adequate for the bacteria to stabilise the BOD to the desired degree. The removal of BOD in the pond follows a unimolecular pattern :

$$L_r = L_a\,(1 - 10^{-P_T \cdot t}) \qquad \text{...(15.10 } a)$$

or

$$t = \frac{1}{P_T} \log_{10}\left(\frac{L_a}{L_a - L_r}\right) = \frac{1}{P_T} \log_{10}\frac{L_a}{L_e} \qquad \text{...(15.10)}$$

where L_a = influent BOD

L_r = BOD removed (say 90% of L_a or 95% of L_a)

L_e = BOD remaining in the effluent

P_T = BOD removal rate costant for the pond, at temp. T° C

t = detention period in days.

The removal rate coefficient at 20°C (*i.e.* P_{20}) is found to be approximately 0.1 per day. The removal rate constant at any other tempurature (T°C) may be determined from the formula :

$$P_T = P_{20}\,(1.047^{T-20}) \qquad ...(15.11)$$

(iii) Depth. Having determined the surface area and detention capacity or volume of the pond, it becomes necessary to consider the depth of the pond only in regard to its limiting values. Shallow depths in facultative ponds will allow growth of aquatic weeds in the pond while excessive depths will cause the entire pond contents to turn anaerobic. *The optimum range of depth for facultative ponds is 1.0 – 1.5 m.* When the depth determined from area and detention period works out lesser than 1.0 m, the depth should be increased to 1 m, keeping surface area unchanged. When the depth works out to be greater than 1.5 m, the depth should be restricted to 1.5 m and surface area increased to give the required detention period.

(iv) Sludge accumulation. The rate of sludge accumulation in facultative ponds depends primarily on the SS concentration in the influent wastes. The reported rate of sludge accumulation in ponds treating municipal sewage ranges from 0.05 to 0.1 m^3/capita/year. *A value of 0.07 m^3/capita/year forms a reasonable assumption in design.* In multiple cell ponds operated in zones, most of the sludge accumulation will be in the primary cells. Continued sludge accumulatioin in ponds over many years will cause (*i*) sludge carry-over into the effluent, (*ii*) development of aquatic weeds on sludge banks, and (*iii*) reduction in pond efficiency due to reduction in the detention period. *Hence, facultative ponds requires periodical desludging at intervals ranging from 6 to 12 years.*

(v) Pretreatment. It is desirable to provide medium screens and grit removal devices before facultative ponds.

Construction Details

1. ***Pond Shape.*** The pond should be such that there are no narrow or elongated portions. Round, square or rectangular ponds with length not exceeding three times the width are acceptable. Maximum basin length of 750 m is generally adopted. The corners should always be rounded to minimise accumulations of floating matter and to avoid dead pockets.

2. ***Embankment.*** The embankment is constructed of the soil obtained from excavation of the pond. It should be compacted well. Top width may vary from 1.5 m (min.) to 3 m for large ponds. A minimum F.B. of 0.5 m should be provided. In larger ponds, the F.B. should be equal to 1.5 times the wave height. Outer slopes may vary from 2 : 1 to 2.5 : 1. The inner slopes may vary from

1:1 to 1.5:1 if fully pitched, or from 2:1 to 3:1 when the face is unprotected.

3. Influent lines. The influent pipeline should extend into the pond at least 15 to 20 m from the water edge. When the sewage is to be pumped into the pond, the pipeline may be laid along the bottom of the pond to discharge through an upward inclined 90° bend. In gravity flows, the outfall sewer should terminate at a manhole located as close to the pond as practicable. The invert of the manhole should be atleast 0.2 m above the MWL of the pond. From the terminal manhole, the influent piping may be carried into the pond grade, supported on masonry pillars. A tee or bend should be provided to discharge the wastes at about 0.3 m depth below the liquid surface. A concrete apron should be provided below the discharge end of the influent end to prevent erosion of the pond bottom.

4. Pond outlets. Multiple outlets are desirable except in small ponds and may be provided at the same rate as for inlets, one for every 0.5 ha pond area. The outlet should be so located with reference to inlets as to avoid short circuiting. The outlet structures may consist of either pipes projecting into the ponds or weir boxes. In the former case vertical tees and in the latter case hanging baffles submerged to a depth of 0.25 m below the water surface should be provided to ensure that floating algal scum is not drawn along with the effluent.

5. Multiple units of ponds. When the required area of pond exceeds 0.5 ha, multiple units should be provided. These units can be either in series or in parallel or in series-parallel system. Multiple cells in *parallel* provide better distribution of settled solids. Also, from operation point of view, one unit at a time can be taken out of operation temporarily for desludging. Multiple cells in *series* enable better coliform removal and reduced algal concentration in the effluent. The series system implies a high BOD loading in the primary cell and to avoid anaerobic conditions in these cells they should have 65 – 75% of the total surface area requirements. A *parallel-series system* possesses the advantages of both parallel and series operations. A convenient arrangement for this system consists of three cells of equal area, of which two are in parallel and serve as *primary ponds* and the third serves as *secondary pond* in series.

6. Pond inter-connections. Pond inter-connections are required when ponds are designed in multiple cells in series. These inter connections should be such that the effluent from one cell withdrawn from the aerobic zone can be introduced at the bottom of next cell. Simple interconnections may be formed by pipes laid through the separating embankments.

7. Performance. Well designed facultative ponds give about 80 – 90% BOD reduction based on *filtered* BOD_5 of the effluent. However, BOD_5 and SS value of *unfiltered* effluent may be in the range of 50 to 100 mg/l each. Facultative ponds also effect high bacterial reduction, the efficiency being particularly high in multicell ponds operated in series. Coliform and faecal streptococci removals are as high as 99.99%. Intestinal pathogens are also removed completely.

Example 15.3. ***Design of oxidation pond***

Design an oxidation pond for the following data :

(*i*)	*Location*	*24°Latitude.*
(*ii*)	*Elevation*	*900 m above MSL*
(*iii*)	*Mean monthly temperature*	*30° max and 10° min.*
(*iv*)	*Population to be served*	*8000*
(*v*)	*Sewage flow*	*160 lpcd*
(*vi*)	*Desired effluent* BOD_5	*30 mg/l*
(*vii*)	*Pond removal constant at 20 °C*	*0.1/d*

Solution

1. ***Total BOD load***

BOD per capita/day $= (160 \times 300)\, 10^{-6} = 0.048$ kg/day

$\therefore$ Total BOD load $= 8000 \times 0.048 = 384$ kg/day

2. ***Permissible areal BOD loading***

Areal BOD loading at 24° latitude = 225 kg/ha/d

$$\text{Correction factor for elevation} = 1 + 0.003 \times \frac{900}{100} = 1.027$$

$$\text{Correction factor for sky clearance} = \frac{100}{100 + 3 \times \frac{15}{10}} = \frac{100}{104.5}$$

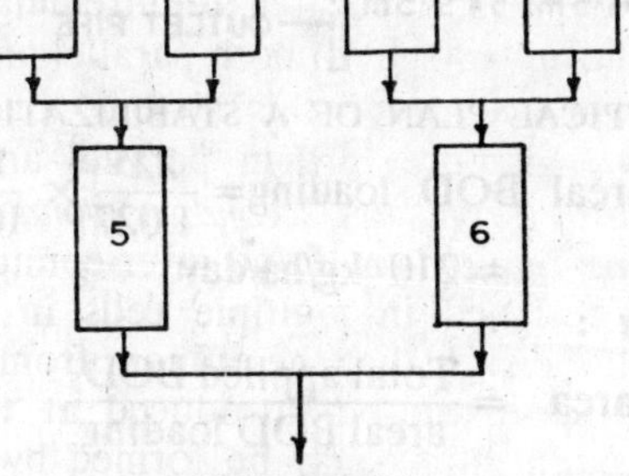

FIG. 15.11. LAYOUT PLAN OF STABILIZATION PONDS

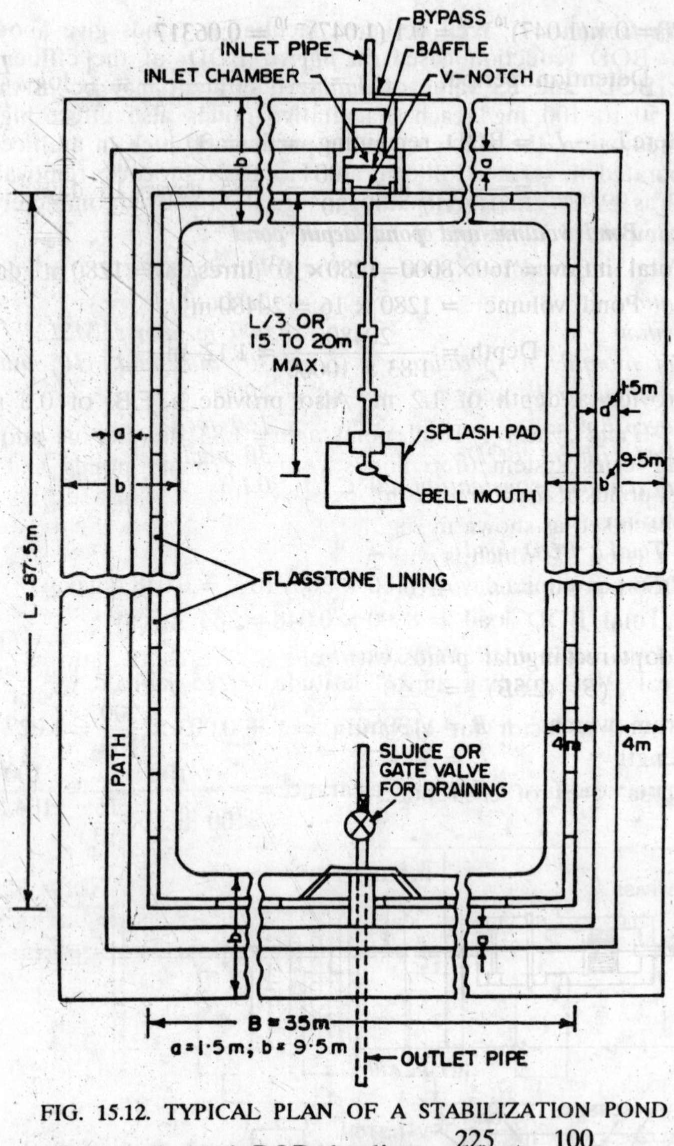

FIG. 15.12. TYPICAL PLAN OF A STABILIZATION POND

$\therefore$ Corrected areal BOD loading $= \dfrac{225}{1.027} \times \dfrac{100}{104.5}$

≈ 210 kg/ha/day.

3. Pond area :

$$\text{Pond area} = \frac{\text{Total applied BOD}_5}{\text{areal BOD loading}}$$

$$= 384/210 \approx 1.83 \text{ ha} = 1.83 \times 10^4 \text{ m}^2$$

4. Detention period. Pond removal constant at 10°C is

$P_{10} = 0.1\,(1.047)^{10-20} = 0.1\,(1.047)^{-10} = 0.06317$

$\therefore$ Detention period, $t = \dfrac{1}{P_T}\log_{10}\dfrac{L_a}{L_a - L_r} = \dfrac{1}{P_T}\log_{10}\dfrac{L_a}{L_e}$

Here $L_a - L_r$ = BOD remaining $= L_e = 30$

$\therefore$ $t = \dfrac{1}{0.06317}\log_{10}\dfrac{300}{30} = 15.85$ days ≈ 16 days

5. ***Pond volume and pond depth***

Total inflow $= 160 \times 8000 = 1280 \times 10^3$ litres/day $= 1280\ m^3/day$

$\therefore$ Pond volume $= 1280 \times 16 = 20480\ m^3$

$\therefore$ Depth $= \dfrac{20480}{1.83 \times 10000} \approx 1.12$ m.

Provide a depth of 1.2 m. Also provide a F.B. of 0.6 m.

6. ***Pond system*** : Total pond area = 1.83 ha. Let us adopt a parallel-series system of 6 ponds with 4 primary ponds and 2 secondary ponds of *equal area*, with 2 primary ponds feeding a secondary pond in each set, as shown in Fig. 15.11. This would give the primary pond area as 66.7% which is within the required range of 65 to 75% of the total pond area. Area of each pond = 1.83/6 = 0.305 ha. $= 3050\ m^2$

Adopt rectangular ponds with length to breadth ratio as 2.5

$\therefore$ $(B)\ (2.5B) = 3050$

From which $B = \sqrt{3050/2.5} \approx 35$ m

Length $L = 35 \times 2.5 = 87.5$ m.

Actual area of each pond $= 35 \times 87.5 = 3062.5\ m^2$

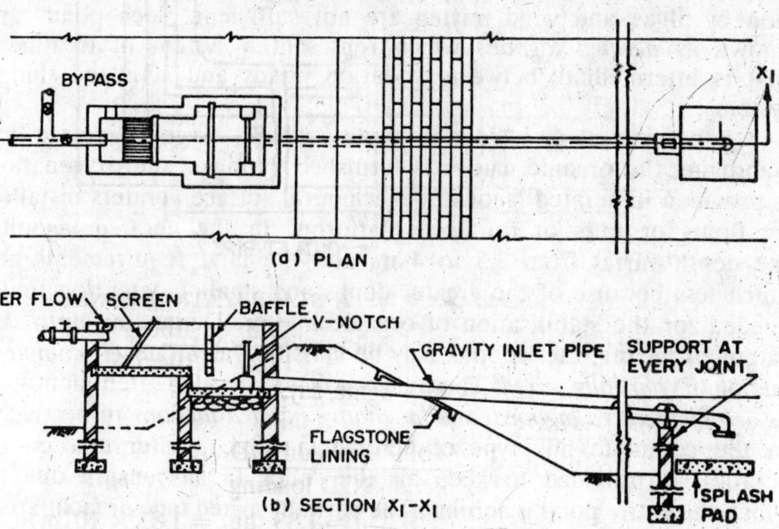

FIG. 15.13. INLET ARRANGEMENT

The details of the pond are shown in Fig. 15.12, 15.13 and 15.14.

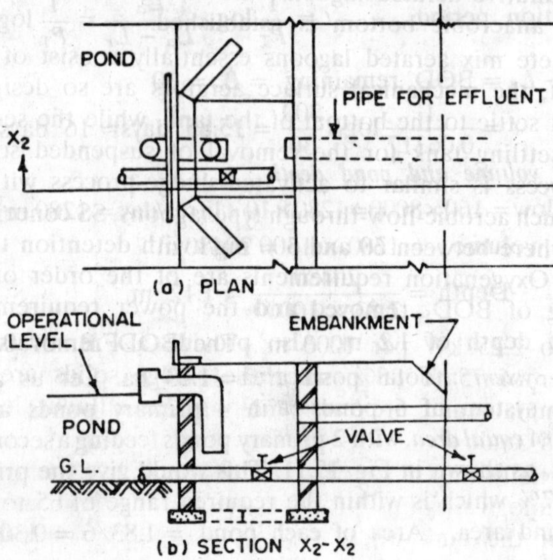

FIG. 15.14. OUTLET ARRANGEMENT

15.6. AERATED LAGOONS

Stabilization ponds overloaded due to industrial wastes or reduced temperature under ice cover often produce odours and insufficient BOD removals. Aeration and mixing provide distribution of dissolved oxygen for decomposition of organic matter when oxygenation by algae and wind mixing are not sufficient. Such ponds are known as *aerated lagoons* which represent a system of treatment that is intermediate between oxidation ponds and activated sludge systems.

In contrast to oxidation ponds, where oxygen required for stabilizing the organic matter is furnished by algae, the oxygenation is provided in aerated lagons by mechanical surface aeraters installed on floats or rafts or on fixed platforms. In the aerated lagoons, the depth varies from 2.5 to 4 m, and the land requirements are much less because of the greater dephs and smaller detention times needed for the stabilisation of organic matter. Depending upto the extent of mixing, the lagoons may be classified as either (*i*) *complete mix* or (*ii*) *partially mixed*. These lagoon types are also often identified as *aerobic areated lagoons*, and *facultative aerated lagoons* respectively. In the complete mix type of aerobic lagoons, greater amount of aeration is provided to keep all the solids in suspension due to which the entire pond is aerobic. The partially mixed type or facultative aerated lagoons are operated at a low rate of aeration not adequate to keep all the solids in suspension but enough to keep top layers aerobic. In a facultative aerated lagoons, the sewage solids tend to settle down and anaerobic bottom is established.

The complete mix aerated lagoons essentially consist of two units. In the first, the mechanical surface aerators are so designed that solids do not settle to the bottom of the tank, while the second unit is used as settling tank for the removal of suspended solids. As such, the process is similar to activated-sludge-process without recirculation. In such aerobic-flow-through type lagoons, SS concentration may be anywhere between 60 and 300 mg/l with detention times of 2 to 10 days. Oxygenation requirements are of the order of 0.7 to 1.3 kg per kg of BOD_5 removed and the power requirements vary from 0.75 to 2.25 kW per 1000 m^3. The BOD removals are only of the order of 75 to 85%, since the sludge solids are not allowed to accumulate and are carried along the effluent led to the sedimentation tank.

In the partially mixed system, a large portion of incoming solids and the biological solids produced within the lagoon settle to the bottom of the tank where anaerobic decomposition takes place. Therefore effluent from this type of tanks is more stable. In such facultative type lagoons, SS concentration varies from 30 to 150 mg/l with detention times ranging from 3 to 5 days. Oxygenation requirements are about 0.8 kg per kg of BOD_5 removed and the power requirements vary from 0.8 to 1.0 kW per 1000 m^3. The BOD removals are of the order of 75 to 90%, but there is need for removal of accumulated sludge after some years.

15.7. ROTATING BIOLOGICAL CONTACTORS

The rotating disc biological contactor (RBC) is one of the relatively new methods of biological treatment of wastes; the method falls under the attached-growth treatment processes. A rotating biological contactor consists of a series of closely spaced circular discs of polystyrene or polyvinyl chloride. The discs are about 3 m in diameter, 10 mm thick and placed 30 to 40 mm on centres along a shaft of variable length. The discs are immersed approximalety 40% in a contour-bottomed tank. The discs are spaced so that during submergence, wastewater can enter between the surfaces. The shaft rotates at 1 to 2 revolutions per minute. The discs provide a large surface area for biological growth. When rotated out of the tank, air enters the voids while liquid (*i.e.* wastewater) trickles out over the fixed films of biological growth attached to the disc media.

Alternating exposure to organics in the wastewater and oxygen in the air is similar to dosing a trickling filter with a revolving distributor. In operations, biological growths become attached to the

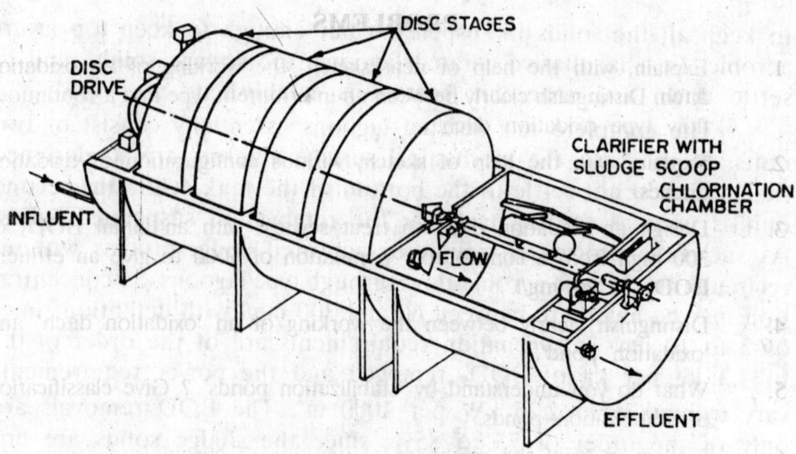

FIG. 15.15. ROTATING-DISC BIOLOGICAL CONTACTOR : PACKAGE PLANT

surfaces of the discs and eventually form a film or slime layer over the entire wetted surface area of the discs. This film adsorbs the organic pollutants during the submerged period of rotation cycle, and the major amount of oxygen transfer occurs when the bio-film is exposed to the atmosphere during the other half of the rotating cycle. However, the substrate utilisation within the microbial film is a continuous process. The disc rotation effects oxygen transfer and maintains the biomass in an aerobic condition. Excess microbial solids slough from the media and are carried out in the process effluent for gravity separation in a final clarifier. Due to the operational limitations of the number of discs on one shaft, *i.e.* in one unit, the process may consist of a number of *stages* in series. The number of *stages* or modules may vary from 2 to 6, depending on the characteristics of the waste and the effluet characteristics required.

The hydraulic loading rates are 0.04 to 0.06 m^3/day/m^2 and organic loading rates 0.05 to 0.06 kg BOD_5/m^2 per day based on disc surface area. Sloughing of biological solids is more or less continuous and the effluent contains a relatively constant concentration. The solids settle well and clarifier surface overflow rates of 33 m/day are reported to be satisfactory. Studies on full scale tests have indicated that bio-disc process provides stable operation under conditions of fluctuating hydraulic and organic loadings. BOD and suspended solids removals proved that it is possible to produce high-quality effluent from secondary treatment of municipal wastewater. Rotating biological contactors generally are more reliable than other fixed-film processes because of large amount of biological mass present (low operating F/M).

PROBLEMS

1. Explain, with the help of neat sketch, the working of an oxidation ditch. Distinguish clearly between an intermittent type and a continuous flow type oxidation ditch.
2. Explain, with the help of sketch, various configuration of oxidation ditches.
3. Design an oxidation ditch to treat sewage with an initial BOD_5 of 300 mg/l from a contributory population of 5000 to give an effluent BOD_5 of 20 mg/l.
4. Distinguish clearly between the working of an 'oxidation ditch' and 'oxidation pond'.
5. What do you understand by 'stabilization ponds' ? Give classification of stabilization ponds.
6. Explain in brief the principles of working of aerobic, anaerobic and facultative type stabilization ponds.
7. Explain the working of an aerobic pond (algae pond).
8. Explain the working of anaerobic type waste stabilization pond.
9. (*a*) Explain the mechanism of purification in facultative ponds.
   (*b*) Explain how do you determine the size and detention period for a facultative type stabilization pond.
10. Design an oxidation pond for treating domestic sewage of 1000 persons supplied with 200 litres per capita water per day. The BOD and the suspended solids are each of 300 mg/l.
    Permissible organic loading for the pond is not less than 500 kg/ha/day. The detention period is not to exceed 6 days. Assume the width of the pond to its length as 1:2 and the operational depth as 1.2. Assume any other suitable data. (UPSC, Engg. Services, 1975)
11. Design an oxidation pono based on the following given data :
    (*i*) Location .... 26° Latitude
    (*ii*) Elevation .... 100 m above sea level
    (*iii*) Mean monthly temperature .... 25° max. and 10° min.
    (*iv*) Population served 10000
    (*v*) Sewage flow .... 150 lpcd
    (*vi*) BOD_5 for raw sewage .... 300 mg/l.
    (*vii*) Desired effluent BOD_5 .... 30 mg/l.
    (*viii*) Sky clearance factor 0.60
    (*ix*) Per capita BOD contribution per day .... 0.045 kg/day.
    (*x*) Pond removal constant at 20° C 0.1/d

16

Treatment and Disposal of Sludge

16.1. INTRODUCTION

There are two end products obtained from various wastewater treatment plants : (*i*) effluent, and (*ii*) sludge. The treated effluent is directly discharged either in the receiving waters, or on land. However, the sludges are to be first processed before their final disposal. Objectives of processing sludge are to extract water from the solids and dispose the dewatered residue through a combination of physical, chemical and biological operations.

The constituents removed by a variety of methods in wastewater treatment plants are important by-producss of the treatment process. They include screenings, grit, scum and sludge. The term sludge is used to designate the solids that settle when sewage is passed through a settling tank. Because the various stages or methods of treatment yield sludges which have some what different characteristics, more specific terms are applied as follows : (*i*) *Raw* or *primary sludge* obtained from primary settling tanks (*ii*) *Secondary*, or *humus, sludge* obtained from secondary settling tanks of trickling filters, and (*iii*) *activated sludge* resulting from activated sludge process. Primary sludge is odoriferous and quickly putrescible. Secondary and activated sludges are less objectionable, but they will decompose quickly.

The sludge resulting from wastewater treatment processes and operation is usually in the form of a liquid or semi-solid liquid which typically contains from 0.25 to 12 percent solids. Sludge is loaded mostly with putrescible organic substances. The high organic content of the sludge demands futher treatment prior to its final disposal either on land or into the sea.

16.2. SLUDGE TREATMENT PROCESSES

Sludge treatment may include all or a combination of the following *unit operations* and processes : 1. Thickening or concentration. 2. Digestion. 3. Conditioning. 4. Dewatering. 5. Drying. 6. Incineration.

Fig. 16.1 shows the flow chart for sludge treatment and disposal. The arrows in the diagram indicate possible flow paths.

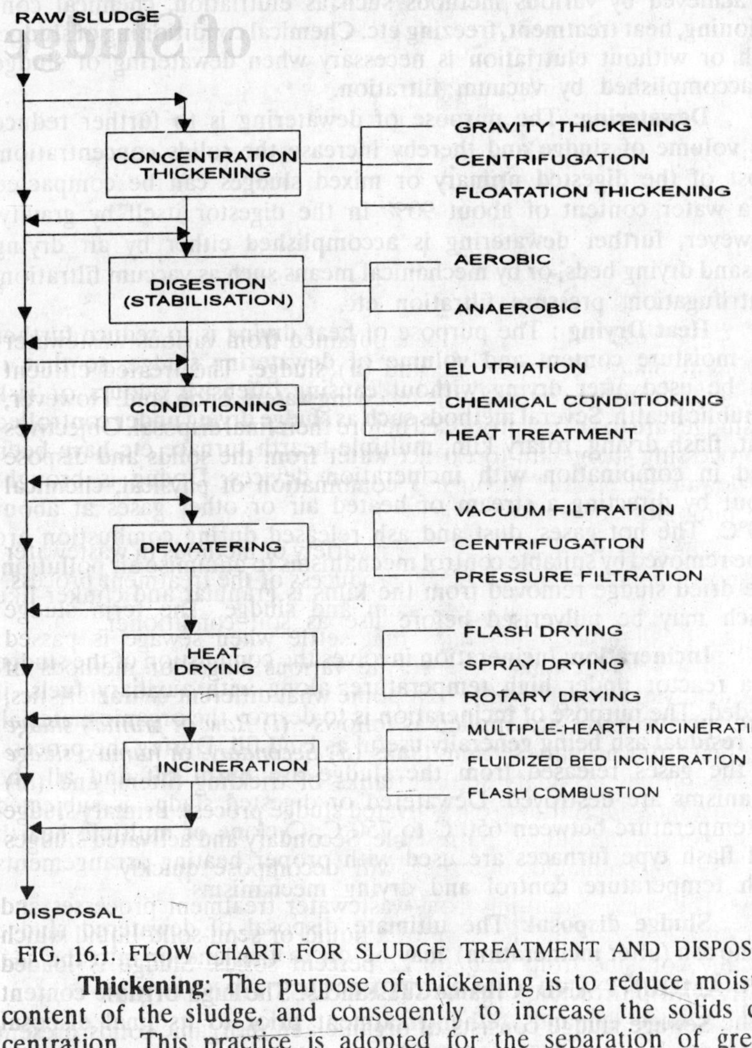

FIG. 16.1. FLOW CHART FOR SLUDGE TREATMENT AND DISPOSAL.

Thickening: The purpose of thickening is to reduce moisture content of the sludge, and conseqently to increase the solids concentration. This practice is adopted for the separation of greater amount of water from the sludge solids than can be attained in settling tanks. Three types of thickening commonly practised are (*i*) gravity thickening, (*ii*) air floatation, and (*iii*) centrifugation.

Digestion. The principal objective of sludge digestion is to subject the organic matter present in the settled sludge to anaerobic or aerobic decomposition so as to make it innocuous and amenable to dewatering on sand beds or mechanical filters before final disposal.

Conditioning: Conditioning improves the drainability of digested sludge. Prior conditioning of sludge before application of dewatering methods renders it more amenable to dewatering. Conditioning can be achieved by various methods such as elutriation, chemical conditioning, heat treatment, freezing etc. Chemical conditioning of sludge, with or without elutriation is necessary when dewatering of sludge is accomplished by vacuum filtration.

Dewatering: The purpose of dewatering is to further reduce the volume of sludge and thereby increase the solids concentration. Most of the digested primary or mixed sludges can be compacted to a water content of about 90% in the digestor itself by gravity. However, further dewatering is accomplished either by air drying on sand drying beds, or by mechanical means such as vaccum filtration, centrifugation, pressure filtration etc.

Heat Drying : The purpose of heat drying is to reduce further the moisture content and volume of dewatering sludge, so that it can be used after drying without causing offensive odours or risk to public health. Several methods such as sludge drying under controlled heat, flash drying, rotary kiln, multiple-hearth furnace etc have been used in combination with incineration devices. Drying is brought about by directing a stream of heated air or other gases at about 350°C. The hot gases, dust and ash released during combustion are to be removed by suitable control mechanisms to minimise air pollution. The dried sludge removed from the kilns is granular and clinker-like which may be pulverised before use as soil conditioner.

Incineration: Incineration involves the combustion of the sludge in a reactor under high temperature, along with auxiliary fuels, if needed. The purpose of incineration is to destroy the organic material, the residual ash being generally useful as land fill. During the process, all the gases released from the sludge are burnt off and all the organisms are destroyed. Dewatered or digested sludge is subjected to temperature between 650°C to 750°C. Cyclone or multiple hearth and flash type furnaces are used with proper heating arrangements with temperature control and drying mechanisms.

Sludge disposal: The ultimate disposal of dewatered sludge or the ash (after incineration) may be on to the land or into the sea.

16.3. AMOUNT AND CHARACTERISTICS OF SLUDGE

Sewage sludge consists of organic and inorganic solids present in the raw sewage and removed in the primary clarifier plus organic solids generated in secondary treatment and removed in secondary clarifier. The quantity and nature of sludge generated relates to character of the raw wastewater and processing units employed. The inorganic fraction of sludge has specific gravity of about 2.5 while organic matter has a specific gravity of 1.01 to 1.06 depending upon its source.

The characteristics of sludge are determined by various tests for the determination of specific gravity, moisture content, volatile and fixed solids, colour, odour etc. The moisture content of sludge is usually determined indirectly by evaporating a given weight of sludge on the water bath and drying the residue to constant weight in an oven at a temperature of about 110°C. The weight of residue gives the percentage of dry solids by weight and the difference in weights is the measure of moisture content. When the dry sludge, obtained above, is ignited, an ash-like residue is obtained which is also *'fixed residue'*, while the loss of weight on ignition represents the organic content (or volatile matter) of the dry solids.

The amount and characteristics of sludge produced in various treatment processes are briefly as follows :

1. Sludge from primary settling tanks : Primary sludge is a gray-coloured, greasy, odourous slurry of settleable solids accounting for about 50 – 60% of the SS applied, and tank skimmings. It contains faecal matter, vegetable parrings and fine silt. The sludge is sticky and does not drain easily. Primary precipitates can be dewatered readily after chemical conditioning because of fibrous and coarse nature. Typical solids concentration in raw primary sludge are 6 – 8%, and the portion of volatile solids varies from 60 to 80%.

2. Chemical precipitation (primary treatment) : It resembles in its general properties with the above except that the odour is less pronounced. It includes precipitated chemicals and entangled solids in more or less stoichiometric amount. It takes its colour from the coagulating chemicals used in the process and is slightly heavier also.

3. Trickling filter humus : This is dark brown in colour, flocculent and relatively less offensive when fresh. However, during unloading period of filters that are operated at low rates, humus will contain many dead worms and its odour becomes quite offensive. The suspended particles are fragments of biological growth washed from the filter media. Although they exhibit good settleability, the precipitate does not compact to a high density. For this reason, and the fact that sloughing is irregular, underflow from the final clarifier containing filter humus is returned to the wet well for mixing with the inflowing raw wastewater. Thus humus is settled with raw organics in the primary clarifier. The combined sludge has a solids content of 4 – 6% which is slightly thinner than residue with raw organics only.

4. Activated sludge: It is golden-brown, flocculent suspension of activate microbial masses, inoffensive when fresh, but turns septic because of its biological activity. It contains the largest percentage of moisture (about 98 – 99%) and is the most voluminous amongst the sludges. Thickens of return activated sludge is 0.5 – 2% suspended solids with a volatile fraction of 0.7 – 0.8. Waste activated sludge can be thickened effectively by floatation or centrifugation, though chemical addition may be needed to ensure high solids capture in the concentrating process.

Volume of solids

The volume of daily sludge varies depending upon the degree of removal of suspended solids in primary and final settling tanks, moisture content and specific gravity of the sludge. In the case of sludge continuously desludged from primary sedimentation tanks, solids do not generally exceed 4 to 5% while in secondary tanks, they range between 0.5 to 1%. When the secondary sludge is combined with the primary sludge, the solids in the mixed sludge range between 2.5 to 5%. The solids content of digested sludge is usually in the range of 6 to 13% when reduction in bulk takes place due to separation of supernatant.

The quantity of SS in raw sewage is typically 90 g/day per capita and its concentration varies from 200 to 250 mg/l depending upon per capita sewage flow. Out of these suspended solids, about 60% solids are removed in primary clarifier while the remaining SS are either oxidized in secondary clarifier or incorporated in the biological mass. The *solids generated* in the biological processes are about 0.4 to 0.5 kg of applied BOD_5 in fixed growth processes and 0.2 to 1.0 kg of applied BOD_5 in suspended growth process depending upon the sludge age.

Typical figures of solids are given in Table 16.1

Volume-mass relationships

If solid matter in a sludge is composed of fixed (or mineral) solids and volatile (or organics) solids, the specific gravity of dry solids matter can be computed from the following expression.

$$\frac{W_s}{S_s} = \frac{W_f}{S_f} + \frac{W_v}{S_v} \qquad \text{...(16.1 } a)$$

$$\text{or} \quad \frac{100}{S_s} = \frac{\%\text{ of mineral matter}}{\text{sp. gr. of mineral matter}} + \frac{\%\text{ of volatile matter}}{\text{sp. gr. of organic matter}} \qquad \text{...(16.1)}$$

where W_s = weight of solids

S_s = specific gravity of solids

W_f = weight of fixed solids (mineral matter)

S_f = specific gravity of mineral solids (which varies from 2.4 to 2.65)

W_v = weight of volatile matter.

S_v = specific gravity of volatile (or organic) matter (which varies from 1.0 to 1.2)

TABLE 16.1 SOLIDS IN SLUDGES (Per capita per day)

Treatment Process	Total (gm)	Volatile (g)	Fixed (g)	Sp. gr. of dry solids	% solids in wet sludge
(1)	(2)	(3)	(4)	(5)	(6)
Raw sewage					
1. Total solids	90	63	27	1.22	
2. Settleable solids (60%)	54	38	16	1.22	
3. Non-settleable solids (40%)	36	25	11	1.22	
I *Primary (Pr.) sedimentation and digestion*					
1. Solids removed as fresh sludge	54	38	16	1.22	4–5
2. Solids digested*	–	–25	+6		
3. Solids remaining in Pr. digested sludge	35	13	22	1.6	10–13
II *Activated sludge Process* (AS)					
1. Non-settleable solids affected	36	25	11	1.22	
2. Solids digested during activation (10%)	–	–2.5	+0.5		
3. Solids to be removed	34	22.5	11.5		
4. Solids removed as fresh activated sludge (90%)	31	20	11	1.25	
5. Combined primary (I) & secondary excess A.S. (II)	54+31 =85	38+20 =58	16+11 =27	1.22	2.5–4
6. Solids digested*	–	–38	+10		
7. Solids remaining in digested combined Pr. & Excess As	57	20	37	1.65	6–7
III *Trickling filter (T.F.) Process*					
1. Non-settleable solids	36	25	11	1.22	
2. Solids digested during filtration (10%)	–	–2.5	+0.5		
3. Solids to be removed in sedimentation tank	34	22.5	11.5		
4. Solids to be removed in S.S. Tank (40%)	14	9	5		
5. Combined Pr. (I) & T.F. humus (III)	68	47	21	1.22	4–5
6. Solids digested in digester*	–	–32	+8		
7. Solids remaining in digested Pr. & T.F. humus	44	15	29	1.66	8–10

(*i*) *It is assumed that 2/3 of the volatile matter is destroyed during digestion and one fourth of the digested matter is converted to non-volatile residual sludge.

(*ii*) Percentage volatile and non-volatile matter is assumed to be 70% and 30% respectively.

(*iii*) Sp. gravity of volatile and non-volatile matter is assumed as 1.0 and 2.5 respectively.

With the known percentage solids or % moisture in wet sludge, the specific gravity of wet sludge (S_{sl}) could be worked out from the relationship :

$$\frac{100}{S_{sl}} = \frac{\%\text{ moisture}}{\text{sp. gr. of water}} + \frac{\%\text{ of solids}}{S_s} \quad ...(16.2)$$

In order to illustrate the method of computation, let us assume that 30% of the solids matter in a sludge containing 90% water is composed of fixed mineral solids with a specific gravity of 2.5, and 70% is composed of volatile solids with a specific gravity of 1.0, then the specific gravity of solids is given by

$$\frac{100}{S_s} = \frac{30}{2.5} + \frac{70}{1} = 82 \quad ...(\text{Eq. } 16.1\ a)$$

From which $\quad S_s = \dfrac{100}{82} \approx 1.22$

Taking specific gravity of water as 1.0, the specific gravity of sludge is given by

$$\frac{100}{S_{sl}} = \frac{90}{1} + \frac{10}{1.2} = 98.2 \quad ...(16.2)$$

From which $\quad S_{sl} = \dfrac{100}{98.2} = 1.018 \approx 1.02$

Once the specific gravity of wet sludge and % of solids in wet sludge are known, volume of wet sludge can be determined both for raw as well as digested sludges, using the relationship :

$$V_{sl} = \frac{W_s}{\rho_w S_{sl} p_s} \quad ...(16.3)$$

where $\quad V_{sl}$ = volume of sludge
W_s = weight of dry solids (kg)
S_{sl} = specific gravity of sludge
ρ_s = percent solids expressed as a decimal
ρ_w = density of water (10^3 kg/m^3 at 5°C)

In order to calculate the solids content *approximately*, the following expression can be used, remembering that volume of sludge varies inversely with the % of solid matter contained in the sludge:

$$\frac{V_{sl1}}{V_{sl2}} \approx \frac{P_2}{P_1} \quad ...(16.4)$$

where $\quad V_{sl1}, V_{sl2}$ = sludge volumes
P_1, P_2 = percent of solid matter

Variation of volume of sludge with its moisture content

Let V_{sl1} be the volume of a given slude at its moistre content m_1. Its volume V_{sl2} at the changed moisture content (m_2) can be easily found as under, noting that the weight of solids (W_s) remains the same.

At any moisture content m (percent), weight of solids will be $(100 - m)\%$. Hence corresponding to weight of solids W_s, the weight of wet slude will be equal to $\frac{100\,W_s}{100 - m}$. If ρ_{sl} is the unit weight of sludge, volume (V_{sl}) of sludge is given by.

$$V_{sl} = \frac{100\,W_s}{100 - m} \cdot \frac{1}{\rho_{sl}} \qquad ...(16.5)$$

Hence corresponding to moisture contents of m_1 and m_2, the volumes of the sludge are given by

$$V_{sl1} = \frac{100\,W_s}{100 - m_1} \cdot \frac{1}{\rho_{sl}}$$

and
$$V_{sl2} = \frac{100\,W_s}{100 - m_2} \cdot \frac{1}{\rho_{sl}}$$

From which,
$$\frac{V_{sl1}}{V_{sl2}} = \frac{100 - m_2}{100 - m_1} \qquad ...(16.6)$$

Comparing Eqs. 16.4 and 16.6, we note that

$$P_2 \approx 100 - m_2$$

and
$$P_1 \approx 100 - m_1$$

Example 16.1. *A trickling filter plant treats 1200 m^3/day of sewage with a BOD_5 of 220 mg/l and SS of 250 mg/l. Estimate the total solids production, assuming that primary clarification removes 30 percent of BOD and 60% of influent solids.*

Solution. Removal of solids in primary clarification :

$$= 0.6\,(250) = 150 \text{ mg/l} \qquad ...(i)$$

BOD of influent entering trickling filters

$$= (1 - 0.3) \times 220 = 154 \text{ mg/l}$$

Assuming that solids production in the trickling filter is @ 0.5 kg/kg of applied BOD, production of solids in the secondary

$$= 0.5 \times 154 = 77 \text{ mg/l} \qquad ...(ii)$$

$\therefore$ Total solids production $= 150 + 77 = 227$ mg/l/day

$$= \frac{227\,(1200 \times 10^3)}{10^3} = \mathbf{272.4 \text{ kg/day}}$$

Example 16.2. *A sedimentation tank treats 6 mLd containing 250 mg/l of suspended solids. The tank removes 60% of the suspended solids. Compute the weight and volume of the sludge produced daily if the moisture content of sludge is (a) 97.5%, (b) 95%.*

Solution

Weight of solids in sewage $= \frac{(6 \times 10^6) \times 250}{10^6} = 1500$ kg/day

Since only 60% of the influent solids are removed, weight of solids removed in sedimentation tank $= 0.6 \times 1500 = 900$ kg/day.

Hence $W_s = 900$ kg/day

Now volume of sludge is given by

$$V_{sl} = \frac{W_s}{\rho_w \cdot S_{sl} \cdot p_s} \qquad \text{...(16.3)}$$

where $\rho_w = 1000\ \text{kg/m}^3$ and $S_{sl} = 1.02$,

(*a*) *At mositure content* $(m) = 97.5$

$$p_s = 1 - 0.975 = 0.025$$

$$V_{sl} = \frac{900}{1000 \times 1.02 \times 0.025} \approx \mathbf{35.3\ m^3}$$

Weight of sludge, W_{sl} = Vol. × unit weight of sludge

$$= 35.3\,(1.02 \times 1000) = \mathbf{36000\ kg}$$

(*b*) *At moisture content of 95%*

$$p_s = 1 - 0.95 = 0.05$$

$$V_{sl} = \frac{900}{1000 \times 1.02 \times 0.05} = \mathbf{17.65\ m^3}$$

and $W_{sl} = 17.65\,(1.02 \times 1000) = \mathbf{18000\ kg}$

The above results show that the volume is reduced to half when its moisture content is lowered from 97.5% to 95%.

Example 16.3. *Determine the liquid volume before and after digestion and percentage reduction for 600 kg (dry basis) of primary sludge having the following characteristics.*

	Primary	*Digested*
(*i*) *Solids* (%)	*6*	*12*
(*ii*) *Volatile matter* (%)	*65*	*65 (destroyed)*
(*iii*) *Specific gravity of fixed solids*	*2.5*	*2.5*
(*iv*) *Specific gravity of volatile solids*	≈ *1.0*	≈*1.0*

Solution

1. ***Computation of average specific gravity of all the solids in primary sludge***

$$\frac{100}{S_s} = \frac{35}{2.5} + \frac{65}{1.0} \qquad \text{...(16.1)}$$

From which $S_s = \mathbf{1.266}$ (primary solids)

2. ***Computation of specific gravity of primary sludge*** (Eq. 16.2)

$$\frac{100}{S_{sl}} = \frac{6}{1.266} + \frac{94}{1}$$

From which $S_{sl} = 1.013$

3. ***Computations of volume of primary sludge*** (Eq. 16.3)

$$V_{sl} = \frac{W_s}{\rho_w\, S_{sl}\, p_s} = \frac{600}{1000 \times 1.013 \times 0.06}$$

$$= \mathbf{9.874\ m^3}$$

4. *Computation of % volatile matter after digestion*

Fixed matter in primary sludge $= 0.35 \times 600 = 210$ kg

Volatile matter in primary sludge $= 0.65 \times 600 = 390$ kg.

Volatile matter after digestion $= 0.35 \times 390$ (since 65% of 390 kg has been destroyed in digestion)

$\therefore$ Total matter after digestion $= 210 + 0.35\,(390)$

$$\%\ \text{Volatile matter} = \frac{0.35\,(390)}{210 + 0.35\,(390)} \times 100 = 39.39$$

5. *Computation of average specific gravity of all the solids in digested sludge*

$$\frac{100}{S_s} = \frac{60.61}{2.5} + \frac{39.39}{1} \qquad ...(16.1)$$

From which $S_s = 1.571$ (digested solids)

6. *Computation of specific gravity of digested sludge* (Eq. 16.2)

$$\frac{100}{S_{dsl}} = \frac{12}{1.571} + \frac{88}{1}$$

From which $S_{dsl} = 1.046$

7. *Computation of volume of digested sludge* (Eq. 16.3)

$$V_{dsl} = \frac{W_{ds}}{\rho_w\, S_{dsl} \cdot p_s}$$

where $W_{ds} = 200 + 0.35\,(390) = 336.5$ kg

$$\therefore \quad V_{dsl} = \frac{336.5}{1000 \times 1.046 \times 0.12} = \mathbf{2.681\ m^3}$$

8. *Reduction of sludge volume after digestion*

$$\%\ \text{reduction} = \frac{9.874 - 2.681}{9.874} \times 100 = \mathbf{72.85\%}$$

Example 16.4. *A waste water plant produces 800 kg of dry solids per day at a moisture content of 96 percent. The solids are 70 percent volatile with a specific gravity of 1.05 and 30 percent non-volatile with a specific gravity of 2.5. Determine the sludge volume:*

(a) after digestion which reduces volatile solids content by 50 percent and decreases the moisture content to 92 percent.

(c) after dewatering to 70% moisture

(d) after drying to 10 percent moisture

(e) after incineration, when only nonvolatile solids remain.

Solution : This example is being solved by a slightly shorter but direct method.

(a) Raw sewage

Total mass of dry solids $= 800$ kg

Moisture content $= 96\%$

$\therefore$ Solids content $= 4\%$

$\therefore$ Mass of sludge $= 800/0.04 = 20000$ kg

Out of this, mass of water= 0.96 × 20000 = 19200 kg

∴ Volume of water = 19200 l ...(*i*)

Mass of volatile solids = 0.7 × 800 = 560 kg

∴ Volume of volatile solids = 560/1.05 = 533.3 l ...(*ii*)

Mass of fixed solids = 0.3 × 800 = 240 kg

∴ Volume of fixed solids 240/2.50 = 96 l ...(*iii*)

Hence total volume= 19200 + 533.3 + 96 = **19828.3 l**

Alternatively, from Eq. 16.1

$$\frac{100}{S_s} = \frac{30}{2.5} + \frac{70}{1.05}$$

or $$S_s = 1.271$$

From Eq. 16.2, $$\frac{100}{S_{sl}} = \frac{4}{1.271} + \frac{96}{1}$$

or $$S_{sl} = 1.0086$$

Hence from Eq. 16.3, $$V_{sl} = \frac{W_s}{\rho_w S_{sl} p_s} = \frac{800}{1000 \times 1.0086 \times 0.04}$$

$$= 19.829 \text{ m}^3 = 19829 \text{ l}$$

which is the same as the one obtained by the direct method.

(b) Volume after digestion

After digestion, volatile solids are reduced by 50%. Hence mass of volatile solids= 0.5 × 560 = 280 kg

Mass of fixed solids= 240 kg (as before)

∴ Total mass of solids= 280 + 240 = 520 kg

Water content = 92 %

∴ Total mass of sludge $= \dfrac{520}{0.08} = 6500$ kg

Mass of water in the sludge= 6500 − 520 = 5980 kg

Hence volume of water = 5980 l

Volume of volatile solids = 280/1.05 = 266.7 l

Volume of fixed solids = 96 l (as before)

∴ Total volume of sludge = 5980 + 266.7 + 96 = **6342.7 l**

(c) Volume after dewatering

Mass of total solids = 520 kg (as before)

Water content = 70%

∴ Total mass of sludge $= \dfrac{520}{0.3} = 1733.3$ kg

Mass of water in the sludge = 1733.3 − 520 = 1213.3 kg

∴ Volume of water = 1213.3 l

Volume of volatile and fixed solids are the same as before.

∴ Total volume = 1213.3 + 266.7 + 96 = **1576 l**

(d) Volume after drying

Mass of total solids = 520 kg (as before)

Moisture content after drying = 10%

∴ Total mass of sludge = 520/0.9 = 577.8 kg

Hence mass of water in sludge = 577.8 − 520 = 57.8 kg

∴ Volume of water = **57.8 l**

Hence total volume = 57.8 + 266.7 + 96 = **420.5 *l***

(e) After incineration

After incineration, only non-volatile (*i.e.* fixed) solids remain. Total mass of non-volatile solids is 240 kg and its volume is 96 l. Hence volume after incineration is ***96 l***

Summary: In terms of percentage of original volume, the volumes computed above are (*a*) 100%, (*b*) 32% (c) 7.9% (*d*) 2.1% and (*e*) 0.5%, respectively.

16.4. SLUDGE THICKENING OR CONCENTRATION

Thickening is a procedure used to increase the solids content of sludge by removing a portion of the liquid fraction. This volume reduction, obtained by sludge thickening or sludge concentration, is done for the following purposes: (*i*) to permit increased loadings to sludge digesters (*ii*) to increase feed solids concentration to vacuum filters, (*iii*) to economise on transport costs as in ocean barging in case of raw sludges, (*iv*) to minimise the land requirements as well as handling costs when digested sludge has to be transported to disposal site, and (*v*) to a save on the auxiliary fuel that may otherwise be needed when incineration of sludge is practised. As an example, if waste activated sludge, which is typically pumped from secondary settling tanks with a content of 0.8 percent solids can be thickened to a content of 4 percent solids, then a five fold decrease in sludge volume is achieved, as is evident from Eq. 16.4, where $P_1/P_2 = 0.8/4 = 1/5$.

Sludge thickening is commonly achieved by the following three methods :

(*i*) Gravity thickening

(*ii*) Air flotation

and (*iii*) Centrifugation

1. Gravity thickening. Gravity thickening is the most common practice for concentration of sludges. Gravity thickening is the simplest and the least expensive. This is adopted for primary sludge or combined primary and activated sludge but is not successful in dealing with activated sludge independently. Gravity thickening of combined sludge is not effective when the activated sludge exceeds 40% of the total sludge weight, and other methods of thickening of activated sludge have to be considered.

Fig. 16.2 shows a typical waste sludge thickener. The tank resembles a circular clarifier except that the depth/diameter ratio is greater and the hoppered bottom has a steeper slope. Dilute sludge is fed to a center feed well. The feed sludge is allowed to settle and compact, and the thickened sludge is withdrawn from the bottom of the tank. A bridge fastened to the tank walls supports a truss-type scraper arm mounted on a pipe shaft equipped with a power lift device thereby opening up channels for water to escape and promoting densification. Also, the slow motion of the scraper dislodges gas bubbles, prevents bridging of solids and moves the sludge towards a central well for withdrawal. The continous supernatant flow that results is returned to the primary settling tank. Feed is provided continuously while the under flow may be extracted intermittently for further processing.

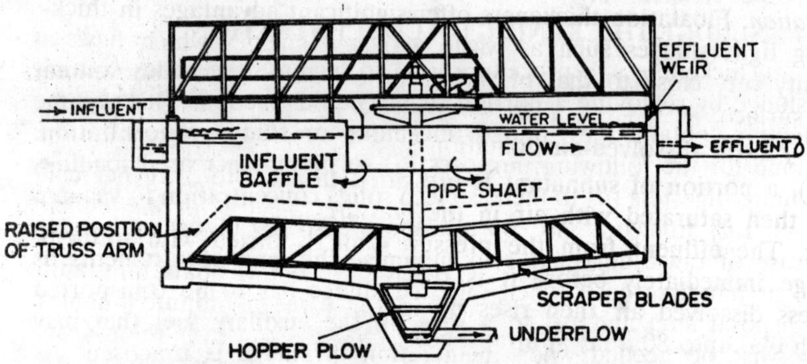

FIG. 16.2. GRAVITY SLUDGE THICKENER

Gravity thickeners are either *'continuous flow'* or *'fill and draw type'*. Continuous flow tanks are designed for a hydraulic loading likely to given rise to odour problems. The normal sludges contain too much with the plant effluent. The recommended solids loading for primary, and activated-sludge (60 : 40 weight ratio) are 100, 60 and 40 kg/day/m^2 respectively. Underflow solids concentration that can be expected with these loadings are about 8 to 10% with primary, 6 to 8% with primary and trickling filters and 2 to 6 % with primary and activated sludge. Better efficiency can be obtained by providing slow revolving stirrers, particularly with gassy sludges.

The continuous flow type circular thickeners have a side water depth of 3 m. Three settling zones occur in a thickener : (*i*) clear supernatant on top, (*ii*) feed zone characterised by *hindered settling* and (*iii*) Compression zone near the bottom where consolidation occurs. Settling data may be collected from batch type laboratory tests conducted on 200 mm dia. cylinder with initial height of atleast 900 mm. Evaluating the performance of a thickener often involves mass-balance calculations.

In operation, a *sludge blanket* is maintained on the bottom of the thickener to aid in concentrating the sludge. Concentration of the underflow solids is governed by the depth of sludge blanket upto 1m, beyond which there is very little influence of the blanket. Underflow solids concentration is increased with increased sludge detention time, 24 hours being required to achieve maximum compaction. Sludge blanket depths may be varied with fluctuation in solids production to achieve good compaction. During peak conditions, lesser detention times will have to be adopted to keep the sludge blanket depth sufficiently below the overflow weirs to prevent excessive solids carryover.

2. Floatation thickening. *Air floatation units* employ floatation of sludge by air under pressure or vacuum. The former process more commonly used, is known as *dissolved air floatation* or *pressure type floatation*. Floatation thickeners offer significant advantages in thickening light sludges such as waste activated sludges which have a density very close to that of water and are thus readily buoyed to the surface.

In the dissolved-air floatation unit (or pressure type floatation unit), a portion of subnatant is pressured from 3 to 5 kg per cm^2 and then saturated with air in the *pressure tank* or *air dissolution tank*. The effluent from the pressure tank is mixed with influent sludge immediately before it is released into the floatation tank. Excess dissolved air then rises up in the form of minute bubbles (with dia. upto 80 μ m) at atmospheric pressure attaching themselves to particles which form the sludge blanket. Thickened blanket (of

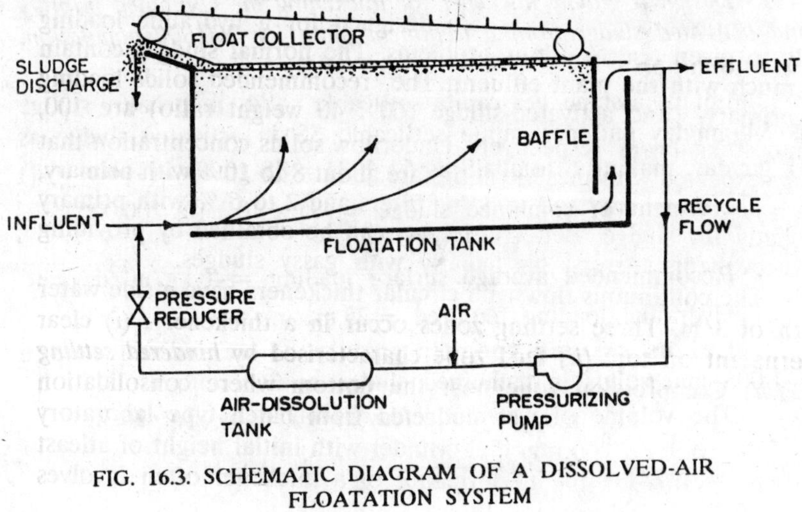

FIG. 16.3. SCHEMATIC DIAGRAM OF A DISSOLVED-AIR FLOATATION SYSTEM

thickness varying from 0.2 m to 0.6 m) is skimmed off from the surface, while the unrecycled subnatant is returned to the plant. Effluent is recycled at a rate of 30–150% of the influent flow through the air-dissolution tank to the feed inlet. Recycle ratio is interelated to with (*i*) feed solids concentration, (*ii*) detention time, and (*iii*) air/solids ratio. Detention time in the floatation unit is not critical, provided that particles rise rapidly enough and the horizontal velocity does not scour the bottom of the sludge *blanket.* An air/solids ratio of 0.01 to 0.03 is *sufficient* to achieve acceptable thickening of waste-activated sludge. Surface overflow rates in floatation thickeners vary from 10 to 45 m/day at retention time of 30 min. to 1 hour. The efficiency of air floatation units is increased by the addition of chemicals like alum and polyelectrolytes. The addition of polyectrolytes does not increase the solids concentration but improves the solids capture from 90 to 98%.

3. Centrifugal thickening. Centrifuges are used both to thicken and to dewater sludges. Their application in thickening is normally limited to waste activated sludge. Thickening by centrifugation involves the settling of sluge particles under the influence of centrifugal forces. Thickening by centifugation is resorted to only when the space limitation or sludge characteristics will not permit the adoption of the other two methods mentioned above. This method involves high maintenance and power costs. The three basic types of centrifuges currently available for sludge thickening are (*i*) nozzle disc, (*ii*) solids bowl and (*iii*) basket centrifuges. Disc centrifuges are prone to clogging while the latter types gives poorer quality effluent.

Example 16.5. Design of gravity thickener

Design a gravity thickener for thickening the combined primary and activated sludge from a treatment plant for 200,000 population.

Solution

Let us assume per capita settleable solids in primary sludge as 54 gm/day and per capita settleable SS in activated sludge as 31 gm/day, making a total SS as 54 + 31 = 85 gm/day (Table 16.1)

$\therefore$ Weight of combined sludge $= (85 \times 10^{-3}) \times 200{,}000$
$= 17000$ kg/day

Recommended average surface loading = 40 kg/m^2/day

Hydraulic loading required = 25 m^3/m^2/day

Let us assume specific gravity of wet mixed sludge = 1.008 and solids in combined sludge as 3%.

The volume of wet sludge/day is given by Eq. 16.3.

$$V_{sl} = \frac{W_s}{\rho_w S_{sl} p_s}$$

$$= \frac{17000}{1000 \times 1.008 \times 0.03} \approx 562 \text{ m}^3/\text{d}$$

$$\text{Surface area needed} = \frac{17000}{40} = 425 \text{ m}^2$$

Flow needed from giving a hydraulic loading of 25 $m^3/m^2/day$

$$= 25 \times 425 = 10625 \text{ m}^3/\text{m}^2/\text{day}$$

$\therefore$ Balance of $10625 - 562 = 10063$ m^3/day is made available by blending with primary or secondary effluent.

Let us provide a side water depth of 3 m.

$\therefore$ Sludge detention period $= V/Q$

$$= \frac{425 \times 3}{562} \times 24 = 54.5 \text{ hours.}$$

This is more than 24 hours, and hence ok.

Let us provide a circular sludge blanket type thickener.

$$\text{Dia. of tank} = \sqrt{\frac{425 \times 4}{\pi}} = 23.26 \text{ m.}$$

Let us provide a 24 m dia. tank

Sludge blanket restricted to 1 m is adopted.

Expected solids in the thickened sludge = 6%.

16.5. ANAEROBIC DIGESTION

Sludge digestion is a biochemical phenomenon involving organisms, enzymes, food and environment. The principal *objective* of sludge digestion is to subject the organic matter present in the settled sludge of the primary and final sedimentation tanks to *anaerobic* or *aerobic* decomposition so as to make it innocuous and amenamable to dewatering on sand beds or mechanical filters before final disposal on land, lagoon or sea. Regardless of the process used, sludge digestion brings about a reduction in its volume. While anaerobic digestion of sludge produces gas which can be utilised wherever feasible, aerobic digestion does not produce any utilisable byproduct other than well stabilized sludge.

Anaerobic digestion is the biological decomposition of organic matter in absence of oxygen. Anaerobic digestion consists of two distinct stages which occurs simultaneously in digesting sluge (Fig. 16.4).

The *first stage*, known as *acid fermentation*, consists of hydrolysis and liquefaction of high molecular-weight organic compounds and conversion to organic acids by *acid forming bacteria*. Acitic acid, propionic acid and butyric acid are the most common end products in the first stage. The second stage known as *methane fermentation*,

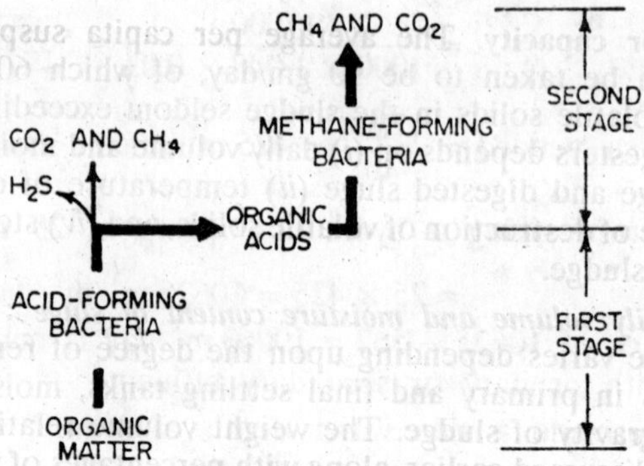

FIG. 16.4.. STAGES IN ANAEROBIC-DIGESTION.

is gassification of the organic acids to methane and carbon dioxide by *acid splitting methane-forming bacteria.* For proper functioning of the process, it is necessary that a good balance between the reactions of the two stages is maintained so that there is no accumulation of excess volatile acids, normal operating values being 200 to 400 mg/l with a maximum of 2000 mg/l. This sequence of reactions brings about a *reduction of volatile solids* present in the sludge thereby permitting easy dewatering on a sand bed or a vacuum filter.

Due to very low *microbial growth rate*, the production of biological sludge in the anaerobe processes is very low. Moreover, a major portion of organic waste is converted into gas. Hence the effluent of an anaerobic process is much more stabilized than from the aerobic process. It should be noted that not all the organic matter in sludge is stabilized under anaerobic condition, since some of the constituents are so complex that they resist biological decomposition. The objective of sludge digestion is therefore, to stabilize only the biodegradable matter that resist dewatering either on sand beds or in mechanical filters.

Types of digesters. Sludge digestors can be of two types : 1. Conventional or low rate digestor, and 2. High rate digestor.

The *conventional* sludge digestion is carried out in a single stage, where digestion, sludge thickening and supernatant formation takes place simultaneously in the same unit. A conventional digestion unit may be *single stage* or in two stages. *In two stage process*, two tanks are provided– first tank meant for digestion and the second stage is used for storage, thickening and supernatant formation. The *high rate digestion* process differs from the conventional single-stage process in that the solids loading rate is much greater, the sludge is intimately mixed and it is heated to achieve optimum digestion rates.

Digestor capacity. The average per capita suspended solids in sewage can be taken to be 90 gm/day, of which 60% settles as sludge with volatile solids in the sludge seldom exceeding 70%. The capacity of digesters depends on (*i*) daily volume and moisture content of input sludge and digested sluge (*ii*) temperature of digestion (*iii*) desired degree of destruction of volatile solids, and (*iv*) storage capacity for digested sludge.

1. Daily volume and moisture content of sluge : The volume of daily sludge varies depending upon the degree of removal of suspended solids in primary and final settling tanks, moisture content and specific gravity of sludge. The weight volume relationships have already been discussed earlier, along with percentages of various solids (Table 16.1).

2. Temperature of digestion. Digestion of sludge is temperature dependent. Table 16.2 gives the time of digestion varying with temperature during digestion. The same is depicted in Fig. 16.5.

TABLE 16.2. VARIATION OF DIGESTION WITH TEMPERATURE

Temperature in °C	*Mesophilic*							*Thermophilic*			
	10	15	20	25	30	35	40	45	50	55	60
Digestion period (days)	75	56	42	33	27	24	25	23	16	14	18

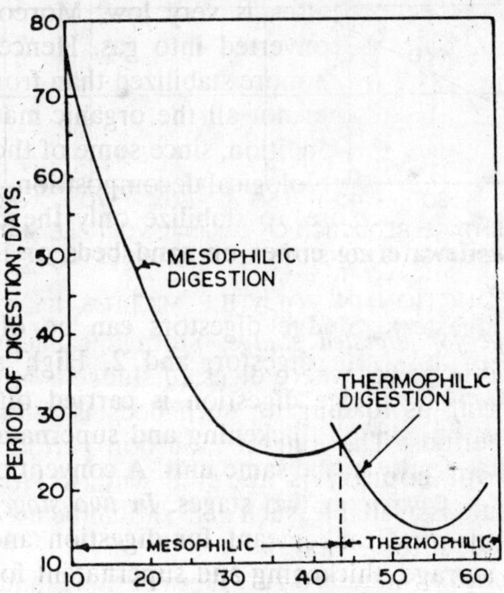

FIG. 16.5. EFFECT OF TEMPERATURE ON DIGESTION PERIOD

From Table 16.2 and Fig. 16.5, it is clear that rate of digestion increases with increase in temperature. Upto a temperature of 40°C the digestion is brought about by a particular type of organisms while a totally different type of oganisms establish in the digester at temperatures higher than 45°C. The former range is called the *mesophilic range* and the latter is known as *thermophilic range*. Normally, digesters are not operated in the thermophilic range as it would warrant special heating and also is beset with operational and maintenance problems. In the mesophilic zone, the range of 25° to 40° gives reasonably short detention period.

3. Desired degree of destruction of volatile solids. The percent volatile matter destroyed is dependent upon the percentage volatile matter in raw sludge. The detention period given in Table 16.2 would accomplish about two-thirds destruction of volatile solids for a 70% volatile solids normally encountered in raw sludge. The different values of volatile solids *destruction* when digestion is considered satisfactory are given in Fig. 16.6.

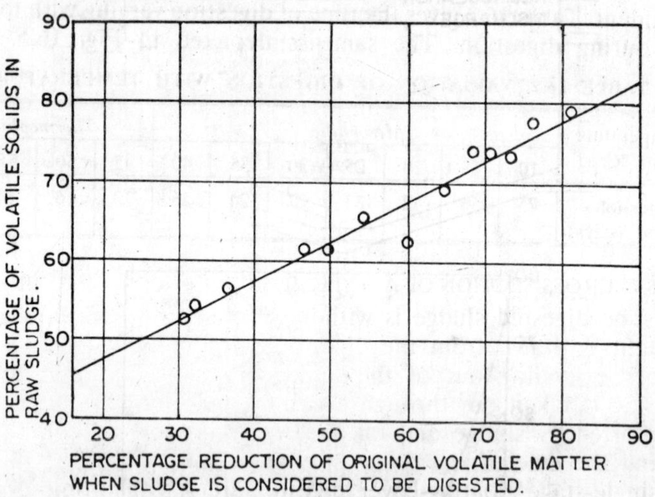

FIG. 16.6. REDUCTION IN VOLATILE MATTER BY DIGESTION.

4. Storage for digested sludge. Storage capacity for digested sludge is required in places where digested sludge is applied to drying beds for dewatering as loading on such beds is interrupted during peak monsoon periods. The longest monsoon period which may be anticipated in our country is about 8 weeks. However, where meteorological data is available, such data should be used to determine the capacity needed for storage.

Conventional or standard-rate digesters. Standard-rate (conventional) sludge digestion is usually carried out as a single-stage process. The functions of digestion, sludge thickening, and supernatant formation are carried out simultaneously. The cross-section of a typical floating-cover digestion tank is shown in Fig. 16.7. The raw sludge is pumped into the digestor through pipes which terminate either near the centre of the tank or in the gas dome. The latter arrangement helps to break up the scum layer which forms on its surface.

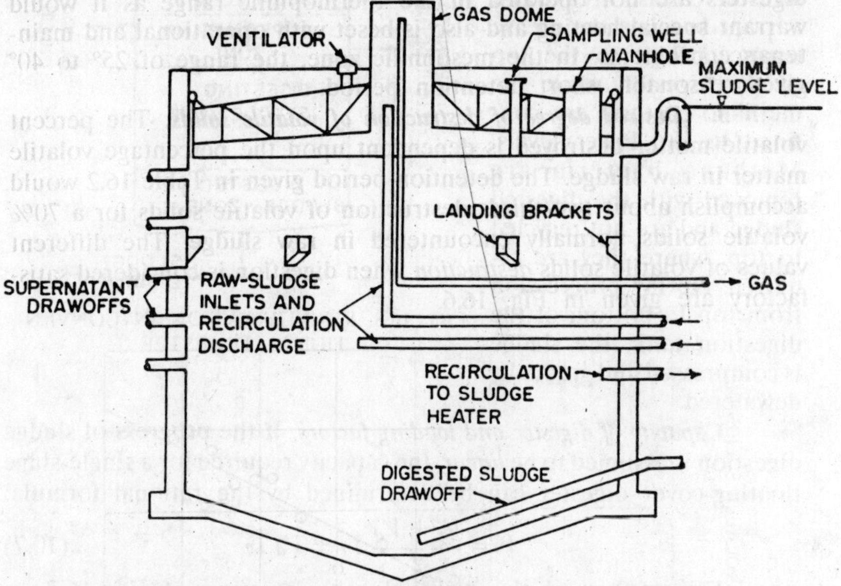

FIG. 16.7. CROSS-SECTION OF A TYPICAL FLOATING COVER DIGESTOR

The digested sludge is withdrawn from the tank bottom. Supernatant is drawn from the digestor through any one of a series of pipes extending out of the tank wall. Digestion gas from the gas dome is taken out through a gas pipe. The weight of the cover is supported by sludge and the liquid forced up between the tank wall and the side of the cover provides a gas seal. three functions of a single-stage floating-cover digester are (*i*) anaerobic digestion of the volatile solids, (*ii*) gravity thickening, and (*iii*) storage of the digested sludge. A floating cover feature of the tank provides for a storage volume equal to approx. one third that of the tank.

Lack of proper mixing in the conventional digesters leads to *stratification*, giving rise to distinct layers of scums, supernatant, actively digesting sludge and digested sludge. (Fig. 16.8). The supernatant is withdrawn periodically and returned to the influent of the treatment plant while the sludge is added at mid-depth and withdrawn from the bottom.

In this type of digesters, much of the digester volume is wasted and sometimes acidification takes place in the top and middle layers while methane fermentation is confined to the lower layers. This can lead to areas of low and high pH in the system, which restrict optimum biological activity. Also, chemicals added for pH control are not dispersed throughout the tank and their effectiveness is limited. Grease breakdown is poor because the grease tends to float to the top of the digester, while the methane bacteria are confined to the lower levels. Methane bacteria are removed with the digested sludge and are not recycled to top, where they are required. During progression from top to bottom of the digestion tank, the sludge is compressed and gradually dewatered.

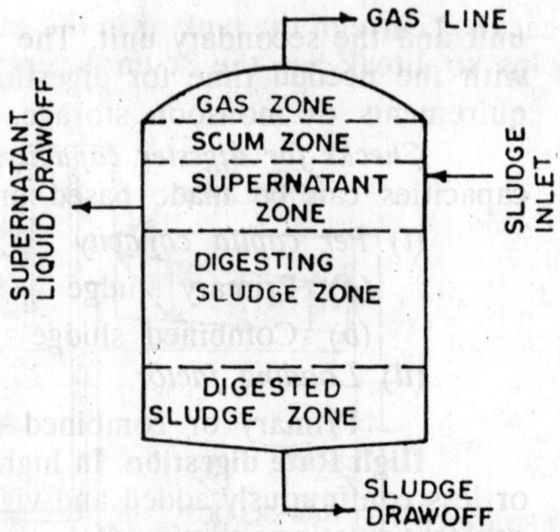

FIG. 16.8. STRATIFICATION IN CONVENTIONAL DIGESTER

Capacity of digester and loading factors. If the progress of sludge digestion is assumed to be *linear*, the capacity required for a single-stage floating-cover digester can be determined by the rational formula:

$$V = \frac{V_f + V_d}{2} T_1 + V_d T_2 \qquad ...(10.7)$$

Fig. 16.9 gives the pictorial representation of Eq. 16.7.

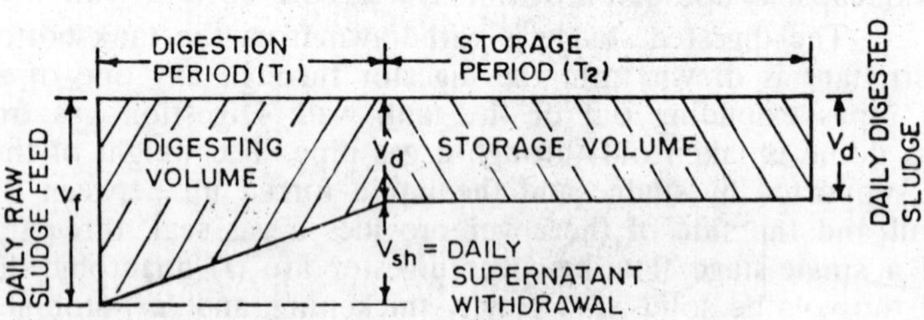

FIG. 16.9. PICTORIAL REPRESENTATION OF EQ. 16.7.

However, since the progress of digestion is parabolic, the capacity of the digester is given by the following expression.

$$V = \left[V_f - \frac{2}{3}(V_f - V_d) \right] T_1 + V_d T_2 \qquad ...(16.8)$$

where V = volume of digester

V_f = volume of fresh sludge added per day

V_d = volume of digested sludge withdrawn per day

T_1 = digestion time in days, and

T_2 = monsoon storage in days.

The digestion can be in a single unit (commonly known as *single stage digestion unit*) or in two units known as the primary unit and the secondary unit. The primary digestion unit is provided with the needed time for digestion and the latter to meet the requirements of monsoon storage.

Checks for digester capacity. Useful checks for the digester capacities can be made based on the following parameters :

(*i*) *Per capita capacity*

(*a*) Primary sludge ..... 0.05 – 0.075 m^3

(*b*) Combined sludge ..... 0.1 to 0.15 m^3

(*ii*) *Loading factor*

Primary or combined sludge .....0.3 to 0.75 $kg/m^3/day$

High Rate digestion. In high rate digestion, the sludge is more or less continuously added and vigorously mixed either mechanically or by recirulating a portion of the digestion gases through a compressor. *Mechanical mixing* is normally accomplished by an impeller suspended from the cover of the digester (Fig. 16.10 *a*). *Gas mixing* may be of three types : (*i*) by injection of compressed gas through a series of small-diameter pipes hanging from cover into the digesting sludge, as shown in Fig. 16.10 (*b*), (*ii*) by use of a draft tube in the centre of the tank, with compressed gas injected into the tube to lift recirculating sludge from the bottom and spill it out on top (Fig. 16.10 *c*), or (*iii*) by supplying compressed gas to a number of diffusers mounted in the centre at the bottom of the tank (Fig. 16.10 *d*). Because of good mixing, there is no stratification and hence loss of capacity does not arise due to supernatant or scum or dead pockets.

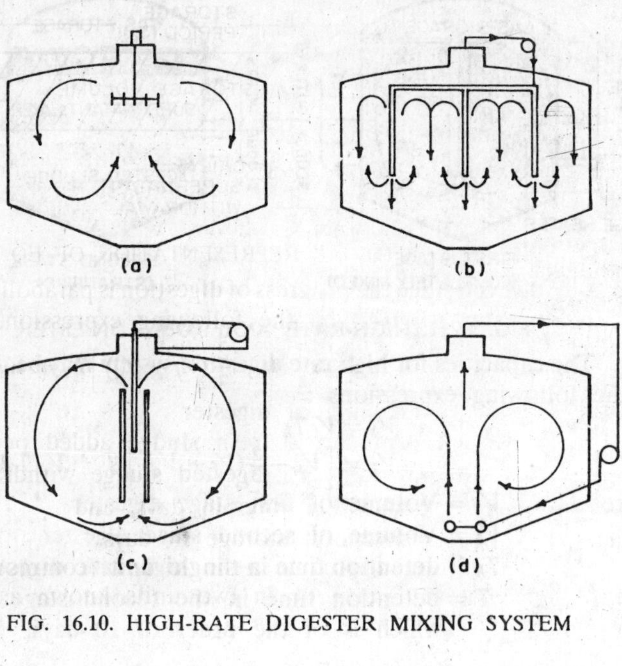

FIG. 16.10. HIGH-RATE DIGESTER MIXING SYSTEM

The digestor is also heated to maintain maximum activity in the mesophilic region.

By adopting more or less continuous addition of raw sludge and resorting to pre-thickening of the raw sludge to a solid content of 6%, the detention time can be reduced to 10 to 15 days. Thickening beyond 6% affects the fluidity of the sludge, hampering its movement in pipes and mixing in the digester. Digesting sludge is displaced when raw sludge is pumped into the digester which may have either a fixed-or a floating cover tank. The homogeneous nature of the digesting sludge in the high-rate digester does not permit formation of supernatant, and only about 50% destruction of VS is achieved. Also, thickening cannot be performed in a complete-mix-digester. Hence high-rate-digestion systems normally consist of two tanks operated in series (Fig. 16.11). The *first stage* is a complete mixing, heated, floating or fixed-cover digester fed as continuously as possible, whose function is anaerobic digestion of the volatile solids. In the second stage digester, which may be heated or unheated, separation of supernatant accompanied with a reduction in volume of sludge due to gravity thickening takes place and digestion is completed. Additional storage capacity needed for the monsoon period has also to be provided in the second stage digester.

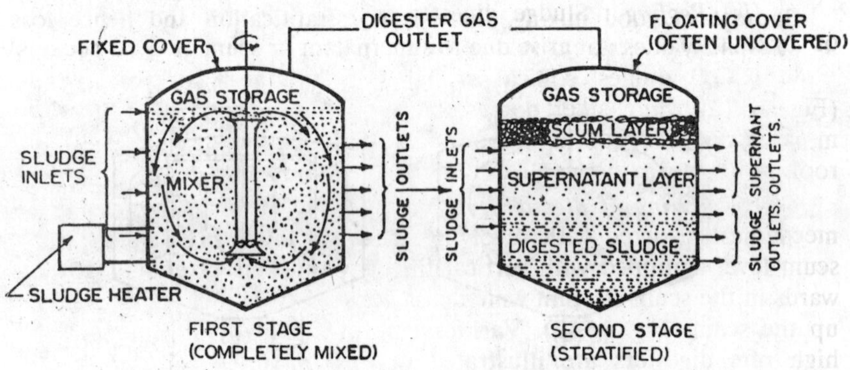

FIG. 16.11. HIGH-RATE ANAEROBIC DIGESTER

The capacities for high rate digestion system may be determined by the following expressions :

$$V_1 = V_f T_h \quad ...(16.9)$$

$$V_2 = \left[V_f - \tfrac{2}{3}(V_f - V_d) \right] T + V_d T_2 \quad ...(16.10)$$

where V_1 = Volume of first stage digester

V_2 = Volume of second stage digester

T_h = detention time in the high rate digester, in days

and T = detention time in the second stage digester which is of the order of 10 days.

Design of digester elements

(*i*) ***Number of units*** : Conventional digesters are designed as a single unit for plants treating upto 4 mLd. For larger plants, units are provided in multiples of two, the individual capacity not exceeding 3 mLd. High rate digesters are designed comprising primary and secondary digestion tanks, each unit generally capable of handling sludge from treatment plants upto 20 mLd.

(*ii*) ***Tank shape and size*** : Circular tanks are most common for sludge digestion and they should preferably be not less than 6 mm nor more than 55 m in dia.

(*iii*) ***Water depth and F.B.*** Side water depth of a digester may be kept between 4.5 to 6.0 m, but should not exceed 9 m even for very large tanks. For fixed dome or conical roofs, F.B. should be not be less than 0.4 m while for floating covers, F.B. should not be less than 0.6 m. For slab roof, F.B. of 0.8 m is recommended.

Note : Before the tank depth and surface area of a digester are worked out, maximum gas production rate should also be determined. On an average, about 0.9 m^3 of gas is produced per kg of volatile matter destroyed. The optimum diameter or depth is calculated such that at twice the average rate of daily gas production, the value of 9 m^3/m^2 of tank area is not exceeded.

(*iv*) ***Roofing*** : Sludge digester may either have floating roof or fixed roof.

R.C.C. domes, conical or flab slabs are used for fixed roof (Fig. 16.12) while steel domes, having vertical travel of 1.2 to 2 m, are used for floating cover. Gas domes are provided in the roof at or near the centre of circular tanks.

(*v*) ***Mixing of digester contents*** : Conventional digesters are mechanically stirred with revolving arms dipping a little below the scum level equipped with vertical fingers or pickets extending down wards in the scum stratum which pass between fixed pickets, breaking up the scum (Fig. 16.12). Various type of mixing arrangements for high rate digesters are illustrated in Fig. 16.10.

Gas composition and collection. Sludge gas is normally composed of about 60 to 70% methane and 25 to 35% carbon dioxide by volume, with smaller quantities of other gases like hydrogen sulphide, hydrogen, nitrogen and oxygen. The combustible constituent in the gas is primarily the methane. Depending upon the sulphate content of the sewage and the sludge, the concentration of hydrogen sulphide in the gas varies. Hydrogen sulphide in addition to its corrosive properties impose a limit on the utilisability or cause nuisance during the burning of the gas. Sludge gas containing 70% methane has a fuel value of about 5800 k cal/m^3. In terms of solids digested, the average gas production is about 0.9 m^3/kg of volatile solids destroyed at a normal operating pressure of 150 to 200 mm of water.

Sludge gas is collected under positive pressure to prevent its mixing with air and causing explosion. The gas may be collected directly from under a floating cover on the digester or from the fixed cover by maintaining a constant water level. A gas dome above the digester roof should be used for gas take off. The velocity in sludge gas piping should not exceed 3.5 m/s. Pressure release valves are provided for controlling the gas in the digester by releasing gas pressure exceeding 200 to 300 mm of water.

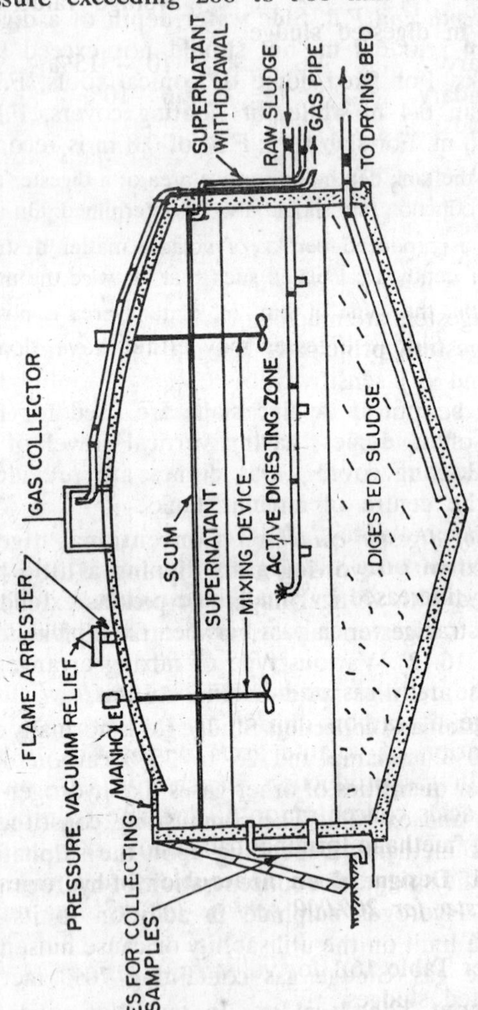

FIG. 16.12 FIXED-ROOF DIGESTER

Performance of digesters

The following parameters are indicative of good digestion :

(*a*) % of volatile solids destroyed : As given in Fig. 16.6.
(*b*) Per capita gas production : 0.025 m^6/day
(*c*) Gas production per kg of volatile matter added : 0.4 m^3
(*d*) Gas production per kg of volatile matter destroyed : 0.9 m^3
(*e*) pH of the digesting sludge : 7 to 8
(*f*) Methane content of gas produced : 60 to 70% with an average value of 65%
(*g*) Dry solids in digested sludge :
  (*i*) Primary ..... : 10 – 15%
  (*ii*) Secondary ..... : 6 –10%
(*h*) Volatile acids ..... : 200 to 400 mg/l
(*i*) Grease ..... : practically absent
(*j*) Colour ..... : black
(*k*) Drainability : quickly drainable
(*l*) Odour : inoffensive

Start-up and monitoring of digesters

Anaerobic digesters are much more difficult to start operating than the aerobic digesters, principally because of two reasons : (*i*) slow growth rate and (*ii*) sensitivity of methane-forming bacteria. A *new digester* can be put in operation within a few weeks if a substantial amount of digesting sludge from an operating digester is used as *seed*. If only raw sludge is available, start up may take months. To startup a digester, only a one-tenth sludge feed rate is applied. Lime may be added with the raw sludge to maintain the pH near 7.0. When once the digestion process is established, the feed is gradually increased by small increments to full loading.

Operation of a digester can be monitored by the following methods :

(*i*) plotting the daily gas production per unit raw sludge fed
(*ii*) percentage of carbon dioxide in the digestion gas.
(*iii*) concentration of volatile acids in the digestion gas.

A reduction in gas production, increase in CO_2 percentage and rise in volatile acids concentration all indicate reduced activity of the acid-splitting methane-forming bacteria.

Example 16.6. Design of an anaerobic sludge digester

Design a digester for 200,000 persons for digesting mixed raw and activated sludge.

Solution Refer Table 16.1 for volumes of mixed raw primary, activated and digested sludges.

1. Volume of mixed raw (undigested) primary plus activated sludge

% of the volatile matter in raw mixed sludge

$$= \frac{58}{85} \times 100 = 68.24\%$$

% of non-volatile (or fixed) matter in sludge

$$= \frac{27}{85} \times 100 = 31.76\%$$

Hence from Eq. 16.1. specific gravity of dry solids in mixed raw sludge is

$$\frac{100}{S_d} = \frac{68.24}{1} + \frac{31.76}{2.5}$$

From which $S_d = 1.235$

Taking percentage solids as 4%, specific gravity of wet-sludge (mixed primary plus activated sludge) is given by Eq. 16.2.

$$\frac{100}{S_{sl}} = \frac{96}{1} + \frac{4}{1.235}$$

From which $S_{sl} = 1.0077$

Hence, from Eq. 16.3, volume of mixed sludge is

$$V_{sl} = \frac{W_s}{\rho_w S_{sl} p_s}$$

$$= \frac{85 \times 200000 \times 10^{-3}}{1000 \times 1.0077 \times 0.04}$$

$$= 421.8 \text{ m}^3/\text{day}$$

2. ***Volume of mixed digested primary plus activated sludge.***

% of volatile matter in digested sludge (Table 16.1)

$$= \frac{20}{57} \times 100 = 35.09\%$$

% of non-volatile (or fixed) matter in digested sludge

$$= \frac{37}{57} \times 100 = 64.91\%$$

Hence $$\frac{100}{S_d'} = \frac{35.09}{1} + \frac{64.91}{2.5}$$

From which $S_d' = 1.638$

Taking percentage solids digested sludge as 7%, specific gravity of wet digested sludge is given by

$$\frac{100}{S_{sl}'} = \frac{93}{1} + \frac{7}{1.638}$$

From which $S_{sl}' = 1.028$

Hence volume of digested mixed sludge is

$$V_{sl}' = \frac{57 \times 200000 \times 10^{-3}}{1000 \times 1.028 \times 0.07} \quad \text{(Eq. 16.3)}$$

$$= 158.4 \text{ m}^3/\text{day}$$

3. Volume of digester. Let us assume average working temperature = 27.5°C. Hence from Table 16.2, digestion period = 30 days. Also assume 60 days storage in monsoon. Assuming a parabolic reduction of volume, the capacity (or volume) of the digester is given by Eq. 16.8 :

$$V = \left[V_f - \tfrac{2}{3}(V_f - V_d)\right] T_1 + V_d T_2$$

$$= [421.8 - \tfrac{2}{3}(421.8 - 158.4)]\, 30 + 158.4 \times 60$$

$$= 7386 + 9504 = 16890 \text{ m}^3$$

(This is within 0.08 to 0.15 m^3/capita of combined sludge)

4. Loading factor. Total loading of volatile solids $= 58 \times 200000 \times 10^{-3} = 11600$ kg/day.

$\therefore$ Volatile solids loading factor = 11600/16890 = 0.687 kg/day^3

(This is within the prescribed range of 0.3 to 0.75 kg/day/m^3)

5. Dimensions of digester. Let us assume that for a cylindrical digester, the average gas production is @ 0.9 m^3/kg of volatile matter destroyed.

From Table 16.2, volatile matter destroyed during digestion of combined sludge = 38 gm/capital.

$\therefore$Total volatile matter destroyed $= (38 \times 10^{-3}) \times 200000 = 7600$ kg

$\therefore$ Gas produced $= 0.9 \times 7600 = 6840 \text{ m}^3$

It is recommended that in order to avoid foaming, the optimum diameter or depth is calculated such that twice the average rate of gas production, the value of 9 m^3/m^2 of tank area is not exceeded.

Hence min. area of digester required (to avoid foaming)

$$= \frac{2 \times 6840}{9} = 1520 \text{ m}^2$$

$$\text{Hence depth of digester} = \frac{16890}{1520} = 11.11 \text{ m.}$$

But the depth should not exceed 9 m. Hence 2 digesters are proposed.

$$\therefore \text{ Volume of each tank} = \tfrac{1}{2} \times 16890 = 8445 \text{ m}^3$$

Adopting depth of 8 m in each tank,

$$\text{Dia. of each tank} = \sqrt{\frac{8445 \times 4}{\pi \times 8}} = 36.7 \text{ m}$$

Provide a free board of 0.6 m (for floating cover).

Hence adopt 2 numbers of digestion tanks, each of 37 m dia. and 8.6 m height.

Typical details of the digestor are shown in Fig. 16.12.

16.6. AEROBIC DIGESTION

The function of aerobic digestion is to stabilize waste sludge by long term aeration, thereby reducing the BOD and destroying volatile solids. Aerobic digestion is generally defined as a process in which micro-organisms obtain energy by endogenous or auto-oxidation of their celluler protoplasm. The biologically degradable constituents of the celluler material are slowly oxidized to carbon dioxide, water and ammonia. Aerobic digestion produces volatile solids reductions, comparable to those in anaerobic digestion, has low BOD in the supernatant, fewer operational problems and lower initial cost. Aerobic digestion is accomplished in one or more tanks mixed by *diffused aeration*. Since aerating solids have a low rate of oxygen demand, the need for effective mixing rather than microbial metabolism usually governs the air supply required. In general, aerobic digesters are similar to conventional activated sludge tanks in that they are not covered or insulated. Thus they are generally more economical to construct than are covered, insulated, and heated anaerobic digesters. However, long-term aeration of waste-activated sludge creates a bulking materal difficult to thicken. Aerobic digestion can be used for secondary tank humus or for a mixture of primary and secondary sludges but not for primacy sludges alone.

The factors that should be considered in designing an aerobic digester include detention time, loading criteria, oxygen requirement, mixing and process operation. The volatile solids destroyed in aerobic digestion in about 10 to 12 days time, at a temperature of 20° C would be 35 to 45%. Higher temperature will result in reduction in the period of digestion. Oxygen requirements normally vary between 1.7 to 1.9 gm/gm of volatile solids destroyed. It is also desirable to maintain the dissolved oxygen between 1 to 2 mg/l in the system.

Advantages

1. Lower BOD concentration in digester supernatant
2. Production of odourless and easily dewaterable biologically stable digested sludge.
3. Recovery of more basic fertiliser value in digested sludge
4. Lower capital cost
5. Fewer operational problems.

Disadvantages

1. Higher power costs generate higher operating costs comparable with anaerobic digestion.
2. Gravity thickening processes following aerobic digestion tend to generate high solids concentration in the supernatant.
3. Some aerobically digested sludges donot dewater easily in vacuum filtration.
4. No methane gas is produced for recovery as a by-product.

The aerobic digestion process appears to be particularly well suited for industrial sludge treatment and for small, municipal activated sludge plants.

16.7. SLUDGE CONDITIONING

Sludge is conditioned expressly to improve its dewatering characteristics. Prior conditioning of sludge before application of dewatering used for sludge conditioning are (*i*) chemical conditioning and (*ii*) heat treatment. *Elutriation*, a physical washing operation, is used to reduce chemical conditioning requirements.

Chemical conditioning. Chemical conditioning may be applied to sludge which have been digested or to raw sludges. Chemical conditioning is the process of adding certain chemicals to enable coalescence of sludge particles facilitating easy extraction of moisture. The chemicals used are ferric and aluminum salts and lime, the more common being ferric chloride with or without lime. The mechansim of chemical conditioning consists of neutralization of charge and formation of a polymeric bridge which incorporates the individual particles into a lattice structure with sufficient rigidity and porosity to permit water to escape. Digested sludge, because of its high alkalinity, exerts a huge chemical demand and therefore the alkalinity has to be reduced to effect a saving on the chemicals. This can be achieved by *elutriation*.

Polyelectrolytes show promise for sludges with finely dispersed solids. The choice of chemical depends on pH, ash content of sludge, temperature and other factors. Optimum pH values and chemical dosage for different sludges has to be based on standard laboratory tests. The dosage of ferric chloride and alum for elutriated digested sludge are of the order of 1.0 kg/m^3 of sludge. Alum when vigorously mixed with the sludge reacts with the carbonate salts and releases CO_2 which causes the sludge to separate and the water drains out more easily. Hence for effective results, alum must be mixed quickly and thoroughly.

Feeding devices are necessary for applying chemicals. Mixing of chemicals with sludge should be gentle but thorough, taking not more than 20 to 30 seconds. Mixing tanks are generally of the vertical type for small plants and of the horizontal type for large plants. They are provided with mechanical agitators rotated at 20-60 rpm.

Elutriation. Elutriation is, literally, a *'washing'* of the sludge. It is a unit operation in which a solid or a solid-liquid mixture is intimately mixed with a liquid for the purpose of transferring certain components to the liquid. A typical example is the elutriation of digested waste water sludge prior to chemical conditioning to reduce the coagulant demand exerted by alkalinity of the digested sludge. This is achieved with dilution with water of lower alkalinity (usually plant effluent) followed by sedimentation and decantation. Some end products of digestion such as ammonium bicarbonate which exert increased demand of chemicals in conditioning are removed in the process. There are three methods of elutriation : (*i*) single stage, (*ii*) multi-stage and (*iii*) counter current washing. The water requirement is dependent upon the method used. For a given alkalinity reduction, single stage elutriation requires 2.5 times as much water as the two stage and 5 times as much water as counter-current washing. Hence single stage washing is used only in small plants. Counter current washing although higher in initial cost, in adopted in all large plants. Water requirement also depends on alkalinity of dilution water, alkalinity of sludge and desired alkalinity of elutriated sludge. Sludge and water are mixed in a chamber with mechanical mixing arrangement, the detention period being about 20 seconds. The sludge is then settled in settling tanks and excess water decanted. A maximum surface loading on settling tank of about 40 m^3/m^2/day and a detention period of about 4 hours are adopted.

Counter current elutriation is generally carried out in twin tanks similar to sedimentation tanks, in which sludge and wash water enter at opposite ends. Piping and channels are so arranged that wash water entering the second stage tank comes first in contact with sludge already washed in the first stage tank. The volume of wash water required is roughly 2 to 3 times the volume of sludge elutriated.

16.8. SLUDGE DEWATERING

Dewatering is a physical unit operation used to reduce the moisture content of the sludge, and thus to increase the solids concentration. Dewatering is accomplished either by *air drying* in *sludge drying beds* or by mechanical means such as vacuum filtration, centrifugation, pressure filtration etc.

Purposes of dewatering

1. Cost of trucking sludge to ultimate disposal site is reduced, because of reduced sludge volume consequent to dewatering.
2. Ease in handling dewatered sludge.
3. Increase in calorific value of sludge by removal of moisture, prior to incineration.
4. Rendering the sludge totally odourless and nonputrescible.
5. Sludge dewatering is commonly required prior to land filling to reduce lechate production at landfill site.

Sludge drying beds. This method of dewatering and drying the sludge is specially suitable for those locations where temperature are higher, similar to the one prevailing in our county. The method consists of applying the sludge on specially prepared open beds of land. A typical section of sludge drying beds is shown in Fig. 16.13. A sludge drying bed usually consists of a bottom layer of gravel *of uniform size* over which is laid a bed of clean sand. Open jointed tile underdrains are laid in the gravel layer to provide positive drainage as the liquid passes through the sand and gravel.

Underdrains are made of vitrified clay pipes or tiles of atleast 10 cm dia. laid with open joint. Underdrains are placed not more than 6 m apart. Graded gravel is placed around the underdrains in layers upto 30 cm with a minimum of 15 cm above the top of the under drains. Atleast 8 cm of the top layer should consist of gravel of 3 to 6 mm size. Clean sand of effective size of 0.5 to 0.75 mm and uniformity coefficient not greater than 4.0 is placed over the gravel. The depth of sand may vary from 15 to 30 cm.

The drying beds are commonly 6 to 8 m wide and 30 to 45 m long. A length of 30 m away from the inlet should not be exceeded with a single point of wet sludge discharge, when the bed slope is about 0.5%. Multiple discharge points should be used with large sludge beds to reduce the length of wet sludge travel. In order to have flexibility in operation, beds should be atleast two in number.

The area needed for dewatering digested sludge is dependent on total volume of sludge, climate, temperature and location. Areas required for drying beds range from 0.1 to 0.15 m^2/capita with dry solids loading of 80 to 120 kg/m^2 of bed per year for digested primary sludge and from 0.175 to 0.25 m^2/capita with dry solids loading of 60 to 120 kg/m^2/year for digested mixed sludge.

Sludge should be deposited evenly to a depth of not greater than 20 cm. When digested sludge is deposited on a well drained bed of sand described above, the dissolved gases tend to buoy up and float the solids leaving a clear liquid at the bottom which drains through the sand rapidly. The major portion of the liquid drains off in a few hours after which drying commences by evaporation. The sludge cake shrinks producing cracks which accelerates evaporation from the sludge surface. With good drying conditions, the sludge

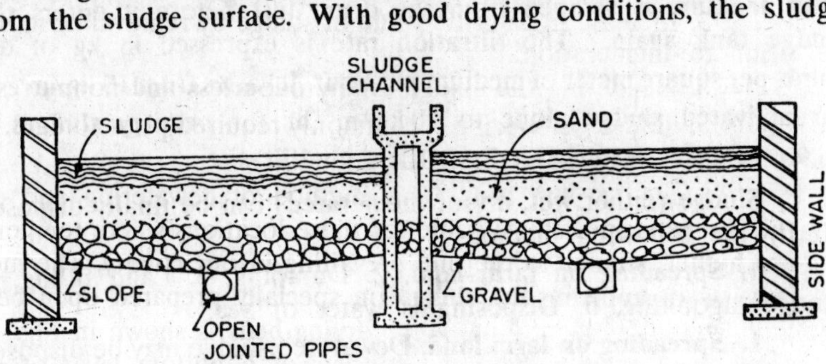

FIG. 16.13. SLUDGE DRYING BEDS.

will dewater satisfactorily and become fit for removal in about 2 to 3 weeks producing a volume reduction of 20 to 40%. Dried sludge cake can be removed by shovel or forks when the moisture content is less than 70%. When the moisture content reaches 40%, the cake becomes lighter and suitable for grinding. Wheel barrows or pick up trucks are used for hauling of sludge cakes.

Mechanical methods. *Vacuum filtration* is the most common mechanical method of dewatering, *filter presses* and *centrifugation* being the other methods. Chemical conditioning is normally required prior to the mechanical methods of dewatering. Mechanical methods may be used to dewater raw or digested sludges preparatory to heat treatment by vacuum filtration because the coarse solids are rendered fine during digestion. Hence filtration of raw primary or a mixture of primary and secondary sludges permits slightly better yields, lower chemical requirements and lower cake moisture contents than filtration of digested sludges. When the ratio of secondary and primary sludge increase, it becomes more and more difficult to dewater the filter. The feed solids concentration has a great influence, the optimum being 8 to 10%. Beyond 10%, the sludge becomes too difficult to pump and lower solids concentration would demand unduly large filter surface. In this method, conditioned sludge is spread out in a thin layer on the filtering medium, the water portion being separated due to the vacuum and the moisture content is reduced quickly.

Vacuum filters : Vacuum filter consists of a cylindrical drum over which is laid a filtering medium of wool, cloth or felt, synthetic fiber or plastic or stainless steel mesh or coil springs. The drum is suspended horizontally so that one quarter of its diameter is submerged in a tank containing sludge. Valves and piping are arranged to apply a vacuum on the inner side of the filter medium as the drum rotates slowly in the sludge. The vacuum holds the sludge against the drum as it continues to be applied as the drum rotates out of sludge tank. This pulls water away from the sludge, leaving a moist cake mat at the outer surface. The sludge cake on the filter medium is scraped from the drum just before it enters the sludge tank again. The filtration rate is expressed in kg of dry solids per square meter of medium per hour. It varies from 10 kg/m^2/hr for activated sludge alone to 50 kg/m^2/hr for primary sludges.

16.9. FINAL DISPOSAL OF SLUDGE

Sludge (either wet, dry or incinerated) can be finally disposed off by the following methods :

1. Spreading on farm land, 2. Dumping, 3. Land filling, 4. Sludge lagooning, 5. Disposing is water or sea.

1. Spreading on farm land. Dewatered sludge may be disposed of by spreading over farm land and ploughing under after it has dried. Wet dewatered sludge can be incorporated into soil directly by injection. Usually a number of shallow trenches, 50 to 90 cm wide and 0.3 to 0.4 m deep are provided about 1 to 1.5 m apart, and wet sludge is discharged into it. After a sludge cake is formed due to evaporation of water, it is covered with dry earth. After about a month, the whole land is ploughed and used for cultivation.

In general, digested sludges are of moderate but definite value as a source of slowly available nitrogen and some phosphate. They are comparable to farmyard manure except for deficiency in potash. They also contain many essential elements to plant life and minor nutrients, in the form of trace metals. The sluge humus also increases the water holding capacity of soil and reduces soil erosion making an excellent *soil conditioner* specially in arid regions by making available needed humus content which results is greater fertility.

2. Dumping. Dumping in an abandoned mine quarry can be resorted to only for sludges and solids that have been stabilized so that no decomposition or nuisance conditions will result. This method can be safely adopted for digested sludge, clean grit and incinerator residue.

3. Disposal by land filling. If a suitable site is convenient, a sanitary landfill can be used for disposal of sludge, grease, grit and other solids, whether stabilised or not. However, dewatering is recommended before such disposal, so that the cost of hauling the sludge is reduced. In a true sanitary landfill, the wastes are deposited in a designated area, compacted in place with a tractor or roller and covered with 30 cm layer of clean soil. The *sanitary land fill method* is most suitable if it is also used for disposal of the other solids wastes of the community. However, drainage from the site that would cause pollution of ground water supplies or surface streams must be guarded against.

4. Sludge Logooning. A lagoon is a shallow earth basin into which untreated or digested sludge is deposited. Untreated-sludge lagoons stabilize the organic solids by anaerobic and aerobic decomposition, which may give rise to objectionable odours. Hence the lagoons should be located away from the town. Fig 16.14 shows a typical section of a lagoon. The depth of the lagoon may vary from 0.5 to 1.5 m. The depth of the lagoon and its area should be about twice that is required for sand drying under comparable conditions. Agricultural tile drains of about 10 cm dia. are placed at 3 m centres at the bottom of the lagoon, and a 15 cm thick layer of ashes or clinker is placed over it to facilitate drain of water from wet sludge. The detention time may vary from 1 to 2 months. After the sludge has been stabilized and the moisture is drained/evaported,

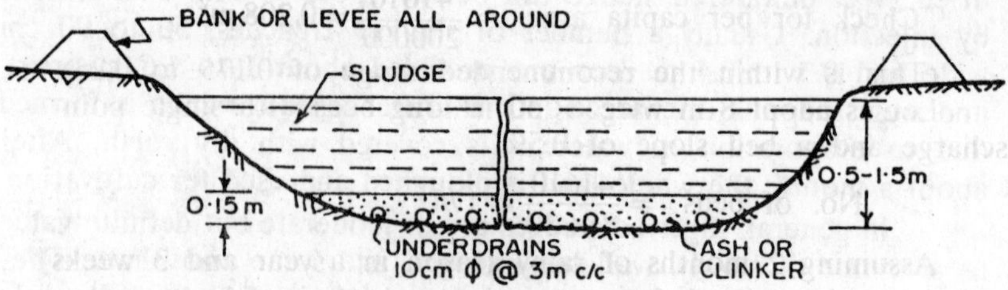

FIG. 16.14. SLUDGE LAGOON

the contents of the lagoon are dug out to about half of its volume and used as manure.

Use of sludge lagoons for storage, digestion , dewatering and final disposal of dried sludge may be adopted in isolated locations where soil is fairly porous and when there is no chance of ground water contamination. Drainage water should not be allowed to enter the lagoon, and hence earth banks are constructed along its perimeter. Lagoons have been used for regular drying of sludge on a fill and draw basis or allowed to fill dry and then levelled out and used as lawns. Lagoons have also been employed as emergency storage when digesters have to be emptied for repairs. As they are less expensive to build and operate, they have been resorted to, particularly for digested sludge in areas where large open land suitably located is available. Use of lagoons is not generally desirable, as they present an ugly sight and cause odour and mosquito breeding.

5. Disposal in water or sea. This is not common method of disposal because it is contingent on the availability of a large body of water adequate to permit dilutions. At some sea coast sites, the sludge either raw or digested may be barged to sea far enough to make available the required dilution and dispersion. The method requires careful consideration of all factors for proper design and siting of outfall to prevent any coastal pollution or interference with navigation.

Example 16.17. Design of sludge drying beds

Design a sludge drying bed for digested sludge from an activated sludge plant serving 200,000 people.

Solution

From Table 16.1, total solids remaining in digested sludge (combined primary and activated) = 57 gm/capita/day

$\therefore$ Daily solids = $200000 \times 57 \times 10^{-3} = 11400$ kg/day

Let us adopt a dry solids loading of 100 kg/m²/year

$$\therefore \quad \text{Area of bed needed} = \frac{11400 \times 365}{100} = 41610 \text{ m}^2$$

Check for per capita area $= \frac{41610}{200000} = 0.208\ m^2$

(This is within the recommended range of 0.175 to 0.25)

Let us adopt 8 m wide × 30 m long beds with single point discharge and a bed slope of 0.5%.

$$\therefore \quad \text{No. of beds} = \frac{41610}{8 \times 30} \approx 174$$

Assuming 2 months of rainy season in a year and 3 weeks for drying and one week for preparation and repair of bed, number of cycles per year $\approx (12 - 2)\, 4/4 = 10$.

Let us assume 7% solids and a specific gravity of 1.025, the volume of digested studge is given by Eq. 16.3 :

$$V_{sl} = \frac{W_s}{\rho_w \cdot S_{sl} \cdot p_s}$$

$$= \frac{11400}{1000 \times 1.025 \times 0.07} \approx 1.59\ m^3/day$$

$$\therefore \text{ Depth of application of sludge} = \frac{159 \times 365}{174 \times 8 \times 30 \times 10} = 0.139\ m.$$

$$\approx 14 \text{ cm (say)}$$

Typical details of the sludge drying beds are given in Fig. 16.13.

PROBLEMS

1. What do you understand by 'digestion' of sludge ? Differentiate between anaerobic and 'aerobic digestion'. Explain the mechanism of anaerobic digestion.
2. Explain, with the help of a flow chart, various processes involved in sludge treatment and disposal.
3. Explain in brief the characteristics of sludge produced in various treatment processes.
4. Explain how do you determine the volume of a sludge if its consti tuents (i.e. volatile matter, fixed matter etc.) are known.
5. The moisture content of a sludge is reduced from 95% to 85%. Show that its volume is reduced to on-third of the volume during this process.
6. The moisture content of a sludge is reduced from 95% to 90%. Find the decrease in the volume of the sludge. **Ans.** [50%]
7. Estimate the solids production from a trickling filter plant treating 1000 m^3/day with a BOD_5 of 200 mg/l and SS of 250 mg/l. Assume that primary clarification removes 30 percent of the BOD and 60 percent of the influent solids. **Ans.** [220 kg/day]
8. A wastewater plant produces 1200 kg of dry solids per day at a moisture content of 95 percent. The solids are 70 percent volatile with a specific gravity of 1.05 and 30 percent nonvolatile with a specific gravity

of 2.5. Determine the sludge volume (*a*) as produced, (*b*) after digestion which reduces the volatile solids content by 50% and decreases the moisture content to 90% (*c*) after dewatering to 75 percent moisture (*d*) after drying to 10% moisture, and (*e*) after incineration where only nonvolatile solids remain.

Ans. [(*a*) 23744 l (*b*) 7564 l (*c*) 2884 l (*d*) 630.7 l (*e*) 144 l]

9. A sedimentation tank treats 8 mLd containing 240 mg/l of suspended solids. The tank removes 65% of the suspended solids. Compute the weight and volume of the sludge produced yearly if the moisture content is (*a*) 97% (*b*) 94%.

Ans. [(*a*) 41596 kg ; 40.78 m^3 (*b*) 20800 kg ; 20.39 m^3]

10. Determine the liquid volume before and after digestion and the percent reduction for 500 kg (dry basis) of primary sludge with the following characteristics.

	Primary	*Secondary*
(*i*) Solids %	5	10
(*ii*) Volatile solids %	60	60 (destroyed)
(*iii*) Specific gravity of fixed solids	2.5	2.5
(*iv*) Specific gravity of volatile solids	1.0	1.0

Ans. [9.9 m^3 ; 3.1 m^3 ; 68.7%]

11. What do you understand by sludge thickening ? Enumenrate various methods. Describe, with the help of sketch, gravity-sludge thickener.

12. Design a gravity thickener for thickening the combined primary and activated sludge from a treatment plant for 1,20,000 population.

13. Write a detailed note on anaerobic sludge digestion. Explain the effect of temperature and pH.

14. Differentiate clearly between conventional and high rate digester. How do you determine the capacity of (*a*) conventional digester (*b*) high rate digestor.

15. Design an anaerobic sludge digester for a 1,20,000 persons for digesting mixed raw and activated sludge.

16. What do you understand by aerobic digestion ? What are its advantages and disadvantages ?

17. Write a note on sludge conditioning. Why elutriation is necessary before chemical conditioning ?

18. Why dewatering of sludge is necessary ? Explain the method of dewatering the sludge on sludge drying beds.

19. Explain vacuum filtration method of dewatering sludge.

20. Explain the construction and working of sludge lagooning.

21. Explain in brief various methods of final disposal of sludge.

22. Design sludge drying beds for digested sludge from an activated sludge plant, serving 1,20,000 persons.

17

Septic and Imhoff Tanks

17.1. SEPTIC TANKS

A septic tank is a special form of primary sedimentation tank with a *longer* detention time, in which digestion of settled sludge also takes place. In other words, a septic tank is a combined sedimentation cum digestion tank. This tank has, therefore, larger capacity than ordinary primary sedimentation tank, so as to accommodate and hold the settled sludge for its subsequent digestion. This unit, therefore, is grouped under the primary treatment units. The digestion of settled sludge is carried out by *anaerobic decomposition process*, giving rise to septicity or septic condition; that's why this unit is known as septic tank. Since the foul gases (such as hydrogen sulphide, methane, carbon dioxide) are evolved during the digestion process, the tank is kept completely covered on the top, with a provision of a high vertical vent shaft for the escape of these gases. The effluent from the septic tank is sufficiently foul in nature, containing considerable amount of dissolved and suspended putrescible organic solids and viable pathogens. Hence it is disposed of either for sub-surface irrigation or in soak pits or through other suitable methods described in §17.3. Due to anaerobic digestion of sludge and consequent release of gases, appreciable reduction in the volume of sludge takes place.

Because of unsatisfactoy quality of the effluent and also difficulty of providing proper disposal system for the effluents, septic tanks are recommended only for individual homes and small communities and institutions (such as schools, hospitals etc.) whose contributory population does not exceed 300 or where suitable wastewater carriage system is not available. For septic tanks to function satisfactorily, a fairly adequate water supply is a prerequisite. Wastes containing excessive detergents and disinfectants are not suited for treatment in septic tanks as they adversely affect the anaerobic decomposition.

17.2. DESIGN AND CONSTRUCTION FEATURES

Since a septic tank is a settling-cum-digestion tank, its rational design is based on the following three functions it is expected to perform : (*i*) *sedimentation* to remove the maximum possible amounts of suspended solids from the sewage, (*ii*) *digestion* of settled sludge resulting in a much reduced volume of dense, digested sludge, and (*iii*) *storage* of sludge and scum accumulating in between successive cleanings thereby preventing their escape. Hence the tank should be large enough to provide for the above three requirements.

1. Sewage flow. The maximum sewage flow into a septic tank is based on the number of plumbing fixtures *discharging simultaneously*, rather than the number of users and per capita wastewater expected to reach the tank. Table 17.1 gives the number of *equivalent* fixture units, wherein a fixture unit is a standard receptacle which gives a discharge of 10 lpm when flushed.

TABLE 17.1. EQUIVALENT FIXTURE UNITS.

Facility	*Equivalent Fixture unit*	*Facility*	*Equivalent Fixtuere unit*
1. Water closet	1	8. Combination fixture	1
2. Bath kitchen	1/2	9. Shower bath	1
3. Wash basin/kitchen sink	1/2	10. Bath tub	2
4. Urinal (autoflush)	1	11. Drinking fountain	1/2
5. Urinal (ordinary)	1/2	12. Ablution tap	1/2
6. Slop tank	1	13. Dish washer	1/2
7. Lab. sink	2		

The estimated number of fixture units and the number of fixture units that contribute to the peak discharge in small installations serving upto 50 persons, for residential housing colonies upto 300 persons and for eating establishments and boarding schools are given in Table 17.2, 17.3 and 17.4 respectively.

Tank dimensions: As stated earlier, the tank capacity, and hence tank dimensions are based on the *three functions* of sedimentation, sludge digestion and sludge storage.

(i) Sedimentation : For Indian conditions, at a temperature of 25°C the surface area required will be 0.92 m^2 for every 10 lpm peak flow rate. This is based on 75% removal of sewage particles of size 0.05 mm and above with a specific gravity of 1.2. A min. depth of 25–30 cm is necessary. The length is mainted 2–4 times the breadth.

(ii) Sludge digestion : The time required for anaerobic digestion is temperature dependant. Fig. 17.1 shows the relation between av. annual temperature and the period (days) or detention time required

TABLE 17.2. ESTIMATED PEAK DISCHARGES FOR SMALL TANKS UPTO 50 USERS.

No. of users	*No. of fixtures*	*Probable No. of fix. units discharging simultaneously*	*Probable peak discharge (lpm)*
5	1	1	10
10	2	2	20
15	3	2	20
20	4	3	30
25	5	4	40
30	6	4	40
35	7	5	50
40	8	6	60
45	9	6	60
50	10	7	70

TABLE 17.3. ESTIMATED PEAK DISCHARGE FOR RESIDENTIAL HOUSING COLONIES

No. of users	*No. of Households*	*No. of fixture units*	*Probable peak discharge (based on 60% fixture units discharging simultaneously) (lpm)*
100	20	40	240
150	30	60	360
200	40	80	480
300	60	120	720

TABLE 17.4. ESTIMATED PEAK DISCHARGE FOR EATING ESTABLISHMENTS, BOARDINGS SCHOOLS ETC.

No. of users	*W.C.*	*Baths*	*Wash basin/ kitchen sinks*	*No. of fixture units*	*Probable peak discharge (based on 70% fixture units discharging simultaneously) (lpm)*
50	6	6	6	12	84
100	12	12	12	24	168
150	19	19	19	38	266
200	25	25	25	50	350
300	37	37	37	74	518

for digestion. The suspended solids per capita may be taken as 70 gm/day. Assuming that 60% of the solids is removed along with fresh sludge, of which 70% is volatile, with a solids content of 5% (or moisture content of 95%), the volume of fresh sludge will work out to be 0.00083 m^3/cap/day. Considering that 2/3 of the volatile matter is destroyed of which 1/4 is mineralised during digestion and solids content of 13% in the digested sludge, the volume of digested sludge

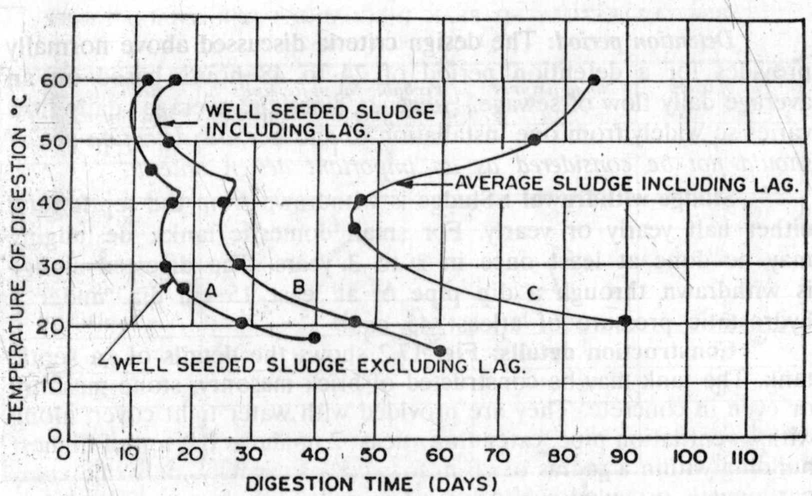

FIG. 17.1. TIME FOR ANAEROBIC DIGESTION OF 10% OF WELL SEEDED SEWAGE SLUDGE

works out to be 0.0002 m^3/day. The digestion zone contains both the fresh sludge, and hence volume of both of these will work out to be 0.000515 m^3/cap/day. From Fig. 17.1, the digestion period at 25°C, for average sludge works out to be 63 days. Hence the capacity for sludge digestion works out to be $= 63 \times 0.000515 = 0.032$ m^3/capita.

(iii) ***Storage of sludge and scum*** : A sludge storage capacity of $0.0002 \times 365 \times 100 = 7.30$ m^3/100 persons for an interval of cleaning of 1 year is provided below the sedimentation zone.

Total Capacity : The total capacity will be equal to the sum of the above three requirements, plus a free board of not less than 30 cm (to accommodate the depth of scum). No additional capacity for seed sludge is provided, if care is taken during desludging to leave about 5 to 10% digested sludge or a min. of 25 to 50 mm depth of sludge on the tank bottom.

Thus, for 10 persons, the total capacity will be as follow :

(i)	*Sedimentation*: Peak flow for 10 persons will be 20 lps, for which area required $= (0.92/10) \times 20 = 1.84$ m^2. Keeping a depth of 30 cm for sedimentation, volume=	1.84×0.3	$= 0.55$ m^3
(ii)	*Digestion*:	0.032×10	$= 0.32$ m^3
(iii)	*Sludge storage* :	(7.3/100) 10	$= 0.73$ m^3
(iv)	*Free board* (including 0.25 m^3 for seed sludge)	$= 1.84 \times 0.3$	$= 0.55$ m^3
	Total for 10 persons		$= 2.15$ m^3

***Detention period*:** The design criteria discussed above normally provides for a detention period of 24 to 48 hours based on an average daily flow of sewage. However, since the average daily flow varies so widely from one installation to another, *the detention period should not be considered as an important design criteria.*

Sludge withdrawal : Sludge is withdrawn from the septic tank either half yearly or yearly. For small domestic tanks, desludging may be done at least once in 2 to 3 years. The digested sludge is withdrawn through a dip pipe of at least 15 cm dia. under a hydrostatic pressure of atleast 45 cm.

Construction details: Fig. 17.2 shows the details of a septic tank. The tank may be constructed of brick masonry, stone masonry or even in concrete. They are provided with water tight cover, along with a ventilation pipe, extending atleast 2 m above the top of highest building within a radius of 20 m for smaller capacity, only one compartment is provided, while for larger capacities, two compartments are provided, with a partition wall at a distance of about 2/3 the length from the inlet. The inlet and outlet are located at the two opposite ends. Baffles are generally provided at both inlet and outlet and should dip 25 to 30 cm into, and project 15 cm above the liquid level. The baffles should be placed at a distance of one fifth of the tank length from the mouth of straight inlet pipe. The invert of outlet pipe should be placed at a level of 5 to 7 cm below the invert level of the inlet pipe. The floor of the tank should be of cement concrete and slope towards the sludge outlet. Both the floor as well as walls should be well plastered to make them water tight.

17.3. EFFLUENT DISPOSAL IN SEPTIC TANKS

The effluent coming out of the tank will be septic and malodorous. Final purification of the effluent and the removal/death of pathogen is affected by the following methods :

1. Soil absorption systems. 2. Upflow filters. 3. Biological filters.

Soil Absorption systems : Soil absorption systems may be of two types : (*a*) seepage pit or soak pit, and (*b*) dispersion trenches. Soakpits or dispersion trenches can be adopted in all porous soils where percolation rate is below 25 minutes per cm and the depth of water table is 180 cm or more from the ground level. The total subsurface soil area required for the soak pits or dispersion trenches is given by the following empirical relation :

$$Q = 130/\sqrt{t} \qquad ...(17.1)$$

where Q = max. rate of effluent application in lpd/m² of *leaching surface.*

t = standard percolation rate for the soil, in minutes per cm.

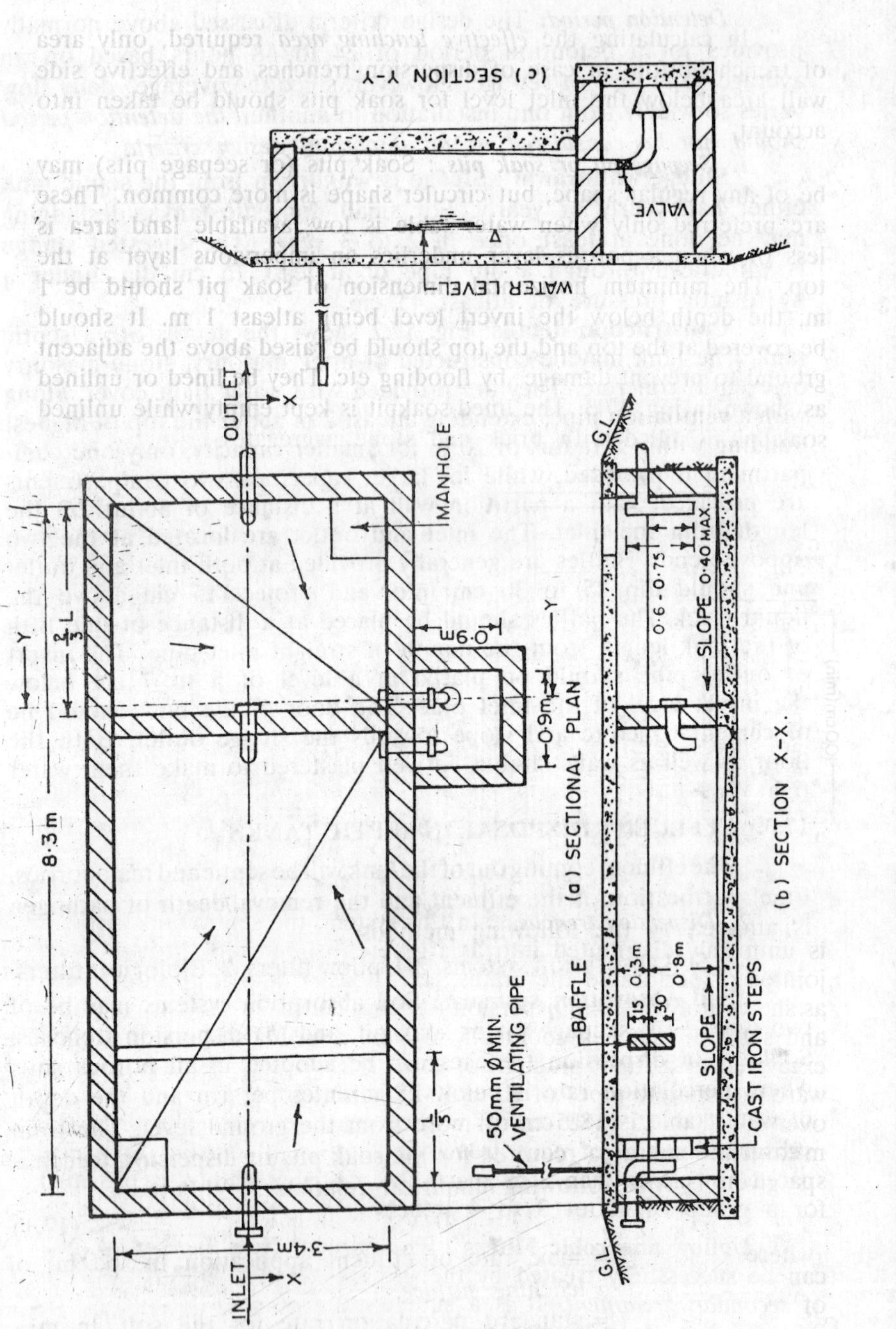

FIG. 17.2. DETAILS OF A SEPTIC TANK

In calculating the *effective leaching area* required, only area of trench bottom in case of dispersion trenches and effective side wall area below the inlet level for soak pits should be taken into account.

1. ***Seepage pit or soak pits*** : Soak pits (or seepage pits) may be of any regular shape, but circuler shape is more common. These are preferred only when water table is low, available land area is less or when a porous layer underlies an impervious layer at the top. The minimum horizontal dimension of soak pit should be 1 m, the depth below the invert level being atleast 1 m. It should be covered at the top and the top should be raised above the adjacent ground to prevent damage by flooding etc. They be lined or unlined as shown in Fig. 17.3. The lined soakpit is kept empty while unlined soak pit is filled with brick and stone aggregates.

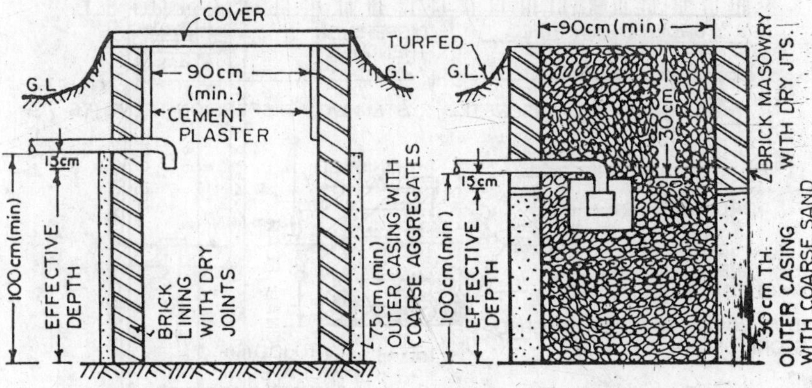

FIG. 17.3. SOAK PITS

2. ***Dispersion trenches*** : In this system, the septic tank effluent is uniformly distributed into a large area of sub-soil through open jointed or perforated tile drains, each housed in a dispersion trench as shown in Fig. 17.4. Dispersion trenches consist of relatively narrow and shallow trenches about 0.5 to 1 m deep and 0.3 to 1 m wide excavated to a slight gradient of about 0.25%. Open jointed earthen ware or concrete pipes of 70 to 100 mm dia. are laid in the trenches over a bed of 15 to 25 cm of washed gravel or crushed stone. The maximum length of each trench is kept as 30 m, and these are spaced not closer than 2 m apart. One distribution box is provided for a group of about 3 to 4 trenches.

Upflow anaerobic Filters : The effluent from the septic tank can be successfully treated by the use of upflow filter in the form of *secondary treatment.* It is a submerged filter with stone media, 0.6 to 1.2 m deep. Septic tank effluent is introduced to it from the bottom. The microbial growth is retained on the stone media

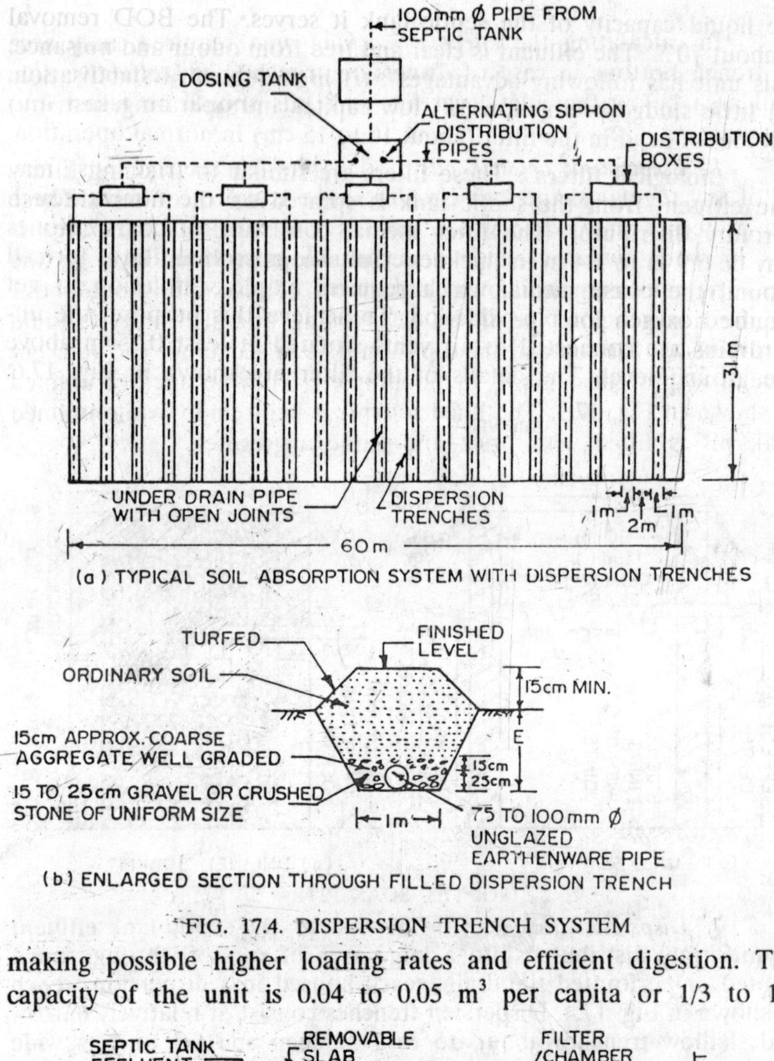

(a) TYPICAL SOIL ABSORPTION SYSTEM WITH DISPERSION TRENCHES

(b) ENLARGED SECTION THROUGH FILLED DISPERSION TRENCH

FIG. 17.4. DISPERSION TRENCH SYSTEM

making possible higher loading rates and efficient digestion. The capacity of the unit is 0.04 to 0.05 m^3 per capita or 1/3 to 1/2

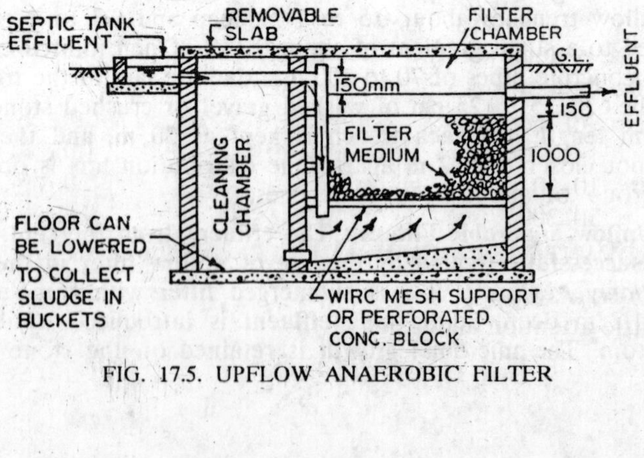

FIG. 17.5. UPFLOW ANAEROBIC FILTER

the liquid capacity of the septic tank it serves. The BOD removal is about 70%. The effluent is clear and free from odour and nuisance. This unit has following advantages : (*i*) high degree of stabilisation, (*ii*) little sludge production (*iii*) low capital and operating cost, (*iv*) low loss of head in the filter (about 10 to 15 cm) in normal operation.

Biological filters : These filters are similar to trickling filters. The effluent from the septic tank is sprayed on the filter through a rotary distributor. The filter media, consisting of coarse stones may be 0.9 m to 1.4 m in thickness. Suitable microbial film is formed around the coase media, which requires ample ventilation to get required oxygen for bio-oxidation. To achieve this purpose, the underdrains are connected to air vents situated at least 0.15 m above the ground level. The details of the filter are shown in Fig. 17.6.

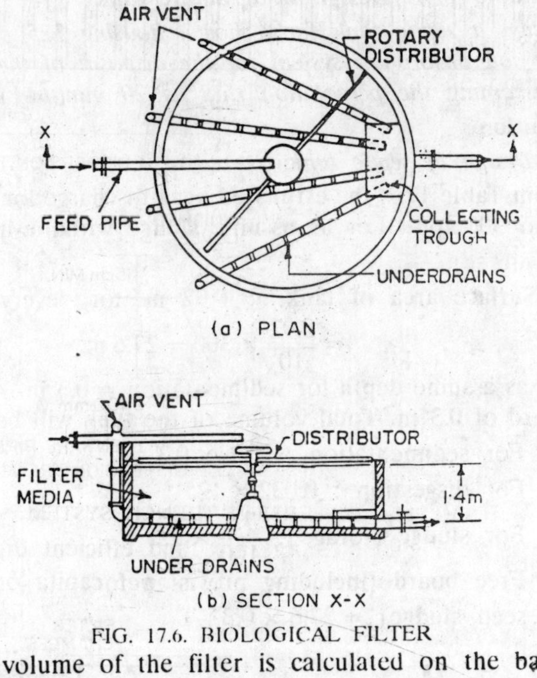

FIG. 17.6. BIOLOGICAL FILTER

The volume of the filter is calculated on the basis of the following data :

No. of persons.	Volume of filter/head m^3
1–10	1.0
10–50	0.8
50–300	0.6

17.4. ADVANTAGES AND DISADVANTAGES OF SEPTIC TANKS

Advantages : (1) Septic tanks can be constructed easily. (2) Very little attention and skiled attendance is required. (3) The sludge is relatively small, most of it being liquefied and digested. The reduction in volume is about 60% and reduction in weight is about 30%. (4) The cost is quite small and within the reach of private householders. (5) There are no moving parts for its operation. (6) The effluent can be disposed of easily, without much trouble. (7) When once installed, they give long, carefree service.

Disadvantages : (1) The size required is large and uneconomical. (2) Their functioning and action is erratic. (3) The effluent is dark and foul smelling with high BOD and is often worse than the influent. (4) Leakage of gases from the top cause bad smell and environmental pollution. (5) Periodic cleaning, removal and disposal of sludge is often tedious.

Example 17.1. Design of a Septic tank

Design a septic tank for a hostel housing 125 persons. Also design the soil absorption system for the disposal of the septic tank effluent, assuming the percolation rate as 20 minutes per cm.

Solution

1. Design of septic tank

From Table 17.3, the estimated peak discharge for 125 persons is equal to 300 lpm. Let us assume sludge withdrawal once in a year.

$\therefore$ Surface area of tank @ 0.92 m^2 for every 10 lpm is

$$= \frac{0.92}{10} \times 300 = 27.6 \text{ m}^2$$

Let as assume depth for sedimentation = 0.3 m. Also provide a free board of 0.3 m. Total volume of the tank will be as follows:

1. For sedimentation : 27.6×0.3 $= 8.28 \text{ m}^3$
2. For disgestion : 0.032×125 $= 4.00 \text{ m}^3$
3. For sludge storage : $\frac{7.3}{100} \times 125$ $= 9.125 \text{ m}^3$
4. Free board (including provision for seed sludge) $= 27.6 \times 0.3$ $= 8.28 \text{ m}^3$

Total $= 29.685 \text{ m}^3$

$$\therefore \quad \text{Total depth of tank} = \frac{29.685}{27.6} = 1.076 \text{ m.}$$

$$\text{Provide depth} = 1.1 \text{ m.}$$

Let us keep L/B ratio as 2.5.

$\therefore$ $(2.5\,B)\,(B) = 27.6$

From which $B = 3.32$. Keep $B = 3.4$ m and $L = 8.3$ m

Hence the tank dimension are 8.3 m × 3.4 m × 1.1 m. The details of the tanks are shown in Fig. 17.2.

2. *Design of absorption system*

$$Q = 130/\sqrt{t} \quad ...(17.1)$$

Here, $t = 20$ min./cm

$\therefore$ $Q = 130\sqrt{20} = 29.07$ lpd/m^2

Let us assume sewage flow @ 150 lpcd.

$\therefore$ Total flow per day= $120 \times 150 = 18000$ lpd.

$\therefore$ Total trench area required= $18000/29.07 = 619.2$ m^2

Let us assume trench width as 1 m, and a separation between trenches as 2 m, total land area required $= 3 \times 619.2 = 1857.6$ m^2

Let us provide length of each trench $= 30$ m. Hence width of land $= 1857.6/30 \approx 62$ m. Hence provide 21 trenches.

Actual percolation area provided $= 21 \times 30 \times 1 = 630$ m^2 Hence OK. Typical section of the asborption system is shown in Fig. 17.4(*b*).

17.5. LEACHING CESSPOOLS

A leaching cesspool is a circular or rectangular underground chamber or tank, in which the top portion acts as an absorption field and the bottom portion acts as a septic tank. As shown in Fig 17.7, the upper portion has an open jointed lining backed by 15 cm of coarse aggregate and 30 cm outer casing of coarse sand. This will permit the upper portion of liquid (or effluent) to be easily dispersed without disturbing the soil around. However, the bottom portion of cesspool is made water tight so as to retain heavier sludge for septic action/digestion.

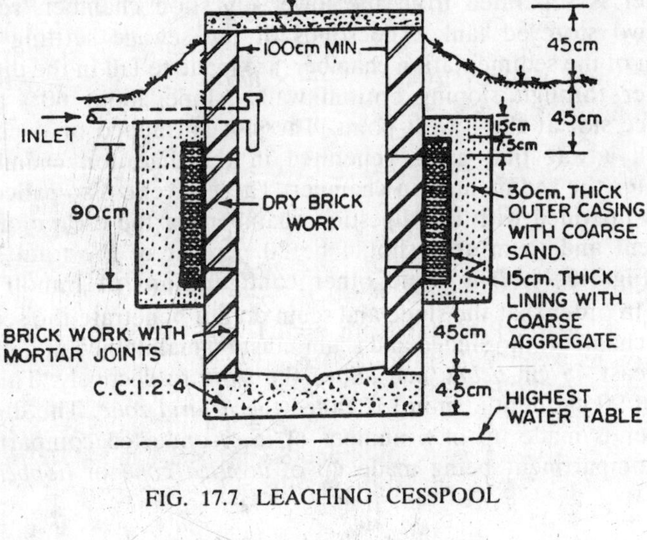

FIG. 17.7. LEACHING CESSPOOL

17.6. IMHOFF TANKS

An Imhoff tank, designed and developed by Karl Imhoff (Germany), is an improvement over the septic tank. The tank, which

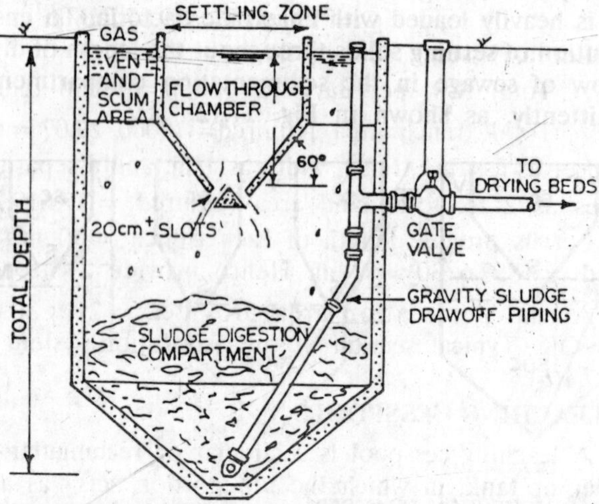

FIG. 17.8. IMHOFF TANK CONFIGURATION

is basically a sedimentation-cum-digestion tank, has two chambers. The *uppper chamber* is called the *sedimentation chamber* or flowing-through chamber, through which sewage flows at a very low velocity so that sedimentation may take place. The *lower chamber* is called the *digestion chamber* in which anaerobic or septic decomposition occurs. The process that takes place in an Imhoff tank is similar, except that the tank is so designed that the flow-through upper chamber is separated from the lower digestion chamber, resulting in a two storeyed tank. The solids of the sewage settling to the bottom of the sedimentation chamber are made to fall in the digestion chamber through sloping bottom walls (slope about 60°) and an entrance slot at the lowest point. The slot is trapped or overlapped in such a way that gases generated in the digestion chamber do not enter the sedimentation chamber. The gas vent, also called *scum area* is provided with the digestion chamber for the escape of gases. Gas vent and scum area should be 20 percent of the total surface area. Fig. 17.9 shows some other configurations of Imhoff tanks.

In order that the slude and scum do not penetrate the sedimentation chamber, the sludge and scum must be maintained at a distance of at least 45 cm *below* and *above* the slots respectively. This clear zone of 90 cm depth. (min.) is called the *neutral zone*. The digestion chamber is made up of a number of interconnected compartments, each compartment being made up of *inverted cone* or *hoppers*, with sides sloping at 1 : 1, so as to concentrate the sludge at the bottom of the hopper (Fig. 17.12). The digested sludge is removed periodically with the help of desludging pipes provided in each compartment. However, some sludge is left out in the bottom to act as *seed sludge* which is heavily loaded with anaerobic bacteria. To ensure uniform distribution of settling solids throughout the length of digestion zone, the flow of sewage in the sedimentation compartment is *reversed* intermittently, as shown in Fig. 17.12.

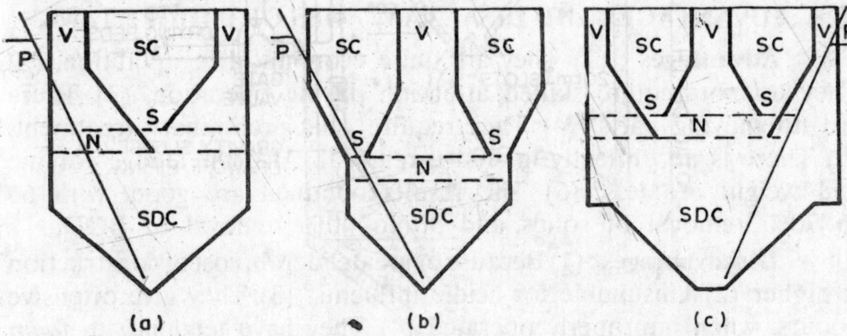

SC : SEDIMENTATION CHAMBER; N : NEUTRAL ZONE
SDC : SLUDGE DIGESTION CHAMBER; S : SLOT
V : VENTS P : SLUDGE WITHDRAWAL PIPE

FIG. 17.9. DIFFERENT FORMS OF IMHOFF TANK

17.7. IMHOFF TANK DESIGN CONSIDERATIONS

***(a)* Sedimentation chamber** : Sedimentation chamber is a rectangular chamber, having the following specifications.

(*i*) Detention period	: 2 to 4 hours; usually 2 hours.
(*ii*) Max. flowing through velocity	: 30 cm/min.
(*iii*) Surface loading	: Max. of 30,000 litres/m^2/day,
(*iv*) Length and width	: Max. length limited to 30 m; length to width ratio between 3 : 1 to 5 : 1
(*v*) Depth	: To be kept shallow to permit particles falling to the slot before reaching the end of sedimentation chamber : General depth 2.5 to 3.5 m, so as to limit the total depth to 9 – 10.5 m.
(*vi*) F.B.	: 45 cm.

***(b)* Digestion chamber :**

(*i*) Capacity	: Capacity of 0.028 – 0.056 m^3 per capita is considered to be

sufficient in warmer climates where shorter period of sludge withdrawals are possible.

(*ii*) Scum chamber : Surface area of scum chamber to be 25–30 percent of the horizontal projection of the top of the digestion chamber. Minimum width of vents should be 60 cm.

17.8. ADVANTAGES AND DISADVANTAGES OF IMHOFF TANKS

Advantages : (1) They are quite economical in operation. (2) They do not require skilled attention during operation. (3) There are no moving parts. (4) They require only preliminary treatment. (5) There is no difficulty in sludge removal. Also the sludge volume and weight are less. (6) The results obtained are good, with 60 to 70% removal of solids and 30 to 40% removal of BOD.

Disadvantages : (1) Because of greater depth, cost of construction is higher (2) Unsuitable for acidic influents. (3) They give oftensive odours, when improperly operated (4) They have tendency to *foam* or *boil*, due to which the sludge particles may enter the sedimentation zone through the slot, thus affecting the quality of the effluent.

Trouble of foaming and their control : Foaming in Imhoff tanks may be *caused* by : (1) Insufficient capacity of the digestion chamber (2) Violent biological action giving rise to large volume of gases (3) Fermentation caused by highly acidic influents (4) Entry of large quantitiy of soap water. Foaming can be *controlled* by (*i*) correcting the acidity by adding hydrated lime (*ii*) withdrawing sludge more frequently (*iii*) increasing capacity of digestion chamber (*iv*) proper seeding, and (*v*) allowing the tank to rest for a few days.

Example 17.2. Design of an Imhoff tank

Design an Imhoff tank to treat the sewage from a small town with a population of 20000 persons, with sewage flow rate of 180 litres per day.

1. *Design of sedimentation chamber*

$$Q = 20000 \times 180 = 3600 \times 10^3 \text{ lpd} = 3600 \text{ m}^3/\text{day}$$

Let us assume a detention period of 2 hours.

$$\therefore \text{ Capacity of chamber } = \frac{3600}{24} \times 2 = 300 \text{ m}^3$$

Let us assume an effective depth of 1.9 m and a width of 4 m. This effective depth includes part of the bottom sloping walls of the chamber.

$$\therefore \text{ Length of chamber } = \frac{300}{1.9 \times 4} = 39.47 \text{ m}$$

Since this length is quite large, let us use two tanks, each of 20 m length, and of width 4 m, giving L/B ratio $= 19/4 = 4.75$, which is between the desirable limit of 3 to 5.

Surface loading

$$= \frac{20000 \times 180}{2 \times 20 \times 4} = 22500 \text{ l/m}^2\text{/day},$$

which is less than 30000 l/m²/day, and hence satisfactory.

$$\textit{Velocity} = \frac{L}{T} = \frac{20}{2 \times 60} = 0.167 \text{ m/min.}$$

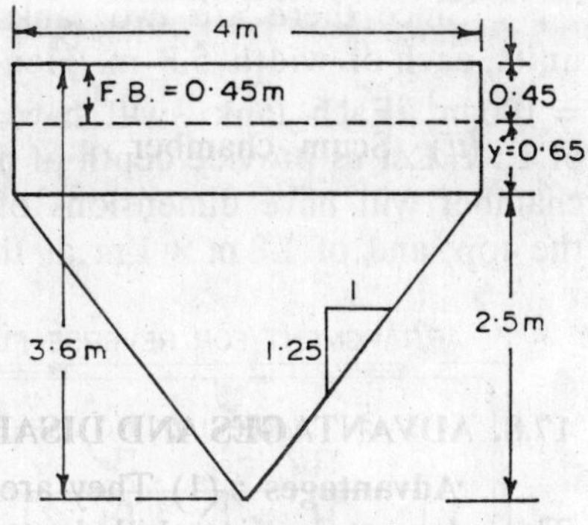

FIG. 17.10

This is less than 0.3 m/min, and hence satisfactory.

Determination of depth : Let us provide bottom slope as 1 H : 1.25 V. Width of chamber = 4 m

$\therefore$ Height $y = 1.25 \times \frac{4}{2} = 2.5$ m. But we have to provide an effective depth of 2 m. Hence height y' of the rectangular portion below the liquid surface is, $y' = 1.9 - \frac{1}{2}(2.5) = 0.65$ m. (Since effective depth of triangular portion is half that of equivalent rectangle). Let us provide a F.B. of 0.45 m.

$\therefore$ Total depth of the sedimentation chamber, upto the bottom of entrance slot $= 0.45 + 0.65 + 2.5 = 3.6$ m.

2. ***Design of gas vent and nutral zone.*** Provide neutral zone of depth 0.45 m *below* the depth of 3.6 m. The width of gas vent should be 25 to 30% of the total width (b).

Assuming 15 cm thick chamber walls to both the sides, we get $b = 4 + (2 \times 0.15) + 0.25\,b$

From which, we get $b = 5.73$ m. Hence providing total width of chamber = 5.8 m, width available for gas vents $= 5.8 - 4 - 2 \times 0.15 = 1.5$ m (*i.e.* 0.75 m width on either side of sedimentation chamber), making of equal to $1.5 \times 100/5.8 = 25.86$ % of the total width.

3. ***Design of digestion chamber.*** Let us further divide the total length L (= 20 m) into four compartments of 5 m each. Required capacity @ 40 litres per capita
$= 20000 \times 40 = 800{,}000$ litres
$= 800 \text{ m}^3$

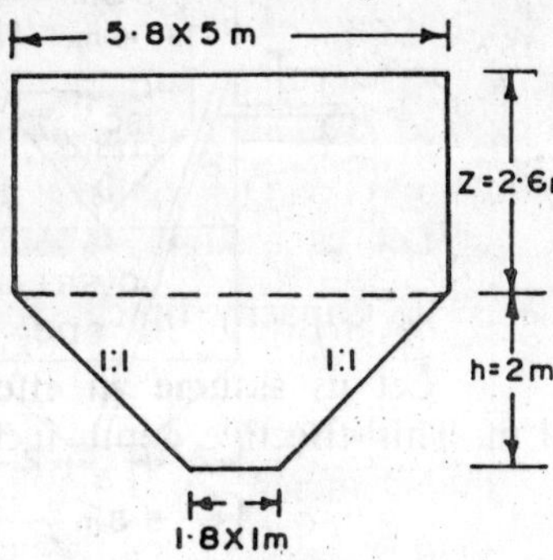

FIG. 17.11

Since there are two tanks, there will be in all 8 digestion units, each of width 5.8 m. Hence capacity of each unit = 800/8 = 100 m^3. Each tank will have hopper bottom, with side slopes of 1 : 1. Let us provide depth of hopper as 2 m. Thus, each digestion chamber will have dimensions of 5.8 m (width) × 5 m (length) at the top, and of 1.8 m × 1 m at the bottom, as shown in Fig. 17.11.

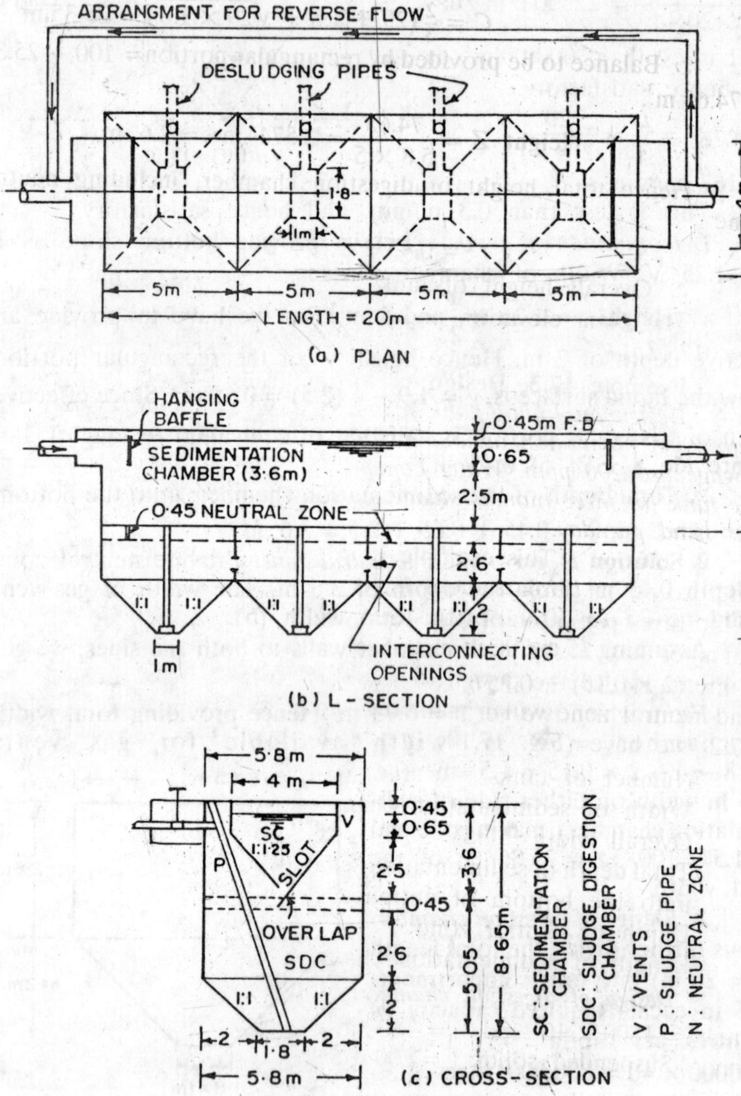

FIG. 17.12. IMHOFF TANK

∴ Capacity of each hopper

$$C = \frac{h}{3}(A_1 + A_2 + \sqrt{A_1 A_2})$$

where $A_1 = 5.8 \times 5 = 29 \text{ m}^2$.

$A_2 = 1.8 \times 1 = 1.8 \text{ m}^2$

$h = 2 \text{ m}$

$$\therefore \quad C = \frac{2}{3}(29 + 1.8 + \sqrt{29 \times 1.8}) = 25.35 \text{ m}^3$$

∴ Balance to be provided by rectangular portion $= 100 - 25.35 = 74.65 \text{ m}^3$.

$$\therefore \quad \text{Height } Z = \frac{74.65}{5.8 \times 5} = 2.574 \text{ m} \quad \approx 2.6 \text{ m}$$

Hence total height of digestion chamber, including neutral zone

$$= 0.45 + 2.6 + 2 = 5.05 \text{ m}.$$

∴ Overall height of tank $= 3.6 + 5.05 = 8.65$ m

The plan, elevation and sections of the tank are shown in Fig. 17.12.

Example 17.3. Design of an Imhoff Tank

Design an Imhoff tank to treat the sewage from a small town with 20000 population, assuming that the suspended solids in the influent sewage are 200 ppm. Water content of sludge is 97 percent. Design the tank for three months sludge storage. Rate of sewage is 180 litres per head per day.

Solution : This example is exactly the same as the previous one, except that the data about determining the capacity was determined @ 40 litres per capita. In this example, the capacity of the digestion chamber has to be determined on the basis of suspended solids content. Hence the design of the sedimentation chamber, gas vent and neutral zone will remain the same. As determined in example 17.2, we have (Fig. 17.10)

Number of tanks	= 2
Width of sedimentation chamber	= 4 m
Overall width	= 5.8 m
Total depth of sedimentation chamber upto the bottom of entrance slot	= 3.6 m.
Depth of neutral zone	= 0.45 m
Length of sedimentation chamber	= 20 m

Design of digestion chamber. Volume of sewage entering per tank $= 1800 \times 10^3$ lpd $= 1800 \text{ m}^3/\text{day}$.

Suspended solids = 200 ppm.

$$\therefore \text{ Volume of sewage solids } = 1800 \times 10^3 \times \frac{200}{10^6}$$

$$= 360 \text{ litres/day}$$

Let us assume that SS removal efficiency of Imhoff tank is 50%.

$\therefore$ Solids in the tank $= \frac{1}{2} \times 360 = 180$ litres per day.

Now water content of sludge = 97%. This means that in 100 litres of sludge, solids are only 3%.

$\therefore$ Total volume (V_1) of sludge $= \dfrac{100}{3} \times 180 = 6000$ litres/day

Again, let us assume that volume of digested sludge is one third of the undigested sludge.

$\therefore$ Volume (V_2) of digested sludge $= \frac{1}{3} \times 6000 = 2000$ litres/day

Storage time (T) = 3 months = 90 days.

$\therefore$ Capacity of digestion chamber $= \frac{1}{2}(V_1 + V_2)\, T$

$$= \frac{1}{2}\left[6000 + 2000\right] \times 90$$

$$= 360 \times 10^3 \text{ litres} = 360 \text{ m}^3.$$

Since there are 4 compartments for each tank, required capacity of each compartment, of length 5 m, will be $360/4 = 90 \text{ m}^3$. Let each tank have hopper bottom, with side slopes of 1 : 1. Let us provide depth of hopper as 2 m. Thus, each digestion chamber will have dimensions of 5.8 m (width) × 5 m (length) at the top and of 1.8 m × 1 m at the bottom, as shown in Fig. 17.13.

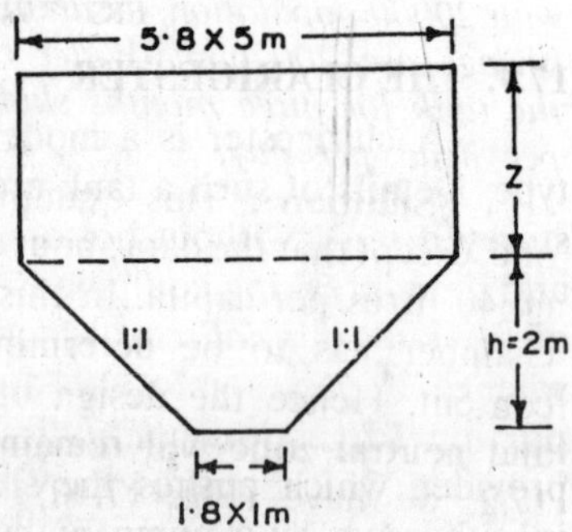

FIG. 17.13

Capacity of each hopper is

$$C = \frac{h}{3}\left(A_1 + A_2 + \sqrt{A_1 A_2}\right)$$

Here $A_1 = 5.8 \times 5 = 29 \text{ m}^2$;

$A_2 = 1.8 \times 1 = 1.8 \text{ m}^2$

and $h = 2$ m

$$\therefore \quad C = \frac{2}{3}\left(29 + 1.8 + \sqrt{29 \times 1.8}\right) = 25.35 \text{ m}^3$$

$\therefore$ Balance to be provided by rectangular portion

$$= 90 - 25.35 = 64.65 \text{ m}^3$$

$$\therefore \quad \text{Height } Z = \frac{64.75}{5.8 \times 5} = 2.233 \text{ m} \approx 2.24 \text{ m}.$$

$\therefore$ Total height of digestion chamber, including neutral zone

$$= 0.45 + 2.24 + 2 = 4.69 \text{ m}$$

Overall height of tank = 3.6 + 4.69 = 8.29 m

The cross section of the tank is shown in Fig. 7.14.

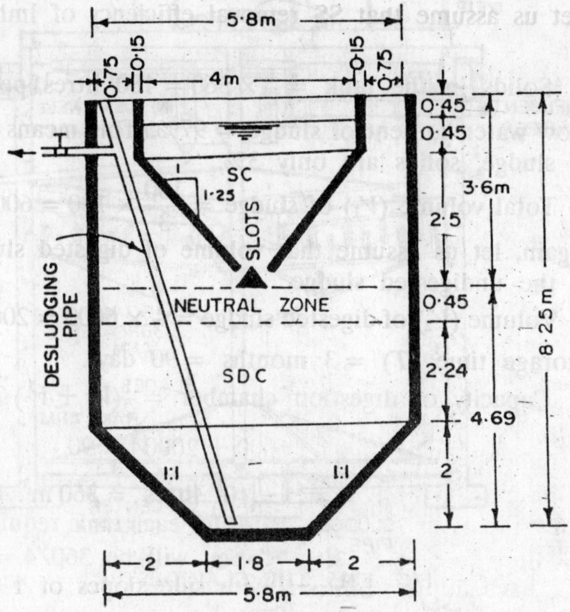

FIG. 7.14. IMHOFF TANK

17.9. THE CLARIGESTER

A clarigester is a modernized mechanical tank of the Imhoff type. Details of such a tank are shown in Fig. 17.15. This is a double storeyed tank, without bottom hopper, fitted with mechanical sludge and scum breaking equipment. In the upper or sedimentation portion of the tank is a *collector* which *concentrates* the sludge to the centre, whence if passes into the digestion chamber below. In the lower portion (digestion compartment), a sludge-stirring mechanism is provided which pushes the sludge towards a hopper from which it is withdrawn by a pump or by hydrostatic pressure. If required, the sludge can be heated by means of heating coils provided in the digestion chamber. Such mechanization reduce the labour requirements. Dorr-Oliver co. of N.Y. manufactures equipment required for clarigesters of diameters 4 to 12 m to treat the sewage for a population of 1500 to 20000 at the rate of 115 litres per capita per day.

Advantages (1) Lesser depth is required because of the elimina tion of hopper bottom. (2) Cost of construction is less (3) Operation is continuous and uniform. (4) Mechanical handling of sludge, and hence there are no labour problems. (5) No scum problem. (6) No foaming problem. (7) Sludge can be heated to improve digestion. (8) Better effluent. (9) Sludge gas can be collected in a dome and used.

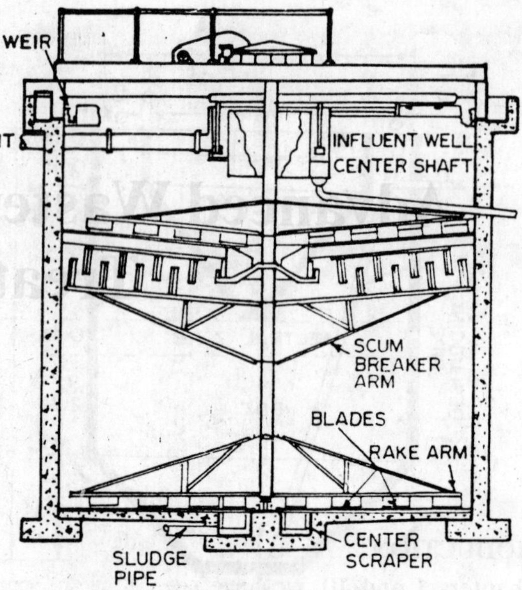

FIG. 17.15. THE CLARIGESTER

PROBLEMS

1. Describe, with the help of a neat sketch, the components of a septic tank, along with the functions of each.
2. Discuss the criterion for design of a septic tank.
3. Discuss the methods of disposal of septic tank effluent.
4. Design a septic tank for the following data :
   No. of Persons : 100
   Sewage/capita/day : 120 litres
   De-sludging period : 1 years.
5. (*a*) Discuss advantages and disadvantages of a septic tank
   (*b*) Discuss advantages and disadvantages of an Imhoff tank.
6. Compare and contrast septic tank with Imhoff tank in scope, function and performance.
7. Write a detailed note on design of Imhoff tank.
8. Design an Imhoff-tank to treat the sewage from a small town with 30000 population, assuming that the suspended solids in the influent sewage are 200 ppm. Water content of sludge is 97 percent. Design the tank for three months sludge storage. Rate of sewage is 135 litres per head per day. (UP.S.C. Engg. Services, 1968)
9. Write notes on the following :
   (*a*) Clarigester (*b*) Biofilter (*c*) Upflow filter
   (*d*) Dispersion trenches. (*e*) Seepage pit

18

Advanced Wastewater Treatment

18.1. INTRODUCTION

In chapters 1 and 10, we have seen that the *conventional (or traditional) wastewater treatment* consists of the following stages :

(*i*) Preliminary treatment (*ii*) Primary treatment, and (*iii*) Secondary (or biological) treatment.

By *conventional treatment* is meant that the treatment processes are directed towards the reduction of the objectional properties of water-carried waste and render it to less dangerous and repulsive to man. Early treatment objectives were concerned with the following: (*i*) Removal of suspended and floatable material (*ii*) Treatment of biodegradable organics and (*iii*) Elimination of pathogenic organics.

Later, major effort was undertaken to achieve more effective and widespread treatment of wastewater. As a consequence, the required degree of treatment was significantly increased, and additional treatment objectives and goals have been added. *The recently added treatment objectives are the removal of* (*i*) nutrients (like plant nitrogen, phosphorous etc), (*ii*) suspended solids (*iii*) BOD (*iv*) dissolved solids and (*v*) toxic substances. This phase of wastewater treatment is known as *advanced wastewater treatment* or *tertiary treatment*.

Thus *advanced wastewater treatment* (AWT) refers to methods and processes that remove more contaminants from the wastewater than are usually taken out by the usual conventional treatment and techniques. Several of the pollutants/contaminants not taken out by secondary biological methods can adversely affect aquatic life in the receiving waters, accelerate eutrophication of lakes and hinder reuse of surface water for domestic needs. *Renovation* and *reuse* of wastewater are becoming more important as water needs cannot be met by the relatively fixed natural supply available. The effluents from the ordinary secondary treatment may contain 300 mg/l of total solids, 200 mg/l of volatile solids, 30 mg/l of suspended solids, 30 mg/l of BOD, 20 mg/l of nitrogen and 7 mg/l of phosphorous, apart from traces of heavy metals and other chemicals. The aim of tertiary treatment is to remove these. Table 18.1 gives a list of popular AWT methods.

TABLE 18.1.
SELECTED ADVANCED WASTEWATER TREATMENT PROCESSES

Contaminant	*Removal methods*
1. Suspended solids	(*i*) Microscreening (*ii*) Diatomaceous-earth filters (*iii*) Ultra filtration (*iv*) Filtration through granular beds (*v*) Chemical coagulation and clarification
2. Organics	(*i*) Adsorption on granular activated carbon (*ii*) Extended biological oxidation
3. Phosphorous	(*i*) Biological-chemical precipitation and clarification (*ii*) Chemical coagulation and clarification (*iii*) Irrigation of crop land
4. Nitrogen	(*i*) Biological nitrification-denitrification (*ii*) Ammonia reduction by air stripping (*iii*) Breakpoint chlorination (*iv*) Ion change extraction (*v*) Irrigation of cropland

18.2. SUSPENDED SOLIDS REMOVAL

Suspended solids removal is one of the most important and common application of advanced wastewater treatment. The effluent from the secondary treatment plant may contain suspended solids concentration ranging from 20 to 40 mg/l, depending upon the type of method adopted. The suspended solids in treated wastewater are in part colloidal and in part discrete ranging from 10^{-3} to 100 microns. None of this material settles readily. Techniques used for reduction of suspended solids concentration are (*i*) microscreening, (*ii*) diatomaceous earth filters (*iii*) ultrafiltration (*iv*) granular media filtration and (*v*) chemical coagulation followed by sedimentation.

Microscreening. Microscreening utilizes a special woven metallic or plastic filter fabric which is mounted on the periphery of a revolving drum provided with continuous backwashing. The drum rotates about a horizontal axis at a variable low-speed (up to 4 rpm). The size of openings of the filter fabric may range from 23 to 60 μ. The drum operates submerged in the flowing wastewater to approximately two thirds of its depth. Wastewater enters through the open upstream end of the drum and flows radially outward through the microfabric, leaving beind the suspended solids. Back washing is accomplished by high pressure water jets placed at the highest point of the drum. The solids which are retained on the fabric are washed through a trough, which recycle the solids to the sedimentation tank. Water for back flushing is drawn from the filtered water effluent. Drum rotation and backwash are continuous and adjustable. The pressure head develops due to intercepted solids which build up on the inside of the microfabric and create a filtration mat capable of removing particles that are smaller than the aperture size of the mesh.

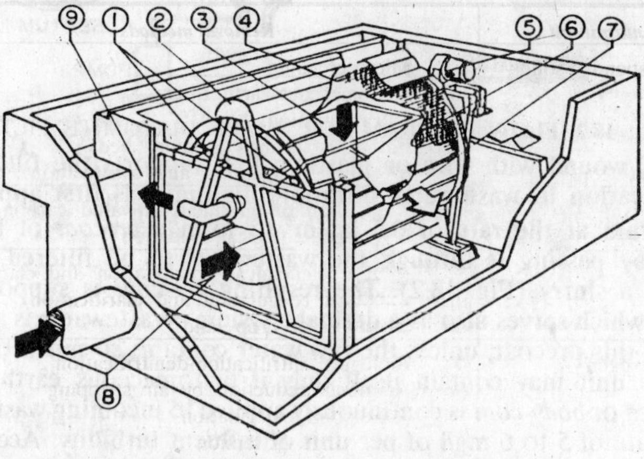

FIG. 18.1 MICROSTRAINER

The design parameters include mesh size, submergence, allowable head losses and peripheral speed. Typical hydraulic loading rate varies from 0.06 to 0.44 $m^3/m^2/min$, based on the submerged surface area of drum. Backwash water requirement may vary from 3 to 5% of the product water flow. Effluent suspended solids from microstrainers following activated sludge and trickling filter plant can be expected to range from 2 to 20 mg/l.

18.3. DIATOMACEOUS-EARTH FILTERS

Diatomaceous earth filters are available both as a vacuum and as a pressure filter. The diatomaceous earth consists of skeletons of diatoms 0.5 to 12 μ in size mined from deposits laid down in ancient seas. These filters are designed with either septums or leafs covered with screenings or fine slots having openings of 0.08 to 0.12 mm. The septum consists of a hollow, flueted and perforated

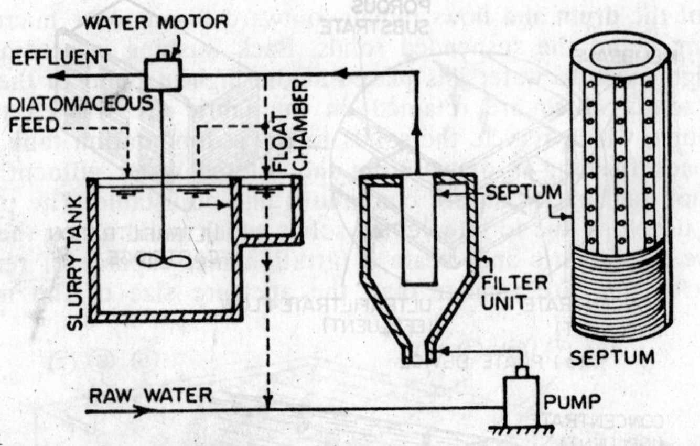

FIG. 18.2. FLOW DIAGRAM FOR SMALL DIATOMITE FILTER

cylinder, wound with wire or plastic which supports the filter cake. In application to wastewater filtration, diatomite is first applied on the septum at the rate of 0.5 kg/m^2 to form a *precoat* of filtering media, by passing it through the wastewater to be filtered, in the form of a slurry (Fig. 18.2). The resulting *pre-coat* is supported by septum which serves also as a drainage system. Wastewater is strained through this precoat, unless the raw water contains so much turbidity that the unit may contain itself only if diatomaceous earth, called *body-feed* or *body-coat* is continuously applied to incoming wastewater, at the rate of 5 to 6 mg/l of per unit of influent turbidity. Acceptable operation is reported to be possible only with influent suspended solids less than 13 mg/l and process-coats are high. Diatomaceous earth filtration produces clear effluents with removal efficiencies of over 90%. However, the costs are also high. Colloidal substances are usually not removed, unless coagulants are added. Coagulants are seldom added to diatomaceous earth filter influents because this drastically shortens the runs.

18.4. ULTRAFILTRATION

Ultrafiltration is a system similar in operation to reverse osmosis save that the membrane is far coarser and the pressure lower. Ultrafiltration membranes are thin film cast from organic polymer solutions, the film thickness being 5 to 10 mils (0.005 to 0.01 inches). The film is anisotropic *i.e.* it has an extremely thin separation layer (of thickness 0.1 to 10 μ) on a relatively porous structure, consisting of pores of closely controlled size ranging from 3 to 100 Å. Ultrafiltration membranes may be packed either as a *plate device* or as a *tube device*, as shown in Fig. 18.3 (*a*) and (*b*) respectively.

Ultrafiltration technique is used in conjunction with biological oxidation processes such as activated sludge process. When the effluent

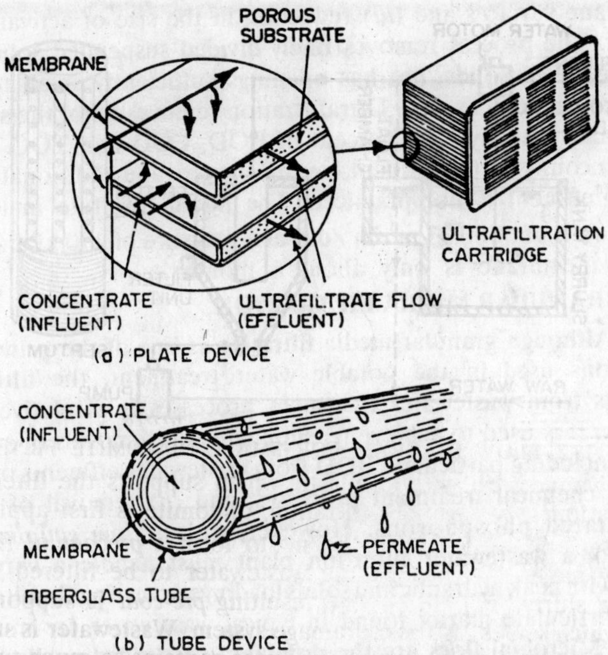

FIG. 18.3. ULTRAFILTRATION MEMBRANE PACKINGS

from the activated sludge process is passed through ultrafiltration system (Fig. 18.4), the filter membranes filter out the biological cells while allowing passage of treated effluent. The retained solids are returned to the activated sludge reactor. This results in high concentration of biological solids in the activated sludge reactor, resulting in (*i*) quick degradation of organics, (*ii*) prevention of fouling of

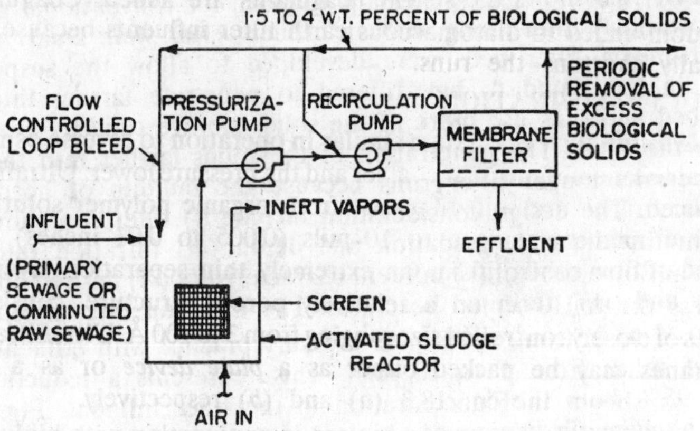

FIG. 18.4. COMBINATION OF UILTRAFILTRATION AND BIODEGRADATION BY ACTIVATED SLUDGE.

membrane surfaces and (*iii*) reduction in the size of activated sludge reactor. The process removes finely divided suspended solids, rather than dissolved solids, but has openings sufficiently small to remove colloidal particles as well. Ultrafiltration removes all of the suspended solids and virtually 100 percent of BOD, COD and TOC. Coliform bacteria counts in the effluent have been zero, and additional stabilization of effluent is not practiced. The liquid flux rate in long term operation is about 0.33 m^3/m^2/day at a pressure of 170 kPa. However, life of membrane is only about 6 months.

18.5. GRANULAR MEDIA FILTRATION

Although granular media filtration is one of the principal unit operations used in the potable water treatment, the filtration of effluents from wastewater treatment processes is a relatively recent practice. It is used to achieve (*i*) supplemental removals of suspended solids, including particulate BOD from wastewater effluents of biological and chemical treatment processes, and (*ii*) removal of chemical precipitated phosphorous. However, unless *flow equalization* is provided, a wastewater filtration plant must handle a varying rate of flow with peak hydraulic and solids loadings occurring simultaneously. Also, particulate matter found in typical wastewater are much more 'sticky', Microbial flocs are the dominant suspended solids following secondary biological treatment, while carry over from biological-chemical and physical-chemical methods contain a significant quantity of coagulant residue.

Apart from conventional downflow filters, various other filter configurations that can be used for this purpose include, (*a*) upflow filter with surface grid, (*b*) biflow filter (*c*) dual-media, downward-flow filter and (*d*) mixed media (triple media). These are shown in fig. 18.5. Fig. 18.6 shows the schematic diagram for conventional gravity (or downflow) filter along with an effluent clear well. Dual media and tri-media filter beds were developed to allow the suspended solids in the liquid to be filtered to penetrate farther into the filter bed and thus use more of the solids-storage capacity available within the filter. The penetration of the solids farther into the bed also permits longer filter runs because the buildup of head loss is reduced. The design consideration include (*i*) filter configuration, (*ii*) filter media size, gradation and depth (*iii*) filtration rate (*iv*) method of flow control (*v*) influent characteristics (*vi*) desired effluent quality and (*viii*) terminal head loss. The filtration rates may range from 0.06 to 0.50 m^3/m^2/min in ordinary practice with effluent suspended solids from 1 to 10 mg/l. This represents a reduction of 20 to 95% from the concentration in the filter influent. Fig. 18.7 shows a schematic diagram of a biological treatment plant with tertiary filtration. For operational flexibility, atleast two and usually four filter cells are provided. Length of filter runs should be minimum 6 hours and maximum 40 hours.

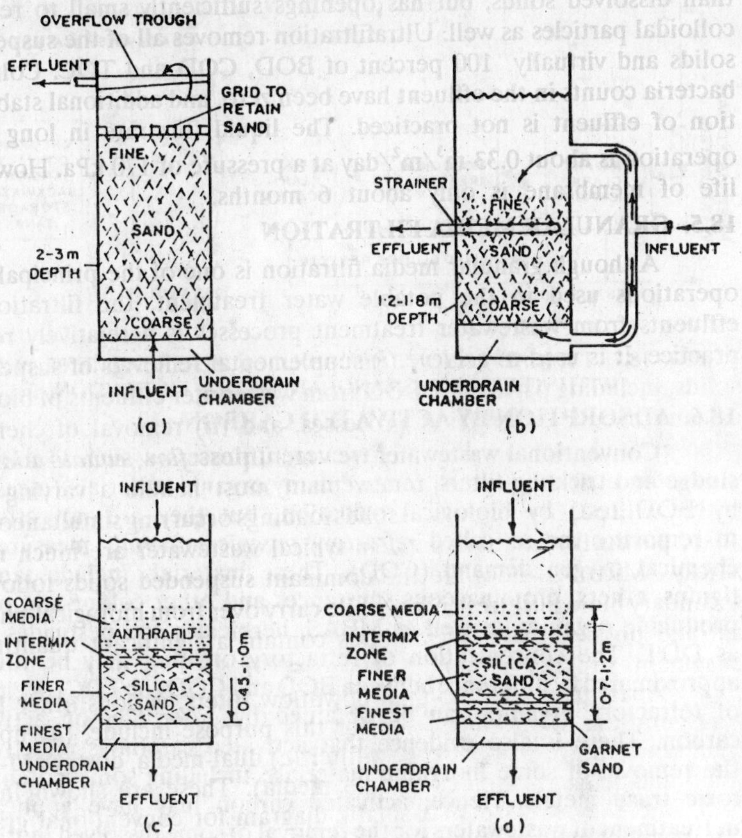

FIG. 18.5. GRANULAR-MEDIA FILTER CONFIGURATIONS.

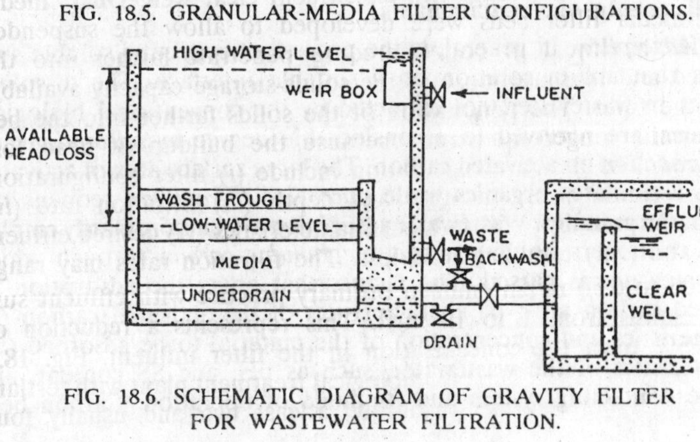

FIG. 18.6. SCHEMATIC DIAGRAM OF GRAVITY FILTER FOR WASTEWATER FILTRATION.

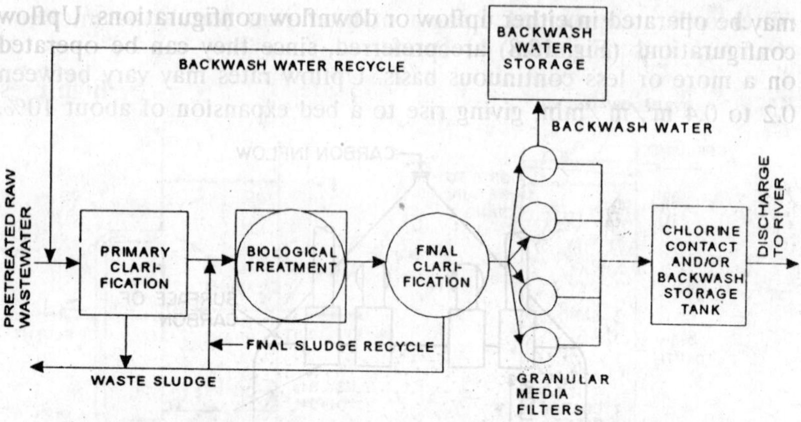

FIG. 18.7. TYPICAL LAYOUT OF BIOLOGICAL TREATMENT PLANT WITH TERTIARY GRANULAR MEDIA FILTRATION.

18.6. ADSORPTION BY ACTIVATED CARBON

Conventional wastewater treatment processes, such as activated sludge and trickling filters, remove many organic materials (measured by BOD test) by biological oxidation, but they are not effective in removing the so called *refractory organic materials* measured by chemical oxygen demand (COD). These materials include tannins, lignins, ethers, proteinaceous substances and other colour and odour producing organics, as well as MBAS, herbicides and pesticides such as DDT. The concentration of refractory organics may be roughly approximated by difference between BOD and COD (or TOC). Removal of refractory organics can be obtained by *adsorption* on activated carbon. There is also evidence that activated carbon is effective in the removal of some inorganic materials, including some potentially toxic trace metals. Hence, activated carbon may have application in treatment of wastewaters for the removal of some dissolved inorganic as well as organic contaminants.

Adsorption, in general, is the process of collecting soluble substances that are in solution on a suitable interface. The dissolved organics in wastewater, not extracted by the conventional biological treatment, are removed to a considerable degree by *adsorption* and *bio-degradation* on activated carbon. The large surface area of activated carbon assimilates organics while *microbial degradation* reopens the pores in the granules. Wastewater adsorption systems ordinarily employ closed containers holding a bed of *granular carbon* through which the wastewater is passed. The factors that affect the adsorption by activated carbon are : (*i*) characteristics of the activated carbon (*ii*) characteristics and concentration of the material to be adsorbed (*iii*) characteristics of the wastewater, such as pH, and SS content, and (*iv*) the contacting system and its mode of operation. Carbon beds may be operated in either upflow or downflow configurations. Upflow configurations (Fig. 18.8) are preferred, since they can be operated on a more or less continuous basis. Upflow rates may vary between 0.2 to 0.4 $m^3/m^2/min$, giving rise to a bed expansion of about 10%.

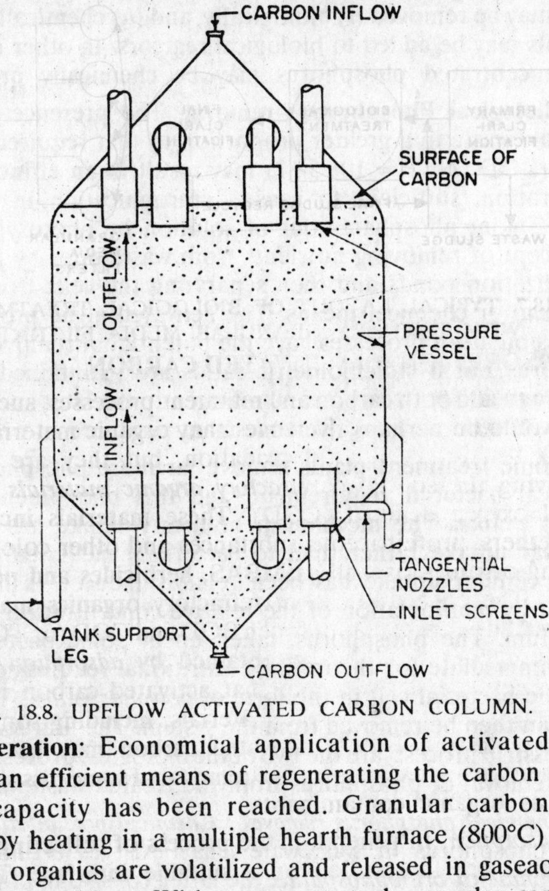

FIG. 18.8. UPFLOW ACTIVATED CARBON COLUMN.

Regeneration: Economical application of activated carbon depends on an efficient means of regenerating the carbon after its adsorptive capacity has been reached. Granular carbon can be regenerated by heating in a multiple hearth furnace (800°C) wherein the adsorbed organics are volatilized and released in gaseaous form from the carbon surface. With proper control, granular carbon can be restored to a *near virgin* adsorptive capacity with only 5-10% weight loss. Thermal generation involves drying, backing (pyrolysis of adsorbates), and activating by oxidation of the remaining residue.

18.7. PHSOPHORUS REMOVAL

Principal sources of phosphorus in wastewater are man-generated wastewater, agricultural return water and land runoff. Phosphorus, like nitrogen contributes to the eutrophication of surface waters. It is often cited as culprit responsible for the stimulation of aquatic plants, since the concentration of phosphorus necessary to support algal blooms is only 0.005 to 0.05 mg/l as P. Phosphorus in wastewater may be present in three forms : (*i*) orthophosphate, (*ii*) polyphosphate and (*iii*) organic phosphorus. Typically, phophorus enters the domestic wastewater from human body wastes, from food wastes discharged to the sewers, from kitchen grinders, and from condensed inorganic phosphate compounds used in various household detergents. Phosphorus may be removed (*i*) biologically, and (*ii*) chemically. Sometimes, chemicals may be added to biological reactors. In other cases, biologically concentrated phosphorus may be chemically precipitated.

Biological Phosphorous removal. The presence of phosphate in the raw waste in a greater proportion to that required for bacterial growth (BOD : N : P = 10 : 5 : 1) may result in an effluent phosphate concentration, sufficient to cause eutrophication in the receiving stream. Taking phosphates out of solution by photosynthesis led to the concept of removing nutrients from wastewater by growing algae in stabilization ponds and then separating the cells from suspension by physical or chemical means. The activated algae process and bacterial-assimilation processes have the potential of removing additional phosphorus, but if stoichiometric ratios are maintained, it would be necessary to add both carbon and nitrogen to the later and the resultant sludge would be perhaps five times that produced in ordinary plants.

Some treatment plants remove far more phosphorus than the theoretical bacterial requirement, and this phonomenon is known as *luxury intake*. The mechanism of luxury intake has been related to cellular uptake rather than to bacterially mediated precipitation, and the cellular uptake has been reliably induced in laboratory experiment by manipulation of the relative concentration of potassium and sodium. The phosphorus, taken up by some bacteria in excess of their immediate (or theoretical) nutritional requirements, is stored within the bacterial cell in the form of *polyphosphates*. The polyphosphates can then be removed from the system. The 'Bardenpho process' and 'Phosstrip process' are the two of the biological processes developed for the removal of phosphorus from the treated wastewater effluents.

Chemical phosphorus removal : Chemical precipitation. As stated earlier, phosphorus in wastewater may exist as *organic phosphate*, *polyphosphate* or *orthophosphate*, the latter consisting of four different ionic forms. For simplicity, phosphorus is considered to be present as phosphate ion (PO_4^{3-}). Phosphate can be removed by *chemical precipitation*, and principal chemicals used for this purpose are lime, alum and ferric chloride. Reactions with calcium, aluminium and iron ions are as follows :

(*i*) *Calcium ion*

$$10\,Ca^{++} + 6\,PO_4^{3-} + 2\,OH^- \Leftrightarrow \underset{\text{Hydroxyapatite}}{Ca_{10}(PO_4)_6(OH)_2} \quad \text{...(18.1)}$$

(*ii*) *Aluminum ions*

$$Al^3 + H_n\,PO_4^{3-n} \Leftrightarrow \underset{\text{Aluminum phosphate}}{Al\,PO_4} + n\,H^+ \quad \text{...(18.2)}$$

(*iii*) *Iron ions*

$$Fe^{3+} + H_n PO_4^{3-n} \Leftrightarrow \underset{\text{Ferric phosphate}}{Fe PO_4} + n H^+ \quad ...(18.3)$$

Fig. 18.9 shows a variety of different points at which the chemicals can be added in the treatment process. However, since polyphosphates and organic phosphorus are less easily removed than orthophosphorus, adding aluminum or iron salts *after secondary treatment* (where organic phosphorus and polyphosphorus are transformed into orthophosphorus) gives best removal.

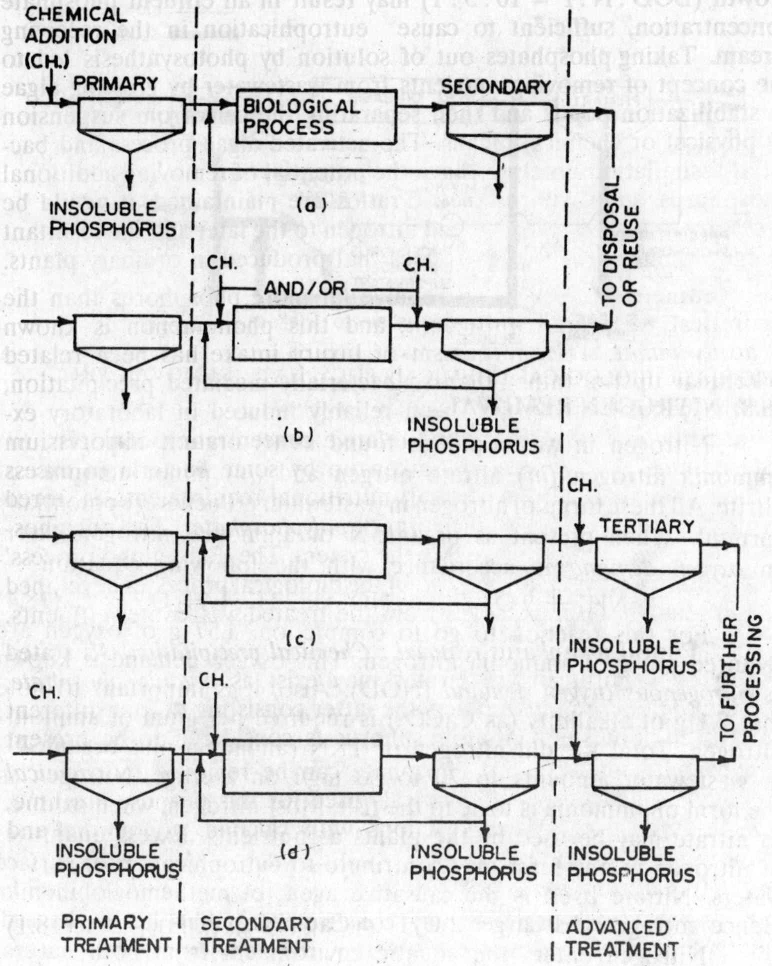

FIG. 18.9. ALTERNATIVE POINTS OF CHEMICAL ADDITION FOR PHOSPHORUS REMOVAL

Biological-Chemical Phosphate-removal. Fig. 18.10 shows the schematic of a novel biological-chemical process. It utilizes an activated sludge system and an anaerobic cell in which phosphorus taken up in the aeration tank (or activated sludge reactor) is released to the liquid phase producing a concentrated phosphate stream and a phosphate deficient sludge which is returned to the aeration tank where the uptake of phosphorus is repeated. Phosphorus is precipitated by dosing the concentrated phsophate stream with lime. The precipitated phosphorus is then removed from the liquid stream in the primary clarifier.

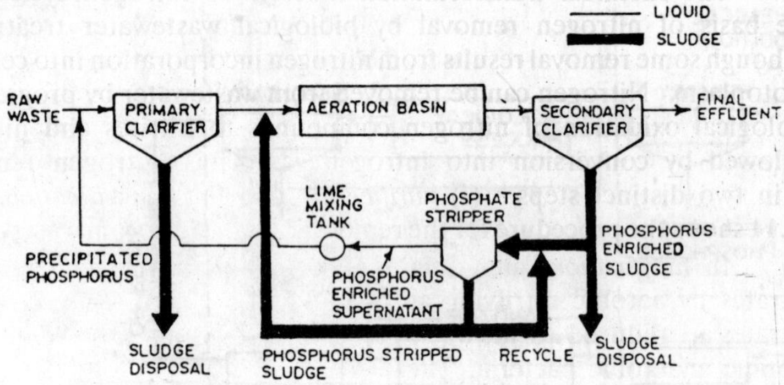

FIG. 18.10. BIOLOGICAL-CHEMICAL PHOSPHATE REMOVAL PROCESS.

18.8. NITROGEN REMOVAL

Nitrogen in wastewater is found as (*i*) organic nitrogen, (*ii*) ammonia nitrogen (*iii*) nitrate nitrgen and (*iv*) minor amounts of nitrite. All these forms of nitrogen in wastewater effluents are potentially harmful. When present as organic-N or ammonia, nitrogen exerts an *oxygen demand* in accordance with the following equation :

$$NH_4^+ + 2\,O_2 \longrightarrow NO_3^- + 2\,H^+ + H_2O \qquad ...(18.4)$$

For this reaction to go to completion, 4.57 g of oxygen are required per g of ammonia nitrogen. This oxygen demand is known as *nitrogenous oxygen demand* (NOD). Also, it is important to note that 7.1 g of alkalinity (as $CaCO_3$) is required per gram of ammonia nitrogen. Total Kjeldhl nitrogen or TKN (ammonia plus organic-N) in wastewater amounts to 20 to 30 mg/l on average. Nitrogen in the form of ammonia is toxic to the fish. Most nitrogen, when oxidized to nitrate may be used by the plants as nutrients. Excess quantities of nitrogen in any form may contribute to eutrophication of surface waters. Nitrate itself is the causative agent of methemoglobinemia. Hence nitrogen discharges may constitute a public health hazard.

Nitrogen enters the aquatic environment from both natural and man made sources. Natural sources include precipitation, dustfall, nonurban run off and biological fixation. Man made sources, which contribute nitrogen in domestic wastewater are faeces, urine and food processing discharges. The largest single source of nitrogen is urea, which together with ammonia, comprises approximalety 85% of the nitrogen excreted by man.

Following are the popular methods of removal of nitrogen from the waste water :

1. Biological nitrification-denitrification 2. Ammonia stripping

18.9. BIOLOGICAL NITRIFICATION-DENITRIFICATION

The chemical transformation inherent to nitrogen cycle form the basis of nitrogen removal by biological wastewater treatment, although some removal results from nitrogen incorporation into cellular protoplasm. Nitrogen can be removed from wastewater by progressive biological oxidation of nitrogen compounds to nitrites and nitrates followed by conversion into introgen gas. Thus, nitrogen removal is in two distinct steps : (*i*) *nitrification* and (*ii*) *denitrification*. Fig. 18.11 shows the procedure for the removal for nitrogen from wastewater.

In *nitrification*, ammonia is oxidised to nitrites and then to nitrates by aerobic nitrifying autotrophic bacteria. In *denitrification*, nitrates are reduced to nitrogen gas by either autotrophic or heterotrophic anaerobic bacteria. The nitrogen forms of interest in AWT are (*i*) organic (*ii*) inorganic and (*iii*) gasesous nitrogen. As a first step, *bacterial decomposition* releases ammonia by *deamination* of nitrogenous organic compounds :

$$\underset{\text{(organic)}}{N} \xrightarrow[\text{(deamination)}]{\text{Bacterial Decomposition}} NH_3 \quad ...(18.5)$$

Continued aerobic oxidation results in *nitrification* as under:

$$NH_3 + O_2 \xrightarrow[\text{(nitrification)}]{\text{Nitrifying bacteria}} NO_3^- \quad ...(18.6)$$

As a further step, *biochemical denitrification* occurs with heterotrophic metabolism in an anaerobic environment :

$$NO_3^- \xrightarrow{\text{Bacterial denitrification}} N_2 \uparrow \quad ...(18.7)$$

Thus, Eqs. 18.5, 18.6 and 18.7 give three reactions in sequence defining biological nitrification-denitrification process.

Biological nitrification. Nitrification is the process in which ammonia is first converted into nitrate form, thereby eliminating problems of toxicity to fish and reducing to nitrogen oxygen demand (NOD) in streams. Ammonia oxidation to nitrate is a diphasic process performed by autotrophic bacteria with nitrite as intermediate product (Eq. 18.6). These aerobic reactions yield energy for metabolic functions such as synthesis of carbon dioxide into new cell growth. The oxidation of ammonia to nitrate is a *two-step process* if carried to completion. In the *first step*, ammonia is oxidized to nitrite by the *Nitrosonomas*,

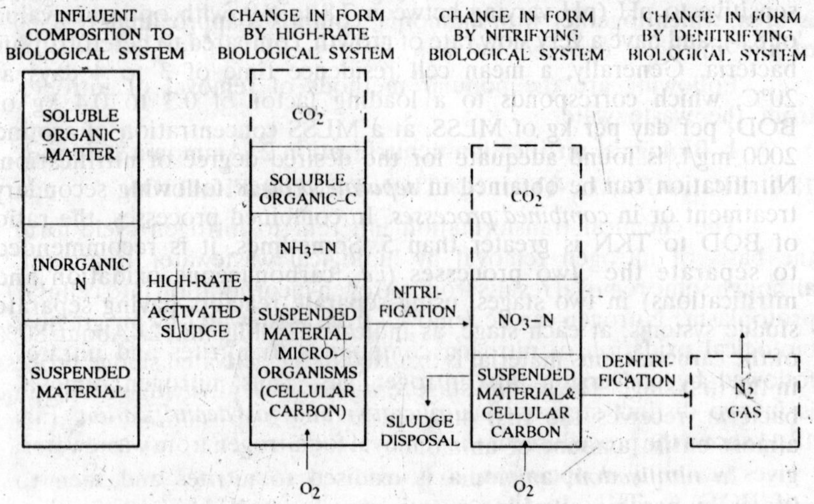

FIG. 18.11. REMOVAL OF ORGANIC AND NITROGENOUS POLLUTANTS FROM WASTEWATER

a genera of strict aerobic autotrophic bacteria that utilize ammonia as their sole source of energy. This is indicated by the following reaction equation :

$$NH_4 + 1.5\,O_2 \xrightarrow{\text{Nitrosonomas}} NO_2^- + 2\,H^+ + H_2O + \text{Energy} \quad ...(18.8)$$

The *second step*, the conversion of nitrite to nitrate, is accomplished by the *Nitrobacter*, which is a genera of autotrophic bacteria which utilize nitrite as their sole energy source. This is indicated by the following reaction equation :

$$NO_2^- + 0.5\,O_2 \xrightarrow{\text{Nitrobacter}} NO_3^- + \text{energy} \quad ...(18.9)$$

The important parameters in the *nitrification kinetics* are (*i*) temperature, (*ii*) pH and (*iii*) DO concentration. The rate of nitrification in wastewater, being essentially linear, is a function of time and independant of ammonia-nitrogen concentration.

The biological processes used for nitrification are identified as aerobic *suspended growth* and aerobic *attached growth*. Nitrification can be achieved by any of the suspended-growth activated sludge processes by the maintenance of conditions suitable for the growth of nitrifying organisms. The two attached-growth processes that can be used for combined carbon oxidation nitrification processes are the trickling filter and rotating biological contactor.

In the suspended growth process, nitrification is achieved in an activated sludge process along with the usual carbonaceous oxidation, with some modifications in the design. Nitrifying bacteria are very sensitive to pH (pH varying between 7.8 to 8.9, with optimum value of 8.4), and have a very slow rate of growth, compared to heterotrophic bacteria. Generally, a mean cell residence time of 3 to 4 days at 20°C, which corresponds to a loading factor of 0.3 to 0.4 kg of BOD_5 per day per kg of MLSS, at a MLSS concentration of around 2000 mg/l, is found adequate for the desired degree of nitrification. Nitrification can be obtained in *separate process* following secondary treatment or in *combined processes.* In combined processes, the ratio of BOD to TKN is greater than 5. Sometimes, it is recommended to separate the two processes (*i.e.*, carbonaceous oxidation and nitrifications) in two stages, using separate reactors having separate sludge systems, at each stage, as indicated in Fig. 18.12. About 80% of the carbonaceous material is oxidised in the activated sludge process in the first stage. The second stage, enriched with nitrifying autotrophic bacteria, receives the first stage effluent with residual carbon and almost entire amount of ammonia, for nitrification. Such a scheme gives flexibility in operation, to give optimum performance. The ratio of BOD to TKN in the second stage is less than 3.

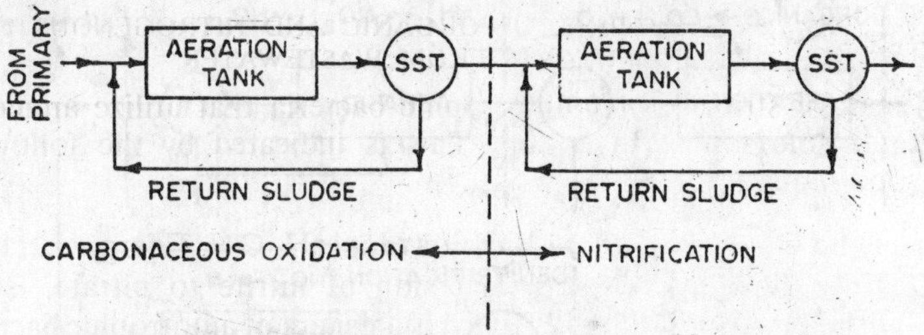

FIG. 18.12. TWO STAGE BIOLOGICAL NITRIFICATION

Biological Denitrification. *Denitrification* is a biological process which may be applied to nitrified water in order to convert nitrate to nitrogen. Biological denitrification is achieved under anaerobic conditions by heterotrophic micro-organisms that utilize nitrate as a hydrogen acceptor. In this process, the nitrate present in the waste process essentially requires a *fully nitrified influent*, an anaerobic condition and a supply of proper substrate. In the process of biological denitrification, a balanced amount of substrate is supplied to reduce the nitrate in the process of stabilization of the supplied substrate under anaerobic conditions.

In the denitrification process, the heterotrophic bacteria found in activated sludges reduce NO_2^- and NO_3^- to nitrogen gas. However, these organisms require an organic carbon source for cell growth. *Methanol* (CH_3OH) has been used as a carbon source in denitrification processes. The organisms metabolize mathanol according to the following reaction :

$$NO_3^- + \frac{5}{6} CH_3OH \rightarrow \frac{1}{2} N_2 + \frac{5}{6} CO_2 + \frac{7}{6} H_2O + OH^- \quad ...(18.10)$$

From the above equation we note that 5/6 mole of methanol is required to reduce completely one mole of nitrate to molecular nitrogen, or 1.9 mg/l of CH_3OH for 1 mg/l of nitrate on weight basis. The overall reaction requires methanol for the reduction of nitrite, nitrate and dissolved oxygen (DO) in the nitrified wastewater. St. Amant and McCarty have suggested the following equation for *methanol demand* :

$$C_m = 2.47\,(NO_3) + 1.53\,(NO_2) + 0.87\,(DO) \quad ...(18.11)$$

in which C_m is the methnol required in mg/l, and NO_3, NO_2 and DO are the concentrations, respectively of nitrate, nitrite and DO in mg/l as N and O_2.

Fig. 18.13 shows a three stage nitrification-denitrification process.

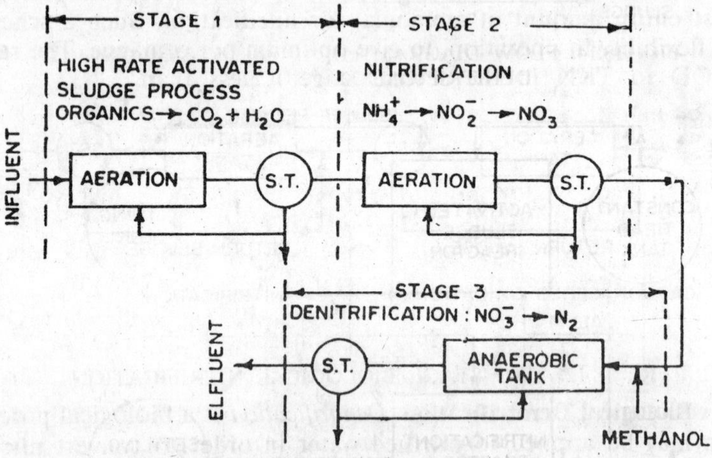

FIG. 18.13. THREE STAGE NITRIFCATION-DENITRIFICATION PROCESS.

Both suspended-growth systems as well as fixed-growth systems have been tried for denitrification. The anaerobic (or submerged) upflow filters are found to be more effective in denitrification of secondary effluents. Pilot plant studies indicate that a mean cell residence time of 3 to 4 days is required for 90% reduction of nitrate in complete mix suspended growth systems while a little more than 2 hours of detention time is adequate for similar efficiency in an upflow anaerobic filter–probably due a very high mean cell residence time established in those filters. Upflow filter beds appear to have the advantages of efficient operation in cold temperature and greater resistance to shock loadings than suspended-growth systems, but clogging of the media with biological growth is an unresolved problem.

The process of denitrification by suspended growth system consists of a complete-mixing basin followed by a clarifier for sludge separation and return as shown in Fig. 18.14. Underwater stirrers mix the contents sufficiently to keep the microbial floc in suspension

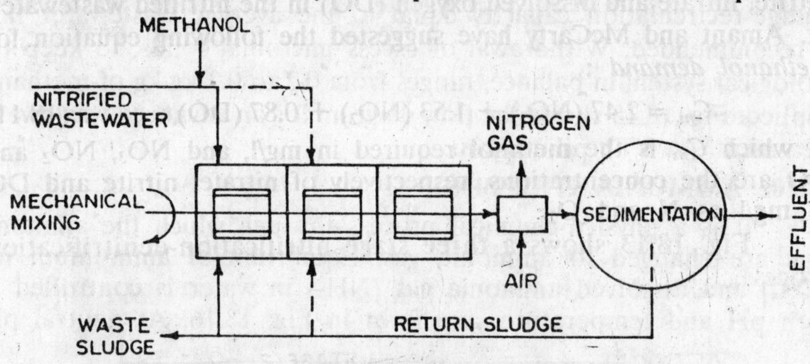

FIG. 18.14. FLOW DIAGRAM FOR COMPLETE MIXING DENITRIFICATION PROCESS.

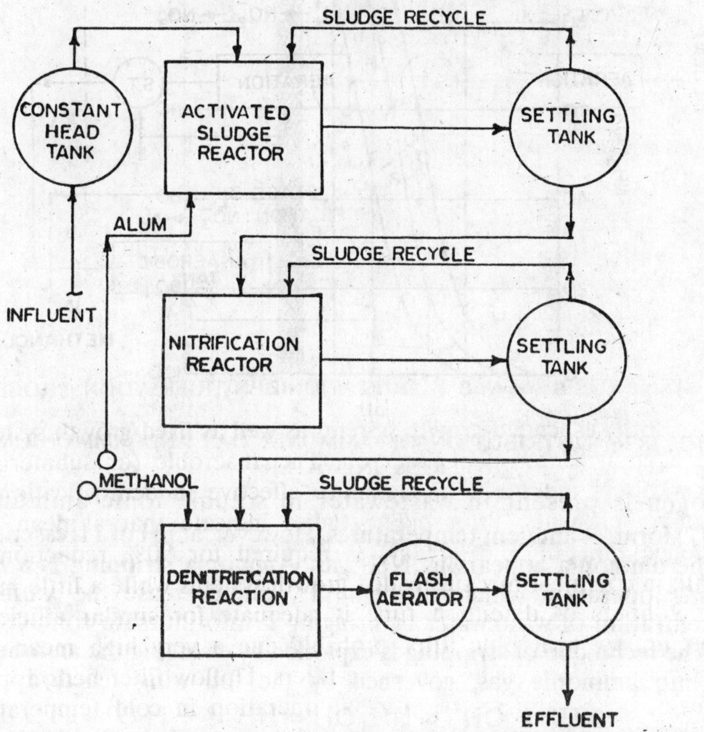

FIG. 18.15. ADVANCED WASTEWATER TREATMENT PILOT PLANT

without producing undue aeration. The reactions during denitrification process form CO_2 and N gas bubbles that inhibit gravity settling by adhering to the biological floc. Hence an open channel or tank known as flash aerator is provided between the denitrification basin and final clarifier, to provide short term aeration as to relieve the supersaturation of the mixed liquor with gases. Due to such *air-stripping*, the settleability of sludge solids in the sedimentation tank increases. Sludge-recirculation capacity equal to the average wastewater flow is recommended. Withdrawal of excess microbial solids, to keep the biological system in balance, ranges from 0.2 to 0.3 kg/kg of methanol applied. Fig. 18.15 shows the flow diagram for an advanced wastewater treatment (AWT) pilot plant.

18.10. AMMONIA STRIPPING

It is a physico-chemical process through which the ammonia ions are changed to ammonia gas. Equibrium of ammonium ion (NH_4) and dissolved ammonia gas (NH_3) in water is controlled by both pH and temperature, as shown in Fig. 13.16. At neutral pH,

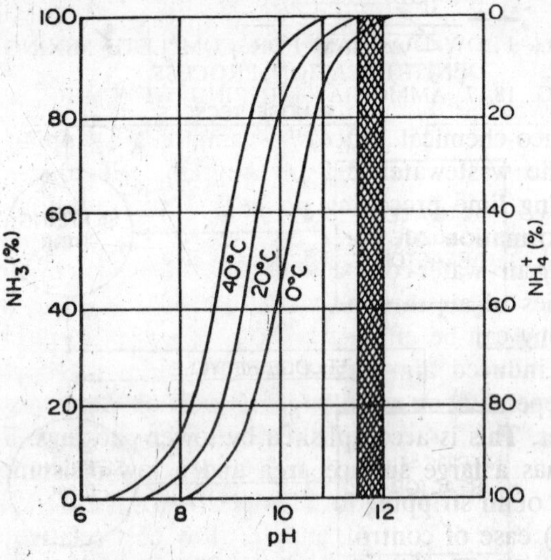

FIG. 18.16. DISTRIBUTION OF AMMONIA GAS AND AMMONIUM IONS IN WATER

nitrogen is present in wastewater in soluble ionic ammonium (NH_4^+) form, at ambient temperatures. However, at pH of 11, essentially all the ammonia appears as NH_3 gas. Ammonia stripping is a *mass transfer* operation which, at high pH, can decrease the ammonia concentration of wastewater by bringing it into intimate contact with air. The technique of stripping is exploited by changing the ammonium ion into ammonia gas, governed by the following reaction:

$$NH_4^+ + OH^- \Leftrightarrow NH_4\,OH \Leftrightarrow NH_3 \uparrow + H_2O \qquad ...(18.12)$$

The increase in the pH (*i.e.* the OH^- concentration) shifts the reaction to the right to create more free ammonia (NH_3). The free ammonia (*i.e.* ammonia gas) is air stripped in a stripping tower shown in Fig. 18.17.

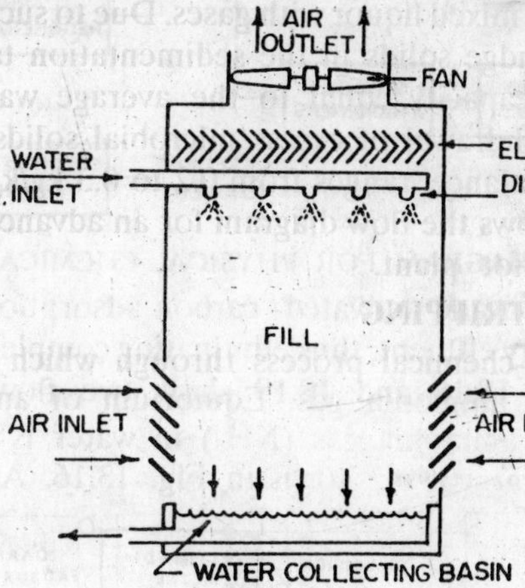

FIG. 18.17. AMMONIA STRIPPING WITH AIR

The physico-chemical process of ammonia stripping consists of (*i*) raising the wastewater pH in the range of 10.8 – 11.5, by means of applying lime preceding phosphate precipitation, (*ii*) formation and reformation of water droplets in the stripping tower and (*iii*) providing air- water contact and droplet agitation by circulation of large quantities of air through the tower. The cooling tower for ammonia stripping can be either of the crossflow or counter current flow type with induced air circulation. The efficient performance of the tower is dependent on a prolonged and intimate contact between the air and water. This is accomplished by tower packings. The ideal tower packing has a large surface area and a low resistance to air flow. *Advantages* of air stripping for nitrogen removal are (*i*) simplicity of operation, (*ii*) ease of control, and (*iii*) low cost relative to other extraction processes. The *disadvantage* are (*i*) inability to operate below at ambient temperature below 32°F, and (*ii*) deposition of calcium carbonate scale on the packing.

18.11. PHYSICO CHEMICAL WASTEWATER TREATMENT

When large quantities of industrial effluents are added to the domestic wastewater, the resulting wastewater mixture is rendered *untreatable* by biological means. In such situations, an *independent* physical-chemical treatment is an alternative approach. In physico-chemical treatment the most important operations are chemical

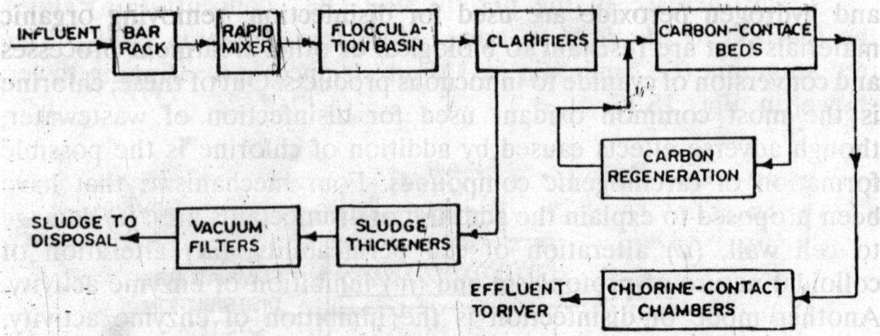

FIG. 18.18. FLOW DIAGRAM FOR PHYSICAL-CHEMICAL TREATMENT

coagulation, filtration and activated- carbon adsorption of either raw wastewater or primary effluent, thus eliminating completely the biological treatment. Fig. 18.18 and 18.19 show two flow sheets.

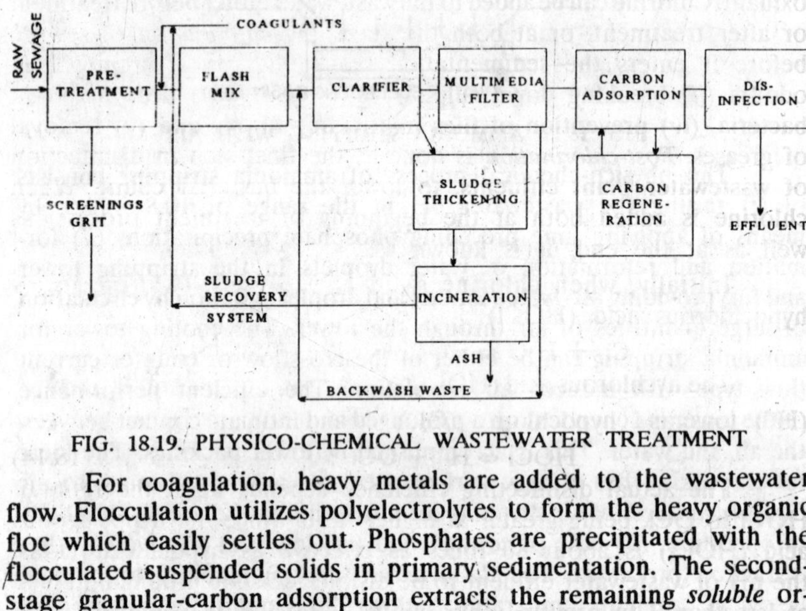

FIG. 18.19. PHYSICO-CHEMICAL WASTEWATER TREATMENT

For coagulation, heavy metals are added to the wastewater flow. Flocculation utilizes polyelectrolytes to form the heavy organic floc which easily settles out. Phosphates are precipitated with the flocculated suspended solids in primary sedimentation. The second-stage granular-carbon adsorption extracts the remaining *soluble* organics, and filtration either by the carbon column, or by separate multimedia beds (Fig. 18.19), to clarify the plant effluent. The treated effluent from the carbon column is usually chlorinated before discharge to the environment. The principal *drawback* to this method of treatment, which has limited its widespread use, is the handling and disposal of the great volumes of sludge resulting from the addition of chemicals.

18.12. DISINFECTION OF SEWAGE : CHEMICAL OXIDATION

Disinfection of treatment plant effluent is done to further reduce the pathogens. Chemical oxidants such as chlorine, ozone and hydrogen peroxide are used for disinfection, removing organic materials that are resistant to biological or other treatment processes and conversion of cyanide to innocuous products. Out of these, chlorine is the most common oxidant used for disinfection of wastewater, though adverse effects caused by addition of chlorine is the possible formation of carcinogenic compounds. Four mechanisms that have been proposed to explain the addition of disinfectants are : (*i*) damage to cell wall, (*ii*) alteration of cell permeability, (*iii*) alteration of colloidal nature of protoplasm and (*iv*) inhibition of enzyme activity. Another mode of disinfection is the inhibition of enzyme activity. Oxidizing agents such as chlorine, can alter the chemical arrangement of enzymes and inactivate the enzymes.

Use of Chlorine. *Chlorine* is very effective in destroying pathogenic bacteria on contact at low concentrations and it remains effective for many days. It is a practical and relatively inexpensive oxidant. Chlorine can be added to the wastewater either before treatment or after treatment, or at both the stages. *Prechlorination* of sewage before it enters the sedimentation tanks helps in (*i*) removal of odours (*ii*) breaking down sulphur compounds, (*iii*) destruction of bacteria, (*iv*) prevention of flies in trickling filters, and (*v*) removal of greases. *Post-chlorination* is done as the final step in disinfection of wastewater plant effluents, to lower the bacterial count. When chlorine is added both at the beginning of treatment process, as well as at the end, it is known as *split chlorination.*

Initially, when chlorine is added to water, chlorine forms hypochlorous acid (HOCl) :

$$Cl_2 + H_2O \Leftrightarrow HOCl + H^+ + Cl^- \qquad ...(18.13)$$

The hychlorous acid (HOCl) *ionizes* or dissociates into hydrogen (H^+) ions and hypochlorite (OCl^-) ions :

$$HOCl \Leftrightarrow H^+ + OCl \qquad ...(18.14)$$

The actual disinfecting efficiency depends upon the ratio of HOCl to OCl, being greater at higher ratio. Since the hypochlorous acid (HOCl) is about 80 times as effective as hypochlorite ions, the pH of wastewater effluent to be disinfected should be maintained at less than 7 to prevent ionization of HOCl. Also, if any reducing agents such as ferrous ions or H_2S are present in wastewater, chlorine reacts with them :

$$H_2S + 4\,Cl_2 + 4\,H_2O \rightarrow H_2SO_4 + 8\,HCl \qquad ...(18.15)$$

Due to this, the concentration of chlorine available to destroy pathogenic bacteria is reduced. Wastewater usually contain ammonia, due to which HOCl reacts to form sequentially : monochoramine (NH_2Cl), diachloramine ($NHCl_2$) and trichloramine (NCl_3).

Typical chlorine dosages required for disinfection are given in Table 18.2.

TABLE 18.2. TYPICAL CHLORINE DOSAGES FOR DISINFECTION

Effluent from	*Chlorine dosage (mg/l)*
1. Untreated waste water	6–25
2. Primary sedimentation	5–20
3. Chemical precipitation	2–6
4. Trickling filter	3–15
5. Activated sludge plant	2–8
6. Multi-media filter following activated sludge plant	1–5

PROBLEMS

1. What do you understand by advanced wastewater treatment ? How it is different from the conventional treatment ? Give, in a tabular form, important AWT processes.
2. Discuss in brief various methods used for suspended solids removal through AWT process.
3. Write a detailed note on ultrafiltration.
4. Describe in detail the granular media filtration methods of suspended solids removal during advanced wastewater treatment.
5. Describe the method of adsorption by activated carbon. How granular carbon is regenerated.
6. Discuss in brief the biological and chemical methods of removal of phosphorous from wastewater.
7. Why it is necessary to remove nitrogen from the secondary plant effluents. Mention different methods used for this. Describe ammonia stripping method.
8. Discuss in brief biological nitrification-denitrification for removal of nitrogen from wastewater.
9. Discuss, with the help of a neat flow diagram, physico-chemical method of wastewater treatment.

19

Industrial Wastewater Treatment

19.1. INTRODUCTION

Industrial wastes are the one which result from industrial processing operation. Though industrial wastes include liquid, solids and gases, we are presently concerned with the liquid part, which is commonly known as industrial wastewater. Industrial wastewaters are either discharged directly into the receiving water (if they meet the necessary standards) or else are discharged into municipal sewers. Some times, industrial wastewaters are *treated* partially before their discharge into sewers, or else are treated *separately* through suitable treatment processes so that the treated effluent is safe. From the standpoint of characteristics of their wastes, various industries can be divided into following four groups :

(*i*) Food and food-processing industries like sugar, canning, dairy, fermentation etc.

(*ii*) Industries of mineral products, *i.e.* coal coke and gas, iron and steel, oil, petroleum etc.

(*iii*) Industries processing chemicals, chemical manufacturing, metal finishing, paper and pulp, rubber, tanning, textiles etc.

and (*iv*) Miscellaneous industries like atomic power, radio active waste etc.

Industrial wastes are varied and complex. These wastes affect, in some way or the other, the normal life of a stream or normal functioning of sewerage and sewage treatment plants, unless pretreated at the source point itself. If they are discharged directly in the receiving waters, it may result in discolouring, foul smell and killing of aquatic life, apart from making the water unfit for various other purposes. Industrial wastewater disposal needs *proper* considerations from the points of view of manufacturer, public and the sanitary engineer alike. From the public point of view, industrial wastes cause pollution to stream making it unfit for domestic, recreational and commercial purposes, deteriorate sewers and treatment plant, and increase cost of treatment. From the point of view of manufacturer or industrialist, the waste is a liability and it has to be disposed of with least possible cost. Finally, from the point of view of sanitary engineer, knowledge of chemistry and bio-chemistry has to be judiciously applied to reduce the volume and polluting strength of the waste.

As mentioned above, there are *three alternatives* for the disposal of industrial wastes :

(*i*) Direct disposal into the receiving waters, without any treatment

(*ii*) Disposal into municipal sewers for combined treatment

and (*iii*) Separate treatment of industrial wastes, before their disposal into receiving water/land.

The factors that affect the choice of any of the above alternatives are : (*i*) self purification capacity of streams (*ii*) tolerance limits for the inland surface waters (*iii*) economic aspects of both the manufacturers as well as municipalities, and (*iv*) technical advantages of mixing the industrial wastes with domestic sewage.

19.2. CHARACTERISTICS OF INDUSTRIAL WASTEWATER

Although the characteristics of the industrial wastewater usually vary from industry to industry, and also vary from process to process even for the same industry, yet industrial effluent as a group differs widely from normal domestic sewage obtained from residences and commercial establishments. They have either too high a proportion of suspended solids, dissolved organic and inorganic solids, BOD, alkalinity or acidity, and their different constituents will not be in the same proportions as they exist in a normal domestic sewage. The pollutants in the industrial wastewater include the raw materials, process chemicals, final products, process intermediates, process by-products and impurities in raw materials and process chemicals. Such industrial wastewaters cannot always be treated easily by the normal methods of treating domestic wastewaters, and certain specially designed methods or sequence of methods may be necessary.

Pollutants in industrial wastewater and their effects : Following are the various types of pollutants that may be present in industrial wastewaters.

1. Organic substances : These deplete DO of streams and impose great load on secondary treatment units.

2. ***Inorganic substances:*** These include carbonates, chlorides, nitrogen, phosphorous etc. They are undesirable for micro-plants of

receiving water body. They render the water body unfit for further use.

3. Acids and alkalies : These greatly affect the aquatic life of receiving water body. They also cause serious problems in the operation of treatment units.

4. Toxic substances : These include cyanides, sulphides, acetylene, alcohol, etc, due to which flora and fauna of receiving waters is greatly affected. They cause problems in the treatment units apart from endangering the life of workmen. Table 19.1 gives the list of toxic substances from some selected industries.

5. Colour-producing substances : They impart objectionable colour in the receiving water bodies.

6. Oil etc. : These hinder self-purification process of stream. They also cause operational problems in the treatment plants.

TABLE 19.1. TOXIC POLLUTANTS IN INDUSTRIAL WASTEWATERS

Industry	*Total Pollutants*
1. Fertilizers	Ammonia, Arsenic
2. Coke ovens	Phenols, Cyanide, Thiocyanate, Ammonia
3. Metallurgicals	Heavy metals (i.e. Copper, Cadmium, Zinc)
4. Electroplating	Hexa-valent Chromium, Cadmium, Copper, Zinc
5. Synthetic wool	Acrylonitrite, Acetonitrite, HCN
6. Petrochemicals	Phenol, Heavy metals, Cyanides.

19.3. METHODS OF TREATMENT

As pointed out earlier, industrial wastewater have either too high a proportion of suspended solids, dissolved organic and inorganic solids, BOD, alkalinity or acidity, and their different constituents will not be in the same proportion throughout, as they exist in domestic wastewater. The important factors which affect the planning for an industrial wastewater treatment plant are : (*i*) discontinuous flow and sometimes seasonally discharged wastewater, (*ii*) high concentration of pollutants, and (*iii*) non-biodegradability and toxicity of some pollution parameters. Treatment of industrial wastewater generally consists of one or more of the following *processes* :

(*i*) Equalization, (*ii*) Neutralization, (*iii*) Physical treatment, (*iv*) Chemical treatment, and (*v*) Biological treatment.

1. Equalization : When the characteristics of the wastewater varies during the day and also when the discharge is either not uniform or else is discontinuous, equalization is necessary. The process of equalization consists of holding the wastewater for some predetermined time, in continuously mixed holding tanks/basins so as to get wastewater of uniform character and at uniform rate. Thus, equalization needs adequate storage.

2. Neutralization : This is necessary when the wastewater contains either excess alkali or exess acid, and is achieved by the addition of either acid or alkali respectively. This may be done either in the equalization tank, if the conditions so permit, or else in a separate neutralization tank.

3. Physical treatment : Physical treatment is similar to primary treatment of domestic wastewater. Various processes that fall under this head are : (*i*) Screening (*ii*) Sedimentation, (*iii*) Floatation and (*iv*) Filtration. Primary sedimentation becomes essential when the wastewater contains high percentage of settleable solids. Floatation is provided to remove the finer particles. In floatation, the rising air bubbles lift the finer particles to the surface from where these are removed by skimming.

4. Chemical treatment : Chemical treatment is one of the essential part in the treatment of industrial wastewater, specially for those which are amenable to biological treatment. Sometimes, physico-chemical treatment is applied to the industrial wastewaters, either in absence of biological treatment or in conjunction with biological treatment. Various chemical and physico-chemical processes that are used for industrial wastewater treatment are : (*i*) Coagulation, (*ii*) Chemical precipitation, (*iii*) Hyper-filtration or reverse-osmosis, (*iv*) Chemical oxidation, (*v*) Adsorption, (*vi*) Ion exchange, (*vii*) Air stripping, (*viii*) Electrodialysis, (*ix*) Thermal reduction.

Most of the above processes have already been described and explained earlier, and hence these are not repeated here.

5. Biological treatment : This is resorted to only when the industrial wastewater contains large quantities of biodegradable substances. If the BOD/COD ratio of the industrial wastewater is more than 0.6, it is biologically treatable. If the ratio is less than 0.6 but is upto 0.3, acclimatisation is necessary before biological treatment. If, however, the ratio is less than 0.3, biological treatment is not necessary. *Acclimatisation* is a process of seeding or raising initial microbial population under a controlled condition, by gradual exposure of the wastewater in increasing concentration. Various biological treatment methods that can be used are (*i*) activated sludge process, (*ii*) trickling filter (*iii*) oxidation pond (*iv*) aerated lagoon (*v*) oxidation ditch, and (*vi*) oxidation lagoon followed by aerated lagoon. Amongst these, trickling filters prove better than activated sludge process, since filters can handle a number of toxic wastes quite satisfactorily. When BOD of wastewater approaches 1000 mg/l, recirculation of effluent will prove more useful.

We shall now consider treatment procedures for the effluents from some important industries.

19.4. DAIRY INDUSTRY

Large quantities of wastewater originates due to different operations of dairy industry. Various operations in a dairy industry may include pasteurisation, bottling, filling in cans, and preparation of butter, creame, cheese, milk powder etc. Wastewater from dairy industry consists primarily of dilutions of milk and its products which impart a very high BOD, sometimes upto about 1000 mg/l. The wastewater may also contain detergents, germicides and other chemicals. Table 19.2 shows composition of wastewater of a typical dairy. Table 19.3 shows various pollution characteristics of dairy industry wastewater, along with suggested treatments. Fig. 19.1 shows a flow chart for treating dairy industry wastewater.

TABLE 19.2. TYPICAL COMPOSITION OF DAIRY WASTEWATER

Characteristic	*Typical values*
1. pH	7.2
2. Alkalinity	580 mg/l as $CaCO_3$
3. Suspended solids	720 mg/l
4. Total dissolved solids	1020 mg/l
5. BOD	1190 mg/l
6. COD	90 mg/l
7. Total nitrogen	90 mg/l
8. Phosphorous	12 mg/l
9. Oil and grease	320 mg/l
10. Chlorides	115 mg/l

TABLE 19.3. POLLUTION CHARACTERISTICS OF DAIRY INDUSTRY WASTE

Pollution characteristics	*Suggested Treatment methods*
1. High dissolved solids 2. High suspended solids 3. High BOD 4. Phosphorous 5. Nitrogen 6. Oil and grease	Physical treatment followed by aerobic biological treatment in the form of high rate trickling filters or activated sludge process. Equalization tank is commonly provided. Sludge can be dried on sludge drying beds. Oxidation ponds and oxidation ditches can also be provided.

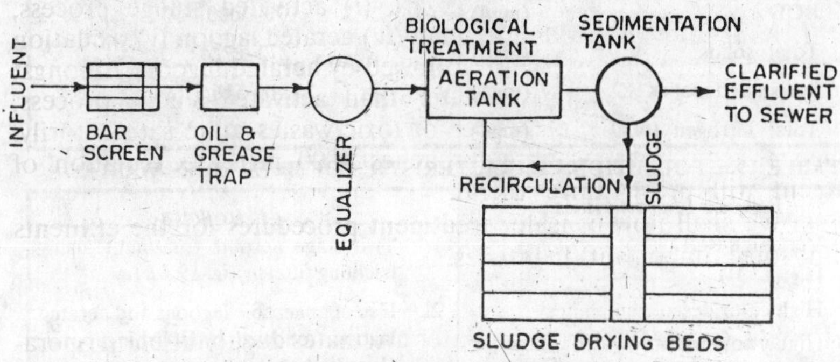

FIG. 19.1. TYPICAL FLOW CHART FOR TREATING DAIRY WASTEWATER.

19.5. CANE SUGAR INDUSTRY WASTEWATER

The average quantity of liquid waste produced from a cane sugar industry wastewater is of the order of about 3000 litres/tonnes of cane crushed. The wastewater is quite strong, having SS of 400 to 500 mg/l and BOD of 500 to 600 mg/l. Because of low pH (5 – 7), disposal becomes a difficult problem. Table 19.4 gives the pollution characteristics along with suggested treatment methods. Fig. 19.2 and 19.3 shows a flow chart for treating the wastewater.

TABLE 19.4. POLLUTION CHARACTERISTICS OF SUGAR INDUSTRY WASTE

Pollution characteristics	*Suggested Treatment methods*
1. High BOD 2. High volatile solids 3. Low pH.	Using digestion in stages with detention periods of one and six days for the primary and secondary digester followed by oxidation pond (Alternatives : See Fig. 19.2 and 19.3)

19.6. DISTILLERY AND BREWERY WASTEWATER

Various products that are manufactured in distilleries include industrial alcohols, rectified spirit, silent spirit, absolute alcohol, beverage alcohol etc. These products are obtained through the high bio-chemical processes of fermentation by yeast, using carbohydrates as raw material. Also, these products contain ethyl alcohol in different proportions. The unwanted residue obtained at the time of preparation of the medium, contain very high BOD, which makes it non-amenable to aerobic biological treatment. Table 19.5 gives the composition of combined bewery wastes. Table 19.6 gives the pollution characteristics and suggested treatment methods. Fig. 19.4 gives the flow diagram for treatment of wastewater obtained from breweries.

TABLE 19.5. COMPOSITION OF COMBINED BREWERY WASTES

Parameter		*Values*
1. pH		4.0–7.0
2. COD	(mg/l)	30–1225
3. BOD	(mg/l)	70–3000
4. Total solids	(mg/l)	272–2724
5. Suspended solids	(mg/l)	16-516
6. Total nitrogen (N)	(mg/l)	7–42

TABLE 19.6. POLLUTION CHARACTERISTICS OF BREWERIES WASTES

Pollution Characteristics	*Suggested treatment*
1. Very high BOD 2. High COD 3. High chlorides and sulphates 4. Highly coloured-brownish yellow	1. Two stage aerobic treatment, using trickling filters (Fig. 19.4 *a*) 2. Use of anaerobic lagoons and aerated lagoon, followed by polishing ponds (Fig. 19.4 *b*)

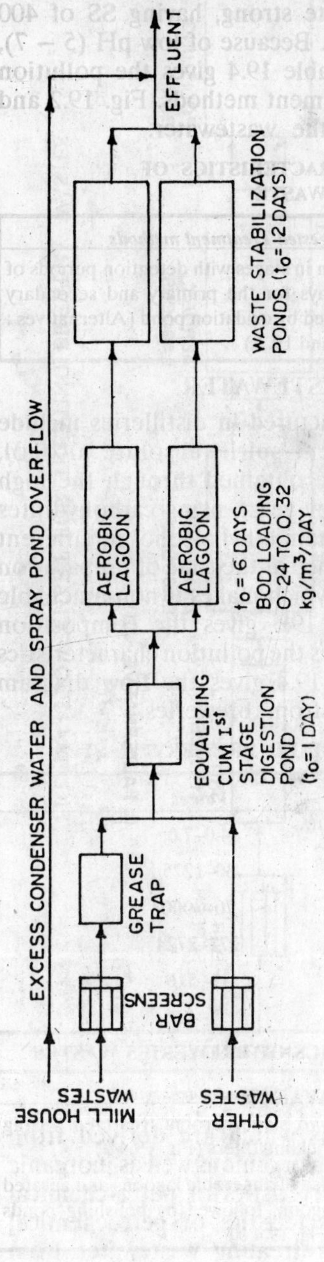

FIG. 19.2. FLOW CHART FOR TREATMENT OF SUGAR MILL WASTE

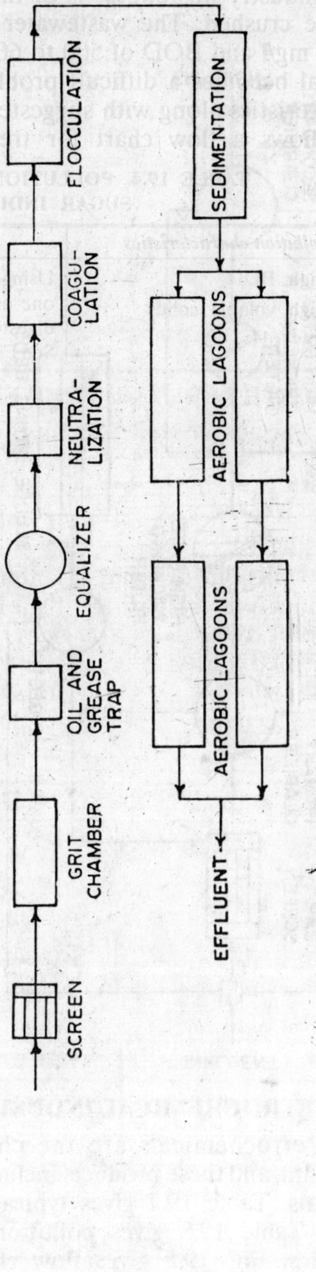

FIG. 19.3. FLOW CHARTS FOR TREATMENT OF SUGAR MILL WASTE

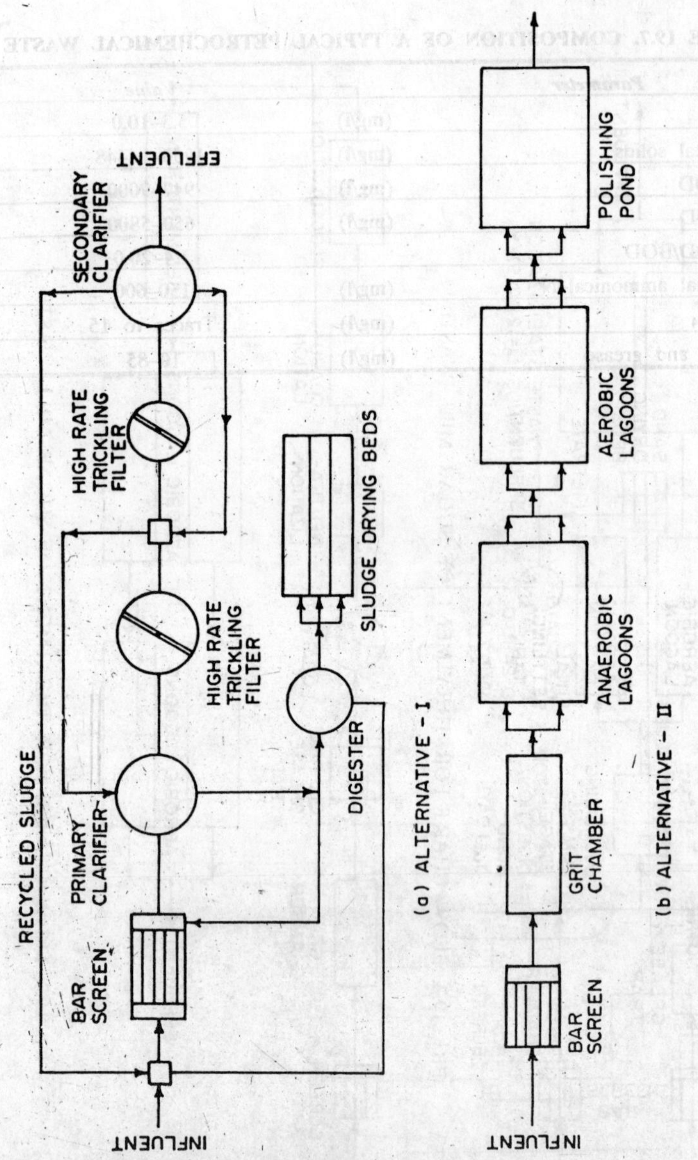

FIG. 19.4. FLOW CHART FOR TREATING BREWERY WASTES

19.7. PETROCHEMICAL INDUSTRIES WASTE

Petrochemicals are the chemicals which are derived from petroleum, and these products includes both organic as well as inorganic chemicals. Table 19.7 gives typical characteristics of petro-chemical wastes. Table 19.8 gives pollution characteristics of petrochemical industries. Fig. 19.5 gives flow chart for treating wastewater from petrochemical industries.

TABLE 19.7. COMPOSITION OF A TYPICAL PETROCHEMICAL WASTE

Parameter		Value
1. pH	(mg/l)	5.3–10.0
2. Total solids	(mg/l)	1956–14448
3. COD	(mg/l)	947–9000
4. BOD	(mg/l)	650–5800
5. COD/BOD		1.4–20.0
6. Total ammonical–N	(mg/l)	150–600
7. PO_4	(mg/l)	Traces to 15
8. Oil and grease	(mg/l)	10–85

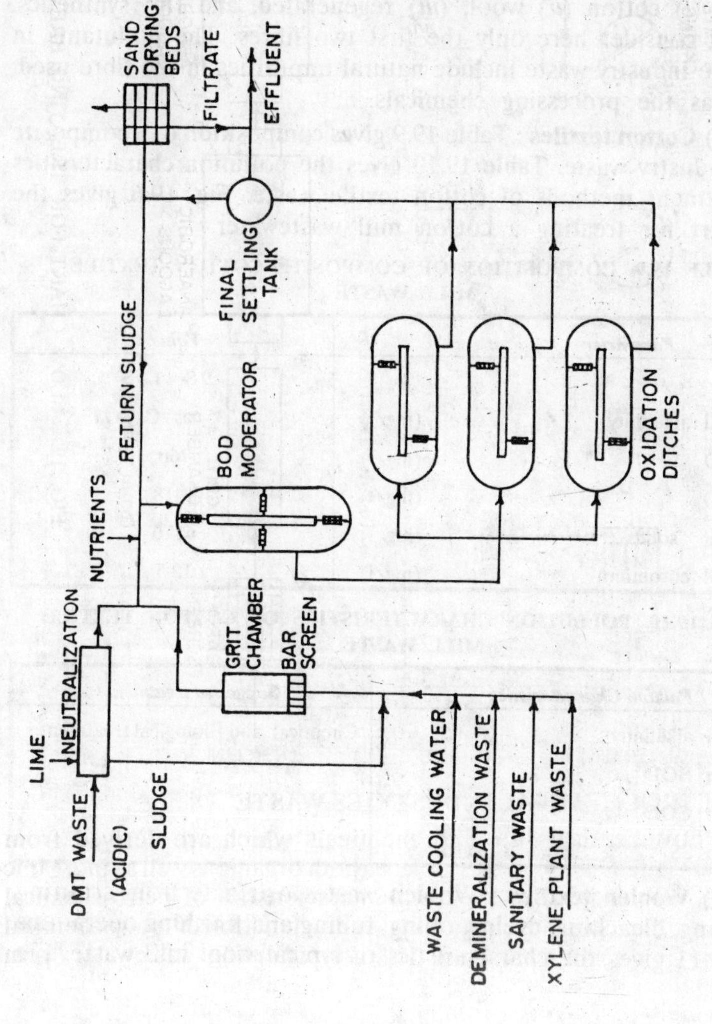

FIG. 19.5. FLOW CHART FOR AN AROMATIC WASTE TREATMENT PLANT

TABLE 19.8. POLLUTION CHARACTERISTICS OF PETROCHEMICAL WASTE

Pollution Characteristics	*Suggested treatment*
1. Oil 2. High BOD and COD 3. High total solids 4. Hydrocarbons and their products	Chemical treatment and Biological treatment.

19.8. TEXTILES INDUSTRIES WASTE

The characteristics of waste from a textile industry depends upto the type of fibre used. These fibres may be classified into four groups : (*i*) cotton (*ii*) wool, (*iii*) regenerated, and (*iv*) synthetics. We shall consider here only the first two fibres. The pollutants in the textile industry waste include natural impurities in the fibre used, as well as the processing chemicals.

(*a*) **Cotton textiles** : Table 19.9 gives composition of a composite cotton industry waste. Table 19.10 gives the pollution charactersitics and treatment methods of cotton textile waste. Fig. 19.6 gives the flow chart for treating a cotton mill wastewater.

TABLE 19.9. COMPOSITION OF COMPOSITE COTTON TEXTILE MILL WASTE

Parameter		*Value*
1. pH		9.8–11.8
2. Total alkalinity	(mg/l)	17.35 as $CaCO_3$
3. BOD	(mg/l)	760
4. COD	(mg/l)	1418
5. Total solids	(mg/l)	6170
6. Total chromium	(mg/l)	12.5

TABLE 19.10. POLLUTION CHARACTERISTICS OF COTTON TEXTILE MILL WASTE.

Polltion Characteristics	*Suggested treatment*
1. High alkalinity 2. High BOD 3. High COD 4. High SS	Chemical and Biological treatment.

(*b*) **Woolen textiles** : Woolen wastes originate from scouring, carbonizing, bleaching, dyeing, oiling, fulling and finishing operations. Table 19.11 gives the characteristics of typical wool mill waste. Fig.

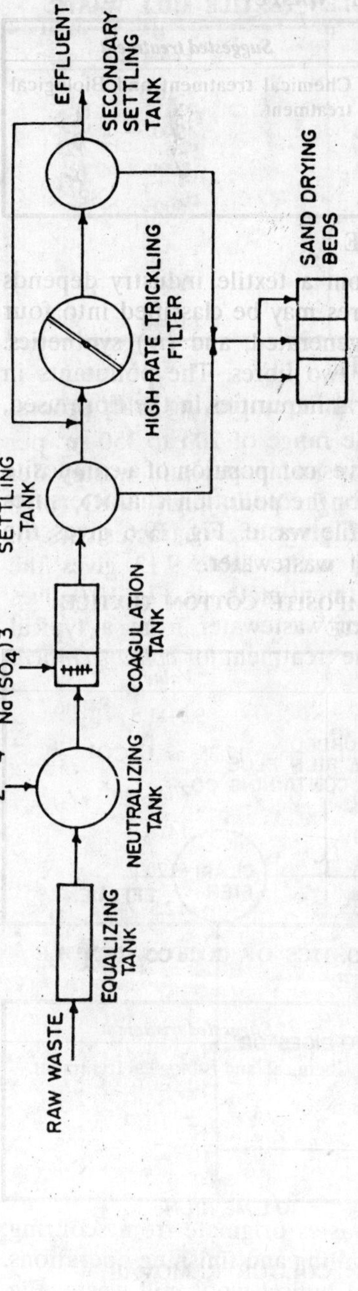

FIG. 19.6 FLOW DIAGRAM FOR TREATMENT OF COTTON TEXTILE INDUSTRIES WASTE

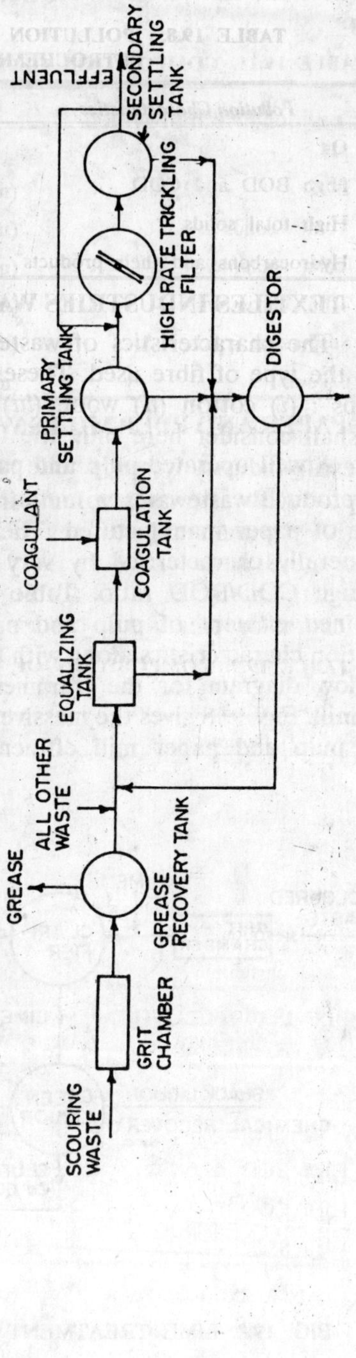

FIG. 19.7. FLOW DIAGRAM FOR TREATMENT OF WOOLEN TEXTILE INDUSTRIES WASTE

19.7 gives the flow diagram for treating the woolen textile mill waste water.

TABLE 19.11. COMPOSITION OF WOOLEN TEXTILE MILL WASTE.

	Parameter		*Value*
1.	pH		9–10.5
2.	BOD	(mg/l)	900
3.	Total solids	(mg/l)	3000
4.	Total alkalinity	(mg/l)	600
5.	Total chromium	(mg/l)	4
6.	Suspended solids	(mg/l)	100
7.	Colour	(mg/l)	Brown

19.9. PAPER AND PULP MILLS WASTE

A well operated pulp and paper mill, employing kraft process may produce wastewater volume in the range of 200 to 350 m^3 per tonne of paper manufactured. The pulp and paper mill wastewater is generally characterised by very strong colour, high BOD, high SS, high COD/BOD ratio. Table 9.12 gives the characteristics of combined effluent of pulp and paper mills. Table 9.13 gives the pollution characteristics along with treatment methods. Fig. 19.9 gives the flow diagram for the treatment of wastewater from a typical pulp mill. Fig. 19.8 gives the massive lime treatment for *colour removal* from pulp and paper mill effluent.

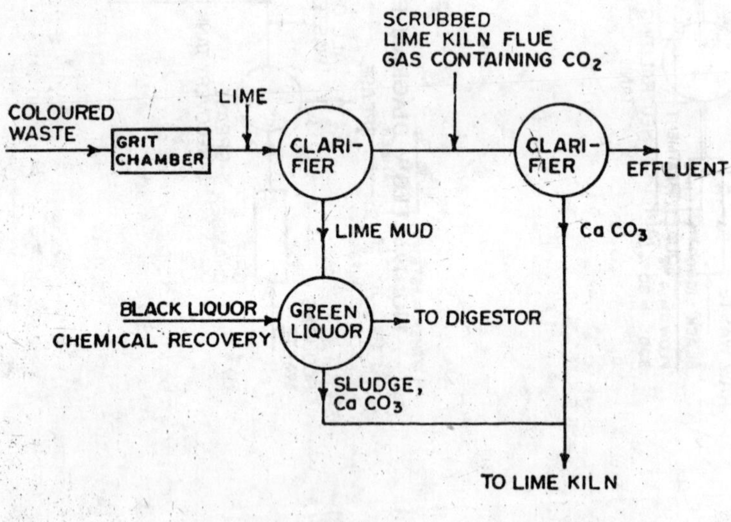

FIG. 19.8. LIME TREATMENT FOR COLOUR REMOVAL.

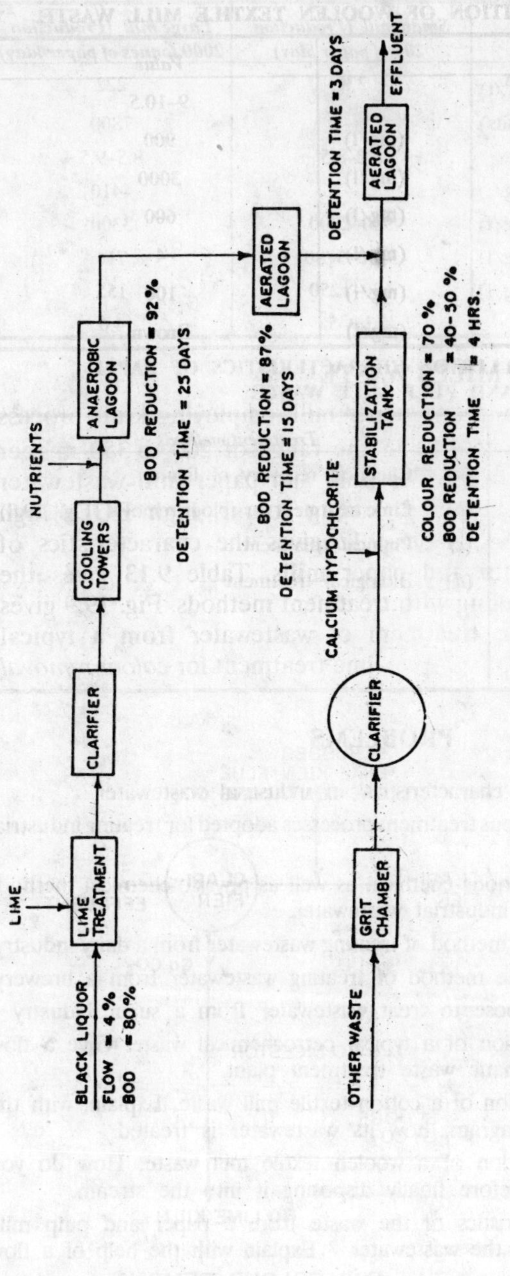

FIG. 19.9. FLOW DIAGRAM FOR TREATMENT OF WASTE FROM A PULP MILL.

TABLE 19.12. CHARACTERISTICS OF COMBINED EFFLUENT FROM PULP AND PAPER MILL.

	Parameter		*Small mill (Production 20 t of paper /day)*	*Large mill (Production 2000 tonnes of paper /day)*
1.	Flow per day	(m^3/t)	330	222
2.	Colour	(units)	–	7800
3.	pH		8.2–8.5	8.5–9.5
4.	Total solids	(mg/l)	–	4410
5.	Suspended solids	(mg/l)	900–2000	3300
6.	COD	(mg/l)	3400–5780	716
7.	BOD	(mg/l)	680–1250	155
8.	COD/BOD		3.9–5	4.6

TABLE 19.13. POLLUTION CHARACTERISTICS OF PAPER AND PULP MILL WASTE.

	Pollution characteristic		*Treatment methods*
1.	High BOD	(*a*)	Chemical recovery of lignin
2.	High BOD/COD rate	(*b*)	Lime treatment for colour removal (Fig. 19.8)
3.	Strong colour	(*c*)	Physical treatment
4.	High alkalinity	(*d*)	Biological treatment
5.	High sodium content		
6.	Lignin, high resistant to biological treatment		

PROBLEMS

1. Write a note on 'characteristics' of industrial wastewater.
2. Discuss in brief various treatment processes adopted for treating industrial wastewater.
3. Explain in brief various chemical as well as physico-chemical methods used for treating industrial wastewater.
4. Discuss in brief the method of treating wastewater from a dairy industry.
5. Discuss in brief the method of treating wastewater from a brewery.
6. How do you propose to treat wastewater from a sugar industry ?
7. Give the composition of a typical petrochemical waste. Give a flow chart for an aromatic waste treatment plant.
8. Give the composition of a cotton textile mill waste. Explain, with the help of a flow diagram, how its wastewater is treated.
9. Give the composition of a woolen textile mill waste. How do you treat this waste before finally disposing it into the stream.
10. Give the characteristics of the waste from a paper and pulp mill. How do you treat the wastewater ? Explain with the help of a flow diagram.

20

House Drainage

20.1. GENERAL PRINCIPLES

The arrangement provided in a house or building, for collecting and conveying wastewater through drain pipes, by gravity, to join either a public sewer or a domestic septic tank, is termed as *house drainage* or building drainage.

Aims of house drainage. House drainage is provided

(*i*) to maintain healthy conditions in the building

(*ii*) to dispose off waste water as early and quickly as possible

(*iii*) to avoid the entry of foul gases from the sewer or the septic tank

(*iv*) to facilitate quick removal of foul matter (*e.g.* human excreta)

(*v*) To collect and remove waste matters systematically

20.2. PRINCIPLES OF HOUSE DRAINAGE

The following principles are adopted for the efficient drainage system :

1. The lavatory blocks should be so located that the length of drainage line is minimum. In the case of multistoreyed building they should be located one above the other. At least one wall of the lavatory block should be an outside wall, to facilitate the fixing of soil and vent pipes.

2. The drainage pipes should be laid by the side of the building rather than below the building.

3. All the drains should be aligned straight between successive inspection chambers. All sharp bends and junctions should be avoided except through chambers.

4. The slope of the drains should be sufficient to develop self cleansing velocity.

5. The size of drain should be sufficient, so that flooding of the drain does not take place while handling the maximum discharge.

6. The drainage system should contain enough number of traps at suitable locations.

7. The house drain should be disconnected to the public sewer by the provision of an intercepting trap. This will avoid the entry of foul gases from entering the house drainage system. It should be seen that the public sewer is deeper than the housedrain.

8. Rain water pipes should drain out rain water directly into the street gutters from where it is carried to the storm water drain.

9. All the connections should be water tight.

10. The entire drainage system should be properly ventilated from the starting point to the final point of disposal. It should permit free circulation of air.

11. All the materials and fittings of the drainage system should be hard, strong and resistant to corrossive action. They should be non-absorbent type.

12. The entire system should be so designed that the possibilities of formation of air locks, siphonage, under deposits etc. are minimised.

20.3. PIPES AND TRAPS

Pipes. In a house drainage system, a pipe may have the following designations, depending upon the function it carries :

1. ***Soil pipe.*** A soil pipe is a pipe through which human excreta flows.

2. ***Waste pipe.*** It is a pipe which carries only the liquid waste. It does not carry human excreta.

3. ***Vent pipe.*** It is a pipe which is provided for the purpose of the ventilation of the system. A vent pipe is open at top and bottom, to facilitate exit of foul gases. It is carried at least 1 m higher than the roof level.

4. ***Rain water pipe.*** It is a pipe which carries only the rain water.

5. ***Anti-siphonage pipe.*** It is pipe which is installed in the house drainage to preserve the water seal of traps.

The following sizes of pipes are commonly used in house drainage :

Soil pipe	: 100 mm
Waste pipe : horizontal	: 30 to 50 mm
Waste pipe : vertical	: 75 mm
Rain water pipe	: 75 mm
Vent pipe	: 50 mm

Anti-siphonage pipe :

(*i*) Connecting soil pipe : 50 mm

(*ii*) Connecting waste pipe : 40 mm

Traps. A trap is a depressed or bent fitting which, when provided in a drainage system, always remains full of water, thus maintaining a *water seal.* It prevents the passage of foul air or gas through it, though it allows the sewage or waste water to flow through it. The depth of water seal is the vertical distance between the crown and dip of a trap (Fig. 20.1). The depth of water seal represents its *strength* or effectiveness. Greater the depth of water seal, more effective is the trap. The depth of water seal varies from 25 mm to 75 mm.

Causes of breaking of seal . Water seal may break due to the following reasons :

(*i*) faulty joints

(*ii*) crack in the bottom of seal

(*iii*) creation of partial vacuum in the sewer fittings

(*iv*) increase in the pressure of sewer gases, and

(*v*) non-use for a prolonged period.

The breaking of the water seal can be prevented by (*i*) connecting the portion between the soil pipe and trap by a vent pipe, and (*ii*) use of anti-siphonage pipe in the building.

Characteristics of traps. A trap should possess the following characteristics :

1. It should possess adequate water seal at all times, to fulfill the purpose of its installation. However, it should retain minimum quantity of water for this purpose.
2. It should be of non-absorbent material.
3. It should be free from any inside projections, angles or contractions, so that flow is not obstructed or retarded.
4. It should be simple in construction, cheap and readily available.
5. It should be self cleansing.
6. It should be provided with suitable access for cleaning.
7. Its internal and external surfaces should have smooth finish so that dirt etc. does not stick to it.

20.4. CLASSIFICATION OF TRAPS

Traps are classified as follows :

(a) Classification according to shape (Fig. 20.1)

(*i*) *P-Trap* (Fig. 20.1 *a*). This resembles the shape of letter *P*, in which the legs are at right angles to each other.

(*ii*) *Q-trap* or *half-S-trap* (Fig. 20.1 *b*). This resembles the shape of letter *Q*, in which the two legs meet at an angle other than a right angle.

(*iii*) *S-trap* (Fig. 21.6 *c*). This resembles letter-*S*, in which both the legs are parallel to each other, discharging in the samc direction. Fig. 20.1 (*d*) shows the development of all the three types of traps.

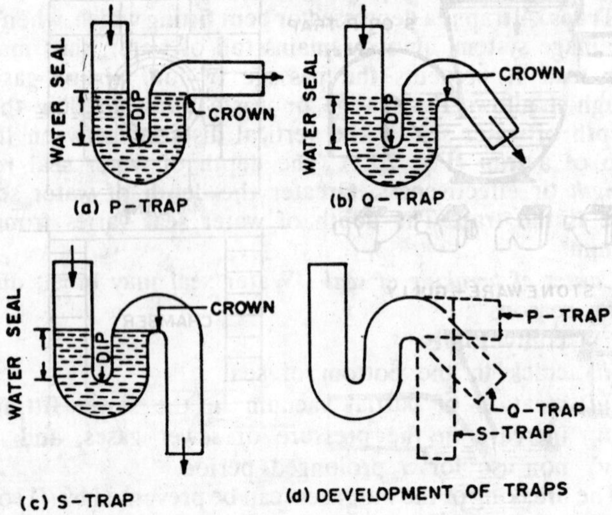

FIG. 20.1. TRAPS ACCORDING TO SHAPES.

(b) Classification according to use:

(*i*) Floor trap or nahni trap

(*ii*) Gully trap

(*iii*) Intercepting trap.

20.5. FLOOR TRAP OR NAHNI TRAP

A floor trap, commonly known as a *nahni trap* is used to collect wash water from floors, kitchens and bath rooms. It forms the starting point of waste water floor. It is made of cast iron, with a gravity at top, to exclude entry of solid matter of big size. This cover can be removed to do frequent cleaning of the trap. These traps have small water seal.

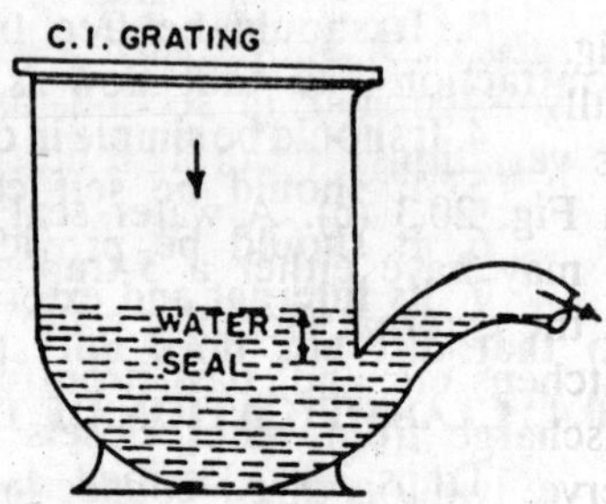

FIG. 20.2. FLOOR TRAP.

20.6. GULLY TRAP

These are special types of traps which disconnect sullage drain (collected from baths, kitchen etc.) from the main drainage system. It is either made of stone-ware or of cast iron. Stone ware gully trap is of square section at the top on which C.I. grating is fitted.

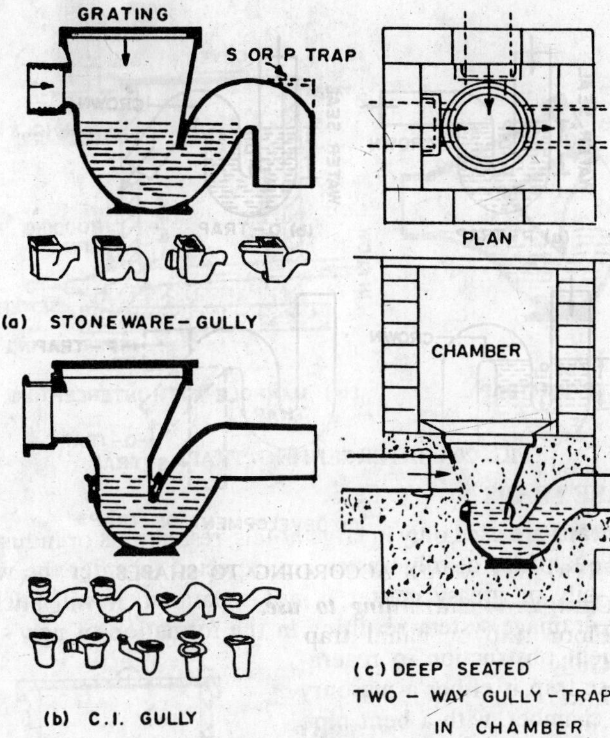

FIG. 20.3. VARIOUS FORMS OF GULLY TRAPS.

Fig. 20.3 (*a*) shows such a gully along with its variations. A C.I. gully is circular in section, as shown in Fig. 20.3 (*b*), along with its variations. It can also be fitted in a masonry chamber as shown in Fig. 20.3 (*c*). A water seal of 60 to 70 mm is usually provided. It may have either a *S*-trap and *P*-trap. A gully trap, is provided at the external face of a wall. It thus receives wastewater from baths, kitchens etc. and pass it on to the house drain carrying excremental discharge from water closets etc. A well designed gully trap may serve two or three connections from nahni traps.

20.7. INTERCEPTING TRAPS

This is a special type of trap provided at the junction of house drain with the public sewer or septic tank. It is thus provided in the last manhole of the house drainage system. It has a deep water seal of 100 mm, so as to effectively prevent the entry of sewer gases from public sewer line into the house drain.

The trap has an opening at the top, called the *cleaning eye* or *rodding arm*, having a tight fitting plug, for frequent cleaning of the trap.

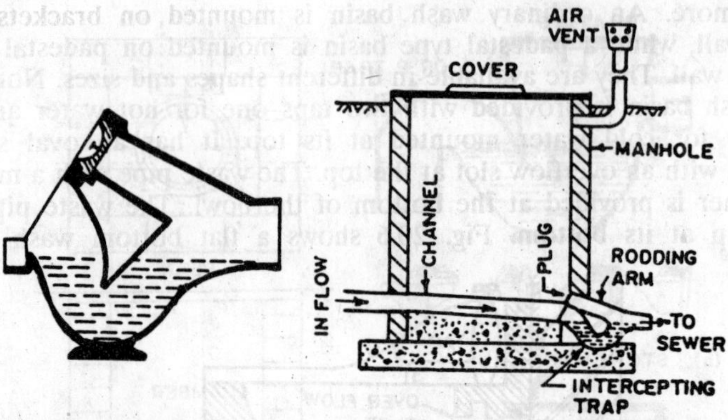

(a) INTERCEPTING TRAP (b) MANHOLE WITH INTERCEPTING TRAP

FIG. 20.4 INTERCEPTING TRAPS

20.8. GREASE TRAPS

Such traps are used only in large hotels, restaurants or industries where large quantities of oily wastes are expected to enter the water flow. If the oily or greasy matter is not separated, it will stick to the building drainage system resulting in the formation of ugly scum and consequent obstruction to reaeration. A grease trap is either a masonry or cast iron chamber, with a bent pipe or Tee-pipe at the outlet end. Because of sudden increase in the area of flow at entry, the velocity of flow is reduced, resulting in the separation of oily and greasy matter from the wastewater. This greasy matter, floating on the top can be removed later.

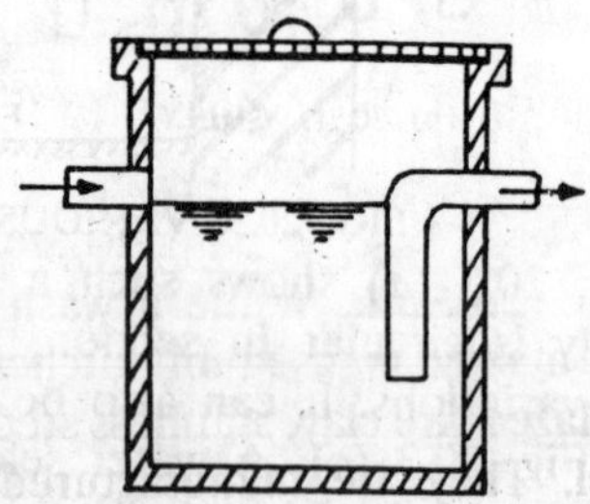

FIG. 20.5 GREASE TRAP

20.9. SANITARY FITTINGS

The following fittings are commonly used in buildings, for efficient collection and removal of wastewater to the house drain:

(*i*) Wash basins
(*ii*) Sinks
(*iii*) Bath tubs
(*iv*) Water closets
(*v*) Urinals
(*vi*) Flushing cisterns.

1. Wash Basin. Wash basins are usually made of pottery or porcelain ware. Sometimes, they are also made of porcelain enamelled cast iron, pressed steel or plastic, specially where number of users are more. An ordinary wash basin is mounted on brackets fixed on wall, while a padestal type basin is mounted on padestal rising from wall. They are available in different shapes and sizes. Normally, a wash basin is provided with two taps–one for hot water and the other for cold water mounted at its top. It has an oval shaped bowl, with an overflow slot at the top. The waste pipe with a metallic strainer is provided at the bottom of the bowl. The waste pipe has a trap at its bottom. Fig. 20.6 shows a flat bottom wash basin.

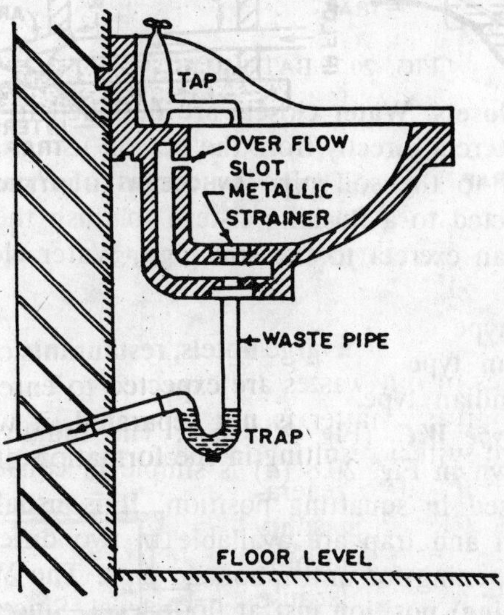

FIG. 20.6. WASH BASIN.

2. Sink. While a wash basin is used for washing hands, face etc. a sink is used in kitchen or laboratory. These may be made of glazed fire clay, stainless steel, metal porcelain or enamelled pressed steel. They are manufactured in various sizes and shapes, though rectangular shape is quite common in kitchens. It may also have a drain board attached to it. A sink may also be constructed of cast-in-situ concrete, with suitable finishing surface such as marble, terrazo etc. The out-let pipe, provided with a grating of brass or nickel, may discharge over a floor trap or nahni trap.

3. Bath tub. Bath tubs, are usually made of iron or steel coated with enamel, enamelled porcelain or of plastic material. They may also be made of cast-in-situ concrete finished with marble chips or terrazzo, or else may be made of marble slabs properly jointed at the side. It has a length varying from 1.7 to 1.85 m, width between 0.7 m to 0.75 m and depth near waste pipe varying from 0.43 m to 0.45 m. The overall height may vary between 0.58 to 0.6 m. It is provided with outlet and overflow pipes, usually of 40 mm diameter. A trap with proper waterseal is used at the outlet.

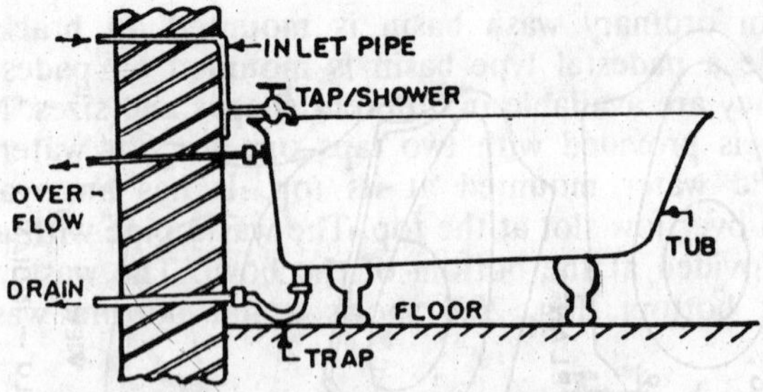

FIG. 20.7. BATH TUB.

4. Water closets. Water closets are designed to receive and discharge human excreta directly from the person using it. The appliance is connected to the soil pipe by means of a suitable trap. It is usually connected to a flushing cistern to flush the closet and discharge the human excreta to the soil pipe. Water closets are of three types.

(*i*) Indian type
(*ii*) European type
(*iii*) Anglo-Indian type.

1. Indian type W.C. (Fig. 20.8 *a*) : The Indian style water closet (W.C.), shown in Fig. 20.8 (*a*) is simple in construction and working, but is used in squatting position. It is usually made of porcelain. The pan and trap are available in two different pieces. The trap has an opening for antisiphonage pipe. The W.C. is fixed in squatting (or siting) position just at floor level. Since the excreta does not directly fall into the trap, therefore, there are chances for excreta to become foul. The excreta may stick to the surface of the pan if the flushing is not proper. The flush water enters the rim of the pan through the opening provided in the front of the pan. The flushing cistern is normally kept 2 m above the closet. Indian type closet requires greater quantity of water (atleast 10 litres) for flushing.

2. European type W.C. : Fig. 20.8 (*b*) shows a typical European type water closet. It is usually made of porcelain. It is a wash down water closet, provide with a seat and a cover. The pan has flushing rim to spread the flush water. The excreta directly falls in the trap, and therefore there are less chances of excreta becoming foul. The padestal type European W.C. also known as *commode* is commonly used. The closet is fitted with either a *P*-trap or *S*-trap. It can also be used at upper floors, while in case of Indian type W.C., the upper floor has to be depressed to receive the pan fixed at floor level. Generally, a low level flushing cistern is used with the European type W.C.

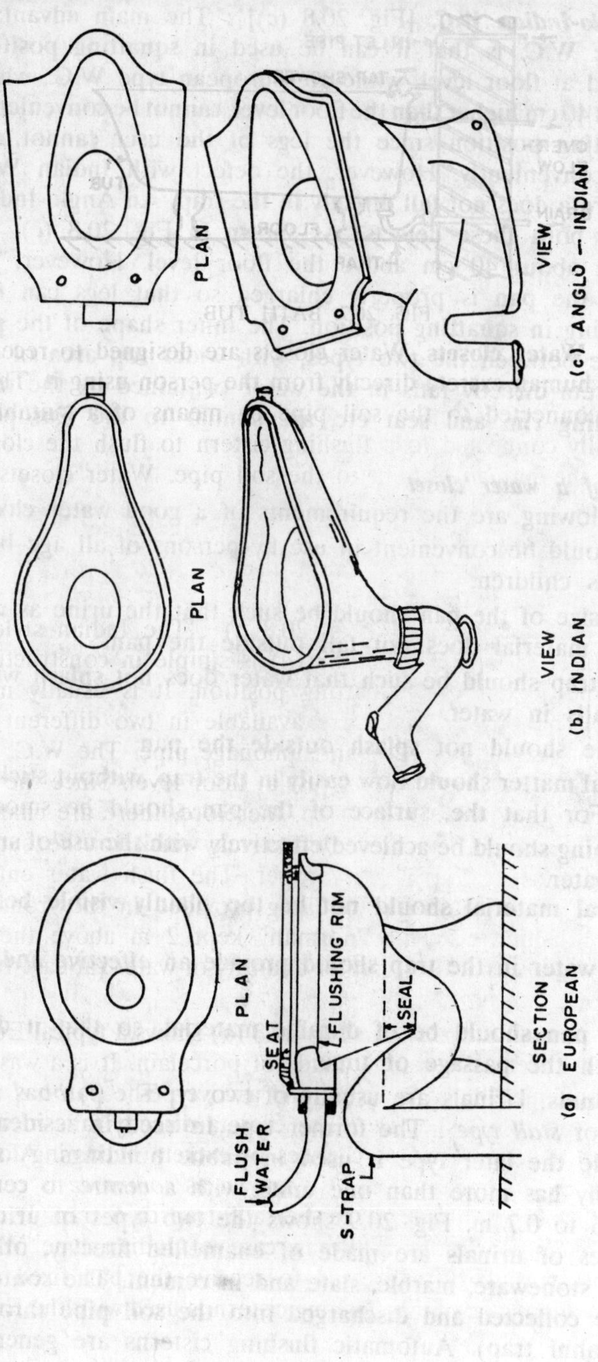

FIG. 20.8. WATER CLOSETS

3. ***Anglo-Indian W.C.*** [Fig. 20.8 (*c*)] : The main advantage of Indian type W.C. is that it can be used in squatting position since it is fixed at floor level, while an European type W.C., which is fixed at about 40 cm higher than the floor level, cannot be conveniently used in squatting position since the legs of the user cannot rest on thin rim conveniently. However, the defect with Indian W.C. is that the excreta does not fall directly in the trap. An Anglo-Indian W.C. removes both these defects. As shown in Fig. 20.8 (*c*), the closet is fixed about 40 cm above the floor level. However, the upper rim of the pan is properly enlarged so that legs can rest on it while using in squatting position. The inner shape of the pan is intermediate between the two types, with wider top area of the trap. The excreta directly falls in the water contained by the trap. The top flushing rim and seat etc. are similar to the European type.

Requirements of a water closet

The following are the requirements of a good water closet:

1. It should be convenient in use by persons of all age–both old as well as children.

2. The size of the pan should be such that the urine as well as the faecal material does not fall outside the pan.

3. The trap should be such that water does not splash when the excreta falls in water.

4. Urine should not splash outside the pan.

5. Faecal matter should flow easily in the trap, without sticking to the pan. For that the, surface of the pan should be smooth.

6. Flushing should be achieved effectively with the use of small quantity of water.

7. Faecal material should not be too plainly visible before flushing.

8. The water in the trap should provide an effective and air tight seal.

9. The pan should be of durable material, so that it does not crack with the passage of time.

5. **Urinals.** Urinals are usually of two types : (*i*) *bowl type and* (*ii*) *slab* or *stall type* . The former type in used in residential buildings while the later type is used in public buildings. A stall urinal normally has more than one units, with a centre to centre spacing of 0.6 to 0.7 m. Fig. 20.9 shows the two types of urinals. The best types of urinals are made of enamelled fireclay, others of salt glazed stoneware, marble, slate and in cement. The contents of urinals are collected and discharged into the soil pipe through floor trap (nahni trap). Automatic flushing cisterns are generally

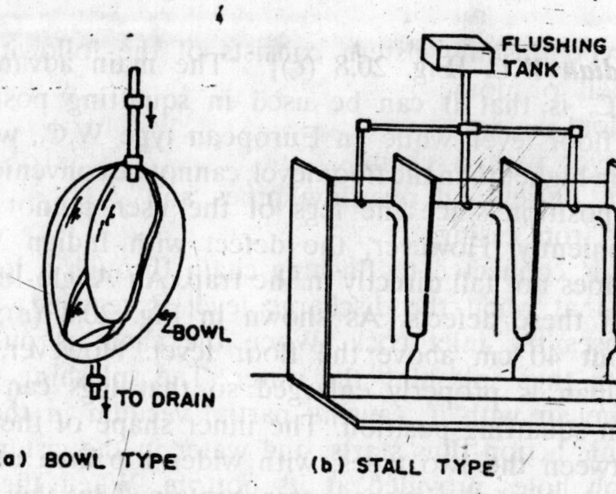

FIG. 20.9. URINALS

provided for stall type urinals, which operates, at regular interval of 10 to 15 minutes.

6. Flushing cisterns. Flushing cisterns are used for flushing out water closets and urinals. These are made of either cast iron or of porcelain. For Indian type W.C., cast iron flushing cistern is normally used, fixed at about 2 m above the floor level. For European type and Anglo-Indian type closets, porcelain cisterns are normally used, fixed at about 60 cm above floor levels. The low level flushing cisterns, made of porcelain, are decent in look, and operate very easily by simply turning a handle.

Flushing cisterns are of two types : (*i*) *valveless* siphonic type and (*ii*) *valve fitted* siphonic type. *Bell type* flushing cistern, commonly used with Indian type closets, is the typical example of valveless siphonic cistern, shown in Fig. 20.10.

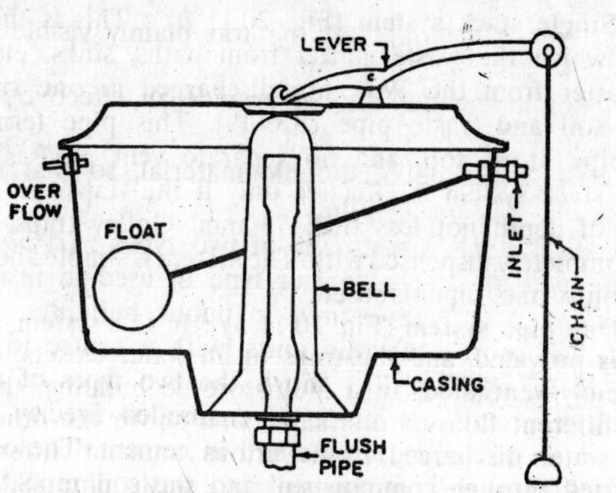

FIG. 20.10. BELL TYPE FLUSHING CISTERN

A bell type flushing cistern consists of the following parts:

1. A bell or dome
2. A float
3. A lever with a chain
4. Inlet, outlet and overflow pipes and
5. Cast iron casing.

The bell is connected to flushing chain through a lever. The float is so set that when the discharge level is reached, the float rod slightly closes the inlet cock. When the chain is pulled, the bell is lifted up, thus splashing the water. The splashing of water takes away some air with it, causing partial vacuum in the top of the bell. Siphonic action thus starts, and water in the cistern enters the bell through holes provided at its bottom. When the tank is emptied, air enters from the bottom and siphonic action is broken. The lowering of the float results in the opening of the inlet cock, and water thus enters the cistern. It should be noted that the chain should be released immediately after the pull, otherwise the partial vacuum caused by splashing water may be destroyed by the entry of air from the flush pipe. The capacity of a bell type flushing cistern may vary between 5 to 15 litres.

20.10. SYSTEMS OF PLUMBING

There are four principal systems of plumbing for drainage of buildings :

(*i*) Single stack system
(*ii*) One pipe system
(*iii*) Partially ventilated single stack system and
(*iv*) Two pipe system.

All the four systems are shown diagrammatically in Fig. 20.11.

1. Single stack system (Fig. 20.11 *a*) : This is the simplest system, in which the waste matter from baths, sinks, etc., as well as foul matter from the W.C. are discharged in one single pipe, called the soil and waste pipe (S.W.P.). This pipe terminates as the vent pipe at its top, and no separate vent pipe is provided. The single stack system is effective only if the traps are filled with water seal of depth not less than 75 mm. Gulley traps and waste pipes are completely dispensed with. The system is simple and economical since only one pipe is used.

2. One pipe system (Fig. 20.11 *b*) : In this system, a separate vent pipe is provided, and the traps of all water closets, basins etc. are completely ventilated. In a multistoreyed building, the lavatory blocks of different floors are situated one above the other, so that the waste water discharged from various units at different floors can be carried through common soil and waste pipe (S.W.P.). The system is costlier than the single stack system.

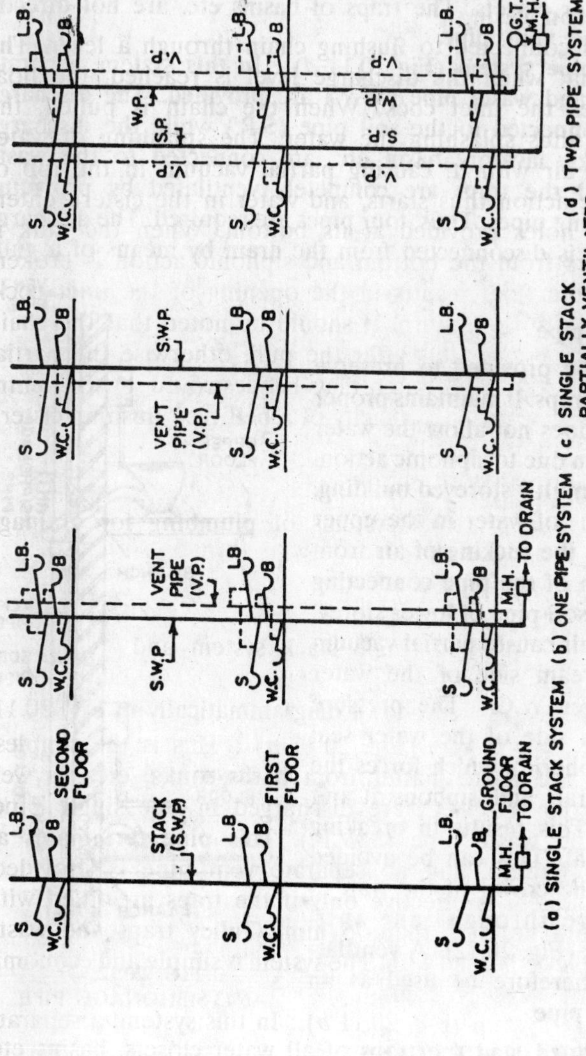

(a) SINGLE STACK SYSTEM (b) ONE PIPE SYSTEM (c) SINGLE STACK PARTIALLY VENTILATED (d) TWO PIPE SYSTEM

W.C.= WATER CLOSET ; B= BASIN ; L.B.= LAVATORY BASIN ; S= SINK ; S.P. = SOIL PIPE
V.P.= VENTILATING PIPE ; S.W.P.= SOIL PIPE AND WASTE PIPE ; S.P. = SOIL PIPE
W.P. =WASTE PIPE ; M.H. = MAN HOLE ; G.T.= GULLEY TRAP.
NOTE. ALL PIPES TERMINATE AS VENT PIPES AT THE TOP.

FIG. 20.11. PLUMBING SYSTEMS.

3. Single stack system partially ventilated (Fig. 20.11 *c*) : This is modified form of the single stack system and one pipe system. In this system, the waste from W.C., basins, sinks etc. is discharged into one common soil and waste pipe (S.W.P.) However, in addition, a relief vent pipe is also provided which provides ventilation to the traps of water closets. The traps of basins etc. are not directly connected to the vent pipe.

4. Two pipe system (Fig 20.11 *d*) : In this system, separate soil pipe (S.P.) and waste pipe (W.P.) are provided. The discharge from W.C. is connected to the soil pipe (S.P.) while the discharge from baths, sinks, lavatory basin etc. are connected to the waste pipe (W.P.). All the traps are competely ventilated by providing separate ventilating pipes. Thus, four pipes are required. The discharge from waste pipe is disconnected from the drain by means of a gully trap.

Anti-siphonage pipe

It is a pipe provided to preserve the water seal of traps. It maintains proper ventilation and does not allow the water seal to get broken due to siphonic action. In the case of a multi- storeyed building, the sudden flush of water in the upper storey results in the sucking of air from the short branch of the pipe connecting the W.C. to the soil pipe of lower storey. This sucking of air causes partial vacuum on the downstream side of the water seal of the lower W.C. The pressure at the upstream side of the water seal is more (atmospheric), which forces the water up the trap and siphons it out in the branch. This results in breaking of the water seal. This can be avoided by connecting the crown of the trap to the atmosphere through an anti-siphonage pipe (Fig. 20.12). A ventilating pipe can therefore be used as an anti-siphonage pipe.

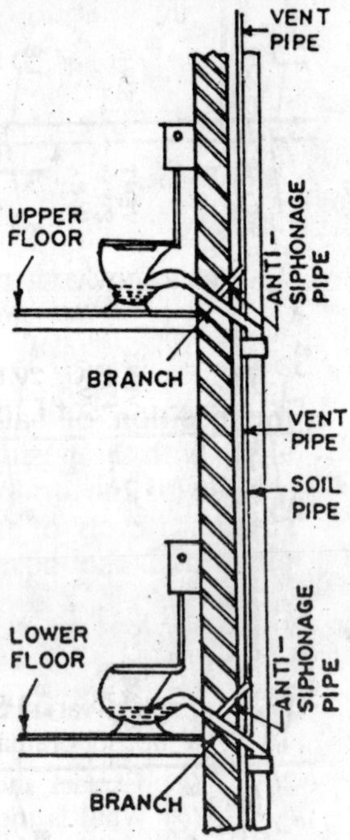

FIG. 20.12.
ANTI-SIPHONAGE PIPE.

20.11. HOUSE DRAINAGE PLANS

For efficient drainage, it is always better to prepare house drainage plan. In some cities, it is statuatory to submit such plans. Fig. 20.13 shows a typical plan for drainage of a small house.

The site plan is drawn to a suitable scale, showing onto it

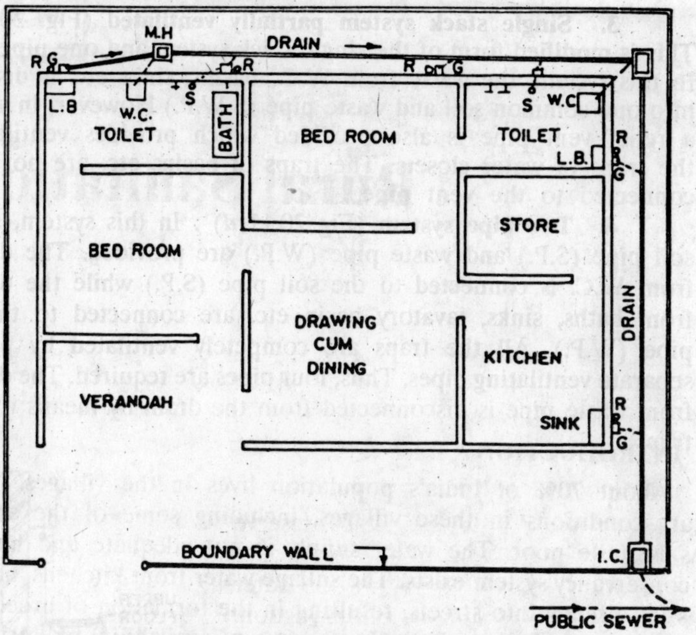

R= RAINWATER PIPE ; G= GULLEY TRAP ; L.B.= LAVATORY BASIN ;
W.C.= WATER CLOSET ; S= SINK M.H.= MANHOLE;
I.C.= INTERCEPTING CHAMBER.

FIG. 20.13. DRAINAGE PLAN OF A BUILDING.

the position of baths, W.C., urinals, wash basins and other units, along with the position of gully traps and floor traps. The longitudinal section of the drain is also drawn, showing distances, invert levels, size and levels of inspection chambers and man holes, gradient of pipes and the position and level of the public sewer.

PROBLEMS

1. Explain various principles that should be kept in mind while designing a house drainage system.
2. (*a*) Explain the terms : soil pipe ; waste pipe ; vent pipe
   (*b*) What is meant by anti-siphonage pipe ? Explain its function with a neat sketch.
3. Explain various types of traps commonly used.
4. (*a*) Explain in brief various types of water closets.
   (*b*) Enumerate the requirements of a good water closet.
5. Explain, with the help of a diagram, the working of a flushing cistern.
6. Explain, with the help of diagrams, various systems of plumbing used for house drainage.
7. Draw a typical house drainage plan of a residential building.

21

Rural Sanitation

21.1. INTRODUCTION

About 70% of India's population lives in the villages. The sanitary conditions in these villages, including some of the small towns is quite poor. The water supply is not adequate and hence only conservancy system exists. The sullage water from kitchens, baths etc. is let directly into streets, resulting in the formation of breeding place for flies and mosquitoes. In absence of availability of latrines, children defecate in front of houses. Unless this faecal matter is quickly removed, phathogenic bacteria thrive on this faecal matter, resulting in spread of epidemics. The rural areas, consisting of villages and small towns are mostly unsewered because of two main reasons: (*i*) adequate water supply, necessary for efficient functioning of water carriage system is not available, and (*ii*) population is not concentrated. This is true even for the outskirts of urban areas. The most important part of the work, which a sanitary engineer is called upon to do in such *unsewered* rural areas, in this connection, consists of the efficient collection, removal and disposal of the following :

(*i*) *Night Soil* from privies on the conservancy system.

(*ii*) *Sullage*, including urine and ablution water from previes

and (*iii*) All *refuse* solid matter.

21.2. RURAL SANITATION

Rural sanitation comprises of the following four essentials:

1. Provision of safe and potable water for domestic purposes.
2. Collection and disposal of dry refuse.
3. Collection and disposal of sullage.
4. Disposal of excretal waste.

1. Provision of safe and potable water for domestic purposes

Usual sources of water in rural areas are : rivers, lakes, ponds and wells. River water is generally not considered safe for domestic use since it may be polluted at the upstream.

Lakes and ponds may be considered as safe sources, provided these are properly protected against contamination caused due to daily defecation by people or animals. Village tanks, which are mostly the source of water, are generally unsafe because animals, mostly buffalos, swim in these throughout the day and some people also swim and take bath, in these tanks. Due to this, epidemics in the form of some 'thread disease' is quite common. However, well water is considered to be the safest of all the available sources. Though limited quantity of water may be available from these wells, this quantity may be sufficient for rural population. Sometimes, water from the nearby canals may be collected in village tanks and subsequently used. Such tanks may have storage for 1 to 4 weeks. Tanks, wells and other sources of water used by the villagers for the drinking purposes should be *disinfected* from time to time by potassium permanganate ($KMnO_4$).

Alternatively, the available source of water should be properly *chlorinated using a pot chlorinator.* In this method, an earthen pot, having 2 to 10 mm holes in the sides (Fig. 21.1) is used. About 1.5 kg of bleaching powder, mixed in 3 kg of 2 mm size coarse sand is filled in the pot. This pot, after covering with a polythelene cover at the top, is suspended 1 m below the water level to be disinfected. This device has a capacity of disinfecting/chlorinating about 9000 to 13000 litres, with a withdrawal rate of 800 to 1200 litres per day. Thus, the unit can serve 20 to 30 people per day. The pot is refilled every week.

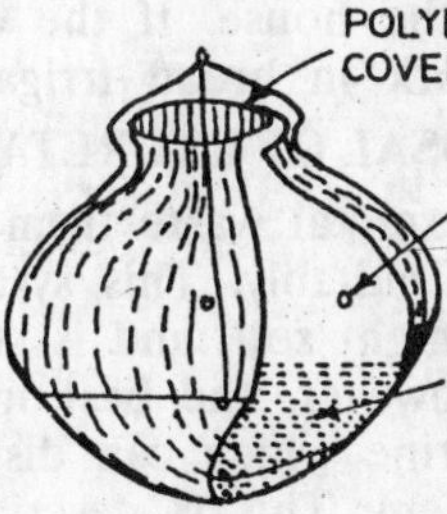

FIG. 21.1. SINGLE POT CHLORINATOR.

21.3. COLLECTION AND DISPOSAL OF DRY REFUSE

Dry refuse in rural areas consists of (*i*) house sweepings, vegetable wastes, paper waste, rags etc. In fact it consists of all sorts of dry waste except excretal waste from latrines. Such a waste does not give any bad or foul smell, and does not create any nuisance. It can be collected and disposed off by any of the following methods:

(*i*) Keeping dust bins in front of every house, and collecting dry refuse from these bins daily and regularly. Sweepers may collect dry refuse from the whole area, at one central place, from where it is lifted periodically in a bullock-cart or a lorry.

(*ii*) The dry refuse so collected may be dumped in a low lying area.

(*iii*) Alternatively, the collected dry refuse may be burnt, and the ash can be dumped in low lying area.

(*iv*) The animal dung is collected and dumped either in a pit and covered. After some time (about 1 – 2 months), this will be converted into manure for agricultural field.

21.4. COLLECTION AND DISPOSAL OF SULLAGE

As stated earlier, sullage is merely the wastewater from bath rooms, kitchens, house washings, washing of clothes etc. It does not contain any excretal refuse, and hence it is neither very dirty nor create bad smell. It is collected and disposed off by the following methods :

(*i*) The wastewater may be led to pits in which animal dung and dry refuse is also dumped. Wastewater will help easy decomposition of waste refuse and convert it into refuse by composting.

(*ii*) Wastewater may be carried outside the village in open drains and admitted into a nallah or a natural water course.

(*iii*) Wastewater may be used for watering trees or kitchen gardens in the house. If the quantity is more, it can be utilized for farm work in broad irrigation.

21.5. DISPOSAL OF EXCRETAL WASTE

The excretal waste from house is collected by sweepers and disposed off suitably. This system is called conservancy system of disposal of night soil, and is still prevalent in most of the villages in India. However, the best method is to construct any type of privy or latrine which can dispose off without the help of water carriage system. This is described in subsequent articles. There is generally a lot of open space available in the rural area. Hence construction of such previes or latrines, with self-disposal system is not a problem. Many times, *soak pits* are used for individual houses. The excretal waste as well as sullage from bathrooms, kitchen etc. is led directly into the soak pit. Such soak pits are usually of 3 m dia. and 3 to 5 m deep, constructed of masonry walls with open joints, and are covered at the top by rigid stone or concrete slab. Such a soak pit, seving 4 to 8 people in a house has a life of 5 to 15 years. When these soak pits get filled up, these are emptied and the material is disposed off. Such a system of soak pit is quite common and popular in the unsewered urban areas.

21.6. NIGHT SOIL DISPOSAL WITHOUT WATER CARRIAGE : PREVIES

Following are the *general requirements* of a good privy :

(1) It should be located away from the building, on the leeward side. (2) Its floor should be atleast 1.2 m above the general level.

(3) Its floor should be of impervious material. (4) It should be well ventilated. (5) The receptable to receive excreta should be of sufficient capacity so as to hold at least 24 hours the excreta, urine and ablution water. (6) It should remain closely covered so that flies donot reach there. (7) Contents should not be visible to the user. (8) It should have moderate cost.

Previes may be of the following types :

(1) Trench or earth privy, (2) Pit privy, (3) Bore hole privy, (4) Pale or can privy, (5) Concrete vault privy, (6) Aqua privy, (7) Chemical toilet, (8) PRAI latrine.

The *choice* of a system depends on (*i*) the availability of water, (*ii*) position of ground water, and (*iii*) available resources and human behaviour.

The prevy *design* depends upon : (*i*) number of persons and quantity of excreta, (*ii*) method of decomposition (*iii*) availability of water (*iv*) permeability of soil (*v*) humidity and temperature conditions, (*vi*) ground slope (*vii*) types of squatting plates available (*viii*) availability of funds (*ix*) availability of man power (*xi*) cultural and religious customs, and (*vii*) location of excreta disposal site.

1. Trench or Earth Privy : These are used for camps etc. A shallow pit of either square or rectangular shape, 30 cm in width and 50 to 100 cm in depth is excvated, for the accumulation of excreta. The excavated earth is piled round the trench. Each user, after use, has to throw equal quantity of earth on the faecal matter. The urine falls outside the privy. One privy can be used for about 100 faecal droppings. After that, another trench is excavated nearby.

2. Pit Privy : It consists of a pit, 90 to 120 cm in diameter (or 1 m × 1 m in plan) and 2 to 3 m deep. It may be lined or unlined, but its bottom is not lined. The pit is covered with a squatting slab with a hole or squatting plate. A cover may also be provided to cover the hole. The user discharges excreta into the pit by sitting on the squatting plate. Faeces are stored in the pit while the urine seeps into the soil. A super structure may be built over the privy. A capacity of 40 – 60 litres/person/year is provided. Storage period may be 4 to 6 years. A pit capacity of 1.5 m^3 can serve an average sized family for several years.

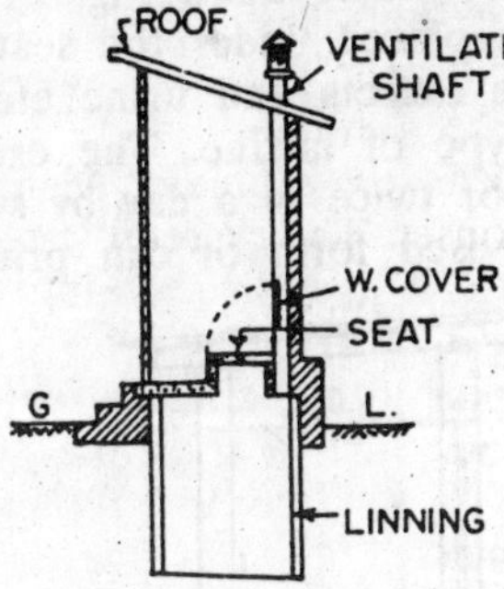

FIG. 21.2. PIT PRIVY

3. Bore Hole Privy : Bore hole privy is similar to pit privy except that size of bore hole privy is smaller than the pit privy while its depth is very much larger. It consists of a bore hole of 30 to 40 cm in diameter and 4 to 8 m in depth. Cover plate and super-strucutre are similar to the pit privy. The upper portion of the borehole is lined with clay [Fig. 21.3 (*a*)].

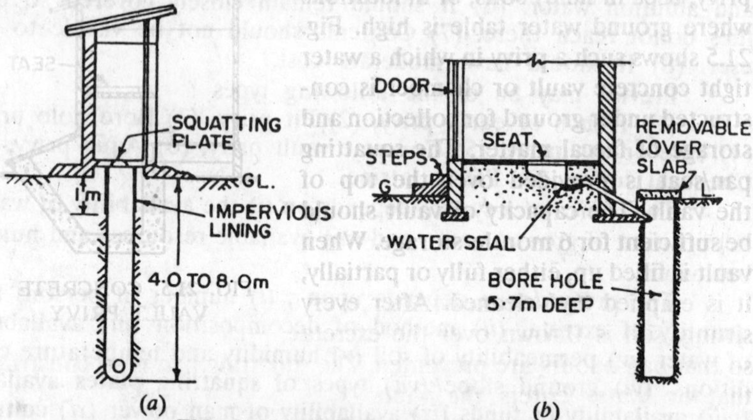

FIG. 21.3. BOREHOLE LATRINE

Fig. 21.3 (*b*) shows another form of borehole latrine with a pan and trap arrangement. When the bore hole gets filled up with excreta, another bore hole is drilled in the vicinity and the compartment and the seat are shifted. The old hole is covered with thick layers of soil and abandoned.

4. Pale or Can Privy : This is also known as removable receptable privy. This is the oldest and primitive type of privy, which is still in use in rural areas. In this privy, excreta is collected in iron pales, tubs or cans. These cans are 40 cm × 30 cm × 20 cm in size and are water tight. Fig. 21.4 (*a*) shows such a privy, where the can is placed under the seat or squatting plate for the collection of the excreta and urine etc. There is odour and fly nuisance in this type of latrine. The excreta is collected from this container once or twice in a day by sweepers, for further disposal. Another improvised form of can privy is shown in Fig. 21.4 (*b*).

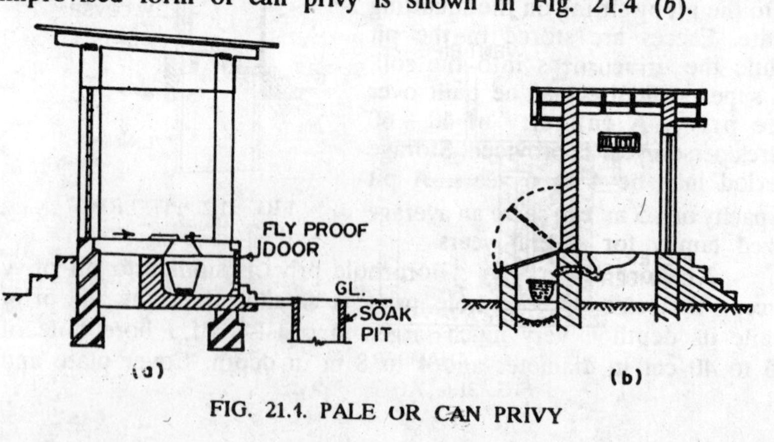

FIG. 21.4. PALE OR CAN PRIVY

5. Concrete Vault Privy : This is an improved form of pit or bore hole privy, used in sandy soils, or in locations where ground water table is high. Fig. 21.5 shows such a privy in which a water tight concrete vault or chamber is constructed under ground for collection and storage of faecal matter. The squatting pan/seat is provided over the top of the vault. The capacity of vault should be sufficient for 6 months storage. When vault is filled up, either fully or partially, it is emptied and cleaned. After every sitting, soil is thrown over the excreta so that bad odours are prevented. The squatting pan is so constructed that no water enters the vault.

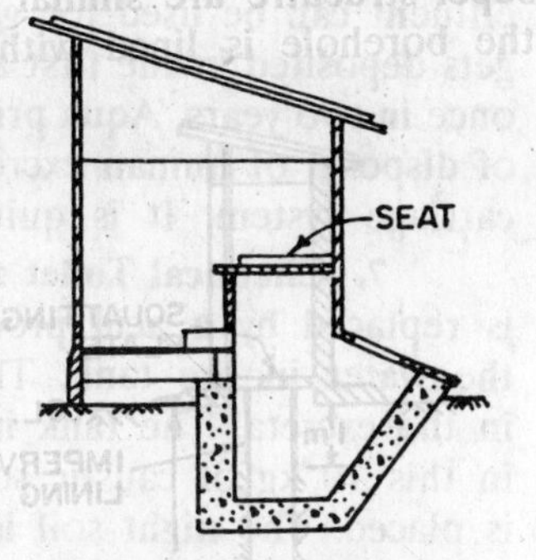

FIG. 21.5. CONCRETE VAULT PRIVY

6. Aqua Privy : This is also known as *wet latrine*. As shown in Fig. 21.6, an aqua privy consists of three chambers : (*i*) the anaerobic digestion chamber, (*ii*) aerobic digestion/clarifying chamber and (*iii*) filter chamber. The latrine pan is kept on the top of the anaerobic chamber and its discharge pipe is taken 10 cm deep below the water level in the first chamber (anaerobic chamber) to maintain water-seal. Night soil, urine and ablution water falls smoothly into the first chamber and is decomposed by anaerobic bacteria. The gases produced escape through the vent pipe. In the second chamber (aerobic chamber), further decomposition takes place by the action of aerobic bacteria. The effluent from the second chamber is discharged into the third chamber called the filter chamber containing brick bats/clinker as the filter media. As the effluent rises in this chamber, it is purified. Finally, the effluent is led out through the effluent chamber. This

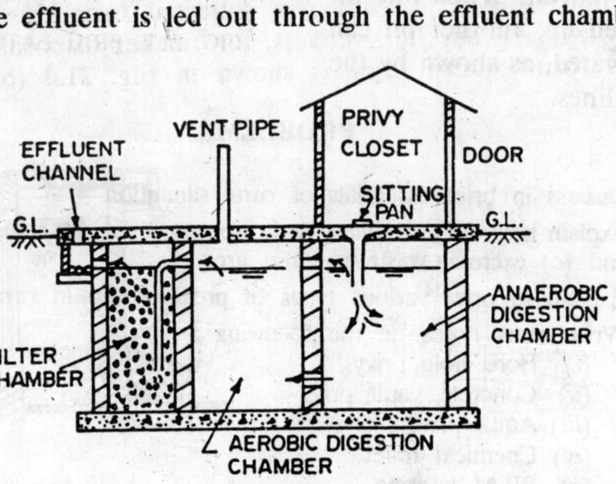

FIG. 21.6. AQUA PRIVY

effluent can be used for gardening or farming. The digested sludge gets deposited in the first and second chamber, which can be cleared once in two years. Aqua privy is considered to be quite useful method of disposal of human excreta, specially in absence of suitable water carriage system. It is quite economical and hygienic.

7. Chemical Toilet : In the chemical toilet, the concrete vault is replaced by a acid proof metal tank. Caustic soda is added to the water in the tank. This stabilises the organic matter present in the excreta. The tank has generally a capacity of 600 litres, and in this 10 kg of caustic soda, dissolved in 40 to 60 litres of water is placed. The night soil is liquified and sterilised in the tank, and pathogens are also destroyed. The spent chemical and accumulating liquids and solids are removed at frequent intervals and drained out in a cesspool. Such type of toilets are also used in aircrafts. There is no odour problem. However, its cost is more.

8. PRAI Laterine : Planning, Research and Action Institute (PRAI), Lucknow (U.P.) has developed a latrine, the details of which are shown in Fig. 21.7. It is a hand-flushed, water seal, lateral pit laterine used for disposal of night soil in rural area.

The laterine essentially consists of a pan which discharges the excreta etc. into a pit which has honey comb masonry in mud mortar. When one pit gets filled up, another pit can be excavated, as shown by the dotted lines.

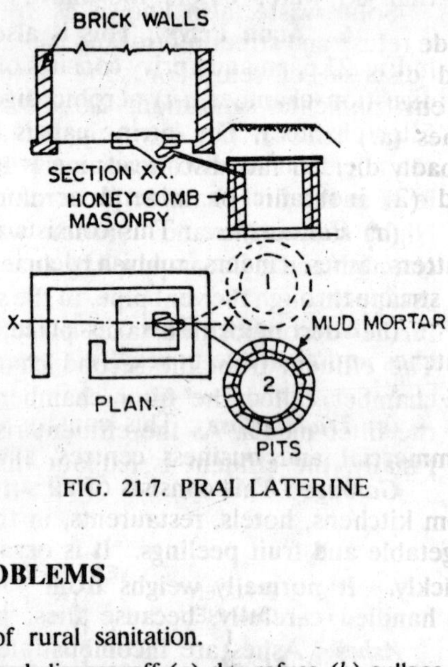

FIG. 21.7. PRAI LATERINE

PROBLEMS

1. Discuss in brief essentials of rural sanitation.
2. Explain how do you collect and dispose off (*a*) dry refuse (*b*) sullage, and (*c*) excretal waste in rural area.
3. Explain in brief various types of previes used in rural area.
4. Write short notes on the following :
   (*i*) Bore hole privy
   (*ii*) Concrete vault privy
   (*iii*) Aqua privy
   (*iv*) Chemical toilet
   (*v*) PRAI laterine.

22

Solid Waste Disposal

22.1. INTRODUCTION :

Solid waste, also known as *dry refuse* includes house refuse, trade refuse and street refuse, and is practically in a dry state. Removal and disposal of refuse or solid waste is a very important aspect of environmental sanitation. Solid waste consists of (*i*) garbage (*ii*) ashes (*iii*) rubbish (*iv*) dust etc. Solid waste or dry refuse can be broadly divided into two heads : (1) Organic or combustible matter, and (2) inorganic or mineral or non-combustible matter.

(*a*) ***House refuse*** : This consists of vegetable and animal waste matters, ashes, cinders, rubbish, debris from cleaning and demolition of structures.

(*b*) ***Street refuse*** : This consists empty packets and bottles, empty matches and cigarette boxes, fruit peels, tree leaves, street sweepings etc.

(*c*) ***Trade refuse*** : This consists of solid wastes from factories, commercial and business centres, slaughter houses etc.

Garbage : This consists of all sorts of putrescible organic waste from kitchens, hotels, restaurents, in the form of waste food articles, vegetable and fruit peelings. It is organic in nature and decomposes quickly. It normally weighs from 450 to 900 kg/m^3. It should be handled carefully, because flies, insects, rats etc. breed in it.

Ashes : Ashes are incombustible waste products from houses, industries, hearths and furnaces. With the introduction of kerosene oil and cooking gas, its quantity is now gradually decreasing. It weighs between 700 to 850 kg/m^3.

Rubbish : It consists of all non-putrescible wastes, excluding ashes. Common items that fall under this category are : rags, paper pieces, paper packets, glass and plastic bottles, brocken pieces of glass, broken crockery, broken furniture and stationary items, card boards. It thus includes a wide variety of combustible and non combustible wastes. It is lighter and its weight varies from 50 to 400 kg/m^3.

Organic waste : It includes dry animal and vegetable refuse, cow dung, excreta of birds, tree leaves, sticks, plastic bottles, paper waste, rags. This waste is subject to decay with time and evolve highly offensive odour and gases which are highly detrimental to health.

Inorganic waste : This consists of non-combustible materials such as grit, dust, mud, metal pieces, metal containers, broken glass and crockery, tiles, waste building material. It is not subject to decay and is therefore not harmful to public health.

22.2. QUANTITY AND COMPOSITION OF REFUSE

The quantity and composition of refuse varies from place to place and also varies from season to season. Ashes increase in winter and in Northern latitudes while these are less in summer and in hot tropical countries. The quantity of garbage depends upon the food available, food habits and standard of living. It also depends upon whether the town is residential, commercial or industrial. For an average Indian city, quantity of refuse varies from 0.4 to 0.5 kg per capita per day. The average composition of refuse by weight is about 25% cinder, 27% fine dust, 15% ashes, 4% empty tins and cans, 14% putrescible matter, 2% glass and crockery, 2% rags, 1% bone and 8% miscellaneous matter. Organic wastes are generally 55% while inorganic wastes are 45%. Table 22.1 gives an average composition of refuse (by weight) for an average Indian city and an average American city, at the disposal site.

TABLE 22.1 AVERAGE COMPOSITION AND PROPERTIES OF REFUSE

S.N.	*Constituent*	*Average composition % by weight at disposal site*	
		Typical Indian city	*Typical U.S.A. city*
1.	Garbage	55	25
2.	Rubbish	15	55
3.	Ashes	10	10
4.	Dust, silt, sand	20	10
(a)	Unit wight (kg/m^3)	350–600	250—300
(b)	Calorific value (kilo. cal/kg)	1500–1800	3000

22.3. COLLECTION AND REMOVAL OF REFUSE

The frequency of collection of solid waste depends upon the quantity of the refuse and the season. Generally, refuse is collected in individual houses in small containers or cans kept outside the premises of the house, from where it is removed daily by sweepers. Sometimes, public dust-bins are provided by the municipalities/local boards at convenient places by the sides of roads. Dry refuse fallen on the public streets and roads, along with road sweepings, are usually collected once or twice a day by the sweepers employed by the local authority. For this purpose, a portable galvanized iron receptable with a closely fitting lid, and having a capacity of 0.02 to 0.1 m^3 is generally used. The containers or cans used for refuse collection should be clean, without any dirt left in, otherwise fresh refuse gets seeded with the putrefactive organisms and start giving foul smell. The frequency of refuse collection is kept such that the refuse may not start giving nuisance by odour and fly breeding. The collection of refuse from the business areas should be done during non-working hours.

22.4. TRANSPORT OF REFUSE

The refuse collected in the public dust bins located by the sides of roads, is transported to the disposal site by means of following vehicles : (*i*) Auto-rickshaws, (*ii*) Trailers, and (*iii*) Trucks.

Autorickshaws : These, having three or four wheels, have covered bodies. Since their capacity is limited to $\frac{1}{2}$ to $\frac{3}{4}$ tonnes, these are used only for those narrow localities where other heavy vehicles cannot go.

Trailors : Trailors have slightly larger capacity (2 to 3 tonnes). They are also used for localities where trucks cannot go. Loading of trailors is done manually. However, they are of tilting-tipping type and hence their unloading is done automatically with the help of hydraulically operated jacks.

Trucks : Trucks have larger capacity (5-10 tonnes). They are generally of tilting-tipping type so that unloading is automatic. Special types of trucks, capable of bodily lifting covered skip boxes (in place of ordinary dust bins) are now available, and should be used so as to avoid nuisance of flies.

The vehicles employed for the transport of refuse should be of such pattern and design that collected garbage does not fall once again on the road during the transport. The transport vehicle should be strong durable and water tight. They should be made of steel with smooth interior surface and round edge and corners, so that they can be kept clean.

22.5. DISPOSAL OF REFUSE

Refuse or solid waste can be finally disposed of by the following methods.

1. Controlled tipping.
2. Filling of low lying areas. (Land filling).

3. Dumping into sea.
4. Pulverisation.
5. Incineration.
6. Composting.

1. Controlled tipping : This method is useful where adequate site for redevelopment is available. The method consists of tipping the refuse in hollows to a depth of 1 to 2 m. While tipping, coarse material is tipped at the bottom while fine material is tipped on the top. These tips are covered with soil, so as to provide a seal under which bacterial decomposition takes place. At the end of about 12 months period, the decomposition is complete, due to which the tip settles down to a height of 30 cm only. Normally, an area of 0.3 to 0.5 m^2 per capita per year is required.

2. Filling of low lying area (landfilling) : This method is quite common. The garbage is dumped into low lying areas or depressions available nearby. Dumping is done in layers of 1 to 2 m, and each layer is covered by 0.2 m thickness of good earth. A rest of 2 to 3 weeks is given before dumping the second layer. If dry refuse is loosely packed, it may give rise to health hazards. Hence each layer should be compacted by movement of dumping vehicles, for its settlement, before starting filling the second layer of refuse.

The *advantages* of the method are : (*i*) It is simple and economical (*ii*) No plant/equipment is required (*iii*) Separation of various materials of the refuse is not required (*iv*) There are no by-products and hence there is no problem of the disposal of the by-products (*v*) The low lying areas can be reclaimed and put to better use by this method. The *disadvantages are* (*i*) proper site may not be available nearby, (*ii*) wind direction may not be favourable (*iii*) large land areas are required (*iv*) It may be difficult to get large quantities of covering material and (*v*) dumped garbage containing carcinogenic non-biodegradable matter (such as plastics, unused medicines, paints, insecticides, sanitary napkins etc) may cause trouble later because of *leachate* coming out of the dump during rainy season (*vi*) leachate from the dumped garbage may pollute surface water as well as ground water.

3. Trenching : This method is generally adopted when low lying areas are not available. Trenches of size 4 to 10 m long, 2 to 3 m wide and 1 to 2 m deep are excavated with a clear spacing of 2 m. These trenches are then filled with refuse/garbage in layers of 15 cm. On the top of each layer, 5 cm thick sandwiching layer of night soil/animal dung is spread in semi liquid form. On the top layer protuding 0.3 m above ground surface, 10 cm layer of good earth or other non-combustible material is spread to act as a seal, so that flies do not get access and wind does not blow the refuse off. The dumped garbage is converted into a type of compost by the fermentation carried out by anaerobic bacteria, within a period of 6 months. The compost available from trenching is of low agricultural value ; on the contrary, it may sometimes have elements which may be harmful to the soil.

4. Dumping into the sea : Solid waste/refuse can also be disposed of by bargeing out into the sea, after carrying it at reasonable distance (say 15 to 20 km) into the sea. This is necessary to prevent the shores from refuse nuisance, because sea waves can carry back the refuse to the shores. The depth at such disposal point should not be less than 30 m. The following are the *defects* of this method : (*i*) bulky and lighter matter in the refuse may float, spread out and tend to return to the shores during high tides, (*ii*) during stormy weather and monsoons, it is not possible to send barges out into the sea, and (*iii*) inspite of best care, some portion of refuse may return the shores and spoil them. Also, the method is possible only in cases of coastal cities.

5. Pulverisation : In this method, the dry refuse is pulverised into powder form, without changing its chemical form. The powder can either be used as a poor qualitity manure, or else be disposed of by land filling. In some countries the pulverised refuse is discharged in the sewer. The method is quite costly and hence not commonly used in India.

6. Incineration : This consists of burning the refuse in the incinerator plant. This is commonly used in disposing of garbage from hospitals and industrial plants. Before incineration, non-combustible and inert material like earth, broken glass, chinaware, metal etc are separated, so as to reduce the load on the hearth. The by product of this method is ash and clinker which can be easily disposed of by land filling. The heat generated by burning the dry refuse may be utilized for raising steam power. The quantity and quality (moisture and calorific value) of refuse is, however, changing and hence the power generated will fluctuate. Emission of air pollutants from incinerators includes particulates such as flyash, unburnt fuel and others. Permissible level of particulate emissions from large incinerators is 0.23 g per standard cubic metre of exhaust gas corrected to 12% CO_2. Smoke includes all liquid and solid matter in the exhaust that hinders visibility. Smoke can be eliminated by mixing the exhaust with hot air to complete combustion, but black smoke requires fairly high air temperatures.

The following points should be carefully observed during incineration :

1. The refuse charging should be thorough, rapid and as nearly continuous as possible.

2. Each batch of refuse entering furnace should be well mixed.

3. Auxiliary burners are usually installed above the refuse to ignite it and to establish the draft at the beginning of the cycle. This is all the more necessary when the moisture content of air is high.

4. Minimum temperature in the combustion chamber should be sufficient (> 670°C) so that all the organic matter is incinerated and foul smelling gases are oxidised.

5. After burners are sometimes required, together with particulate removal devices such as settling chambers or scrubbers.

Furnaces can be (*i*) vertical, (*ii*) circular (*iii*) rotary multicell and (*iv*) rectangular in their design. In the *vertical circular furnace*, the refuse is charged through a door in the ceiling and drops into a central cone grate surrounded by a circular grate. Primary combustion is supported by underfire air. In the *rotary kiln furnace*, the wastes are partially burnt in a rectangular furnace and then fed, via grates, to a rotating kiln. The rotary action exposes the unburnt material for combustion. Final combustion takes place in the chamber after the kiln discharge point. The *multicell furnace* (Fig. 22.1) has cells side by side. Each cell has grates for moving the refuse across them. Several cells have a common combustion chamber and residue hopper. In *rectangular furnace* unit, two or more grates are arranged in tiers so that the refuse is agitated as it falls from one level to the next. Secondary combustion is usually employed. *Pathological waste incinerator* handle organic wastes of human or animal origin and crematory furnaces. Such incinerators

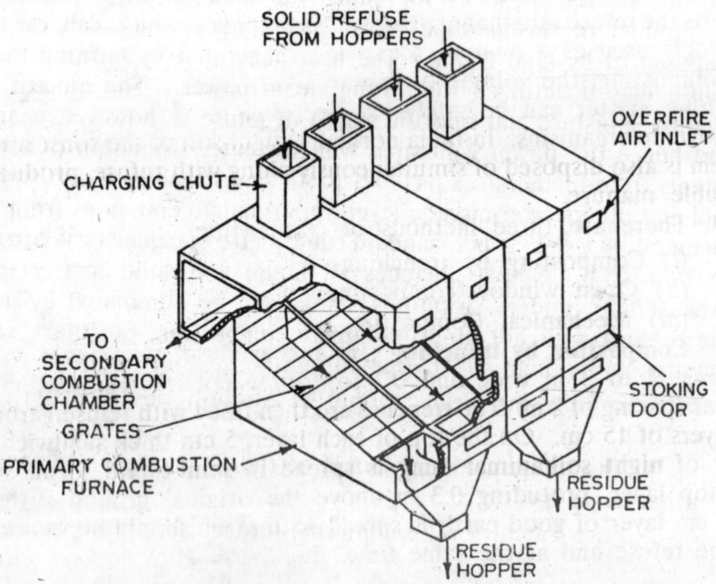

FIG. 22.1. MULTICELL INCINERATION FURNACE.

are multiple chamber units which release fluids as the material is being destroyed. These fluids do not evaporate quickly and, therefore, solid hearth rather than grating is required.

Advantages of incineration : (*i*) This is most hygienic method, since it ensures complete destruction of pathogens (*ii*) There is no odour trouble or dust nuisance (*iii*) The heat generated can be used for raising steam power (*iv*) Clinker produced can be used for road purposes (*v*) The disposal site (i.e. incineration site) can be located at a convenient distance (*vi*) Lesser space is required for disposal of residues (*vii*) Modern incineraters can burn a great variety of refuse materials which are otherwise not biodegradable (*viii*) Adverse weather condition have no effect on the incinerator's operation.

Disadvantages : (*i*) Large initial expenditure (*ii*) Improper operation results in air pollution problems and incomplete reduction of the waste materials (*iii*) Disposal of the remaining residue is required (*iv*) High stacks needed for natural draft chimneys present safety problems.

It should be clearly noted that municipal incineration of solid waste or refuse is a volume reduction process and not one of complete or ultimate disposal. Safe disposal of remaining residue is an essential requirement. Also, the plant need be operated properly so that the gases are completely burned and a stable residue is produced.

22.6. COMPOSTING

Composting is a method in which putrescible organic matter in the solid waste/refuse is digested anaerobically and converted into *humus* and stable mineral compounds. It is a hygienic method which coverts the refuse into manure through the bacterial agencies. Compost is widely used as a manure which is rich in nitrogen content. Due to composting, the volume of refuse is very much reduced, and the resulting matter can be safely handled since it becomes free from pathogenic organisms. In India normally, night soil of the conservancy system is also disposed of simultaneously along with refuse, producing valuable manure.

There are three methods of composting.

(*i*) Composting by trenching
(*ii*) Open window composting
(*iii*) Mechanical Composting

Composting by trenching : In this method, trenches 4 to 10 m long, 2 to 3 m wide and 0.7 to 1 m deep are excavated with a clear spacing of 2 m. The trenches are then filled with refuse/garbage in layers of 15 cm. On the top of each layer, 5 cm thick sandwiching layer of night soil/animal dung is spread in semi-liquid form. On the top layer, protuding 0.3 m above the original ground surface, a 10 cm layer of good earth is spread so that flies do not get access to the refuse and at the same time, the refuse does not get blown off by wind. Within 2-3 days, intensive biological action starts to destroy/reduce organic matter present in the refuse. In this process, considerable heat is generated and the temperature of the composting mass rises to about 75°C. Due to this reason the breeding of flies does not take place. The refuse gets stabilised in about 4-5 months period, and gets changed into a brown coloured odourless innocuous powdery form known as *humus*, which has high manure value because of its nitrogen content. The stabilised mass is removed from the trenches, sieved through 12.5 mm sieve to exclude coarse inert materials like stones, brick bats, broken stone etc. The sieved material is then sold out as a manure.

Open Window Composting : In this method, a large proportion of mineral matter like dust, stone, broken glass pieces etc. are first removed from the refuse. The refuse is then dumped on the ground in the form of 0.6 to 1 m high, 6 m long and 1 to 2 m wide piles at about 60% moisture content. The pile is then covered with night soil, cowdung, cattle urine etc. through which the organisms or germs that are necessary for fermentation are added. Due to biological activity through aerobic bacteria, heat starts developing upto about 75°C in the refuse piles. Due to this, the microbial reaction shifts from mesophilic to thermophilic stage. After this, the pile is turned up for cooling and aeration to avoid anaerobic reactions. The temperature of pile again rises to 75°C, and the process of turning, cooling and aeration is repeated. The complete process may take about 4-6 weeks, after which the compost is ready for use as manure when the temperature falls considerably.

Mechanical Composting : The open window method of composting is very laborious and time consuming process. Also, it requires large area of land which may not be available in big cities. These difficulties are overcome by adopting *mechanical compositing* in which the process of stabilization is expedited by mechanical devices of turning the compost. The mechanical method stabilizes the refuse compost only within 3-6 days. The operations involved in a large scale composting plant, shown in Fig. 22.2 are as follows : (1) Reception of refuse, (2) Segregation, (3) Shredding or pulverising, (4) Stabilisation, (5) Marketing the humus.

The refuse is *received* at the plant site in quantities of 2 to 6 tonnes per vehicle. Hence the plant site must have a storage capacity of about 25 to 50% of total daily arrival, before it can be segregated and shreded/pulverised. *Segregation* is done by hand picking on smaller plants and by mechanical devices on large plants, to remove paper, rags, non-ferrous metals and large objects. Ferrous metals are removed by magnetic separators. Finer material such as ash, particles of garbage etc. are removed by passing the refuse

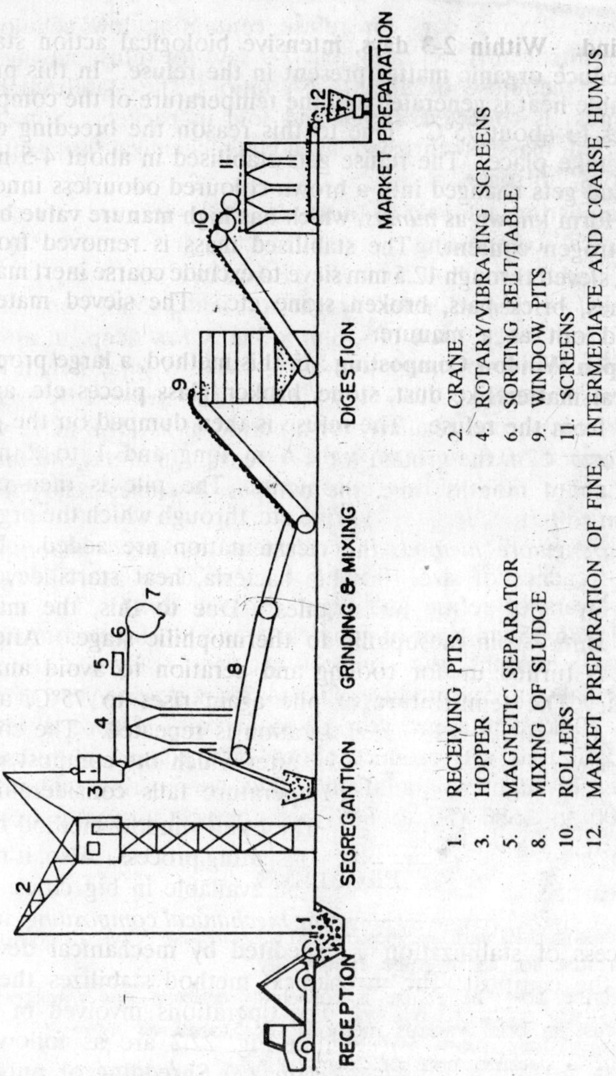

FIG. 22.2. PROCESSING OF REFUSE BY MECHANICAL COMPOSTING

over shaker-screens. The remaining refuse is then shreded and pulverized mechanically.

The *prepared refuse* is then decomposed or stabilized under controlled conditions of temperature and moisture content. Mechanical digestors of various types such as (*i*)) Pits or cells (*ii*) Windows or stacks and (*iii*) Vertical cylinder, horizontal cylinder, or silo type closed digestors. Closed digestors are most hygienic and occupy less space. In these digestors, the refuse is digested and converted into humus and stable mineral compounds. The digestion period vary between 2 to 5 days for refuse containing low cellulose or low carbon-nitrogen (C/N) ratio, and 7-9 days for refuse having more quantities of cellulose or high (C/N) ratio. The stabilized brown mass (humus) is collected, sieved and sold in packets. Sometimes, the stabilized mass is enriched by adding chemical nutrients like phosphorous, nitrogen.

Composting adopted in India : In India, there are two methods of mechanical composting :

(*i*) Indore method and (*ii*) Bangalore method

In *Indore method*, refuse, night soil and animal dung etc. are placed in small brick lined pits, 3 m × 3 m × 1 m deep, in alternate layers of 7.5 to 10 cm height, so as to make a total height of 1.5 m. Chemicals (such as DDT etc.) are added to prevent fly breeding. The material is turned regularly for a period of about 8 to 12 weeks, and then stored on the ground for 4-6 weeks. In about 6-8 turnings and in about 4 months time, the compost becomes ready for use as manure.

In *Bangalore method*, the refuse is stabilized anaerobically. Earthern trenches of size 10 × 1.5 × 1.5 m deep are filled up in alternate layers of refuse and night soil/cow dung. The material is covered with 15 cm layer of good earth and left for decomposition. In about 4-5 months, the compost becomes ready for use.

Normally, a city produces 200 to 250 kg/capita/year of refuge and 8 to 10 kg/capita per year of night soil. Hence a town of 10,000 population will produce about 2000 tonnes of refuse and 800 tonnes of night soil annually. The composting will produce about 1400 to 1680 (50 to 60%) of compost annually from the above waste.

PROBLEMS

1. Discuss in brief various types of solid waste/refuse. Give the composition of refuse for an average Indian city.
2. Describe how the refuse is collected, removed and transported.
3. Discuss in brief various methods of disposal of refuse.
4. Write a detailed note on 'composting'.
5. Write a detailed note on incineration of refuse. What are its advantage and disadvantage ?
6. Write short notes on the following :
   (*i*) Mechanical method of composting
   (*ii*) Incineration,
   (*iii*) Open window composting
   (*iv*) Bangalore method of composting
   (*v*) Indore method of composting.

23

Air Pollution

23.1. INTRODUCTION : DEFINITIONS

Air is defined as the elastic, invisible and tasteless mixture of gases that surrounds the earth. *Air pollution* is an environmental evil. Under ideal conditions, the air that we inhale has a qualitative and quantitative balance that maintains the well being of a man. *When the balance among the air components is disturbed, air is said to be polluted. Some* specific definitions of air pollution are given below.

According to Indian Standards Institution, air pollution is the presence of ambient atmosphere of substances, generally resulting from the activity of man, in sufficient concentration, present for a sufficient time and under circumstances which interfere significantly with the comfort, health or welfare of persons or with the full use or enjoyment of property. As per American Medical Association, 'air pollution is the excessive concentration of foreign matter in the air which adversely affects the well being of individual or causes damage to the property. As per Engineer's Joint Council (U.S.A.), 'air pollution means the presence of outdoor atmosphere of one or more contaminants such as dust, fumes, gas, mist, odour, smoke or vapour, in quantities, of characteristics and of duration as to be injurious to human, plant or animal life or to property or which reasonably interferes with the comfortable enjoyment of life and property'.

Air pollution is an increasing problem in all countries where urban growth and population increases are accompanied by rapid and extensive industrial development and extensive use of automobiles. Air pollution is attributed as the '*price of industrialisation*.' Air pollution can impair health and sometimes cause death. It can cause respiratory problems and eye irritation. It also affects plants, animals and buildings.

23.2. THE BIOSPHERE AND ENVIRONMENT

The solid earth and its interior is known as *lithosphere*. The entire collection of water over the earth, as well as inside the earth is called the hydrosphere. The gaseous layers surrounding the earth upto a distance of about 700 km is called *atmosphere*. The atmosphere is therefore a life supporting gas mantle, surrounding the earth, which decreases in temperature and pressure with altitude (Fig. 23.1). Based on this, atmosphere is further subdivided into (*i*) troposphere, (*ii*) stratosphere and (*iii*) ionosphere. The atmospheric layer blow tropopause is known as *troposphere.* It is the region of turbulent and large-scale vertical motions of air masses where the prevailing westerlies (in the middle latitudes) blow in both Northern and Southern hemispheres. North-South flow also occurs by a number of processes (including stratosphere) and thus leads to air exchange between hemispheres. Most of the heat in the troposphere is received from the ground rather than directly from the sun. In the stratosphere, little vertical mixing takes place. Troposphere contains about 90% of the mass of the atmosphere.

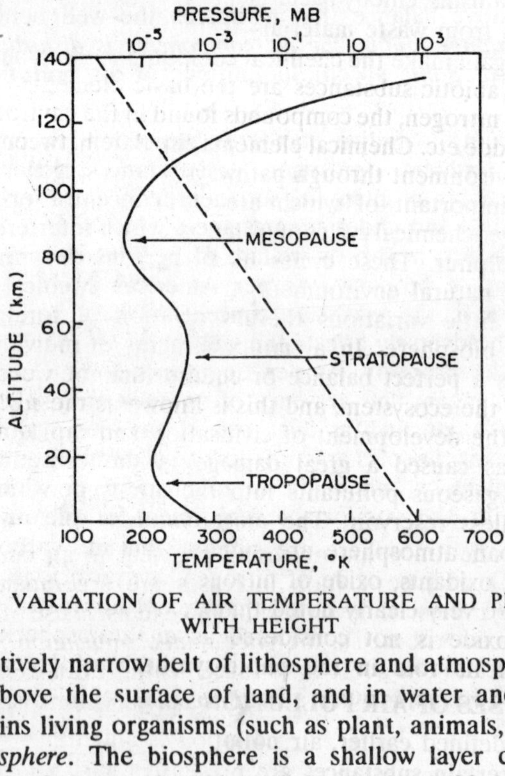

FIG. 23.1. VARIATION OF AIR TEMPERATURE AND PRESSURE WITH HEIGHT

A relatively narrow belt of lithosphere and atmosphere, a little below and above the surface of land, and in water and air, which largely contains living organisms (such as plant, animals, human life) is called *biosphere*. The biosphere is a shallow layer compared to the total size of the earth and extends to about 20 km from the bottom of the ocean to the highest point in the atmosphere at which life can surivie without man-made protective devices. In other words, biosphere is the environment which supports life and sustains various human activities. The essential requisites of life for all the living species (*i.e.* light, heat, water, food and habitats) are supplied by the biosphere. The four major elements *i.e.* (*i*) land (*ii*) water (*iii*) air and (*iv*) living organisms (both plants as well as animals) together constitute what is known as *environment* or *ecosystem*. The ecosystems of bioshpere can be divided into two components : the *biotic* (or living) components and the *abiotic* (or non-living) components. The biotic category can further be subdivided into the following three functional groups : (*i*) *The producers* : These are the autotrophic (self nourishing) organisms, largely the green plants and algae. (*ii*) *The consumers*: These are the heterotrophic (other-nourishing) organisms which ultimately depend upon consumption of producers to sustain life. These includes species of animal life, including mammals, fish, insects and birds, and (*iii*) *The decomposers* : These are heterotrophic organisms, chiefly bacteria and fungi, that break down complex compounds from waste materials–including dead producers and consumer–to again make the chemical components available to producers.

The abiotic substances are the basic elements– such as phosphorus and nitrogen, the compounds found in the environment, oxygen, carbon dioxide etc. Chemical elements circulate between the organisms and the environment through pathways comprising the natural cycles, the most important of which are : (1) the hydrologic cycle, and (2) the biogeochemical cycles of (*i*) carbon, (*ii*) nitrogen (*iii*) phosphorus and (*iv*) sulphur. These cycles have been described earlier. In the unpolluted natural environment, these cycles operate in a balanced state with little variations thereby contributing to the stability of the whole biosphere. In a natural environment, in the biosphere, there exists a perfect balance or equilibrium between various components of the ecosystem, and this is known as the *ecological balance*. However, the development of civilisation and rapid industrialisation by man has caused a great damage to the ecosystem, by way of release of gaseous pollutants into the atmosphere, presuming it to be an endless reservoir. The most common gaseous contaminants in the urban atmosphere are sulphur dioxide, carbon-mono-oxide, ozone and oxidants, oxide of nitrogen, hydrocarbons and aldehydes. It should be very clearly noted that except for special circumstances, carbon dioxide is not considered as an atmospheric contaminant, because of its role in the photo-synthesis of green plants.

23.3. CAUSES OF AIR POLLUTION

As defined earlier, air pollution is any atmospheric condition in which certain substances are present in such concentrations that they produce undesirable effects on man and his environment. Following are the main *causes* of air *pollution* :

*1. **High rate of population growth*** : Consequent to high population density, there is higher rate of fuel consumption.

*2. **Rapid industrialisation*** : This is one of the major cause. The smokes from the factories, coke oven and furnaces, steam engines; exhauset fumes from power plants etc; chemical fumes from oil refineries, zinc refineries, chemical industries, metallurgical plants, iron and steel plants, incineration plants etc. are the common sources of atmospheric pollution. Industrial pollution is most complex, because of the large number of chemicals emitted by it in the atmosphere, and the modifications which these emitted pollutants can undergo by reaction with one another in presence or absence of sun light. Some, critics comment on air pollution as the *'price of industrialisation'*.

*3. **Transportation facilities*** : Intensive increase in transportation services such as motor vehicles, rail-trains, aeroplanes etc, throughout the world is another *major* contributing factor to air pollution. The *exhaust fumes* from the automobiles pollutes the atmosphere considerably in the urban area. Air pollution caused by automobiles is sometimes described as *'disease of wealth'*.

*4. **Radioactive substances*** : Evolution of radioactive gases and suspended radioactive dusts from atomic explosions and accidental discharges from nuclear reactors are very dangerous sources of air pollution.

*5. **Natural causes*** : The natural sources which cause air pollution include the organic compounds from vegetation, ground dust, salt spray from oceans, cosmic dust and evolution of hydrogen sulphide from natural sources. These natural sources are beyond the control of man.

23.4. COMPOSITION OF ATMOSPHERIC AIR

Clean dry air contains 78.09% nitrogen by volume and 20.94% oxygen. The remaining 0.97% consists of gaseous mixture of carbon dioxide, helium, argon, neon, krypton, nitrous oxide etc, including very small amounts of some other organic and inorganic gases whose content varies with time and place. Table 23.1 gives the composition of clean, dry atmospheric air, including their residence time. The values mentioned in the table indicate only the order of magnitude and are not the specific and universally accepted concentrations.

23.5. CLASSIFICATION AND SOURCES OF AIR POLLUTANTS

1. Categories of air pollution. Air pollution exists in three distinct categories : (*i*) Personal air pollution (*ii*) Occupational air pollution and (*iii*) Community air pollution.

TABLE 23.1. COMPOSITION OF CLEAN, DRY ATMOSPHERIC AIR.

Component	*Concentration*		*Estimated Residence Time*
	(ppm)	*Percentage*	
1. Nitrogen	78.09×10^4	78.09	Continuous
2. Oxygen	20.94×10^4	20.94	Continuous
3. Argon	93×10^2	0.93	Continuous
4. Carbon dioxide	3.2×10^2	0.032	2–4 years
5. Neon	18	0.0018	Continuous
6. Helium	5.2	0.00052	≈ 2 million years
7. Krypton	1.0	0.0001	Continuous
8. Xenon	8×10^{-2}	0.000008	Continuous
9. Carbon monoxide	1×10^{-1}	0.00001	0.5 years
10. Methane	1.2	0.00012	4–7 years
11. Nitrous oxide	25×10^{-2}	0.000025	4 year
12. Nitric oxide	6×10^{-4}	0.00000006	5 days
13. Ammonia	6×10^{-3}	0.0000006	7 days
14. Hydrogen Sulphide	2×10^{-4}	0.00000002	2 days
15. Sulphur dioxide	2×10^{-4}	0.00000002	4 days
16 Hydrogen	5×10^{-1}	0.00005	Not known
17. Ozone	2×10^{-2}	0.000002	≈ 60 days

*(i) **Personal air pollution*** : This represents the exposure of an individual to dust, fumes and gases. The common example of personal air pollution is when a person indulges in smoking of cigarette, cigar or pipe.

*(ii) **Occupational air pollution*** : This represents the exposure of individuals to aerosols, vapours and gases in harmful concentrations, in their working or occupational environment.

*(iii) **Community air pollution*** : It represents pollution from a variety of sources and factors which cause adverse social, economic and health effects. Apart from affecting individuals the community air pollution exerts a significant impact on man's total environment including plants, animals, property and the weather.

2. Sources of air pollution. There are two main *sources* of air pollution : (*i*) Natural air pollution sources and (*ii*) Man made or anthropogenic air pollution sources.

*(i) **Natural air pollution sources*** : The natural air pollution sources consists of the following : (1) Wind blown dust. (2) Salt particles from sea water. (3) Dust of meteoric origin. (4) Micro-organisms (bacteria, spores, pollen). (5) Gases and odours from swamps and

marshes (6) Fog (7) Volcanic ash and gases (hydrogen fluoride and hydrogen sulphide) (8) Oxides of nitrogen from electric storm.

(ii) ***Man made sources*** : Table 23.2 shows the main sources of air pollution due to man's activities.

TABLE 23.2. MAN MADE SOURCES OF AIR POLLUTION

Class	*Aerosols*	*Gases and vapours*
1. *Combustion processes* (Domestic gas burning, thermal power plants, automobiles, aeroplanes, railways, refuse incineration)	Dust, fume, smoke	SO_2, NO_2, CO_2 organic vapours, odours
2. *Chemical Processes* (Paper mills, cement plants, fertilisers, other chemical plants etc.)	Dust, fume, mist	Process dependant (SO_2, CO_2, NH_3, NO_2 organic vapours, odours)
3. *Petroleum operations*	Dust, mist	SO_2, H_2S, NH_3, CO, Hydro-carbons, mercaptans
4. *Metallurgical processes* (Alluminium industries, steel pla-nts / mills, smelters, non-ferrous industries)	Dust, fume,	SO_2, CO, fluorides, organic vapours
5. *Mineral Processing*	Dust, fume,	Process dependant (SO_2, CO, fluourides, organic vapours)
6. *Food and feed operations*	Dust, mist	Odourous materials
7. *Agricultural activities* (*a*) Crop spraying (*b*) Field burning	Dust, mist	Organic phosphates, chlorinated hydrocarbons, Arsenic, lead.
8. *Nuclear energy programmes*		
(*a*) Fuel preparation		Fluorides
(*b*) Ore preparation	Dust,	Iodine–131 and Argon 41
(*c*) Nuclear device explosion/testing		Radiactive gases (Sr.–90, Cs–137, C–14 etc)

3. Classification of pollutants Air pollutants can be classified under three heads as under : (*a*) *Natural contaminants* (such as natural fog, pollen grains, bacteria and products of volcanic eruption) (*b*) *Aerosols* (*particulates*) (such as dust, smoke, mist, fog, fume) (*c*) *Other contaminants* (such as gases, vapours and compounds)

4. Primary and secondary air pollutants. Air pollutants can be broadly classified into two groups : (*i*) *primary air pollutants* and (*ii*) *secondary air pollutants.*

Primary air pollutants are those which are emitted directly from identifiable sources. The atmosphere contains hundreds of air pollutants from natural or from anthropogenic sources. All such pollutants are called *primary pollutants*. Important primary pollutants are :

(i) Particulate matter

(a) Coarse particles (> 100 μ)

(b) Finer aerosols including particles of metal, carbon, tar, resin, pollen, bacteria etc.

(ii) Sulphur oxides and sulphur compounds

(iii) Nitrogen oxides

(iv) Carbon monoxide

(v) Hydrocarbons including photochemically reactive aliphatic hydrocarbons and carcinogenic substances consisting of aeromatic hydrocarbons.

(vi) Halogen compounds.

(vii) Organic compounds.

(viii) Radioactive compounds.

(ix) Photochemical oxidants.

Secondary air pollutants are those which are formed in the atmosphere as a result of interaction between two or more primary pollutants or by reactions with the normal atmospheric constituents, with or without photoactivation. These are therefore chemical substances which are often more harmful than the original basic chemicals that produce them. Examples of secondary pollutants are :

(i) Sulphuric acid (H_2SO_4) or acid mist.

(ii) Ozone (O_3)

(iii) Formaldehyde

(iv) Peroxy-acetyl-nitrate (PAN)

(v) Photochemical smog.

5. Stationary and mobile sources of air pollution. Air pollutants may also be classified on the basis of its *source* as : (i) *stationary source* and (ii) *mobile source*. Alternatively, emitting sources can be classified as (a) *point sources* (b) *area sources* and (c) *line sources*.

The automobiles moving on roads form the *mobile sources* of emissions of undesirable gases and particulates into the atmosphere. It is found that two and three wheelers emit the highly toxic gases (such as carbon monoxide) in quantities that are about twice the amount emitted by other vehicles.

The industrial installations form the *stationary sources* of air pollution. Important examples are : (i) asphaltic plants (ii) boilers and heating installations (iii) cement manufacturing units, (iv) steel plants (v) fertilizer manufacturing plants. (vi) mineral acid manufacturing units (vii) paper and pulp plants (viii) power plants (ix) municipal incinerators (x) sewage treatment plants.

Point sources are the large stationary sources mentioned above. *Area sources* are small stationary sources (such as residential heating, institutional and commercial heating, on site incinerations, open burnings etc.) and mobile sources with indefinite routes (such as air crafts, port vessels, motor vehicles, rail yard locomotives etc.). *Line sources* are the mobile sources with definite routes, and these include highway vehicles, railroad locomotives and channel vessels.

23.6. CHARACTERISTICS OF IMPORTANT AIR POLLUTANTS.

Natural Contaminants : The natural contaminants consist of natural fog, pollen grain, bacteria and products of volcanic eruptions. Out of these, *pollen grains* are most important since these can be carried away by wind to a great distance. These are discharged into the atmosphere from natural vegetation such as trees, plants, grasses, weeds etc. These area borne pollutants produce allergic responses in sensitive individual, causing asthma, hay fever, bronchitis, dermatitis etc.

Aerosols (particulates). *Aerosols* refer to the dispersion of solids or liquid particles of microscopic size in gaseous media, such as smoke, fog or mist. An aerosol can also be defined as a colloidal system in which the dispersion media is a gas and the dispersed phase is a solid or liquid. *In general, the term particulate represents to all atmospheric substances that are not gases.* They can be suspended droplets or solid particles or the mixture of the two. The terms *aerosol* is used during the time it is *suspended* in the air. Thus, a particulate matter is an air pollutant only when it is an aerosol. A particulate matter is nuisance both as an aerosol causing visibility reduction and as settled or deposited matter resulting in soiling of surfaces, corrosions etc. Particulates can be composed of inert or extremely reactive materials ranging in size from 0.01 μ to 100 μ. The various types of aerosols are (*i*) dust, (*ii*) smoke, (*iii*) mists, (*iv*) fog, (*v*) fumes.

(*i*) ***Dust***: Dust is a commonly used term applied to solid particles larger than colloidal particles and capable of temporary suspension in air or other gases. These are formed by natural disintegration of rock and soil or by the mechanical process grinding and spraying, with their size varying from 1 to 200 μ. Dusts do not tend to flocculate except under electrostatic forces. They also do not diffuse but settle under the influence of gravity. Fine dust particles act as centres of catalysts for many of the chemical reactions taking place in the atmosphere.

(*ii*) ***Smoke*** : Smoke consists of finely divided particles resulting from incomplete combustion or other chemical processes. They consist predominantly of carbon particles and other combusible materials, with their size less than 1 μ.

(*iii*) ***Mists*** : Mist is made up of liquid droplets which are formed by condensation in the atmosphere or are released from industrial operations. Natural mist particles formed from water vapour in the atmosphere are rather large, ranging from, 40–500 μ in size. Mist particles may coalesce.

(*iv*) ***Fog*** : Fog is a loose term applied to liquid dispersed aerosols in air by condensation. Fog refers to visible aerosols in which dispersed phase size of particles ranges from 1.0 to 40 μ.

(*v*) ***Fumes***: Fumes are solid particles generated by condensation from the gaseous state, generally after volatilisation from melted substances and often accompanied by a chemical reaction such as oxidation. These are released from chemical or metallurgical processes. The particle size varies from 0.1 to 1 μ. Fumes flocculate and sometimes coalesce.

Other contaminants : Theses include gases, vapours and compounds, which may be derived from solids or liquids such as gasoline. Table 23.3 gives various groups of air contaminants.

TABLE 23.3. GROUPS OF AIR CONTAMINANTS

	Group	*Examples*
1.	Solids	Carbon flyash (Soot), zinc oxide, lead chloride
2.	Sulphur compounds	SO_2, SO_3, H_2S, mercaptans
3.	Nitrogen compounds	NO, NO_2, NH_3
4.	Oxygen compounds	O_3, CO, CO_2
5.	Halogen compounds	HF, HCl
6.	Organic compounds	Aldehydes, hydrocarbons
7.	Radioactive compounds	Radioactive gases.

Following are the important gases in air pollutants.

1. ***Sulphur dioxide*** : The most important oxide emitted by pollution sources is sulphur dioxide (SO_2) ; it is nations major air pollution source. Although hydrogen sulphide oxidation is a major source of sulphur dioxide, the man made sources (*i.e.* combustion of sulphur-containing fossil fuels, specially coal) can lead to undesirable local concentrations which may exceed 1 ppm. The sulphur content in coal may vary from 1 to 4% while crude petroleum products contain less than 1% sulhur. Another common sources of sulphur dioxide in the atmosphere are the metallurgical operations, sulphuric acid plants and paper manufacturing plants. The open burning of refuse as well as municipal incinerators also contribute some amount of SO_2 to the atmosphere. However thermal power houses/plants are the main contributors to SO_2 emissions and accounts for 40% of the total emission. SO_2 is a colorless gas with a characteristic, sharp, pungent odour. It is moderalety soluble in water forming weakly acidic sulphurous acid (H_2SO_3). It is oxidized slowly in clean air to sulphur trioxide (SO_3). Sulphur oxides are transformed to sulphate aerosols by more than one mechanism. They can be removed via rainfall and coagulation. In a polluted atmosphere, SO_2 reacts photo-chemically or catalytically with other pollutants or normal atmospheric constituents to form sulphur trioxide, sulphuric acid and salts of sulphuric acid.

2. *Hydrogen sulphide* : It is a foul smelling gas which is principally produced from the refining of petroleum, coking of coal, refining of natural gas, manufacturing of viscose rayon and evaporation of black liquor in craft pulping process. The sources of its natural emission are anaerobic biological decay processes on land, in marshes and in the ocean. Volcanoes and natural water springs emit hydrogen sulphide to some extent. Other sulphur compounds of interest in air pollution are methyl mercaptan (CH_3SH) dimethyl sulphide (CH_3SCH_3), dimethyl disulphide (CH_3SSCH_3) and their higher molecular homologs. Atmsospheric concentration of hydrogen sulphide and other reduced forms of sulphur can create objectionable odour and cause direct effects on materials, plants, animals and humans. The repeated exposures to even low concentrations has irritating effects on the mucous membrane, eyes and respiratory tract.

3. *Hydrogen fluoride and other fluorides* : Fluorides may be found in the air as gases or as particulate matter and very often industrial emissions will contain fluoride compounds in both states. The industries which emit fluoride compounds include the mining and processing of phosphate rock and other fluoride-containing minerals, the fertilizer-manufacturing industry, the aluminum and steel industry, the glass making industry and the chemical industry. Various fluoride compounds include hydrogen fluoride (HF), silicon tetrafluoride, hydrofluosilicic acid, sodium fluoride, aluminum fluorite etc. Out of these, hydrogen fluoride is an important air contaminant even in extremely low concentration of 0.001–0.10 ppm by volume. It is extremely irritant and corrosive. It is more important is terms of injury to vegetation and animals than in terms of injury to humans.

4. *Chlorine and hydrogen chloride* : Chlorine is found in polluted atmosphere in various forms such as (*i*) elemental chlorine, (*ii*) hydrogen chloride and (*iii*) other organic as well as inorganic chlorides. Various sources of chlorine and its compounds are the manufacturing processes in which chlorine is used to produce other chemicals, water purification plants and wastewater treatment plants. Elemental chlorine produce respiratory irritation and damage to vegetation while hydrogen chloride cause corrosion as well as damage to vegetation.

5. *Oxides of Nitrogen* : Oxides of nitrogen are the second most abundant atmospheric contaminants in many cities ranking next to sulphur dioxide. From the air pollution standpoint, the most important nitrogen oxides in the atmosphere are nitric oxide (NO) and the nitrogen dioxide (NO_2), and these two oxides are often referred to jointly in the air pollution literature as NO_x. At high concentrations, these oxides are very toxic and many persons have been killed by accidental emissions. They are also important because of the role they play in the chemistry of *photochemical smog*. Other oxides of nitrogen are : N_2O, N_2O_3, N_2O_4 and N_2O_5. The main source of man-made NO_x are coal, oil, natural gas and motor-vehicle fuel combustion. However, the highest concentration of nitrogen oxides in gaseous emissions occurs in effluents from industries where nitric acid is produced or used in chemical reactions. Also, effluent from large power plants and exhaust from low heat burners and furnace contribute the oxides of nitrogen to the atmosphere. Recent studies have shown that NO_x, emitted by jet engine exhaust gases, would be expected to decrease the concentration of stratospheric ozone. Since ozone serves to prevent much of the ultraviolet radiation, which is harmful to mankind, from reaching the earths surface, even a small decrease in stratospheric ozone is potentially harmful.

Carbon monoxide (CO) : Carbon monoxide is an odourless and colorless gas which is highly poisonous and is generally classified as asphyxiant. The major sources of carbon monoxide in the environment are the man made sources arising from incomplete combustion of carbonaceous fuels and the *natural sources* such as atmospheric oxidation of methane and formaldehyde derived from biological activity and decay of chlorophyll in plants and algae and other biological sources in oceans, and the photochemical oxidation of terpenes. It is estimated that about 420 million tonnes of carbon monoxide is discharged into the atmosphere per year through the human activities, out of which 99.5% is contributed by transportation alone. Global emissions from natural sources amount to an estimated 3400 million tonnes per year which are approximately eight times more than the emission from all human activities. About 88 percent of this total of 3400 million tonnes comes from atmospheric oxidation of CH_4 and formaldehyde and the remaining from other sources. Carbon monoxide, when inhaled, passes through the lungs and diffuses directly into the blood stream where it combines with hemoglobin forming carboxyhemoglobin, COHb. The affinity of CO for hemoglobin is 210 times greater than that of oxygen and as a result the amount of hemoglobin available for carrying oxygen for body tissue is considerably reduced. This may result in death by asphyxiation (lack of oxygen).

6. Hydrocarbons and Photochemical oxidants : Hydrocarbon are compounds containing only hydrogen and carbon. They are generally divided into two categories (*i*) *aliphatic group* of Hydrocarbons which include alkanes (methanes) alkenes (olefins) and alkynes and (*ii*) *aromatic group* of hydrocarbons such as benzene etc. These are chiefly released into atmosphere by automobile exhausts. Hydrocarbons are also released into the atmosphere by smokes of incinerators, through fumes of oil refineries and also by evaporation of gasoline at service stations. However, at the concentration usually found in urban air, the hydrocarbons cause no adverse effect on human health. Aromatic hydrocarbons are more reactive than aliphatic ones and cause irritation of mucous membranes. The major oxidant produced in photochemical smog is ozone which is poisonous and smelly. It exists in great abundance under natural conditions in the upper atmosphere. It is likely that combustion and sunlight are involved in the production of ozone. Many other oxidants produced in photochemical smog, particularly the peroxyacyl nitrates, cause eye irritation. Oxidants such as peroxyacetyl nitrate and peroxybenzonyl nitrate irritate the nose and throat, and cause chest constriction.

Aldehydes : Although not present in as high a concentration as hydrocarbons, aldehydes are important constituents of smog because they combribute to eye irritation and possibly to other physiological effects and participate in the chemical reactions that produce unpleasant products. At higher concentrations than what usually occurs in the atmosphere, the mucous membranes of the nose and throat are also irritated. Aldehydes are produced by the combustion of gasoline, diesel oil, fuel oil, and natural gas. Automobiles is by far the major source of aldehydes in urban pollution. Not only does it emit aldehydes directly, but it also emits hydrocarbons which are converted to aldehydes in smog by photochemical reactions. However, essentially all burning of organic fuel (such as in municipal incinerations, home heating, power plants etc.) also introduces aldehydes into the atmosphere. Aldehydes are also introduced in the atmosphere by nature in the form of forest and brush fire initiated by lighting, and in the form of essential oils emitted by plants.

Carcinogens : These are cancer producing agents. Increase in lung cancers in cities have been reported due to aromatic hydrocarbons emitted by automobile emissions. Common examples of such *organic carcinogens* are Benzon(a)pyrene, Benzo(b)Phenanthrene and Benzo(e) pyrene of the polynuclear aromatic hydrocarbons group. Another class of organic carcinogens leading to bladder cancers are *aromatic amines.* Possible *inorganic carcinogens* include arsenic, beryllium, cadmium, chromium, cobalt, lead, mercury nickel, selemium, silicon, silver, zinc and asbestos. Exposure to these can occur in industrial situations, in water, in insecticide residues on food, as airborne gases or dusts.

Redioactive gases : The radioactive materias from explosions of nuclear devices are the main causes of radioactivity in the environment. Another major source of radioactive gases and particulates is the nuclear power reactor and related fuel handling facillities and fuel reprocessing plants. To date, the most serious effects of gaseous releases have been in the underground mining of uranium ore. The major generation of radioactive gaseous wastes occurs in the nuclear reactor, though these gases are not necessarily released directly to the environment.

Smog : Smog is a synchronym of two words – smoke and fog. Smog can be of two types : photochemical and coal induced. The conditions for the formation of photochemical smog are (*i*) air stagnation, (*ii*) abundant sunlight and (*iii*) high concentrations of hydrocarbon and nitrogen oxides in atmosphere. It is restricted to highly motorised areas in metropolitan cities. Photochemical smog was first observed in Los Angeles (U.S.A.) in mid-1940 and since then the phenomenon has been detected in most major metropolitan cities of the World. Smog is caused by the interaction of some hydrocarbons and oxidants under the influence of sunlight giving rise to dangerous peroxy-acetyl-nitrate (PAN). Its main constituents are nitrogen oxide, PAN, hydrocarbons, carbon monoxide and ozone. *Photochemical smog* has following illeffects ; (*i*) It causes coughing and chest soarness (*ii*) It is irritant to eyes (*iii*) It kills leaf tissue (*iv*) It degrades the rubber and cellulose and (*v*) It reduces visibility. The fog from burning coal covers urban areas at night or on cold days when the temperature is below 10° C and when calm meteorological conditions prevail. Such a fog consists of smoke, sulphur compounds and fly ash. Prolonged exposure of such smog may result in high mortality rate especially among the elderly and those who have histories of chronic bronchities, asthma broncho-pheumonia and other lung or heart deseases.

A typical London's smog incidence took place between 5th and 8th December 1952 in which 4000 people died due to suffocation. It is understood that the incident occured due to a combination of fog and pollution, mainly by sulphur dioxide released when sulphur containing coal and oil were burnt domestically or in fossil fuel power plants. At that time, an anticyclone reached London from North west and became stationary, with the result that the whole London was under the pool of cold stagnant air with a very effective lid overhead.

23.7. EFFECTS OF AIR POLLUTION

In general the effects of the polluted atmosphere can be classified under the following four heads : (*i*) Effect on certain materials (*ii*) Effect on plants (*iii*) Effect on animals (*iv*) Effect on human health (*v*) Effect on physical features of the atmosphere.

1. Effect of air pollutants on certain materials. Air pollutants affect certain materials by the following five ways : (*i*) abrasion, (*ii*) deposition and removal, (*iii*) direct chemical attack, (*iv*) indirect chemical attack and (*v*) corrosion. Table 23.4 shows the effect of air pollution on various materials.

2. Effect of air pollutants on plants : Air pollution has long been known to have adverse effect on plants. Air pollutants affecting plants are (*i*) sulphur dioxide (*ii*) hydrogen fluoride (*iii*) hydrogen chloride (*iv*) chlorine (*v*) ozone (*vi*) oxides of nitrogen (*vii*) ammonia (*viii*) mercury (*ix*) ethylene (*x*) hydrogen sulphide (*xi*) hydrogen cynide, (*xii*) PAN, (*xiii*) herbicides and (*xiv*) smog. The most obvious damage caused by air pollutants to plants and vegetation occurs is the leaf structure. The stomata of leaf gets clogged thereby reducing intake of CO_2 and thus affecting photosynthesis. The advice effects range from reduction in growth rate to death of the plant.

TABLE 23.4 EFFECT OF AIR POLLUTION ON CERTAIN MATERIALS

	Materials	*Air pollutants*	*Other factors*	*Effects*
1.	Building materials	SO_2, acid, gases and adhering substances	Moisture	Discolouration
2.	Metals	SO_2, acid, gases	Moisture, temperature	Tarnishing of surfaces; loss of metal
3.	Textiles	SO_2, acid, gases	Moisture, Sunlight	Reduction in tensile strength
4.	Rubber	Oxidants	Sunlight	Cracking
5.	Paints	SO_2, H_2S and particulates	Moisture, fungus	Discolouration
6.	Paper	SO_2, acid, gases	Sunlight	Embrittlement.

3. Effect of air pollutants on animals. The effect of pollutants on farm animals takes place in two steps : (*i*) accumulation of air pollutants in the vegetation, plants and forage, and (*ii*) subsequent poisoning of the animals when they eat the contaminated vegetation. Important contaminant that affect the live stock are (*a*) fluorine, (*b*) arsenic and (*c*) lead. These pollutants originate either from the industries situated nearby, or from dusting and spraying. Out of the these contaminants, fluorine contamination is most prominent since cattle and sheep are found to be more susceptible to it. Symptoms of advanced fluorosis include lack of appetite, general ill health due to malnutrition, lowered fertility, reduced milk production and growth retardation. Arsenic in dusts or sprays on plants can cause to poisoning of cattle leading to salivation, thirst, vomiting, uneasiness, feeble and irregular pulse and respiration. Lead contamination of atmosphere takes place on account of various industries such as smelters, coke ovens, and other coal based industries. Prostration, staggering and inability to rise are the prominent symptoms of lead poisoning. There is complete loss of appetite, paralysis of digestive tract and diarrhoea.

4. Effect of air pollutants on human health. The inhalation of undesirable gases from the atmosphere has marked adverse effects on human health . There adverse effect can be divided into two classes: acute effects and chronic effects. Acute effects manifest themselves *immediately* upon short term exposure to air pollutants of high concentrations while chronic effects become evident only after continuous exposure to low levels of air pollution. Following is a list of health effects of air pollutants : (*i*) Ear, nose and throat irritation. (*ii*) Irritation of respiratory tract. (*iii*) Odour nuisance due it hydrogen sulphide, ammonia merceptans, even at low concentrations. (*iv*) Chronic pulmonary deseases (such as bronchitis, asthma) etc. are aggravated by high concentrations of SO_2, NO_2, particulate matter and photochemical smog.(*v*) Pollens etc. initiate asthamatic attacks. (*vi*) Carcinogenic agents cause cancer (*vii*) Respiratory disease is caused by dust particles. *Silicosis* is caused by silica dust of cement factories and *asbestosis* is caused by asbestos plants. (*viii*) Lead poisoning is caused due to entry of lead throu^n the lungs.(*ix*) Bone fluorosis and mottling of teeth is caused by hydrogen fluoride. (*x*) Carbon monoxide may cause death by asphyxiation. It also increases stress on persons suffering from cardiovascular and pulmonary diseases. (*x*) Air pollution in general cause increase in mortality rate and morbidity rate. (*xii*) Radio-active fallout may cause (*a*) cancer, (*b*) shortening of life span and (*c*) genetic effects.

Table 23.5 gives a list of some major air pollution disasters.

TABLE 23.5 MAJOR AIR POLLUTION DISASTERS

S.N.	*Period and Location*	*Conditions and causes*	*Symtoms and Effects*
1.	Dec. 1930. Meuse valley, Belgium	Inversion, smoke, SO_2 and H_2SO_4 mist	63 deaths ; thousands ill ; Irritation of eyes and nose; cough
2.	Oct. 1948. Donora, (Pennsylvania) U.S.A.	Inversion and Fog ; SO_2 ; smoke ; zinc particles ; H_2SO_4 mist	20 deaths ; 45% of population ill ; Irritation of eyes, nose and respiratory track; breathlessness, nausea.
3.	Nov. 1950. Poza Rica, Maxico	Shallow inversion, fog and calm conditions ; H_2SO_4 released due to burner failure	22 deaths ; 320 hospitalised ; irritation of respiratory tract.
4.	December 1952 London, England	Low temperature inversion, thick fog, stagnant air, smoke and SO_2 accumulation	4000 deaths ; thousands hospitalised for respiratory and heart diseases ; chronic bronchitis, broncopneumonia
5.	Jan. 1956. London, England	Extended fog conditions similar to 1952 episode; particulates and SO_2	1000 deaths
6.	Dec. 1962. London, England	Shallow inversion, thick fog, smoke, SO_2	700 deaths ; increased illness.
7.	Nov. 1966. New york, U.S.A.	SO_2 ; particulates	168 deaths.
8.	Dec. 1984. Bhopal, India.	Release of 30 tonnes of deadly methyl isocynate gas due to failure of system	> 2500 deaths ; 100,000 people severaly affected ; vomiting ; violent coughing ; chemical conjunctivitis; suffocation ; cardiac failure.

5. Effect of air pollutants on physical features of the atmosphere

The physical effects of pollutants on the atmosphere can be classified under three heads:

(*i*) Effects on visibility: (*ii*) Effects on urban atmosphere and weather conditions, and (*iii*) Effects on atmospheric constituents.

1. Effects on visibility: The visibility is reduced due to the concentration and physical properties of particulate pollutants present in the atmosphere. The measurement of prevailing visibility is a standard meteorological practice. The stormy wind raises dust particles resulting in decrease in the visibility. In unsaturated humidity conditions, the hygroscopic particles pick up moisture and as they increase in size, the visibility is affected. Due to the angle of the sun, visibility observation in polluted areas show strong directional variations. Other meteorological factors such as inversion, height and wind speed, presence of hygroscopic particles and relative humidity also affect the visibility. Fog and photochemical smog reduce the visibility considerably.

2. Effects of urban atmosphere and weather conditions : Urban, air pollution is mainly caused due to smoke, dust, fog and other aerosols, and all of these affect the weather conditions. Polluted area becomes more cloudy, more foggy, resulting in reduction of solar radiation to an extent of about 30% . The area may have 5 to 10% more precipitation, since air pollutants can add to the condensation of nuclei of the cloud system.

3. Effects on atmospheric constituents: Due to air pollution, the balance between various constituents of air (Table 23.1) is disturbed. Atmospheric carbon dioxide is the main source of organic carbon in the biosphere. It has been noted that there is steady increase in the percentage of CO_2 in the atmosphere due to combustion and other factors causing air pollution. CO_2 is interpreted as a factor responsible for rise in ambient temperature. Due to continued air pollution, the lead aerosol concentration is now 30 times more than that in pre-industrial days. Freezing nuclei are formed in large numbers when automobile exhaust gases are exposed to minute quantities of iodine vapours. These nuclei are the main basis of weather modification cloud seeding operations.

23.8. METEOROLOGICAL ASPECTS OF AIR POLLUTION DISPERSION

Air pollutants discharged from anthropogenic sources gets dispersed into the atmosphere into various directions depending upon prevailing winds (direction as well as speed), temperature and pressure conditions. It is essential that these air pollutants emitted from the source must first be transported and diluted in the atmosphere before these undergo various physical and photochemical transformations and ultimately reach their receptors. Otherwise, the pollutant concentrations will reach dangerous levels near the source of emission. Following are the meteorological parameters that influence air pollution dispersion:

(a) Primary Parameters

1. Wind direction and speed.
2. Temperature.
3. Atmospheric stability
4. Mixing height.

(b) ***Secondary Parameters*** :

1. Precipitation
2. Humidity
3. Solar radiation
4. Visibility.

1. Wind direction and speed : Morning air is known as *wind*. The movement of air is caused by the unequal distributions of atmospheric temperature and pressure over earth's surface, and it is largely influenced by the rotation of the earth . The temperature and pressure gradients are caused due to differential solar heating of the Earth's surface. The motion of air near the surface of the Earth is retarded by friction, which varies with surface roughness. The planetary boundary layer, in which friction is significant, extends to about 1 km above the Earth's surface. The wind speed is usually measured by an *anemometer*, at a test altitude (Z_0). Knowing the wind speed u_0 at this height, the wind velocity u at any other altitude Z can be found from the following simple empirical *power law*

$$\frac{u}{u_0} = \left[\frac{Z}{Z_0}\right]^{\alpha} \quad ...(23.1)$$

where α is the exponent which varies between 0.14 to 0.40 depending upon the roughness of the ground as well as *temperature stability* of the atmosphere. If the environmental lapse rate is dry adiabatic and the terrain is fairly smooth, $\alpha = 1/7$. The value of α increases with increasing stability or with increasing surface roughness. Eq. 23.1 can also be expressed in the following alternative form given by Sutton :

$$u = u_0\left[\frac{Z}{Z_0}\right]^{n/(2-n)} \quad ...(23.1b)$$

where n the turbulence parameter $0 < n < 1$. The approximate value of n for a large lapse rate is 0.20 and for marked inversion is 0.5.

The direction and speed of the surface winds govern the drift and diffusion of air pollutants discharged near the ground level. The higher the wind speed at or near the point of emission, the more rapidly the pollutants would we carried away from the source. The concentration of the pollutants, so dispersed, will go on decreasing with the distance. However, if the wind speed's are low, the dispersion will not be effective, and the pollutants will tend to concentrate near the area of their emission. *Gustiness*, which is directly proportional to wind speed, determines the extent to which the pollutants are mixed and diluted with the surrounding area. Other factors which influence the mixing and dilution are atmospheric turbulence and stability, topographical features such as hills or valleys and the presence of moisture in the environment. Surface topography can modify the general pattern of wind speed and direction. Hills may deflect the air flow either horizontally, vertically or both while in valleys, the winds carrying a pollutant tend to flow either up or down the valley, following its meanderings.

2. Atmospheric stability and temperature inversions. The effective dispersion of pollutants in the atmosphere depends primarily on the degree of stability of the atmosphere, which in turn depends upon the rate of change of ambient temperature with altitude. In well mixed air which is dry, for every 1000 m increase in altitude, the temperature decreases by about 65°C. This *vertical temperature gradient* is known as *lapse rate* expressed by the equation.

$$\frac{dT}{dZ} = -\frac{n-1}{n} \cdot \frac{g}{R} \quad ...(23.2a)$$

where R is the gas constant for air.

Based on the meterological data in the troposphere up to about 10 km, the *environmental lapse rate* (or normal lapse rate) is found to be 6.5°C/1000 m. Hence for polytropic model, we have

$$\left(\frac{dT}{dZ}\right)_{env} = -\frac{6.5°C}{1000 \text{ m}} = -\frac{n-1}{n}\frac{g}{R} \quad ...(23.2)$$

From the above equation, n comes out to be 1.23, and this value is applicable only for the *troposphere*. Above troposphere, we has *stratosphere* the lower region (upto 25 km) of which, has a constant temperature (*i.e* $n = 1$) while in the upper region the temperature increases with the altitude as a result of ozone formation. Fig 23.2 shows the temperature-altitude profile.

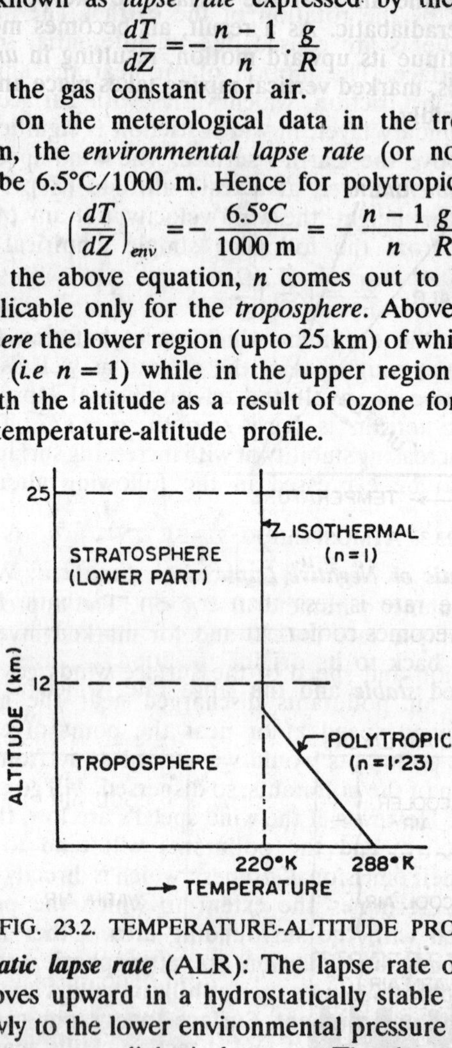

FIG. 23.2. TEMPERATURE-ALTITUDE PROFILE.

Adiabatic lapse rate (ALR): The lapse rate of a parcel of dry air, as it moves upward in a hydrostatically stable environment and expands slowly to the lower environmental pressure without exchange of heat, is known as *adiabatic lapse rate*.The *dry adiabatic lapse rate* (DALR) is given by

$$\left(\frac{dT}{dZ}\right)_{ad} = -\frac{9.86°C}{1000\ m} \qquad ...(23.3)$$

Super-adiabatic lapse rate (SALR) : When the *prevailing lapse rate* or *ambient lapse rate* or *environmental lapse rate* (ELR) is greater than the dry adiabatic lapse rate, the atmosphere is said to be *super-adiabatic*. For such a case, the rising parcel of air, cooling at the adiabatic rate, will be warmer and less dense than the surrounding environment. On a clear summer day, rapid heating of the earth by the sun warms the air near the surface, to the point where the lapse rate is superadiabatic. As a result, air becomes more buoyant and tends to continue its upward motion, resulting in *unstable equilibrium*; due to this, marked vertical mixing takes place and pollutants are dispersed rapidly.

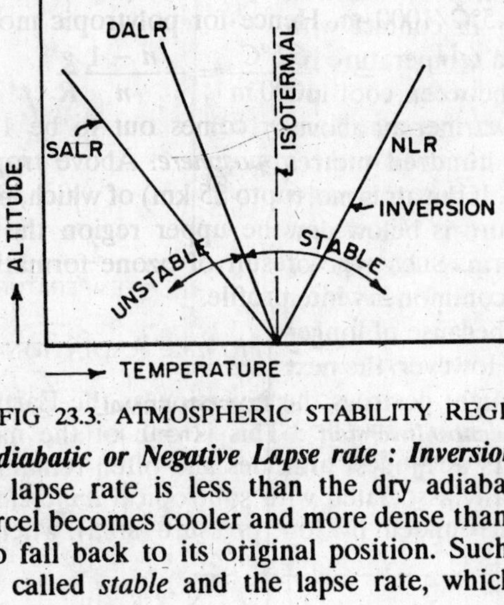

FIG 23.3. ATMOSPHERIC STABILITY REGIONS

Sub-adiabatic or Negative Lapse rate : Inversion: When the environmental lapse rate is less than the dry adiabatic lapse rate, a rising air parcel becomes cooler and more dense than its surroundings and tends to fall back to its original position. Such an atmospheric condition is called *stable* and the lapse rate, which is subadiabatic

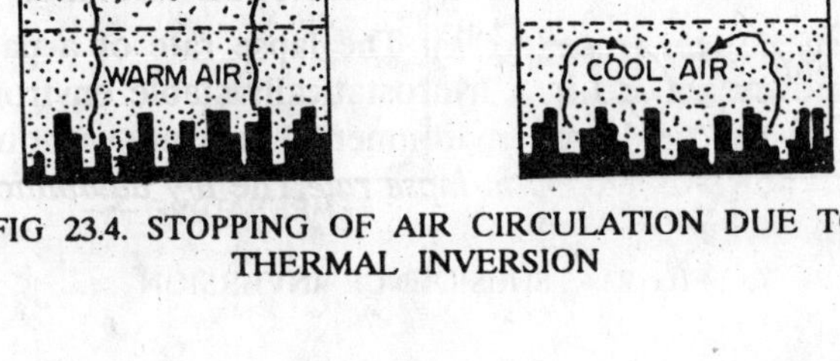

FIG 23.4. STOPPING OF AIR CIRCULATION DUE TO THERMAL INVERSION

is called *negative lapse rate.* In such a condition, a dense cold stratum of air at ground level gets covered by lighter warmer air at higher lever. *This phenomenon is called inversion.*

During *inversion* vertical air movement is stopped and the pollution is concentrated beneath the inversion layer, *i.e.* the denser air at ground level. Thus, atmospheric inversion influence the dispersion of pollutants by restricting vertical mixing ; hence pollutants in the air do not disperse. Inversion is a frequent occurrence in the autumn and winter months. Inversions may be of the following types: *(i)* Radiation inversion *(ii)* Subsidence inversion *(iii)* Double inversion.

(i) ***Radiation Inversion*** : Such type of inversion usually occurs at night, resulting from the normal diurnal cooling cycle. After sunset, the ground looses heat by radiation and cools the air in contact with it. Consequently a temperature inversion is setup between cool low level air and the warmer air above, in the first few hundred metres above the surface. If the air is moist and its temperature is below dew point, *fog* will form. Such type of inversion is more common in winter than in summer, because of longer nights in winter. However, the next morning, the sunlight destroys the inversion as the Earth is warmed.

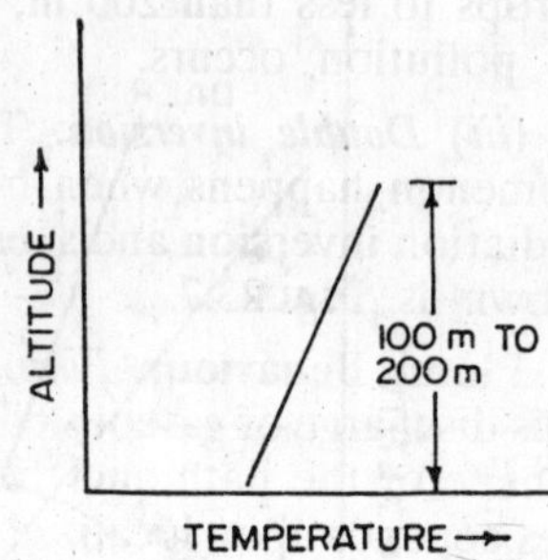

FIG 23.5. RADIATION INVERSION

(ii) ***Subsidence inversion*** : This is one of the most common types which occurs at modest altitudes and often remains for several days. This is usually associated with subtropical *anticyclone* (i.e. high pressure area surrounded by low pressure area) where the air is

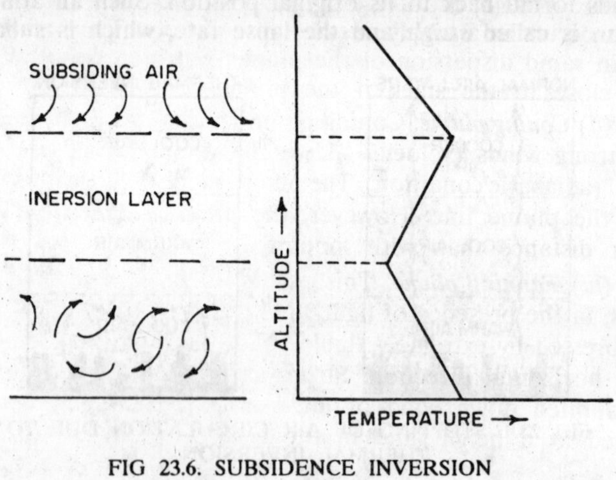

FIG 23.6. SUBSIDENCE INVERSION

warmed by compression as it descends in a high pressure system and achieves temperature higher than that of the air underneath. An inversion will result if the temperature increase is sufficient. The air circulating around the area descends slowly at a rate of about 1000 m per day. This acts as a lid to prevent the upward movement of the contaminants. The inversion height may vary from the ground surface to 1600 m, When this drops to less than 200 m, extreme pollution occurs.

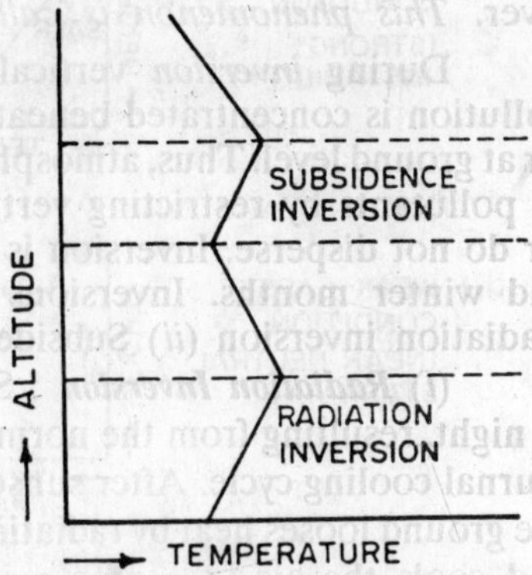

FIG 23.7. DOUBLE INVERSION.

(*iii*) ***Double inversion***: This phenomenon happens when both the radiation inversion and subsidence inversion occur simultaneously, as shown is Fig 23.7.

Plume Behaviour. *Plume* is defined by the path taken by continuous discharge of gaseous effluents emitted from a stack or chimney. The shape of the path and the concentration distribution of gaseous plumes depends upon localized air stability. Typical situations generally encountered in the lower atmosphere (< 300 m above ground), illustrated in Fig 23.8 are (*i*) looping plume (*ii*) coning plume (*iii*) fanning plume (*iv*) lofting plume (*v*) fumigating plume and (*vi*) trapping plume.

(*i*) ***Looping plume***: This is common type of plume behavior which occurs under superadiabatic lapse rate (SALR) conditions with light to moderate wind speeds on a hot summer afternoon when large scale thermal eddies are present. The plume has wavy behaviour since it occurs in a highly *unstable* atmosphere. The high turbulence helps in rapid dispersion of the plume, but high concentration may occur close to the stack if the plume touches the ground.

(*ii*) ***Coning plume***: Coning plume occurs on cloudy day or nights with strong winds (velocity 32 km/h) when the lapse rate is near neutral (adiabatic condition). The plume shape is vertically symmetrical about the plume line. However, the plume reaches the ground at greater distance than with looping.

(*iii*) ***Fanning plume***: This occurs under extreme inversion conditions, in the presence of light wind. Most of the vertical dispersion is suppressed by extremely stable condition, and the plume fans out in the horizontal direction. Strong concentrations at plume height are exhibited down wind of the stack. A fanning plume is often observed at a height and in the early morning in all seasons.

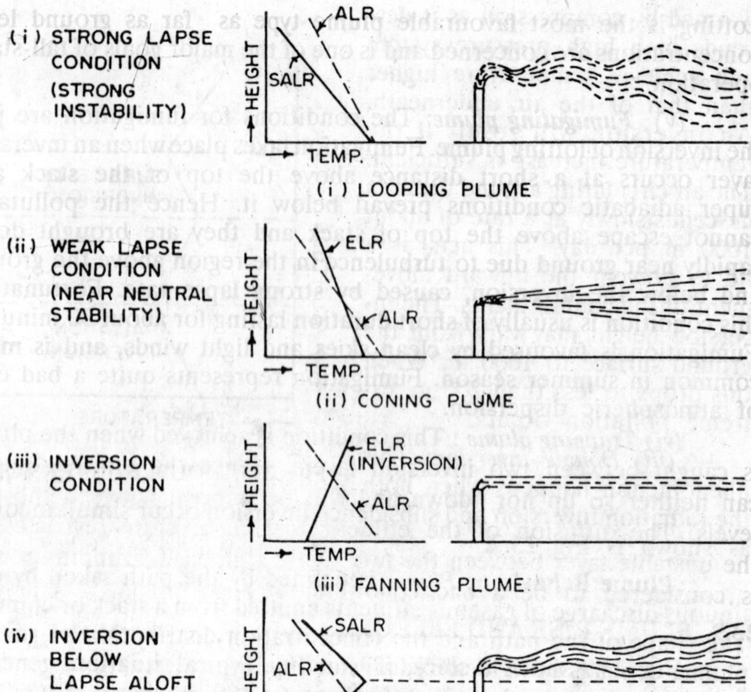

FIG. 23.8. TYPES OF PLUME BEHAVIOUR

(*iv*) ***Lofting plume***: plume is said to be lofting when there exists a strong super adiabatic lapse rate (SALR) above a surface inversion. In such a condition, downward motion and mixing is prevented by surface inversion but the upward mixing will be quite turbulent and rapid. The emission will therefore not reach the ground surface.

Lofting is the most favourable plume type as far as ground level concentrations are concerned and is one of the major goals of tall-stack operation.

(*v*) ***Fumigating plume***: The conditions for fumigation are just the inversion of lofting plume. Fumigation takes place when an inversion layer occurs at a short distance above the top of the stack and super adiabatic conditions prevail below it. Hence the pollutants cannot escape above the top of stack and they are brought down rapidly near ground due to turbulence in the region above the ground and below the inversion, caused by strong lapse rate. Fortunately, this condition is usually of short duration lasting for about 30 minutes. Fumigation is favoured by clean skies and light winds, and is more common in summer season. Fumigation represents quite a bad case of atmospheric dispersion.

(vi) Trapping plume : This condition is achieved when the plume is caught between two inversion layers. Hence the emitted plume can neither go up nor down and will be trapped between the two levels. The diffusion of the effluent will be severely restricted to the unstable layer between the two stable regions. A trapping plume is considered to be a bad condition for dispersion.

23.9. PLUME RISE AND EFFECTIVE HEIGHT OF STACK

Fig. 23.9 shows a stack of *actual height h*. The *effective height H* will be equal to the sum of the actual height of stack (h) and the *rise of plume* (Δh) beyond the stack exit.

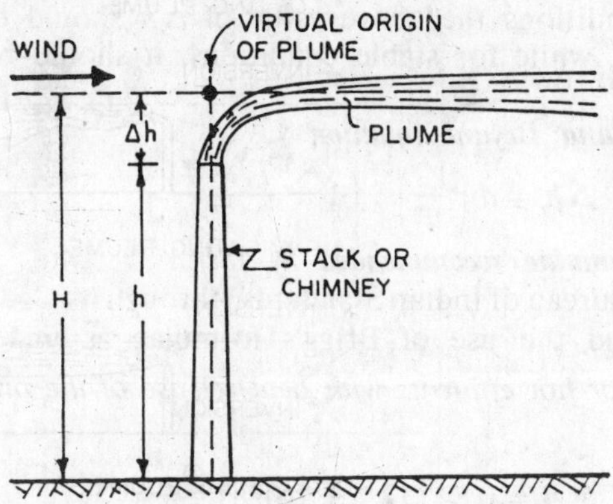

FIG 23.9. PLUME RISE AND EFFECTIVE HEIGHT OF STACK

The plume rise (Δh) depends upon many factors such as exit velocity, wind speed, diameter of stack, temperature of plume, lapse rate etc, and several empirical formulae are available to predict this.

(i) General equation

In general, plume rise can be expressed by :

$$\Delta h = K\frac{Q_H^{\alpha}}{u^{\beta}} \qquad ...(23.4)$$

where α, β and K are dimensional constants, Q_H is the heat emission rate from the stack, and u is the average wind velocity.

According to Canadian Combustion Research laboratory formula of the above form, $\alpha = 0.25$, $\beta = 1$ and $K = 66.4$, where K is expressed in kcal/s and u in m/s.

2. *Moses and Carson equation.*

$$\Delta h = C_1\frac{V_s . d}{u} + C_2\frac{Q_H^{1/2}}{u} \qquad ...(23.5)$$

Where Δh = plume rise (m)
V_s = stack exit velocity (m/s)
u = wind speed (m/s)
d = diameter of stack at exit (m)
Q_H = heat emission rate(kcal/s)
C_1, C_2 = Plume rise regression coefficients which depend on atmospheric stability.

3. Holland's equation

$$\Delta h = \frac{V_s . d}{u}\left[1.5 + 2.68 \times 10^{-3} pd\left(\frac{T_s - T_a}{T_s}\right)\right] \qquad ...(23.6)$$

where p = atmospheric pressure in millibars
T_s = stack gas temperature (°K)
T_a = air temperature (°K)

The above equation is applicable for neutral conditions. For unstable conditions the above value of Δh should be increased by 10 to 20 %, while for stable conditions, it should be decreased by 20 to 10 %

4. Davidson and Bryant equation

$$\Delta h = d\left(\frac{V_s}{u}\right)^{1.4}\left(1 + \frac{T_s - T_a}{T_s}\right) \qquad ...(23.7)$$

5. Brigg's formulae recommended by ISI

The Bureau of Indian Standards, through its Code IS : 8829-1978 has suggested the use of Brigg's formulae as under :

(*a*) *For hot effluents with heat release of the order of* 10^6 *cal/s or more*

$$\Delta h = 0.84\,(12.4 + 0.09h)\,\frac{Q_H^{1/4}}{u} \qquad ...(23.8)$$

Where Q_H = heat release in calories per second
h = height of stack (m)
u = wind speed (m/s)

(*b*) *For not very hot releases, and which can be counted as momentum sources above*:

$$\Delta h = \frac{3 V_0 d}{u} \qquad ...(23.9)$$

V_0 = afflux velocity (m/s)
u = wind speed (m/s)
d = stack exit diameter.

23.10. DISPERSION EQUATIONS

Pollutants emitted into the atmosphere are mixed thoroughly with the surrounding air and diluted by *atmospheric dispersion*. This dispersion is primarily due to turbulent diffusion and bulk air flow. The *turbulent diffusion models* are based on *Fick's law* of mulecular diffusion . Several equation/ models have been develepod to estimate the concentration of pollutants in the plume at any distance x, y and z, in horizontal down-wind direction, horizontal cross-wind direction and vertical direction respectively. If K_x, K_y and K_z are eddy *diffusion coefficients* in the three directions, C is concentration of pollutant and v_x, v_y and v_z are the respective velocity components, then from *continuity principle*, one obtains the following diffusion equation in terms of time t :

$$\frac{\partial C}{\partial t} + v_x \frac{\partial C}{\partial x} + v_y \frac{\partial C}{\partial y} + v_z \frac{\partial C}{\partial z} = \frac{\partial}{\partial x}\left(K_x \frac{\partial C}{\partial x}\right) + \frac{\partial}{\partial y}\left(K_y \frac{\partial C}{\partial y}\right) + \frac{\partial}{\partial z}\left(K_z \frac{\partial C}{\partial z}\right) \qquad ...(23.10)$$

On the basis of the above Ficks law and based on the knowledge of shape of the concentration distribution for a continuous single emission source, Gaussian developed the following simple statistical equation to compute turbulent transport of gas or aerosols (particulates < 20 μ)

$$C_{x,y,z,H} = \underset{\text{(Centre line)}}{\frac{Q}{2 \pi u \sigma_y \sigma_z}} \underset{\text{(Horizontal)}}{e^{-\frac{1}{2}(y/\sigma_y)^2}} \underset{\text{(Vertical)}}{\left[e^{-\frac{1}{2}(z-H)^2/\sigma_z^2} + e^{\frac{1}{2}(z+H)^2/\sigma_z^2} \right]} \qquad ...(23.11\ a)$$

For concentration calculated at ground level (*i.e.* $z = 0$),

$$C_{x,y,0,H} = \frac{Q}{\pi u \sigma_y \sigma_z} e^{-\frac{1}{2}(y/\sigma_y)^2} . e^{-\frac{1}{2}(H/\sigma_z)^2} \qquad ...(23.11)$$

where
C = Concentration of pollutant (g/m^3)
Q = Pollutant emission rate (g/s)
u = mean wind speed (m/s) at height h_s of the stack
H = effective height of stack (m)
x, y = down-wind and cross-wind horizontal distances (m)
z = level of computation of concentration (m)
σ_y = plume's standard deviation in cross-wind direction (m)
σ_z = plume's standard deviation in vertical direction (m)

Ground level centre line concentration

If the ground level concentration is required only in the down wind horizontal direction (*i.e.* *x*-direction), $y = 0$. Hence Eq. 23.11 reduces to the following from

$$C_{x,0,0,H} = \frac{Q}{\pi\,\sigma_y\,\sigma_z\,.\,u}\,e^{-\frac{1}{2}(H/\sigma_z)^2} \qquad ...(23.12)$$

Again, when the effluent is discharged at ground level , the effective stack height H is zero and Eq. 23.12 is further reduced to the form

$$C_{x,0,0,0} = \frac{Q}{\pi\,\sigma_y\,.\,\sigma_z\,.\,u} \qquad ...(23.13)$$

In the above equations, σ_y and σ_z are functions of down wind distance x as well as of atmospheric stability. Fig 23.10 and 23.11 give the values of σ_y and σ_z respectively, for various types of atmospheric stabilities.

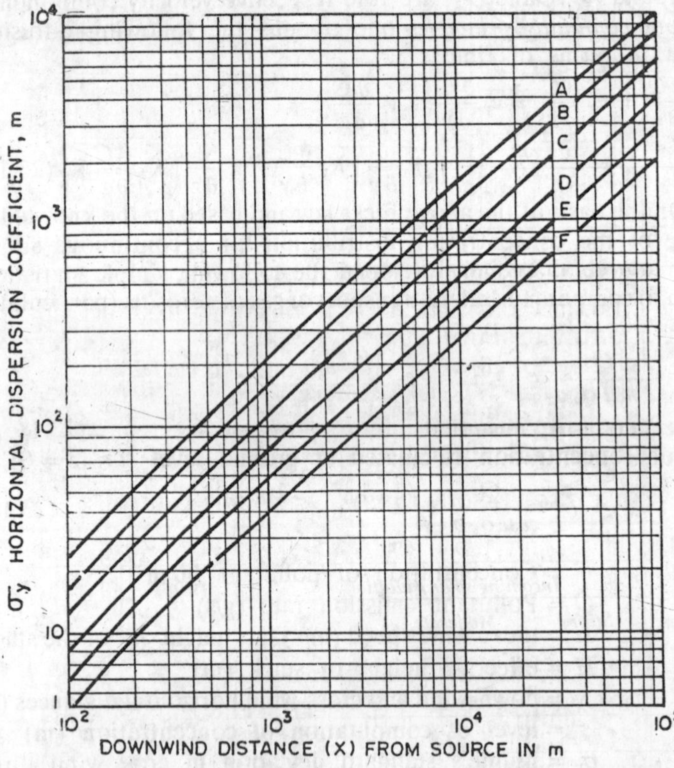

A. EXTREMELY UNSTABLE B. MODERATELY UNSTABLE
C. SLIGHTLY UNSTABLE D. NEUTRAL
E. SLIGHTLY STABLE F. MODERATELY STABLE

FIG. 23.10. VALUES OF σ_y FOR DIFFERENT ATMOSPHERIC STABILITIES.

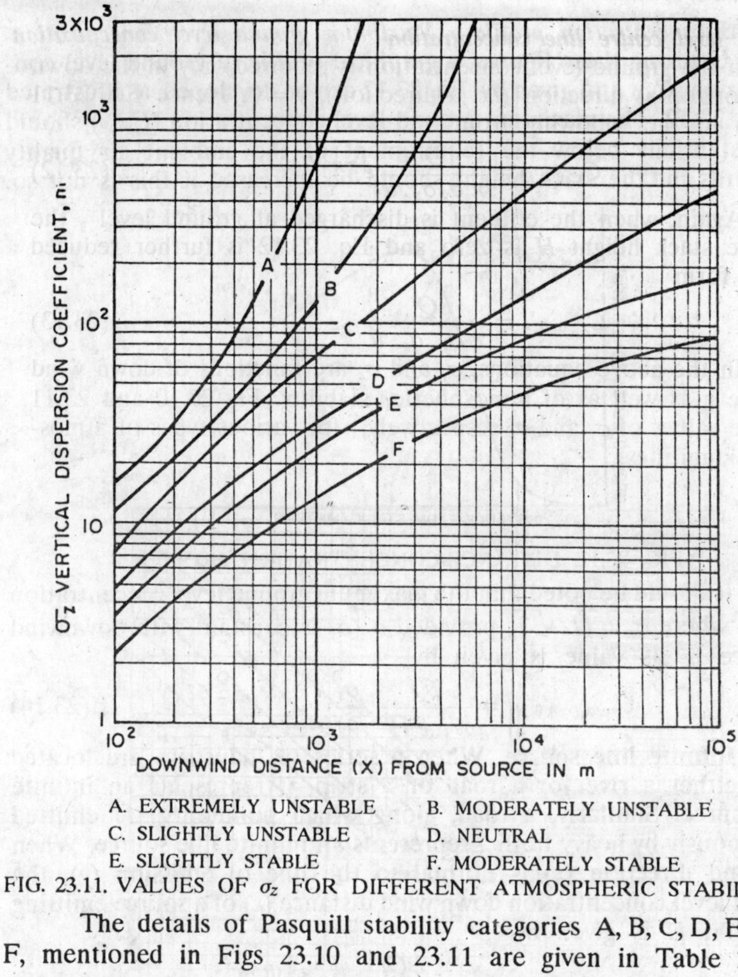

FIG. 23.11. VALUES OF σ_z FOR DIFFERENT ATMOSPHERIC STABILITIES

The details of Pasquill stability categories A, B, C, D, E, and F, mentioned in Figs 23.10 and 23.11 are given in Table 23.6.

TABLE 23.6. PASQUILL'S STABILITY CATEGORIES

Surface wind speed (m/sec) at z = 10 m	*Day*			*Night*	
	Incoming solar radiation			*Thin overcast or 4/8 Low cloud*	$\leq \frac{3}{8}$ *cloud*
	Strong	*Moderate*	*Weak*		
< 2	A	A–B	B	E	F
2–3	A–B	B	C	E	F
4–5	B	B–C	C	D	E
5–6	C	C–D	D	D	D
>6	C	D	D	D	D

Distance-concentration profile : Maximum ground level concentration

Utilizing these figures, a graphical profile of ground level concentration ($C_{x,0,0,H}$) given by Eq. 23.12 can be developed, as illustrated in Fig. 23.12. The maximum ground level concentration (C_{max}) should be sustantially below the requirement of the ambient air quality standards and the stack height should be increased if this is not so.

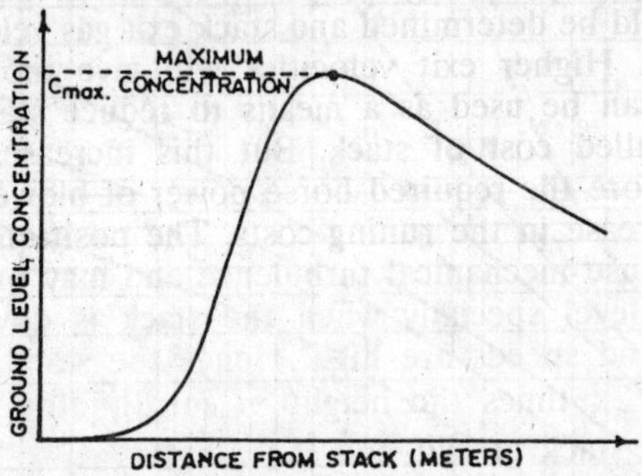

FIG. 23.12. DISTANCE-CONCENTRATION PROFILE

It should be noted that the maximum ground level cencentration occurs where $\sigma_z = H/\sqrt{2}$, provided σ_z/σ_y is constant with downwind distance x. Its value is given by

$$C_{max\,(x,\,0,\,0,\,H)} = \frac{2Q}{\pi\, u\, e\, H^2} \cdot \frac{\sigma_z}{\sigma_y} \qquad \text{...(23.14)}$$

Infinite line source. When a series of industries are located along either a river or a road or a strip, it represents an infinite line source. Similarly, a road, along which pollutants are emitted continuously by heavy traffic, represents an infinite line source. When the wind direction (x) is normal to the line of emission (y), the ground level concentration down wind distance (x) of a source emitting as height H is given by

$$C_{x,\,y,\,0,\,H} = \frac{2\,Q_l}{(2\pi)^{1/2}\,\sigma_z\,.\,u}\, e^{-\frac{1}{2}(H/\sigma_z)^2} \qquad \text{...(23.15)}$$

where, Q_l = Strength of source per unit distance.

If the source is at ground level, (*i.e.* $H = 0$), the above equation reduces to

$$C_{x,\,y,\,0,\,0} = \frac{2\,Q_l}{(2\,\pi)^{1/2}\,\sigma_z\,.\,u} \qquad \text{...(23.15 a)}$$

In both the equations, y does not appear because the concentration is *uniform* along y direction at a given downwind distance x.

23.11. DESIGN OF STACK HEIGHT

From Eq 23.14 it is clear that maximum ground level concentration is inversely proportional to the square of the height of the chimney. This clearly explains why the pollutants are emitted into the atmosphere through tall stacks. Historically, the basic function of a stack was to provide natural draft for combustion. Stacks are designed to fulful two additional requirements (*i*) that the pollutants should be sufficiently dispersed, and (*ii*) that the smoke should not re-enter the building under anticipated wind conditions. The ratio of the stack exit velocity to wind velocity should be greater than 1.5 to allow the effluent to break cleanly from the stack and to eliminate downwash. For design purposes, maximum local wind velocity conditions should be determined and stack exit gas velocities selected proportionately. Higher exit velocities will give more plume rise, and therefore can be used as a means to reduce stack height, and hence the installed cost of stack. But this increases the pressure drop and therefore the required horse power of blowers is increased resulting in increase in the runing costs. The position of the nearby building may cause mechanical turbulence and may bring the plume to the ground level specially when the stack is downwind of the building and wind speeds are high. Hence the stack height should be atleast 2 to $2\frac{1}{2}$ times the height of surrounding buildings. The diameter of the stack at exit end can be determined on the basics of exhaust gas flow rate and on the exit velocity required.

Central Board for prevention and control of water pollution (New Delhi), through its publications : (*i*) Emission Regulation (July 1984) *Part I* and (*ii*) *A method to determine minimum chimney height has* given the following recommendations for the height of chimney :

(a) For Chimney emitting particulate matter :

$$h = 74\,(Q_p)^{0.27} \qquad ...(23.16)$$

where h = height of chimney

Q = particulate matter emission (tonnes/hour)

(b) For chimney emitting SO_2

$$h = 14\,(Q_s)^{1/3}$$

where $Q_s = SO_2$ emission (kg/hour)

(c) Minimum values

The Board has further recommended that the height of the chimney calculated on the basis of particulate matter and SO_2 emissions by the above two equations should be subject to the following minimum values :

(*i*) For chimney adopted for industries in general (except thermal power plants) --- 30 m

(*ii*) For thermal power plants above 200 MW and below 500 MW capacity --- 220 m

(*iii*) For thermal power plants having greater than 500 MW capacity --- 275 m

(d) Relaxation in the minimum values

According to the regulations (*c*) above, the minimum height of the chimney was to be 30 m, but these recommendations were relaxed by the Board later, for the chimneys of boilers and diesel generating sets, as per its latest guide lines contained in their publication *'Emission Regulations part IV.* Following are the revised provisions for minimum stack height.

(*i*) For a boiler, generating steam @ less than 2t/hr. --- 9 m

(*ii*) For a boiler generating steam @ 30 t/hr --- 30 m

(*iii*) For a boiler of intermediate capacity --- 9-30 m

(*iv*) For *diesel generating sets* of different capacities (KVA), the minimum height of stack should be kept only 1.5 to 3.5 m more than the height of the building, and is to be worked out using the following equation :

$$h = h' + 0.2 \sqrt{\text{kVA}} \quad ...(23.17)$$

where h = Stack height

h' = Height of the building.

Example 23.1. *A thermal power plant burns coal at the rate of 8 tonones per hour and discharges the flue gases through a chimney having effective height of 90 m. The coal has a sulphur content of 4.5%. The wind velocity at the top of the stack is 7.5 m/s. The atmospheric conditions are slightly unstable. Determine the maximum ground level concentration of* SO_2 *and the distance from the stack at which this occurs.*

Solution

1. Rate of emission of SO_2

Coal burnt per hour = 8 t = 8000 kg.

∴ Sulphur produced per hour = 8000 × 0.045 = 360 kg.

Now $S + O_2 = SO_2$

Since the molecular mass of both S and O_2 is 32, they combine on a one to one mass basis to produce SO_2.

∴ SO_2 produced per hour = 360 kg S + 360 kg of O_2

= 720 kg/hour.

∴ Rate of emission of SO_2 $= \dfrac{720 \times 1000}{3600} = 200$ g/sec.

2. *Location of point where maximum ground level concentration occurs*

Maximum ground level concentration of pollutant occurs at a point where σ_z is equal to $H/\sqrt{2}$, provided σ_z/σ_y is constant with x.

$$\sigma_z = 90/\sqrt{2} = 63.64 \text{ m}$$

For slightly unstable condition, we have stability class C.

From Fig. 23.11, we observe that the value of $\sigma_z = 63.64$ m is obtained at $x \approx 900$ m for C class stability provided σ_z/σ_y remains constant upto this distance. From Figs. 23.10 and 23.11, we observe that σ_z/σ_y remains contant upto a distance $x = 1000$ m. Hence $x_{max} \approx 900$ m, for which we get $\sigma_y \approx 100$ m for C class stability.

3. *Maximum ground level concentration*

Now from Eq. 23.12,

$$C_{max} = C_{900,0,0,H} = \frac{Q}{\pi \sigma_y . \sigma_z u} e^{-\frac{1}{2}(H/\sigma_z)^2}$$

$$= \frac{200}{\pi (100)(63.64)\, 7.5} (e)^{-\frac{1}{2}(90/63.64)^2} \text{ g/m}^3$$

$$= \mathbf{4.9 \times 10^{-4}\ g/m^3}$$

Alternatively, from Eq. 23.14,

$$C_{max} = \frac{2Q}{\pi u e H^2} \frac{\sigma_z}{\sigma_y}$$

$$= \frac{2 \times 200}{\pi \times 7.5\,(e)\,(90)^2} \times \frac{63.64}{100} \quad \text{(where } e = 2.7183)$$

$$= 4.9 \times 10^{-4} \text{ g/m}^3$$

Example 23.2. *For the data of Example 23.1, determine the ground level concentration at a distance 2000 m downwind at (a) centre line of plume, and (b) at a crosswind distance of 400 m on either side of the centre line.*

Solution

(*a*) For this case, we have $x = 2000$ m and $y = 0$

From Fig. 23.10, we get $\sigma_y \approx 210$ for $x = 2000$ m

From Fig. 23.11, we get $\sigma_z \approx 130$ for $x = 2000$ m.

Both the above values are for C class stability

For this case, Eq. 23.12 is applicable.

$$\therefore \quad C_{x,0,0,H} = \frac{Q}{\pi u \sigma_y \sigma_z} e^{-\frac{1}{2}(H/\sigma_z)^2}$$

$$\therefore \quad C_{2000,0,0,H} = \frac{200}{\pi \times 7.5 \times 130 \times 210} e^{-\frac{1}{2}(90/130)^2}$$

$$= 2.45 \times 10^{-4} \text{ g/m}^3$$

(*b*) For this case, $x = 2000$ m and $y = 400$ m

Hence from Eq. 23.11.

$$C_{x,y\ 0,H} = \frac{Q}{\pi u \sigma_y \sigma_z} e^{-\frac{1}{2}(y/\sigma_y)^2} . e^{-\frac{1}{2}(H/\sigma_z)^2}$$

$$\therefore \quad C_{200,400,0,H} = \frac{200}{\pi \times 7.5 \times 130 \times 210} e^{-\frac{1}{2}(400/210)^2} . e^{-\frac{1}{2}(90/130)^2}$$

$$\approx 0.4 \times 10^{-4} \text{ g/m}^3$$

Example 23.3. *For the data of example 23.1, compute the concentration a $x = 1000$ m, $y = 100$ m and $z = 20$ m.*

Solution. For this case, Eq. 23.11 (*a*) will be applicable.

$$C_{x,y,z,H} = \frac{Q}{2\pi u \sigma_y \sigma_z} e^{-\frac{1}{2}(y/\sigma_y)^2} [e^{-\frac{1}{2}(z-H)^2/\sigma_z^2} + e^{-\frac{1}{2}(size7z + H)^2/\sigma_z^2}]$$

From Fig. 23.10, we get $\sigma_y \approx 115$ m for $x = 1000$ m and C class.

From Fig. 23.11, we get $\sigma_z \approx 68$ m for $x = 1000$ m and C class.

Now $$e^{-\frac{1}{2}(y/\sigma_y)^2} = e^{-\frac{1}{2}(100/115)^2} = 0.6852$$

$$e^{-\frac{1}{2}(z-H)^2/\sigma_z^2} = e^{-\frac{1}{2}(20-90)^2/(68)^2} = 0.5887$$

$$e^{-\frac{1}{2}(z+H)^2/\sigma_z^2} = e^{-\frac{1}{2}(20+90)^2/(68)^2} = 0.2703$$

$$\therefore \quad C_{1000,\,100,\,20,\,H} = \frac{200}{2\pi\,(7.5)115\times 68}\times 0.6852\times 0.5887\times 0.2703$$

$$= \mathbf{0.59\times 10^{-4}\ g/m^3}$$

Example 23.4. *If for the data of Example 23.1, the value of* $C_{2000,y,0,H}$ *is* 1×10^{-4} g/m^3, *what is the value of y ?*

Solution

Dividing Eqs. 23.11 and 23.12, we get

$$\frac{C_{x,y,0,H}}{C_{x,0,0,H}} = e^{-\frac{1}{2}(y/\sigma_y)^2}$$

$$\therefore \quad y = \left[2\log_e \frac{C_{x,0,0,H}}{C_{x,y,0,H}}\right]^{1/2}\sigma_y \qquad ...(23.12\ a)$$

From example 23.2, $C_{x,0,0,H} = 2.45\times 10^{-4}$ g/m^3 and $\sigma_y = 210$ m

Also $C_{x,y,0,H} = 1\times 10^{-4}$ g/m^3 (given)

$$\therefore \quad y = \left[2\log_e \frac{2.45\times 10^{-4}}{1\times 10^{-4}}\right]^{1/2}\times 210 \approx \mathbf{281}\ \text{m}$$

Example 23.5. *The traffic density along a busy highway is 1600 vehicles/hour and the average vehicle speed is 70 m/hour. The average carbon monoxide emission per vehicle is 50 g/s. Estimate the carbon monoxide at 100 m, 500 m and 1000 m down wind distances of the highway if the wind speed normal to highway is 2.5 m/s. Assume neutral conditions.*

Solution

Strength of source per unit length (m) is

$$Q_l = 1600\left(\frac{\text{vehicles}}{\text{hr}}\right)\frac{1}{70}\left(\frac{\text{hr}}{\text{km}}\right)\frac{1}{1000}\left(\frac{\text{km}}{\text{m}}\right)50\left(\frac{g}{s}\frac{1}{\text{vehicle}}\right)$$

$$= 1.14\ \text{g/m–s}$$

From Fig. 23.11, we get $\sigma_z \approx 4.7$ m, 18 m and 32 m for $x = 100$, 500 and 1000 m respectively, for neutral condition (D).

Now, taking $H = 0$, we get from Eq. 23.15 (a),

$$C_{x,y,0,0} = \frac{2\,Q_l}{(2\pi)^{1/2}\,\sigma_z\,.\,u}$$

$$= \frac{2\times 1.14}{\sqrt{2\pi}\,(2.5)\,\sigma_z} = \frac{0.3638}{\sigma_z}$$

$$\therefore \quad C_{100,\,y,\,0,\,0} = \frac{0.3638}{4.7} = 77.4\times 10^{-3}\ g/m^3$$

$$C_{250, y, 0, 0} = \frac{0.3638}{18} = 20.2 \times 10^{-3}\ g/m^3$$

$$C_{1000, y, 0, 0} = \frac{0.3638}{32} = 11.4 \times 10^{-3}\ g/m^3$$

Example 23.6. *Determine the plume rise and the effective height of the stack for the following data :*

(i) *Physical stack height : 200 m.*
(ii) *Inside dia. of stack at exit : 1 m*
(iii) *Wind velocily : 3 m/s*
(iv) *Air temperature : 25° C.*
(v) *Barometric pressure : 1000 millibars*
(vi) *Stack gas exit velocity : 12 m/sec.*
(vii) *Stack gas exit temperature : 150°C.*

Solution. Here, $h = 200$; $d = 1$ m; $u = 3$ m/s ; $V_s = 12$ m/sec $p = 1000$ millibars.

Also, $T_s = 150 + 273 = 423°$ K

$T_a = 25 + 273 = 298°$ K

Hence from Eq. 23.6 (Holland's equation)

$$\Delta h = \frac{V_s . d}{4}\left[1.5 + 2.68 \times 10^{-3} pd \left(\frac{T_s - T_a}{T_s}\right)\right]$$

$$= \frac{12 \times 1}{3}\left[1.5 + 2.68 \times 10^{-3} \times 1000 \times 1\left(\frac{423 - 298}{423}\right)\right]$$

$= \mathbf{9.17\ m}$

$\therefore \quad H = h + \Delta h = 200 + 9.17 = \mathbf{209.17\ m}$

Example 23.7. *A factory uses 1.5 Ml of fuel oil per month. The exhaust gases from the factory contain the following quantities of pollutants per Ml per year :*

(i) *Particulate matter : = 4 t/year*
(ii) *SO_2 : 20 t/year*
(iii) *NO_x : 5 t /year*
(iv) *HC, CO and others : 3 t/year.*

Determine the safe height of the chimney required for the safe dispersion of the pollutants.

Solution.

The concentrations of NO_x, HC, CO and others are generally very much less than the concentration of SO_2 in various industries. Hence the Board has made only SO_2 as the criterion for design, along with the particulate matter.

(a) Height of chimney on the basis of particulate matter

$$h = 74\,(Q_p)^{0.27} \qquad \text{...(23.16)}$$

Particulate matter emission $= 4\,(1.5 \times 12) = 72$ t/yr.

Assuming 300 working days in a year, and 24 hours working a day,

$$Q_p = \frac{72}{300 \times 24} = 0.01 \text{ t/hour.}$$

$$\therefore \quad h = 74\,(0.01)^{0.27} = 21.34 \text{ m}$$

(*b*) ***Height of chimney on the basis of*** SO_2

$$h = 14\,(Q_s)^{1/3} \quad ...(23.17)$$

Now, SO_2 emission $= 20\,(1.5 \times 12) = 360$ t/year

$$\therefore \quad Q_s = \frac{360}{300 \times 24} = 0.05 \text{ t/hour}$$

$$= 50 \text{ kg/hr.}$$

$$\therefore \quad h = 14\,(50)^{1/3} = 51.58 \text{ m.}$$

$\therefore$ Reguired height of chimney $\approx$ **52 m**

23.12. AIR POLLUTION CONTROL METHODS AND EQUIPMENT

There are following four methods of controlling air pollution:

1. Proper planning of industrial area like zoning.
2. Dilution of source discharge by use of tall stacks
3. Using source correction methods through (*i*) raw material changes (*ii*) process changes, and (*iii*) equipment modification or replacement.
4. Reduction of pollutant discharge at source by use of controlling equipments.

1. Air pollution control by zoning. Air polution can be effectively controlled by adopting the zoning system at the planning stage itself. '*Cumulative zoning*' in the past has resulted in less availability of land for industries. This system has been now modified as 'permissible system', though this system has also resulted in crowding of industrial zones with other uses also besides industry. The next improved system is the '*exclusive zoning system*' which provides for compatible uses for each zone, excluding other uses. In this sytem a separate zone or area is set aside for industries (known as *industrial area* or *industrial zone*), thereby reducing the ill effects of air pollution on urban dwellers. By proper zoning, the planning of the city should be done such that residential areas and heavy industries are not located too close to each other. This is achieved by providing a *green belt* between the industries and the township. Zoning for the industrial areas can be done on the two aspects : (*a*) functional requirement and (*b*) performance characteristics. The *functional requirements* of the industries include inter-industries linkages, railway sidings, groupings, land traffic generation utilities etc. The *performance characteristics* of industries include the traffic congestion, obnoxious and hazardous character of the industry, industrial nuisances such as dust, noise, smoke, odour, heat, fire, hazardous gases etc. Separate areas must be earmarked according to their performances so that neat industries are placed away from obnoxious industries. In India, zoning system varies from city to city. For example, in the industrial estate of

Bangalore, only three zones are provided for light, medium and large industries. Bombay plan has listed four zones : (*i*) small repairing and light manufacturing units, permissible in residential areas, (*ii*) service industrial zone, (*iii*) special industrial zone and (*iv*) general industrial zone.

2. Dilution of source discharge by use of tall stacks : Various aspects of atmospheric dispersion of pollutants have already been discussed earlier. The atmosphere, like natural stream, possesses self cleansing properties which continuously clean and remove the pollutants from the atmosphere under natural conditions, provided the pollutants are discharged in the atmosphere judiciously so that effective dispersion takes place. If the pollutants are carried away to some distance or taken to high attitudes, they are reduced in concentration by diffusion and dilution. The pollutants are taken to high altitudes by means of tall stacks. The height of the stack should be such that the maximum ground level concentration, which varies inversely with the square of the stack height, is within the permissible limits.

3. Control by using source correction methods : This method is known as 'air pollution prevention at source'. This can be achieved through (*a*) change in raw materials (*b*) process changes, and (*c*) equipment modification or replacement.

(a) Change in raw material : If one type of raw material, currently in use, results in air pollution problem, while a substitute material, which may be of purer grade, does not, the substitution will be more desirable. A typical example in this context is the use of low sulphur fuel in place of high-sulphur fuel. The raw material in current use may contain certain ingredient which is not essential for the process but which contributes to pollution ; the non-essential ingredient should be removed through prior processing so that pollution can be minimised.

(*b*) ***Process modification*** : Atmospheric pollutants emissions can some times be reduced by adopting *modified* or *new* processes. A typical example is the use of exhaust hoods and ducts over several types of industrial ovens have not only reduced pollutants but also have resulted in the recovery of valuable solvents that could have become air pollutants. Similarly, volatile substances can be *recovered* by condensation and the non-condensable gases can be recycled for additional reactions.

(*c*) ***Equipment modification or replacement*** : Old equipment, which contribute to greater degree of air pollution can be modified or completely replaced. For example, basic oxygen furnaces, that are replacing the open-hearth furnaces in the steel industry, pose lesser pollution problems. In many cases, newer types of equipment are less pollution prone. Newer types of equipment in paper and pulp industry also cut down the quantity of pollutants emitted. Also, air pollutant emissions from industrial operations can be reduced by proper equipment maintenance, house keeping and cleanliness in the facilities and premises.

4. **Reduction of pollutant discharge at source by use of controlling equipment** : This is a method of reducing the pollutants at source by use of *air pollution control equipment.* In the subsequent articles, we will discuss the pollutant control methods, both for aerosals (particulates) as well as gaseous emissions separately.

23.13. EQUIPMENT FOR CONTROL OF PARTICULATE POLLUTANTS

Following are the equipment or devices which are used to control the emission of particulate pollutants :

(*a*) Gravity settling chambers, (*b*) Inertial or impact separators (other than cyclones), (*c*) Cyclones, (*d*) Fabric filters, (*e*) Electrostatic precipitaters and (*f*) Scrubbers or wet collectors

(*a*) Gravity settling chambers : A settling chamber consists of a large rectangular expansion chamber in which dust is separated from the gas by reducing the velocity of the gas. Due to this, the dust particles settle down under gravity, in the chamber. Gravity settling chamber are set horizontally, often on the ground, and can be constructed in brick and concrete. In order to reduce the size of the chamber, the gas velocity is kept between 0.5 m/s (to obtain good results) to 3 m/s to prevent re-entrainment of settled particles. Hence such chambers are capable of removing only the large size particles between 40 to 50 μ. Their efficiency is quite low for smaller particles, and hence these devices are normally used as precleaners prior to passing the gas stream through high efficiency collection devices. Settling chambers are used widely for the removal of large solid particulate from natural draft furnaces, kilns etc. They are also used in process industries, particularly the food and metallurgical industries, as a first step in dust control. The efficiency of a settling chamber can be improved if the height to be travelled by the particles is less. This is achieved by incorporating horizontal trays or shelves, as in Howard settling chamber shown is Fig. 23.13 (*b*), in which

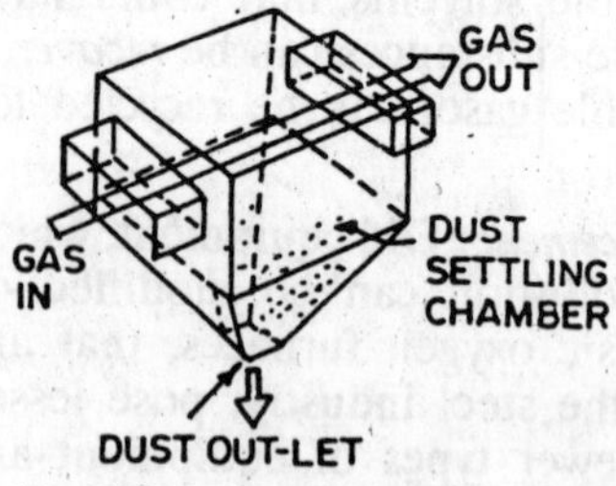

(*a*) SIMPLE CHAMBER

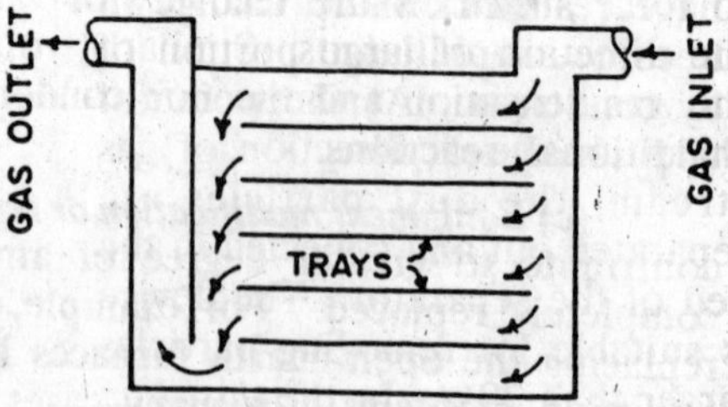

(*b*) HOWARD CHAMBER

FIG. 23.13. GRAVITY SETTLING CHAMBER.

particles upto $10\,\mu$ size can be removed. However, such modified chambers present cleaning difficulties, and the horizontal plate tend to warp at higher temperatures. The advantages of gravity collectors are (*i*) low initial cost (*ii*) simple construction (*iii*) low maintenance cost (*iv*) low pressure drop (*v*) dry and continuous disposal of solids. The disadvantages are (*i*) large space requirements and (*ii*) only comparitively larger size particles can be collected.

2. Inertial or impact Separators : Such separation devices cause sudden change in the direction of gas stream and thereby separate the particles by inertia, by impingement on a target or by centrifugal force. Two types of equipment utilising this principle are : (*a*) inertial or impact separaters and (*b*) cyclonic separators. The inertial or impact separators employ *incremental changes of direction* of the carrier gas stream to exert the greater inertial effects of the dispersoid while the cyclonic separators produce a continuous centrifugal force as a means of exerting the greater inertial effects of the dispersoid.

There are three fundamental types of inertial or impact separaters : (*a*) Baffle type separator (*b*) Louver type collector, and (*c*) Dust traps.

In the *baffle type separater*, the gas stream is made to follow a tortuous flow path which is obtained by insertion of staggered plates into the path of the gas stream. Such a device removes particles of greater than $20\,\mu$ size and is widely used for particulate removal in power plant and rotary kilns. A *louver type separator* (or collector) mainly consists of several blades set at an angle to the flow path of the gas stream (Fig. 23.14). The blades are set to force a quick sharp change in the direction of large portion of the gas stream. At the point of sharp change of direction of gas stream, the dust particles are separated out and collected in the bed of the separator. The device is suitable for removing particles larger than $30\,\mu$. In the *dust trap* type separator, the dust laden gas is introduced into a chamber through a central pipe and is made

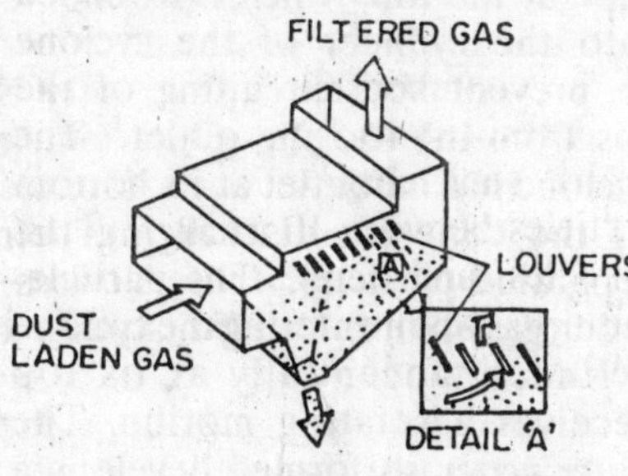

FIG. 23.14. LOUVER SEPARATOR

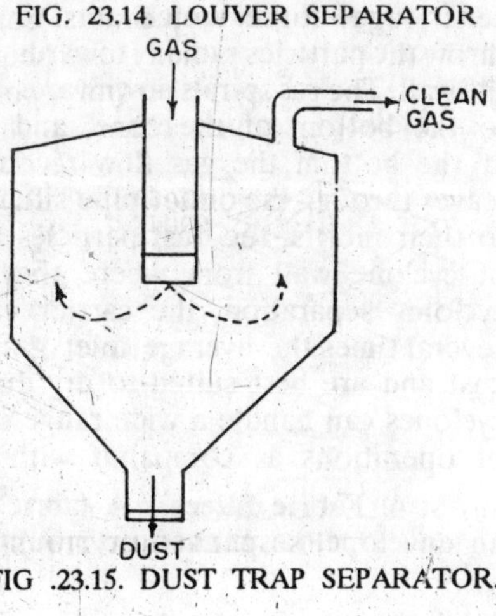

FIG .23.15. DUST TRAP SEPARATOR.

to undergo a change in direction by 180° (Fig. 23.15). The dust falls down in the conical chamber. The collection efficiency of particles of 30 μ size is about 70%. Hence it is used mostly as a precleaner.

3. **Cyclone Separators** : Such a separator depends upon centrifugal force for its action. It is a structure, without moving parts, where the velocity of an inlet gas stream is transformed into a confined vortex from which the centrifugal forces tend to drive the suspended particles to the wall of the cyclone body. The centrifugal force on particles in a spinning stream is much greater than gravity and hence cyclones are effective in the removal of much smaller particles than gravitational settling chambers. They also require much less space to handle the same gas volume. A simple reverse flow type cyclone, shown is Fig. 23.16 consists of a vertically placed cylinder having an inverted cone attached at its bottom, and fitted with a tangential inlet located near the top. The outlet pipe for the purified gas is a central cylindrical pipe at the top, which is extended into the cylinder of the cyclone to prevent shortcircuiting of the gas from inlet to the outlet. The cyclone has an outlet at its bottom of the cone for discharging the separated particles. The particle-laden gas upon entering the cyclone cylinder tangentially at its top receives a rotating motion. The *outer vertex* so formed develops a centrifugal force which acts to throw the particles radially towards the wall. The gas spirals downwards to the bottom of the cone, and at the bottom the gas flow *reverses* to form an *inner vortex* which leaves through the outlet pipe situated at the top of the cyclone. Due to their inertia, the dust particles tend to concentrate on the surface of cyclone wall from where they are led to the receiver. During cyclonic separation, the carrier gas rotational velocity may exceed several times the average inlet gas velocity. Cyclones are cheaper in cost and are best suited to dry dust particles of size 10 – 40 μ. The cyclones can handle a wide range of physical and chemical conditions of operations as compared with other collection equipments.

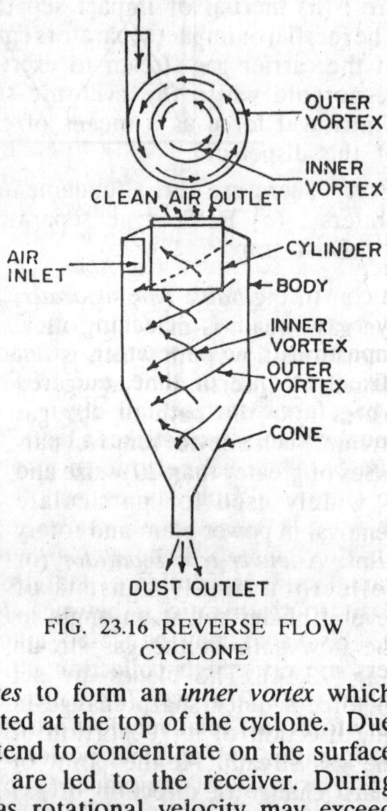

FIG. 23.16 REVERSE FLOW CYCLONE

4. **Fabric filters** : A fabric filter consits of a tubuler bag or an envelope, suspended or mounted with open ends in such a way

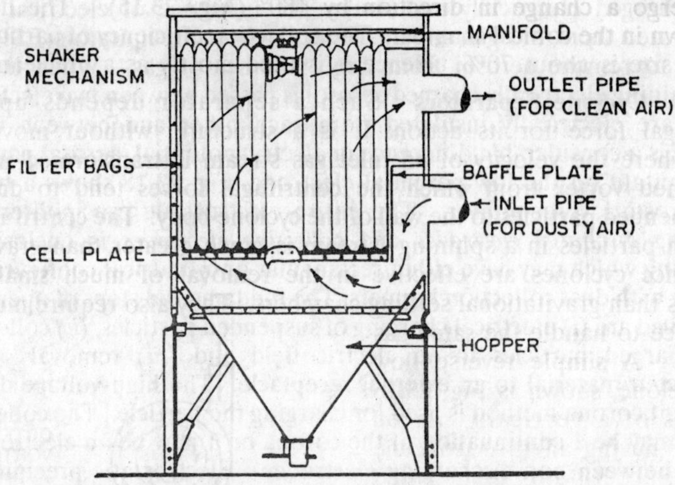

FIG. 23.17. BAG HOUSE FILTER

that when the dirty gas passes through it, the particulate matter is arrested on the inside fabric surface of the bag. The whole system, generally known as a *bag house* consists of a number of such bags, and constructed with several compartments so that one compartment may be isolated for cleaning as and when needed while the other compartments are operating. Cleaning is accomplished by shaking at fixed intervals of time, due to which the finer particles collected on bag fabric surface fall down in the hopper. The gas entering through the inlet pipe strikes a baffle plate which causes larger particles to fall into the hopper due to gravity. The gas then flows upwards into the cloth fabric tubes, and then outwards through the fabric. The normal velocities at which the gas is passed through the bags is 0.4 to 1 m/min. The air to cloth ratio is one of the important parameters in the design of fabric filters. In the *low ratio* bag filter the air to cloth ratio is between 7-25 mm/sec, while in *high ratio* bag, this value may be 25-50 mm/sec. The *advantages* of fabric filters are : (*i*) high collection efficiencies, especially for particles smaller than 10 μ size (*ii*) collection and disposal of particulates in dry form (*iii*) relatively low pressure drop (*iv*) simple construction and operation (*v*) nominal power consumption. The disadvantages are : (*i*) large size of equipment (*ii*) operational limitations for process temperatures higher than 285°C (*iii*) high construction cost (*iv*) operational limitations at high humidity and (*v*) problems in handling dusts which may abrade, corrode or blind the cloth fabric. These are extensively used in metallurgical industry, foundry, cement industry, chalk and lime plants, ceramic industry, flour mills etc.

5. Electrostatic precipitaters : Electrostatic precipitation is a physical process by which particles suspended in a gas stream are charged electrically and, under the influence of electrical field, separated from the gas stream. Electrostatic precipitators(ESP) are used to remove aerosol from large volumes of gas. When the gas containing electrically charged aerosol is passed between two electrodes that are electrically insulated from each other and between which there is a considerable difference in electric potential, aerosal particles precipitate on the low potential electrode. Fig. 23.18 shows a typical electrostatic precipitator. The entire electrostatic precipitater containing *discharge electrodes* and *collecting electrodes* is enclosed in a casing which may have either rectangular or cylindrical configuration, along with dust collection hoppers. The fundamental steps of operation involved are (*i*) electrical charging of suspended particles, (*ii*) collection of charged particles in an electric field and (*iii*) removal of the collected material to an external receptacle. The high-voltage direct-current corona method is used for charging the particle. The collecting field may be a continuation of the corona or it may be an electrostatic field between non-discharging electrodes. Electrostatic precipitators can be classified as *dry precipitators* and *wet precipitators*, depending upon whether the mode of operation is a dry process or wet one. Though wet precipitators are more efficient collectors, dry process plate type precipitators are predominantly used. The *advantages* of ESP are (*i*) precipitators are unique in that the collecting force is applied only to the material to be collected and not to the total gas stream resulting in low operating costs (*ii*) pressure drop through the system is less (0.25-1.25 cm of water) (*iii*) power requirement is less

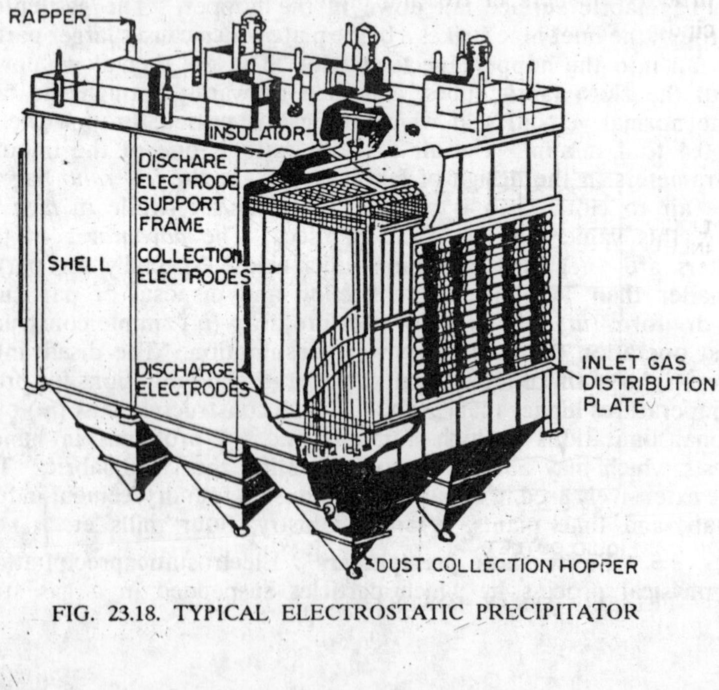

FIG. 23.18. TYPICAL ELECTROSTATIC PRECIPITATOR

(*iv*) large gas columns can be handled (*v*) high efficiency to remove very small particles, as small as 0.1 μ (*vi*) ability to operate over a wide range of conditions, temperatures upto 800°C and pressures of 50 atmosphere (*vii*) can even remove tar and acid mists which are difficult to be removed by any other method (*viii*) treatment time is negligible (0.1 to 10 s), (*ix*) cleaning is easy. The *disadvantage* are : (*i*) high initial cost (*ii*) precipitators are not easily adapted to charging process conditions (*iii*) high voltage equipment necessitates special safety precautions (*iv*) large space is required (*v*) there are possible explosion hazards during collection of combustible particulates or gases.

6. Scrubbers or wet collectors : A scrubber is a device which utilises a liquid to assist in the removal of particulates from the carrier gas stream. They are mixed phases of gas and liquid. The object of the scrubber is to transfer sus-pended particulate matter in the gas to the scrubbing liquid which can be readily removed by the gas cleaning device. Following are the commonly used types of scrubbers : (*i*) spray towers (*ii*) packed bed scrubbers (*iii*) venturi scrubbers (*iv*) cyclone scrubbers (*v*) mechanical scrubbers.

(*i*) *Spray Towers* : It consistes of a rectangular or round tower with a number of spray nozzles. The carrier gas flows upward and particle collection results because of inertial impaction and interception on the droplets. Only moderate contact between the phases is effected and hence these are used for removing coarse dusts (> 1 – 2 μ), where high efficiency is not required. They are therefore used as primary cleaners in treating blast furnace gas and for fly ash and cinder removal.

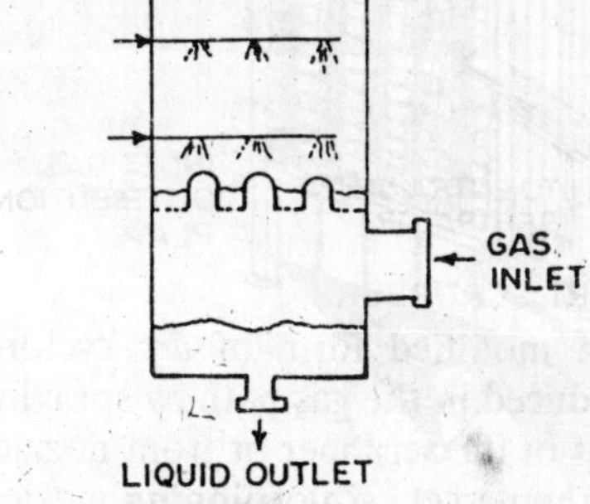

FIG. 23.19. SPRAY TOWER

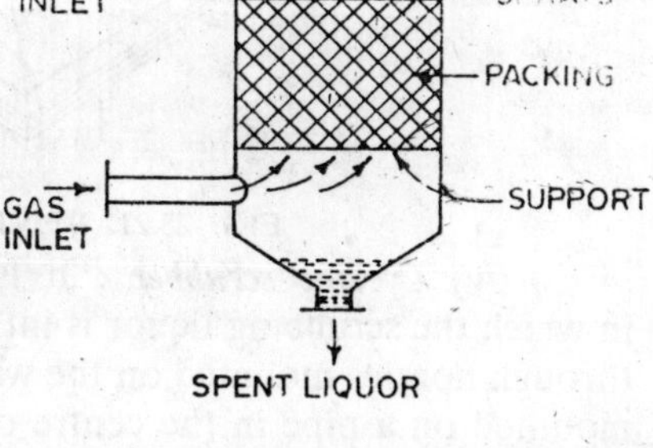

FIG. 23.20. PACKED BED TOWER

(ii) Packed bed towers : Fig. 23.20 shows a packed bed tower. They use fibre glass (fine glass) packing or other packing. The polluted gas stream moves upward and comes in contact with the scrubbing liquid stream which is moving downward over the packing in a film. The particles are captured by inertial impaction while the gas stream takes a curved path through pore space of the packing.

(iii) Venturi Scrubber : In such a scrubber dust laden gas is accelerated to a high velocity (60 to 100 m/s) while passing through the converging section and approaching the throat section. Water sprays are introduced just ahead of the venturi throat (Fig. 23.11). The high-velocity gas impinges (impacts) upon the liquid stream in the throat, thereby atomizing the liquid into large number of fine droplets. The high-differential velocity between the gas and the atomized droplets causes liquor droplets and dust particles to collide, impact and agglomerate. After the particulates have been trapped within the liquor droplets, the resulting agglomerates are readily removed from the gas stream in the separater. Such scrubbers are used in removing mists and dust from gases from pulp and paper industries, steel and non-ferrous metal industries and chemical industries.

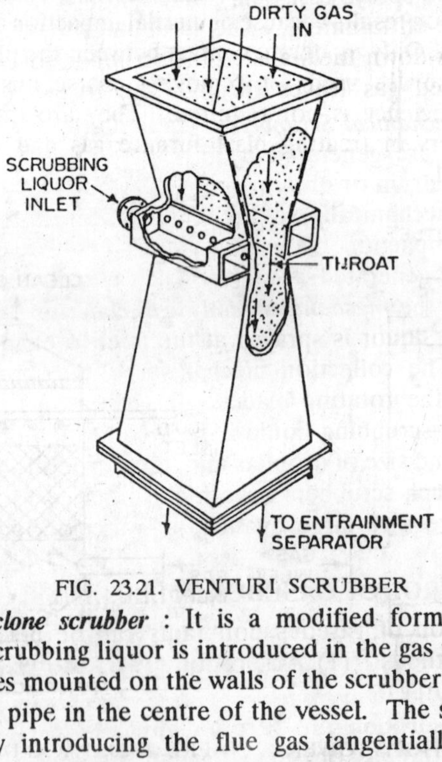

FIG. 23.21. VENTURI SCRUBBER

(iv) Cyclone scrubber : It is a modified form of *dry* cyclone in which the scrubbing liquor is introduced in the gas path by spraying through nozzles mounted on the walls of the scrubber or from nozzles mounted on a pipe in the centre of the vessel. The spinning motion is initiated by introducing the flue gas tangentially or by having

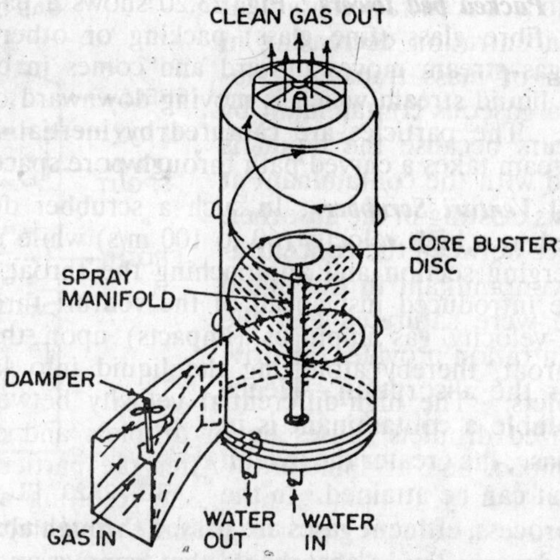

FIG. 23.22. CYCLONIC SCRUBBER

the scrubbing liquid introduced (or sprayed) tangentially inward from the wall. The sprays assist in collection of the dispersoid and tend to prevent re-entrainment. The scrubbing liquor with the collected particles runs down the walls to the scrubber bottom and out, where as the cleaned gas leaves through the top.

(*V*) ***Mechanical Scrubbers*** : These are high emerging scrubbers. Here, sprays are generated by mechanical means such as a rotary paddle, cage, drum or disc, or are generated by spray nozzles. Liquid is sheared mechanically to break into fine droplets for collection by inertial impaction between droplets and dust particles. Sprays are normally generated at right angles to the direction of the flue gas flow. In the *mechanical-centrifugal collector* (also called *dynamic wet scrubber*), liquor is sprayed at the inlet to enhance the collection efficiency. The collection mechanism is the impingement of dust particles on the rotating blades. In the case of *high-pressure spray scrubbers*, the scrubbing liquid is sprayed at high pressure to increase the number and size of droplets and, thereby to increase the collection efficiency. Such scrubbers have high initial cost and high operating cost. They require large quantity of water and considerable maintenance.

23.14. CONTROL OF GASEOUS POLLUTANTS

Emission of gaseous pollutants can be be controlled by the following methods : (1) Absorption (2) Adsorption, and (3) Combustion

1. Absorption : *rption involves the transfer of the pol lutants from the gas phase to the liquid phase across the interfa in response to a concentration gradient with the concentration decreasing in the direction of mass transfer. Absorption of a gaseous contaminant by a liquid occurs because the liquid is not saturated with the contaminant at the conditions existing in the absorber. The difference between the concentration of the contaminant in the liquid if the liquid were saturated and the actual concentration provides the driving force for the absorption. Hence the more soluble a contaminant is in the liquid phase, the greater the overall efficiency that can be attained. In the absorption process, effluent gases are passed through absorbers (scrubbers) which contain liquid absorbents that remove one or more pollutants from the gas stream. Absorbants are being used to remove SO_2, H_2S, SO_3, F and oxides of nitrogen. The absorbents may be either reactive or non-reactive with the pollutants removed by them. The reactive type absorbents may be either regenerative type or non-generative type. Various equipments which use the principle of absorptions are : (*i*) spray towers (*ii*) packed towers (*iii*) plate towers (*iv*) bubble plate tower (*v*) ventury scrubbers. Out of these, spray towers, packed power and ventury scrubbers can be simultaneously used for removing particulate pollutants, and have already been described in the previous article. Fig. 23.23 shows a plate tower while Fig 23.24 shows a bubble plate tower. A plate tower consists of a vertical shell in which are mounted a large number of equally spaced circular perforated plates. Gases and vapours bubble upward through the liquid seal above each plate. A bubble plate tower consists of a vertical shell in which are mounted a

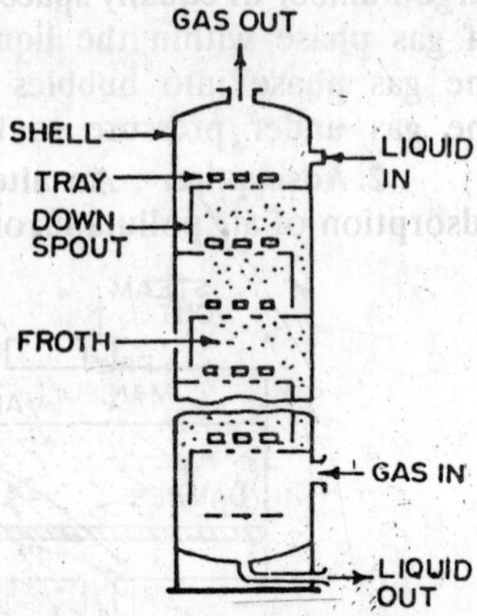

FIG. 23.23. PLATE TOWER

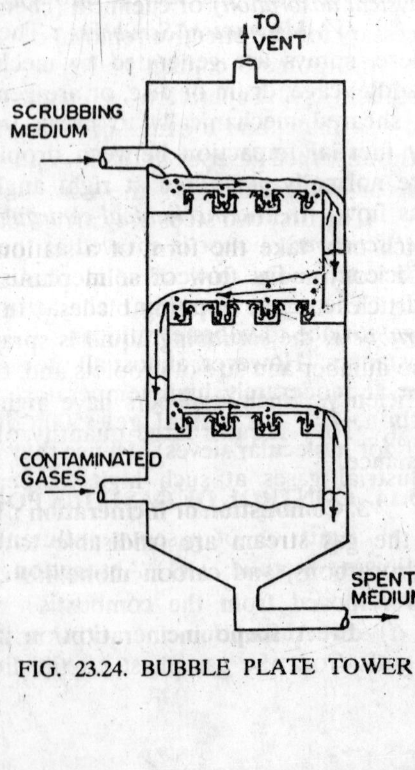

FIG. 23.24. BUBBLE PLATE TOWER

large number of equally spaced circular bubble plates. The dispersion of gas phase within the liquid phase is accomplished by breaking the gas phase into bubbles within the liquid phase by supplying the gas under pressure to bubble plates.

2. Adsorption : An alternative to absorption by liquids is the adsorption of air pollutants on solids. The commonly used adsorbers

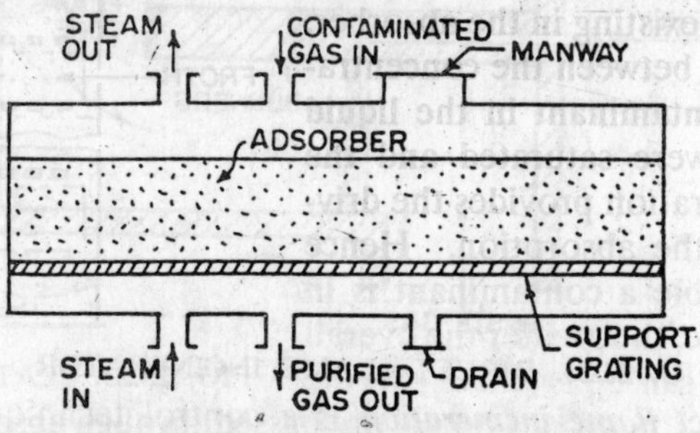

FIG. 23.25. FIXED-BED ADSORBER.

include activated carbon, molecular sieves (such as dehydrated zeolites) silica gel, activated alumina, lithium, chloride, bauxile etc. *Adsorption* is a surface phonomenon by which gas or liquid molecules are captured by and adhere to the surface of solid adsorbent. The attractive forces holding the molecules at the surface may be either physical *(physical adsorption)* or chemical (chemisorption) in nature. The steps necessary for the effective removal of gaseous pollutants by adsorbents are : *(i)* Intimate contact between gaseous pollutants and the solid adsorbent. *(ii)* Separation of unadsorbed gases from the adsorbant and adsorbate, and for final disposal. *(iii)* Separation (or desorption) of the adsorbed gaseous pollutant with the solid adsorbant by regeneration or replacement of the adsorbent.

The first two steps are carried out in the adsorption equipment, which may take the form of a stationary or fixed bed of adsorbent or a continuous flow of solid phase through a series of fluidized bed. The most important characteristics of an adsorbent are its large surface-to-volume ratio and its preferential affinity to specific substances. However, almost all the adsorbents are subject to destruction at moderately high temperatures (such as 150°C for activated carbon, 400° for silical gel, 500°C for activated aluminium and 600° for molecular sieves). Hence they are very inefficient for purifying industrial gases at such high temperatures.

3. Combustion or incineration : This is used when the pollutants in the gas stream are oxidisable to an inert gas. Pollutants like hydrocarbons, and carbon monoxide can be easily burned, oxidised and removed from the combustion equipment. This is achieved by *(i)* direct flame incineration, and *(ii)* catalytic incincration.

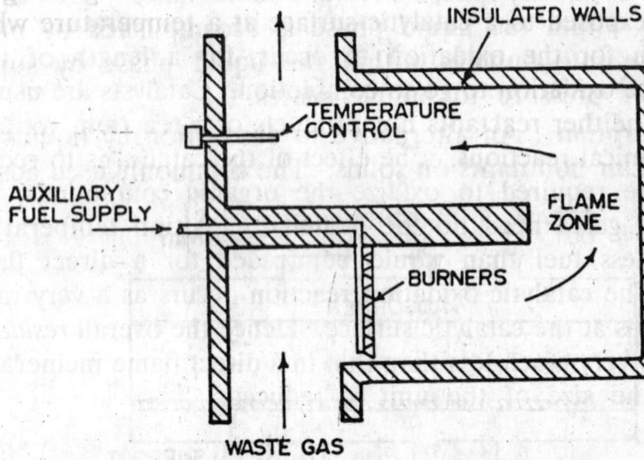

FIG. 23.26. DIRECT FLAME INCINERATOR

The *direct flame incineration* is a control technique for combustible organic air pollutants. This is accomplished in the presence of a flame and sufficient oxygen by raising the temperature of the gases above their ignition temperature and then maintaining this temperature until the oxidation reactions are complete.

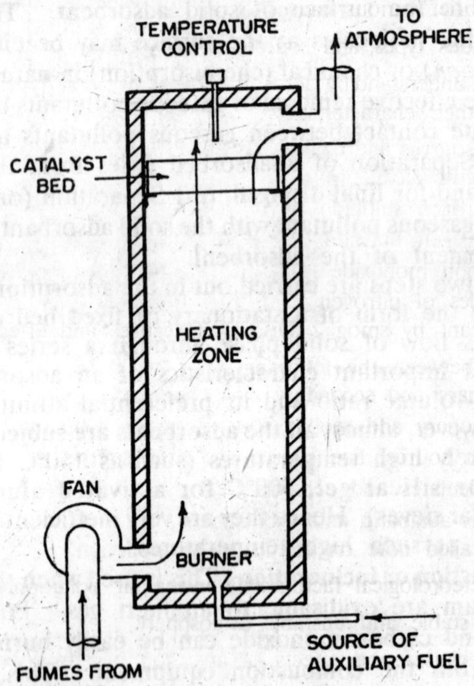

FIG. 23.27. CATALYTIC INCINERATOR.

In the *catalytic incineration*, a mixure of dilute organic gases and oxygen is exposed to a catalytic surface at a temperature which is high enough for the oxidation to occur, for a length of time sufficient for the oxidation to go to completion. Catalysts are usually solids that are neither reactants nor products of a reaction, yet alter the rate of chemical reactions. The effect of the catalyst is to reduce the temperature required to oxidize the organic compounds, and hence the inlet gases need not be heated to ignition temperature. This requires less fuel than would be needed for a direct flame incineration. The catalytic oxidation reaction occurs as a very rapid sequence of steps at the catalytic surface. Hence the overall *residence time* required is very much less than that in a direct flame incinerator. Due to this, the size of the unit is reduced.

PROBLEMS

1. Write a note on biosphere and environment. How temperature and pressure vary with altitude?
2. Write a note on ecosystems of biosphere and ecological balance.
3. Explain in brief various causes of air pollution.
4. Classify various types and sources of air pollution.
5. What do you understand by 'aerosols'? Describe various types of aerosols along with their characteristics.
6. Explain the sources and characteristics of the following air pollutants?
   (*i*) Sulphur dioxide
   (*ii*) Hydrogen sulphide
   (*iii*) Hydrogen fluoride
   (*iv*) Carbon monoxide
   (v) Oxides of nitrogen
7. What is meant by smog ? Discuss its causes and effects.
8. Distinguish between the following :
   (*i*) Primary and secondary air pollutants.
   (*ii*) Stationary and mobile sources of air pollutants.
   (*iii*) Point source, line source and area source.
   (*iv*) Particulates and aerosols.
   (*v*) Super-adiabatic and subadiabatic.
9. Write a detailed note on effects of air pollution.
10. List out meteorological factors influencing air pollution.
11. (*a*) Explain stable and unstable atmosphere.
    (*b*) Explain causes and effects of inversion of atmosphere.
12. Expiain, with the help of diagrams, various types of plume behavior.

13. What do you understand by 'plume rise' ? Give ISI formulae for determining plume rise.
14. What do you understand by atmospheric dispersion ? Write down the equation for determining ground level concentration of pollutant. Where does the max. ground level concentration occur ?
15. Explain how do you select stack height.
16. (*a*) Enumerate various types of equipments used for the control of emission of particulate pollutants.

    (*b*) Write a note on 'scrubbers'.
17. Write notes on the following :

    (*a*) Cyclone separators

    (*b*) Bag house filter

    (*c*) Electrostatic precipitators
18. Write a note on control of 'gaseous pollutants'

Appendix

TABLE A–1 VISCOSITY AND DENSITY OF WATER

Temperature, C	*Density, (grams/cm^3)*	*Absolute Viscosity μ, centipoises*	*Kinematic Viscosity ν, centistokes*	*Temperature F*
0	0.99987	1.7921	1.7923	32.0
2	0.99997	1.6740	1.6741	35.6
4	1.00000	1.5676	1.5676	39.2
6	0.99997	1.4726	1.4726	42.8
8	0.99988	1.3872	1.3874	46.4
10	0.99973	1.3097	1.3101	50.0
12	0.99952	1.2390	1.2396	53.6
14	0.99927	1.1748	1.1756	57.2
16	0.99897	1.1156	1.1168	60.8
18	0.99862	1.0603	1.0618	64.4
20	0.99823	1.0087	1.0105	68.0
22	0.99780	0.9608	0.9629	71.6
24	0.99733	0.9161	0.9186	75.2
26	0.99681	0.8746	0.8774	78.8
28	0.99626	0.8363	0.8394	82.4
30	0.99568	0.8004	0.8039	86.0

Notes : (*i*) 1 Centipoise = 10^{-2} (gram mass)/(cm)(sec).

(*ii*) To convert to (lb force) (sec)/(sq.ft), multiply centipoise by 2.088×10^{-5}.

(*iii*) To convert to (kg f) (sec)/m^2, multiply centipoise by 1.019×10^{-4}

(*iv*) 1 (N-s)/m^2 = 10 poise = 1000 centipoises.
1 centipoise = 10^{-3} (N-s)/m^2

(*v*) 1 centistoke = 10^{-2} cm^2/sec. To convert to (sq. ft)/(sec), multiply centistoke by 1.075×10^{-5}

TABLE A–2 VAPOUR PRESSURE OF WATER AND SURFACE TENSION OF WATER IN CONTACT WITH AIR

Temperature, C	0	5	10	15	20	25	30
Vapour pressure, mm Hg	4.58	6.54	9.21	12.8	17.5	23.8	31.8
Surface tension, dyne/cm	75.6	74.9	74.2	73.5	72.8	72.0	71.2

TABLE A–3 PROPERTIES OF WATER (FPS UNITS)

Temperature (°F)	*Specific Weight* γ (lb/ft^3)	*Mass Density* ρ (lb–sec^2/ft^4)	*Dynamic Viscosity* $\mu \times 10^5$ (lb–sec/ft^2)	*Kinematic Viscosity* $\nu \times 10^5$ (ft^2/sec)	*Surface Energy** $\sigma \times 10^3$ (lb/ft)	*Vapour Pressure* p_v (lb/in.^{2})	*Bulk Modulus* $E \times 10^{-3}$ (lb/in.^{2})
32	62.42	1.940	3.746	1.931	5.18	0.09	290
40	62.43	1.938	3.229	1.664	5.14	0.12	295
50	62.41	1.936	2.735	1.410	5.09	0.18	300
60	62.37	1.934	2.359	1.217	5.04	0.26	312
70	62.30	1.931	2.050	1.059	5.00	0.36	320
80	62.22	1.927	1.799	0.930	4.92	0.51	323
90	62.11	1.923	1.595	0.826	4.86	0.70	326
100	62.00	1.918	1.424	0.739	4.80	0.95	329
110	61.86	1.913	1.284	0.667	4.73	1.24	331
120	61.71	1.908	1.168	0.609	4.65	1.69	333
130	61.55	1.902	1.069	0.558	4.60	2.22	332
140	61.38	1.896	0.981	0.514	4.54	2.89	330
150	61.20	1.890	0.905	0.476	4.47	3.72	328
160	61.00	1.896	0.838	0.442	4.41	4.74	326
170	60.80	1.890	0.780	0.413	4.33	5.99	322
180	60.58	1.883	0.726	0.385	4.26	7.51	318
190	60.36	1.876	0.678	0.362	4.19	9.34	313
200	60.12	1.868	0.637	0.341	4.12	11.52	308
212	59.83	1.860	0.593	0.319	4.04	14.70	300

* In contact with air.

TABLE A–4 CONVERSION FACTORS FROM CUSTOMARY UNITS TO SI METRIC

Customary Units		Multiplier	Metric Units	
Description	Symbol		Symbol	
	Multiply	By	To Obtain	Reciprocal
Acre	ac	0.4047	ha	2.471
British thermal unit	Btu	1.055	kJ	0.9470
British thermal units per cubic foot	Btu/ ft^3	37.30	J / l	0.02681
British thermal units per pound	Btu/lb	2.328	kJ / kg	0.4295
British thermal units per square foot per hour	Btu/ ft^2/hr	3.158	J / m^2/sec	0.3167
Cubic foot	ft^3	0.02832	m^3	35.31
Cubic foot	ft^3	28.32	l	0.03531
Cubic feet per minute	cfm	0.4719	l /sec	2.119
Cubic feet per minute per thousand cubic feet	cfm /1000 ft^3	0.01667	l / m^3/sec	60.00
Cubic feet per second	cfs	0.02832	m^3/sec	35.31
Cubic feet per second per acre	cfs / ac	0.06998	m^3/sec/ ha	14.29
Cubic inch	$in.^3$	0.01639	l	61.01
Cubic yard	yd^3	0.7646	m^3	1.308
Fathom	f	1.839	m	0.5467
Foot	ft	0.3048	m	3.281
Feet per hour	ft / hr	0.08467	mm / sec	11.81
Feet per minute	fpm	0.00508	m / sec	196.8
Foot-pound	ft-lb	1.356	J	0.7375
Gallon, U.S.	gal	3.785	l	0.2642
Gallons per acre	gal / ac	0.00935	m^3/ ha	106.9
Gallons per day per linear foot	gpd/ lin. ft	0.01242	m^3/m /day	80.53
Gallons per day per square foot	gpd/ft^2	0.04074	m^3/m^2/day	24.54
Gallons per minute	gpm	0.06308	l / sec	15.85
Grain	gr	0.06480	g	15.43

TABLE A–4 CONVERSION FACTORS FROM CUSTOMARY UNITS TO SI METRIC (CONTD.)

Grains per gallon	gr / gal	17.12	mg / l	0.05841
Horsepower	hp	0.7457	kW	1.341
Horsepower-hour	hp-hr	2.684	MJ	0.3725
Inch	in.	25.4	mm	0.03937
Knot	knot	1.852	km / h	0.5400
Knot	knot	0.5144	m / sec	1.944
Mile	mi	1.609	km	0.6215
Miler per hour	mph	1.609	km / h	0.6215
Million gallons	mil gal	3785.0	m^3	0.0002642
Million gallons per day	mgd	43.81	l /sec	0.02282
Million gallons per day	mgd	0.04381	m^3/ sec	22.82
Ounce	oz	28.35	g	0.03527
Pound (force)	lb	4.448	N	0.2248
Pound (mass)	lb	0.4536	kg	2.205
Pound per acre	lb / ac	1.121	kg / ha	0.8921
Pound per cubic foot	lb/ft^3	16.02	kg / m^3	0.06242
Pound per foot	lb / ft	1.488	kg / m	0.6720
Pound per horsepower-hour	lb / hp-hr	0.1690	mg / J	5.918
Pound per square foot	lb / ft^2	4.882	kgf / m^2	0.2048
Pound per square inch	psi	703.1	kg f / m^2	0.001422
Pound per square inch	psi	6.895	kN / m^2	0.1450
Pound per thousand cubic feet per day	lb / 1000 ft^3/ day	0.01602	kg / m^3/ day	62.43
Square foot	ft^2	0.09290	m^2	10.76
Square inch	$in.^2$	645.2	mm^2	0.001550
Square mile	mi^2	2.590	km^2	0.3861
Square yard	yd^2	0.8361	m^2	1.196
Ton, short	ton	0.9072	t	1.102
Yard	yd	0.9144	m	1.094

Acceleration of gravity $g = 32.174\ ft/s^2 = 9.80665\ m/s^2$

TABLE A–5 SATURATION VALUES OF DISSOLVED OXYGEN IN WATER EXPOSED TO WATER-SATURATED AIR CONTAINING 20.90% OXYGEN UNDER A PRESSURE OF 760 mm OF MERCURY*

Temperature (°C)	*Chloride Concentration in Water (mg/l)*			*Difference per 100 mg Chloride*	*Temperature (°C)*	*Vapour Pressure (mm)*
	0	*5000*	*10,000*			
	Dissolved Oxygen (mg/l)					
0	14.6	13.8	13.0	0.017	0	5
1	14.2	13.4	12.6	0.016	1	5
2	13.8	13.1	12.3	0.015	2	5
3	13.5	12.7	12.0	0.015	3	6
4	13.1	12.4	11.7	0.014	4	6
5	12.8	12.1	11.4	0.014	5	7
6	12.5	11.8	11.1	0.014	6	7
7	12.2	11.5	10.9	0.013	7	8
8	11.9	11.2	10.6	0.013	8	8
9	11.6	11.0	10.4	0.012	9	9
10	11.3	10.7	10.1	0.012	10	9
11	11.1	10.5	9.9	0.011	11	10
12	10.8	10.3	9.7	0.011	12	11
13	10.6	10.1	9.5	0.011	13	11
14	10.4	9.9	9.3	0.010	14	12
15	10.2	9.7	9.1	0.010	15	13
16	10.0	9.5	9.0	0.010	16	14
17	9.7	9.3	8.8	0.010	17	15
18	9.5	9.1	8.6	0.009	18	16
19	9.4	8.9	8.5	0.009	19	17
20	9.2	8.7	8.3	0.009	20	18
21	9.0	8.6	8.1	0.009	21	19
22	8.8	8.4	8.0	0.008	22	20
23	8.7	8.3	7.9	0.008	23	21
24	8.5	8.1	7.7	0.008	24	22
25	8.4	8.0	7.6	0.008	25	24
26	8.2	7.8	7.4	0.008	26	25
27	8.1	7.7	7.3	0.008	27	27
28	7.9	7.5	7.1	0.008	28	28
29	7.8	7.4	7.0	0.008	29	30
30	7.6	7.3	6.9	0.008	30	32

*Saturation at barometric pressures other than 760 mm (29.92 in.), C_s' is related to the corresponding tabulated values, C_s, by the equation:

$$C_s' = C_s \frac{P - p}{760 - p} \quad \text{where}$$

C_s' = solubility at barometric pressure P and given temperature, mg/litre

C_s = saturation at given temperature from table, mg/litre

P = barometric pressure, mm.

p = pressure of saturated water vapour at temperature of the water selected from table, mm

TABLE A–6 TABLE OF RELATIVE ATOMIC WEIGHTS

Name	Symbol	Atomic Number	Atomic Weight	Name	Symbol	Atomic Number	Atomic Weight	Name	Symbol	Atomic Number	Atomic Weight
Actinium	Ac	89	–	Gold	Au	79	196.967	Praseodymium	Pr	59	140.907
Aluminium	Al	13	26.9815	Hafnium	Hf	72	178.49	Promethium	Pm	61	–
Americium	Am	95	–	Helium	He	2	4.0026	Protactinium	Pa	91	–
Antimony	Sb	51	121.75	Holmium	Ho	67	164.930	Radium	Ra	88	–
Argon	Ar	18	39.948	Hydrogen	H	1	1.00797	Radon	Rn	86	–
Arsenic	As	33	74.9216	Indium	In	49	114.82	Rhenium	Re	75	186.2
Astatine	At	85	–	Iodine	I	53	126.9044	Rhodium	Rh	45	102.905
Barium	Ba	56	137.34	Iridium	Ir	77	192.2	Rubidium	Rb	37	85.47
Berkelium	Bk	97	–	Iron	Fe	26	55.847	Ruthenium	Ru	44	101.07
Beryllium	Be	4	9.0122	Krypton	Kr	36	83.80	Samarium	Sm	62	150.35
Bismuth	Bi	83	208.980	Lanthanum	La	57	138.91	Scandium	Sc	21	44.956
Boron	B	5	10.811	Lead	Pb	82	207.19	Selenium	Se	34	78.96
Bromine	Br	35	79.904	Lithium	Li	3	6.939	Silicon	Si	14	28.086
Cadmium	Cd	48	112.40	Lutetium	Lu	71	174.97	Silver	Ag	47	107.868
Calcium	Ca	20	40.08	Magnesium	Mg	12	24.312	Sodium	Na	11	22.9898
Californium	Cf	98	–	Manganese	Mn	25	54.9380	Strontium	Sr	38	87.62
Carbon	C	6	12.01115	Mendelevium	Md	101	–	Sulphur	S	16	32.064
Cerium	Ce	58	140.12	Mercury	Hg	80	200.59	Tantalum	Ta	73	189.948
Cesium	Cs	55	132.905	Molybdenum	Mo	42	95.94	Technetium	Tc	43	–
Chlorine	Cl	17	35.453	Neodymium	Nd	60	144.24	Tellurium	Te	52	127.60
Chromium	Cr	24	51.996	Neon	Ne	10	20.183	Terbium	Tb	65	158.924
Cobalt	Co	27	58.9332	Neptunium	Np	93	–	Thallium	Tl	81	204.37
Copper	Cu	29	63.546	Nickel	Ni	28	58.71	Thorium	Th	90	232.038
Curium	Cm	96	–	Niobium	Nb	41	92.906	Thulium	Tm	69	168.934
Dysprosium	Dy	66	162.50	Nitrogen	N	7	14.0067	Tin	Sn	50	118.69
Einsteinium	Es	99	–	Nobelium	No	102	–	Titanium	Ti	22	47.90
Erbium	Er	68	167.26	Osmium	Os	76	190.2	Tungsten	W	74	183.85
Europium	Eu	63	151.96	Oxygen	O	8	15.9994	Uranium	U	92	238.03
Fermium	Fm	100	–	Palladium	Pd	46	106.4	Vanadium	V	23	50.942
Fluorine	F	9	18.9984	Phosphorus	P	15	30.9738	Xenon	Xe	54	131.30
Francium	Fr	87	–	Platinum	Pt	78	195.09	Ytterbium	Yb	70	173.04
Gadolinium	Gd	64	157.25	Plutonium	Pu	94		Yttrium	Y	39	88.905
Gallium	Ga	31	69.72	Polonium	Po	84	–	Zinc	Zn	30	65.37
Germanium	Ge	32	72.59	Potassium	K	19	39.102	Zirconium	Zr	40	91.22

INDEX

INDEX